Far-Infrared Spectroscopy

HEINRICH RUBENS
(born March 30, 1865, died July 17, 1922)
Professor of Experimental Physics at the University of Berlin from 1906 to 1922

Far-Infrared Spectroscopy

KARL DIETER MÖLLER

Professor of Physics
Fairleigh-Dickinson University
Teaneck, New Jersey

WALTER G. ROTHSCHILD

Department of Chemistry
Scientific Research Staff, Ford Motor Company
Dearborn, Michigan

WILEY-INTERSCIENCE
A Division of John Wiley & Sons, Inc.
New York . London . Sydney . Toronto

Preface

In the last ten to fifteen years far-infrared spectroscopy has been used at an increasing rate in the investigations of a large variety of phenomena in various fields of science. We therefore felt that there was a need for assembling the major applications of this, in some instances unique, research technique and for attempting to integrate the available and relevant research material into one book. Thus this book is a source of information about the state of far-infrared spectroscopy for anyone interested in inquiring about the recent developments in this field or its research potentialities for solving problems in physics, chemistry, and biochemistry or biophysics. Furthermore, the book will help the student or researcher in other fields who is interested in applying far-infrared spectroscopic techniques to the solutions of his specific research problems.

Since far-infrared spectral techniques are applicable to such a large variety of different phenomena, the theoretical discussions in this book are oriented to the particular subject under consideration rather than to basic, purely infrared principles (for which the reader may consult the standard books and treatises). In this sense we have restricted ourselves to short and pertinent theoretical treatments within the context of the discussions and have elaborated on only a few theoretical principles that are less commonly known.

Although this book is not intended as an exhaustive review of the entire field of far-infrared spectroscopy, we have illustrated the various techniques and subjects of research with thoroughly discussed, representative examples. The first four chapters of the book deal with instrumentation. They are followed by nine chapters on problems in chemical physics, divided into seven chapters on vapor- and liquid-phase spectra and two chapters on crystal spectra. In order to achieve a more comprehensive picture, we have included six specialized chapters written by researchers in several branches of solid state physics as appendices and a bibliography from 1892 to the present compiled by E. D. Palik.

We consider the far-infrared spectral region, somewhat arbitrarily, to be the wavelength range between 50 and 1000μ, or the frequency range from 200 to $10\ \mathrm{cm}^{-1}$. (These units for wavelengths and frequencies, respectively, are used in this book.) In various instances we found it necessary to extend this range to shorter or longer wavelengths for a more complete discussion of a particular phenomenon.

We are grateful for advice and helpful comment to D. R. Bosomworth, R. I. Bryant, J. W. Brasch, E. Burstein, D. M. Dennison, H. L. Friedman, T. M. Hard, R. J. Jakobsen, R. Kaplan, H. B. Levine, E. V. Loewenstein, F. A. Miller, W. J. Moore, E. D. Palik, K. N. Rao, H. Sakai, I. F. Silvera, D. R. Smith, H. L. Strauss, R. Ulrich, and G. A. Vanasse.

Teaneck, New Jersey   K. D. MÖLLER
Dearborn, Michigan   W. G. ROTHSCHILD
1970

Contents

1 Grating Spectrometers

A. INTRODUCTION

Ever since the work of Rubens and coworkers in the early days of far-infrared spectroscopy, monochromators employing a grating as dispersion element have been used for obtaining far-infrared spectral data. The grating spectrometer was almost the only instrument for investigations in the 50 to 1000 μ* spectral region until the recent introduction of equipment using the principles of Fourier transform spectroscopy.

In the wavelength region from the visible to about 50 μ, prism materials with high transmission and satisfactory dispersion characteristics are available. Prisms of CsI (cesium iodide), for instance, are useful in the 25 to 50 μ wavelength range. The long-wavelength cut-off inherent in convenient prism spectrometers has created the term "far-infrared" for wavelengths longer than 50 μ.

The long-wavelength limit of the far-infrared is considered to be at about 1000 μ because of the energy limitations of continuously-emitting sources. Although intense monochromatic sources such as laser- and maser-type oscillators as well as higher harmonics of millimeter waves are available for the long-wavelength region, they are presently of little use since they can be swept, at best, through only a relatively narrow frequency range. Because of this, we shall not discuss them here.

The use of a grating as the dispersion element for continuous radiation requires filtering devices for the separation of the different orders of the diffracted light. These filters have to be very efficient since the spectral radiance in the far-infrared is rather limited. Various types of such filters have been developed in recent years; one of the most efficient, an interference-type transmission filter, will be discussed in detail in Chapter 3.

In the nineteen-sixties, several commercially built far-infrared spectrometers became available. This development was, in great part, a consequence

---

* We will use μ (microns) for wavelength and cm^{-1} (reciprocal cm) for frequency. Note: $1 \mu = 10^{-4}$ cm; $10 \mu \leftrightarrow 1000$ cm^{-1}, $100 \mu \leftrightarrow 100$ cm^{-1}, $1000 \mu \leftrightarrow 10$ cm^{-1}.

1

of the commercial availability of the Golay detector. We remark here that helium-cooled detectors are becoming increasingly popular and, furthermore, that vacuum spectrometers are gradually displacing those using dry air or dry nitrogen for the removal of water vapor.

B. Optical Layout of Spectrometers

a. General

The optical system of a spectrometer consists of a number of curved and flat front-surface mirrors and the grating. (Reflection filters will be considered here as flat front-surface mirrors.) The purpose of the optical system is to transfer the available radiant power from the source to the dispersion element and then to the detector. The location of the sample compartment is dictated by the particular application as well as by convenience: it might be placed between the radiation source and the monochromator or between the monochromator and the detector.

The first basic requirement for the design of the optical system is to maximize the radiant power falling onto the detector within the given spectral resolution width. The radiant power is proportional to the minimum value of the product formed by the solid angle Ω and the aperture area A, taken at each optical element in the spectrometer. In almost all cases the minimum value of $(A\Omega)$ is given by the used area of the detector and by the solid angle under which the detector is able to accept radiation. The $(A\Omega)$ value of all the other optical components is then equal to or greater than this $(A\Omega)$. Once the desired resolution (see Section D) is set, the size of the grating, the widths of the slits, and the characteristics of the mirrors can be found. It is clear that the sample area must also provide an $(A\Omega)$ value which is not smaller than that at the detector, $(A\Omega)_{Det}$, if full use of the radiant power is to be made. This procedure is, of course, not different from that employed in the design of spectrometers for use in the other regions of the infrared; therefore we will mention only a few salient features. This discussion will be divided according to the three most frequently used types of monochromators employing reflection gratings: (1) the Littrow mount, (2) the Czerny-Turner mount, and the (3) Ebert–Fastie mount. In all three mounts the angle between the incident and the diffracted light does not vary during the turning of the grating.

b. Littrow Mount

As an example of the optical layout of a spectrometer using a Littrow mount monochromator, let us consider the Perkin-Elmer Model 301, a commercially available double-beam spectrometer (see Fig. 1.1). The light from the source I, magnified by two by the toroid mirrors M_1 and $M_{1'}$,

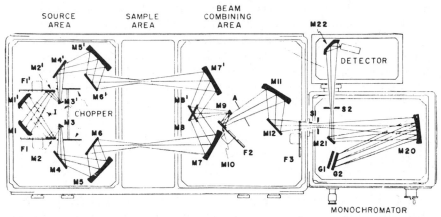

Fig. 1.1   Schematic diagram of the Perkin-Elmer 301 far-infrared double-beam spectrometer.
[C. C. Helms, H. W. Jones, A. J. Russo, and E. H. Siegler, Jr., *Spectrochim. Acta* **19**, 819 (1963),
by permission of Pergamon Press.]

forms images A_1 and $A_{1'}$ on the field mirrors M_3 and $M_{3'}$. Following—for
simplicity's sake—only one beam, we see that the toroid mirror M_5 forms
a 1:1 image A_2 of A_1 in a plane in the sample area. The toroid mirror M_7
forms a 1:1 image A_3 of A_2 at the field mirror M_9. Here, M_8 and $M_{8'}$
designate the halves of a beam-combining mirror in which the upper half
is used by one beam and the lower half by the other beam. Toroid M_{11}
then images A_3 onto the entrance slit S_1. The off-axis parabolic mirror M_{20}
serves simultaneously as collimator and as telescope. Finally, the 90° off-
axis ellipsoid mirror M_{22} concentrates the light from the exit slit onto the
Golay detector window with a demagnification ratio of 6:1.

We make here a few remarks on the particular characteristics of the various
mirrors: Toroid mirrors are used for the off-axis imaging in order to minimize astigmatism, and field mirrors are used to reduce the divergence of
the beam along its path. The paraboloid mirror M_{20} is employed (*1*) to
change the divergent beam which emerges from the entrance slit into a
parallel beam and (*2*) to converge the beam after it is diffracted by the grating.
The ellipsoid mirror is the convenient choice for the necessary demagnification of the cross section of the light beam emerging from the exit slit before
reaching the detector window. (With respect to the geometry of paraboloid
and ellipsoid mirrors, the interested reader is referred to p. 163 of Ref. 1.)
Two gratings (blaze angle 26°45′, ruled area 68 × 68 mm²) are mounted
back-to-back on one holder for easy interchange, thus eliminating the need
of breaking the dried atmosphere of the housing by having to open it.

For completeness' sake we now list the remaining "optical" components
of the Perkin-Elmer 301. There are two choppers, each modulating one light

beam at 13 Hz. The choppers consist of various materials, depending on the frequency range under investigation. The wheels F_1, $F_{1'}$, and F_2 provide for easy insertion of reflection filters made of several materials whereas the wheel F_3 is a holder for transmission filters. The grating drive provides a linear "wavenumber" (cm^{-1}) recording. A Golay cell is used as detector. Transmission measurements can be performed on liquids and solids (for instance, powder samples embedded in polyethylene) without changing this arrangement of mirrors. For investigations of vapors and gases, a cell of maximum length of 15 in. (37 cm) can be accommodated in the sample compartment under special provisions.

c. Czerny-Turner Mount

A monochromator employing the Czerny-Turner mount uses two spherical mirrors in a symmetrical arrangement in order to reduce aberration. Genzel and Eckhardt[2] as well as Yoshinaga et al.[3] have constructed spectrometers using this type of monochromator. Figure 1.2 shows a schematic diagram of the optical layout similar to that used by Genzel and Eckhardt. The size of the gratings was approximately 20×25 cm^2; wavelengths as long as 1 mm could be measured. We see from Fig. 1.2 that the light from the source S is imaged by the spherical mirror M_1 onto the slit S_1 after passing the chopper Ch. The collimator-mirror M_2 and the telescope-mirror M_3, together with the grating G, form the Czerny-Turner mount. Furthermore, S_2 is the exit slit whereas TF designates a holder for transmission filters. The mirror M_4 finally concentrates the radiation onto the detector D.

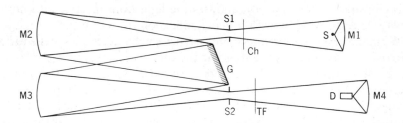

Fig. 1.2   Schematic diagram of a spectrometer using a Czerny-Turner mount.

We mention here that the symmetrical arrangement of the two spherical mirrors* of the Czerny-Turner mount cancels out the "coma."[4] The question of coma and astigmatism has been reinvestigated by Rosendahl.[5] With respect to the coma, he studied a Czerny-Turner mount as shown in Fig. 1.3— a mount which is suitable for transmission gratings (as were used by Czerny).

* The two mirrors do not have to be arranged confocally.

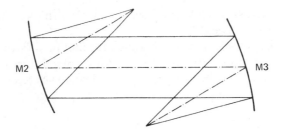

Fig. 1.3   Czerny-Turner mount compensating for coma. [G. R. Rosendahl, *J. Opt. Soc. Am.* **52**, 412 (1962).]

Rosendahl then showed that, in addition, one could compensate for the astigmatism of the Czerny-Turner mount by incorporating two spherical-convex mirrors. This is shown in Fig. 1.4, where M_1 and M_4 are the convex mirrors and M_2 and M_3 are the original mirrors of the Czerny-Turner mount (see Fig. 1.3). The interested reader will find the necessary mathematical theorems to compute the required radius of the mirrors M_1 and M_4 in Ref. 5.

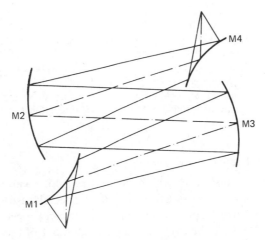

Fig. 1.4   Czerny-Turner mount compensating for coma and astigmatism. [G. R. Rosendahl, *J. Opt. Soc. Am.* **52**, 412 (1962).]

As we mentioned above, Yoshinaga et al.[3] also constructed a large-grating spectrometer based on the Czerny-Turner principle. The optical layout of their spectrometer is shown in Fig. 1.5. The instrumental arrangement provides for reflection filters. These are mounted for easy interchange on the rotating holders D and J in a manner which guarantees correct positioning of each filter in the optical beam. Furthermore, the angles of incidence and reflection are kept small: we shall see below that this is required for the efficient use of some types of filters.

Figure 1.5 also shows that two gratings are mounted back-to-back as in the Perkin-Elmer Model 301 (see above). An elliptical mirror concentrates the light emerging from the exit slit onto the detector by using a Pfund-type arrangement.

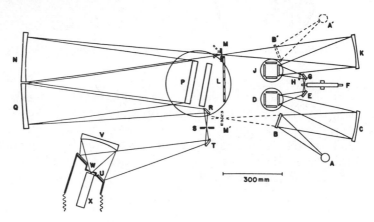

Fig. 1.5   Optical arrangement of a Czerny-Turner mount: *A*, source; *B*, plane mirror; *C* and *K*, spherical mirrors; *D* and *J*, rotating holders; *E* and *G*, vertical holders; *F*, chopper; *H*, shutter; *L*, horizontal holder; *M*, entrance slit; *N* and *Q*, main spherical mirrors; *P*, grating; *R* and *T*, plane mirrors; *S*, exit slit; *U*, plane mirror; *V*, ellipsoidal mirror; *W*, window; *X*, Golay detector; *A′*, *B′*, and *M′* are components corresponding to, respectively, *A*, *B*, *M* for double-beam use. [H. Yoshinaga, S. Fujita, S. Minami, A. Mitsuishi, R. A. Oetjen, and Y. Yamada, *J. Opt. Soc. Am.* **48**, 315 (1958).]

d. Ebert-Fastie Mount

The Ebert-Fastie mount, in which the two reflecting surfaces have coincident centers of curvature (see Fig. 1.6), is a special case of the Czerny-Turner mount. Consequently, only one large mirror need be used. The coma is canceled out by the symmetrical arrangement of the light beams as in the Czerny-Turner mount. Astigmatism is reduced considerably by the use of curved slits,[7] even if these slits are long. [Long slits are desirable in the far-infrared for a large ($A\Omega$) value at the exit slit. We shall discuss this in more detail in Section D.]

Figure 1.6 shows a layout of a spectrometer using an Ebert-Fastie mount (Robinson[6]). The source is mounted on-axis with its own condensing mirror; an image transformer (see Chapter 2) is placed in front of the Golay detector, and reflection filter plates in conjunction with a sample cell for gases are employed.

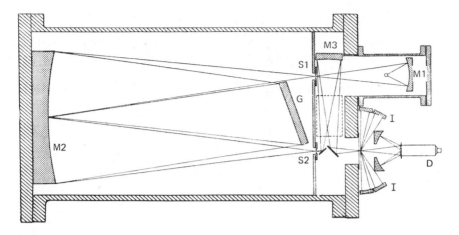

Fig. 1.6   Ebert-Fastie arrangement: M_1 to M_3 are spherical mirrors; S_1 and S_2 are slits; G, grating; I, image transformer; D, detector. [D. W. Robinson, *J. Opt. Soc. Am.* **49**, 966 (1959).]

Figure 1.7 shows a schematic of the optical arrangement in a direction looking directly at the ruled surface of the grating. We see that the entrance and exit slits are on a common circle centered on an axis which goes through the middle of the large mirror and the center of the grating surface.

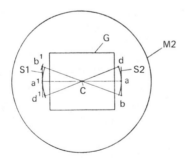

Fig. 1.7   End-on view of the optical system showing the relationship between the line image with straight and curved slits. [W. G. Fastie, *J. Opt. Soc. Am.* **42**, 641 (1952); **42**, 647 (1952).]

From what has been reported in the literature, practical as well as optical considerations have dictated the choice of a mount. For instance, parabolic mirrors are costlier to obtain than spherical mirrors. One spherical mirror fits better into tube-shaped vacuum tanks (such tanks have, from the point of view of the vacuum techniques, several advantages over tanks welded into other shapes).

e. f-Number of the Monochromator

The f-number of the optical system in the monochromator is usually defined as the ratio between the focal length F and the diameter D of the collimator and telescope mirror, $f = F/D$. (For the Czerny-Turner mount, we assume here that both mirrors have the same dimensions.) If aberration is neglected, the resolution depends on the ratio b/f of the slit height b and the focal length f (see Section D). Therefore, a system using a large focal length will have a large aperture area A and a small solid angle Ω in order to satisfy the constancy of the product $(A\Omega)$—and vice versa for a system of small focal length. We note that systems with large focal length reduce aberrations resulting from the small magnitude of Ω but, on the other hand, increase aberration because of the size of the image. (The reversed situation holds in systems with small f values.)

The spectrometers discussed in the previous sections all use monochromators with $f/4$ optics. A spectrometer with $f/1$ optics has been constructed by McCubbin,[8] small-grating spectrometers employing $f/2$ optics have been built by Lecomte and Hadni[9] as well as by Delorme,[10] whereas large-grating monochromators with $f/2$ optics have been described by Lorenzelli[11] and Möller et al.[12] A commercial instrument, the Jarrell Ash 78-900 far-infrared double-beam vacuum spectrometer, uses $f/10$ optics. For other cases, the interested reader should consult the Bibliography (Appendix VII).

With respect to the large physical dimensions of spectrometers using large focal length optics, the choice of $f/4$ seems a favorable compromise. Spectrometers employing large focal lengths have large vacuum housings but problems of aberration are less important. Small-focal length spectrometers can be built very compactly but the corrections for aberrations are more elaborate and placing the sample area often presents a problem. This latter difficulty can be circumvented, in particular cases, by the use of "light-pipes" (see Refs. 12–14 and Chapter 2, Section C).

C. GRATINGS

a. General

In the early days of far-infrared spectroscopy, transmission wire gratings ("amplitude gratings," see p. 232 of Ref. 15) were used as dispersion elements in the monochromator; for instance, Rubens and Czerny employed such devices (see Bibliography). Such gratings can be fabricated by winding wire in a vertical direction on a frame. The size of the frame determines the grating area. The distance between the wires, the "grating constant," can be determined by the use of two threaded rods in the frame. Similar electroformed versions (on plastic substrates) are still finding important applications as polarizers (see Chapter 2, Section E).

The gratings used in modern infrared spectrometers are echelette reflection gratings ("phase gratings," see p. 232 of Ref. 15). Their dispersion characteristics are a consequence of their ruled metal surface; that is, a surface consisting of a periodic array of parallel grooves of triangular shape. Such surfaces diffract the light and act like a mirror which concentrates radiation of a certain wavelength into a certain direction in space.* Figure 1.8 shows the cross section of the grating surface in a direction perpendicular to the grooves.

Ruling engines are employed for the fabrication of echelette gratings for use in the shorter wavelength regions (to about 200 μ). Since less accuracy is required for the 200 to 1000 μ range, milling machines can be employed on metals such as soft brass or mixtures of lead and tin. Also, plastic plates which are plated with a metal surface following the cutting process can be used.[2] We note that the step angle (see Fig. 1.8) is between 15 and 30° in most cases.

In the far-infrared the appearance of "ghosts" has not been reported. Diffraction anomalies,[16,17] however, are observed. Gratings are frequently used only in the first order; since each grating covers about one octave of wavelength, a set of about five gratings is necessary to scan the 50 to 1000 μ region.

Because of this, two or more gratings are mounted on a common table in many spectrometers (see Figs. 1.1 and 1.5). This guarantees ready interchange without the necessity of breaking the vaccum or the dried atmosphere of the instrument. The axis of rotation of the grating when scanning the spectrum is, of course, not identical with that about which grating interchanges are effected. It is clear that the grating drive must be mechanically accurate; however, the requirements are not so stringent as those for near-infrared applications. Depending on operational conditions, scanning speeds of the grating drive between 0.04 and 0.4°/min have been reported.

b. Grating Formula and Intensity Distribution

The echelette grating can be described in terms of two superimposed features. The first is the periodicity of the arrangement of the grooves, the second is the regular orientation of the grooves (see Fig. 1.8).

The first feature is responsible for the directions of diffraction, the second feature is responsible for the distribution of the intensity in these directions. We will discuss the mathematical expressions which describe these two features further below; we mention here that pages 182 and 232 of Ref. 15

---

* For monochromatic incident light the diffracted light is concentrated into one certain order of diffraction.

give a two-dimensional treatment based on the diffraction theory by Kirchhoff. The same reference contains a critical discussion of the applicability of Kirchhoff's theory. The assumption is made that the wavelength λ of the light is small compared to the grating constant d (see Fig. 1.8) and that the step angle ξ is small. Since both assumptions are not valid in the far-infrared, a more elaborate treatment is required. The interested reader is referred to Appendix P of Ref. 18 which gives a three-dimensional calculation (including shadowing effects generated by the shape of the grooves) and a comparison with experimental data employing polarized light. Studies concerning these problems have also been contributed by Hadni and coworkers.[19,20]

We begin now with the discussion of the first feature; namely, the directions of constructive interference of the light. These are, as we pointed out above, a consequence of the periodicity of the grooves. The diffraction features of the grating can be described in an elementary way by considering a reflecting surface possessing a periodic structure of M elements, where M is a large number. The incident light is assumed to be parallel and the diffracted light is observed in the focal plane of the telescope mirror ("Fraunhofer case"). Only monochromatic light is considered at the present. We will use the coordinate system depicted in Fig. 1.8. The grating constant is designated by d, $\mathbf{N}_0(\theta)$ is a vector normal to the plane of the grating in its zero-order position, $\theta = 0$, where zero-order position means that there is no path difference between incident and "diffracted" light. The angle of rotation of the grating, counted from its zero-order position, is θ; i is the angle of the incident light and δ is the angle of the diffracted light. Both i and δ are measured with respect to $\mathbf{N}_0(\theta)$. The angle convention is as follows: All angles measured to the right of $\mathbf{N}_0(\theta)$ are counted as positive, those to the left are taken to be negative. The angle 2ε is constant if the grating is used in the monochromators discussed earlier.

The condition for constructive interference is that the path difference between rays from adjacent grooves must be an integral multiple of λ,

$$m\lambda = d\cos(90 - i) + d\cos(90 - \delta); \qquad m\lambda = 2d\sin\frac{i+\delta}{2}\cos\frac{i-\delta}{2}, \quad (1.1)$$

where m is a positive or negative integer, called the order of diffraction. Employing the angles θ and ε as shown in Fig. 1.8, we obtain the following relation:

$$m\lambda = 2d\cos\varepsilon\sin\theta. \qquad (1.2)$$

From Eq. 1.2 we see that the wavelength for which constructive interference occurs is a function of the rotation angle θ in the form of a sine if the angle 2ε is kept constant. (For example, the angle 2ε in the Czerny-Turner mount shown in Fig. 1.2 is the angle between the centers of the collimator M_2, the

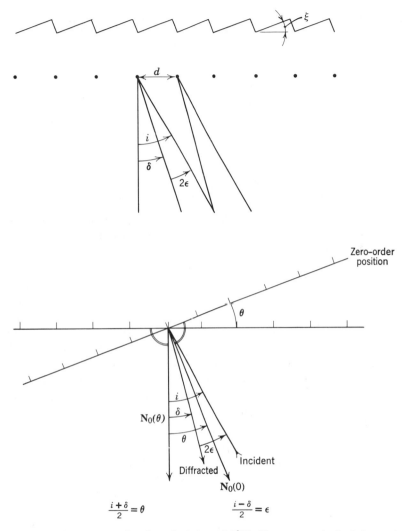

$$\frac{i+\delta}{2} = \theta \qquad\qquad \frac{i-\delta}{2} = \epsilon$$

Fig. 1.8   Coordinate system for the calculation of $S^2(\theta)$. The step angle ξ of the grating is indicated at the top.

grating G, and the telescope mirror M_3.) It is often convenient to scan the angle θ so that $\sin\theta$ is linear with time. The recorded spectrum then appears on a linear wavelength scale.

We now extend our discussion by considering the intensity distribution which is also a consequence of the periodicity of the grooves (see above). Under the above-mentioned assumptions, the intensity distribution of the diffracted light can be described in good approximation as the product of

the blaze position of the grating, λ_B. We furthermore see that all minima of the second factor are placed at $kd \cos \varepsilon \sin (\theta - \xi) = n\pi$ $(n \neq 0)$, which is approximately the location where the other maxima of the first factor occur; namely, $kd \cos \varepsilon \sin \theta = m\pi$ $(m \neq 1)$. Thus, the available energy of the grating is seen to be concentrated in one order of diffraction. This is the great advantage of the echelette grating.

For the actual case of polychromatic light the situation is quite similar. Here, we shall find a certain wavelength range centered at the blaze $\lambda = \lambda_B$ for which almost all energy can be concentrated into the first order. In general, the concentration of the diffracted light of order n takes place at $\lambda = \lambda_B/n$, where λ_B is the blaze angle of the grating as defined above.

We have used the term "higher-order" with respect to monochromatic light, meaning that the path difference between two light beams undergoing constructive interference is $n\lambda$ for the nth order. Consequently, the n orders of diffracted light are observed at n different angles—and as we know, an echelette grating suppresses all but one. On the other hand, for "higher-order spectra" of polychromatic light we understand that, in the absence of filters, light of wavelength $\lambda_1, \lambda_2 = \lambda_1/2, \cdots, \lambda_n = \lambda_1/n$ is diffracted at the *same* angle θ.

The intensity distribution of the second-order spectrum about the blaze angle ξ is narrower than that of the first-order spectrum. This can easily be seen by substituting the value $k = 2\pi/(\lambda/2)$ for $k = 2\pi/\lambda$ into the second factor of Eq. 1.6. The angular difference between the predominant maximum and the first minimum is decreased.

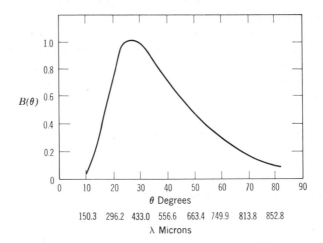

Fig. 1.10   Dependence of the second factor of Eq. 1.6, $B(\theta)$, on θ. Numerical values: $\xi = 26° \ 45'$; $\varepsilon = 15°$; the wavelengths are marked for $d = 500 \ \mu$.

Figure 1.10 shows a plot of the second factor of Eq. 1.6, here designated by the symbol $B(\theta)$, as a function of λ, θ. We notice the maximum of B at the blaze angle ξ (26°45′).

c. Spectral Slit Width

We saw in the previous section that the light rays which are diffracted from the grating have a wavelength distribution (angular distribution) of a certain width. In order to be able to narrow or broaden this distribution at will, adjustable slits are provided, generally at the entrance and exit of the monochromator. We assume in the following that the entrance and exit slits have equal widths. The width of the exit slit corresponds to a certain range $\Delta\lambda$ of wavelengths which pass through it; spectral details narrower than this value of $\Delta\lambda$ will not be resolved. (Rayleigh's criterion for the resolution may be used.*)

The value of the angular resolution $\Delta\lambda$ is obtained from Eq. 1.2 for $m = 1$ as

$$\Delta\lambda = 2d \cos \varepsilon \cos \theta \Delta\theta. \tag{1.7}$$

The variation of λ is here expressed as a function of the variation of θ, the angle of the rotation of the grating. If a is the physical slit width and F the focal length of the telescope mirror, the angular slit width of the exit slit is given by $W = a/F$. This ratio is related to the variation of the rotation of the grating $\Delta\theta$ as $\Delta\theta = a/2F$ since the grating is a reflecting surface. Therefore, we obtain for the resolution

$$\Delta\lambda = d \frac{a}{F} \cos \varepsilon \cos \theta; \qquad \Delta v = v^2 d \frac{a}{F} \cos \varepsilon \cos \theta, \tag{1.8}$$

where v is expressed as frequency (cm^{-1}).

The resolving power is defined by $R = \lambda/\Delta\lambda$ and is thus given here (see Eq. 1.2) by

$$R = \frac{\lambda}{\Delta\lambda} = \frac{2F}{a} \tan \theta. \tag{1.9}$$

We notice that for the case of a grating spectrometer R depends on the angle θ and on the angular slit width.

The theoretical resolving power of a grating is equal to the number of wavelengths per path difference, taken between the rays that are diffracted from the opposite extreme ends of the grating ($= Md$) into the direction θ.[21] Of course, the theoretical value is not reached in practice because of the diffraction limits of the optical system of the monochromator, its aberration, and energy limitations. We shall discuss the latter in the following section.

* See pages 210–215 of Ref. 18 for Rayleigh's criterion and a detailed discussion on principles of resolving power.

D. Energy Limitations on the Resolving Power

The limited available spectral radiance in the far-infrared is more responsible for the restricted resolution than are the diffraction limits of the monochromator. For instance, the slits must be opened sufficiently wide to yield
a signal with an acceptable signal-to-noise ratio. This slit width is larger than
that corresponding to the diffraction limit. Disregarding energy losses in the
optical system, an improvement of the signal-to-noise ratio can be accomplished by improving the brightness (spectral radiance) of the source or by
reducing the noise-equivalent-power (NEP) of the detector. The latter can be
accomplished by employing helium-cooled detectors; this has been reported,
for instance, by Hard and Lord.[22]

Let us now discuss the dependence of the resolution on the flux of light
which passes through the spectrometer and on the geometrical parameters
of the optical setup. Figure 1.11 shows an idealized optical system, drawn

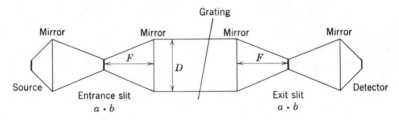

Fig. 1.11   Schematic diagram of the optical system for a consideration of the energy limitations
on resolution.

unfolded for clarity's sake. We assume that the collimator and telescope
mirror of the monochromator have the same focal length F and area S,
and that the grating is square-shaped with grooved area $D^2 \approx S$ of $M = D/d$
grooves (where d is the grating constant). We also assume that the $(A\Omega)$
value at the source is larger than anywhere else (see Section B.a). The flux
P, which passes through the area ab of the exit slit towards the detector, is
then expressed by

$$P = B_\lambda \Delta\lambda \tau ab\Omega, \qquad (1.10a)$$

where B_λ is the brightness of the source in units of watt/(meter^2 × steradian ×
wavelength interval) (see also pp. 55–60 of Ref. 18). The wavelength band
passing through the spectrometer is given by $\Delta\lambda$, and τ is a transmission
coefficient which takes care of losses (for instance, those due to the filtering
action). The solid angle at the exit slit is given by $\Omega \approx S/F^2$. Using the angular slit width $W = a/F$ and the angular slit height $L = b/F$, Eq. 1.10a is

rewritten in the form

$$P \approx B_\lambda L \tau S W \Delta\lambda. \tag{1.10b}$$

Assuming the Rayleigh-Jeans law for the emission of the source at absolute temperature T, $B_\lambda = B_0 T/\lambda^4$ (where B_0 is a constant), and introducing the f-number, $f = F/D$, Eq. 1.10a yields

$$P \approx B_0 T \tau (\Delta\lambda/\lambda^4)(ab/f^2). \tag{1.10c}$$

If we go to frequency units (cm^{-1}) rather than wavelength units, Eq. 1.10c reads

$$P \approx B_0 T \tau v^2 \Delta v (ab/f^2). \tag{1.10d}$$

We will call now the quantities $\Delta\lambda$ and Δv of Eq. 1.8 the "spectral slit width." We thus have the following relations between the spectral slit width and the physical slit width a:

$$\Delta\lambda = (da/F) \cos \varepsilon \cos \theta; \tag{1.11a}$$

$$\Delta v = v^2 (da/F) \cos \varepsilon \cos \theta. \tag{1.11b}$$

Inserting $M = D/d$, we obtain

$$\Delta\lambda = (a/Mf) \cos \varepsilon \cos \theta; \tag{1.11c}$$

$$\Delta v = (v^2 a/Mf) \cos \varepsilon \cos \theta. \tag{1.11d}$$

Eliminating either a or Δv from Eqs. 1.10d and 1.11d, we find for the flux of radiation which passes through the exit slit the expression

$$P \approx B_0 T \tau \frac{\cos \varepsilon \cos \theta}{M} \frac{b}{f} v^4 (a/f)^2 \tag{1.12}$$

or

$$P \approx B_0 T \tau \frac{M}{\cos \varepsilon \cos \theta} \frac{b}{f} (\Delta v)^2. \tag{1.13}$$

Equations 1.12 and 1.13 will now be applied to some special cases of interest.

1. We first assume that the product $(A\Omega)$ at the detector is larger than that elsewhere, $(A\Omega)_{Det} > (A\Omega)_{Spectr}$. This situation is realized in a spectrometer with which one is able to scan over several octaves but where one wishes to keep the flux P constant throughout the spectral region. From Eq. 1.12 we see then that decreasing values of v^4/M (with decreasing v) can be balanced by increasing values of $(a/f)^2$. This would be effective until the point where the detector is fully illuminated.

For such a type of spectrometer, the question of improving the resolution by the use of a larger grating is of interest. From Eq. 1.13 we find for the minimum flux P_m belonging to the minimum acceptable value of the ratio signal/noise

$$P_m \approx \left(\frac{B_0 T \tau}{\cos \varepsilon \cos \theta} \frac{b}{f} \right) M(\Delta v_m)^2. \tag{1.14}$$

Assuming that the angular dependence of the factor contained in the first pair of parentheses is essentially a constant, we find for a given minimum flux P_m that

$$M(\Delta v_m)^2 = \text{constant}. \tag{1.15}$$

Equation 1.15 tells us that a grating with, for instance, four times as many grooves will give twice the resolution. However, we notice that as a consequence the slit width is now twice as large and, therefore, the product $(A\Omega)$ is also twice as large:

$$\frac{a'}{f} = 2\frac{a}{f}; \quad (A\Omega)' = 2(A\Omega). \tag{1.16}$$

We can also think of improving the resolution in a different way; namely, by increasing the ratio b/f. For instance, if f is constant, we might increase the height of the slits. (A difference between the shape of the image at the entrance slit and the shape of the sensitive area of the detector can be reconciled by the use of an image transformer, as we shall indicate in Section G of Chapter 2.) The result is again a larger value of $(A\Omega)$.

We remark (considering the assumptions made above) that other parameters might be changed in order to obtain an increased resolution. But if P_m is considered constant, this will always result in a larger value of $(A\Omega)$.*

2. We assume now the case that $(A\Omega)_{Spectr} = (A\Omega)_{Det}$. In this situation the detector is fully illuminated with respect to solid angle and area. As mentioned above, we can accomplish this with the help of an image transformer or by "over"-illumination alone. The obtainable resolution follows directly from Eq. 1.10a:

$$\Delta\lambda = \frac{P_m}{B_\lambda \tau (A\Omega)_{Det}}. \tag{1.17}$$

Since P_m is the minimum acceptable value for the flux, Eq. 1.17 is only true for a certain wavelength λ. No improvement in $\Delta\lambda$ is possible under the above assumptions.

---

*A similar consideration holds also for multiple-pass dispersion systems. See Nolt et al., Appl. Opt. **8**, 309 (1969).

E. Filters

a. Introduction

A grating spectrometer needs filters for the separation of one particular order of the spectrum from the others. This is a consequence of the use of light sources emitting a broad band of frequencies and the property of the grating to diffract into a specific direction of space radiation of wavelength $\lambda_1, \lambda_2 = \lambda_1/2, \lambda_3 = \lambda_1/3$, and so forth (see Section C).

The separation of a particular order of the spectrum is achieved by the use of a bandpass filter; that is, a filter which passes only a wavelength band from λ_c to λ_f and no others. Bandpass filters can be composed of a long-wavelength cut-on filter and a long-wavelength cut-off filter, as shown in Fig. 1.12. The

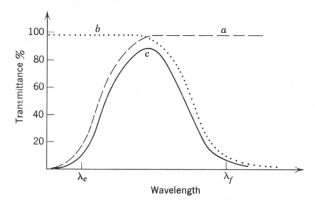

Fig. 1.12   Sketch of the transmission of a bandpass filter c, composed of a long-wavelength cut-on filter a and a long-wavelength cut-off filter b.

cut-on filter attenuates all radiation with wavelengths up to λ_c and lets all radiation with longer wavelengths pass. A long-wavelength cut-off filter lets all radiation with wavelengths up to λ_f pass and attenuates the radiation with wavelengths longer than λ_f. The superposition of these two filters results in a bandpass filter for the wavelength band λ_c to λ_f.

If the first-order spectrum of a grating is used, a long-wavelength cut-on filter is needed to eliminate the second- and higher-order spectra. The grating itself acts as the long-wavelength cut-off filter. This can be seen from the following. Light having wavelengths longer than the path difference in first order (see Eq. 1.1 for $m = 1$) cannot be diffracted by the grating and is reflected from the grating as if it were a plane mirror. Consequently, this portion of the light is not traveling to the exit slit and therefore is eliminated.

This property of the grating; namely, to reflect radiation of wavelengths longer than a certain wavelength (which is related to the grating constant) and to diffract radiation of wavelengths shorter than this certain wavelength, is a useful property for the construction of a long-wavelength pass filter (cut-on filter). (We shall discuss this in Part f of this section).

It is clear that the slope of the long-wavelength cut-on or cut-off should be as steep as possible since the available energy is rather limited in the far-infrared. Also, the filter should have an attenuation as low as possible in its bandpass region.

Considerable attention has been paid to the development and improve-ment of such types of filters for use in the far-infrared. For instance, take the case where a grating is used in such a way that a portion of the spectrum larger than one octave ($\lambda_1 \to 2\lambda_1$) can be scanned by rotation of the grating from position $\theta(\lambda_1)$ to position $\theta(\lambda_2)$, where $\lambda_2 > 2\lambda_1$. A cut-on filter is employed for a cut-on at $\theta(\lambda_1)$ and to serve the useful range; namely, until the grating is in position $\theta(2\lambda_1)$. At that point overlapping would occur. Therefore, another filter with cut-on at $\theta(2\lambda_1)$ has to be inserted for scanning from $\theta(2\lambda_1)$ to $\theta(\lambda_2)$.

b. Attenuation Requirements

If the available radiation had a constant energy distribution, $dE/d\lambda =$ const., a filter having 1 % transmission for wavelengths smaller than the cut-on wavelength λ_c and 95 % transmission for $\lambda > \lambda_c$ would be an acceptable device. The difficulty in the far-infrared, however, is that the amount of radiation in the short-wavelength region is several orders of magnitude higher than that in the long-wavelength range.

We will follow a consideration by Oetjen et al.[23] to estimate the required attenuation which a filter device has to give. A grating of 7.2 lines/mm is considered. It will disperse, at the blaze, radiation of 100 μ wavelength in the first order. Radiation with $\lambda = 50 \mu$, $\lambda = 33.3 \mu$, and so forth will be diffracted into the same direction as second-order spectrum, third-order spectrum, and so forth. The radiation source is assumed to be a black body of temperature 1300°K. By applying Planck's radiation law, we obtain

$$J_\lambda = \frac{C_1}{\lambda^5(e^{C_2/\lambda T} - 1)}, \tag{1.18}$$

where J_λ = radiation power per unit wavelength at wavelength λ and at temperature $T°K$; C_1 is a constant, depending on the units of J_λ and the geometrical conditions, chosen to normalize the radiation intensity at $\lambda = 100 \mu$, and C_2 is a product of universal constants. The intensity reaching the detector is not proportional to J_λ for the higher-order spectra since the dispersion increases proportional to the order number m and, consequently, the spectral interval passing through the exit slit decreases. The radiant power is, therefore, proportional to J_λ/m.

In Table 1.1 the relative intensity for the first seven orders is given. For orders higher than these, contributions of all orders in spectral intervals covering a band 3 μ wide are added and the sum of these is listed as the relative intensity.

TABLE 1.1*

RELATIVE INTENSITIES IN VARIOUS SPECTRAL REGIONS CALCU-LATED ACCORDING TO PLANCK'S RADIATION LAW. AN IDEAL ECHELETTE GRATING WAS ASSUMED, USED AT BLAZE ANGLE

Spectral region in μ	Order of spectrum	Relative intensity
100	1	1
50	2	8
33.3	3	24
25	4	54
20	5	99
16.66	6	161
14.28	7	241
10–13	8–10	1,371
7–10	11–14	3,870
4–7	15–25	25,100
1–4	26–100	145,000
1–100	1–100	175,000

* R. A. Oetjen, W. H. Haynie, W. M. Ward, R. L. Hansler, H. E. Schauwecker, and E. E. Bell, *J. Opt. Soc. Am.* **42**, 559 (1952).

We see from Table 1.1 that the ratio of unwanted to wanted radiation at 100 μ is of the magnitude 1.8×10^5. If only 1 % false radiation can be tolerated, the required attenuation is therefore 1.8×10^7.

For the long-wavelengths regions, the Rayleigh-Jeans radiation law can be used for the computation of the relative intensity of the second, third, or higher orders:

$$\frac{E_{\lambda_1}}{E_{\lambda_m}} = \frac{1}{m^3},$$ (1.19)

where E_{λ_1} is the intensity at wavelength λ_1, E_{λ_m} is the intensity at wavelength λ_m, and m is the order number.

A special application of these principles with respect to the use of a band-pass filter in conjunction with the second-order spectrum can be taken from the data of Table 1.1. If we wish to use the second-order spectrum, the first-order spectrum—which makes about 12.5 % of the second-order energy—

must be blocked out at the long-wavelength side. Of course, the required attenuation of the (large) amounts of radiation of third and higher order is much greater.

All of these considerations treat idealized cases; they do not, for instance, take into account any variation of the grating characteristics with the order of the diffracted spectrum.

c. Elimination of Short-Wavelength Radiation

The short-wavelength radiation can be suppressed by using absorption filters, scatter plates, crystal choppers, or a combination of these. Absorption filters commonly used are black polyethylene and crystal quartz.

Black polyethylene[24] consists of carbon black embedded in polyethylene sheets or films. It is commercially available in sheets of varying thickness (~ 0.1 mm). A transmittance curve of one such sheet in the 3000 to 30 cm^{-1} spectral region is presented in Fig. 1.13. Depending on the other filtering

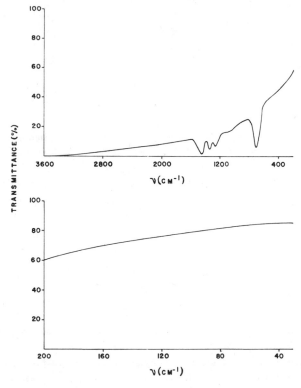

Fig. 1.13   Transmittance of a sheet of black polyethylene in the 3600-30 cm^{-1} spectral region. Thickness 0.1 mm. [K. D. Möller, D. J. McMahon, and D. R. Smith, *Appl. Opt.* **5**, 403 (1966).]

devices in the spectrometer, several sheets are needed to cut down the visible and near-infrared light. In order to avoid reflection losses on the many surfaces, the sheets might be baked together (see Section E.e).

Crystal quartz of several millimeters thickness is a very useful absorption filter for the spectral region from about 5 to 45 μ. The lattice vibrations in this range have strong absorption bands. The cut-on slope at about 40 to 50 μ wavelength depends on the thickness of the plate. However, as is the case for many absorption filters using the lattice vibrations of a single crystal, the cut-on slope is not very steep; therefore, crystal quartz is not a very efficient filter device for the 45 to 90 μ region. Figure 1.14 shows the cut-on slope of some crystal quartz samples of several thicknesses and types.

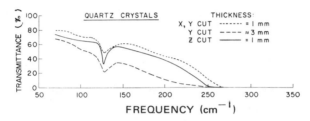

Fig. 1.14   Far-infrared transmittance of natural Brazilian quartz-crystal plates. [R. V. McKnight and K. D. Möller, *J. Opt. Soc. Am.* **54**, 132 (1964).]

Scatter plates are frequently used as reflection filters. Their surface is roughened in such a way that the average size of the rough surface spots is comparable to the wavelength of the unwanted radiation. Longer-wavelength radiation is reflected as by a mirror whereas the shorter-wavelength radiation is scattered in all directions. Some small losses of energy for the long-wavelength radiation and some scattering of the short-wavelength radiation into the "mirror" direction will occur. Scatter plates may be constructed by grinding aluminum plates with emery or carborundum of certain mesh numbers. It is also possible to grind glass plates to the desired roughness and deposit an aluminum layer on the treated surface. Such scatter plates, which show rather steep cut-on slopes, have been used for the 500–240 cm^{-1} region (made with the help of 320 mesh Carborundum) and for the 240–150 cm^{-1} region (using 220 mesh).[6]

Selective modulation of the light which passes through the spectrometer can be used for the elimination of short-wavelength radiation. Since it is at any rate advantageous to modulate the light beam passing through the spectrometer (for convenient amplification with ac amplifiers), one can use the "chopping" process simultaneously for filtering action. The modulator is then manufactured in such a way that it serves as a long-wavelength pass filter. For this purpose the modulator is, for instance, constructed in the form of a rotating half disk consisting of a material which is opaque for the wanted wavelength region and transmissive for the unwanted radiation. The detection system is locked-in to the chopping frequency and thus rejects the unwanted short-wavelength radiation since the latter represents a dc level.

Energy losses due to absorption or reflection of the short-wavelength radiation by the chopper material are compensated for by a series of wire spikes extending into the open section of the chopper. Choppers of NaCl, KBr, and CsI have been successfully used for the partial elimination of short-wavelength radiation up to 15, 25, and 50 μ, respectively.

A periodic interference modulator may also be used as a filtering device (see Appendix VI).

d. Reflection Crystal Filters (Reststrahlen Filters)

Many inorganic crystals exhibit an intense reflectivity in certain wavelength regions of the far-infrared spectrum. The range of strong reflectivity, the so-called reststrahlen band, depends on the material of the crystal. (The theoretical aspects of this phenomenon are discussed in detail in Chapter 12.) Such materials can therefore be used advantageously as convenient reflection filters for the elimination of unwanted radiation. Figure 1.15 gives the reflectivity of some pure alkali halides, of some mixed crystals, and of calcium and barium fluoride. We note that there remains a significant amount of reflectivity beyond the main peak towards longer wavelengths, but that the cut-off towards shorter wavelengths is much sharper (see Chapter 12).

Table 1.2 shows the peak wavelengths and the wavelengths corresponding to a 50% level of the reflectivity on either side of the main peaks for the crystals of Fig. 1.15. The reflectivity for the short-wavelength radiation is about 3–4% but depends on the surface conditions. A very good filter action can be achieved with two of these reflection plates in series. If the two reflection plates are arranged crosswise at the polarizing angle, elimination of almost all the radiation outside the reststrahlen region is accomplished.[25,26] Many far-infrared spectra in various frequency ranges have been obtained with the help of only one of these reststrahlen plates in conjunction with scatter plates, crystal choppers, or black polyethylene. These reflection filters may also be used as bandpass filters.

e. Transmission Crystal-Powder Filters

The strong absorption of the alkali halides and similar compounds can also be utilized in the construction of transmission filters. For thick specimen the absorption band is very broad but with thinner and thinner crystals the band is found to be centered approximately at the wavelength of the reflection

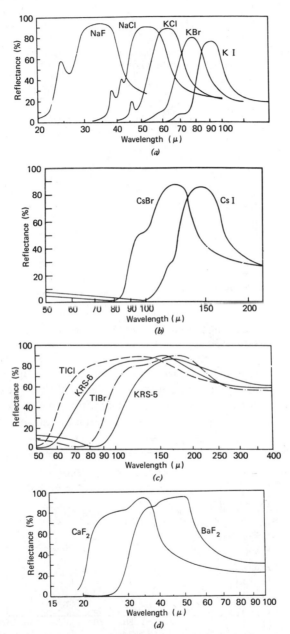

Fig. 1.15   Reflectances of various single crystals. (*a*) NaF, NaCl, KCl, KBr, and KI; angle of
incident light: 12°. (*b*) CsBr and CsI; angle of incidence: 15°. (*c*) TlCl, TlBr, KRS-5, and KRS-6;
angle of incidence: 12°. (*d*) CaF$_2$, BaF$_2$; angle of incidence: 12°. [A. Mitsuishi, Y. Yamada,
and H. Yoshinaga, *J. Opt. Soc. Am.* **52**, 14 (1962).]

TABLE 1.2*

FREQUENCY OF REFLECTION PEAKS AND OF THE 50% REFLEC-
TANCE OF SOME RESTSTRAHLEN CRYSTALS

Reststrahlen crystal	Peak wavelength, μ	Wavelength at 50% level, μ
NaF	34	27–42.5
CaF_2	34	21.3–41.2
BaF_2	45	31–60
NaCl	53	44.3–65
KCl	63	54–72
KBr	79	68.5–88
KI	92	82–101
CsBr	122	95–143
CsI	145	124–170
TlCl	130	63
TlBr	170	95
KSR-6	155	75
KRS-5	170	112

* A. Mitsuishi, Y. Yamada, and H. Yoshinaga, *J. Opt. Soc. Am.*
52, 14 (1962).

peak (see also Chapter 12). Barnes and Czerny[27] investigated thin films of NaCl. From their results one could have predicted that cut-on filters can be constructed by the use of the proper combination of thin films of varying thicknesses and of different materials. The actual fabrication of such types of

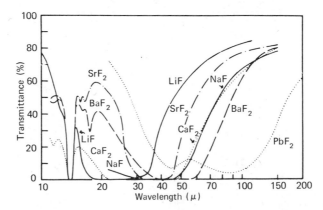

Fig. 1.16 Transmittance of polyethylene sheets containing LiF, NaF, CaF_2, SrF_2, BaF_2, and PbF_2 powder, respectively, as a function of wavelength. [Y. Yamada, A. Mitsuishi, and H. Yoshinaga, *J. Opt. Soc. Am.* **52**, 17 (1962).]

cut-on filters would involve evaporation techniques. Yamada *et al.*[28] introduced the simple method of suspending crystal powders in polyethylene in amounts which would yield the desired transmission characteristics of the filter. For example, Fig. 1.16 shows the transmittance of some of their crystal-powder filters. The filters were 0.3 mm thick and contained the following amounts of crystal powder (in g/2 g polyethylene): LiF (0.25), NaF (0.25), CaF_2 (0.25), SrF_2 (0.2), BaF_2 (0.25), and PbF_2 (0.2). Figure 1.17 shows filters of the same thickness for the following materials: NaCl (0.2), KCl (0.2), KBr (0.4), KI (0.4), TlCl (0.5), TlBr (0.5), and TlI (0.5). Figure 1.18 shows the corresponding plots for some alkali-earth oxides and carbonates: BeO (0.25), MgO (0.25), ZnO (0.25), $MgCO_3$ (0.25), and $CaCO_3$ (0.25).

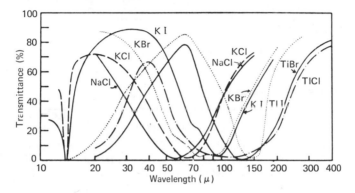

Fig. 1.17   Transmittance of polyethylene sheets containing NaCl, KCl, KBr, KI, TlCl, TlBr, and TlI powders, respectively, as a function of wavelength. [Y. Yamada, A. Mitsuishi, and H. Yoshinaga, *J. Opt. Soc. Am.* **52**, 17 (1962).]

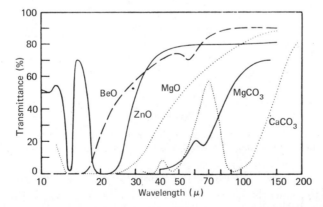

Fig. 1.18   Transmittance of polyethylene sheets containing BeO, ZnO, MgO, $MgCO_3$, and $CaCO_3$ powders as a function of wavelength. [Y. Yamada, A. Mitsuishi, and H. Yoshinaga, *J. Opt. Soc. Am.* **52**, 17 (1962).]

In order to construct a filter which cuts off a broad part of unwanted radiation in the shorter-wavelength region and passes the radiation of wavelength longer than a cut-on wavelength, it is necessary to combine several powders in one filter. Figure 1.19 shows a filter containing BaF_2, SrF_2, and LiF (solid curve) which has a cut-on at 60 μ. The individual transmittance of each powder (in a separate filter) is indicated by the broken curves. Figure 1.20 shows combinations for the 25 to 300 μ region.

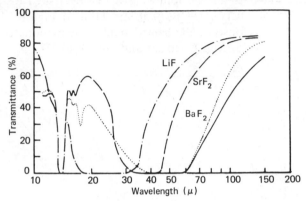

Fig. 1.19   Elimination of the transmittance of BaF_2 in the shorter wavelength region by addition of LiF and SrF_2. The solid curve shows the transmittance of a polyethylene sheet containing BaF_2, SrF_2, and LiF crystal powders. [Y. Yamada, A. Mitsuishi, and H. Yoshinaga, *J. Opt. Soc. Am.* **52**, 17 (1962).]

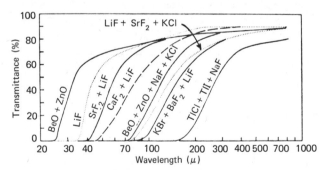

Fig. 1.20   Transmittance of several crystal-powder filters with cut-on wavelength in the 25 to 200 μ region. [Y. Yamada, A. Mitsuishi, and H. Yoshinaga, *J. Opt. Soc. Am.* **52**, 17 (1962).]

The cut-on slope depends on the powder size, the amount of powder contained in a unit thickness of the filter, and the thickness of the filter itself. Figure 1.21 shows the increase of the steepness of the slope as a function of different powder sizes of NaF, whereas Fig. 1.22 displays the dependence of the slope on the amount of crystal powder and on the filter thickness.

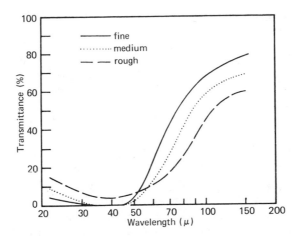

Fig. 1.21   Transmittance curves for NaF powder of different particle sizes as a function of wavelength. [Y. Yamada, A. Mitsuishi, and H. Yoshinaga, *J. Opt. Soc. Am.* **52**, 17 (1962).]

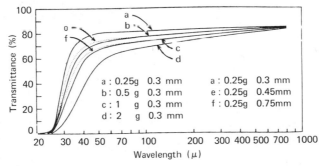

Fig. 1.22   Transmittance curves for various amounts of ZnO powder and thicknesses of polyethylene sheets as a function of wavelength. [Y. Yamada, A. Mitsuishi, and H. Yoshinaga, *J. Opt. Soc. Am.* **52**, 17 (1962).]

Yamada *et al.*[28] have used heated rollers for the construction of these filters. A somewhat different method, using hydraulic presses, was employed by Möller *et al.*[29] The first method offers the possibility of manufacturing large filters (for instance, $16 \times 4 \, cm^2$), whereas the second technique affords a better control of the ratio of the filter components and, in addition, it is faster.

Various recipes for the construction of such filters usable in the 300–150, 160–80, 110–55, 70–35, and 35–18 cm^{-1} regions have been given by Möller *et al.*[29] Similar work was done by Manley and Williams.[30] Zwerdling and Theriault[31] report that low temperatures (77 and 4.2°K) change the cut-on frequency and steepen the slope.

f. Reflection Filter Gratings

As discussed in Section E.a of this Chapter, the echelette grating can only diffract radiation in the first order up to a certain wavelength limit. If the wavelength (at normal incidence) exceeds the grating constant d, the radiation is reflected from the plane of the grating as if there were no grooves present; in other words, all the radiation is reflected into the zero-order spectrum. White[32] has used this property of the grating for a cut-on filter device in the near-infrared region.

The important property of the echelette grating with respect to its use as a filter is that its cut-on characteristics depend only on geometrical parameters. The cut-on point is given by the groove spacing d, the shape of the groove, and the angle of incidence of the radiation. Unfortunately, only infrequent literature reference is made to the dependence of filter characteristics on the grating parameters (for instance, the shape of the grooves) although echelette reflection gratings are frequently used as filters. Peters et al.[33] have investigated the properties of the echelette grating using polarized light, and Genzel[34] has used their measurements to plot the zero-order reflection spectrum from an echelette grating as a function of λ/d, as shown in Fig. 1.23.

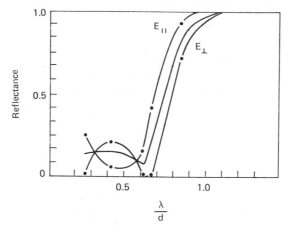

Fig. 1.23   Reflectance of an echelette grating in zero order as a function of λ/d. $E_{\parallel}$ = electric vector parallel to the grooves, $E_{\perp}$ = electric vector perpendicular to the grooves. Solid curve: calculated average. [L. Genzel, H. Happ, and R. Weber, Z. Physik **154**, 1 (1959), by permission of Springer: Berlin-Göttingen-Heidelberg.]

The two curves with dots indicate the reflectance for polarized light (parallel and perpendicular to the grooves, respectively) and the solid curve gives the average; that is, the reflectance for unpolarized radiation.

Hadni[35] has published reflectance vs λ/d data of one of the echelette gratings used in his spectrometer. This is shown in Fig. 1.24. We also refer the interested reader to the recent work of Nelson and Wong[36] (reflection measurements with polarized light; see also Section G.a below). Finally we remark that focusing devices (spherical mirrors) with ruled surfaces have been reported, but their use for filtering purposes has not become widespread.

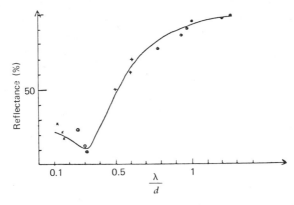

Fig. 1.24   Reflectance in zero order of an echelette grating with grooves in a direction perpendicular to that of the grooves of the diffraction grating. [*A. Hadni, Spectrochim. Acta* **19**, 793 (1963), by permission of Pergamon Press.]

g. Transmission Filter Gratings

The disadvantage of reflection filters is that they must be aligned accurately in the optical system of the spectrometer. The mechanical mechanism for manual or automatic interchange of reflection filters must therefore be constructed very carefully. A type of filter which does not suffer from this drawback is the transmission filter. Its efficiency and characteristics are insensitive to small displacements or tilt; furthermore, it can be placed almost anywhere in the optical path of the spectrometer.

Möller and McKnight[37,38] have shown that transmission filter gratings with a desired cut-on frequency can be fabricated by pressing polyethylene sheets on heated brass plates into which grooves had been machined (using a milling cutter). The heating temperature was between 120–140°C; after

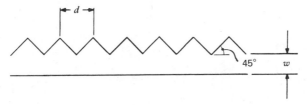

Fig. 1.25   Cross section of a polyethylene transmission filter grating. Grating constant d, step angle 45°, ungrooved thickness w. [K. D. Möller and R. V. McKnight, *J. Opt. Soc. Am.* **53**, 760 (1963).]

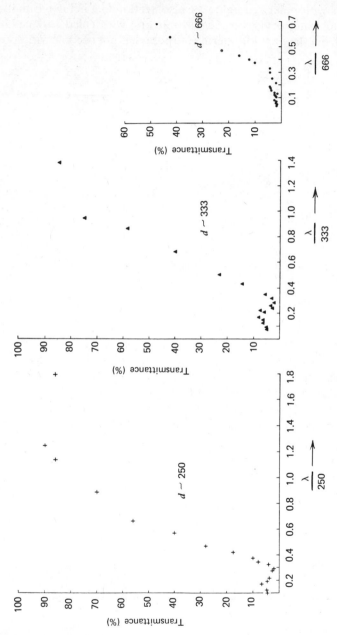

Fig. 1.26 Transmittance of filter gratings with $d \sim 250\ \mu$, $333\ \mu$, and $666\ \mu$ as a function of λ/d. Step angle 45°. [K. D. Möller and R. V. McKnight, *J. Opt. Soc. Am.* **53**, 760 (1963).]

cooling, the filter grating peals easily off its brass template. (For the detailed procedure the reader is referred to Refs. 37 and 38.) Figure 1.25 shows a cross section through such a filter. This filter has a step angle of 45°; the thickness (w) of the ungrooved portion of the filter is between 0.2 and 0.4 mm.

Figure 1.26 shows the transmittance of three such polyethylene transmission filters with spacings of $d \sim 250$, 333, and 666 μ, respectively. The data were obtained by placing the filters into the sample compartment (where the light beam is almost parallel) of a Perkin-Elmer Model 301 spectrometer. The grooves were in a vertical direction, facing the incident light. We see that the cut-on point for all three filters is at $\lambda/d \sim 0.3$.

Möller and McKnight[38] investigated the transmittance of these filters as a function of step angle for asymmetric grooves (Fig. 1.27) and for symmetric

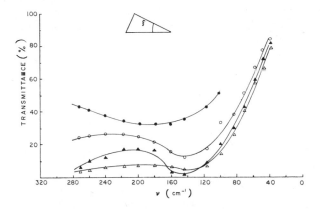

Fig. 1.27 Transmittance as a function of frequency (cm^{-1}) for transmission filter gratings of various asymmetric groove shapes with $\xi = 15°$ (●), 20° (○), 25° (▲), 30° (△). Light is incident on the grooved side and the grooves are in vertical positions, three grooves per millimeter. [K. D. Möller and R. V. McKnight, *J. Opt. Soc. Am.* **55**, 1075 (1965).]

grooves (Fig. 1.28). All filters had 3 grooves/mm; the grooves were aligned vertically and faced the incident radiation. We see that the best filter action is obtained with a symmetric groove of step angle 45°. The transmittance is independent of the relative orientation of the grooves of the filter grating and the grooves of the diffraction grating. Furthermore, it is immaterial whether the light is incident on the grooved or on the flat backside of the filter (see Fig. 1.29). Möller and McKnight did not investigate the polarization effect of these gratings. The filter characteristics are considerably changed if grooves are molded on both sides of the grating at a 90°-angle to each other; the filtering is less efficient.

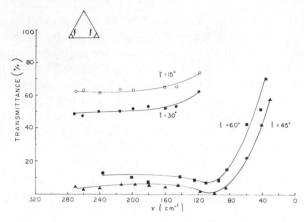

Fig. 1.28 Transmittance as a function of frequency (cm^{-1}) for various transmission filter gratings with symmetrical groove shapes, $\xi = 15°$ ($\bigcirc$), 30° ($\bullet$), 45° ($\blacktriangle$), and 60° ($\blacksquare$); grooves are in vertical positions, three grooves per millimeter. [K. D. Möller and R. V. McKnight, *J. Opt. Soc. Am.* **55**, 1075 (1965).]

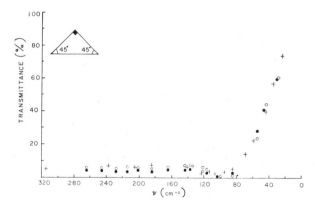

Fig. 1.29 Transmittance as function of frequency (cm^{-1}) for filter gratings in different orientations with symmetrical grooves of $\xi = 45°$ and three grooves per millimeter. Light incident on the flat side, grooves vertical ($\bullet\bullet$) and horizontal ($\bigcirc\bigcirc$). Light incident on the rough side, grooves vertical ($++$). [K. D. Möller and R. V. McKnight, *J. Opt. Soc. Am.* **55**, 1075 (1965).]

A comparison of the efficiency of reflection and transmission filter gratings, published by Möller and McKnight,[38] is shown in Fig. 1.30. The reflectance of one and two reflection filters, taken from the data of Fig. 1.24,[35] is plotted under the symbol R and R^2, respectively (left-side ordinate in Fig. 1.30). On the other hand, the curve with the symbol T gives the transmittance (right-side ordinate) of the transmission filter grating taken from Fig. 1.29.[38]

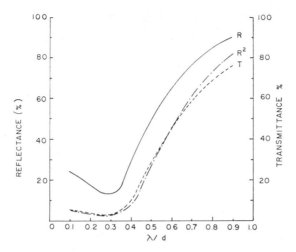

Fig. 1.30   Comparison of the efficiency of reflection and transmission filter gratings: R and R^2 designate the reflectance curves of a single and two reflection filter gratings, respectively, and T is the transmittance of one transmission filter grating. R is taken from Fig. 1.24. [K. D. Möller and R. V. McKnight, *J. Opt. Soc. Am.* **55**, 1075 (1965).]

We note that the curves for R^2 and T correspond closely. This indicates that the filter action of two transmission filter gratings is similar to that of four reflection filter gratings (provided they are of the type and optical mount as reported in Refs. 35 and 38).

In conclusion, we remark that Yoshinaga et al.[3] and Hadni et al.[39] have reported the use of four reflection filter gratings in conjunction with crystal quartz and with black polyethylene for an efficient removal of unwanted radiation in the far-infrared region. The extension of the spectral range of the Perkin-Elmer spectrometer Model 301 from 50 to about 33 cm^{-1} with the help of two polyethylene transmission filter gratings of the type described above (in conjunction with black polyethylene and crystal quartz) has been reported by Rothschild.[40]

h. Metal Mesh Reflection Filters

We now describe another type of reflection filter; namely, metal mesh reflection filters. An early report on metal meshes as filtering devices in the millimeter wavelength region was that by Theissing and Caplan.[41] Renk and Genzel[42] published the transmission characteristics of electroformed metallic meshes used in their Fabry-Perot spectrometer. Ulrich et al.[43] discussed the use of electroformed metal meshes as reflection filters. At about the same time, Mitsuishi et al.[44] reported on the transmission and reflection properties of wire-cloth metal meshes and their applicability as reflection filters.

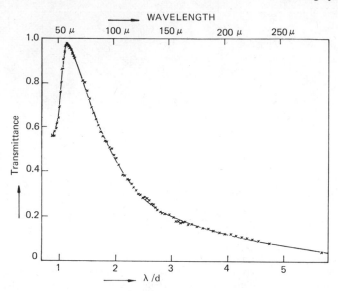

Fig. 1.31 Transmittance of an electroformed (inductive) mesh. The ratio of the grating constant to the strip half-width is 8 : 1, $d = 50.8\,\mu$, $b = 12\,\mu$. [K. F. Renk and L. Genzel, *Appl. Opt.* **1**, 643 (1962).]

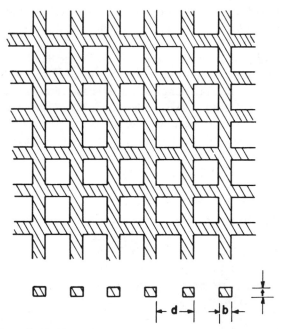

Fig. 1.32 Schematic sketch of an electroformed wire mesh and definition of the parameters d, $b(= 2a)$, and t. [G. M. Ressler and K. D. Möller, *Appl. Opt.* **6**, 893 (1967).]

The transmittance of a nearly parallel beam of light at normal incidence by an electroformed, self-supporting metal mesh in the region of $\lambda/d = 0.8$ to 5.5 is shown in Fig. 1.31.[42] The parameters d, b, and t, as defined in Fig. 1.32, are $d = 50.8\ \mu$, $b = 12\ \mu$, and $t = 6\ \mu$. The maximum transmittance is seen to be about 97 % at $\lambda/d = 1.15$. The transmittance was the same whether the meshes were made of copper or of nickel.

TABLE 1.3*

PARAMETERS OF ELECTROFORMED MESHES d, b, AND t AS EX-
PLAINED IN FIG. 1.32 AND AS PLOTTED IN FIGS. 1.33 AND 1.34
(IN μ)

	#50	#100	#250	#500
d	506	250	99	51
b	60	32	33	18
t	5	10	24.5	4
b/d	0.118	0.128	0.333	0.353
t/d	0.0099	0.0400	0.2480	0.0784

* G. M. Ressler and K. D. Möller, *Appl. Opt.* **6**, 893 (1967).

Ressler and Möller[45] made transmission and reflection measurements in the region $\lambda/d = 0.01$ to 5 on electroformed meshes for several sets of parameters d, b, and t as shown in Table 1.3. The data are depicted in Figs. 1.33 and 1.34. The meshes had been mounted in the diverging beam close to a light-pipe. This particular optical arrangement (in which the light emerges

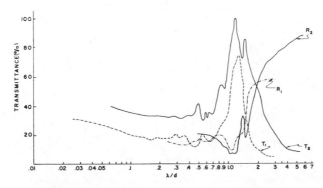

Fig. 1.33   Transmittance and reflectance curves of electroformed meshes #250 (T_1, R_1) and #500 (T_2, R_2) as a function of λ/d. [G. M. Ressler and K. D. Möller, *Appl. Opt.* **6**, 893 (1967).]

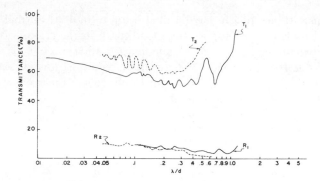

Fig. 1.34   Transmittance and reflectance curves of electroformed meshes #100 (T_1, R_1) and #50 (T_2, R_2) as a function of λ/d. [G. M. Ressler and K. D. Möller, *Appl. Opt.* **6**, 893 (1967).]

under a 30° cross section angle) was different from that used to obtain the data shown in Fig. 1.31. This may account, in part, for the divergence between the transmittance curves depicted in Fig. 1.31 on one hand and Figs. 1.33 and 1.34 on the other hand.

The relative peak transmission of the $d \sim 50\,\mu$ mesh is seen to be almost unity, that of the $d \sim 100\,\mu$ mesh is only 70% (see No. 250 in Fig. 1.33). The b/d value for these meshes is about the same but the t/d value of the $100\,\mu$ mesh is higher than that of the $50\,\mu$ mesh. The thicker mesh has the lower transmission. A 10 to 20% reflectance was observed for the short-wavelength region.

The results of reflection and transmission measurements of wire-cloth meshes[44] of varying d and b values are given in Figs. 1.35 and 1.36. The diameter of the wire, b, and the periodicity constant d are given in Table 1.4.

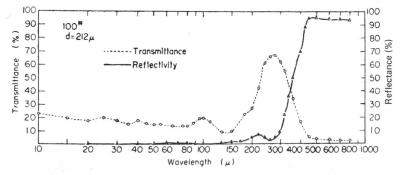

Fig. 1.35   Transmittance and reflectance at 15° incident angle of a #100 metal mesh between 10 and 800 μ. [A. Mitsuishi, Y. Otsuka, S. Fujita, and H. Yoshinaga, *Japan J. Appl. Phys.* **2**, 574 (1963).]

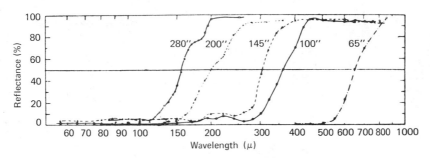

Fig. 1.36   Reflectance of five different meshes at 15° incident angle as a function of wavelength.
[A. Mitsuishi, Y. Otsuka, S. Fujita, and H. Yoshinaga, *Japan J. Appl. Phys.* **2**, 574 (1963).]

TABLE 1.4*

PARAMETERS OF WIRE-CLOTH MESHES AS PLOTTED IN FIGS. 1.35 AND 1.36
(IN μ)

	#280	#200	#145	#100	#65
d	95	127	171	212	384
b	39	46	60	82	177
b/d	0.41	0.36	0.35	0.39	0.46

*A. Mitsuishi, Y. Otsuka, S. Fujita, and H. Yoshinaga, *Japan J. Appl. Phys.*
2, 574 (1963).

(These meshes are manufactured for use as sieves.) Their performance is
comparable to that of electroformed meshes of rectangular cross section if
$t = b$. We see that the transmittance peak of the $d = 212\ \mu$ mesh in Fig. 1.35
is 70%; again, the thicker mesh has lower transmission.

The important application of wire-cloth meshes as reflection filters is
based on their small reflection for the short-wavelength radiation. Figure 1.37
shows this for the $\lambda/d = 0.5$ to 1.5 region. (This holds up to a value of $\lambda/d =
0.05$.[45])

Figure 1.35 shows the steepness of the cut-on slope; a value of 95% reflec-
ance is reached at one-third of an octave after the cut-on point. However,
aside from the type of optical arrangement, such a steep increase of reflectance
depends on the manner in which the wire-cloth is woven and is not encountered in all commercially available materials.[46] Ressler and Möller[45] have
not found cut-on slopes as steep as those reported by Mitsuishi,[44] but (as
mentioned above) their optical arrangement used large-angle converging or

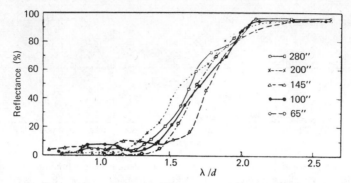

Fig. 1.37   Reflectance of five different meshes at 15° incident angle plotted against λ/d. [A. Mitsuishi, Y. Otsuka, S. Fujita, and H. Yoshinaga, *Japan J. Appl. Phys.* **2**, 574 (1963).]

diverging light at the sample surface. Similarly, Mitsuishi observed a less steep cut-on slope when the angle of the incident light was enlarged from 15 to 52° for a $d = 171\ \mu$ mesh as indicated in Fig. 1.38.

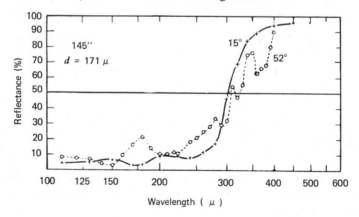

Fig. 1.38   The reflectance of a #145 mesh filter at incident angles of 15 and 52°. [A. Mitsuishi, Y. Otsuka, S. Fujita, and H. Yoshinaga, *Japan J. Appl. Phys.* **2**, 574 (1963).]

In conclusion, the wire-cloth mesh is, in general, a better cut-on filter than the electroformed mesh (see also Vogel and Genzel[47]). However, it has the disadvantages of a reflection filter (which we had outlined earlier in this chapter). On the other hand, adjustment to the desired cut-on point depends only on geometrical parameters. At present, the wire-cloth mesh reflection filter is the most efficient reflection filter in the far-infrared.

i. Bandpass Filters

As we have discussed in Section E.b, the required attenuation for a bandpass filter is considerably higher on the short-wavelength than on the long-wavelength side. A bandpass filter can be constructed from a combination of a high-pass and low-pass filter; in other words, from a filter with a cut-on at a certain wavelength and a filter with a cut-off at a certain longer wavelength. For example, a combination of a crystal quartz and a CsI plate gives a bandpass filter for the 45 to 70 μ spectral region. For the visible and near-infrared light waves, additional filters have to be used as discussed in Section E.c. We note that a bandpass filter can also be constructed from a cut-on filter and a crystal reflection filter (or two reflection filters in series).

The transmission characteristics of metal mesh, as shown in Fig. 1.31, indicate that a bandpass filter can be constructed from a metal mesh used in transmission together with a crystal-powder filter (see Section E.e). Ressler and Möller[45] have made such a bandpass filter for the 70 to 140 μ region, using a crystal-powder cut-on filter for the 8 lines/mm spectral region and two electroformed meshes (of No. 300). The peak transmission of this device was about 50%.

j. Interference Filters

Interference filters of the cut-on, cut-off, and bandpass type have been developed by Ulrich and will be discussed in detail in Chapter 3.

F. CHECKS FOR FALSE RADIATION

If a cut-on filter does not sufficiently attenuate the radiation of wavelength equal to and shorter than one-half of the wavelength which is diffracted in first order by the grating, the radiation of the first-order spectrum is over-lapped with a certain amount of radiation belonging to higher orders. This is usually called the "false" radiation contained in the first-order spectrum.

One way to check false radiation is to use a cut-off filter which passes all short-wavelength radiation up to a certain wavelength λ_c and attenuates all longer-wavelength radiation. A thick CsI plate serves well for checking false radiation in the 70 to 140 μ spectral region of the first-order spectrum of a dispersion grating. No radiation should be recorded when the CsI plate is inserted if the particular cut-on filter satisfactorily attenuates all radiation of $\lambda < 70 \mu$.

Another method, also based on total absorption, is to put a certain alkali halide powder into a matrix (Nujol, polyethylene) and to see if total absorption for a wavelength interval $\Delta\lambda$ can be achieved by using different concentrations of the alkali halide powder.

The same idea of total absorption for a small wavelength interval $\Delta\lambda$ is employed in the absorption of gases. If the pressure in the gas cell can be raised to give total absorption for a spectral width $\Delta\lambda$ at λ, and if no other absorption peaks of width $\Delta\lambda/2, \Delta\lambda/3, \cdots$ are observed at $\lambda/2, \lambda/3, \cdots$, one can conclude that no false radiation is present within the interval $\Delta\lambda$.

Sage and Klemperer[14] have described the use of the pure rotation spectrum of NH_3 as a check for false radiation. Here, if absorption by second-order radiation is present, it shows up as shoulders on the pure rotation lines of the first-order spectrum and as additional lines between the first-order lines.

A different method has been proposed by Möller and McKnight.[37] They checked the false radiation of a first-order spectrum of a mono-chromator (grating constant d) with the help of a grating of grating constant $d/2$. Such a grating diffracts only the second- and higher-even orders of the grating of constant d into the same directions. For a sufficiently effective cut-on filter, a grating with constant $d/2$ should record no radiation if the filters (and slit widths) are set to correspond to those which would be used to record the first-order spectrum of the grating with grating constant d. This method implies that the gratings have the same reflectivities and polarizations. (We saw in Section E that the distribution of the diffracted intensity for the second-order spectrum of the grating with constant d is different from that of the first-order spectrum of the grating with constant $d/2$.) In practice, this method has shown results which agree well with those of the methods mentioned above in regions of the spectrum where a comparison is possible. The method in which the grating of constant $d/2$ is used is particular-ly convenient in the "very" far-infrared where alkali halide crystals and similar materials do not have intense "normal" absorption bands.

We can also check a filter for false radiation with the help of a far-infrared Fabry-Perot interferometer which is tunable over a certain wavelength range. This has been proposed by Ulrich et al.[43] and put into practice by Nelson and Wong.[36] If we were to set the Fabry-Perot to the wavelength λ in first order, radiation of wavelengths $\lambda/2, \lambda/3, \cdots$ can pass the spectrometer in second, third, $\cdots$ order if the filtering action of the filter does not eliminate the short-wavelength regions. However, if we set the Fabry-Perot at $\lambda/2$, $\lambda/3, \cdots$, radiation of wavelength λ will not pass through; therefore the false radiation of $\lambda/2, \lambda/3, \cdots$ can be detected. (This requires that we take the transmission factors of the Fabry-Perot interferometer into account; details on transmission factors will be discussed in Chapter 3.)

G. POLARIZATION OF SPECTROMETER COMPONENTS

A sample can partially polarize the radiation which passes through it. In order to ascertain the magnitude and orientation of this effect, we first need to know how the different optical components of the spectrometer polarize the radiation *per se*. We shall describe the results of studies on the polarization of light by (a) reflection filter gratings, (b) diffraction gratings, and (c) crystal reflection plates (Mitsuishi et al.[48]). The investigation was performed with the help of the spectrometer shown in Fig. 1.5 (Section B.c and Ref. 3), which employs a polarizer made up of a stack of polyethylene plates. (Details on the action of such polarizers will be discussed in Chapter 2, Section E.a.)

a. Reflection Filter Gratings

The results of the polarization measurements are shown in Fig. 1.39. Curve 1 represents the s-component (electric vector parallel to the grooves of the filter grating) and Curve 2 gives the p-component (electric vector perpendicular).* The angle of incidence amounted to 15° and the grating constant of the filter grating was $d = 60\ \mu$. Comparison was performed with the help of an aluminized mirror. Curves 3 and 4 show the same measurements on the same filter grating but with an angle of incidence of 52°. The experimental accuracy of these measurements is claimed to be within 3%.

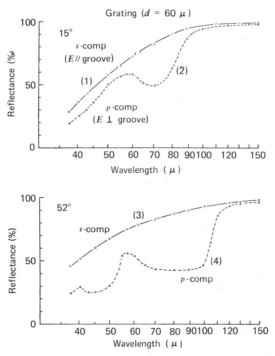

Fig. 1.39   Reflectance of the s-component and p-component of light from a filter grating at incident angles of 15 and 52°. [A. Mitsuishi, Y. Yamada, S. Fujita, and H. Yoshinaga, *J. Opt. Soc. Am.* **50**, 433 (1960).]

* The letters s and p stand for the German "senkrecht" (= perpendicular) and "parallel," respectively. The directions s and p are taken with respect to the plane which is formed by the incident and diffracted light beams.

b. Diffraction Gratings

Two diffraction gratings of constants $d = 79.4$ and $158.8\ \mu$ with a blaze angle of 15° in a Czerny-Turner mount with $2\varepsilon = 15°$ were investigated. Figure 1.40 shows the relative reflectivities $I_s/(I_s + I_p)$ and $I_p/(I_s + I_p)$, where s and p have the meanings defined above. We observe that for wavelengths shorter than the blaze wavelength the s-component is stronger than the p-component, and vice versa for wavelengths longer than the wavelength at the blaze angle. Accuracy of the measurements is claimed to be within 3%.

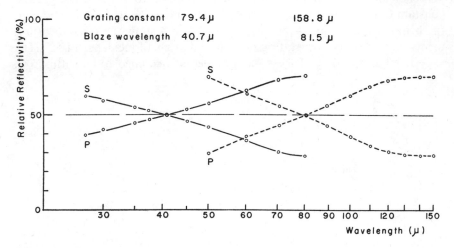

Fig. 1.40   Relative reflectivity of two dispersion gratings in Czerny-Turner mounting. [A. Mitsuishi, Y. Yamada, S. Fujita, and H. Yoshinaga, *J. Opt. Soc. Am.* **50**, 433 (1960).]

c. Crystal Reflection Plates

The polarization of NaCl-plates has been measured for angles of incidence of 12 and 52°. The results are shown in Fig. 1.41. We see that the s-component has a higher reflectivity than the p-component for both angles of incidence. The experimental accuracy is claimed to be within 2%.

H. WAVELENGTH CALIBRATION

a. General

As we have seen, *wavelength* measurements are performed with a grating spectrometer. It is therefore necessary to calibrate the grating spectrometer with respect to this unit. For a rough calibration we can use Eq. 1.2. If we consider $K = 2d \cos \varepsilon$ as an "empirical constant" or calculate it, we have a relation between the radiation wavelength and the angle of rotation of the

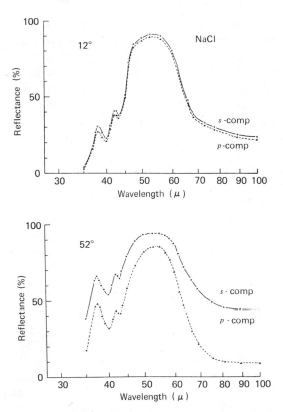

Fig. 1.41   Reflectance of plane-polarized light by an NaCl crystal at an incident angle of 12 and 52°, respectively. [A. Mitsuishi, Y. Yamada, S. Fujita, and H. Yoshinaga, *J. Opt. Soc. Am.* **50**, 433 (1960).]

grating θ (as defined in Section C.b):

$$\lambda = K \sin \theta. \qquad (1.20)$$

The limiting factor for an accurate application of Eq. 1.20 is the accuracy with which the angular scale is known. If the intensity of the radiation is sufficiently large, one can use the $+1$ and -1 order of diffraction and improve the angular scale by making use of its symmetry about the zero position. In a precise calibration the refractive index of the medium and its pressure must be taken into account.

Another possibility is to use absorption lines of known wavelength. The calibration spectrum can be directly superposed on the sample spectrum, and the wavelengths of the sample spectrum are then obtained by interpolation.

An alternative is to use Eq. 1.20 and obtain the K-value with the help of a calibrating gas (or vapor) and the angle θ by observing the rotation of the grating about its axis. If a variety of calibrating substances are used in this method, the value of K should turn out to be the same for all. We note that this procedure is sometimes not without difficulties; for instance, if gratings are to be interchanged for different spectral regions. In this case the requirements on the mechanical grating drive to reproduce a given wavelength at a certain setting are very high (although this is not quite as serious in spectrometers which use a linear wavelength or frequency recording).

A fourth possibility is to employ the higher orders of diffraction of a single strong line (in the visible) of the radiation source—in most cases this is conveniently the mercury lamp; a laser is frequently useful here. Some gratings can give up to the 100th or 200th order of such a line. The line can be detected with the help of a phototube or a solid-state detector mounted at the exit slit. The slit width of the spectrum of the reference line and the sample spectrum may not be the same; the necessary details have been discussed in Refs. 35 and 49.

b. Calibration With Water Vapor

The most commonly employed method of wavelength calibration in far-infrared spectrometers is comparison with a standard. The most frequently used standard is the rotational spectrum of water vapor. The reasons for this are easy to understand. (1) Water is a strongly polar molecule; therefore the intensity of the pure rotational spectrum is large. (2) The intensity distribution of the spectrum of the molecule is such that the far-infrared region is rather well covered by the pure rotational transitions. (3) Because water is an asymmetric top, the number of rotational transitions is large. (4) Preparation and handling of water samples is very easy; in some cases—for instance, in single-beam instruments—one merely admits a small amount of atmosphere into the instrument. Because of the resulting long effective cell path, the partial water vapor pressure can be kept rather low, giving the additional advantage of relatively sharp lines (because pressure broadening is minimized).

The pure rotational spectrum of water vapor in the far-infrared has been mapped by many authors, for instance Benedict,[50] Blaine et al.,[51] Randall et al.,[52] and Rao et al.[49] Figures 1.42 and 1.43 show the data of Rao et al.[49] which are believed to provide wavelength calibrations for the region between 30 and 200 μ (333 to 50 cm^{-1}) with an accuracy within $\pm 0.1\ cm^{-1}$.

Hall and Dowling have recently published the principal rotational constants and a set of centrifugal distortion coefficients of the water molecule in its ground vibrational state with an accuracy higher than reported previously.[53] They obtained these values from a study of the far-infrared

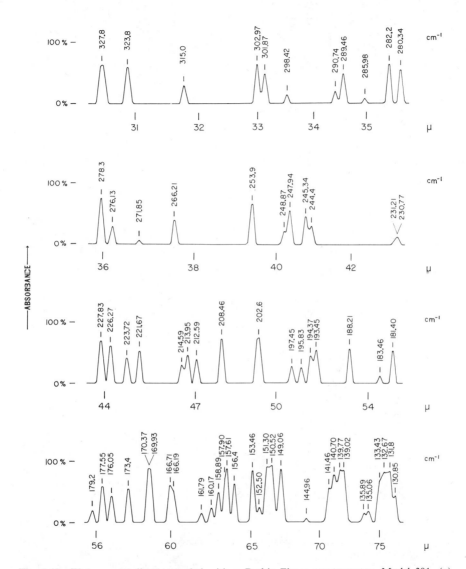

Fig. 1.42   Water vapor lines recorded with a Perkin-Elmer spectrometer, Model 301. (*a*) Spectrometer flushed with dry nitrogen. (*b*) Spectrometer flushed with dry nitrogen; P_2O_5 trays had been kept inside the spectrometer for 12 hours before observations were made. The cm^{-1}-value given for each pure rotational H_2O line refers to vacuum. [K. N. Rao, R. V. de Vore, and E. K. Plyler, *J. Res. Natl. Bur. Stand.* **67A**, 351 (1963).]

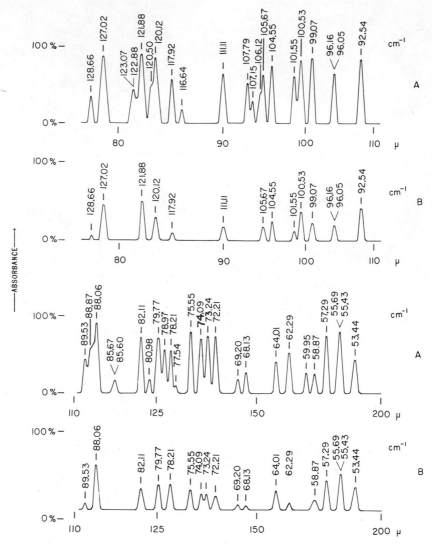

Fig. 1.43   Water vapor lines recorded with a Perkin-Elmer Spectrometer, Model 301 (continued from Fig. 1.42).

rotational transitions in the interval of 5 to 125 cm^{-1} using a high-resolution lamellar-grating interferometer.[54]

For purposes of calibration, the reader will find a table of the frequencies and relative intensities of water vapor lines in the 12 to 305 cm^{-1} region at the end of Chapter 9.

c. Calibration With Methyl Chloride and Other Compounds

Möller *et al.*[12] have shown that the pure rotational spectrum of methyl chloride (CH_3Cl) lends itself well for calibration purposes in the long-wavelength region of the far-infrared (where the pure rotational water transitions are weak). Figure 1.44 shows the $\Delta J = 1$ ($\Delta K = 0$) rotational transitions of CH_3Cl in the region of 9 to $16 \, cm^{-1}$, and Table 1.5 gives the computed transition frequencies for both chlorine isotopic species ($^{35}Cl/^{37}Cl \sim 3$) on the basis of the microwave spectral data of Simmons and Anderson.[55]

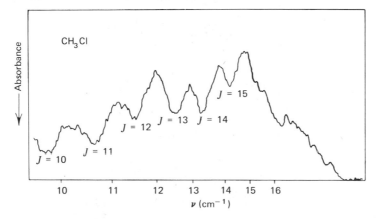

Fig. 1.44   Pure rotational spectrum of methyl chloride obtained with a path length of 120 cm. Pressure: 30 mm Hg. The spectral slit width was $0.21 \, cm^{-1}$ at $10 \, cm^{-1}$. Time constant, 130 sec, scan speed, 0.12°/min. [K. D. Möller, V. P. Tomaselli, L. R. Skube, and B. K. McKenna, *J. Opt. Soc. Am.* **55**, 1233 (1965).]

Methyl chloride is a vapor at room temperature and under atmospheric pressure (bp $-24.2°C$). The compound is easy to handle and is obtainable in high purity (from The Matheson Company, Inc., East Rutherford, New Jersey) in steel flasks under about 5 atm. Since the pure compound is not corrosive, there is *a priori* no objection of filling the spectrometer with it, as was described above for water vapor, unless there are components made of plastic (including cell windows) which could absorb the compound and not release it without prolonged pumping.[56]

Rao *et al.*[49] have recommended the use of the pure rotational transitions of CO, HCN, and N_2O for calibration purposes in the very long-wavelength region of the far-infrared. They calculated (from microwave and near-infrared rotation and rotation-vibration data, respectively) the frequencies of these transitions as given in Table 1.6.

The pure rotational transition of other small molecules such as ammonia,[57] phosphine,[58] hydrogen fluoride,[59] hydrogen chloride,[60] hydrogen bromide,[60] nitrosyl fluoride (NOF) and chloride,[61] nitrogen oxide (NO),[62] and dioxide (NO_2)[63] can also be used for calibration purposes. The smaller of these molecules, for instance hydrogen fluoride and ammonia, show intense rotational transitions far into the 200 to 400 cm^{-1} region whereas the larger molecules of the above series have their lowest transitions at very long wavelengths.

There is, however, little which recommends the molecules just enumerated for calibration purposes. Nearly all of them are reactive, corrosive, and some of them are extremely poisonous when inhaled or brought into contact

TABLE 1.5*

CALCULATED FREQUENCIES FOR THE PURE ROTA-
TION SPECTRUM OF METHYL CHLORIDE

$$\nu = 2B(J + 1) - 4D(J + 1)^3$$

CH$_3{}^{35}$Cl	CH$_3{}^{37}$Cl
$2B = 0.8868$ cm^{-1}	$2B = 0.8731$ cm^{-1}
$4D = 2.4 \times 10^{-6}$ cm^{-1}	$4D = 3.6 \times 10^{-6}$ cm^{-1}

J	ν, cm^{-1}	ν, cm^{-1}
0	0.886_8	0.8731
1	1.77_4	1.746
2	2.660	2.619
3	3.547	3.492
4	4.43_4	4.365
5	5.320	5.238
6	6.20_7	6.11_1
7	7.093	6.983
8	7.979	7.85_6
9	8.86_6	8.72_8
10	9.75_2	9.60_0
11	10.6_4	10.47
12	11.52	11.34
13	12.4_1	12.21
14	13.29	13.0_9
15	14.1_8	13.9_6
16	15.06	14.8_3
17	15.9_5	15.7_0
18	16.83	16.5_7
19	17.7_2	17.43
20	18.60	18.30

* K. D. Möller, V. P. Tomaselli, L. R. Skube, and B. K. McKenna, *J. Opt. Soc. Am.* **55**, 1233 (1965).

TABLE 1.6*

CALCULATED POSITIONS OF THE PURE ROTATIONAL LINES OF CO, N_2O, AND HCN (VACUUM-CM^{-1}). RECOMMENDED CONDITIONS FOR OBSERVATIONS: PATH LENGTH 40 CM; PRESSURE (CM HG) 2–3 FOR HCN AND 40–60 FOR CO AND N_2O

J	$^{12}C^{16}O$	$^{14}N_2{}^{16}O$	$H^{12}C^{14}N$
0	3.84_5	0.83_8	2.95_6
1	7.69_0	1.67_6	5.91_3
2	11.53_4	2.51_4	8.86_9
3	15.37_9	3.35_2	11.82_5
4	19.22_2	4.19_0	14.78_1
5	23.06_5	5.02_8	17.73_6
6	26.90_7	5.86_8	20.69_1
7	30.74_8	6.70_4	23.64_6
8	34.58_8	7.54_2	26.59_9
9	38.42_6	8.38_0	29.55_3
10	42.26_3	9.21_7	32.50_5
11	46.09_8	10.05_5	35.45_7
12	49.93_2	10.89_3	38.40_8
13	53.76_3	11.73_0	41.35_8
14	57.59_8	12.56_8	44.30_7
15	61.42_0	13.40_5	47.25_5
16	65.24_5	14.24_3	50.20_2
17	69.06_8	15.08_0	53.14_8
18	72.88_8	15.91_8	56.09_2
19	76.70_5	16.75_5	59.03_6
20	80.51_9	17.59_2	61.97_7
21	84.33_0	18.42_9	64.91_8
22	88.13_8	19.26_6	67.85_6
23	91.94_3	20.10_3	70.79_3
24	95.74_4	20.94_0	73.72_9
25	99.54_1	21.77_6	76.66_3
26	103.33_4	22.61_3	79.59_5
27	107.12_4	23.44_9	82.52_4
28	110.90_9	24.28_5	85.45_3
29	114.69_0	25.12_2	88.37_9
30	118.46_7	25.95_8	91.30
31		26.79_4	94.22
32		27.62_9	97.14
33		28.46_5	100.06
34		29.30_1	102.98

TABLE 1.6 (contd.)

J	$^{12}C^{16}O$	$^{14}N_2^{16}O$	$H^{12}C^{14}N$
35		30.13_6	105.89
36		30.97_1	108.80
37		31.80_6	111.71
38		32.64_1	114.61
39		33.47_6	117.51
40		34.31_0	120.41
41		35.14_5	
42		35.97_9	
43		36.81_3	
44		37.64_7	
45		38.48_1	
46		39.31_4	
47		40.14_7	
48		40.98_0	
49		41.81_3	
50		42.64_6	

* K. N. Rao, R. V. de Vore, and E. K. Plyler, *J. Res. Natl. Bur. Stand.* **67A**, 351 (1963).

with the skin. Their use is therefore only possible with special handling and filling techniques.

Equations for the computation of the rotational transition frequencies (from the known rotational constants) and their intensities can be obtained from relevant books published on or dealing with pure rotational spectra.[64]

REFERENCES

1. M. R. Holter, S. Nudelman, G. H. Suits, W. L. Wolfe, and G. J. Zissis, *Fundamentals of Infrared Technology*, Macmillan, New York, 1962.
2. L. Genzel and W. Eckhardt, *Z. Physik* **139**, 578 (1954).
3. H. Yoshinaga, S. Fujita, S. Minami, A. Mitsuishi, R. A. Oetjen, and Y. Yamada, *J. Opt. Soc. Am.* **48**, 315 (1958).
4. M. Czerny and A. Turner, *Z. Physik* **61**, 792 (1930).
5. G. R. Rosendahl, *J. Opt. Soc. Am.* **52**, 412 (1962).
6. D. W. Robinson, *J. Opt. Soc. Am.* **49**, 966 (1959).
7. W. G. Fastie, *J. Opt. Soc. Am.* **42**, 641 (1952); **42**, 647 (1952).
8. T. K. McCubbin, Jr., *J. Chem. Phys.* **20**, 668 (1952).
9. A. Hadni and E. Decamps, *Rev. Universelle Mines* **15**, 423 (1959).
10. P. Delorme, Thesis, Paris, 1964.

11. V. Lorenzelli, *Compt. Rend.* **254**, 1017 (1962).
12. K. D. Möller, V. P. Tomaselli, L. R. Skube, and B. K. McKenna, *J. Opt. Soc. Am.* **55**, 1233 (1965).
13. D. R. Smith, B. K. McKenna, and K. D. Möller, *J. Chem. Phys.* **45**, 1904 (1966).
14. G. Sage and W. Klemperer, *J. Chem. Phys.* **39**, 371 (1963).
15. A. Sommerfeld, *Vorlesungen über Theoretische Physik*, Vol. V, Diederich'sche Verlags-buchhandlung, Wiesbaden, 1950, p. 233.
16. J. E. Stewart and W. S. Gallaway, *Appl. Opt.* **1**, 421 (1962).
17. C. H. Palmer, F. C. Evering, Jr., and F. M. Nelson, *Appl. Opt.* **4**, 1271 (1965).
18. J. Strong, *Concepts of Classical Optics*, Freeman, San Francisco, 1958.
19. A. Hadni, C. Janot, and E. Decamps, *J. Phys. Rad.* **20**, 705 (1959).
20. J. M. Munier, J. Claudel, E. Decamps, and A. Hadni, *Rev. Opt.* **41**, 245 (1962).
21. M. Born and E. Wolf, *Principles of Optics*, 2nd ed., Pergamon, London, 1964, p. 406.
22. T. M. Hard and R. C. Lord, *Appl. Opt.* **7**, 589 (1968).
23. R. A. Oetjen, W. H. Haynie, W. M. Ward, R. L. Hansler, H. E. Schauwecker, and E. E. Bell, *J. Opt. Soc. Am.* **42**, 559 (1952).
24. L. R. Blaine, *J. Res. Natl. Bur. Stand.* **67**, 207 (1963).
25. M. Czerny, *Z. Physik* **16**, 321 (1923).
26. J. Strong, *J. Opt. Soc. Am.* **29**, 520 (1939).
27. R. B. Barnes and M. Czerny, *Z. Physik* **72**, 477 (1931).
28. Y. Yamada, A. Mitsuishi, and H. Yoshinaga, *J. Opt. Soc. Am.* **52**, 17 (1962).
29. K. D. Möller, D. J. McMahon, and D. R. Smith, *Appl. Opt.* **5**, 403 (1966)
30. T. R. Manley and D. A. Williams, *Spectrochim. Acta* **21**, 737 (1965).
31. S. Zwerdling and J. P. Theriault, *Appl. Opt.* **7**, 209 (1968).
32. J. U. White, *J. Opt. Soc. Am.* **37**, 713 (1947).
33. C. W. Peters, P. V. Deibel, W. K. Pursley, and T. F. Zipf, *Reports*, U.S. Army Corps of Engineers Contract DA-44-009 ENG 1410.
34. L. Genzel, H. Happ, and R. Weber, *Z. Physik* **154**, 1 (1959).
35. A. Hadni, *Spectrochim. Acta* **19**, 793 (1963).
36. E. D. Nelson and J. Y. Wong, *Appl. Opt.* **6**, 1259 (1967).
37. K. D. Möller and R. V. McKnight, *J. Opt. Soc. Am.* **53**, 760 (1963).
38. K. D. Möller and R. V. McKnight, *J. Opt. Soc. Am.* **55**, 1075 (1965).
39. A. Hadni, E. Decamps, D. Gandjean, and C. Janot, *Compt. Rend.* **250**, 2007 (1960).
40. W. G. Rothschild, *J. Opt. Soc. Am.* **54**, 20 (1964).
41. H. H. Theissing and P. J. Caplan, *J. Opt. Soc. Am.* **46**, 971 (1956).
42. K. F. Renk and L. Genzel, *Appl. Opt.* **1**, 643 (1962).
43. R. Ulrich, K. F. Renk, and L. Genzel, *IEEE Trans.* **MTT-11**, 363 (1963).
44. A. Mitsuishi, Y. Otsuka, S. Fujita, and H. Yoshinaga, *Japan J. Appl. Phys.* **2**, 574 (1963).
45. G. M. Ressler and K. D. Möller, *Appl. Opt.* **6**, 893 (1967).
46. A. Mitsuishi, Private Communication.
47. P. Vogel and L. Genzel, *Infrared Phys.* **4**, 257 (1964).
48. A. Mitsuishi, Y. Yamada, S. Fujita, and H. Yoshinaga, *J. Opt. Soc. Am.* **50**, 433 (1960).
49. K. N. Rao, R. V. de Vore, and E. K. Plyler, *J. Res. Natl. Bur. Stand.* **67A**, 351 (1963).
50. W. S. Benedict, *Theoretical Studies of High-Resolution Spectra of Atmospheric Molecules*, Final Report, AFCRL-65-573, 1965.
51. L. R. Blaine, E. K. Plyler, and W. S. Benedict, *J. Res. Natl. Bur. Stand.* **66A**, 223 (1962).
52. H. M. Randall, D. M. Dennison, N. Ginsburg, and L. R. Weber, *Phys. Rev.* **52**, 160 (1937).
53. R. T. Hall and J. M. Dowling, *J. Chem. Phys.* **47**, 2454 (1967).
54. R. T. Hall, D. Vrabec, and J. M. Dowling, *Appl. Opt.* **5**, 1147 (1966).
55. J. W. Simmons and W. E. Anderson, *Phys. Rev.* **80**, 338 (1950).

56. W. G. Rothschild, *Spectrochim. Acta* **21**, 852 (1965).
57. R. B. Barnes, *Phys. Rev.* **47**, 658 (1935).
58. N. Wright and H. M. Randall, *Phys. Rev.* **44**, 391 (1933).
59. A. A. Mason and A. H. Nielsen, *J. Opt. Soc. Am.* **57**, 1464 (1967); W. G. Rothschild, *J. Opt. Soc. Am.* **54**, 20 (1964).
60. G. Herzberg, *Diatomic Molecules*, Van Nostrand, New York, 1955.
61. J. R. Durig and R. C. Lord, *Spectrochim. Acta* **19**, 421 (1963).
62. R. T. Hall and J. M. Dowling, *J. Chem. Phys.* **45**, 1899 (1966).
63. H. A. Gebbie, N. W. B. Stone, G. R. Bird, and G. R. Hunt, *Nature* **200**, 1304 (1963).
64. G. Herzberg, *Infrared and Raman Spectra of Polyatomic Molecules*, Van Nostrand, New York, 1951; G. Herzberg, *Diatomic Molecules*, Van Nostrand, New York, 1955; C. H. Townes and A. L. Schawlow, *Microwave Spectroscopy*, McGraw-Hill, New York, 1955; J. E. Wollrab, *Rotational Spectra and Molecular Structure*, Academic Press, New York, 1967; T. M. Sugden and C. N. Kenney, *Microwave Spectroscopy of Gases*, Van Nostrand, New York, 1965; W. Gordy, *Microwave Spectroscopy*, Wiley, New York, 1953; E. B. Wilson, Jr., J. C. Decius, and P. C. Cross, *Molecular Vibrations*, McGraw-Hill, New York, 1955; H. C. Allen, Jr. and P. C. Cross, *Molecular Vib-Rotors*, Wiley, New York, 1963.

2 Common Parts of Grating Spectrometers and Interferometers and Some Accessories

A. SOURCES

The mercury-arc lamp and the globar (silicon carbide) are the most frequently used radiation sources in the far-infrared. As early as the beginning of this century, Rubens and coworkers reported that they had employed the mercury-arc lamp as an emitter for "extremely long waves."

A comparison between the emissivity of the two sources was published by Plyler et al.[1] The *mercury-arc lamp* was a Philips HPK-125 W lamp, listed for 3-atm working pressure, 135 V dc, and 0.98 A.* The arc, which emits in the long-wavelength region (probably through the bremsstrahlung of electrons and heavy atoms), burns in a fused quartz envelope of 1 to 2 mm wall thickness. The heating of this envelope generates infrared radiation comparable to black body radiation of 800 to 900°K; the envelope itself is transmissive for long wavelengths of $\lambda > 100\,\mu$. We note that ordinary mercury lamps have an opaque (for the far-infrared) outer envelope which must be removed. The *globar source* used in this comparison was a rod 60 mm long and of 6 mm diameter, operated at 2.5 A and 32.5 V dc. Its temperature, as measured by an optical pyrometer, was found to be 1200°K (which is considerably less than the temperature equivalent of the mercury-arc output).

Figure 2.1 shows the actual data of Plyler et al. for the 50 to 170 cm^{-1} region; the radiant energy is given in arbitrary units. We see that for frequencies smaller than 140 cm^{-1}, the mercury-arc lamp is more efficient in the far-infrared than the globar source.

---

* The commonly used mercury lamp operates on ac current.

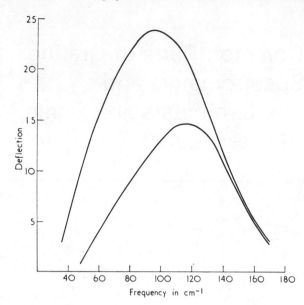

Fig. 2.1   Radiant energy of a quartz-mercury lamp (*upper curve*) and a globar source (*lower curve*) from 50–170 cm^{-1}. [E. K. Plyler, D. J. C. Yates, and H. A. Gebbie, *J. Opt. Soc. Am.* **52**, 859 (1962).]

Hadni *et al.*[2] have investigated the long-wavelength emission peak of mercury lamps as function of the operational conditions; they found that the emission peak shifts under certain conditions.

Despite many attempts to find better continuous radiation sources in the far-infrared, the medium-pressure mercury-arc lamp is still the most useful source. Also, it seems that lamps of a higher wattage rating are not always brighter sources since the emitted radiant energy per area and solid angle is seen to be about the same for lamps of different wattage. However, it might be convenient to illuminate a wider slit width with a larger mercury lamp possessing a broader arc.

To conclude this section we mention the use of "negative light flux" for far-infrared purposes (Arefyev[3]). In this method, used in connection with emission experiments, the detector is at a higher temperature than the radiation source. In the experimental set-up, a mercury-arc lamp and a bismuth bolometer at room temperature were compared with a liquid-nitrogen cooled rod in place of the mercury lamp. (Another convenient detector operating at room temperature is the Golay cell.) At a wavelength of 100 μ, the (negative) signal from the cooled source was only four times weaker than the signal emanating from the hot source. This smaller fraction of useful radiation is believed to be amply counterbalanced by the much smaller effort needed to filter out the unwanted short-wavelength radiation emitted by the cool source compared to the extensive methods required for high-temperature radiation sources (see Chapter 1). For far-infrared emission measurements and techniques, we refer the interested reader to Appendix IV, Section A.b.

B. Detectors

Of the many different methods of detecting far-infrared radiation, we will restrict our discussion to the Golay detector and some helium-cooled solid-state devices. These detectors are commercially available and some of them can be constructed in the laboratory. In the following sections we shall present a short discussion of the far-infrared detectors most commonly used at present. For a more detailed treatment of the underlying basic principles the reader is referred to Appendices IV and V. In Appendix V a discussion of the superconducting bolometer (Martin and Bloor[4]) and the Josephson junction (Grimes et al.[5]), including a description of certain aspects of noise, low-temperature optical, and low-temperature electronic systems, is presented. We also mention that references and discussions regarding the detection of infrared and far-infrared radiation are the subject of a book by Smith et al.[6] and of the June 1965 issue of *Applied Optics* (American Institute of Physics). The field of detection of far-infrared radiation by solid-state devices is constantly expanding, and the exact operating principles (and the specifications) of some of these devices are, at present, not fully understood.

a. Detector Characterization

We would like to recall some definitions used for a characterization of the performance of detectors.[7] *Noise equivalent power (NEP)*. The NEP is defined by

$$\text{NEP} = \frac{V_n}{V_s} H_s A \ [\text{W}], \tag{2.1}$$

where V_n is the voltage observed with no radiation falling on the detector and V_s is the voltage observed with the irradiance H_s [W/cm^2] falling on the detector area A [cm^2]. The noise-equivalent-power "per unit bandwidth" is frequently used in units of W/Hz. *The detectivity (D)* is defined as the reciprocal value of NEP,

$$D = \frac{1}{\text{NEP}} \left[\frac{1}{\text{W}} \right]. \tag{2.2}$$

D-starred (D*) is conveniently defined as

$$D^* = \frac{\sqrt{A\,\Delta f}}{\text{NEP}} \left[\frac{\text{cm}\sqrt{\text{Hz}}}{\text{W}} \right] \tag{2.3}$$

because the detectivity of various detectors inherently contains the factor $\sqrt{A\,\Delta f}$, where Δf is the bandwidth at the observed frequencies.

b. Golay Detector

The Golay pneumatic-cell detector[8,9] is a thermal detector.* Figure 2.2 shows its schematic (The Eppley Laboratory, Inc.) and Fig. 2.3 gives a schematic of the pneumatic cell in the detector head.

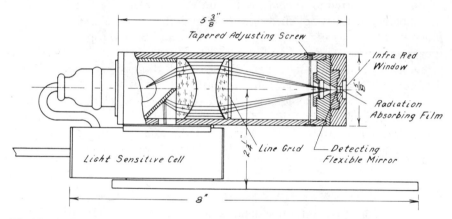

Fig. 2.2   Optical schematic diagram of the Golay infrared detector. [Eppley Laboratory, Inc.]

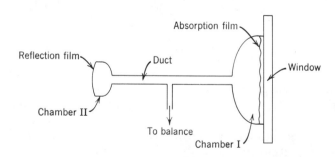

Fig. 2.3   Schematic of the pneumatic cell of the Golay detector head.

* The Golay cell is fabricated in the U.S. by The Eppley Laboratory, Inc. and in Great Britain by Unicam Ltd.

We begin with a description of the detector head. It contains a window, a film to absorb the incident radiation, a reflection film which acts as a flexible mirror, and chambers I and II, filled with xenon. The absorption film consists of a 0.01 μ thick film of collodion on which a thin layer of aluminum has been deposited. The reflection film is similarly constructed, only its aluminum layer is much thicker than that of the absorption film. The working principle is simple: The incident radiation is partially absorbed by the absorption film and the gas in chamber I is heated. The pressure change is conducted through the duct to chamber II; the subsequent movement of the reflection film is sensed by an "optical relay" (see Fig. 2.2) which operates in the following manner: light from a small electric bulb is focused to traverse the upper half of the line grid and then to fall onto the reflection mirror. From there it is reflected through the lower half of the grid and subsequently conducted into the phototube. Let us assume that in the null position of the flexible mirror the lines of the upper half of the grid are imaged between the lines of the lower part: then no light passes from the bulb to the phototube. If radiation now falls onto the detector, the resulting movement of the flexible reflection film will result in a displacement of the image of the upper line grid at the lower line grid. The light of the small bulb can now reach the phototube and the signal is recorded electrically. After the small amount of heat is dissipated, the cell is back in its equilibrium position. This arrangement has a time constant of 10^{-2} sec. Its NEP amounts to 10^{-9} to 10^{-10} W/Hz. Since the deflection of the reflection film can be as small as 1 Å, the NEP is limited by noise due to Brownian motion.

In the Golay detector it is necessary to compensate for slow temperature variations which would change the null position of the mirror-grid arrangement. This is accomplished by attaching a large balance volume, which is not sensitive to rapid pressure variations, to the duct between chambers I and II (see Fig. 2.3). Therefore, the detector is an ac device and the signal must be modulated (and the detector tuned accordingly). Modulation frequencies must be in the range of 5 to 20 Hz.

Since the Golay detector is a thermal detector, its frequency response depends predominantly on the window material. Windows are made of crystal quartz, diamond, or alkali halide crystals of sizes up to 1 cm diameter (the window can also be in the form of a rectangle). By using a diamond window, for instance, radiation from the visible to the millimeter wave region can be detected. Diamond has no first-order allowed infrared-active transitions (see also Chapters 12 and 13).

Figure 2.4 shows an electric schematic. The 2.5 V for the lamp is best supplied by a lead storage battery or a transistorized regulated power supply. The battery for the phototube supplies about 90 V. The signal is proportional to the resistance R, but the capacitance C_L of the connection from the photo-

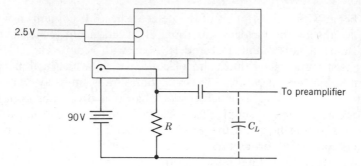

Fig. 2.4   Circuit diagram for the electrical connection of the phototube to the preamplifier.

tube to the preamplifier has to be taken into account. For $R = 1$–10 MΩ, one has to use a high amplification ratio in the amplifier. For $R = 200$ MΩ, a cathode follower must be used but less amplification is required. Proper grounding is essential for the reduction of electrical pick-up.

The Golay detector is used in a lock-in system as is any other detector in the infrared; the radiation falling on the detector is chopped and the ac-output voltage, after amplification, is synchronuously rectified. This type of rectification can be performed with a rectifier operated directly by the chopper, as in the Perkin-Elmer spectrometers (see Chapter 1, Section B.b). But one can also generate, at the chopper, a signal synchronuously chopped with the infrared light and use this signal in the amplifier as a reference for rectification. This phase-sensitive system allows effective discrimination of the signal from the background noise.*

The linearity of the Golay cell has been reported[10] to be better than 1% if the radiation power is not larger than 3×10^{-6} W.

The main advantage of the Golay detector is that it is essentially a frequency-independent far-infrared detector which can be used at room temperature. Its weak point, at present, is that its lifetime is rather limited since the reflection film becomes useless after a certain period of operation and the detector head has to be reconditioned at the factory.

c. Liquid-Helium Temperature Detectors

1. Introduction and General Considerations. Various types of solid-state far-infrared detectors operating at liquid-helium temperatures are now being used at an increasing rate. By their mode of operation, these detectors can be classified into thermal detectors (Section B.c.2) and photoconducting detectors (Sections B.c.3 and B.c.4). This classification is very rough and does

---

* Amplifiers and preamplifiers are offered, for instance, by Brower Lab., Inc., Turnpike Rd., Westboro, Massachusetts 01581, and by Princeton Applied Research Corp., P.O. Box 565, Princeton, New Jersey.

not consider subtle distinctions arising from the various (and sometimes simultaneous) modes and physical principles of operation. However, such a classification is useful for our purposes.

Far-infrared thermal detectors are the carbon resistance bolometer and the single-crystal doped germanium bolometer. A rise in temperature due to the absorption of radiant power alters the electrical resistance of the bolometer material; the change of resistance is converted into a change of potential by the usual electronic apparatus.

Frequently used far-infrared photoconducting detectors are the indium-antimonide detector (Section B.c.3) and the doped germanium photoconducting detector (Section B.c.4).

The indium-antimonide (InSb) detector (Section B.c.3) is particularly useful in the far-infrared region. Its working principle, which is rather complicated, is based on two processes:

1. Because of the high mobility of the electrons and the weak electron-lattice coupling in this semiconductor, the absorbed radiation changes the distribution in energy of the electrons *within* their band. This intraband effect leads to an extremely long wavelength cut-off (microwave region).

2. Because of the presence of very shallow impurity centers in the InSb semiconductor, liquid-helium temperatures alone would not avoid metallic behavior at the technically lowest possible impurity concentrations. However, if (for instance) a magnetic field is applied, the impurity energy levels are separated from the conduction band and the resistance is increased. The InSb detector thus works not only as an "electron bolometer" but also as a photoconducting detector.*

In contrast, in the doped germanium photoconducting detector (Section B.c.4) the incident radiation frees electrons (or holes) from impurity centers into the conduction band of the semiconductor. Depending on the size of the energy gap, there is thus a long-wavelength cut-off in this type of detector (about 120 to 130 μ). It is therefore not of universal use in the far-infrared and we shall discuss it rather briefly.

2. Carbon Resistance and Doped Germanium Bolometers. The characteristics of carbon resistance bolometers operated at liquid-helium temperature have been described by Boyle and Rodgers[11] and Richards.[12] The bolometer of Boyle and Rodgers was a flat slab cut from the core of a 56-Ω resistor. It was mounted directly on a copper backing (with a 0.025 mm Mylar sheet for electrical insulation) in good thermal contact with the helium-bath. The detector assembly, with the relevant filtering devices for far-infrared operation, is shown in Fig. 2.5. To avoid temperature fluctuations of the helium-

* Instead of employing a magnetic field, a helium-cooled step-up transformer can be used (see Section B.c.5).

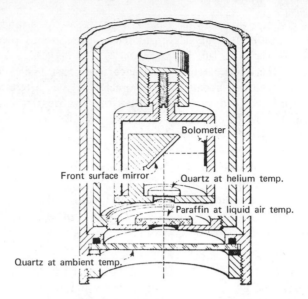

Fig. 2.5   Low-temperature carbon-resistant bolometer assembly for the far-infrared region.
[W. S. Boyle and K. F. Rodgers, Jr., *J. Opt. Soc. Am.* **49**, 66 (1959).]

bath, the temperature of the liquid was suppressed below the lambda point
by pumping; an acoustic filter was installed in series with the pumping line.
The main source of noise in the bolometer was current noise, generally
arising from the contacts. (The lowest amount of such noise was obtained
with gallium amalgam contacts.)

The carbon resistance bolometer described by Richards was a $0.3 \times 5 \times 7$-mm^3 slab, cut from a 50-Ω, 2-W Allen Bradley resistor. The electrodes
consisted of electro-deposited copper. Thermal conductance was obtained
in a manner similar to the bolometer of Boyle and Rodgers. Again, the
current noise of the bolometer was relatively high in comparison to the noise
originating from a good amplifier.

The theory of bolometer performance has been given by Jones,[13] and
optimum operating conditions have been described by Richards.[12] We are
mainly interested in the noise level and the (low-frequency) responsivity, S,
defined as the quotient dE/dQ $[V/W]$, where E is the voltage drop across the
bolometer and Q is the radiant power. S is given by the relation

$$S = \frac{R_L(Z - R)}{2E(Z + R_L)},\tag{2.4}$$

where $Z = dE/dI$ is the dynamic impedance, I is the current, R is the resistance of the bolometer element, and R_L the load resistor. The resistance R at

temperatures below the lambda point of He follows the relation

$$R = A \exp\left(\frac{\Delta}{T}\right), \tag{2.5}$$

where T is the temperature (in °K), and A and Δ are constants describing the characteristics of the bolometer. Values for the carbon resistance bolometer[12] are $\Delta = 8°K$, $R = 30\,k\Omega$, a response time (thermal relaxation time) of $\tau = 10^{-2}$ sec when $I = 10^{-4}\,A$. The temperature of the bath of this bolometer was 1.2°K and the temperature of the bolometer in the absence of the chopped infrared radiation was 2.8°K. Optimum responsivity is obtained with relatively slow bolometers since τ is given by the quotient of the thermal capacity c and the thermal conductance (to the heat bath) κ and it is desirable to have κ small. In practice, c is made as small as possible and κ is subsequently adjusted (by the thickness of the bolometer) to obtain convenient low chopping frequencies ($\omega = 1/\tau$).

Doped germanium (Ge) bolometers are made with the dopents arsenic (As), indium (In),[12] gallium (Ga),[14] or GaAs,[15] among others. The arrangement for the In-doped Ge bolometer is shown in Fig. 2.6. The bolometer was a

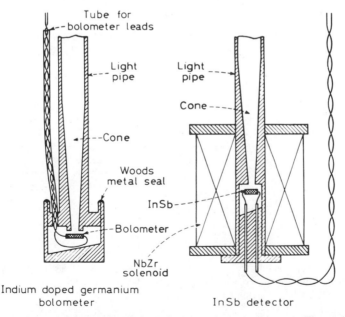

Fig. 2.6 Mounting of far-infrared liquid helium-temperature detectors. The carbon and germanium detectors are operated *in vacuo*, whereas the indium antimonide detector is immersed in liquid helium. [P. L. Richards, International Union of Pure and Applied Chemistry, Molecular Spectroscopy VIII, 1965, p. 535. Copyright and permission by Butterworths, London.]

$0.25 \times 4 \times 4$-mm^3 slab which has a room temperature resistivity of $0.06 \, \Omega$-cm. The 0.18-mm diameter leads were made of copper, attached with indium solder. They also furnished the thermal contact to the bath. The characteristic parameters of the bolometer were $\Delta = 14°K$, $R = 140 \, k\Omega$, $\tau = 6 \times 10^{-3}$ sec $(I = 5 \times 10^{-5}$ A$)$, T_0 (bolometer) $= 1.9°K$, and T_B (He-bath) $= 1.0°K$. The amount of impurity in the bolometer material is rather critical: too much causes Δ to vanish, too little causes the bolometer to absorb the radiation ineffectively.[12] After a slab of Ge with a range of impurities is prepared, the proper piece can be obtained when it has been located by resistivity measurements made at low temperatures.

The carbon resistance and the doped Ge bolometers sometimes show a certain frequency dependence which indicates that not all of the incident radiation is absorbed. This can be reduced by mounting the bolometer into a cavity or on a reflecting backing (see Fig. 2.6).

Low[14] has described a Ga-doped Ge bolometer and compared its performance with the carbon resistance bolometer of Boyle and Rodgers.[11] The measured NEP (W), noise (V), responsivity (V/W), response time (sec), and specific detectivity (defined as the quotient of the square root of the area of the bolometer element and its noise equivalent power value) together with other information are given in Table 2.1.

3. Indium-Antimonide Detector. The InSb detector, whose principle of operation was briefly described above (Section B.c.1), was developed by

TABLE 2.1

SOME CHARACTERISTICS OF THE CARBON RESISTANCE AND THE GALLIUM-
DOPED GERMANIUM BOLOMETERS*

	Detector	
	Doped germanium	Carbon resistance
Noise equivalent power	5×10^{-13}	1×10^{-11}
Noise	2×10^{-9}	1.6×10^{-7}
Responsivity	4.5×10^3	2.1×10^4
Area of bolometer (cm^2)	0.15	0.20
Response time	0.0004	0.01
Specific detectivity[†]	8×10^{11}	4.5×10^{10}
T_0 (°K)	2.15	2.1
Resistance (Ω), R	1.2×10^4	1.2×10^6
Resistance (Ω), R_L	5.0×10^5	3.2×10^6

* *After* F. J. Low, *J. Opt. Soc. Am.* **51**, 1300 (1961).
[†] Of limited significance (see Section B.c.3).

Putley[16] and others. It is a very fast detector, with a response time of 10^{-7} sec. Richards[12] has described a representative InSb detector. It consisted of a $1.8 \times 3 \times 5 \, mm^3$ block of n-type InSb with a carrier density of about $10^{14}/cm^3$. The leads were attached with indium solder and the detector was mounted in a niobium-zirkonide superconducting solenoid as shown in Fig. 2.6.

In general, the responsivity is higher at pumped He temperatures and in a magnetic field for reasons which we have explained in Section B.c.1. The responsivity of the InSb detector relative to the In-doped Ge bolometer (see Section B.c.2), which was set equal to unity, is shown in Fig. 2.7 as a function of frequency and of the magnetic field (in units of kOe). Figure 2.7 also contains the corresponding data on Richards carbon resistance bolometer (see Section B.c.2). We see how an increase in magnetic field and a decrease in

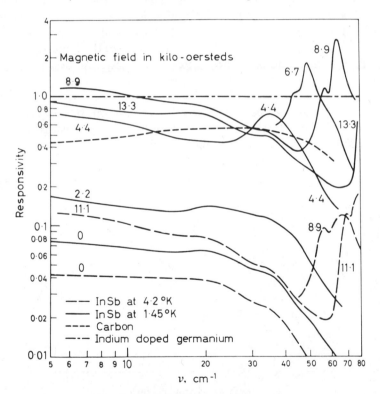

Fig. 2.7 Comparison between the responsivities of the InSb detector and the indium-doped germanium bolometer at various magnetic-field strengths as a function of frequency. The responsivity of the germanium bolometer is set equal to unity. The responsivity of the carbon bolometer is also indicated. [P. L. Richards, International Union of Pure and Applied Chemistry, Molecular Spectroscopy VIII, 1965, p. 535. Copyright and Permission by Butterworths, London.]

temperature increase the responsivity of the InSb detector. The pattern of the curves at lower frequencies is due to transitions between impurity levels with different angular momenta associated with the first and second energy levels of the electrons in the magnetic field (Landau levels).

The performance of the InSb detector is limited by amplifier noise, as was the case for the doped Ge bolometer; quantities such as the specific detectivity as defined in Section B.c.2 are therefore of limited significance. In case of major improvements in detector-amplifier matching, the magnetic field may no longer be advantageous.[12,16]

It should be mentioned that the InSb detector can be used as a tunable infrared detector since the frequency response depends on the magnetic field.[16]

4. Doped Germanium Photoconducting Detectors. Doped photoconducting detectors made from Ge have special applications in the shorter-wavelength region of the far-infrared. As stated in Section B.c.1, the long-wavelength cut-off of these devices is (depending on the nature of the impurity) at 120 to 130 μ. Figures 2.8a and 2.8b show the spectral response of a Ga-doped[17] and B-doped[18] Ge photoconducting detector, respectively. The intrinsic time constant of the detectors are rather small ($\sim 10^{-7}$ sec); however, the effective time constant is usually longer. We explain this further below.

A compilation of doped Ge detectors and their characteristics for the 1 to 120 μ region has been given by Levinstein.[19]

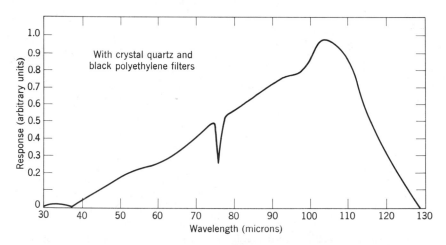

Fig. 2.8a   Spectral response curve, on an equal energy basis, for a gallium-doped germanium photoconducting detector. The behavior at wavelengths shorter than 42 μ is due to absorption by the crystal quartz filter. The same holds for the dips at 76 and 100 μ. [W. J. Moore and H. Shenker, *Infrared Phys.* **5**, 99 (1965), by permission of Pergamon Press.]

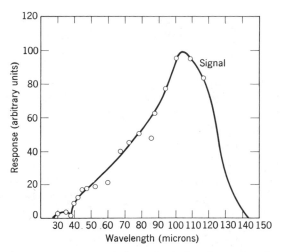

Fig. 2.8*b*   Spectral response, on an equal photon basis, for a boron-doped germanium photo-conducting detector. [H. Shenker, W. J. Moore, and E. M. Swiggard, *J. Appl. Phys.* **35**, 2965 (1964).]

5. Comparison of the Liquid-Helium Temperature Detectors and Problems of Their Use. In this section we compare some of the far-infrared detectors and mention various problems connected with their use.

Obviously, we want a detector of optimum detectivity; that is, one with the highest possible responsivity and the lowest possible noise level. Two additional requirements are sometimes very desirable;[19] namely, a high speed of response and linearity of response over a large range of signal. The latter requirement becomes important in Fourier transform spectroscopy where a wide band of wavelengths falls on the detector. For such applications a thermal detector, such as the doped Ge single-crystal bolometer, has been successfully applied[12,15] (see also Section B.b). In this case it is also necessary to insure uniform absorption over the frequency range, but this is not always so (see Section B.c.2). Improvement may be obtained by covering the bolo-meter element with some form of absorbing black. This increases the thermal capacity c and, therefore, the time constant. An increase of the latter is of no great disadvantage for most spectroscopic purposes.

Concerning the time constant itself, it is possible to use some detectors, notably the indium-antimonide detector, for a study of transient phenomena (response times are about 10^{-6} sec). Although the photoconducting doped Ge detectors (see Section B.c.4) also have a very short intrinsic time response (about 10^{-7} sec), the effective time constant is larger because the capacitance of the high-resistance detectors plus associated circuits makes it difficult to obtain RC times smaller than 10^{-6}; in most cases, the RC times are much longer. Only a reduction of the load resistance below the detector resistance would decrease the RC time; however, there would then be a loss in the detector signal.

With respect to the noise level, we have already said that the amplifier noise is the limiting factor except for the carbon resistance bolometer. It is therefore difficult to make exact comparisons between the merits of the various detectors. In general, the broad-band InSb detector with a magnetic field of 6–7 kOe has been found[16] to be about 40 to 50 times more sensitive than the Golay cell at 1 mm wavelength. This, of course, is little more than a very rough indication since there is a wide variation in the detectivity of Golay cells (besides that among the InSb detectors). Similar values for comparing the NEP among Golay detectors and He-cooled detectors have been given by Smith.[20]

If speed is not the object, the detectivity of the InSb detector can be raised about an order of magnitude by operating the device without a magnetic field; in this case the resistance of the detector is stepped up by means of a He-cooled transformer.[21] This possibility has also been discussed by Richards[12] who showed that the resistance of the detector drops faster than the responsivity when the magnetic field is decreased. The operation of the InSb detector in this mode, however, has only been done with an experimental device. In general, as can be seen from Fig. 2.7, the InSb detector is less useful in the shorter wavelength region.

In conclusion, Fig. 2.9 shows the spectrum of water vapor in the region of 140 to 250 μ scanned with a Perkin-Elmer spectrometer Model 301 (single beam) with (1) a hand-picked Golay detector (NEP $\sim 3 \times 10^{-11}$ W) and (2) a Texas Instruments Ge bolometer system (NEP $= 5 \times 10^{-13}$ W, described in operating manual No. 183522).[22] The bolometer is operated at 4.2°K. The test was performed under the same gain settings and signal levels; the wider slits necessary for the Golay detector are apparent in the loss of resolution. The comparison shows that an improvement in the signal-to-noise ratio of at least three-to-one is easily obtained in the far-infrared under ordinary conditions with this doped Ge detector. Much higher signal-to-noise ratios are to be expected under optimum conditions.[22]

C. LIGHT-PIPES

Light-pipes of uniform diameter have been used in far-infrared investigations to conduct light into a Dewar,[15,23] through a magnetic field,[12] to serve as low-volume long-path gas cell,[24] or to circumvent some of the difficulties arising in the adaptation of sample optical arrangements (as mentioned in Chapter 1, Section B.e). The use of light-pipes for low-temperature solid-state work offers the opportunity to place the detector, the sample, and the filters

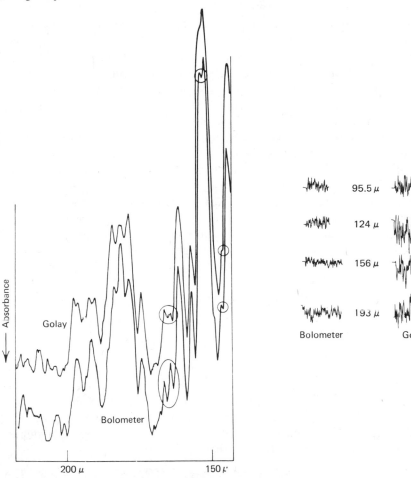

Fig. 2.9 Water vapor spectrum between 150 and 250 μ, taken with a Perkin-Elmer spectrometer, Model 301, equipped with a Golay detector (Eppley Laboratories, Inc.) and with a gallium-
doped germanium detector (Texas Instruments, Inc.), respectively. A comparison of the signal/
noise ratio at various fixed wavelengths is also given. [C. E. Jones, Jr., A. R. Hilton, J. B. Damrel,
Jr., and C. C. Helms, *Appl. Opt.* **4**, 683 (1965).]

(and any magnet) into one helium-cooled Dewar. An example is presented
in Fig. 2 of Appendix III.

It will be helpful to give a few definitions and explanations. The rays which
are reflected through a light-pipe in such a way that they remain in a fixed
plane which goes through the central axis of the pipe are called "nonskew
rays." All other rays passing through the pipe are designated as "skew
rays." The path lengths of the skew rays are longer than those of nonskew

rays (which can be seen by considering light conducted from an image at one side of the pipe to the other). Obviously, light-pipes are not image-forming.

The use of light-pipes as low-volume absorption cells depends on the large contribution from skew rays; the effective absorption length exceeds the physical length of the tube. However, light-pipes do not lend themselves easily to photometric measurements (where an exact path length is needed) since the theoretical treatment of the effect of skew rays is difficult.

Light-pipes are advantageously employed for temperature studies. They are conveniently heated and there is no displacement of components as in long-path cells using the principle of multiple reflections by mirrors (see Section F).

We next give a résumé of a simplified theoretical study on the attenuation of far-infrared nonskew light rays passing through a light-pipe (Ohlmann et al.[25]). It is assumed that light of wavelength λ passes under m reflections with grazing angle α (α small) through a light-pipe of length L and uniform diameter d. It can then be shown[26] that the coefficient of reflection per reflection is given approximately by

$$R_s \approx 1 - 2x\alpha \qquad (2.6)$$

for light polarized perpendicularly to the plane of incidence, and by

$$R_p \approx \frac{2\alpha^2 - 2x\alpha + x^2}{2\alpha^2 + 2x\alpha + x^2} \qquad (2.7)$$

for light polarized in a direction which is parallel to the plane of incidence. Here $x = 0.18(\rho/\lambda)^{1/2}$, ρ is the resistivity in Ω-cm, and λ is in cm.

For a large number of grazing reflections, we can recast Eqs. 2.6 and 2.7 into continuous functions of m by setting $m = L\alpha/d$. In the region of interest, $x \ll \alpha \ll 1$, we find for $R_s{}^m$ and $R_p{}^m$ that

$$R_s{}^m \approx e^{-2(xL/d)\alpha^2} \approx e^{-2xm\alpha}, \qquad (2.8)$$

$$R_p{}^m \approx e^{-2(xL/d)} \approx e^{-2(xm/\alpha)}. \qquad (2.9)$$

We see that both coefficients decrease exponentially with the number of reflections, but that $R_p{}^m$ decreases faster with m than $R_s{}^m$.

For unpolarized radiation, the reflection coefficient is defined, per m reflections, by[25]

$$R(\alpha, m) = \frac{R_s{}^m + R_p{}^m}{2}. \qquad (2.10)$$

In order to find the total transmission of the pipe, we average $R(\alpha)$ over the solid angle. We thus obtain

$$T = \frac{2}{\alpha_m{}^2} \int_0^{\alpha_m} \alpha R(\alpha) \, d\alpha, \qquad (2.11)$$

where α_m is the maximum grazing angle of the incident radiation.

In a system with f-number $f = 1/\gamma$, we have $\alpha_m = (\tfrac{1}{2})\gamma$ and, using the abbreviation $q = xL/d$, we arrive at[25]

$$T = \tfrac{1}{2}e^{-2q} + (1 - e^{-q/2\gamma^2})\frac{\gamma^2}{q} \approx \tfrac{1}{2}(1 + e^{-2q}) - \frac{q}{8\gamma^2}. \qquad (2.12)$$

In Fig. 2.10 we show a plot of T as a function of q for $f/3$ and $f/1.5$, respectively. Ohlmann *et al.*[25] compared their computations with experimental observations on light-pipes of brass, copper, aluminum, and silvered glass at 70 and 140 μ. The results, in terms of transmittance, are presented in Fig. 2.11.

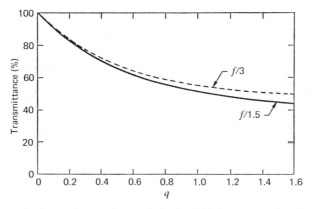

Fig. 2.10   Theoretical transmittance characteristic of metal light-pipes as a function of $q = xL/d$. [R. C. Ohlmann, P. L. Richards, and M. Tinkham, *J. Opt. Soc. Am.* **48**, 531 (1958).]

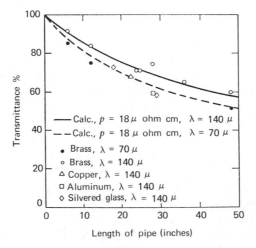

Fig. 2.11   Observed transmittance of brass, copper, aluminum, and silvered glass pipes. The input radiation had an $f/1.5$ aperture. [R. C. Ohlmann, P. L. Richards, and M. Tinkham, *J. Opt. Soc. Am.* **48**, 531 (1958).]

The brass pipes with an internal diameter of 0.43 in. (about 11 mm) were taken as standards. For pipes with different diameters, the length were scaled according to $L^* = (0.43/d)L$.

The resistivity of a brass pipe which led to the best fit between calculations and measurements was found to be 18 $\mu\Omega$-cm; the resistivity of the bulk material is only 6.4 $\mu\Omega$-cm. Part of this difference is certainly a consequence of neglecting the skew rays. Other sources leading to disagreement between calculations and experiments arise from the neglect of the wavelength distribution of the light and the imperfect surface of the inside wall of the pipes. The importance of the latter effect was demonstrated by a comparison of a badly oxidized pipe with an acid-polished pipe; the oxidized pipe was 10 to 20 % less transmissive.

In most cases the Dewar which cools the sample or detector is in an upright position whereas the light beam which leaves the spectrometer is horizontal. In this case one has to conduct the light around a corner. According to Ref. 25, it is more advantageous to use a 90° corner made out of two light-pipes and a reflecting surface (reflecting the light from one light-pipe into the other) than to bend the pipe, even with a large radius. The former arrangement conducts about 95 % of the light around the corner, the latter only 50 % (see Figs. 1 and 2 in Appendix III).

D. Conical Light-Pipes (Channel Condensers)

A conical light-pipe (or "cone channel condenser") instead of an elliptical or toroid mirror can be used as an image reducer. Conical light-pipes are especially convenient for helium-cooled detectors.[12,15,16,23-25,27-29] Williamson[29] has reported an investigation on the nonskew rays in a conical light-pipe and has derived a relationship between the length of the conical light-pipe x, the radii of the larger opening s and the smaller opening c, and the angle of the incident light V. These parameters are illustrated in Fig. 2.12. It is assumed that the light is focused to an image at the larger opening of the conical light-pipe, that the detector is placed in the smaller opening, and that the emerging light-cone fills the solid angle 2π.

If the shape of the conical light-pipe is too steep, not all the incident light passes through it. If the conical light-pipe is not steep enough, the light leaves with a solid angle smaller than 2π. This latter condition may be of some use in a Golay detector if we wish to have space between the detector window and the smaller opening of the cone. We shall now outline Williamson's results and the general manner of their derivation by using Fig. 2.13. Instead of drawing the light rays at various angles, the position of the light-pipe is displaced as shown. We see that the points c, b, e, and so forth define a circle. Consider the ray AA. We note that it will never leave the conical light-pipe

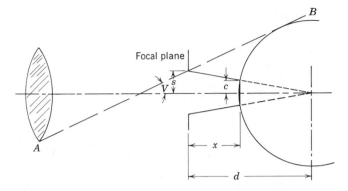

Fig. 2.12 Design layout for a cone channel condenser system. [D. E. Williamson, *J. Opt. Soc. Am.* **42**, 712 (1952).]

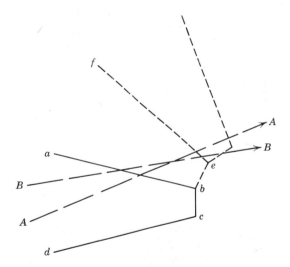

Fig. 2.13 Geometrical method of ray tracing in a channel condenser. [D. E. Williamson, *J. Opt. Soc. Am.* **42**, 712 (1952).]

at the smaller opening but will travel backwards after having reached a certain point. The ray *BB* will leave the conical light-pipe after two reflections. Let these rays represent the extreme rays of a cone of light incident on the larger opening of the conical light-pipe. The extreme ray has the largest value of *V*. Only a fraction of the cone of light limited by the *AA* ray as the extreme ray will pass through the conical light-pipe. The cone of light limited by the *BB* ray as the extreme ray will leave the conical light-pipe with a solid angle smaller than 2π. On the other hand, if an incident cone of light has extreme rays according to the ray AB of Fig. 2.12, the cone of light leaving the smaller opening of the conical light-pipe will fill the solid angle 2π. If the detector is placed at the smaller opening of the conical light-pipe, optimum conditions are achieved with respect to low reflection losses and total light gathering capability with a minimum length of the conical light-pipe.

By using the geometrical parameters defined in Fig. 2.12, we arrive at the following formula for the minimum length of the conical light-pipe:

$$X = \left(1 - \frac{c}{s}\right)\frac{s\cos V}{(c/s) - \sin V}. \tag{2.13}$$

We have mentioned that the emerging light cone should have a solid angle smaller than 2π if the detector is not placed at the smaller opening of the conical light-pipe but at a small distance from it. Equation 2.13 is then modified to[27]

$$X = \left(1 - \frac{c}{s}\right)\frac{s\cos V}{(c/s)\sin\alpha - \sin V}, \tag{2.14}$$

where α is shown in Fig. 2.14. For $\alpha = 90°$ (see Fig. 2.12), Eq. 2.14 reduces to Eq. 2.13, as it should. Williamson[29] also treats the case where a field lens is positioned at the larger opening of the conical light-pipe.

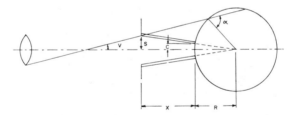

Fig. 2.14   Conical light-pipe. V, angle between the extreme ray and center ray of incoming light; α, angle between the extreme ray and center ray of outgoing light; s and c, larger and smaller radii of cone opening; X, length of the cone. [K. D. Möller, V. P. Tomaselli, L. R. Scube, and B. K. McKenna, *J. Opt. Soc. Am.* **55**, 1233 (1965).]

The conical light-pipe may be fabricated by an electroforming process. A core of stainless steel of the proper shape is plated first with cadmium and then with a silver deposit about 0.25 mm thick. Copper is deposited on this layer to form the body of the conical light-pipe. After the electroforming process is completed, the cone is heated and the stainless steel core is removed. The cadmium layer is dissolved with acid and the inner silver surface is polished. Another technique is simply to obtain a suitable reamer and, after having removed enough material from the inside of a brass or copper rod, use the reamer to machine the cone.

Ohlmann[25] reports that only 40% of the incident light was conducted into the Golay detector when he used a cone of brass of the following dimensions: $x = 2$ in. (~ 5 cm), $2s = 0.43$ in. (~ 11 mm), and $2c = 0.125$ in. (~ 3.13 mm).* He established these values by permitting the light to pass directly onto the Golay detector head via a 0.43 in. (~ 11 mm) uniform diameter light-pipe and, on the other hand, by interspersing the conical light-pipe between the light-pipe and the Golay detector head.

Loewenstein and Newell[30] have reported on ray tracing through straight light-pipes, conical light-pipes, and (mitered) right-angle corners. Skew rays as well as nonskew rays were taken into account.

For conical light-pipes it was found that the sine condition is obeyed to a good approximation, regardless of the condenser length. The conical light-pipes should be kept as short as possible to minimize reflection losses. Right-angle corners were shown to be superior over ideal toroidal right-angle bends in all cases of practical interest.

E. POLARIZERS

Polarized light is required for many far-infrared investigations; for instance, in the study of the polarization which is produced by the different components of a spectrometer or interferometer, in the magneto-optical studies of semiconductors[31] (see Appendix IV), in the reflection or transmission measurements of the optical constants under an oblique angle of incidence (see Chapter 12), and for the assignment of vibrational modes in crystals (see Chapter 13). Most polarizers employed in the far-infrared are in principle similar to corresponding devices in the visible or near-infrared spectral regions. Of course, it is necessary to find suitable materials which are transmissive in the far-infrared. The manipulation of polarized light is similar to that in the shorter-wavelength regions and will not be discussed here.

a. Linear Polarizers

1. Pile-of-Plates Polarizer. Mitsuishi et al.[32] have published a report on the fabrication and the use of a pile-of-plate polarizer made of polyethylene sheets based on the following working principle. If light is incident on one polyethylene sheet at Brewster's angle, the electric vector of the reflected light has a reduced component in the plane of incidence. Therefore, the transmitted light is partially polarized. Since this polarization process is repeated on each sheet, linear-polarized light can be obtained. The index of refraction of polyethylene in the far-infrared is about 1.46; the required angle of incidence is then $i_B = 55°$ according to the well-known relation $\tan i_B = n$.

---

* The beam opening was estimated to be about $f/1.5$.

One polarizer was made up of 15 sheets, each sheet having a thickness of about 50 μ. Through the use of such a device, a degree of polarization of 97 % was observed in the wavelength region around 34, 47, and 80 μ; the corresponding transmission was higher than 75 % of the incident plane-polarized light. Between these wavelength regions, at 40 and 60 μ, second- and third-order interference peaks in the polyethylene sheets were observed. This reduced the polarization to approximately 35 and 15 %, respectively.

Considerably more successful was the construction of a pile-of-plate polarizer using sheets of two different thicknesses. This is demonstrated in Fig. 2.15. The polarizer was constructed of 24 sheets, 12 each of 20 and 30 μ

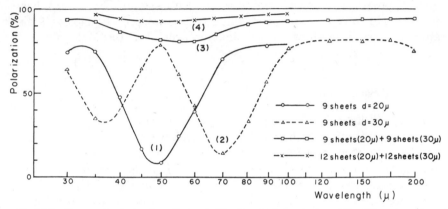

Fig. 2.15   Degree of polarization obtained with piles-of-plate polarizers. The different numbers and thicknesses of polyethylene sheets are indicated. [A. Mitsuishi, Y. Yamada, S. Fujita, and H. Yoshinaga, *J. Opt. Soc. Am.* **50**, 433 (1960).]

thickness (see Curve 4 in Fig. 2.15). The resulting polarization was about 95 % over most of the wavelength region. The transmittance of the incident plane-polarized light again was more than 75 %. This polarizer is easy to construct; however, sometimes its physical dimensions may be too large for easy application.

2. Wire-Grating Polarizer. A wire-grating polarizer (used by Rubens, see Bibliography) transmits only components of the incident electric vector which are oriented perpendicularly to the direction of the wires. Modern forms of grating polarizers are made from gold lines deposited on Mylar film.*

Rowntree[33] has published a report on the use of such a polarizer. The "wires" had a width of 0.01 mm; their centers were 0.025 mm apart. The

---

* Buckbee Mears Company, 245 6th Street, St. Paul, Minnesota. Mylar is polyethylene terephthalate.

polarizer transmitted more than 80% of the perpendicularly oriented component of the incident radiation at a frequency of 50 cm^{-1}. The transmittance dropped to 40% at 100 cm^{-1}, probably due to absorption by the Mylar film. The transmittance of the parallel component of the incident light was reported to be less than 2% at frequencies below 150 cm^{-1}.

This "sheet" polarizer has the advantage that it can be located at almost any place in the optical path and that even two polarizers will not take up too much space. A disadvantage is that Mylar is birefringent.

Another type of grating polarizer can be constructed, according to Bird and Parrish,[34] from an echelette transmission grating as follows. The edges of an echelette transmission grating are covered by metal by means of vacuum sputtering (see Figs. 2.16 and 2.17). As discussed in Section E.g of Chapter 1,

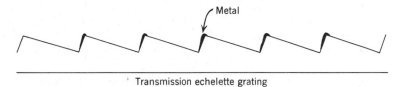

Transmission echelette grating

Fig. 2.16 Schematic of a transmission echelette grating, with edges covered with metal, for use as wire-grating polarizer.

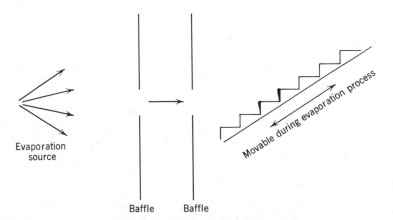

Fig. 2.17 Schematic of the metal-coating procedure for the edges of an echelette grating.

if the wavelength of the light traversing the grating is much longer than the groove spacing, the edges have no effect on the transmission and can be used for the deposition of the metal, thus forming the "wires."

Hass and O'Hara[35] have fabricated and studied this type of polarizer in the far-infrared. Two different substrates were used for the echelette gratings.

One was made of polymethyl methacrylate (DP) of 0.051 mm thickness and an area of $5.1 \times 5.1 \ cm^2$ with 2160 grooves/mm and a blaze angle of 20°.* The other was made of polyethylene (NRL) of 0.152 mm thickness with 600 grooves/mm and produced as replica of a reflection echelette diffraction grating.

Figure 2.17 shows schematically how the metal is deposited on the transmission grating by a metal-sputtering process. Collimators, made out of razor blades, assured a narrow and guided stream of metal vapor. The grating itself was placed at near-grazing incidence.

The experimental measurements using these devices were evaluated with proper consideration of the imperfect polarization characteristics (Rupprecht et al.[36]). The results obtained are shown in Table 2.2 in terms of the degree of polarization (100% = perfect, 0% = none) and the transmittance of the linearly polarized light.

TABLE 2.2*

COMPARISON OF THE PERCENTAGE OF POLARIZATION AND THE TRANS-
MITTANCE FOR TWO POLARIZERS AT VARIOUS FREQUENCIES (CM^{-1})

	Degree of polarization		Transmittance	
cm^{-1}	DP	NRL	DP	NRL
2.5	99.0		0.985	
49.5	97.8	96.4	0.86	0.87
83	98.8	97.9	0.86	0.88
160	98.9	98.0	0.86	0.83
300	98.1	96.6	0.65	0.84

* M. Hass and M. O'Hara, *Appl. Opt.* **4**, 1027 (1965).

3. Pyrolytic Graphite Transmission Polarizer. Rupprecht et al.[36] have fabricated a transmission polarizer from a piece of pyrolytic graphite. The piece was cut from a thin foil of pyrolytic graphite containing the c-axis of the crystal. The foil was then ground to a thickness of about 10 μ. The polarization action is due to the fact that an electromagnetic wave with its electric vector oriented parallel (perpendicular) to the c-axis has a minimum (maximum) of reflection. (For details on the performance of this device, the interested reader may refer to Ref. 36.) This polarizer has the advantage that it can be used in highly convergent or divergent beams. On the other hand, it is very brittle (since the foil must be thin) and thus difficult to handle.

* Diffraction Products, Inc., 344 Lathrop Ave., River Forest, Illinois.

Furthermore, its small diameter (about 13 mm) may be disadvantageous in some applications.

b. Circular and Elliptical Polarizers

1. Soleil Compensator. For the production and analysis of circular- or elliptical-polarized light, a linear polarizer and a retardation plate can be used. It is convenient if the retardation plate can be adjusted to give various phase retardations. A Soleil compensator using crystal quartz has been fabricated and investigated in the far-infrared by Palik.[37] Figure 2.18 shows

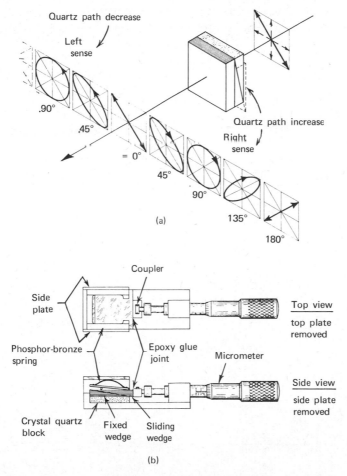

Fig. 2.18 Soleil compensator: (*a*) left-hand and right-hand senses of elliptically polarized radiation, produced by a Soleil compensator; (*b*) top and side view. [E. D. Palik, *Appl. Opt.* **4**, 1017 (1965).]

the optical arrangement. (The optical axis of the birefringent quartz crystal is indicated as a line if it is to be considered to lie in the plane of the figure, and as a dot if it lies perpendicular to it.)

The compensator is constructed from two blocks of crystal quartz of dimensions $2 \times 2 \times 0.477$ cm^3. One block is cut into two wedges. The wedge on the top may be moved by a micrometer, thus varying the thickness of the resulting parallel quartz plate. The total thickness of the compensator may be varied from 0.759 to 0.954 cm (null position) to 1.024 cm. The plates have to be flat to slightly better than one sodium-light fringe and plane-parallel to within less than 20 sec of angle.

If d_1 is the effective thickness of the variable first plate (in a certain position) and d_2 the thickness of the second plate, the phase retardation is obtained as

$$\delta = \frac{2\pi}{\lambda} d |n_0 - n_e|, \tag{2.15}$$

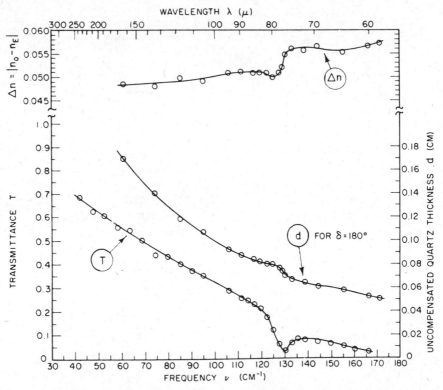

Fig. 2.19   Transmittance of a Soleil compensator in null position; the uncompensated quartz thickness d to produce a 180°-phase shift and the difference in the index $\Delta n = |n_0 - n_e|$ as a function of frequency in the far-infrared are also indicated. [E. D. Palik, *Appl. Opt.* **4**, 1017 (1965).]

where $d = d_1 - d_2$ and n_0 and n_e are the indices of refraction of the ordinary and extraordinary ray, respectively. If $d_1 > d_2$, the elliptically polarized light is polarized in the right-hand sense; if $d_1 < d_2$, it is polarized in a left-hand sense (see Fig. 2.18).

The refractive indices of crystal quartz at zero frequency are $n_0 = 2.102 \pm 0.002$ and $n_e = 2.153 \pm 0.002$ (Roberts and Coon[38]).

For higher frequencies, the indices of refraction and the value $n_e - n_0$ have been determined by Russell and Bell.[39] Figure 2.19 shows a plot of the $(n_0 - n_e)$ values.[37] The dispersion of the 128 cm^{-1} absorption line of quartz is clearly seen. Figure 2.19 also shows the transmittance of the Soleil polarizer in its null position (T) and in the position where a 180°-phase shift is produced (d).

The transmittance in arbitrary units of the compensator as a function of the thickness d of uncompensated quartz (counted from the null position) is shown in Fig. 2.20. In these transmittance measurements two linear polarizers were used; one located before and one placed behind the compensator. They were placed uncrossed at a 45° angle to the optical axes. Figure 2.20 presents the transmittance maxima and minima for light of the frequencies $v = 60.7$, 85.0, 116.6, and 166.6 cm^{-1}. The total quartz thickness is also indicated.

2. Polarizer Using a Magnetic Field. Richards and Smith[40] have described the possibilities of constructing circular polarizers, isolators, and circulators as solid-state devices in conjunction with a magnetic field. A circular polarizer

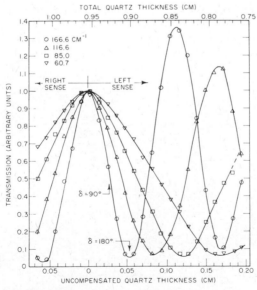

Fig. 2.20 Transmission (arbitrary units) of a compensator at several wavelengths as a function of the uncompensated quartz thickness. [E. D. Palik, *Appl. Opt.* **4**, 1017 (1965).]

made from n-type indium antimonide is considered here. The polarizer action is based on the principle that electrons undergoing cyclotron motion absorb left-hand circular-polarized radiation if the direction of propagation is parallel to the magnetic field. Thus the sample transmits right-hand circularly polarized radiation at the resonant frequency $\omega_c = eH/m^*c$, where $m^* = 0.014m_0$ is the effective mass of electrons in InSb (see also Appendix IV, Section C.a).

As we have shown, the various types of linear and circular polarizers have inherent advantages and disadvantages. It is necessary to keep these in mind if a polarizer is to be chosen for a particular experimental set-up or spectrometer.

F. MULTIPLE-PATH CELL

In the spectroscopy of vapors and gases it is sometimes desirable or necessary to use long-path absorption cells, either because the spectrum is inherently very weak and vapor pressures at convenient temperatures are too low or because pressure broadening of the absorption lines has to be minimized.

White[41] has designed a folded light-path absorption cell which permits the use of large angular apertures off the optical axis. It can be used for all types of gases and vapors which do not corrode the mirrors. The system is built of three spherical concave mirrors, all having the same radius of curvature. The two smaller-diameter mirrors, A and A', are on one side and the larger one, B, is at the other side, as shown in Fig. 2.21. The centers of

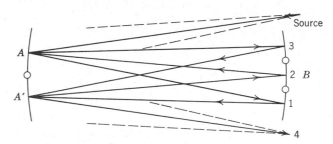

Fig. 2.21   Schematic of a White cell set for eight light passes. [J. U. White, *J. Opt. Soc. Am.* **32**, 285 (1942).]

curvature of A and A' are on the surface of mirror B, and the center of curvature of B is midway between the mirrors A and A' as indicated by the circles in Fig. 2.21. The light enters at the "source point" and an image at Point 1 on B is formed by A. From this point, A' forms an image on B at

Point 2 and, subsequently, the image at Point 3 is formed again by *A*. Finally the light emerges at the "exit point" 4 which is located at the same distance from point 2 as the "source point." The positions of the mirrors *A* and *A'*, and correspondingly the separation of the centers of curvature of *A* and *A'* on *B*, determine the number of times the light passes through the cell. If there is only one image on *B*, the light passes four times the distance between the mirrors. (Source point and exit point are positioned close to the mirror *B*.) For three images on *B*, the light traverses the cell eight times, for five images it passes twelve times, and so forth. The path length can be adjusted only to these distances; intermediate distances are not possible.

The optical adjustment of this system is not critical. Figures 2.22 to 2.24 show smoke photographs of four, eight, and twelve passes of the light through the cell. The numbers in Figs. 2.22 to 2.24 indicate the order of the images. It was possible to pass the light ninety times through the cell and thus obtain a path length of 56 m. The radii of curvature of the mirrors in this cell were 63.5 cm and their area was 8×11 cm^2 and 5.5×15 cm^2, respectively.

A modification of the White cell has been introduced by Bernstein and Herzberg.[42] The front mirror *B* is modified as shown in Fig. 2.25. By using the lower and the upper half of this mirror, twice as many images may be placed on it and therefore twice the path length is obtainable.

White has mentioned that the principal aberration is astigmatism. Edwards[43] has considered the astigmatism appearing in the normal use of the

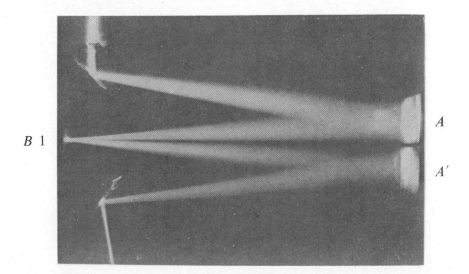

Fig. 2.22   Smoke photograph of four optical passes. [J. U. White, *J. Opt. Soc. Am.* **32**, 285 (1942).]

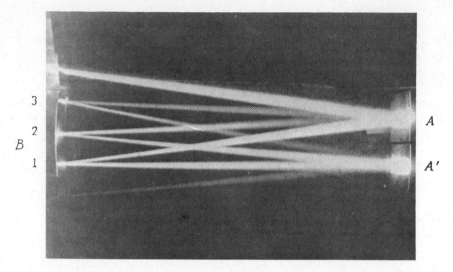

Fig. 2.23   Smoke photograph of eight optical passes. [J. U. White, *J. Opt. Soc. Am.* **32**, 285 (1942).]

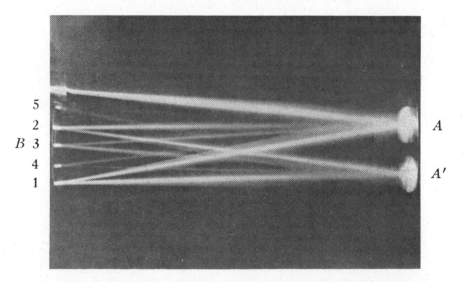

Fig. 2.24   Smoke photograph of twelve optical passes. [J. U. White, *J. Opt. Soc. Am.* **32**, 285 (1942).]

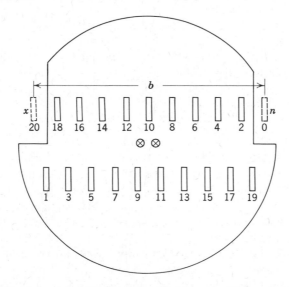

Fig. 2.25 Modified front mirror showing image positions of a White cell when it is adjusted for 40 traversals (20 images). The images are numbered. The crossed circles indicate centers of curvature of the back mirrors, n and x indicate the entrance and exit images, and b is the total separation of the entrance and exit images. [H. J. Bernstein and G. Herzberg, *J. Chem. Phys.* **16**, 30 (1948).]

White cell and has derived an expression for the resulting image elongation. For a point source he found the vertical (tangential) image length ΔL_T after N traversals as

$$\Delta L_T = \frac{hb^2}{12R^2}\left(N - \frac{4}{N}\right) = \frac{b^2}{12Rf_v}\left(N - \frac{4}{N}\right), \tag{2.16}$$

where h is the height of the back mirrors A and A', $f_v = R/h$ is the vertical f-number, b is the separation of the entrance and exit images, and R is the common radius of curvature for the three spherical mirrors.

If a White cell is constructed for a certain spectrometer, it is necessary to match the f-numbers of the two systems. It is advantageous to make the back mirror slightly larger to have some tolerance for the positioning of the cell. The distance b should be as small as possible. The minimum b that will give distinct images on the front mirror and will simultaneously allow the entrance and exit beams to pass the front mirror B without losses is given by $b_{min} = (N/4)w_s$, where w_s is the width of the entrance aperture. After the specification of f_v and b, the remaining dependence of ΔL_T is $\Delta L_T \approx N/R$; for small f-number systems, as they are generally used in the far-infrared, R should be chosen to be 30 to 50 cm.

This type of multiple-path cell has been shown to be very useful in the far-infrared. Furthermore, energy losses do not seem to be excessive; with a cell of 1 m physical length and based on a $f/10$-system, 75% of the radiation was transmitted with 40 transversals.

G. Image Transformer

The receiver is sometimes not fully illuminated in an optical system (see Chapter 1, Section D). We shall now discuss briefly how this can be avoided through the use of an image transformer in the case of a grating spectrometer.

Detector windows and the orifices of light-pipes are in most cases circular. However, the slits of the monochromator form straight or curved rectangular areas and, if they are imaged on a circular detector entrance, only part of the detector area is illuminated. In order to utilize the full detector area, longer slits may be used. They can then be optically divided into several sections and imaged side-by-side onto the detector window.

This has been shown by Benesch and Strong[44]; Fig. 2.26 shows their system designed for this purpose. The rectangular slit S_1 is imaged by M_1

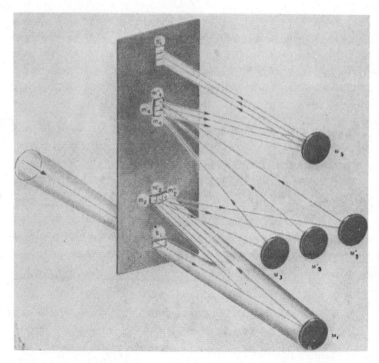

Fig. 2.26   Image transformer. [W. Benesch and J. Strong, *J. Opt. Soc. Am.* **41**, 252 (1951).]

onto three small mirrors m_2, m_2', and m_2'' which, in turn, slice the image of the slit into three segments. Each segment is imaged by one of the spherical mirrors M_3, M_3', and M_3'' onto one of the small mirrors m_4, m_4', and m_4'' which are oriented rectangular to the set m_2 to m_2''. The mirror M_5 combines the three images from the mirrors m_4 to m_4'' into the final image S_1', which forms a rectangular area in a perpendicular orientation to S_1. Theoretically, the resolution can be improved by a factor which is proportional to the square root of the increased flux falling on the detector. In practice, however, a smaller amount is gained. The Jarrell-Ash 78-900 spectrometer uses an image transformer with $f/10$ optics which slices the slit into three parts. The gain in flux is about 2 to 2.5.*

H. Vacuum System and Cryogenic Equipment

Atmospheric water vapor absorption is very strong in the far-infrared; it is therefore necessary to dry or to evacuate the atmospheres of spectrometers and interferometers. The use of nitrogen, evaporating from liquid-nitrogen storage flasks, or dried air from other generating sources is not completely satisfactory in some cases. Therefore, many spectrometers and interferometers are enclosed in vacuum tanks. Tanks of sufficient mechanical strength are not difficult to manufacture and the required vacuum of 50–100 μ pressure can easily be maintained by standard mechanical pumps. The necessary openings of the tank can be covered with flanges and sealed by O-rings. In locations where it is desirable to look into the instrument, windows can be made out of plastic. Plastic plates of 5 cm thickness can conveniently cover an area as large as $50 \times 50 \text{ cm}^2$.

The different parts of the instrument need not be enclosed into one tank. They can be placed into separate tanks and the optical connection between the tanks can be accomplished by the use of bellows or light-pipes. Electrical feed-throughs are commercially available. Movable parts of the spectrometer or interferometer can be operated from the outside through quick-couplings which can be soldered into the walls of the tank or glued on with epoxy.

The optical parts of the instrument should be placed on a separate base plate. This plate should be mounted in the tank in a way which guarantees that the distortion of the tank upon evacuation will not cause displacements and maladjustments of the optical components.

Cryogenic equipment is necessary for operation of many solid-state detectors (liquid-helium temperatures) and for studies of various phenomena

* Image transformers built with spherical mirrors and employing f-numbers smaller than or equal to four have been shown not to be efficient.

at low temperatures. An example of the operation of a detector at this low temperature was given in Section B. Discussions on far-infrared absorption and reflection measurements on solids at low temperatures are given in Chapter 12 and Appendices III and V. Appendix V contains descriptions and other pertinent information on refrigeration systems and related cryogenic apparatus.

REFERENCES

1. E. K. Plyler, D. J. C. Yates, and H. A. Gebbie, *J. Opt. Soc. Am.* **52**, 859 (1962).
2. A. Hadni, E. Decamps, and J. M. Munier, *Rev. Opt.* **42**, 584 (1963).
3. I. M. Arefyev, *Opt. Spektr.* **17**, 300 (1964); *Opt. Spectry.* **17**, 157 (1964).
4. D. H. Martin and D. Bloor, *Cryogenics* **1**, 159 (1961).
5. C. C. Grimes, P. L. Richards, and S. Shapiro, *Phys. Rev. Letters* **17**, 431 (1966).
6. R. A. Smith, F. E. Jones, and R. P. Chasmar, *The Detection and Measurement of Infrared Radiation,* Clarendon Press, Oxford, 1960.
7. M. R. Holter, S. Nudelman, G. H. Suits, W. L. Wolfe, and G. J. Zissis, *Fundamentals of Infrared Technology,* Macmillan, New York, 1962.
8. H. Zahl and M. Golay, *Rev. Sci. Instr.* **17**, 511 (1946).
9. M. Golay, *Rev. Sci. Instr.* **18**, 357 (1947).
10. K. Hennerich, W. Lahmann, and W. Witte, *Infrared Phys.* **6**, 123 (1966).
11. W. S. Boyle and K. F. Rodgers, Jr., *J. Opt. Soc. Am.* **49**, 66 (1959).
12. P. L. Richards, *International Union of Pure and Applied Chemistry, Molecular Spectroscopy,* Vol. VIII, Butterworths, London, 1965, p. 535.
13. R. C. Jones, *J. Opt. Soc. Am.* **43**, 1 (1953).
14. F. J. Low, *J. Opt. Soc. Am.* **51**, 1300 (1961).
15. R. G. Wheeler and J. C. Hill, *J. Opt. Soc. Am.* **56**, 657 (1966).
16. E. H. Putley, *Appl. Opt.* **4**, 649 (1965).
17. W. J. Moore and H. Shenker, *Infrared Phys.* **5**, 99 (1965).
18. H. Shenker, W. J. Moore, and E. M. Swiggard, *J. Appl. Phys.* **35**, 2965 (1964).
19. H. Levinstein, *Appl. Opt.* **4**, 639 (1965).
20. R. A. Smith, *Appl. Opt.* **4**, 631 (1965).
21. M. A. Kinch and B. V. Rollin, *Brit. J. Appl. Phys.* **14**, 672 (1963).
22. C. E. Jones, Jr., A. R. Hilton, J. B. Damrel, Jr., and C. C. Helms, *Appl. Opt.* **4**, 683 (1965).
23. P. L. Richards, *J. Opt. Soc. Am.* **54**, 1474 (1964).
24. D. R. Smith, B. K. McKenna, and K. D. Möller, *J. Chem. Phys.* **45**, 1904 (1966).
25. R. C. Ohlmann, P. L. Richards, and M. Tinkham, *J. Opt. Soc. Am.* **48**, 531 (1958).
26. J. A. Stratton, *Electromagnetic Theory,* McGraw-Hill, New York, 1941, p. 507.
27. K. D. Möller, V. P. Tomaselli, L. R. Scube, and B. K. McKenna, *J. Opt. Soc. Am.* **55**, 1233 (1965).
28. *Nimbus Wide Fieldradiometer,* Final Report 4257, NASA-Goddard Space Flight Center, Contract No. NAS 5-1016.
29. D. E. Williamson, *J. Opt. Soc. Am.* **42**, 712 (1952).
30. E. V. Loewenstein and D. C. Newell, *J. Opt. Soc. Am.* **59**, 407 (1969).
31. E. D. Palik, *Appl. Opt.* **2**, 527 (1963).
32. A. Mitsuishi, Y. Yamada, S. Fujita, and H. Yoshinaga, *J. Opt. Soc. Am.* **50**, 433 (1960).
33. R. F. Rowntree, *Scientific Report No. 4,* Air Force Cambridge Research Laboratory, Bedford, Massachusetts, Contract No. AF 19(604)-4119.

34. G. R. Bird and M. Parrish, *J. Opt. Soc. Am.* **50**, 886 (1960).
35. M. Hass and M. O'Hara, *Appl. Opt.* **4**, 1027 (1965).
36. G. Rupprecht, D. M. Ginsberg, and J. D. Leslie, *J. Opt. Soc. Am.* **52**, 665 (1962).
37. E. D. Palik, *Appl. Opt.* **4**, 1017 (1965).
38. S. Roberts and D. D. Coon, *J. Opt. Soc. Am.* **52**, 1032 (1962).
39. E. E. Russell and E. E. Bell, *J. Opt. Soc. Am.* **57**, 341 (1967).
40. P. L. Richards and G. E. Smith, *Rev. Sci. Instr.* **35**, 1535 (1964).
41. J. U. White, *J. Opt. Soc. Am.* **32**, 285 (1942).
42. H. J. Bernstein and G. Herzberg, *J. Chem. Phys.* **16**, 30 (1948).
43. T. H. Edwards, *J. Opt. Soc. Am.* **51**, 98 (1961).
44. W. Benesch and J. Strong, *J. Opt. Soc. Am.* **41**, 252 (1951).

3 Fabry-Perot Interference Filters and Tunable Fabry-Perot Etalon

a. General

The use of Fabry-Perot interferometers and interference filters, which we mentioned briefly in Sections A and E of Chapter 1, will now be discussed in detail since these devices are of great importance in far-infrared applications. For instance, Fabry-Perot interferometers have been employed in low-order as spectrometers, and interference filters are of general use as narrow- and broad-bandpass filters and cut-on or cut-off filters.

In the ultraviolet, visible, and near-infrared regions the Fabry-Perot and interference filters have been used extensively.[1] The Fabry-Perot, constructed from two parallel and highly reflecting plates, found its application as a high-resolution high-order multiple-beam interferometer in narrow spectral regions or in the construction of lasers. The interference filters have been constructed, in general, from a number of dielectric layers of different indices of refraction according to well-known design procedures to give the desired optical characteristics.

The reflectors for the Fabry-Perot and meshes for the interference filters for use in the far-infrared are constructed from various types of metallic meshes. Most of the work on far-infrared Fabry-Perot interferometers and interference filters has been published by Ulrich, Renk, and Genzel[2-5]; we will follow their papers closely in the following discussions.

b. Airy Formula and Finesse

An interference filter or a Fabry-Perot spectrometer employs a pair of identical plane-parallel reflectors, separated by a distance d in a medium of refractive index n. The power transmittance, P_t, of the device for normally incident light of wavelength λ is given by Airy's formula[1-3,6,7]

$$P_t(\lambda) = \left(1 - \frac{A}{1 - R}\right)^2 \left(1 + \frac{4R}{(1 - R)^2} \sin^2 \frac{\delta}{2}\right)^{-1}, \qquad (3.1)$$

where $\sqrt{R}e^{i\phi}$ is the complex amplitude of reflectance of the *single* reflector, ϕ is the phase shift upon reflection on one reflector, A is the power absorptance, and δ is the phase difference of two successive interfering beams, given by[3]

$$\delta = \frac{4\pi nd}{\lambda} - 2(\phi - \pi). \qquad (3.2)$$

The maximum of transmittance which occurs when $\delta = 2\pi q$ with $q = 1, 2, 3, \cdots$ is called the qth-interference order. The value of the maximum transmissivity is given by

$$P_{t_0} = \left(1 - \frac{A}{1 - R}\right)^2 = \left(1 + \frac{A}{T}\right)^{-2}, \qquad (3.3)$$

where T is the power transmissivity of the *single* reflector,

$$T = 1 - R - A. \qquad (3.4)$$

The resolving power, Q, is defined by the ratio of the wavelength (λ_q) and the bandwidth $(\Delta\lambda)$ at half peak height of the qth-order transmittance maximum:

$$Q = \frac{\lambda_q}{\Delta\lambda} = q\mathscr{F}. \qquad (3.5)$$

(Perfectly flat and parallel reflector plates are assumed here and the influence of the finite aperture is neglected.) The dimensionless number $\mathscr{F}$ is called the "finesse"; it is equal to the Q-value in the first order $(q = 1)$. The quantities A, R, T, ϕ, Q, and $\mathscr{F}$ are, in general, wavelength-dependent but can be considered to be constant over the region $\Delta\lambda$ in most cases.

A convenient way to determine $\mathscr{F}$ is to measure P_t as a function of the spacing d using monochromatic light. From such measurements, $\mathscr{F}$ is found (Eqs. 3.1 and 3.3) to be given by the relation

$$\mathscr{F} = \frac{\text{separation of peaks}}{\text{width at half height}} = \frac{\pi}{2 \arcsin\left[(1 - R)/2\sqrt{R}\right]}. \qquad (3.6)$$

For $R > 0.6$, Eq. 3.6 can be approximated by

$$\mathscr{F} \approx \frac{\pi\sqrt{R}}{1 - R}. \qquad (3.7)$$

The finesse, which depends on the reflectance of a single reflector, gives a measure of the resolution. The finesse is large if $1 - R \ll 1$; in other words, if $R \sim 1$. We see from Eq. 3.3 that the power transmittance at the peak of the interference orders approaches unity if A is small compared to $1 - R$.

B. Reflectors for the Far-infrared

For most applications we wish to construct a Fabry-Perot interferometer which offers high resolution (large value of Q) in the lowest possible order together with a sufficiently high peak transmission P_{t_0}. Inspection of Eqs. 3.3 and 3.7 indicates that these two desirable properties of the Fabry-Perot interferometer require

$$A \ll (1 - R) \quad \text{and} \quad (1 - R) \ll 1. \quad (3.8)$$

These conditions cannot be satisfied simultaneously by the employment of thin homogeneous metal layers[8] in the far-infrared region (in contrast to conditions in the visible spectrum); a homogeneous metal layer thick enough to allow the desired reflection characteristics would be so thick it would absorb too much of the radiation. In microwave optics, perforated metal plates[9] and metal gratings have been used for resonant cavities, and very large Q-values were obtained. Similar plates can be used in the far-infrared; because of the much shorter wavelength, diffraction effects due to the finite size of the plates are less pronounced and can usually be neglected.

The plates are made in the form of thin perforated metal layers, either self-supporting or on a substrate, and are aptly called "metal meshes." A type of mesh made by an electroforming process has been investigated intensively; we have discussed their use as reflection and transmission filters (see Chapter 1, Section E.h–i).

Gratings made of wire strips (with and without a substrate) must also be listed in this connection. They are, however, only applicable for one direction of polarization (see Chapter 2, Section E, and Refs. 2 and 10). Because of this, they are not very important for practical far-infrared Fabry-Perot interference devices. However, a discussion of their properties will be very helpful for an understanding of the properties of the metal meshes. In the following section we shall describe the reflection, the transmission, and the absorption of electromagnetic waves by a "grating" and a "mesh." We shall understand the concept of a "grating" to mean an array of parallel metal strips separated by empty space or by a dielectric medium, and a "mesh" shall be understood to designate an array formed by the fusion of two such gratings laid across each other. (The designation "1-dimensional grid" for gratings and "2-dimensional grids" for meshes is sometimes used in the literature.)

a. Metal Gratings in Empty Space

A theoretical treatment of the reflection, transmission, and absorption characteristics of an electromagnetic wave by a metal strip grating has been given by several authors.[10,11] The strips of the grating are of either circular, elliptical, or rectangular cross section. As mentioned before, the grating is essentially transparent to the component of the electric vector which is polarized in a normal direction with respect to the direction of the wires. We shall call the grating constant g.* The component of the electric vector which is polarized in a parallel direction is partly reflected, partly absorbed, and partly transmitted under this condition; for $\lambda \gg g$, it is nearly reflected totally.

Under the restrictions that the electric vector and the wires are in parallel directions, that $\lambda > 2g$, that the skin depth is much smaller than the radius of the wire (a), and that $a \ll \lambda$, Casey and Lewis[10] solved the boundary value problem approximately and arrived at the following expression for the absorptance of the grating:

$$A = \frac{g}{\pi a} \left(\frac{c}{\sigma \lambda} \right)^{1/2} R, \qquad (3.9)$$

where R is the reflectance, σ the conductivity of the wire, and c the velocity of light. In the limiting case $\lambda \gg g$, the absorptance is very small; the corresponding equations describing the reflectance and transmittance (T) are then given by[2,10]

$$R = 1 - \left(\frac{2g}{\lambda} \ln \frac{g}{2\pi a} \right)^2 = 1 - \tan^2 \phi$$

$$T \approx 1 - R = \tan^2 \phi = \left(\frac{2g}{\lambda} \ln \frac{g}{2\pi a} \right)^2 \qquad (3.10)$$

$$A = \frac{g}{\pi a} \left(\frac{c}{\sigma \lambda} \right)^{1/2} R,$$

where we have written down once more the expression for the absorptance A. With these relations we can check Eq. 3.8 for the example of a copper wire-grating with the numerical values $g/a = 8$, $R = 0.88$, $\lambda = 200\,\mu$, and $\sigma = 5 \times 10^{17}$ sec.$^{-1}$. We obtain $A = 0.004$ and $T = 0.116$; Eq. 3.8 is therefore satisfied.

We now digress somewhat and describe briefly some topics of transmission line theory[12,13] because this theory lends itself advantageously to a description of the effects of more complex interference filters, such as a combination

---

* In Chapter 1 we employed the letter d for the grating constant. Here we will use the symbol g. The grating or mesh constant g is the reciprocal of the "mesh number" (a commercially used designation, usually in units of lines/inch).

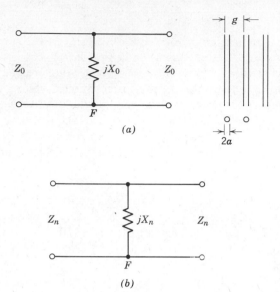

Fig. 3.1 Equivalent circuit for the transmission characteristics of metal gratings (or meshes):
(a) grating in empty space; (b) grating in a medium. Grating constant: g; wire diameter: $2a$.
[R. Ulrich, K. F. Renk, and L. Genzel, *IEEE Trans.* **MTT-11**, 363 (1963), by permission of The
Institute of Electrical and Electronics Engineers, Inc.]

of three or more gratings or meshes. The simplest equivalent circuit is shown
in Fig. 3.1(a). We shall show that its power transmittance is the same as that
of Eq. 3.10 if X_0, Z_0 are chosen in a certain way. We begin by writing down
the reflection coefficient of the voltage amplitude for a wave incident from
the left (see p. 16 of Ref. 12):

$$\Gamma = \frac{Z - Z_0}{Z + Z_0}, \tag{3.11}$$

where Z is the total impedance terminating the transmission line at the
reference point F (see Fig. 3.1(a)) and Z_0 is the characteristic impedance of the
transmission line. The total impedance Z is given by the impedance of the
parallel connection of the shunt reactance jX_0 (representing the mesh) and
by the impedance Z_0 of the line to the right of F:

$$Z^{-1} = (jX_0)^{-1} + (Z_0)^{-1}, \quad \text{or} \quad Z = \frac{jX_0Z_0}{Z_0 + jX_0}. \tag{3.12}$$

(The symbol $j = \sqrt{-1}$, as commonly used in the theory of electricity.) We
first assume that the dielectric constant of the medium surrounding the mesh
is equal to unity (empty space).

Inserting Eq. 3.12 into Eq. 3.11, we obtain[†]

$$\Gamma = -\frac{Z_0}{Z_0 + j2X_0} = \frac{Z_0}{\sqrt{Z_0^2 + 4X_0^2}} \exp j[\pi - \arctan(2X_0/Z_0)]. \qquad (3.13)$$

Writing $\Gamma = |\Gamma|e^{j\phi}$ for the amplitude reflectance, we arrive at

$$|\Gamma| = \frac{Z_0}{\sqrt{Z_0^2 + 4X_0^2}} \quad \text{and} \quad \phi = \pi - \arctan\frac{2X_0}{Z_0}. \qquad (3.14)$$

The sine of the phase angle ϕ is given by

$$\sin\phi = \frac{2X_0}{\sqrt{4X_0^2 + Z_0^2}}, \qquad (3.15)$$

and the power reflectance R is given by[‡]

$$R = \Gamma\Gamma^* = |\Gamma|^2 = \frac{Z_0^2}{Z_0^2 + 4X_0^2}. \qquad (3.16)$$

Since the shunt impedance in Fig. 3.1 is purely imaginary, the element represents a lossless grating ($A = 0$) and one obtains the power transmittance

$$T = 1 - R = \frac{4X_0^2}{Z_0^2 + 4X_0^2} = \sin^2\phi. \qquad (3.17)$$

The most interesting situation is when $R \sim 1$; that is $T \ll 1$. Then we can approximate Eq. 3.17 by

$$T = 1 - R \approx \frac{4X_0^2}{Z_0^2}. \qquad (3.18)$$

The quantity X_0/Z_0 is, in general, a function of the wavelength. It completely describes the amplitudes and phases of the transmission and the reflection; for a wire grating (wire radius a), under the conditions that $\lambda \gg g$, $a \ll g$, it is given (p. 280 of Ref. 12) by

$$\frac{X_0}{Z_0} = \frac{g}{\lambda}\ln\frac{g}{2\pi a}. \qquad (3.19)$$

With Eqs. 3.18 and 3.19, we obtain for T

$$T = \left(\frac{2g}{\lambda}\ln\frac{g}{2\pi a}\right)^2. \qquad (3.20)$$

The expression 3.20 for T is exactly that quoted in Eq. 3.10.

[†] The time dependence of the field strength in this representation is assumed to be of the form $\exp(j\omega t)$.
[‡] In the following we will use both notations: R and $|\Gamma|^2$, T and $|\tau|^2$. The letter $\bar{R}$ will designate a resistance.

b. Grating and Mesh in the Vicinity of a Dielectric Medium

If we consider a grating affixed to a not too thin substrate, the effect of the dielectric medium has to be taken into account. This leads to a modification of the expressions of the transmittance and reflectance.

In the following we neglect absorption and assume that $\lambda \gg g$. We first treat the case where the grating is imbedded in a dielectric medium of dielectric constant $\varepsilon = n^2$ and of unit magnetic susceptibility. The normalized impedance of the medium is then $Z_n/Z_0 = 1/n$; the wavelength in the medium is $\lambda_n = \lambda_0/n$, where λ_0 is the wavelength in the vacuum. Since the refractive index n is contained neither in the spatial part of the wave equation at the grating nor in the boundary conditions, we can use Eq. 3.19 in the form

$$\frac{X_n}{Z_n} = \frac{g}{\lambda_n} \ln \frac{g}{2\pi a} \tag{3.21}$$

if circular wires are assumed. From Eq. 3.21 it follows that

$$X_n = Z_0 \frac{g}{\lambda_0} \ln \frac{g}{2\pi a} = X_0. \tag{3.22}$$

This means that the absolute impedance of the grating is independent of n. Therefore, the influence of the dielectric medium merely causes Z_0 to be replaced by $Z_0/n = Z_n$ (see Fig. 3.1(b)); we thus must use Z_n in Eqs. 3.11 to 3.19.

As a second case, we shall consider the grating (or mesh) to be placed on the surface of the dielectric. This case is often encountered in practice; for instance, when the grating or mesh is glued or deposited on a substrate in order to maintain the required flatness. We keep the above assumptions that the absolute impedance is independent of n, which is only valid for $\lambda \gg g$.* The physical picture is then as follows. The grating on the substrate "sees" a transmission line with characteristic impedance Z_0 on the "free" side and a line of impedance Z_n on the "bounded" side. The power transmittance T' and the power reflectance R' in this case are related to the corresponding quantities T and R of the grating in a vacuum by the relations[3]

$$\frac{T'}{T} = \frac{1 - R'}{1 - R} = \frac{n}{R + \frac{1}{4}(n + 1)^2 T}. \tag{3.23}$$

Experimental verification of Eq. 3.23 was made[3] with the help of a metal mesh affixed to the surface of a crystal quartz plate ($n = 2.12$). The mesh ($g = 50\ \mu$) consisted of copper strips of rectangular cross section ($12.5 \times 6\ \mu^2$). At a wavelength of $\lambda = 200\ \mu$, T' was found to be 1.77 times greater than T (the transmittance in vacuum). This experimental ratio should be com-

---

* Balakhanov maintains that this assumption is poor (see Ref. 14).

pared to the value of 1.81, computed from Eq. 3.23 (absorption losses in the quartz plate and reflections on the second surface had been taken into account).

A more exact theoretical treatment of the case where the grating is on the surface of a dielectric medium has been given by Balakhanov.[14] He found, as we have assumed, that the equations discussed in this context are only valid for wavelengths which are much longer than the grating constant.

In this connection we note here that there is a close relation between transmission gratings and meshes: The properties of meshes can be derived qualitatively from those of the gratings (under our assumption that $\lambda \gg g$) since we can imagine that a mesh is made up of two crossed gratings. We had mentioned above that a grating is nearly completely transparent to radiation with the electric vector polarized perpendicularly to the wires; this configuration will be termed a capacitive grating. On the other hand, a configuration with the electric vector parallel to the wires is called an inductive grating.[12] With respect to meshes, we shall also encounter a classification of inductive and capacitive meshes as we will discuss in the next section.

In conclusion, we mention that the optical properties of an inductive grating or mesh at the interface of two different dielectric media (n_1, n_2) are obtained (for $\lambda \gg g$) in the form

$$\frac{T'}{T} = \frac{1 - R'}{1 - R} = \frac{n_1 n_2}{R + \frac{1}{4}(n_1 + n_2)^2 T}. \tag{3.24}$$

c. Metal Meshes and Their Equivalent Circuit Representation

We first discuss some general properties. Metallic meshes are considered, as we mentioned above, to exist in two complementary structures. Figure 3.2(a) shows the inductive mesh and Fig. 3.2(b) the capacitive mesh. The parameters g, a, and t, which determine the structure, are indicated. The meshes are made of copper, nickel, silver, or gold. The inductive meshes are unsupported in most cases. The capacitive meshes must be supported by a thin dielectric film; for instance, Hostaphan of 2.5 μ thickness.* The designations "inductive" and "capacitive" enumerate the electric elements of the equivalent electrical circuit representation which are used to describe the transmission, reflection, and phase characteristics of these meshes. The solutions of the boundary problem in terms of the electromagnetic theory of light are, at present, not yet known, but numerous measurements of the reflectivity (R) and transmissivity (T) of such meshes have been reported.[2-5,15-18] Ulrich,[4] in particular, has shown that the infrared properties of such meshes can be represented in good approximation by a simple resonant circuit

* Mylar is the trade name for polyethylene terphthalate in the United States. Hostaphan is the trade name used by Kalle A.G., Wiesbaden-Biebrich, Germany.

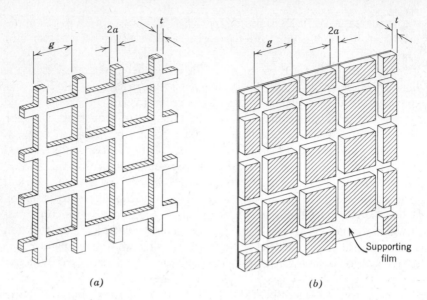

Fig. 3.2   Definition of mesh parameters: (a) inductive mesh; (b) capacitive mesh. Thickness: t. [R. Ulrich, *Infrared Phys.* **7**, 37 (1967), by permission of Pergamon Press.]

treatment as long as the meshes are thin and the wavelength λ is larger than the mesh constant g.

In the following we shall outline the theory and compare its results with some of the experimental data. We assume that the mesh is very thin and made of a perfectly conducting metal. Losses, for instance those due to a substrate, are usually small and can be neglected. However, we shall show below how they can be represented in the equivalent circuit.

We will employ the "normalized" frequency $\omega = g/\lambda$ in agreement with the literature. This definition of ω is convenient since we see, for instance from Eq. 3.10 or 3.20, that the ratio λ/g rather than λ enters into the relevant mathematical expressions.

In quite general terms, we first describe what happens when a plane-polarized electromagnetic wave of unit amplitude is considered to be incident normally to the plane of the mesh. The scattered electric field is symmetric to the plane of the mesh. If $\omega < 1$, no diffraction orders can propagate. Behind the mesh, where the field consists of the superposition of the incident wave and the scattered wave $\Gamma(\omega)$, the amplitude is given by

$$\tau(\omega) = 1 + \Gamma(\omega). \tag{3.25}$$

(We note that $|\Gamma(\omega)|$ equals the reflected amplitude). Both τ and Γ depend also on a/g. Since losses are neglected, the energy conservation law requires that

$$|\tau(\omega)|^2 + |\Gamma(\omega)|^2 = 1. \tag{3.26}$$

Equations 3.25 and 3.26 restrict Γ to a circle in the complex plane as is indicated in Fig. 3.3. Therefore, the following relations hold:

$$\sin^2 \varphi_\Gamma(\omega) = |\tau(\omega)|^2$$
$$\sin^2 \varphi_\tau(\omega) = 1 - |\tau(\omega)|^2 \tag{3.27}$$

where φ_Γ* is the phase of Γ and φ_τ is the phase of τ. Figure 3.3 shows that the

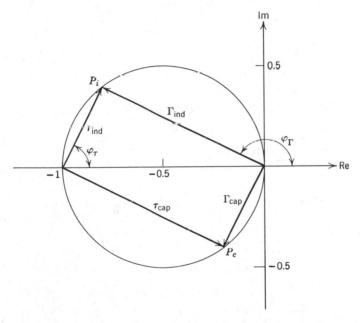

Fig. 3.3   Representation of τ and Γ in the complex plane. [R. Ulrich, *Infrared Phys.* **7**, 37 (1967), by permission of Pergamon Press.]

point P moves around the circle when the frequency varies but that at all frequencies the reflected and the transmitted wave have a constant phase difference of 90°.

Since a capacitive and an inductive mesh have complementary structures, Ulrich[4] has used the electromagnetic equivalent of Babinet's principle (see Ref. 6, p. 559) to arrive at the following relations between the various

* The φ_Γ is identical with the ϕ of Section A and B.

quantities of a capacitive (c) and its complementary inductive (i) mesh:

$$\tau_i(\omega) + \tau_c(\omega) = 1, \tag{3.28a}$$

$$|\tau_i(\omega)|^2 + |\tau_c(\omega)|^2 = 1, \tag{3.28b}$$

$$\tau_i(\omega) = -\Gamma_c(\omega) \quad \text{and} \quad \tau_c(\omega) = -\Gamma_i(\omega). \tag{3.28c}$$

The relations Eq. 3.28c, which follow from Eqs. 3.25 and 3.28a, are depicted in Fig. 3.3.

We now discuss briefly the equivalent circuit representation. The general equivalent circuit for a thin mesh can be taken to be that of a four-terminal transmission line with the lumped admittance $Y(\omega)$, representing the mesh at the reference plane F (see Fig. 3.4).* The characteristic admittance of the

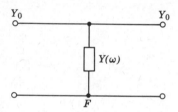

Fig. 3.4   Transmission line shunted by the lumped admittance $Y(\omega)$.

transmission line is $Y_0 = 1/Z_0$ and only the region $\omega < 1$ is considered. The results of transmission line theory then show that the shunt $Y(\omega)$ gives rise to a voltage reflectance, $\Gamma(\omega)$, and a voltage transmittance, $\tau(\omega)$:

$$\Gamma(\omega) = -\frac{Y(\omega)/2Y_0}{1 + Y(\omega)/2Y_0}; \quad \tau(\omega) = \frac{1}{1 + Y(\omega)/2Y_0}. \tag{3.29}$$

(We see that $\tau(\omega) = 1 + \Gamma(\omega)$ in accordance with Eq. 3.25). For a lossless mesh, Y/Y_0 is purely imaginary: $Y/Y_0 = jB(\omega)/Y_0$, where $B(\omega)/Y_0$ is normalized and real. The quantity $B(\omega)/Y_0$ is obtained empirically from measurements of $\Gamma(\omega)$ and $\tau(\omega)$. For the capacitive susceptances, $B > 0$; therefore, τ_c and Γ_c are located in the lower half of the circle in Fig. 3.3. For inductive susceptances, $B < 0$; therefore τ_i and Γ_i are found in the upper half.

Equations 3.28 express quantitatively the relations between complementary inductive and capacitive meshes. We can rewrite this in the equivalent circuit

---

* The susceptance is chosen in this book as $Y(\omega)/Y_0$ compared to $2Y(\omega)$ in the original paper (Ref. 4). This brings the formulas into accord with standard microwave notations (see Ref. 12) but distracts somewhat their simplicity and symmetry. The relations between the quantities used here (left-hand side) and those of the original reference (right-hand side) are: $Y(\omega)/Y_0 = 2Y(\omega)$, $B(\omega)/Y_0 = 2B(\omega)$, $2\Xi = Z_0$, $R = 2R$, $C = 2C$, and $L = L/2$.

representation as

$$\frac{B_c(\omega)}{Y_0} \frac{B_i(\omega)}{Y_0} = -4, \tag{3.30}$$

where $B_c(\omega)/Y_0$ is the susceptance of the capacitive mesh and $B_i(\omega)/Y_0$ that of the complementary inductive mesh.

All the information we can obtain on $|\tau(\omega)|^2$, $|\Gamma(\omega)|^2$, $\varphi_\tau(\omega)$, or $\varphi_\Gamma(\omega)$ is contained in $B(\omega)/Y_0$. Conversely, we can obtain $B(\omega)/Y_0$ if we know any one of these four quantities. We would proceed by synthesizing an equivalent circuit from constant inductances and capacitances (and sometimes resistances) in order to obtain the representation of the susceptance $B(\omega)/Y_0$ and to compare it with values of $B(\omega)/Y_0$ obtained from the measured values of, say, $|\tau(\omega)|^2$. We show an example in the following.

We synthesize our circuit using one element for $Y(\omega)/Y_0$. For a capacitive mesh a capacitance C is used, and $Y(\omega)/Y_0 = j\omega C/Y_0$. For an inductive mesh we have $Y(\omega)/Y_0 = -jZ_0/\omega L$ (C and L have their usual meaning).* The dependence of the quantities $|\Gamma(\omega)|^2$, φ_Γ, $|\tau(\omega)|^2$, and φ_τ on ωC and ωL for both types of meshes has been compiled in Table 3.1 for $\omega < 1$ and $\omega \ll 1$. For complementary meshes we find, from Eq. 3.30, that $C_c/2Y_0 - 2L_i/Z_0$. The dashed line in Fig. 3.5 shows the transmissivity $|\tau(\omega)|^2$ as a

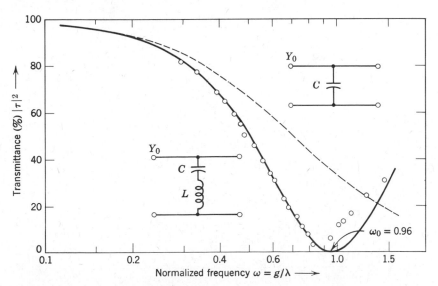

Fig. 3.5 Transmittance $|\tau|^2$ of a single mesh. Measurement: ○○○. Calculated with one element for Y: −−−−. Calculated with two elements for Y: ———. [R. Ulrich, *Infrared Phys.* **7**, 37 (1967), by permission of Pergamon Press.]

* This is identical with the Lewis and Casey approximation (see Section B.a).

TABLE 3.1*

EQUIVALENT CIRCUITS EMPLOYING ONE ELEMENT FOR THE REPRESENTATION OF CAPACITIVE AND INDUCTIVE MESHES

Type of mesh	Capacitive	Inductive
Equivalent circuit		
Normalized admittance[†]	$\dfrac{Y(\omega)}{Y_0} = j\dfrac{\omega C}{Y_0}$	$-j\dfrac{Z_0}{\omega L}$
Reflectance	$|\Gamma|^2 = \left(\dfrac{\omega C}{2Y_0}\right)^2\left[1 + \left(\dfrac{\omega C}{2Y_0}\right)^2\right]^{-1}$	$\left[1 + \left(\dfrac{2\omega L}{Z_0}\right)^2\right]^{-1}$
	$\varphi_\Gamma = \pi + \arctan\left(\dfrac{2Y_0}{\omega C}\right)$	$\pi - \arctan\left(\dfrac{2\omega L}{Z_0}\right)$
Transmittance	$|\tau|^2 = \left[1 + \left(\dfrac{\omega C}{2Y_0}\right)^2\right]^{-1}$	$\left(\dfrac{2\omega L}{Z_0}\right)^2\left[1 + \left(\dfrac{2\omega L}{Z_0}\right)^2\right]^{-1}$
	$\varphi_\tau = -\arctan\left(\dfrac{\omega C}{2Y_0}\right)$	$\arctan\left(\dfrac{Z_0}{2\omega L}\right)$
For complementary meshes	$\dfrac{C_c}{2Y_0} = \dfrac{2L_i}{Z_0}$	

Low frequency approximations ($\omega \ll 1$)[†]

Reflectance	$|\Gamma|^2 = \left(\dfrac{\omega C}{2Y_0}\right)^2$	$1 - \left(\dfrac{2\omega L}{Z_0}\right)^2$
	$\varphi_\Gamma = \dfrac{3\pi}{2} - \left(\dfrac{\omega C}{2Y_0}\right)$	$\pi - \left(\dfrac{2\omega L}{Z_0}\right)$
Transmittance	$|\tau|^2 = 1 - \left(\dfrac{\omega C}{2Y_0}\right)^2$	$\left(\dfrac{2\omega L}{Z_0}\right)^2$
	$\varphi_\tau = -\left(\dfrac{\omega C}{2Y_0}\right)$	$\dfrac{\pi}{2} - \left(\dfrac{2\omega L}{Z_0}\right)$

* R. Ulrich, *Infrared Phys.* **7**, 37 (1967), by permission of Pergamon Press.
† For most practical applications of these formulas (and those in the rest of the text) it is not important whether the frequency ω is considered here to be the actual infrared frequency (in rad/sec) or the dimensionless normalized frequency g/λ, defined above. Both quantities differ only by a factor of proportionality which can be absorbed in the definition of Y_0. Here, preference is given to the second alternative since then C/Y_0 and L/Z_0 become dimensionless.

TABLE 3.2*

EQUIVALENT CIRCUITS EMPLOYING MULTIPLE ELEMENTS FOR THE REPRESENTATION OF CAPACITIVE AND INDUCTIVE MESHES

Type of mesh	Capacitive	Inductive
Equivalent circuit		
Resonance frequency[†]	ω_0	
Characteristic impedance of empty space	$Z_0 = 1/Y_0$	
Normalized impedance of L and C at resonance[†]	$\Xi = \dfrac{\omega_0 L}{Z_0} = \dfrac{Y_0}{\omega_0 C}$	
Generalized frequency	$\Omega = \dfrac{\omega}{\omega_0} - \dfrac{\omega_0}{\omega} = \dfrac{\lambda_0}{\lambda} - \dfrac{\lambda}{\lambda_0}$	
Normalized admittance	$\dfrac{Y(\omega)}{Y_0} = \dfrac{1}{\dfrac{\bar{R}}{Z_0} + j\Omega\Xi}$	$\dfrac{1}{\dfrac{\bar{R}}{Z_0} - j\dfrac{\Xi}{\Omega}}$
Reflectance	$|\Gamma|^2 = \dfrac{1}{\left(1 + \dfrac{2\bar{R}}{Z_0}\right)^2 + (2\Xi\Omega)^2}$ $\varphi_\Gamma = \pi - \arctan \dfrac{2\Xi\Omega}{1 + \dfrac{2\bar{R}}{Z_0}}$	$\dfrac{1}{\left(1 + \dfrac{2\bar{R}}{Z_0}\right)^2 + \left(\dfrac{2\Xi}{\Omega}\right)^2}$ $\pi + \arctan \dfrac{2\Xi/\Omega}{1 + \dfrac{2\bar{R}}{Z_0}}$
Transmittance	$|\tau|^2 = \dfrac{\left(\dfrac{2\bar{R}}{Z_0}\right)^2 + (2\Xi\Omega)^2}{\left(1 + \dfrac{2\bar{R}}{Z_0}\right)^2 + (2\Xi\Omega)^2}$ $\varphi_\tau = \arctan \dfrac{2\Xi\Omega}{\left(\dfrac{2\bar{R}}{Z_0}\right)\left(1 + \dfrac{2\bar{R}}{Z_0}\right) + (2\Xi\Omega)^2}$	$\dfrac{\left(\dfrac{2\bar{R}}{Z_0}\right)^2 + \left(\dfrac{2\Xi}{\Omega}\right)^2}{\left(1 + \dfrac{2\bar{R}}{Z_0}\right)^2 + \left(\dfrac{2\Xi}{\Omega}\right)^2}$ $-\arctan \dfrac{\dfrac{2\Xi}{\Omega}}{\left(\dfrac{2\bar{R}}{Z_0}\right)\left(1 + \dfrac{2\bar{R}}{Z_0}\right) + \left(\dfrac{2\Xi}{\Omega}\right)^2}$
Absorptance	$A = 4\left(\dfrac{R}{Z_0}\right)|\Gamma|^2$	
For complementary meshes	$\dfrac{C_c}{2Y_0} = \dfrac{2L_i}{Z_0}$	$\dfrac{C_i}{2Y_0} = \dfrac{2L_c}{Z_0}$

* R. Ulrich, *Infrared Phys.* **7**, 37 (1967), by permission of Pergamon Press.

† Resonance at $\omega = \omega_0$ (losses are taken into account). Numerical values of Ξ and ω_0 can be taken from Fig. 3.7 and Eqs. 3.31 and 3.32. $\lambda_0 = g/\omega_0$ is the wavelength corresponding to the frequency ω_0.

function of g/λ with $C/Y_0 = 2.85$. The agreement with the experimental values (dots) is only good for low values of ω.

We now synthesize $Y(\omega)/Y_0$ employing two elements. The results, based on the values $C/Y_0 = 2.85$ (see above) and $L/Z_0 = 0.318$, are represented in the form of the solid line in Fig. 3.5. We see that the measured values are rather well represented over nearly the whole region of $\omega < 1$. Ulrich has therefore attempted to represent the capacitive and inductive meshes by resonant circuits given in Table 3.2, using a resistance $\bar{R}$ to take care of losses. Besides $\bar{R}$, the two parameters representing the equivalent circuits are $\Xi = \omega_0 L/Z_0 = Y_0/\omega_0 C$ and ω_0; Ξ is called the normalized impedance (see above) and ω_0 is called the normalized resonant frequency. In terms of the generalized frequency $\Omega = \omega/\omega_0 - \omega_0/\omega$, Table 3.2 gives the formulas for the normalized admittance Y/Y_0, the reflectance Γ, the transmittance τ, the absorptance A, and the relevant phase angles.

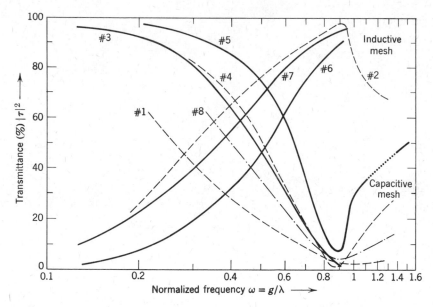

Fig. 3.6 Transmittance $|\tau|^2$ of the meshes given in Table 3.3. [R. Ulrich, *Infrared Phys.* **7**, 37 (1967), by permission of Pergamon Press.]

We now briefly discuss some comparisons between theory and experiment. Figure 3.6 shows the measured transmittance curves for a number of inductive and capacitive meshes, as specified in Table 3.3. The calculated curves (not shown) for the capacitive meshes represented well the experimental points in

TABLE 3.3*

Compilation of Mesh Parameters

Mesh			Dimension, in μ				Equivalent circuit parameters[†]		
No.	Type	Material	g	a	t	a/g	2Ξ	ω_0	$2\bar{R}/Z_0$
1	Cap	Cu on Mylar	368	17.7	5	0.051	0.274	1.0	$\leqslant 0.02$
2	Ind	Ni[‡]	368	17.0	14	0.046	3.00	1[§]	0.02[‖]
3	Cap	Cu on Mylar	250	40.5	5	0.161	0.70	0.96	$\leqslant 0.02$
4	Cap	Cu on Mylar	473	72.5	5	0.153	0.73	0.96	0.001[‖]
5	Cap	Cu on Mylar	342	68.5	6	0.200	1.19	0.945	0.002[‖]
6	Ind	Cu[‡]	216	28.8	7	0.133	1.33	1[§]	0.004[‖]
7	Ind	Ni[‡]	216	13.5	12	0.0625	2.42	1[§]	$\leqslant 0.04$
8	Cap	Cu on Mylar	368	35	7	0.095			—

* R. Ulrich, *Infrared Phys.* **7**, 37 (1967), by permission of Pergamon Press.

† The equivalent circuit parameters refer to the equivalent circuits given in Table 3.2. They were determined by adaption of the calculated value of $|\tau|^2$ to the measurements.

‡ Product of Buckbee Mears Comp., St. Paul, Minnesota.

§ Chosen arbitrarily.

‖ Ohmic losses only, calculated from Eq. 3.34 for $\lambda = 0.5$ mm and $\sigma = 0.25\sigma_{bulk}$.

the region $\omega < 1$. In the neighborhood of $\omega = 1$, the agreement was poorer, similar to the example shown in Fig. 3.5. With respect to the inductive meshes, the fit between experiment and theory was not so good, presumably as a consequence of the finite thickness of the mesh.

From Fig. 3.6 we can see that the capacitive meshes are generally low-pass filters. The transmittance minima or maxima, respectively, are in the range $\omega = 0.9 - 1$. The slope of T increases with the ratio a/g for both types of mesh.

Complementary meshes should obey Eq. 3.28b for $\omega < 1$ if absorption can be neglected (see above). This has been confirmed experimentally for a broad wavelength region.[4]

Let us now describe briefly the dependance of Ξ and ω_0 on the ratio a/g. Figure 3.7 shows a plot of Ξ *vs* a/g for all capacitive and inductive meshes of Table 3.3. For the capacitive meshes the Ξ_c-scale is at the left edge, and for the inductive meshes the Ξ_i-scale is at the right edge of the figure. Both scales are related, according to Babinet's theorem, by $\Xi_i \cdot \Xi_c = \frac{1}{4}$. The solid line in Fig. 3.7 represents the impedance of a "capacitive grating" of thin

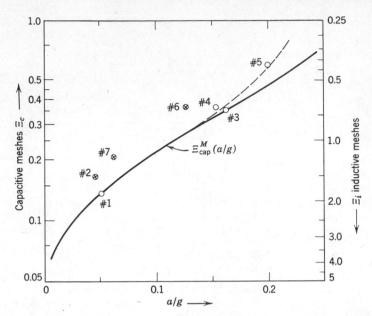

Fig. 3.7   Dependence of 2Ξ on a/g. Experimental values for the capacitive mesh: $\bigcirc\bigcirc\bigcirc$; for the inductive mesh: $\otimes\otimes\otimes$. [R. Ulrich, *Infrared Phys.* 7, 37 (1967), by permission of Pergamon Press.]

strips, $\Xi_c{}^M(a/g)$, calculated according to[12]

$$\Xi_c{}^M\left(\frac{a}{g}\right) = \left[2\ln\operatorname{cosec}\left(\frac{a\pi}{g}\right)\right]^{-1*}. \qquad (3.31)$$

The good agreement between the curve calculated for the grating and the values measured for the capacitive mesh at low values of a/g is not too surprising because the capacitive mesh is obtained from the grating by cutting equidistant gaps of width $2a$ (see Fig. 3.2(b)) into the strips of the grating. Since these gaps are oriented parallel to the direction of the incident electric vector, they exert little influence on the characteristics of the surface currents (see the beginning of Section B.c).

The impedance Ξ_i of the inductive meshes can be obtained with the help of $\Xi_i \cdot \Xi_c = \frac{1}{4}$ and Eq. 3.31. Most commercially available self-supporting inductive meshes, however, are too thick for our description to be valid. This is probably the reason why the values of the inductive meshes are not in good agreement with the solid line[12] in Fig. 3.7. (For higher values of a/g the dashed curve should be used since deviations become apparent.)

* This formula was not correctly given in Refs. 4, 5, and 22.

The dependence of the resonance frequency ω_0 on a/g has been established by Ulrich[4] in terms of the approximate expression

$$\omega_0\left(\frac{a}{g}\right) = 1 - 0.27\frac{a}{g}. \tag{3.32}$$

Equation 3.32 was obtained by matching the transmittance of a number of capacitive meshes of different a/g values.

We now take into account the effects of losses (see also the beginning of Section B.c) by adding a small ohmic resistor $\bar{R}$ to the equivalent circuit. This is shown for the meshes Nos. 1 to 7 in the last column of Table 3.3. An approximate analytical expression is given by[4]

$$\frac{\bar{R}}{Z_0} = \left(\frac{c}{\lambda\sigma}\right)^{1/2}\frac{\eta}{4}, \tag{3.33}$$

where η is a form-factor. For the capacitive mesh η is roughly given by $\eta = 1/[1 - 2(a/g)]$, for inductive meshes $\eta = g/2a$. We see that $\bar{R}$ does not depend very strongly on the frequency. We note here that for some interference filters it is important to take $\bar{R}$ into account since losses reduce the peak transmissivity.

d. Phase Measurements

If we wish to check the relations given by Eq. 3.27, phase measurements have to be performed as will be discussed below. We only mention here that for this purpose Russell and Bell[15a] have employed Fourier transform spectroscopy using a Michelson interferometer operated in the asymmetric mode. This technique allows calculation of the transmissivity and the phase from the interferogram as discussed in Chapter 4. Figure 3.8 shows the power transmittance and the phase of the amplitude transmittance at normal incidence for a self-supporting nickel mesh (inductive) with the parameters $g = 50.8\,\mu$ and $2a = 14\,\mu$.

e. Fabrication of Meshes

1. Inductive Meshes. A large variety of inductive meshes are available from Buckbee Mears Co., St. Paul, Minnesota. They can also be made by a photoetching process as described below.

2. Capacitive Meshes. Ulrich[4,15b] has reported a photoetching process using a substrate of 2.5 μ thick Hostaphan sheets. He deposited copper on the Hostaphan film *in vacuo* and coated the copper layer with Kodak Photo Resist. An inductive mesh of the desired parameters was used as a mask, and

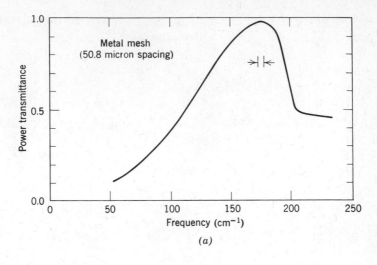

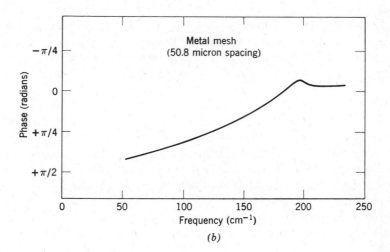

Fig. 3.8   Optical properties of a nickel mesh of $g = 50.8\ \mu$; (a) power transmittance at normal incidence; (b) phase angle of the (complex) amplitude transmittance at normal incidence. [E. E. Russell and E. E. Bell, *Infrared Phys.* **6**, 75 (1966), by permission of Pergamon Press.]

after exposure and development, the shaded areas could be etched with FeCl$_3$ solution and rinsed away.

Small capacitive meshes on a quartz substrate have been made by Ressler and Möller[18] by direct vacuum deposition of aluminum using an inductive mesh as a mask. This technique also works well on Mylar or high-density polyethylene substrates.

C. INTERFERENCE FILTERS

a. Theory of Two-Mesh Filters

We now discuss the far-infrared properties of interference filters. These filters use metal meshes as reflector plates. Their theoretical principles have been established by Ulrich[4] on the basis of their transmission line representation (see Fig. 3.9(a)) following p. 10 of Ref. 12. He gives the following relation between the frequency at maximum transmittance of a filter, ω_q, and the susceptances $B_1(\omega)/Y_0 = B_2(\omega)/Y_0$ of the two lossless, thin meshes serving as reflector plates:

$$\cot\left(2\pi\omega_q\frac{d}{g} \pm q\pi\right) - \frac{B(\omega_q)}{2Y_0} = 0. \tag{3.34}$$

The quantity $q = 1, 2, \cdots$ denotes the order of interference, and d is the distance between the meshes. We remark that Eq. 3.34 is essentially identical with the condition that $\delta = 2\pi q$ (derived in Section A.b, Eq. 3.2).

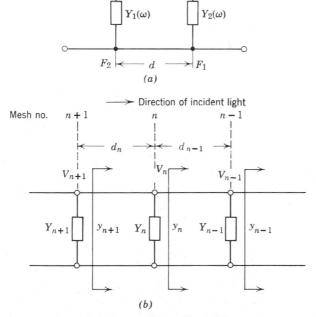

Fig. 3.9 Equivalent circuits: (a) for a two-mesh filter, distance between the two meshes, d; (b) for a multiple-mesh filter, V_n = voltage across the shunt admittance Y_n, d_n = spacing between two plates, y_n = admittance of the filter to the right of the nth layer.

We begin with filters made from *inductive meshes*. Considering thin and lossless inductive meshes, we can replace $B(\omega)/Y_0$ in Eq. 3.34 by $-\omega_0/\omega\Xi$ (Lewis and Casey low-frequency approximation; see Tables 3.1, 3.2, and the definition of $B(\omega)$ in Section B.c). Applying the small-angle approximation for the cotangent, we then obtain

$$\omega_q = \frac{q}{2d/g + 2\Xi/\pi\omega_0} \qquad (3.35)$$

for the position of the interference maxima. From Eq. 3.35 we can see that in this low-frequency approximation the transmittance maxima are found at equally spaced frequency intervals of ω_1 ("harmonic"). If this type of filter is used as the dispersion element in a spectrometer, the removal of higher-order radiation would meet the same difficulties as those encountered in grating spectrometers.

For thin and lossless *capacitive* meshes, we replace $B(\omega_0)/Y_0$ in the low-frequency approximation by $\omega/\omega_0\Xi$ (see Table 3.1 and 3.2). In the small-angle approximation, from Eq. 3.34 we obtain

$$\omega_q = \frac{q - \frac{1}{2}}{2d/g + 1/2\pi\omega_0\Xi}. \qquad (3.36)$$

We see that in the case of thin, lossless capacitive meshes the frequencies of the interference orders q_1, q_2, q_3 is $\omega_1, 3\omega_1, 5\omega_1$, and so forth; there are no even harmonics of the fundamental passband. However, at frequencies where the above approximate expression is applicable, the reflectance of the capacitive mesh, $|\Gamma|^2$, is small and therefore good interference filters with high finesse cannot be constructed. At higher frequencies the finesse increases, but Eq. 3.36 is no longer applicable. Instead, a more complex formula applies, yet it still follows that there will be no even, harmonically lying higher orders. This type of filter, especially in an improved version with more than two meshes (see below), is useful as a prefilter for a grating spectrometer or tunable Fabry-Perot interferometer using inductive meshes.

Ulrich[4] has also investigated a combination consisting of one inductive and one capacitive mesh. This combination yields a bandpass filter of a width (at half of the peak transmittance) which is less than one octave but has wide side wings.

We conclude this section with a few remarks on the effects of unevenness of the meshes on the optical properties of the interference filters (for more details, see Ulrich et al.[3]). It turns out that the unevenness Δd should be small in comparison with the quantity $\lambda/4n\mathscr{F}$ (n is the refractive index, $\mathscr{F}$ the finesse). The filter properties remain almost unchanged as long as Δd is less than one-tenth of this quantity. If $\Delta d \approx \lambda/4n\mathscr{F}$, the maximum transmittance (P_{t_0}) and $\mathscr{F}$ will have dropped to about one-half of their ideal values. To give a numerical example, let $\lambda = 100\,\mu$, $\mathscr{F} = 100$, and $n = 1$; for $\Delta d \approx 0.1(\lambda/4n\mathscr{F})$, we find $\Delta d \sim 0.025\,\mu$. This value, which is considerably smaller than the thickness of inductive or capacitive meshes and their substrates (if any), gives an indication of the limits of the applicability of this consideration to the evaluation of such metal meshes and the construction of efficient interference filter plates.

With respect to the spacing (d) between the filter plates, it is always assumed that d is made sufficiently large $(d > g/3,\ \lambda \gg g)$ so that interactions between diffraction modes of higher orders are negligible.

b. Formulas for the Calculation of the Transmittance of Multi-Mesh
   Filters (according to R. Ulrich)

We will give some relations which are useful for calculations on multi-mesh filters. The filter is assumed to consist of meshes $n + 1, n, n - 1, \cdots$ separated by the distances $d_{n+1}, d_n, d_{n-1}, \cdots$ as indicated in Fig. 3.9(b). The equivalent circuit consists of shunt admittance Y_n (we set $Y_0 = 1$) and a voltage of V_n across the shunt admittance Y_n. The part of the filter to the right side of the nth layer is represented by the admittance y_n; Y_n adds to y_n. The resulting recursion formula is given by

$$y_{n+1} = \frac{y_n + Y_n + j\tan(kd_n)}{1 + j(y_n + Y_n)\tan(kd_n)}, \tag{3.37}$$

where k is the quantity $(2\pi/\lambda)$ and y_1 is set equal to unity. The ratio of the voltage across the $(n + 1)$th and the nth shunt admittance is given by

$$V_{n+1}/V_n = \cos(kd_n) + jy_n\sin(kd_n). \tag{3.38}$$

Using Eqs. 3.37 and 3.38 repeatedly, and starting at grid No. 1 (right end of the filter), we need to evaluate the two quantities y_M and $V_M/V_1 = (V_M/V_{M-1})(V_{M-1}/V_{M-2})\cdots(V_2/V_1)$, where M indicates the last layer on the left side. From this, the amplitude transmission of the whole filter is obtained as

$$\tau = \frac{2}{1 + y_M + Y_M}\frac{V_1}{V_M}. \tag{3.39}$$

The power transmission is given by $|\tau|^2$.

c. Construction of Fabry-Perot Interference Filters

1. Self-Supporting Meshes. The meshes have to be stretched in order to be sufficiently flat. Figure 3.10 shows such an arrangement. The meshes are fixed on rings which can be pressed down by spring-loaded screws. The edges, indicated by B and C, stretch the mesh (like the rims of a drum stretch the skin). Ring 1, which holds only Mesh 1, can be turned against Ring 2;

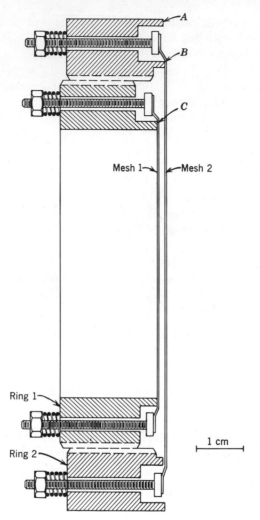

Fig. 3.10 Mount for a two-mesh interference filter. [R. Ulrich, *Infrared Phys.* **7**, 37 (1967), by permission of Pergamon Press.]

consequently, the distance between the two meshes can be changed. Ulrich *et al.*[3] also constructed a device in which the distance d between the meshes could be varied in a more reproducible and precalibrated manner. Such a device can serve as a tunable interference filter. It will be discussed in more detail in Section D.

2. Meshes on the Surface of a Dielectric. The mesh can be glued to the surface of an optically flat crystal quartz plate with thinned collodion. The adjustment of parallelism can be done interferometrically by observing Haidinger fringes[6] with visible light (these fringes are caused by reflection from the surfaces of the quartz plate). The fringes can be greatly enhanced by using semitransparent silvered annular areas around the interference filter area on both plates, as shown in Fig. 3.11. To obtain a mesh on the surface of a quartz or of a plastic film, direct vacuum deposition or a photoetching process can be employed as mentioned in Section B.e.

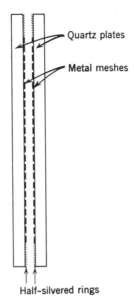

Fig. 3.11   Schematic arrangement of meshes glued on quartz plates. Half-silvered rings serve for the adjustment.

3. Meshes on Both Sides of a Dielectric. The mesh can be vacuum-deposited, glued, or photoetched at both sides of a crystal quartz plate or a plastic film. This method gives, of course, a fixed-spaced Fabry-Perot interference filter.

d. Measurements on Interference Filters and Comparison with Theory

1. Two Inductive Meshes. Filters of two meshes ($g = 50\,\mu$) have been investigated by Renk and Genzel.[2] The meshes were cemented on crystal quartz plates with thinned collodion. Figure 3.12 shows the interference maxima ($q = 1$ to 6) in the 100–600 μ wavelength region. The maxima turned out to be harmonically spaced by equal frequency intervals in agreement with Eq. 3.35. (Note that the abscissa in Fig. 3.12 is in wavelength units.) The experimental slit widths are indicated in the figure, and the dotted

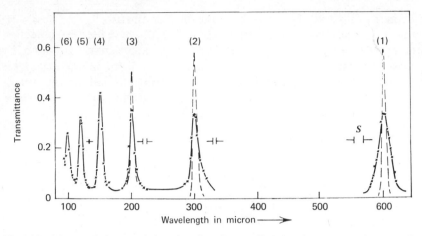

Fig. 3.12   Measured characteristics of an interference filter (meshes on crystal quartz plates, each 5 mm thick): x, measured, – – – – corrected to zero spectrometer slit width and parallel beams; S is the equivalent spectrometer slit width. The orders of interference are given in parentheses. [K. F. Renk and L. Genzel, *Appl. Opt.* **1**, 643 (1962).]

lines designate the line shapes corrected to zero slit width and parallel light.[19,20] At 600 μ the finesse $\mathscr{F}$ turns out to be 60 and the power transmittance $|\tau|^2$ is about 0.6.

Using higher resolution, Rawcliffe and Randall[16] investigated an interference filter made of self-supporting meshes (mesh No. 750, $g \sim 34\,\mu$) at a fixed distance of 0.807 mm (see Section c above). The finesse, computed from the transmittance data on one mesh with the help of Eqs. 3.1 to 3.7, is presented in Fig. 3.13 by the line of open squares. Correction for the convergence angle leads to the curve with the open circles whereas additional correction for the instrumental resolution results in the curves of open triangles. The solid dots show the measured values of the finesse of the interference filter (obtained from determinations of the half-width). We see that these measured values are somewhat higher than the fully corrected computed results but, on the whole, agree satisfactorily with them.

2. Two Capacitive Meshes. An interference filter consisting of two capacitive meshes was investigated by Ulrich.[4] The meshes had the dimensions $g = 368\,\mu$, $a = 17.7\,\mu$, $t = 5\,\mu$, and were photoetched in copper on a Hostaphan substrate of 2.5 μ thickness. The distance between the reflector plates was 1.250 mm. The full line in Fig. 3.14 is the filter transmittance calculated for these meshes with $2\Xi = 0.274$, $\omega_0 = 1$, and $2\bar{R}/Z_0 \sim 0.02$ (see Table 3.3, No. 1). The dotted line represents the experimental values (open circles). We see that the interference maxima are not harmonically positioned (confirming the expectation from Eq. 3.36).

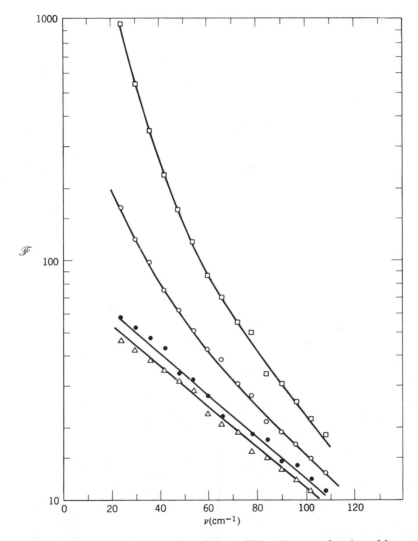

Fig. 3.13   Finesse of an interference filter of two #750 meshes as a function of frequency. Measured values: ●●●. [R. D. Rawcliffe and C. M. Randall, *Appl. Opt.* **6**, 1353 (1967).]

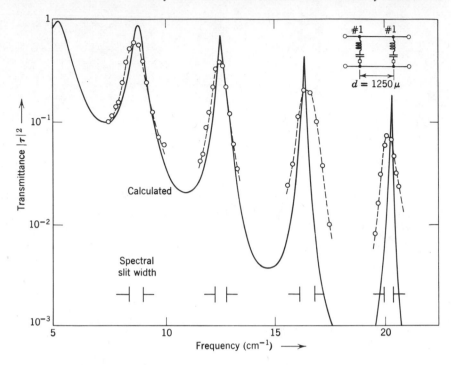

Fig. 3.14   Transmittance $|\tau|^2$ of an interference filter constructed from two capacitive meshes. [R. Ulrich, *Infrared Phys.* **7**, 37 (1967), by permission of Pergamon Press.]

An investigation of some interference filters constructed of meshes of different g values was reported by Varma and Möller.[21] Their data are shown in Fig. 3.15. In general, the design of their filters followed the principles given by Ulrich,[4] but note that the same spacing $d^* = 37\,\mu^\dagger$ was used for all g values between 78 μ (mesh No. 325) and 423 μ (mesh No. 60) (Ulrich used $d = g/2\omega_0$).

3. Four Capacitive Meshes. Ulrich[5,22] published a theoretical and experimental study on the optical properties of interference filters consisting of four capacitive meshes. The meshes were put on a Hostaphan substrate of 2.5 μ thickness. The four layers were glued to metal rings and mounted in equidistant positions on a holder similar to the one shown in Fig. 3.10, but at fixed spacings d. The theoretical design curves are shown in Fig. 3.16. After selecting a suitable filter and using the values of 2Ξ and ω/ω_0, the geometrical parameters can be found with the help of Eqs. 3.31 and 3.32 and the relation

---

† The thickness of the spacer actually used is given by the symbol d^*. In the theory, the spacing between the meshes is given by d. In the case of inductive meshes of thickness t, one may use $d \approx d^* + t$.

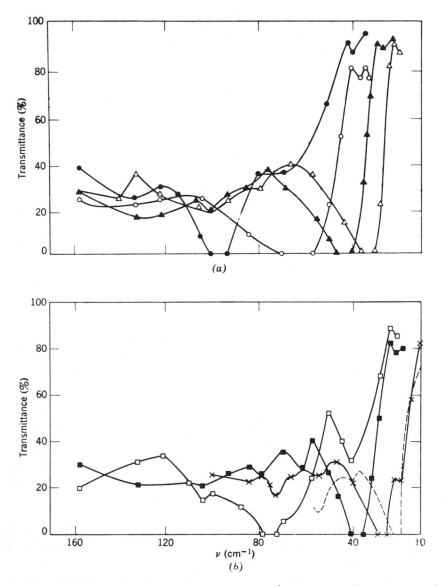

Fig. 3.15 Transmittance as a function of frequency (cm⁻¹) for two-mesh filters constructed from capacitive meshes of the following dimensions: All meshes have an a/g value of about 0.2. The value for the spacing is the same for all filters: $d^* = 37\,\mu$ and $g_1 = g_2 = g$. (a) ●—●—●—●, $g = 78\,\mu$ (#325); ○—○—○—○, $g = 127\,\mu$ (#200); ▲—▲—▲—▲, $g = 181\,\mu$ (#140); △—△—△—△, $g = 254\,\mu$ (#100). (b) □—□—□—□, $g = 101\,\mu$ (#250); ■—■—■—■, $g = 211\,\mu$ (#120); ×—×—×—×, $g = 317\,\mu$ (#80); ----, $g = 423\,\mu$ (#60). [S. P. Varma and K. D. Möller, *Appl. Opt.* **8**, 1663 (1969).]

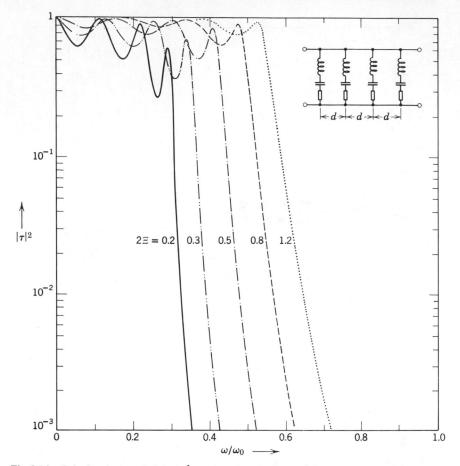

Fig. 3.16   Calculated transmissivity $|\tau|^2$ as a function of ω/ω_0 and the parameter 2Ξ of an interference filter made from four capacitive meshes. [R. Ulrich, *Infrared Phys.* **7**, 65 (1967), by permission of Pergamon Press.]

$d = g/2\omega_0$. Some of the experimental results are shown in Fig. 3.17 for three of such four-mesh filters ($g = 51\ \mu$, $a/g = 0.18$) as a function of the spacing d^* as indicated in the figure.

Ulrich[22] also studied four-layer filters having different mesh constants and different spacings; an example is shown in Fig. 3.18. By comparing Fig. 3.18 with Fig. 3.17, we see that undesirable maxima in the stop-band are suppressed. The transmittance measured in the 10^{-3} to 10^{-4} power range is a consequence of noise and was not reproducible.

Varma and Möller[21] have used two interference filters with different g values and constructed a four-layer filter from them by putting a spacer

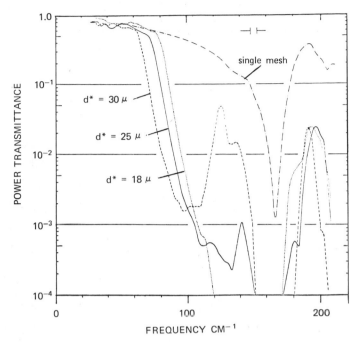

Fig. 3.17   Power transmittance as function of frequency (cm^{-1}) for a four-layer filter of capacitive meshes of $g = 51 \mu$, $a/g = 0.18$, and equal spacers. The thickness of the spacers was varied for the three filter assemblies as indicated. [R. Ulrich, *Appl. Opt.* 7, 1987 (1968).]

between the two filters. The resulting device is a "cut-on" filter with a stop-band of about one octave width. Examples are shown in Fig. 3.19. The transmittance of the two-layers filters employed for the construction of the four-layer filter is shown in Fig. 3.15.

4. Eight Capacitive Meshes. Varma and Möller[21] have also constructed an eight-layer filter by using two four-mesh filters separated by a spacer. Four of the meshes with their spacers were glued, one after the other, onto a metal ring with Eastman Kodak Adhesive 910. The other four meshes were then stretched on the package of the first four meshes as described above. (It is possible to glue all meshes and spacers together into one package.) The transmittance of this filter is shown in Fig. 3.19.

5. Capacitive and Inductive Crosses. Figure 3.20 shows the power transmission of capacitive meshes with mesh patterns in the form of crosses (Ulrich[22]). The dotted line shows the transmittance for a single mesh, the solid line for a double-mesh filter.

We see from the figure that the two-mesh device acts as a band-stop filter. Its complementary structure, called "inductive crosses," acts as a bandpass

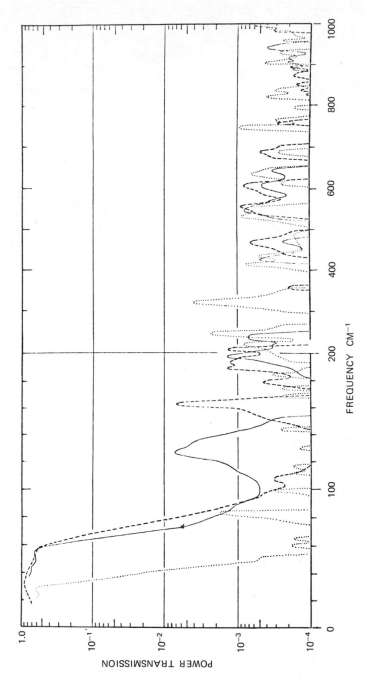

Fig. 3.18   Power transmittance of four-mesh filters as function of frequency (cm^{-1}) for capacitive meshes of the following dimensions: ······ $g_1 = g_2 = g_3 = 102$ μ, $a/g = 0.08$ and $g_4 = 51$ μ, $a/g = 0.18$, with spacers $d_1^* = d_2^* = 50$ μ, $d_3^* = 40$ μ. ——— $g_1 = g_4 = 25$ μ, $a/g = 0.10$; $g_2 = g_3 = 51$ μ, $a/g = 0.18$, with spacers $d_1^* = d_3^* = 28$ μ, $d_2^* = 20$ μ. ------- $g_1 = g_4 = 25$ μ, $a/g = 0.10$; $g_2 = g_3 = 51$ μ, $a/g = 0.18$, with spacers $d_1^* = d_2^* = d_3^* = 20$ μ. [R. Ulrich, *Appl. Opt.* **7**, 1987 (1968).]

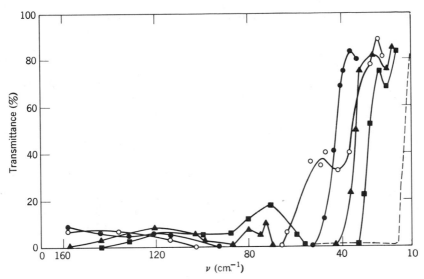

Fig. 3.19   Transmittance as function of frequency (cm^{-1}) of four-mesh and one eight-mesh filter constructed from capacitive meshes with the following dimensions. All capacitive meshes have an approximate a/g value of 0.2. For all four-mesh filters $d_1^* = d_3^* = 37\,\mu$, $d_2^* = 50\,\mu$, and $g_1 = g_2$, $g_3 = g_4$: ●—●—●—● $g_1 = 78\,\mu$; $g_3 = 127\,\mu$; ○—○—○—○ $g_1 = 78\,\mu$; $g_3 = 101\,\mu$; ▲—▲—▲—▲ $g_1 = 127\,\mu$; $g_3 = 181\,\mu$; ■—■—■—■ $g_1 = 181\,\mu$; $g_3 = 211\,\mu$. Eight-mesh filter: – – – – – – $g_1 = g_2 = 211\,\mu$; $g_3 = g_4 = 182\,\mu$; $g_5 = g_6 = 425\,\mu$; $g_7 = g_8 = 317\,\mu$; $d_1^* = d_3^* = d_5^* = d_7^* = 37\,\mu$; $d_2^* = d_6^* = 50\,\mu$ and $d_4^* = 100\,\mu$. [S. P. Varma and K. D. Möller, *Appl. Opt.* **8**, 1663 (1969).]

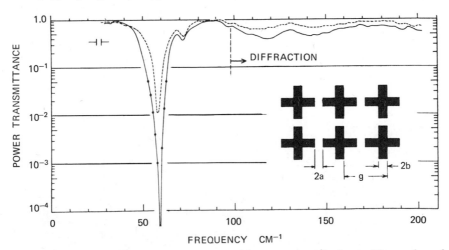

Fig. 3.20   Power transmittance as a function of frequency (cm^{-1}) of capacitive meshes of crosses of the following dimensions: $g = 102\,\mu$, $a/g = 0.13$, $b/g = 0.06$; – – – – – – single mesh; ———— two-mesh filter with spacing $d^* = 38\,\mu$. [R. Ulrich, *Appl. Opt.* **7**, 1987 (1968).]

filter. These types of filters, although difficult to make at present, will probably have important future applications.

e. Phase Measurements with Interference Filters

Measurements of the phase of the complex reflectance (see also Section A.b) for meshes of both the inductive and the capacitive type (mesh No. 3 and 7 of Table 3.3) have been reported,[4] and the transmittance characteristics of an interference filter, constructed from two of these meshes, was recorded as a function of the filter spacing d at several fixed wavelengths. Application of Eq. 3.2 then yields the phase. In Fig. 3.21 the measured phase angles are plotted in the complex plane, using $\omega = g/\lambda$ as parameter. The distance of the points from the origin is given by $(1 - |\tau|^2)^{1/2}$, which is equal to $|\Gamma|$ if absorption is neglected (see Eq. 3.25). We see that the experimental data taken for the capacitive mesh are located close to the theoretically

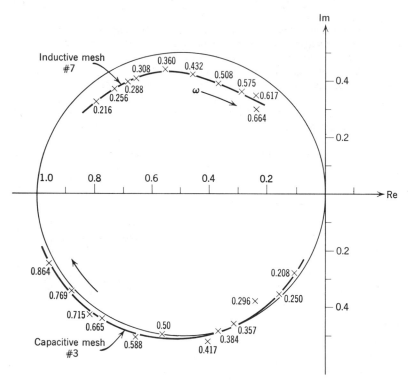

Fig. 3.21   Complex amplitude transmission of an inductive and a capacitive mesh. The numbers on the curves designate the normalized frequency ω. [R. Ulrich, *Infrared Phys.* 7, 37 (1967), by permission of Pergamon Press.]

predicted circle but that the agreement for the inductive meshes is less satisfactory.

D. Tunable Fabry-Perot Interferometer

Ulrich *et al.*[3] have constructed a Fabry-Perot interferometer using self-supporting inductive meshes ($g = 50\,\mu$, $a = t = 6\,\mu$) as the reflecting plates with a continuously variable plate distance d. Figure 3.22 shows a cross section of the reflector assembly of their Fabry-Perot interferometer. The

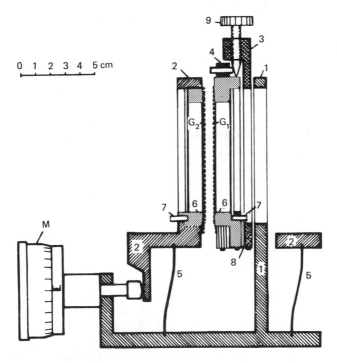

Fig. 3.22   Tunable Fabry-Perot étalon constructed from inductive meshes (without substrate). [R. Ulrich, K. F. Renk, and L. Genzel, *IEEE Trans.* **MTT-11**, 363 (1963), by permission of The Institute of Electrical and Electronics Engineers, Inc.]

meshes are fixed to the frames (No. 2 and 4 in Fig. 3.22) and stretched by semitransparent silvered glass rings pressed against them. Frame 2 is mounted on a table which is held by the blade springs 5. By means of the micrometer screw M, it is possible to position the whole assembly to any small distance with respect to the fixed part holding Frame 4. With two smaller screws it is possible to adjust Frame 4 in a parallel position to the fixed holder 1. The parallelism can be checked by observing interference fringes (of a visible monochromatic source) generated between the two inner surfaces of the two glass rings (see also Section C.c).

This tunable Fabry-Perot interferometer has been used as the dispersion element in a spectroscopic arrangement which is shown schematically in Fig. 3.23: M_1 and M_4 designate the elliptical and M_3 a spherical mirror. (The rays are drawn unfolded at the mirrors.) Ch is a chopper (12.5 Hz modulation), B is the detector (bolometer), and P is a 1.2-mm thick filter made of black polyethylene. In addition, the filtering of higher orders was accomplished by a metal mesh of $g = 108 \, \mu$ and $g/a = 4.8$, which served as a reflection filter (see Section E.h of Chapter 1). Furthermore, since the atmosphere in the instrument was not removed, some filtering action by the stronger absorption lines of water vapor (at wavelengths shorter than $200 \, \mu$) was obtained (effective path length was 5 m).

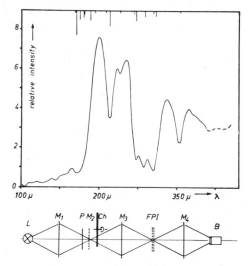

Fig. 3.23   Absorption spectrum of water vapor obtained with a tunable Fabry-Perot. Source: mercury lamp. The optical schematic diagram is drawn unfolded. [R. Ulrich, K. F. Renk, and L. Genzel, *IEEE Trans.* **MTT-11**, 363 (1963), by permission of The Institute of Electrical and Electronics Engineers, Inc.]

The usable diameter of the opening of the Fabry-Perot was 7.5 cm, of which only an opening of 2 cm was used, thus reducing the requirements for perfect parallelism. Figure 3.23 shows the spectrum of water vapor between 200 and $400 \, \mu$ obtained with this Fabry-Perot interferometer. The resolution is about the same as that obtainable with a grating spectrometer which employs the full area ($6 \times 6 \, cm^2$) of its grating (4 grooves/mm).

E. Fabry-Perot Interferometer in Higher Orders

In order to obtain high resolution, Grisar et al.[23] have used a Fabry-Perot interferometer in higher orders in conjunction with a grating spectrometer. This approach is similar to that which makes use of a Fabry-Perot interferometer in the visible spectral region. However, in order to obtain a sufficiently large flux of far-infrared radiation through the Fabry-Perot interferometer, the slits had to be opened rather wide. Since the Fabry-Perot interferometer has axial symmetry, this does not decrease its resolution, but the grating spectrometer then acts as a tunable bandpass filter.

The optical schematic diagram of this assembly is given in Fig. 3.24. It was possible to obtain a resolution of 0.4 cm^{-1} at 90 cm^{-1} in twentieth order with inductive meshes of $g = 35 \mu$. [A sensitive (He-cooled) detector turned out to be a prerequisite for the success of the experiment.]

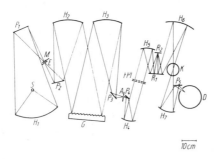

Fig. 3.24   Fabry-Perot interferometer (FPI) used in higher orders in conjunction with a grating spectrometer: S, source, H, spherical mirrors, P, plane mirrors, M, modulator, E and A, entrance and exit slits, R, reflection filters, K, cryostat, and D, detector. [R. G. J. Grisan, K. P. Reiners, K. F. Renk, and L. Genzel, *Phys. Stat. Sol.* **23**, 613 (1967).]

F. Comparison of a Grating and a Fabry-Perot Spectrometer under Energy-Limited Conditions

We now compare a grating spectrometer (as considered in Chapter 1) with a Fabry-Perot tunable interference filter or Fabry-Perot spectrometer (as presented in Section D). We assume that the area of the source is the same in both spectrometers and is sufficiently large to illuminate fully the optical components. The flux which passes through the grating spectrometer is (see Chapter 1, Eq. 1.10b)

$$P_G \approx B_\lambda \Delta\lambda_G \tau_G S_G W L, \tag{3.40}$$

where B_λ is the spectral brightness of the source, $\Delta\lambda_G$ the wavelength band passing through the spectrometer, τ_G the transmission factor, S_G the grating area, $W = a/F$ the angular slit width (a = physical width, F = focal length of the telescope mirror), and $L = b/F$ the angular slit height (b = slit height).

Similarly, the flux passing through the Fabry-Perot spectrometer is given by

$$P_{FP} \approx B_\lambda \Delta\lambda_{FP} \tau_{FP} S_{FP} (c/F)^2, \tag{3.41}$$

where c^2 is the area of the aperture and S_{FP} the area of the Fabry-Perot étalon. The other symbols have the same meaning as in Eq. 3.40. A schematic drawing is shown in Fig. 3.25.

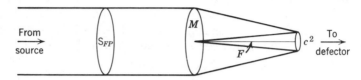

Fig. 3.25   Optical arrangement for the comparison of a Fabry-Perot and a grating spectrometer: S_{FP}, area of Fabry-Perot; F, focal length of telescope mirror M; c^2, aperture area.

Furthermore, we will assume that: (*1*) the detectors of both spectrometers have the same NEP (see Chapter 2, Section B); (*2*) the product of the area and the solid angle for both spectrometers is the same and equal to that at the detectors, that is to say $(A\Omega)_{Det} = (A\Omega)_G = (A\Omega)_{FP}$; and (*3*) we work with the same minimum acceptable signal-to-noise level, $P_{min} = P_G = P_{FP}$. We then find from Eq. 3.40 and 3.41 that

$$\Delta\lambda_G \tau_G = \Delta\lambda_{FP} \tau_{FP}. \tag{3.42}$$

Under ideal conditions (no reflection losses on reflector plates and filters, and so forth) in both types of spectrometers, the transmission coefficients are the same, $\tau_G = \tau_{FP}$. We thus obtain the same resolution in both spectrometers. However, these ideal conditions are certainly never realized, and τ_G and τ_{FP} are different in most cases.

Jacquinot[24] has shown that for the same resolving power a considerably larger flux can pass through a Fabry-Perot étalon than through a grating spectrometer. The flux was calculated for this comparison by the use of the solid angle containing all radiation of bandwidth $\Delta\lambda$, which is given by $\Omega' = 2\pi\Delta\lambda/\lambda$. This solid angle Ω' is larger than the solid angle used for the Fabry-Perot spectrometer in our above comparison because of the limited $(A\Omega)_{Det}$ we had to employ (see also Vanasse and Sakai[25]).

Yet we can realize an advantage of practical importance in using a Fabry-Perot rather than a grating spectrometer: We can make the area of the Fabry-Perot plates, S_{FP}, considerably smaller than the grating area. Therefore, using the same $(A\Omega)_{Det}$, the minimum area of the Fabry-Perot étalon is given by

$$S''_{FP} > \frac{(A\Omega)_{Det}}{\Omega''}, \tag{3.43}$$

where Ω'' is the solid angle emerging from the area S''_{FP} (the comparison mentioned at the end of Section D makes use of this relation; see also Fig. 3.23). The advantage of being able to reduce S_{FP} not only results in a more compact device but also improves the finesse since the plates can be kept in a more even position.

Another practical advantage of the Fabry-Perot spectrometer over the grating spectrometer in the far-infrared is the circular shape of the étalon which allows full illumination of circularly shaped detector windows, thereby dispensing with the need of image transformers (see Chapter 2, Section G).

REFERENCES

1. J. Strong, *Concepts of Classical Optics*, Freeman, San Francisco, 1958.
2. K. F. Renk and L. Genzel, *Appl. Opt.* **1**, 643 (1962).
3. R. Ulrich, K. F. Renk, and L. Genzel, *IEEE Trans.* **MTT-11**, 363 (1963).
4. R. Ulrich, *Infrared Phys.* **7**, 37 (1967).
5. R. Ulrich, *Infrared Phys.* **7**, 65 (1967).
6. M. Born and E. Wolf, *Principles of Optics*, Pergamon, New York, 1964.
7. P. Jacquinot, "New Developments in Interference Spectroscopy," in *Reports Progr. Phys.* **23**, 267 (1960).
8. L. N. Hadley and D. M. Dennison, *J. Opt. Soc. Am.* **37**, 451 (1947).
9. W. Culshaw, *IRE Trans.* **MTT-8**, 182 (1960).
10. J. P. Casey and E. A. Lewis, *J. Opt. Soc. Am.* **42**, 971 (1952).
11. T. Larsen, *IRE Trans.* **MTT-10**, 191 (1962).
12. N. Marcuvitz, *Waveguide Handbook*, MIT Radiation Lab. Series 10, McGraw-Hill, New York, 1951.
13. W. L. Weeks, *Electromagnetic Theory for Engineering Applications*, Wiley, New York, 1964.
14. V. Y. Balakhanov, *Sov. Phys. Doklady* **10**, 788 (1966).
15a. E. E. Russell and E. E. Bell, *Infrared Phys.* **6**, 75 (1966).
15b. R. Ulrich, *Appl. Opt.* **8**, 319 (1969).
16. R. D. Rawcliffe and C. M. Randall, *Appl. Opt.* **6**, 1353 (1967).
17. P. Vogel and L. Genzel, *Infrared Phys.* **4**, 257 (1964).
18. G. M. Ressler and K. D. Möller, *Appl. Opt.* **6**, 893 (1967).
19. G. Koppelmann and K. Krebs, *Z. Physik* **157**, 592 (1960).
20. G. Koppelmann and K. Krebs, *Z. Physik* **158**, 172 (1960).
21. S. P. Varma and K. D. Möller, *Appl. Opt.* **8**, 1663 (1969).
22. R. Ulrich, *Appl. Opt.* **7**, 1987 (1968).
23. R. G. J. Grisar, K. P. Reiners, K. F. Renk, and L. Genzel, *Phys. Stat. Sol.* **23**, 613 (1967).
24. P. Jacquinot, *J. Opt. Soc. Am.* **44**, 761 (1954).
25. G. A. Vanasse and H. Sakai, in *Progress in Optics*, Vol. 6, North-Holland, Amsterdam, 1967, p. 307.

4 Fourier Transform Spectroscopy

A. Introduction

An interferogram is obtained by the superposition of two light beams and the variation of their optical path difference. In particular, the interferogram may be recorded in terms of the intensity as a function of distance, and by Fourier analysis converted into an intensity *vs* frequency distribution.

The most frequently used method at present is the so-called aperiodic Fourier transform spectroscopy which employs two-beam interferometers such as the Michelson and the lamellar-grating interferometer. The reason that this type of spectroscopy has become widely used recently lies in the necessity of performing a large number of numerical calculations, a task which is only possible with the help of a computer.

The principle advantage of Fourier transform spectroscopy over the conventional spectroscopy is its higher signal-to-noise ratio. This may be understood by considering an analogy in the visible spectral region. We might use (*1*) a phototube as detector and study with a conventional spectrometer one spectral element after the other, each for a time T/N, where N is the number of spectral elements and T the total time of observation. We also could use (*2*) a photographic plate and study all N spectral elements for the total time T. It is clear that the second method yields a higher signal-to-noise ratio since all spectral elements are observed during the total time. The Fourier analysis of an interferogram, compared to grating spectroscopy, offers the same advantage. (We will discuss this in more detail in Section F.)

The method of periodic Fourier transform spectroscopy is discussed in Appendix VI.

In the literature, the publications by Loewenstein[1] (many references and historical notes), Strong[2] and Mertz[3] (Fourier transformation in optics), Papoulis[4] (mathematical formulation and theorems), Connes[5] (detailed description of Fourier transform spectroscopy), and Vanasse and Sakai[6] are particularly useful. We will frequently use the ideas expressed in these articles and books.

B. INTERFEROGRAM FUNCTION AND FOURIER TRANSFORMATION

For the following description, we choose a Michelson-type two-beam interferometer. Its schematic is shown in Fig. 4.1. The light from the source is divided by the beam splitter into Beams I and II. Each beam is reflected at the Mirrors I and II, respectively, and divided again at the beam splitter. We are only interested in the parts of Beams I and II which overlap and form Beam III, traveling toward the detector. Let us assume first that the source emits only monochromatic light of frequency v_0 and that the mirror M_{II} is displaced by $x/2$ from the zero point position at $x = 0$ (see Fig. 4.1).*

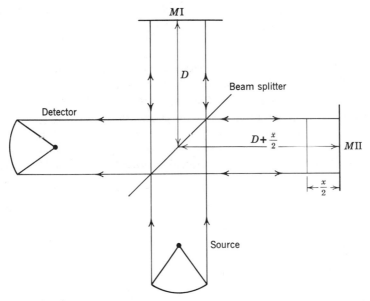

Fig. 4.1   Optical schematic of a Michelson interferometer.

The superposition of the two parts of Beams I and II which form Beam III can then be expressed by

$$g(x) = 4S \cos^2 (\pi v_0 x) = 2S(1 + \cos [2\pi v_0 x]), \tag{4.1}$$

where $\sqrt{S}$ is the amplitude of the equal fractions of Beams I and II. The function $g(x)$ goes through minima and maxima as a function of the displacement x of M_{II} along the optical path. If we now consider a source that emits light of a continuous spectrum of frequencies $0 \leqslant v \leqslant \infty$, S becomes a

---

* In this chapter the frequency v is given in cm^{-1} and the displacement x in cm, unless otherwise specified.

function of v and we have to integrate from $v = 0$ to $v \to \infty$:

$$g(x) = 2 \int_0^\infty S(v)(1 + \cos [2\pi vx])\, dv$$

$$= \int_0^\infty 2S(v)\, dv + \int_0^\infty 2S(v) \cos [2\pi vx]\, dv. \tag{4.2}$$

For $x = 0$, the total illumination, we have

$$g(0) = 2 \int_0^\infty 2S(v)\, dv. \tag{4.3}$$

Inserting Eq. 4.3 into Eq. 4.2, we can write

$$s(x) = g(x) - \tfrac{1}{2}g(0) = \int_0^\infty 2S(v) \cos [2\pi vx]\, dv. \tag{4.4}$$

$s(x)$ is called the interferogram function; that is, the intensity as a function of distance (as mentioned in the Introduction). Figure 4.2 shows a schematic chart recording of $g(x)$ or $s(x)$. We notice that $s(x)$ has positive and negative values. It is a convenient function for the calculation of the Fourier transform $S(v)$ which gives us the desired intensity-frequency distribution.

Considering $S(v)$ to be an even function in the entire frequency range, we can write the following pair of Fourier transforms (see p. 271 of Ref. 6):

$$s(x) = \int_{-\infty}^{+\infty} S(v) \cos [2\pi vx]\, dv, \tag{4.5}$$

$$S(v) = \int_{-\infty}^{+\infty} s(x) \cos [2\pi vx]\, dx. \tag{4.6}$$

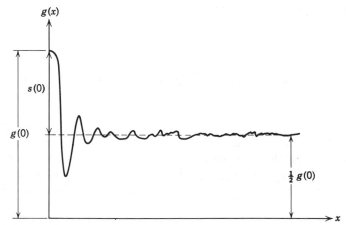

Fig. 4.2   Scheme of an interferogram.

We note that in this formulation we would have to use physically unrealizably large values of x for the calculation of the spectral distribution $S(v)$. We shall return to this point in the next section.

For monochromatic light, $s(x)$ has the form of a cosine function. Its Fourier transformation, which can be directly obtained from a knowledge of the scan speed, recorder speed, and the distance between the minima and maxima, is a function of the type $\delta(2\pi v - 2\pi v_0) + \delta(2\pi v + 2\pi v_0)$ (see Ref. 4, p. 37): $S(v)$ is only different from zero for $v = \pm v_0$. The peak at $-v_0$ is the consequence of the mathematically advantageous extension of $S(v)$ to negative frequencies. Since $S(v)$ in this case is in the form of δ-functions, the peak at $-v_0$ does not stretch into the physically interesting spectral region $v \geqslant 0$ and thus can be disregarded.

If we have two monochromatic waves coming from the source, for $s(x)$ we will observe an interferogram similar to the "phenomenon of beats." The Fourier transform of $s(x)$, $S(v)$, is in the form of δ-functions with peaks at $v = \pm v_1$ and $v = \pm v_2$. If now $S(v)$ is composed of more and more frequencies, the interferogram $s(x)$ will, in general, show less and less modulation at longer and longer displacements x of the mirror M_{II}.

C. FINITE PATH DIFFERENCE, APODIZATION, AND RESOLUTION

In the preceding section we dealt with limits of x between $-\infty$ and $+\infty$. In reality, x is restricted to $-L \leqslant x \leqslant L$. This finite movement of the mirror can be expressed by multiplying $s(x)$ in Eq. 4.6 with a symmetric "window function,"

$$p(x) = \begin{cases} 1 & \text{for } -L \leqslant x \leqslant L \\ 0 & \text{for other values.} \end{cases} \tag{4.7}$$

With it we can calculate an approximate value of the integral in Eq. 4.6:

$$S_1(v) = \int_{-\infty}^{+\infty} s(x)p(x) \cos [2\pi vx] \, dx. \tag{4.8}$$

$S_1(v)$ is the Fourier transform of the product of $s(x)$ and $p(x)$. This Fourier transform $S_1(v)$ can be expressed by using the convolution theorem (see Ref. 4, p. 25) which states: The Fourier transform of the product of the two functions $s(x)$ and $p(x)$ is equal to the convolution of the Fourier transform of $s(x)$ and of $p(x)$:

$$S_1(v) = \int_{-\infty}^{+\infty} S(\tau)P(v - \tau) \, d\tau, \tag{4.9}$$

where

$$S(\tau) = \int_{-\infty}^{+\infty} s(x) \cos [2\pi\tau x] \, dx \qquad (4.10)$$

and

$$P(\tau) = \int_{-\infty}^{+\infty} p(x) \cos [2\pi\tau x] \, dx. \qquad (4.11)$$

The integral of Eq. 4.11 is easily computed with the help of Eq. 4.7:

$$P(\tau) = 2L \frac{\sin [2\pi\tau L]}{[2\pi\tau L]}. \qquad (4.12)$$

The function $(\sin y)/y$ is sometimes designated by sinc y or by dif y. Insertion of Eq. 4.12 into Eq. 4.9 then gives

$$S_1(v) = \int_{-\infty}^{+\infty} S(\tau) 2L \frac{\sin [2\pi(v - \tau)L]}{[2\pi(v - \tau)L]} \, d\tau \qquad (4.13)$$

or

$$S_1(v) = \int_0^{\infty} S(\tau) 2L \left\{ \frac{\sin [2\pi(v - \tau)L]}{[2\pi(v - \tau)L]} + \frac{\sin [2\pi(v + \tau)L]}{2\pi[(v + \tau)L]} \right\} d\tau. \qquad (4.14)$$

Equation 4.13 or 4.14 is interpreted as the convolution of the actual spectrum $S(v)$ with the "theoretical apparatus function"

$$A_p(v, \tau) = 2L\{\text{sinc} [2\pi(v - \tau)L] + \text{sinc} [2\pi(v + \tau)L]\}. \qquad (4.15)$$

For the simple example of a monochromatic input, the interferogram function $s(x)$ in Eq. 4.10 has the form of a cosine-function of frequency v_0. Consequently the Fourier transform $S(\tau)$ (see Eq. 4.10) is a δ-function. Inserting this $S(\tau)$ into Eq. 4.14, we see that the output of a monochromatic input for finite path differences is represented by

$$A'_{p_1}(v, v_0) = 2L\{\text{sinc} [2\pi(v - v_0)L] + \text{sinc} [2\pi(v + v_0)L]\}. \qquad (4.16)$$

The quantity $A'_{p_1}(v, v_0)$ has the appearance of the sinc-functions with two peaks centered at $v = +v_0$ and $v = -v_0$. The peaks at $-v_0$ and $+v_0$ have wings; therefore we cannot neglect here *a priori* the peak at $-v_0$ as we did in the case of simple δ-functions (see Section B) since it may reach into the region $v > 0$. However, if $v_0 \gg 1/L$, this influence can be neglected. In the far-infrared, say for frequencies as low as 10 cm^{-1}, one has to choose $L \gg 0.1$ cm, a condition which is easy to fulfil.

Neglecting the second term in Eq. 4.16, the apparatus function is given by

$$A''_{p_1} = 2L \, \text{sinc} [2\pi(v - v_0)L]. \qquad (4.17)$$

A relation of the type of Eq. 4.17 exists for every spectrometer and expresses its resolving properties. For the Fourier spectrometer, considering a finite interferogram, Eq. 4.17 is plotted in Fig. 4.3 as curve a. The line width Δv (at half peak height) is about $1/2L$ and therefore depends on L; with increasing L it would become smaller and smaller. Consequently, we see that the resolution will be improved by taking larger and larger values for L.

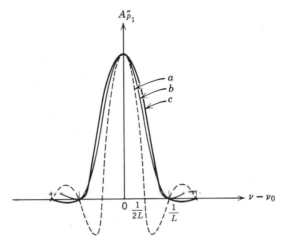

Fig. 4.3   Theoretical apparatus functions: (a) $2L$ sinc $[2\pi(v - v_0)L]$; (b) L sinc^2 $[\pi(v - v_0)L]$; (c) $\mathscr{I}_{5/2}[2\pi(v - v_0)L]/(2\pi(v - v_0)L)^{5/2}$.

The side lobes of the function A_{p_1}'' arise from the abrupt termination of the interferogram at $x = L$ through $p(x)$. These side lobes, some of them negative, may cause distortion in the analyzed spectrum. Removing or suppressing them can be accomplished by replacing the window function $p(x)$ by other functions. This process is called "apodization." The following two apodization functions (see Fig. 4.4) are frequently mentioned and used:

$$(1) \qquad h_1(x) = \begin{cases} 1 - \left|\dfrac{x}{L}\right| & \text{for } -L \leqslant x \leqslant L \\ 0 & \text{for all other values of } x. \end{cases} \qquad (4.18)$$

This is a triangular function as shown in Fig. 4.4. The Fourier transform of $h_1(x)$ is $H_1(v) = L$ sinc^2 $[\pi v L]$. It has only positive values and theoretically presents the same apparatus function as that of a grating spectrometer.

By using $h_1(x)$ instead of $p(x)$ and following the same considerations which lead to Eq. 4.17, we obtain for monochromatic input the following output function:

$$H_1(v) = L \text{ sinc}^2 [\pi(v - v_0)L]. \qquad (4.19)$$

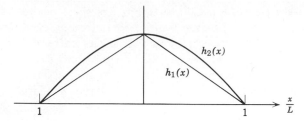

Fig. 4.4   Plot of $h_1(x)$ and $h_2(x)$ as a function of x/L.

Equation 4.19 is plotted in Fig. 4.3 as curve b. We see that the line width Δv is broadened by a factor of about two and consequently the resolution obtainable will be decreased by the same amount. But, on the other hand, we notice that the undesirable side wings are all positive and considerably smaller than in the case of the window function $p(x)$. As a consequence, the spectrum will be smoother.

$$(2) \qquad h_2(x) = \begin{cases} \left(1 - \dfrac{x^2}{L^2}\right)^2 & \text{for } -L \leqslant x \leqslant L \\[2mm] 0 & \text{for all other values of } x. \end{cases} \qquad (4.20)$$

The Fourier transform of $h_2(x)$ is $H_2(v) = \mathcal{I}_{5/2}(2\pi v L)/(2\pi v L)^{5/2}$ (Bessel function, see Ref. 6, p. 285). For a monochromatic input, we obtain for the output

$$H_2(v) = \frac{\mathcal{I}_{5/2}[2\pi(v - v_0)L]}{(2\pi v - v_0 L)^{5/2}}. \qquad (4.21)$$

The function H_2 is plotted in Fig. 4.3 as curve c.

For the general case of a nonmonochromatic input, we substitute $p(x)$ in Eq. 4.8 by the apodization function; that is, $h_1(x)$ or $h_2(x)$. The effect of the apodization on the computed spectrum may be seen by comparing the unapodized spectrum with the apodized one. Apodization is effective if information about unresolved or poorly resolved bands is required or for the evaluation of band contours.[7] An example is given in Fig. 4.5 according to Ref. 8.

From Eqs. 4.9–4.11 we see that the spectrum $S_1(v)$ is calculated through the convolution of the Fourier transform of the interferogram function and the Fourier transform of the window function $p(x)$. If we substitute the window function $p(x)$ by an apodization function $h(x)$, we see that the apodized spectrum is obtained by the convolution of the calculated unapodized spectrum with the Fourier transform of the apodization function:

$$S_1'(v) = \int_{-\infty}^{+\infty} S(\tau)H(v - \tau)\,d\tau. \qquad (4.22)$$

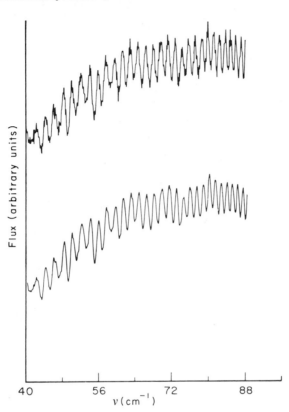

Fig. 4.5   Comparison of an unapodized (*top*) and an apodized (*bottom*) spectrum. Resolution is 0.2 and 0.3 cm^{-1}, respectively. [R. G. Wheeler and J. C. Hill, *J. Opt. Soc. Am.* **56**, 657 (1966).]

Filler[9] has discussed this procedure of *a posteriori* apodization in detail.

One common feature of all three window functions $p(x)$, $h_1(x)$, and $h_2(x)$ is that the central maximum of the resulting apodization functions has a width between $\approx 1/2L$ and $\approx 1/L$. This width, which represents the resolution of the Fourier transform spectrometer, is inversely proportional to the maximum path difference. The same general relation is true for a grating spectrometer, where the corresponding relation, $\Delta v \approx 1/L$, in terms of L, is the path difference between the two rays which originate from the outermost lines of the grating.

In the Fourier transform spectrometer the finite path difference limits the resolution to $\approx 1/2L$ and the apodization decreases it further by a factor of two. Other causes for a further deterioration of the resolution, such as the finite aperture, are discussed below.

D. SAMPLING

We now discuss some aspects of the "sampling" procedure.

Consider a digital computer used for the calculation of the spectrum from the interferogram function. In this case, the interferogram has to be digitized into a number of discrete values which are the input data of the computer. $S_1(v)$, as given in Eq. 4.8, will then be expressed as a sum of discrete values at equal spacings $0, l, 2l, \cdots nl$,

$$S_{11}(v) = l[s(0)p(0) + 2s(l)p(l) \cos [2\pi vl]$$

$$+ \cdots 2s(nl)p(nl) \cos [2\pi vnl] + \cdots]. \qquad (4.23)$$

All the $p(nl)$ considered here are equal to unity (see Eq. 4.7). We can cast Eq. 4.23 into the form of an integral if we multiply $s(x)p(x)$ by $lr_l(x)$ [$r_l(x) =$ Dirac comb function]:

$$S_{11}(v) = \int_{-\infty}^{+\infty} ls(x)p(x)r_l(x) \cos [2\pi vx] dx. \qquad (4.24)$$

The comb function $r_l(x)$ is shown in Fig. 4.6.

$S_{11}(v)$ can also be expressed as the convolution of the Fourier transform of $s(x)$ and $p(x)lr_l(x)$:

$$S_{11}(v) = \int_{-\infty}^{+\infty} S(\tau)F(v - \tau) d\tau, \qquad (4.25)$$

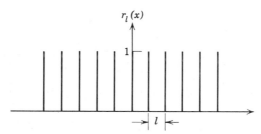

Fig. 4.6   Grating function $r_l(x)$ (Dirac comb).

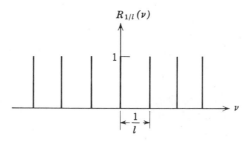

Fig. 4.7   Grating function $R_{1/l}(v)$ (Dirac comb).

where $F(\tau)$ is obtained by the convolution of the Fourier transforms of $p(x)$ and of $lr_l(x)$. These Fourier transforms are $P(v)$ (see Eq. 4.11) and $R_{1/l}(v)$, where $R_{1/l}(v)$ is a Dirac comb function presenting an infinite sequence of Dirac distributions separated by $1/l$ (see Fig. 4.7 and Ref. 4, p. 44). Therefore, for $F(\tau)$ we obtain, remembering Eq. 4.12,

$$F(\tau) = \int_{-\infty}^{+\infty} P(\beta) R_{1/l}(\tau - \beta)\, d\beta$$

$$= 2L \sum_{n=-\infty}^{+\infty} \text{sinc}\left[2\pi L\left(\frac{n}{l} - \tau\right)\right]. \tag{4.26}$$

Inserting $F(\tau)$ as $F(v - \tau)$ into Eq. 4.25, we obtain

$$S_{11}(v) = \int_{-\infty}^{+\infty} S(\tau) 2L \sum_{n=-\infty}^{n=+\infty} \text{sinc}\, 2\pi L\left[\frac{n}{l} - (v - \tau)\right] d\tau. \tag{4.27}$$

We notice the similarity between Eqs. 4.27 and 4.14: The theoretical apparatus function of Eq. 4.14 (as given in Eq. 4.15) is replaced in Eq. 4.27 by $F(v - \tau)$. If we consider a monochromatic input (see Section C), the output is given by $F(v - v_0)$. This function consists of an infinite sum of sinc-functions with peaks at $\pm v_0$, $\pm v_0 \pm 1/l$, $\cdots$, $\pm v_0 \pm n/l$, $\cdots$, as shown in Fig. 4.8; in other words, the computed spectrum is repeated at multiples of $1/l$. This is in contrast to the results of Eq. 4.15 where there were only peaks at $\pm v_0$. Since the side loop of the many peaks would overlap if we did not restrict v to some highest value v_M, we have to eliminate all frequencies higher than v_M in order to avoid disturbance of the calculated spectrum. This can be done by optical or by electronical filtering.

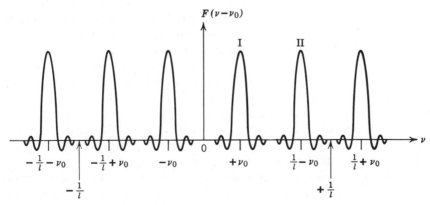

Fig. 4.8   Plot of the function $F(v - v_0)$.

Let us find v_M in the special case of the overlap of the first and second peaks (I and II in Fig. 4.8), centered at v_0 and $(1/l) - v_0$, respectively. Overlap starts if $v_M = v_0 + |\Delta v| = (1/l) - v_0 - |\Delta v|$. Since there are two symmetrically placed peaks in the interval 0 to $1/l$, we obtain $v_M = 1/2l$. We see that the highest frequency now permitted is directly related to the sampling interval. This result is true in general and follows from the sampling theorem of information theory (see Ref. 4). The number of sampling points in the interferogram is then $2v_M L$; in other words, this is twice the number of cycles which the highest frequency undergoes while the path difference is changed from 0 to L.[6]

For an interferogram which is continuously taken (not in steps), the above statement (that the interval of sampling is related to the highest frequency allowed) applies also to the noise frequencies. Therefore noise frequencies higher than v_M must be filtered out in order not to disturb the spectrum in the same way as the signal frequencies would disturb it.

Up to this point we have considered the frequency range $0 \leqslant v \leqslant v_M$. Because of the periodic appearance of the analyzed spectrum, we can also restrict the investigation to a smaller interval $v_m \leqslant v \leqslant v_M$. In this case, a similar consideration as presented above yields the sampling interval $l' = 1/[2(v_M - v_m)]$. We see that by restricting the spectral region to a smaller bandpass, we have increased the sampling interval (fewer sampling points). This may be illustrated by the following example: (a) $0 \leqslant v \leqslant 300 \text{ cm}^{-1}$. It follows that $v_M = 1/(2l) = 300 \text{ cm}^{-1}$, and $l = 1/600 \text{ cm} = 16.7 \mu$. (b) $200 \leqslant v \leqslant 300 \text{ cm}^{-1}$. It follows that $l' = 1/(2 \cdot 100) \text{ cm} = 50 \mu$.

In order to separate out the above interval of the spectrum, a bandpass filter is necessary (see Chapters 1 and 3). The use of a bandpass filter, which allows us to study only a certain frequency interval $v_M - v_m$ at the larger sampling interval $1/[2(v_M - v_m)]$, does not take full advantage of the fact that in Fourier transform spectroscopy we may obtain information over a large spectral region with one interferogram. It might, however, be of some practical importance if the spectrum itself consists of only a few lines and if it is desirable to obtain a high-resolution spectrum of these few lines.

Since the digital computer accepts the interferogram as a sequence of discrete points, we may record the interferogram point by point, corresponding to the sampling intervals l; that is, $M + 1$ points for $0 \leqslant l \leqslant L = Ml$. Provisions have to be made to set the movable mirror accurately at the sampling points. Over a length of $0 \leqslant x \leqslant L = 20 \text{ cm}$, Richards[7] has used a precision lead screw (micrometer screw) with a stepping motor for applications in the 5 to 100 cm^{-1} spectral region. For longer displacements or for operation in the higher-frequency region, we have to use an interferometrically controlled servo system (for instance, by monitoring the system through the interference fringes produced by a laser). We may advance the mirror between the sampling points with high speed and wait at the points for integration of the signal. The time constant for smoothing out the noise has to be chosen with respect to the time between two sample points and the gate time of the digital voltmeter (for details see Ref. 7).

We saw above that the number of sampling points in the interferogram is $M = L/l = 2v_M L$. The number of calculated points in the spectral domain for an unapodized spectrum with $\Delta v = 1/2L$ is

$$N = \frac{v_M}{\Delta v} = \frac{M}{2L} \bigg/ \frac{1}{2L} = M.$$

We see that one point per resolution-width is obtained. Some authors recommend taking more points in order to make full use of the resolution capability of the interferometric setup.

From Eq. 4.23 we have for the sampled interferogram

$$S'_{11}(v) = l\left[s(0) + 2 \sum_{n=1}^{M=L/l} s(nl) \cos\left[2\pi vnl\right]\right]. \tag{4.28}$$

Since the interferogram function is defined by $s(x) = g(x) - (\frac{1}{2})g(0)$ (see Eq. 4.4) and since $g(x)$ is obtained from the observed data, we have to find the value of $(\frac{1}{2})g(0)$. It is difficult to obtain an exact value for $(\frac{1}{2})g(0)$ in many instances since unmodulated radiation may displace the zero line. Richards[7] recommends taking $g(x)$ for a number of points around $x = L$ and averaging to $\overline{g(L)}$. This has also an advantage with respect to a lowering of the drift of the zero line; we see that $g(L)$ is equal to $(\frac{1}{2})g(0)$ if the last integral in Eq. 4.2 can be neglected. In Eq. 4.28 this average is assumed to be subtracted out, yielding $s(nl)$. Of course, this procedure may be incorporated into the computer program.

We can rewrite the dimensionless expression $2\pi vnl$ in Eq. 4.28 by using $x = nl = nL/M$ and $\Delta v = 1/2L$, with $v_m = m/2L$, in the form

$$2\pi vnl = 2\pi \frac{m}{2L} \frac{nL}{M} = \frac{nm\pi}{M}. \tag{4.29}$$

Computer programs for the direct Fourier transformation as well as for the fast Fourier transformation (Cooly-Tukey algorithm[10]) are available in most computer libraries.

E. PHASE ERROR AND ITS CORRECTION. MALADJUSTMENT

The interferogram function $s(x)$ (Eq. 4.4) has its maximum value at $x = 0$. The sampling of the interferogram function in steps of $\Delta x = l$ should start exactly at $x = 0$. If this is not so, all sampling points are displaced by a small amount. This results in a phase error $\varphi(v)$, which makes the interferogram function $s(x)$ asymmetric with respect to $x = 0$. For the calculation of the spectrum, in this case we must use the complete cosine- and sine-Fourier transformation. The application of the cosine Fourier transformation alone would result in a distortion of the line shape, creation of spurious lines, and a displacement of the zero line of the analyzed spectrum. Some of these features are demonstrated in Ref. 11. On the other hand, one would like to avoid a complete cosine- and sine-Fourier transformation just from the point of view that twice as many points have to be recorded; consequently, the calculation of the spectrum takes longer.

Phase errors and their correction have been discussed; for instance, by Connes,[5] Loewenstein,[11] and Forman et al.[12] For the discussion of phase errors and methods for their correction, we will follow the presentation by Forman et al.[12]

Let us go back to Eq. 4.5 for the unapodized cosine-transform of $S(v) = S(-v)$:

$$s(x) = \int_{-\infty}^{+\infty} S(v) \cos\left[2\pi v x\right] dv \tag{4.5}$$

or

$$s(x) = \int_{-\infty}^{+\infty} S(v) e^{-i2\pi v x} dv. \tag{4.30}$$

Let us now assume that a phase error $\varphi(v)$ is present in the interferometer or the recording equipment. This modifies Eq. 4.5 to

$$s_1(x) = \int_{-\infty}^{+\infty} S(v) \cos\left[2\pi v x + \varphi(v)\right] dv \tag{4.31}$$

or

$$s_1(x) = \int_{-\infty}^{+\infty} S(v) e^{-i\varphi(v)} e^{-i2\pi v x} dv \tag{4.32}$$

where we define

$$\varphi(v) = \varphi(-v).$$

We see that the interferogram is now no longer symmetric about $x = 0$. The Fourier transform of $S(v) e^{-i\varphi(v)}$ is $s_1(x)$, and therefore

$$S(v) e^{-i\varphi(v)} = \int_{-\infty}^{+\infty} s_1(x) e^{i2\pi v x} dx. \tag{4.33}$$

Let us now write $S(v) e^{-i\varphi(v)}$ as a sum of a real and imaginary part:

$$S(v) e^{-i\varphi(v)} = P(v) + iQ(v). \tag{4.34}$$

The functions $P(v)$ and $Q(v)$ are obtainable from the complete Fourier transformation, and consequently:

$$S(v) = \sqrt{P(v)^2 + Q(v)^2} \qquad (4.35)$$

and

$$\varphi(v) = -\arctan \frac{Q(v)}{P(v)}. \qquad (4.36)$$

It is possible to avoid the computation of the complete Fourier transformation as we shall show in the following: Under practical conditions the phase error $\varphi(v)$ is a smooth function of the frequency and therefore $\cos \varphi(v)$ and $\sin \varphi(v)$ are also smooth functions of v. A smooth function of v requires only a few sampling points of $s(x)$ on each side of $x = 0$. This means that we take some sampling points for $x < 0$ in addition to the interferogram for $x \geqslant 0$ and use the center part of the interferogram for the calculation of $\varphi(v)$.

Having obtained $\varphi(v)$, the correction of the interferogram is accomplished as follows. The Fourier transform of $e^{i\varphi(v)}$ is calculated:

$$f(x) = \int_{-\infty}^{+\infty} e^{i\varphi(v)} e^{\,i2\pi vx}\, dv. \qquad (4.37)$$

We denote the right side of Eq. 4.33 by $S'(v)$; hence

$$S'(v) = \int_{-\infty}^{+\infty} s_1(x) e^{i2\pi vx}\, dx. \qquad (4.38)$$

According to $S(v) = S'(v)e^{i\varphi(v)}$, the corrected interferogram function $s(x)$ is obtained through the convolution of the Fourier transforms of $S'(v)$ and that of $e^{i\varphi(v)}$; in other words,

$$s(x) = \int_{-\infty}^{+\infty} s_1(\beta) f(x - \beta)\, d\beta. \qquad (4.39)$$

We have obtained a phase-corrected interferogram function and with the use of a symmetric Fourier transformation we can calculate the spectral distribution $S(v)$.

The complete process is demonstrated in Figs. 4.9 and 4.10. The uncorrected interferogram is shown in Fig. 4.9(a) and the double-sided Fourier transform of this interferogram is displayed in Fig. 4.10(a); this is the correct spectrum. The cosine-transform of the uncorrected interferogram of Fig. 4.9(a) is shown in Fig. 4.10(b). We see how much the spectrum is distorted by comparing Fig. 4.10(b) with Fig. 4.10(a). The corrected interferogram (Fig. 4.9(b)) is not completely symmetric, but the spectrum calculated using only the cosine-transform (Fig. 4.10(c)) agrees well with Fig. 4.10(a).

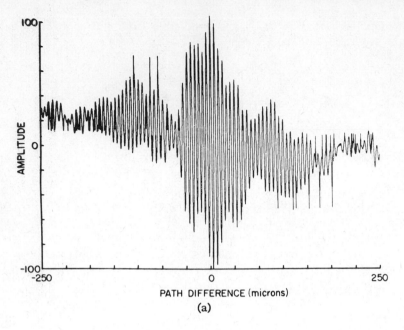

(a)

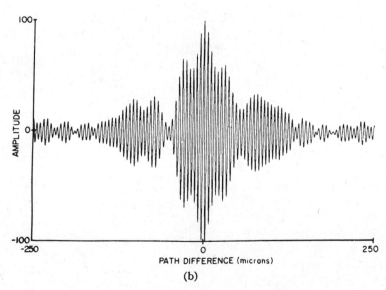

(b)

Fig. 4.9   (a) Uncorrected interferogram; (b) corrected interferogram. [M. L. Forman, W. H. Steel, and G. A. Vanasse, *J. Opt. Soc. Am.* **56**, 59 (1966).]

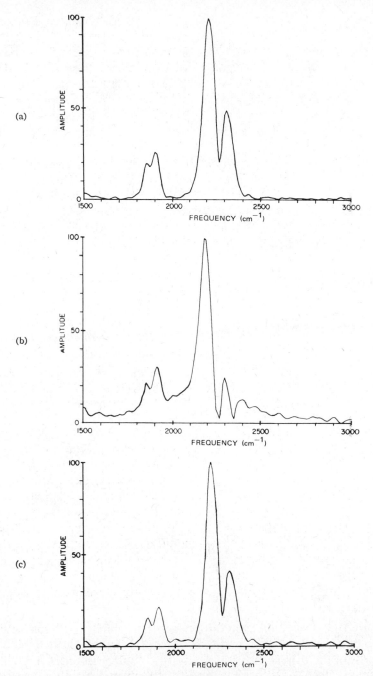

Fig. 4.10 (*a*) Spectrum obtained by a double-sided transform of the interferogram of Fig. 4.9*a*; (*b*) cosine-transformation of the interferogram of Fig. 4.9*a*; (*c*) cosine-transformation of the interferogram of Fig. 4.9*b*. [M. L. Forman, W. H. Steel, and G. A. Vanasse, *J. Opt. Soc. Am.* **56**, 59 (1966).]

A difficulty arises from the fact that we do not know precisely the point $x = 0$ and therefore cannot calculate $\varphi(v)$ with respect to this point. The practical approach is to choose the center where the linear part of $\varphi(v)$ is zero for some mean frequency v within the center part of the interferogram. If the phase error is too large, successive applications of the above procedure will improve the symmetry of the interferogram. This is demonstrated in the paper by Forman.[12]

The adjustment of the movable mirror in the Michelson interferometer should be accurate at least to $\frac{1}{10}$ of the smallest wavelength used in vertical and horizontal deviations. A corner mirror or a "cat eye"[13] has been recommended for large L. The accuracy of the positioning of the mirror through the lead screw should also be within at least $\frac{1}{10}$ of the smallest wavelength considered. For more details of maladjustment, see Ref. 5.

F. Advantages and Disadvantages of Fourier Transform Spectroscopy Compared to Grating Spectroscopy

The principal advantage of Fourier transform spectroscopy is a consequence of the fact that the interferogram contains information about all spectral elements (Fellgett or multiplex advantage). We touched upon this in the Introduction to this chapter, and now we will be more specific. Let us designate the number of the spectral elements by N. Each particular spectral element is therefore studied for the total time T which is required to record the interferogram. In a grating spectrometer, in contrast, one studies each spectral element only during the time interval T/N, thereafter switching to the next spectral element. Since the accuracy of spectroscopy in the far-infrared is generally limited by the noise from the detection system, we realize a gain in the signal-to-noise ratio by using Fourier transform spectroscopy compared to grating spectroscopy. For equal total times of observation this gain amounts to $\sqrt{N}$. Consequently, for equal signal-to-noise ratios in both spectroscopic techniques, the measuring time is considerably shorter in Fourier transform spectroscopy; if, in addition, the measuring time is the same in both methods, a higher resolution is obtainable using Fourier transform spectroscopy. (We refer the interested reader to Ref. 5 and 6, which give detailed discussions on the influence of noise in Fourier transform spectroscopy.)

There are some less important advantages. It is simpler to improve the resolution of a Fourier transform spectrometer since we only have to double the length L of the displacement of the movable mirror in order to double the resolution. The measuring time is then also doubled. In a grating spectrometer, in contrast, we have to narrow the slit width by a factor of two. According to what we said in Chapter 1 (Section C), the signal is then decreased by a factor of four. If it is assumed that the system is noise-limited due to the detection system, it is then necessary to scan the spectrum of the grating spectrometer sixteen times slower to obtain a comparable spectrum with that obtained with the Fourier transform spectrometer.

With respect to the necessity of filtering out unwanted radiation, the requirements for the Fourier transform spectrometer to remove all frequencies $v > v_M$ are much more easily achieved than those for a certain grating region, mainly because the cut-on slope of the filter need not be extremely steep.

A disadvantage of Fourier transform spectroscopy is that we cannot immediately observe the spectrum because of the computer processing time of the whole interferogram. Yoshinaga et al.[14] have circumvented the waiting time, as we shall show below.

G. Finite Aperture, Sensitivity, and Dynamic Range of the Recording System

So far we have considered a point entrance aperture in the interferometer with the light entering in a direction which is parallel to the optical axis. For a practical Fourier transform spectrometer, we have to open the aperture to a certain extent. The relationship between the solid angle Ω of the aperture and the resolving power $\mathcal{R}$ of the interferometer is $\mathcal{R}\Omega = 2\pi$. We see that a finite aperture opening, which corresponds to a finite solid angle, will limit the obtainable resolving power. If b is the radius of the aperture and F the focal length of the telescope mirror, it follows (see Fig. 4.11) that $\Omega \approx \pi\alpha_0^2 \approx \pi(b^2/F^2)$ and therefore $\mathcal{R} = 2F^2/b^2$. For example: If we are interested in the aperture opening having a radius b for a resolution of $\Delta v = 0.01\ \mathrm{cm}^{-1}$ at $300\ \mathrm{cm}^{-1}$ and assume $f/4$ optics, we find $\mathcal{R} = 30000$, $b^2/2F^2 = 1/30000$. Using telescope mirrors of $2B = 7.5\ \mathrm{cm}$, we obtain $F = 30\ \mathrm{cm}$ and $2b \sim 5\ \mathrm{mm}$.

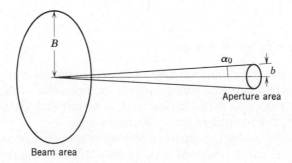

Fig. 4.11   Area and solid angle for an interferometer.

Because a light beam of a certain width is admitted by the finite aperture of the interferometer, there will be a frequency variation between the outermost and center rays due to their path differences. The corrected frequency v_c is obtained by averaging the path differences of all rays of the solid angle Ω,

$$v_c = v\left(1 - \frac{\Omega}{4\pi}\right). \tag{4.40}$$

The application of Eq. 4.40 assumes an equal illumination of the aperture, which is not easy to achieve. The Fourier transformation using the spacings l between the sampling points can then be calculated with

$$l_c = l\left(1 - \frac{\Omega}{4\pi}\right). \tag{4.41}$$

(More details are given in Refs. 5 and 6.)

If a wide spectral region in the far-infrared is examined, a large dynamical range for the recording of the interferogram has to be used. (Dynamical range is understood to mean the ratio of the maximum signal to the noise). The dynamical range of the calculated spectrum is, in general, smaller. Therefore, we have to find which spectral details in the spectrum should be recovered and choose the dynamical range of the recording system for the interferogram accordingly. (Examples are given in Refs. 7 and 16; for more details Refs. 4 and 5 should be consulted.)

The problems resulting from a large dynamical range can be circumvented by "double-beam" operation as used by Hall et al.[20] in a lamellar-grating Fourier transform spectrometer. We shall discuss this in more detail below, but here it will be sufficient to mention that the light from the source passes through a sample beam and a reference beam and is recombined by a chopper. Therefore, only those spectral details which originate from the absorption of the sample beam are modulated. In addition to the considerable reduction of the dynamical range, compensation of instabilities of the source is achieved by this method.

H. MICHELSON INTERFEROMETER

In the foregoing sections we have discussed some principles of Fourier transform spectroscopy and based our considerations on a Michelson two-beam interferometer. In the following we shall describe the operation of such an instrument, following a paper by Richards.[7]

The optical layout of the far-infrared interferometer is presented in Fig. 4.12. The instrument is enclosed in a vacuum housing. The path difference through the displacement of the movable mirror is $-0.5 \leqslant x \leqslant 20$ cm. The

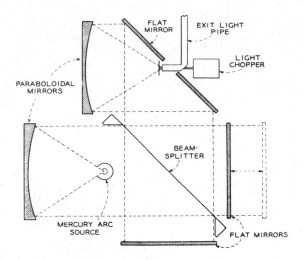

Fig. 4.12   Schematic of a Michelson interferometer. [P. L. Richards, *J. Opt. Soc. Am.* **54**, 1474 (1964).]

mirror is controlled by a precision traveling microscope drive. The optical system is effectively of $f/1.3$ aperture and has collimating mirrors of 18 cm diameter. Figure 4.13 shows the block diagram of the interferometer and the recording system.

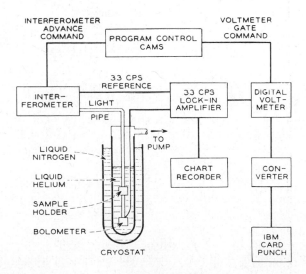

Fig. 4.13   Block diagram of an interferometric spectrometer. [P. L. Richards, *J. Opt. Soc. Am.* **54**, 1474 (1964).]

The beam splitter is the weakest point in a Michelson interferometer if a spectrum extending over one octave or more is to be studied. An ideal beam splitter requires $T = R = 0.5$, where T is the transmissivity and R the reflectivity. The quantity $4RT$ may be used as an indicator of the efficiency of the beam splitter; in the ideal case it is equal to unity. Figure 4.14 shows the efficiency of a Mylar beam splitter of 3 mil (about 75 μ) thickness. We see that due to interference in the film there are regions where the beam splitter cannot be used, whereas at some optimal points its efficiency reaches 70 %.

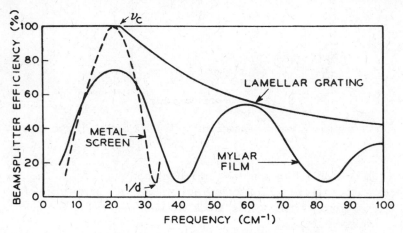

Fig. 4.14   Comparison between the beam splitting efficiencies of a lamellar-grating interfero- meter and a Michelson interferometer using a Mylar beam splitter ($\sim 75\,\mu$) or a metal wire grid (~ 250 lines/cm). The frequency at which cancellation begins to occur in the lamellar-grating spectrometer is ν_c. The frequency at which the wavelength is equal to the grating constant of the wire grids is $1/d$. [P. L. Richards, *J. Opt. Soc. Am.* **54**, 1474 (1964).]

Important developments and detailed determinations of the character- istics of beam splitters have been reported by Gebbie,[15] who first suggested Mylar as an efficient beam splitter. Figure 4.15 shows data on the relative energy *vs* frequency relations of beam splitters of varying thickness between 6 and 125 μ.[16] We see that none of the beam splitters can be used effectively over the whole spectral region of 400 to 10 cm^{-1}.

Loewenstein[17] showed that a Mylar beam splitter polarizes the light considerably and, if a beam splitter with an angle of about 60° (Brewster angle) is used, the beam traveling toward the detector would be almost completely polarized.

Beam splitters can also be constructed from metallic meshes. The efficien- cies and the polarizing properties of such devices for an angle of incidence of 45° and for different directions of polarization have been investigated and reported by Vogel and Genzel.[18]

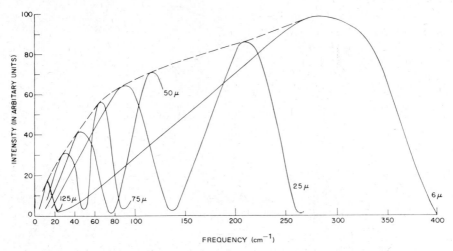

Fig. 4.15   Relative Mylar beam splitter efficiencies as a function of frequency (cm^{-1}). [C. H. Perry, R. Geick, and E. F. Young, *Appl. Opt.* **5**, 1171 (1966).]

We now illustrate some interferograms and spectra of water vapor. Figure 4.16 shows interferograms of the background of a singlet line, a doublet line, and a part of the spectrum of water vapor which contains many lines. In the interferogram of the singlet and doublet we can see the periodic appearance and the beat phenomena, respectively. On the basis of such observations direct interpretations of the interferogram may be attempted; this is naturally only fruitful in very specialized cases.[19] Figure 4.17 shows part of the water vapor pure rotation spectrum calculated from an interferogram[7]; we see that the spectrum covers more than two octaves.

We return now to our remark made at the end of Section F; namely, a disadvantage of Fourier transform spectroscopy is that we have to wait until the transformation is performed by the computer. To eliminate this, Yoshinaga *et al.*[14] have constructed a special "on-line" computer which performs the Fourier transformation while the interferogram is being recorded. Thus, the computed spectrum is available immediately at the completion of the run. The principle is as follows. Usually, the Fourier transform is performed by computing the set of $S(v)$ according to Eq. 4.23,*

$$S(v_1) = s(0) + 2s(l) \cos [2\pi v_1 l] + \cdots 2s(nl) \cos [2\pi v_1 nl] + \cdots$$

$$S(v_2) = s(0) + 2s(l) \cos [2\pi v_2 l] + \cdots 2s(nl) \cos [2\pi v_2 nl] + \cdots$$

$$\vdots \tag{4.42}$$

$$S(v_m) = s(0) + 2s(l) \cos [2\pi v_m l] + \cdots 2s(nl) \cos [2\pi v_m nl] + \cdots,$$

* The constant l can be included in $S(v_m)$.

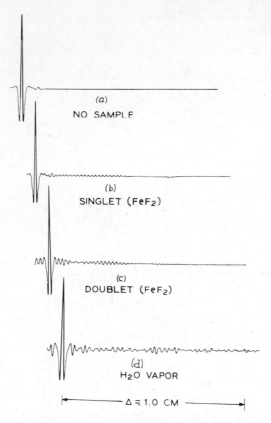

Fig. 4.16   Examples of interferograms measured with a Michelson interferometer: (a) interferogram of the mercury source with filters cutting off at 100 cm^{-1}; (b) singlet, (c) doublet, (d) H$_2$O vapor spectrum. [P. L. Richards, *J. Opt. Soc. Am.* **54**, 1474 (1964).]

after all values of n are available; that is, after the interferogram is taken. In the system described by Yoshinaga *et al.*, all $S(v_m)$ are calculated immediately for each value of $2s(nl) \cos [2\pi v_m nl]$, as these data come from the interferogram, and are added successively to the sum

$$s(0) + 2s(l) \cos [2\pi v_m l] + \cdots 2s[(n-1)l] \cos [2\pi v_m(n-1)l]. \quad (4.43)$$

The sum for each v_m is stored in the computer. The $S(v_m)$, which are computed until the limit $Ml = L$ is reached, can be called from storage and displayed on an oscilloscope. We can thus follow the spectrum along with the interferogram for any multiple of l displacements. This is shown on the example of the pure rotational spectrum of water vapor in Fig. 4.18 which gives the calculated spectrum as a function of path length. We see that the structure in the spectral

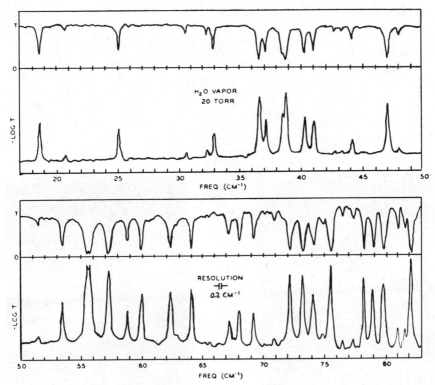

Fig. 4.17 Transmittance T and absorption coefficient $-\log T$ of a 1.5-m path of H_2O vapor at a pressure of 20 mm Hg measured with a Michelson interferometer. Computed points at intervals of 0.1 cm^{-1}. [P. L. Richards, *J. Opt. Soc. Am.* **54**, 1474 (1964).]

region of 40 to 60 cm^{-1} is obtained in finer and finer detail as the path difference, indicated at the top of Fig. 4.18, is increased. The final spectrum was obtained with a path difference of about 14.5 mm.

I. LAMELLAR-GRATING INTERFEROMETER

Another type of two-beam interferometer for Fourier transform spectroscopy is the lamellar-grating interferometer, first used for this purpose by Strong and coworkers.[2] (Reference 2 contains an Appendix on interferometric spectroscopy and a description of such an instrument.) The two interfering beams are made by wavefront division at the lamellar grating. This is in contrast to the amplitude division of the Michelson interferometer. Figure 4.19 shows a scheme of such lamellar grating with its interpenetrating sets of facets.

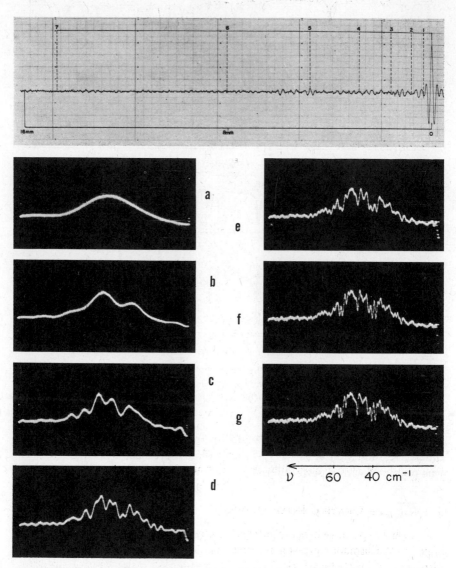

Fig. 4.18 Interferogram of water vapor (top) and spectra at different points of the interferogram displayed on an oscilloscope. Spectra of (a) to (g) are obtained by using the interferogram from 0 to 1 to 0 to 7, respectively. [H. Yoshinaga, S. Fujita, S. Minami, Y. Suemoto, M. Inoue, K. Chiba, R. Nakano, S. Yoshida, and H. Sugimori, *Appl. Opt.* **5**, 1159 (1966).]

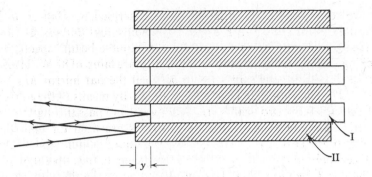

Fig. 4.19 Scheme of lamellar grating with the two sets of interpenetrating facets I and II. The path difference of the two beams is accomplished by the displacements of the two sets of facets by y.

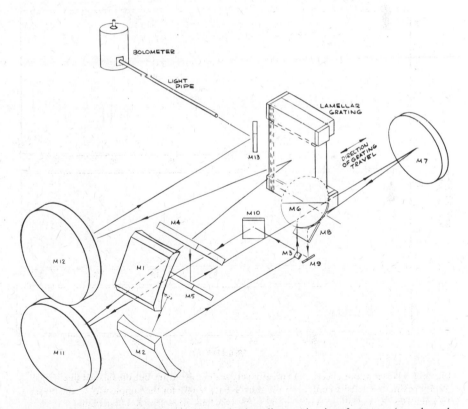

Fig. 4.20 Schematic of the optical system of a lamellar-grating interferometer (sample and reference beam system is shown on the right in the figure). [R. T. Hall, D. Vrabec, and J. M. Dowling, *Appl. Opt.* **5**, 1147 (1966).]

We will discuss here the interferometer described by Hall *et al.*[20] The optical diagram is shown in Fig. 4.20. The ellipsoidal mirrors M_1 and M_2 form the sample and reference beam (of this "double-beam" spectrometer*) by picking up light from the source, a mercury-arc lamp of 85 W. The sample can be inserted into the upper beam between the flat mirror M_5 and the chopper M_6. The chopping frequency is 19 Hz. By means of the mirrors M_7 and the chopper, the two beams are recombined and pass through the instrument. Iris diaphragms in equivalent positions are used for matching the intensities of the two beams. The receiver is a liquid-helium-cooled, gallium-doped, germanium detector from Texas Instruments Inc., operated at 4.2°K (see Chapter 2, Section B.c.). The sensitive area of the detector element is $5 \times 5 \text{ mm}^2$.

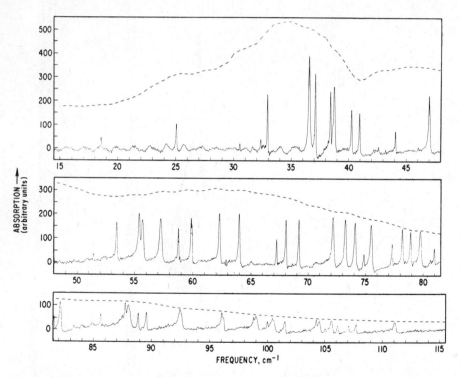

Fig. 4.21 Water vapor spectrum. The intensity scale is arbitrary but the dashed line indicates the approximate intensity contour of the source. Path length for the sample was 15 cm, pressure 15.5 mm Hg. [R. T. Hall, D. Vrabec, and J. M. Dowling, *Appl. Opt.* **5**, 1147 (1966).]

* The main purpose of "double-beam" operation, as we have discussed in Section G, is to reduce the dynamical range considerably and to compensate for fluctuations of the source.

The lamellar grating consists of two sets of 24 facets. Each facet is 0.635 cm wide and 30.5 cm long, the total area being 30.5×30.5 cm². The path difference is achieved by displacing one of the two interpenetrating sets of facets from the other. The maximum path difference is 8.3 cm. The efficiency of the lamellar grating as a beam splitter is shown in Fig. 4.14 (Richards[7]). Above a certain frequency ν_c, the efficiency drops due to diffraction effects.[21] The undesired high frequencies are removed by the use of reflection filter gratings, crystal quartz, and black polyethylene (see Chapter 1, Section E).

The interferogram is recorded in steps and punched on paper tape. The sample intervals are multiples of 10 μ of the optical path difference according to the highest frequency (cm⁻¹) under investigation.

Figure 4.21 shows a water vapor spectrum obtained under the conditions given in Table 4.1 and a maximum optical path difference of about 11 cm.

TABLE 4.1*

EXPERIMENTAL PARAMETERS FOR THE INTERFEROGRAM OF THE
WATER VAPOR SPECTRUM SHOWN IN FIG. 4.21

	H_2O
Maximum path difference, cm	11.112
Theoretical resolution, cm⁻¹	0.09
Sampling interval, μ, optical path difference	30.0
Number of points recorded	3705
Run time, hr	6.0
Number of output points computed	6251
Computer time, min	23.22

* R. T. Hall, D. Vrabec, and J. M. Dowling, *Appl. Opt.* **5**, 1147 (1966).

The highest frequency which was passed through the instrument was $\nu_c = 125$ cm⁻¹. Corresponding to this, the sampling intervals should be 40 μ but somewhat smaller intervals, only 30 μ, were actually used. The authors did not apodize the spectrum since the interferogram function was already smoothed out by appropriately large damping (Ref. 20, p. 1157). With this instrument a resolution of better than 0.094 cm⁻¹ has been demonstrated on chlorine isotopic splitting of the $J = 2$ to $J = 3$ rotational transition of DCl.

Since the mechanical construction of a lamellar-grating interferometer is more complicated than a Michelson interferometer, it has not found as many applications in Fourier transform spectroscopy as has the Michelson interferometer.

J. Interferometry in the Asymmetric Mode

a. Introduction

In the foregoing sections of this chapter we have assumed that the interferogram is symmetric about the zero path difference and that therefore the cosine transformation could be applied. We have discussed the importance of starting the recording of the interferogram exactly at the zero point, and we have mentioned that the complete Fourier transformation is useful if corrections due to asymmetry are necessary.

In this section we discuss a Fourier transform interferometer which allows calculation of the optical constants (index of refraction and extinction coefficient) from the interferogram by the use of the complete Fourier transformation. The measurements can be made in reflection or transmission. This method has been described by Bell[22] and Chamberlin et al.[23] Our discussion will closely follow the papers by Russell and Bell[24] and Bell.[25]

The optical diagram of the Michelson interferometer used for this purpose is shown in Fig. 4.22. The source is a 100-W mercury lamp. The beam

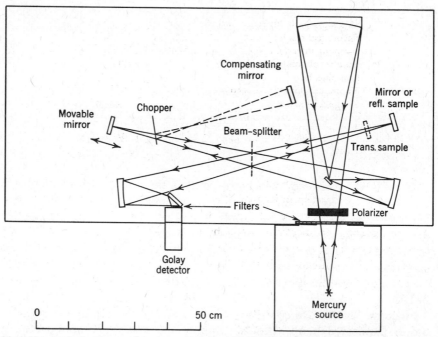

Fig. 4.22   Optical scheme of vacuum far-infrared Michelson interferometer for use in the asymmetric mode. [E. E. Russell and E. E. Bell, *Infrared Phys.* **6**, 75 (1966), by permission of Pergamon Press.]

splitter is made of Mylar ($\sim 13\ \mu$) or electroformed metal mesh with periodicity constants of 25.4, 50.8, and 101.6 μ, respectively (see Chapters 1 and 3 and Ref. 18). The use of a particular beam splitter depends on the spectral region under investigation, as discussed above for the Mylar beam splitter in Section H. The angle of the incident light at the beam splitter is 10° to increase efficient filtering of high-frequency components and for the reduction of polarization effects. One arm of the interferometer contains the transmission sample or the sample as a reflection plate. The other arm holds the movable mirror. The compensation mirror is used to introduce an out-of-phase signal to reduce the average signal level and to compensate for fluctuations of the sources. For transmission measurements a light-pipe is used for the same purpose. In both types of measurements, therefore, the dynamical range of the detection system is limited and it is easier to achieve linearity of response. The source is water-cooled and provisions are made to keep the instrument and its alignment stable with regard to temperature changes. The optical system is enclosed in a vacuum tank. The optical filtering is accomplished with various transmission filters in conjunction with the beam splitter. A Golay cell serves as detector.

Sampling of the interferogram is performed with a step motor, allowing intervals which are multiples of 1.19 μ. A "sample-in, sample-out" technique for transmission measurements is employed for elimination of the influence of instabilities in the system. (A similar technique is described in Chapter 12, Section B.d.) The resolution for the recording of most interferograms was about 5 cm^{-1}.

b. Theory and Experimental Results

We will consider only one direction of polarization and, therefore, the signals traveling through the two equally long arms of the interferometer are scalar functions of the time: $u_1{}^T(t)$.* The superscript T indicates that the function is zero outside the observation time $-T \leqslant t \leqslant T$ and, consequently, the integrations (see below) can be taken between $-\infty \leqslant t \leqslant \infty$. If the optical path of the arm of the interferometer which contains the movable mirror is longer by $x = \tau/c$, where c is the velocity of light, the signal at the detector will have originated at the source time τ earlier ($\tau \ll T$). Therefore the signal from that arm of the interferometer is given by $u_1{}^T(t - \tau)$.

The Fourier transform of $u_1{}^T(t)$ is

$$\hat{U}_{1T}(f) = \int_{-\infty}^{+\infty} u_1{}^T(t) e^{i2\pi ft}\, dt, \tag{4.44}$$

where f is frequency in Hz (cycles per second).

* In order to be consistent with the notation in earlier sections of this chapter, we use small letters for the x- and t-domains and capital letters for the v- and f-domains. This is the opposite of the notation in Ref. 24 and 25.

The inverse transformation is

$$u_1^T(t) = \int_{-\infty}^{+\infty} \hat{U}_{1T}(f) e^{-i2\pi f t} \, df. \qquad (4.45)$$

The subscript on $\hat{U}_{1T}(f)$ indicates that it is the Fourier transform of a truncated function and the circumflex shows that it is a complex function; since $u(t)$ is real we have $\hat{U}_T(-f) = \hat{U}_T^*(f)$, where the asterisk indicates the complex-conjugate. The pair of Eqs. 4.44 and 4.45 can also be expressed, according to common usage, by $\hat{U}_{1T}(f) \rightleftarrows u_1^T(t)$.

If we place a sample in the fixed arm of the interferometer, the wave $u_1^T(t)$ is modified by the impulse response function $g(t)$. The effect of the sample, for instance on $u_1^T(t)$, can be expressed by the convolution of $g(t)$ and $u_1^T(t)$,†

$$u_2^T(t) = \int_{-\infty}^{+\infty} g(t') u_1^T(t - t') \, dt'. \qquad (4.46)$$

The power at the detector without a sample is proportional to the mean square of the superposition of the two waves $u_1^T(t - \tau) + u_1^T(t)$. If a sample is in one arm, $u_1^T(t)$ is replaced by $u_2^T(t)$ and the average power $p_{12}(\tau)$ at the detector is given by

$$p_{12}(\tau) = \lim_{T \to \infty} \frac{1}{2T} \int_{-\infty}^{+\infty} u_1^T(t - \tau) u_2^T(t) \, dt. \qquad (4.47)$$

(Only cross products of the two waves have been considered here; the constant contributions and uninteresting constants are omitted.)

Inserting Eq. 4.46 into Eq. 4.47 we have

$$p_{12}(\tau) = \lim_{T \to \infty} \frac{1}{2T} \int_{-\infty}^{+\infty} u_1^T(t - \tau) \left[\int_{-\infty}^{+\infty} g(t') u_1^T(t - t') \, dt' \right] dt. \qquad (4.48)$$

Equation 4.48 may be written in the equivalent form

$$p_{12}(\tau) = \int_{-\infty}^{+\infty} g(t') \left[\lim_{T \to \infty} \frac{1}{2T} \int_{-\infty}^{+\infty} u_1^T(t) u_1^T(t + \tau - t') \, dt \right] dt'. \qquad (4.49)$$

In the special case when there is no sample in the interferometer arms, the impulse response function is a δ-function: $g(t') = \delta(t')$ and we have

$$p_{11}(\tau) = \lim_{T \to \infty} \frac{1}{2T} \int_{-\infty}^{+\infty} u_1^T(t) u_1^T(t + \tau) \, dt. \qquad (4.50)$$

By comparison with Eq. 4.49, we see that we can write for $p_{12}(\tau)$:

$$p_{12}(\tau) = \int_{-\infty}^{+\infty} g(t') p_{11}(\tau - t') \, dt'. \qquad (4.51)$$

† See Eqs. 4.58 and 4.59.

Therefore the sample interferogram can be expressed as the convolution of the impulse response function and the background interferogram.

The Fourier transform of Eq. 4.51 is

$$\hat{P}_{12}(f) = \hat{G}(f)\hat{P}_{11}(f) \tag{4.52}$$

$$\hat{P}_{12}(f) \rightleftarrows p_{12}(\tau) \tag{4.53}$$

$$\hat{P}_{11}(f) \rightleftarrows p_{11}(\tau). \tag{4.54}$$

In the foregoing section we have used the variables x and v instead of f and t, but since $f \cdot t$ is dimensionless, as is $x \cdot v$, we can write

$$\hat{P}_{12}(v) = \hat{G}(v)\hat{P}_{11}(v) \tag{4.55}$$

$$\hat{P}_{12}(v) \leftrightarrows p_{12}(x) \tag{4.56}$$

$$\hat{P}_{11}(v) \leftrightarrows p_{11}(x) \tag{4.57}$$

and

$$\hat{U}_{2T}(v) = \hat{G}(v)\hat{U}_{1T}(v). \tag{4.58}$$

We see that $\hat{G}(v)$ can be obtained from the interferogram taken with the sample and from the interferogram taken without the sample. The optical constants of the sample can now be calculated from $\hat{G}(v)$ since this is a complex function of the frequency and contains both the amplitude and the phase information. Both quantities are needed to calculate a pair of optical constants at each frequency.

We now discuss some reflection measurements (for transmission measurements, see Ref. 25). The back of the sample might be shaped in such a way that there will be no reflected light passing from that surface back to the beam splitter. On the other hand, the light from the front surface of the sample passes back to the beam splitter under the same geometric optical conditions as the incident light. We will compare the incident and the reflected light at the mirror and, for a sample at the position of that mirror, relate this to the function $\hat{G}(v)$: At the mirror, we have $U(\text{mirror}) = e^{i\pi}U(\text{incident})$ and for the sample $U(\text{sample}) = \hat{R} \cdot U(\text{incident})$, where $\hat{R} = \sqrt{R}e^{i\psi_R(v)}$ is the complex reflectivity. $\hat{R}$ is related to the refractive index n and the extinction coefficient κ as given in Chapter 12, Eqs. 12.12 and 12.13.* Since $U(\text{sample}) = \hat{G}(v)U(\text{mirror})$, it follows:

$$\hat{G}(v) = \sqrt{R(v)}e^{i(\psi_R(v)-\pi)} = \hat{G}(v)e^{i\phi_G(v)}. \tag{4.59}$$

---

* In this chapter we write the complex index of refraction in the form $\hat{n} = n + i\kappa$, and the complex reflectivity as $\hat{R} = (1 - n - i\kappa)/(1 + n + i\kappa) = \sqrt{R}e^{i\psi_R}$. In Chapter 12 we employ a slightly different sign convention $\rho = (n + i\kappa - 1)/(n + i\kappa + 1) = \sqrt{R}e^{i\psi}$. The real quantities are, of course, the same in both cases.

The power spectrum $\hat{P}_{11}(v)$ is real (because the background interferogram is symmetric about the zero point) and, consequently, $\hat{G}(v)$ and $\hat{P}_{12}(v)$ have the same phase. It follows with the relations

$$\hat{P}_{12}(v) = P_{12}(v)e^{i\phi_{12}(v)} \tag{4.60}$$

and

$$\hat{P}_{11}(v) = P_{11}(v)e^{i\phi_{11}(v)} \tag{4.61}$$

that

$$\sqrt{R(v)} = \frac{P_{12}(v)}{P_{11}(v)}$$

and that

$$\psi_R(v) = \phi_{12}(v) - \phi_{11}(v) + \pi = \phi_G(v) + \pi. \tag{4.62}$$

Using these equations, we can calculate the quantities $\sqrt{R(v)}$ and $\psi_R^{(v)}$ from $\hat{G}(v)$. For n and κ we then obtain:

$$n = \frac{(1 - R)}{1 + 2\sqrt{R}\cos\psi_R + R}, \tag{4.63}$$

and

$$\kappa = \frac{-2\sqrt{R}\sin\psi_R}{1 + 2\sqrt{R}\cos\psi_R + R}. \tag{4.64}$$

Figure 4.23 shows an example where $R(v)$ and $\psi_R(v) - \pi$ have been calculated for KBr in reflection. This method is extremely useful for obtaining optical constants since these are calculated directly from the complete Fourier transformation of the interferogram. Both n and κ are obtained simultaneously in the same experiment (see Chapter 12). The method is sensitive to maladjustments and other error sources such as these discussed in Ref. 24. No low-temperature measurements seem to have been performed with this method to the present time, and it appears that such measurements would present a severe instrumental problem.

REFERENCES

1. E. V. Loewenstein, *Appl. Opt.* **5**, 845 (1966).
2. J. Strong, *Concepts of Classical Optics*, Freeman, San Francisco, 1958.
3. L. Mertz, *Transformations in Optics*, Wiley, New York, 1965.
4. A. Papoulis, *The Fourier Integral and Its Applications*, McGraw-Hill, 1962.
5. J. Connes, *Rev. Opt.* **40**, 45, 116, 171, 231 (1961). English Translation: NAVWEPS Report No. 8099, NOTS TP3157, U.S. Naval Ordnance Test Station, China Lake, California.

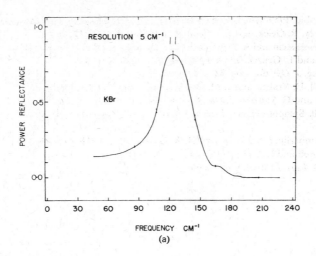

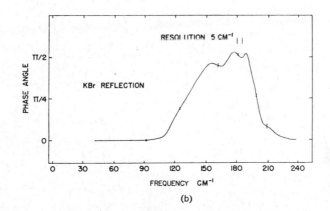

Fig. 4.23 Spectrum of KBr: (a) power reflectance $R(v)$; (b) phase spectrum $\psi_R(v) - \pi$. [E. E. Russell and E. E. Bell, *Infrared Phys.* **6**, 75 (1966), by permission of Pergamon Press.]

6. G. A. Vanasse and H. Sakai, "Fourier Spectroscopy," in *Progr. Opt.* Vol. 6, North-Holland, Amsterdam, 1967, p. 261.
7. P. L. Richards, *J. Opt. Soc. Am.* **54**, 1474 (1964).
8. R. G. Wheeler and J. C. Hill, *J. Opt. Soc. Am.* **56**, 657 (1966).
9. A. S. Filler, *J. Opt. Soc. Am.* **54**, 762 (1964).
10. M. L. Forman, *J. Opt. Soc. Am.* **56**, 978 (1966).
11. E. V. Loewenstein, *Appl. Opt.* **2**, 491 (1963).
12. M. L. Forman, W. H. Steel, and G. A. Vanasse, *J. Opt. Soc. Am.* **56**, 59 (1966).
13. R. Beer and D. Marjaniemi, *Appl. Opt.* **5**, 1191 (1966).
14. H. Yoshinaga, S. Fujita, S. Minami, Y. Suemoto, M. Inoue, K. Chiba, K. Nakano, S. Yoshida, and H. Sigimori, *Appl. Opt.* **5**, 1159 (1966).

15. H. A. Gebbie, 1959 Symposium on Interferometry, Teddington, England, unpublished.
16. C. H. Perry, R. Geick, and E. F. Young, *Appl. Opt.* **5**, 1171 (1966).
17. E. V. Loewenstein and A. Engelsrath, AFCRL-66-536, IP 115, Project 7670.
18. P. Vogel and L. Genzel, *Infrared Phys.* **4**, 257 (1964).
19. J. Dowling, *J. Opt. Soc. Am.* **54**, 663 (1964).
20. R. T. Hall, D. Vrabec, and J. M. Dowling, *Appl. Opt.* **5**, 1147 (1966).
21. J. Strong and G. Vanasse, *J. Opt. Soc. Am.* **50**, 113 (1960).
22. E. E. Bell, Symposium on Molecular Structure and Spectroscopy, Columbus, Ohio, June 1962.
23. J. E. Chamberlin, J. E. Gibbs, and H. A. Gebbie, *Nature* **198**, 874 (1963).
24. E. E. Russell and E. E. Bell, *Infrared Phys.* **6**, 75 (1966).
25. E. E. Bell, *Infrared Phys.* **6**, 57 (1966).

5 Low-Frequency Stretching and Bending Fundamentals

A. INTRODUCTION

Although the greater number of the fundamental modes of a molecule lies in the infrared range of 3500–600 cm^{-1}, there are often various low-lying bending and—in cases where heavy atoms or weak bonds are involved—stretching fundamentals below 600 cm^{-1} (16.7 μ). Many of these low-lying deformation and stretching modes are found within the 500–200 cm^{-1} range, a range which is experimentally less difficult than the spectral regions towards longer wavelengths. Within the last few years various small but nearly fully automatic grating infrared spectrometers, which scan out to 200 cm^{-1} (50 μ), have appeared on the market. It is therefore reasonable to expect that the spectral range between 600 and 200 cm^{-1} will be thoroughly searched and mapped in the coming years, particularly with the aim of completing the "fingerprinting" of chemical compounds.

The nature of the fundamental vibrational modes encountered in the 600–200 cm^{-1} region is somewhat different compared to that of the fundamentals in the infrared to shorter wavelengths. The low-lying modes, on account of their weak force constants, are to a relatively great extent influenced by interaction with other vibrations, by changes in the structure or conformation of the molecule, and so forth. Furthermore, isotopic substitution is not of as great value as the near-infrared range since simple square-root shifts are generally not observed due to the complicated nature of the vibrations.

We feel it would be beneficial to discuss some of the phenomena which influence or determine the vibrational modes in this frequency range. We do not intend to give coverage of the relevant literature but we will concentrate on a variety of molecules of more than routine interest, molecules which are representative examples of some of the effects which show up in the wavelength region between 16 and 50 μ.

163

B. Disulfur Dichloride and Disulfur Dibromide

The structure of disulfur dichloride, disulfur dibromide, and related molecules with the formula S_2X_2 is of considerable interest because these molecules are simple prototypes containing the $-S=S-$ group. This group is important in chemistry. Like the $-C=C-$ group, it can occur in large molecules and in polymeric systems. In addition, the $-S=S-$ group is a reactive site; for instance, it can easily be reduced or oxidized.

On purely spectroscopic grounds, S_2Br_2 and S_2Cl_2 are good examples for the valuable information which is often furnished by a study of up-to-now unrecorded far-infrared fundamentals. For the disulfur dihalides the presence or absence of the far-infrared torsional mode helps in deciding between the two proposed molecular structures, the planar *cis* molecule of point group symmetry C_{2v} and the nonplanar molecule of point group symmetry C_2, respectively. Since the torsional mode has very recently been observed in the far-infrared at $106 \, \text{cm}^{-1}$ for S_2Cl_2 and at $70 \, \text{cm}^{-1}$ for S_2Br_2, there is now no doubt that the structure of the molecule is C_2 (nonplanar) since in the planar configuration (C_{2v}) the torsional (a_2) mode would only be Raman-active. The assignments of the fundamentals of the two molecules are therefore based on the C_2 structure, which is shown in Fig. 5.1.

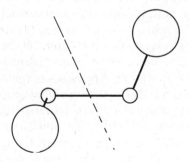

Fig. 5.1   Nonplanar structure of S_2X_2. The dotted line represents the C_2-symmetry axis.

Table 5.1 gives a collection of the assignments of the fundamentals of S_2Cl_2 and S_2Br_2. The spectra of S_2Br_2 had been taken using the pure liquid[1-3,4b] and carbon disulfide solutions;[1] the spectra of S_2Cl_2 had been scanned with the liquid phase.[4a] The descriptions of the nature of the vibrational modes, shown in the last column of Table 5.1, are based on the latest Raman data (quantitative depolarization data), on the observation of the torsional mode in the far-infrared spectrum (see Fig. 5.2), and the near constancy of the frequency of the highest fundamental regardless of the

TABLE 5.1

FREQUENCIES (CM^{-1}) AND ASSIGNMENTS OF THE FUNDAMENTALS OF DISULFUR DICHLORIDE AND DISULFUR DIBROMIDE

S_2Cl_2		S_2Br_2		Species of C_2	Assignment	Nature of vibration
IR	R	IR	R			
540[‖]	540[‖]	531[*,†]	356[‡]	a	ν_1	S—S stretch
537[§]	449[‡]		534[#]			
448[‖]	446[‖]	302*	302[‡]	a	ν_2	S—X stretch
443[§]	436[‡]		357[#]			
206[§]	205[‖]	175*	172[‡]	a	ν_3	S—S—X angle deformation
	206[‡]	176[†]	170[#]			
106[§,‖]	102[‖]	70[#]	66[‡]	a	ν_4	torsion
	102[‡]		68[#]			
434[‖]	434[‖]	355*	531[‡]	b	ν_5	S—X stretch
	540[‡]	354[†]	304[#]			
245[§]	238[‖]	198*	200[‡]	b	ν_6	S—S—X angle deformation
	240[‡]	196[†]	199[#]			

* See Ref. 3.
† See Ref. 1.
‡ See Ref. 6.
§ See Ref. 2.
‖ See Ref. 4a.
See Ref. 4b.

IR = infrared
R = Raman

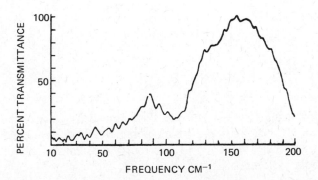

Fig. 5.2   Far-infrared spectrum of S_2Cl_2, $200\,cm^{-1} - 10\,cm^{-1}$, obtained on an interferometer employing a $\frac{1}{2}$-mil Mylar beam splitter. The path difference was 1 cm and yielded a resolution of $1\,cm^{-1}$. The computer was programmed to plot three points per resolution. [E. B. Bradley, M. S. Mathur, and C. A. Frenzel, *J. Chem. Phys.* **47**, 4325 (1967).]

nature of the substituent on the sulfur atoms. Since S_2Br_2 is a red-colored liquid, previous Raman depolarization data were only qualitative; quantitative data were recently obtained with the help of a laser source. They indicate that greater credence should be put into the assignment of the symmetric S—Br stretch to a frequency of 357 cm^{-1} and that of the antisymmetric S—Br stretch to 304 cm^{-1},[4b] a reversal from previous assignments.[3]

A further reassignment is that of the S—S stretch v_1 of the bromide compound. There is now little doubt that the assignment of v_1 to 540 and 531 cm^{-1} for S_2Cl_2 and S_2Br_2, respectively, is correct; a frequency of around 520 cm^{-1} occurs in all S_2X_2 molecules investigated. The corresponding values for X = H, CH_3, F, Cl, and Br are 510, 510, 526, 540, and 531 cm^{-1}, respectively. This strongly hints at the possibility of considering the S=S stretching frequency as a "group frequency"—if it can be shown that the distribution of the potential energy among the coordinates is about the same in these compounds during v_1.*

Bradley *et al.* have computed thermodynamic functions with the help of the observed frequencies.[4a] However, since the frequencies of the liquid phase were used, the computed values should be used with caution on account of the often considerable solvent shifts between the vapor and the liquid-phase frequencies (which are expected to have positive as well as negative signs, depending on the frequency range).[5]

In conclusion, we give the liquid-phase infrared frequencies of the fundamentals of S_2Cl_2 and S_2Br_2 (in cm^{-1}), taken from Table 5.1, according to the most recent determinations:[2,4a,4b]

	cm^{-1}		Species
v_1,	540,	531	a
v_2,	448,	355	a
v_3,	206,	175	a
v_4,	106,	70	a
v_5,	434,	302	b
v_6,	245,	198	b

(5.1)

In the following two sections we shall consider two molecules, carbon suboxide and 1,4-dioxadiene, which possess far-infrared bending modes whose characteristics have led to interesting speculations about resonance forms in these molecules.

---

* It is our opinion that the concept of "group frequency" has often been used indiscriminately in the literature. Of course, this may be a consequence of the rather diffuse definition that underlies many uses of the group frequency concept. In this respect we refer the reader to the pertinent pages of Herzberg's book (Ref. 11), pp. 194–201.

C. CARBON SUBOXIDE AND CARBON SUBSULFIDE

Investigations of the far-infrared spectrum of carbon suboxide, C_3O_2, a linear molecule, have led to a proposal concerning the existence and importance of certain resonance structures of the molecule. In the following we shall discuss this in "historical" order; namely, the experimental observations, the rationalization of the data in terms of resonance structures, and finally the quantum-mechanical treatment of the motion.

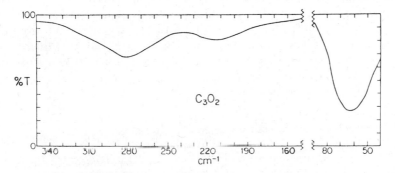

Fig. 5.3 Low-frequency infrared bands. (Retraced) C_3O_2 gas. Pressure 80 mm and 10-m path. Resolution 1.8 cm^{-1} at each of the bands. The band at 280 cm^{-1} is the v_2-v_6 difference band. [F. A. Miller, D. H. Lemmon, and R. E. Witkowski, *Spectrochim. Acta* **21**, 1709 (1965), by permission of Pergamon Press.]

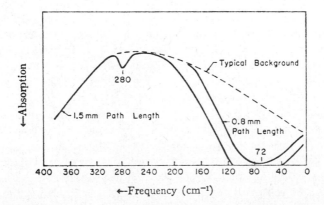

Fig. 5.4 Far-infrared spectrum of liquid carbon suboxide. Fourier transformation of single-beam interferograms taken at 193°K. The band at 280 cm^{-1} is the v_2-v_6 difference band. [Wm. H. Smith and G. E. Leroi, *J. Chem. Phys.* **45**, 1767 (1966).]

In C_3O_2, at a frequency of $63\,cm^{-1}$ for the vapor and $72\,cm^{-1}$ for the liquid, an extremely weak and broad band has been observed and assigned to the central-carbon bending fundamental $v_7(\pi_u)$. Figure 5.3 shows the spectrum of C_3O_2 vapor, taken by Miller et al.[7] with a Beckman IR-11 spectrometer between 40 and $350\,cm^{-1}$. Figure 5.4 shows the spectrum of liquid C_3O_2, measured by Smith and Leroi[8] with a RIIC-FS-520 Michelson interferometer, in the spectral range of 10 to $400\,cm^{-1}$. The corresponding fundamental v_7 of linear carbon subsulfide, C_3S_2, lies at $94\,cm^{-1}$ (vapor), with a clearly defined PQR structure, and at $107\,cm^{-1}$ in CS_2-solution. We notice that the fundamental of the sulfur compound is at higher frequencies than that of C_3O_2 although sulfur is heavier than oxygen.[9]

A rationalization for the lowness of v_7 in C_3O_2 has been given by Pitzer and Strickler.[10] Their arguments, which were extended by Miller et al.[7] to explain the weak intensity of v_7 as well, are as follows (it should be noted that the assignment of v_7 seems now to be beyond doubt): If one writes down, a priori, ionic resonance forms of C_3O_2 of the form

$$O{=}C{=}C{=}C{=}O, \qquad O^+{\equiv}C{-}C^-{=}C{=}O \leftrightarrow O{=}C{=}C^-{-}C{\equiv}O^+ \quad (5.2)$$

it is seen that the group $-C^-{\equiv}$ is isoelectronic with the group $-N{=}$. Since a configuration such as $X{-}N{=}Y$ is nonlinear, bending at the central carbon atom would be made easier if the above resonance forms constitute an important contribution to the structure. As the outer $C{\equiv}O^+$ group is always linear, bending at the outer carbon atoms is not affected by these resonance structures; in other words, the other two bending fundamentals (v_5 and v_6) remain near their "usual" frequencies. The three bending fundamentals are given in Fig. 5.5.[11]

The low intensity of v_7 is a consequence of the sharply reduced bond dipole moment of the $C^-{-}C{\equiv}O^+$ group in the resonance structures compared to its value in the classical form, which possesses a considerable amount of ionic character and (negative) charge concentration on the oxygen atom. (See, for instance, L. Pauling, The Nature of the Chemical Bond, Cornell Univ. Press, Ithaca, New York, 1960, p. 142.)

Another perplexing characteristic in the spectrum of carbon suboxide is that difference (as well as sum) combination bands involving v_7 disappear when the compound is solidified in a CCl_4 matrix at a temperature of 170–190°K. This cannot be a consequence of a significant depopulation of the upper levels of v_7 due to the temperature decrease since, first, the particular combination bands are observable in the liquid phase of C_3O_2 at 190°K and, second, the relative population of v_7 in its, say, fourth excited level is still 0.45 at 190°K.[8] Smith and Leroi propose that this phenomenon is simply a consequence of the transition liquid phase → solid phase; in other words, the low-frequency, large-amplitude motion of v_7 is constricted in the solid phase

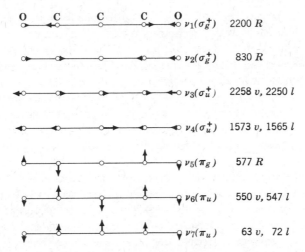

Fig. 5.5 The normal modes (schematic) for the linear model of carbon suboxide C_3O_2. The assigned frequencies are given in cm^{-1}, taken from F. A. Miller and W. G. Fateley, *Spectrochim. Acta* **20**, 253 (1964), F. A. Miller, D. H. Lemmon, and R. E. Witkowski, *Spectrochim. Acta* **21**, 1709 (1965), Wm. H. Smith and G. E. Leroi, *J. Chem. Phys.* **45**, 1778 (1966); v = vapor at room temperature, l = liquid at 193°K, R = Raman (liquid). The mode v_5 has also been observed in the infrared (liquid) by relaxation of the selection rules. [From Herzberg's *Infrared and Raman Spectra of Polyatomic Molecules*, Copyright 1962, D. Van Nostrand Company, Inc., Princeton, New Jersey.]

by the forces of the lattice, with a subsequent loss of intensity. Ultimately, it is the quantity $\partial\mu/\partial Q$, the derivative of the dipole moment with respect to the normal coordinate, which must then decrease with decreasing amplitude. How this comes about does not seem definitely established. We have seen that bending of the molecule at the central atom has been suggested as a cause of the weak intensity of v_7 *per se*; we now have experimental evidence that there is an additional loss of intensity if the compound is solidified.

TABLE 5.2*

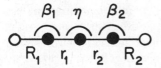

FORCE CONSTANTS OF C_3O_2 AND C_3S_2 (IN PARENTHESES) IN UNITS OF 10^5 DYNE/CM

$f_R = 15.4$ (7.74)	$f_\beta = $ 0.355 (0.359)
$f_r = 11.8$ (10.2)	$f_\eta = $ 0.031 (0.122)
$f_{Rr} = $ 1.27 (0.313)	$f_{\beta\beta} = $ 0.095 (0.083)
$f_{rr} = $ 2.49 (1.02)	$f_{\beta\eta} = $ −0.006 (0.042).

* *After* Wm. H. Smith and G. E. Leroi, *J. Chem. Phys.* **45**, 1767 (1966).

In Table 5.2 we give the force constants obtained from the solutions of the vibrational analysis[8,9] with the help of a harmonic valence force field. All values are in units of 10^5 dyne/cm; those of the sulfur compound C_3S_2 are given in parentheses. Note the larger bending force constant of v_7 (f_n) for C_3S_2; Smith and Leroi have proposed that this is partly a consequence of the smaller degree of overlap of the spacial distribution of the valence $3p$ electrons of sulfur with the $2p$ orbital of carbon compared to the $2p$–$2p$ overlap between oxygen and carbon. Hence, resonance structures such as those proposed for C_3O_2 are not so important in carbon subsulfide.[12]

Inspection of Fig. 5.3 indicates a broad and very weak absorption centered near $214 \, cm^{-1}$. This absorption seems to belong to C_3O_2 but cannot be assigned on the basis of the known transitions of the molecule.[7] There are other unexplained inconsistencies in the mid-infrared and near-infrared spectrum of this molecule. Redington[13] has been able to explain the various aspects of the spectrum of C_3O_2 by considering the molecule to be "quasi-linear." We shall discuss this now in some detail.

First we define a quasi-linear molecule as a molecule which has properties intermediate to those of a bent and a linear molecule. In the theoretical treatment of the quasi-linear molecule, developed by Thorson and Nakagawa,[14] the vibrational motions are separated into fast varying, small-amplitude vibrations (for instance, the bond stretches) and slowly varying, large-amplitude modes (the bending motions). It is assumed that the bond stretches are essentially harmonic and at significantly higher frequencies than the bending modes. The rotational motion of the molecule is described, according to common practice, by the three Eulerian angles θ, ψ, and ϕ. For a strictly linear molecule, there would be no rotation about the long axis (z-axis). For a slightly bent molecule, the moment of inertia about this axis is small and consequently the rotational spacings are large. We therefore retain only the rotation about the long axis (Euler's angle $= \phi$) in the expression for the kinetic energy and neglect the fine structure arising from the other two (large) moments of inertia of the slightly bent molecule. In Fig. 5.6 we give a qualitative energy correlation diagram. On the left, we have the strictly linear molecule. Its bending vibrational motion is expressed in terms of the doubly degenerate (the x and y-directions are equivalent) harmonic oscillator with the levels $E(v) = \hbar\omega(1 + v)$ and quantum numbers $v = 0, 1, 2, \cdots; l = v, v - 2, v - 4, \cdots, -v$.[15] At the other limit, on the right side of the diagram, we have the permanently bent molecule with (single) harmonic oscillator vibrational-rotational levels of energy $E(n,l) = \hbar(n + \frac{1}{2})\omega + (\hbar^2/2I)(l + \frac{1}{2})^2$, where $n = 0, 1, 2, \cdots; |l| = 0, 1, 2, \cdots$. The form of the rotational energy term given above is seen to be equal to $(\hbar^2/2I)\{l(l + 1) + \frac{1}{4}\}$; the term $\hbar^2/8I$ originates from the quantum conditions in the Brillouin-

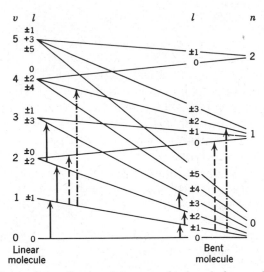

Fig. 5.6   Schematic correlation diagram of the energy levels and of some of the transitions for the linear molecule bending mode and the bent molecule vibration–rotation mode. The sets of n-levels generally overlap and have been pulled apart here for clarity's sake. [See W. R. Thorson and I. Nakagawa, *J. Chem. Phys.* **33**, 994 (1960), (Fig. 2).]

Wenzel-Kramer approximation which was used to obtain the above expression for $E(n, l)$ of the bending mode.[16]

We now give the selection rules. We first note that $\Delta l = \pm 1$ is rigorous for all transitions polarized in a direction perpendicular to the z-axis; that is, those involving a change in the angular momentum about the z-axis.[14] Starting with the linear molecule on the left of the energy correlation diagram (Fig. 5.6) and moving to the bent molecule, we see that the transition $\Delta v = +1$, $\Delta l = +1$, goes over, in the limit of the bent molecule, into $\Delta l = +1, \Delta n = 0$. The resulting frequencies in the bent molecule are those of a "pure rotation spectrum," $v = v_0 + 2B(l + 1)$ (B is the effective rotational constant), except that the first line appears at $v_0 + 2B$ instead of $2B$. Next we consider the transitions $\Delta v = +1, \Delta l = -1$ of the linear molecule (dashed arrow in Fig. 5.6); these become the $\Delta l = -1, \Delta n = +1$ transitions ("P" branch) in the limit of the bent molecule. Finally, the corresponding "R" branch rotational-vibrational transitions of the bent molecule, $\Delta l = +1, \Delta n = +1$ (dot-dash arrow), are seen to have originated from the weakly allowed $\Delta v = +3, \Delta l = +1$ transitions of the linear molecule.

If the variables in the Schrödinger equation for the slow bending mode (α, in radians) and the rotation about the z-axis (Euler's angle ϕ) are separated in the usual fashion,[15] the effective potential energy can be written in the form:[13]*

$$\tilde{V}(\alpha, l) = V(\alpha) + \frac{\hbar^2 l^2}{2\mu\alpha^2}. \tag{5.3}$$

Note that the term $\hbar^2 l^2/2\mu\alpha^2$ fulfills the function of a central (high) barrier for all $l \neq 0$. Hence, the equilibrium position of the molecule in energy levels $n, l = 0$ is linear; for $l \neq 0$ the molecule is slightly bent.

Using the quartic potential function $V(\alpha) = 1.50 \times 10^3 \alpha^2 + 4.50 \times 10^4 \alpha^4$ (cm^{-1}), Redington was able to predict the spectrum of v_7 and the curious aspects observed in the mid-infrared of C_3O_2. The results of his calculations involving the far-infrared transitions of v_7 are collected in Table 5.3. Notice that the hitherto unassignable band near 214 cm^{-1} (see above and Fig. 5.3) is now characterized as the $n = 0, l \rightarrow n = 1, l + 1$ band. From the energy

TABLE 5.3*

ENERGY LEVELS AND PARTIAL SPECTRUM ($V(\alpha) = 1500\alpha^2 + 45000\alpha^4$ CM^{-1})

l	$E_{0,l}$	$E_{1,l}$	$E_{2,l}$	$(0, l \rightarrow 0, l + 1)$		$(0, l \rightarrow 1, l + 1)$	
				$h\nu$	Spacing	$h\nu$	Spacing
0	52.1	189.1	353.0	62.3		214.8	
					5.9		18.7
1	114.4	266.9	442.5	68.2		233.5	
					4.5		15.9
2	182.6	347.9	532.4	72.7		249.4	
					4.3		15.1
3	255.3	432.0	625.7	77.0		264.5	
					4.2		12.6
4	332.3	519.8	721.0	81.2		277.1	
					2.9		11.5
5	413.5	609.4		84.1		288.6	
					3.9		11.1
6	497.6	702.1		88.0		299.7	
7	585.6	797.3					

* R. L. Redington, *Spectrochim. Acta* **23A**, 1863 (1967). By permission of Pergamon Press.

* The Hamiltonian H for the bending mode contains two kinetic energy terms that describe the kinetic energy of the bending motion and the pure rotational energy about the z-axis. The pure rotational term is $T = \frac{1}{2}\mu\alpha^2\dot{\phi}^2$ (where $\mu\alpha^2$ is the moment of inertia about the z-axis) or $T_\phi = p_\phi^2/2\mu\alpha^2 = \hbar^2 l^2/2\mu\alpha^2$ in terms of the conjugate momentum p_ϕ; the kinetic energy of the pure vibrational motion is $T_\alpha = p_\alpha^2/2\mu$. Hence $H \approx T_\alpha + T_\phi + V(\alpha)$, where a small interaction term has been neglected.

level correlation diagram in Fig. 5.6 we see that this corresponds to the $\Delta v = +3$ transition for the linear molecule. The extreme weakness of the $214\,\mathrm{cm}^{-1}$ absorption is therefore not surprising.

TABLE 5.4*

ENERGY LEVELS AND BENDING OF THE QUASI-LINEAR MOLECULE C_3O_2

	Root mean square α $\langle\alpha^2\rangle^{\frac{1}{2}}_{n,l}$			Minimum in effective potential energy
l	$n = 1$	$n = 2$	$n = 3$	
0	5.9°	8.8°	10.6°	0°
1	7.7	9.8	11.3	6.0
2	9.1	10.7	12.0	8.0
3	10.1	11.6	12.7	9.3
4	11.1	12.3		10.4
5	11.9	12.9		11.3
6	12.6			12.0
7	13.2			12.8

* R. L. Redington, *Spectrochim. Acta* **23A**, 1863 (1967). By permission of Pergamon Press.

Finally in Table 5.4 we give the root-mean-square of the angle α and the angle at the minimum of $\tilde{V}(\alpha, l)$ for various states of excitation of the bending mode of C_3O_2. The average nonlinearity of the molecule, obtained by averaging the minima over the Boltzmann factors, turns out to be about 11°. In conclusion we mention that this value agrees acceptably well with results of electron diffraction measurements.

D. 1,4-DIOXADIENE

1,4-Dioxadiene is another example of a molecule where a far-infrared study (Connett *et al.*[17]) has helped in answering questions on the importance of contributions of resonance structures to the total structure. It had been suggested from electron diffraction measurements that nonaromatic conjugation presumably makes some contribution to the total energy, and a resonance structure had been written to this effect:[18]

$$\text{(I)} \qquad \text{(II)} \tag{5.4}$$

The ring deformation vibration b_{3u} of the planar molecule (D_{2h}) was observed at $124\,\text{cm}^{-1}$. The molecule assumes the boat conformation through this mode; the low frequency of the fundamental indicates that ionic resonance structures such as (II) make only a small contribution to the structure of the molecule since the conjugated bonds would strive for maximum overlap by keeping the ring planar. The much higher frequency of the b_{2g} bending mode $(516\,\text{cm}^{-1})$, a mode through which the molecule can assume the chair conformation, has been proposed to be a consequence of torsional strain induced in the ethylenic bonds during the vibration.[17]

E. Molecules Containing Fluorine

The element fluorine plays a special role in chemistry because of its great reactivity and the strength of its bonds with other elements. For instance, the bond energies in the sequence of the halogen molecules F_2, Cl_2, Br_2, and I_2 are 38, 57, 45, and 36 kcal/mole, respectively, whereas the electronegativities decrease monotonically in the series. In other words, about the same energy is required to break the F—F bond and the bond of the much larger I—I molecule but the reactivity of the F atom is much larger than that of the I atom. The spectroscopic characterization of compounds containing fluorine, particularly if the bonding seems to contain unusual features, deserves particular attention.

a. Xenon Difluoride

The preparation of various xenon–fluorine compounds in the last few years had aroused great excitement, particularly with respect to the nature of the xenon–fluorine bonds.[19] Of these xenon fluorides, only the infrared spectra of the xenon difluoride and the xenon tetrafluoride have at the present been unambiguously assigned.

We first concentrate on the spectrum of xenon difluoride. The following observations and assignments of the spectrum of XeF_2 vapor were reported by Smith and Mason.[20] One band without central Q branch, centered at $557\,\text{cm}^{-1}$; one band with a strong, sharp central Q branch at $213\,\text{cm}^{-1}$; and one weak combination, without Q branch, at $1070\,\text{cm}^{-1}$. The absence of a central Q branch of the $557\,\text{cm}^{-1}$ band and the lack of observing more than two strong infrared absorptions are very convincing evidence that xenon difluoride is a symmetric linear molecule. Consequently, the observed transitions are assigned as the infrared-active stretch v_3 $(\Sigma_u^+ \leftarrow \Sigma_g^+) = 557\,\text{cm}^{-1}$, the degenerate bending mode v_2 $(\Pi_u \leftarrow \Sigma_g^+) = 213\,\text{cm}^{-1}$, and the combination $v_1 + v_3 = 1070\,\text{cm}^{-1}$, where v_1 is the infrared-inactive stretching fundamental $(\Sigma_g^+ \leftarrow \Sigma_g^+)$. Figures 5.7 and 5.8 show copies of tracings of v_2 and v_3. From the spacing between the R and P branches of v_3, which corresponds to a maximum of the rotational population at quantum number $J \sim 25$,[21] one calculates an approximate Xe—F bond distance of 1.7 Å. This distance is shorter than that determined from a crystal structure analysis of XeF_2[20] by about 0.3 Å. Smith has proposed that this difference is a consequence of insufficient pressure broadening in the vapor band[22] and has cited evidence that pressures between 0.4 and 10 atm of admixed foreign gases—which is a factor of 10^2 to 2×10^3 above the vapor pressure of the pure compound at room temperature—reduce the P–R separation of v_3 to a constant value which is then equivalent to

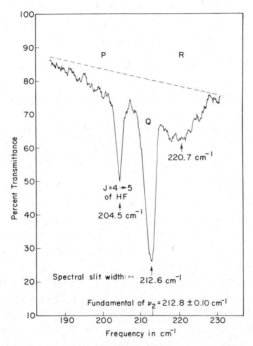

Fig. 5.7   The bending fundamental v_2 of saturated xenon difluoride vapor at room temperature (Unpublished data, W. G. Rothschild and B. Weinstock). The frequencies are the averages of various scans. The band envelope shows incipient resolution of the rotational fine structure (spacing of rotational lines, apart from the different isotopes of Xe, is 0.3 cm^{-1}; the nuclear statistical weights of adjacent rotational lines are 3:1). The band head at 212.8 cm^{-1} has been computed by using the estimated value of the rotational constant. The peak absorption in the R branch, equivalent to $J \sim 25$ and bond distance Xe—F 1.7 Å, is at 220.7 cm^{-1}. Perkin-Elmer 301, grating 20 lines/mm, 10-cm Monel cell with polyethylene windows. HF is a decomposition product. The bending force constant amounts to 0.197×10^5 dyne/cm.

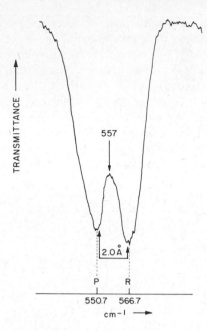

Fig. 5.8   The $\Sigma_u^+ \leftarrow \Sigma_g^+$ stretching fundamental ν_3 of XeF_2 vapor, 40 lines/mm grating. The *P–R* spacing which is equivalent to a Xe—F bond distance of 2.0 Å has been entered in the figure. (Unpublished data, W. G. Rothschild and B. Weinstock).

1.9 Å. On the other hand, Gellings has published some arguments* based on simple electrostatic considerations which point out that a slightly larger bond distance ($\sim 10\%$) in the solid may be real. A recent high-resolution infrared investigation of XeF_2 vapor has yielded a bond distance of 1.977 Å for the vibrational ground state.[23]

It has been assumed that the bonds in XeF_2 have considerable ionic character; Jortner *et al.*[24] give some experimental and theoretical evidence to this effect. On the other hand, Smith has claimed that the relatively low value of the bond stretching interaction constant, $k_{rr} = 0.11$,[20] compared to the value of the principal stretching constant, $k_r = 2.85$ (in units of 10^5 dyne/cm), indicates covalent bonding in XeF_2. This excludes major contributions of structures such as $F-Xe^+F^- \leftrightarrow F^-Xe^+-F$ since in essence

* From the known crystal structure of XeF_2, Gellings estimates that the Xe—F bond is stretched in the solid over its value in the isolated molecule since there is a net force pushing a F atom away from its "own" Xe atom due to the fluorine–fluorine repulsion by the neighboring F atoms (which lie in a line somewhat midway between the Xe and F atoms of a Xe—F bond). Furthermore, an additional contribution to the bond stretching arises from the attraction of the F atom by the neighboring Xe atoms of the other XeF_2 molecules. [See *Z. Phys. Chem.* (Frankfurt), **43**, 123 (1964).]

TABLE 5.5*

OBSERVED AND CALCULATED FREQUENCIES $\nu(\text{cm}^{-1})$ OF THE
FUNDAMENTAL VIBRATIONS OF XENON TETRAFLUORIDE

	Activity	Observed[†]	Calculated	Vibrational mode
$\nu_1(a_{1g})$	R	543[‡]	543	
$\nu_2(a_{2u})$	IR	291[§]	291	
$\nu_3(b_{1g})$	R	235	235	
$\nu_4(b_{1u})$	Inactive	(221)[‖]	(231)	
$\nu_5(b_{2g})$	R	502[‡]	502	
$\nu_6(e_u)$	IR	586	587	
$\nu_7(e_u)$	IR	168**	182	

* W. A. Yeranos, *Mol. Phys.* **9**, 449 (1965).

[†] The infrared spectra were taken with a Beckman spectrometer IR-7 with CsI prism and a Perkin-Elmer spectrometer 421 and 301 (see Ref. 27).

[‡] No polarization measurements could be made. The symmetric stretch was assigned to the higher frequency because of the expected larger degree of fluorine–fluorine repulsion compared to that during the asymmetric stretch ν_5.

[§] The P-Q spacing of $11\,\text{cm}^{-1}$ yields an Xe—F bond distance of 1.9 Å. The Q branch is very sharp and intense as expected from this C-type band.

[‖] Estimated from $2\nu_4 = 442\,\text{cm}^{-1}$, a weak band in the Raman spectrum. The assignment is considered questionable.

** Observed in the solid (vapor condensed at a polyethylene window kept near liquid-nitrogen temperature, H. H. Claassen, private communication, December 28, 1966).

IR = infrared R = Raman

"stretching of one bond tends to make the resonance structure in which this bond is ionic more prominent, changing the nature of the bonding more than normally"—thereby leading to a large value of k_{rr}. He has been disputed by Pimentel and Spratley.[25] It is difficult, in the absence of a reasonably complete series of structurally very similar molecules, to draw conclusions with respect to chemical bonding from the values of force constants of only one member. This, by the way, is also reflected in the scarcity and complexity of *ab initio* force constant calculations from structural data and bond parameters.[26]

b. Xenon Tetrafluoride

The vibrational spectrum of xenon tetrafluoride vapor, XeF_4, and the Raman spectrum of the solid compound have been assigned by Claassen et al.[27] and a normal coordinate analysis in the Urey-Bradley field approximation has been performed by Yeranos.[28] Since the molecule is a planar square with Xe in the center (point group D_{4h}), the mutual exclusion rule between Raman and infrared activity holds and there should be three infrared-active (e_u, e_u, a_{2u}), three Raman-active (a_{1g}, b_{1g}, b_{2g}), and one inactive fundamental (b_{1u}). Table 5.5 shows the frequencies of the observed transitions, the assignments, the calculated frequencies (cm^{-1}), and the description of the vibrational motion. The lowest fundamental, the doubly degenerate mode $v_7(e_u)$, had previously been assigned to $123 \, cm^{-1}$, a sharp band which turned out to be the $J = 2 \rightarrow 3$ pure rotational transition of hydrogen fluoride, a decomposition product.[29] It should be noted that such a low frequency for the v_7 mode would, at any rate, be at great variance to the

TABLE 5.6*

THE POTENTIAL ENERGY DISTRIBUTION IN
XENON TETRAFLUORIDE[†]

	K	H_α	H_γ	F
v_1	0.867_7	0.000_0	0.000_0	0.132_3
v_2	0.000_0	0.000_0	1.000_0	0.000_0
v_3	0.000_0	0.223_1	0.000_0	0.776_9
v_4	0.000_0	0.000_0	1.000_0	0.000_0
v_5	1.015_5	0.000_0	0.000_0	-0.015_5
v_6	0.954_6	0.000_7	0.000_0	0.044_7
v_7	0.021_0	0.231_9	0.000_0	0.747_1

* W. A. Yeranos, *Mol. Phys.* **9**, 449 (1965).
† The force constants are in units of 10^5 dyne/cm: $K = 2.86$, $H_\alpha = 0.034$, $H_\gamma = 0.30$, $F = 0.22$.

calculated frequencies since the force constants of the Urey-Bradley potential, namely the stretch $K(Xe{-}F)$, the two angle deformations $H_\alpha(F_1{-}Xe{-}F_2)$ and $H_\gamma(F_1{-}Xe{-}F_3$, out-of-plane), and the fluorine–fluorine repulsion $F(F{\cdots}F)$, are fully determined by the fundamentals v_1, v_2, v_3, and v_5. Indeed, the fundamental v_7 was subsequently observed by Claassen[30] at a frequency of $168\ cm^{-1}$ in the form of vapor condensed at a polyethylene window kept near liquid-nitrogen temperature.

Finally, in Table 5.6, we give the potential energy distribution in XeF_4 in terms of the contributions of the four Urey-Bradley constants.

In the following section we discuss the low-frequency vibrational motions of AsF_5 and PF_5. These particular motions are very interesting since they conceivably offer a pathway to a postulated mechanism for an intramolecular exchange of fluorine atoms in these molecules.

c. Phosphorous Pentafluoride and Arsenic Pentafluoride

The structures of PF_5 and AsF_5 are thought to be trigonal bipyramids with the P (or As) atom in the center (point group D_{3h}). This has been established for PF_5 by electron diffraction measurements and postulated for AsF_5 by Hoskins and Lord.[31] The interesting aspects of these molecules are that nuclear magnetic resonance studies have indicated the "equatorial" and "axial" fluorine atoms, that is, the three F atoms on the basis and the two at the apexes of the bipyramid, to be equivalent—although the P—F distances in the two sets are different (equatorial: 1.534 Å; axial: 1.577 Å in PF_5). By "equivalence" we mean that an equatorial F atom may change place with an axial F atom (in the same molecule) at a rate which is faster than the observable difference in the chemical shifts (expressed in frequency units).[32] We shall not dwell further on the aspects of the NMR experiments.

In the following we first indicate the proposed pathway of the fluorine atom exchange in terms of vibrational displacements of the F atoms (Berry[33]), and thereafter discuss the character of the normal coordinates of the relevant far-infrared modes (Hoskins and Lord[31]).

Figure 5.9 shows Berry's mechanism: two equatorial F atoms, for instance 2, 3, and the axial F atoms (5, 6) move in the direction of the arrows through an intermediate structure of C_{4v} point group symmetry until they have regained the original D_{3h} structure, but now with the set of F atoms 2, 3 in the axial and the set 1, 5, 6 in the equatorial positions. Hoskins and Lord[31] have investigated the spectra of PF_5 and AsF_5 and calculated the normal coordinates of the three e' fundamentals v_5, v_6, and v_7. These fundamentals are involved in the motions which might conceivably have appreciable displacement vectors in the directions required by Berry's mechanism. Restricting ourselves to AsF_5, the three e' modes were observed

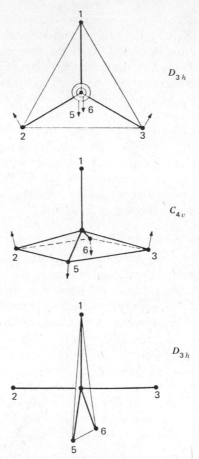

Fig. 5.9   Intramolecular exchange of fluorine atoms between equatorial (1, 2, 3) and axial (5, 6) atoms in PF_5 and AsF_5. In the uppermost configuration the threefold symmetry axis is perpendicular to the plane of the paper and the fluorine atoms 5, 6 are above and below the paper plane. [L. C. Hoskins and R. C. Lord, *J. Chem. Phys.* **46**, 2402 (1967).]

to be centered at 811, 369, and 123 cm^{-1}, respectively. Figure 5.10 shows a tracing of the v_6 mode at 369 cm^{-1}. The assignment of these modes was performed with the help of Raman depolarization values and observed infrared band contours.

Essentially, v_5 is the degenerate stretching mode, v_6 the degenerate bending mode of the equatorial, and v_7 the degenerate bending fundamental of the axial fluorine atoms. Of course, the exact nature of the vibrations must be determined by the normal coordinate analysis. The coordinate system used for this purpose is shown in Fig. 5.11; Table 5.7 gives the

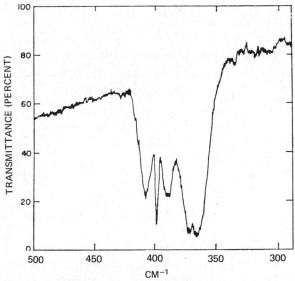

Fig. 5.10   The $\nu_6(e')$ fundamental of AsF$_5$, measured at 15-mm vapor pressure, cell length 10 cm, Perkin-Elmer Model 521. The band at 400 cm^{-1} represents the ν_4 (a_2'') mode, the bending of the equatorial F atoms in a direction normal to the basis of the bipyramid. [L. C. Hoskins and R. C. Lord, *J. Chem. Phys.* **46**, 2402 (1967).]

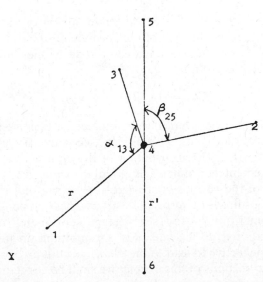

Fig. 5.11   Coordinates of the XY$_5$ molecule. [L. C. Hoskins and R. C. Lord, *J. Chem. Phys.* **46**, 2402 (1967).]

TABLE 5.7*

SYMMETRY COORDINATES FOR THE e' SPECIES FOR PF_5 AND AsF_5.

Species	Symmetry coordinate
e'	$S_{5a} = 6^{-\frac{1}{2}}(2\Delta r_1 - \Delta r_2 - \Delta r_3)$
	$S_{5b} = 2^{-\frac{1}{2}}(\Delta r_2 - \Delta r_3)$
	$S_{6a} = 6^{-\frac{1}{2}}(2\Delta\alpha_{23} - \Delta\alpha_{12} - \Delta\alpha_{13})$
	$S_{6b} = 2^{-\frac{1}{2}}(\Delta\alpha_{13} - \Delta\alpha_{12})$
	$S_{7a} = 12^{-\frac{1}{2}}(2\Delta\beta_{15} - \Delta\beta_{25} - \Delta\beta_{35} + 2\Delta\beta_{16} - \Delta\beta_{26} - \Delta\beta_{36})$
	$S_{7b} = \frac{1}{2}(\Delta\beta_{25} - \Delta\beta_{35} + \Delta\beta_{26} - \Delta\beta_{36})$

* After L. C. Hoskins and R. C. Lord, J. Chem. Phys. 46, 2402 (1967).

symmetry coordinates based on D_{3h} symmetry and a simple valence force field. We turn now to the kernel of our discussion and give, for the modes v_5, v_6, and v_7, the normal coordinates Q_5, Q_6, and Q_7 as obtained from the analysis of AsF_5:

$$Q_5 = 3.6S_5 + 0.3S_6 - 0.1S_7$$

$$Q_6 = -1.4S_5 + 3.8S_6 - 0.2S_7 \qquad (5.5)$$

$$Q_7 = 1.0S_5 + 1.1S_6 + 5.5S_7.$$

We notice at once that there is considerable mixing of the symmetry coordinates S_5, S_6, and S_7 in each of the normal coordinates. We must now compare the Q's with the exchange coordinate of the fluorine exchange (see Fig. 5.9), which is given by

$$Q_{ex} = (1/\sqrt{2})(S_{6a} + S_{7a}), \qquad (5.6)$$

where S_{6a} and S_{7a} are the symmetry coordinates for bending two equatorial and two axial As—F bonds, respectively (see Table 5.7), in accordance with Berry's proposal. The distribution of the coordinates Q_i among the original displacement coordinates is given in Table 5.8. On comparing the coefficients of the various coordinates, we notice that the distribution of the normal coordinates in terms of bond stretches and bond deformations is rather different from that of the exchange coordinate. In other words, the actual vibrational displacements in the molecule in its ground state would be very ineffective to lead to the exchange path proposed by Berry.[33]

In the last two sections of this chapter we shall be concerned with a brief discussion of the vibrational characteristics of some of the metal carbonyls and with a more detailed description of the carbon–halogen vibrations

TABLE 5.8

DISTRIBUTION OF THE NORMAL COORDINATES Q_5, Q_6, AND Q_7 OF AsF$_5$ AND OF THE EXCHANGE COORDINATE Q_{ex} AMONG THE DISPLACEMENT COORDINATES*

Displacement coordinate,[†] Δ_i	Normal coordinate			
	Q_5	Q_6	Q_7	Q_{ex}
Δr_1	0.58	−0.20	0.10	0
Δr_2	0.21	−0.07	0.04	0
Δr_3	−0.78	0.27	−0.14	0
$\Delta \alpha_{12}$	−0.06	−0.74	−0.15	−0.29
$\Delta \alpha_{13}$	0.02	0.20	0.04	−0.29
$\Delta \alpha_{23}$	0.04	0.54	0.11	0.58
$\Delta \beta_{15}$	−0.01	−0.02	0.40	0.41
$\Delta \beta_{16}$	−0.01	−0.02	0.40	0.41
$\Delta \beta_{25}$	−0.004	−0.01	0.14	−0.20
$\Delta \beta_{26}$	−0.004	−0.01	0.14	−0.20
$\Delta \beta_{35}$	0.02	0.03	−0.54	−0.20
$\Delta \beta_{36}$	0.02	0.03	−0.54	−0.20

* Computed from the data of Table 5.7 and Eq. 5.5.
† See Fig. 5.11 and Table 5.7.

of halogenated alkyl compounds. The low-lying modes of the metal carbonyls may conceivably be of use in adsorption studies and catalysis, to name two examples. The characterization of the carbon–halogen modes of the halogenated alkyl compounds is not only required for purposes of identification of a very important class of organic chemicals, but it is even more important for an understanding of the conformational behavior of aliphatic chain molecules.

F. METAL CARBONYLS

The volatile, monomolecular metal carbonyls are a very intriguing class of molecules. They possess a number of carbon–metal–carbon bending and metal–carbon stretching fundamentals in the far-infrared. The example we wish to discuss here is ironpentacarbonyl, a trigonal bipyramid (symmetry point group D_{3h}). Figure 5.12 shows the far-infrared spectrum of this compound [Fe(CO)$_5$] between 94 and 125 cm^{-1} as measured by McDowell and Jones[34] and Table 5.9 gives their most recent assignments of the four lowest fundamentals. These assignments are an improvement over previous assumptions[34,35] and agree better with Raman studies[36] and the apparently

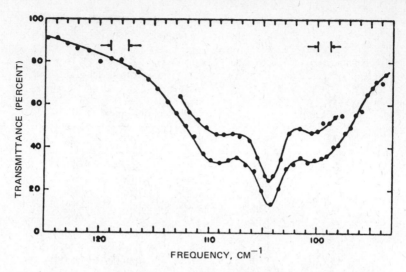

Fig. 5.12   Infrared spectrum of Fe(CO)$_5$, 94–125 cm^{-1}. One-meter absorption cell, pressure $\sim$ 25 mm Hg. Spectral slit widths as indicated. [R. S. McDowell and L. H. Jones, *J. Chem. Phys.* **36**, 3321 (1962).]

TABLE 5.9

ASSIGNMENTS OF THE FOUR CARBON–METAL–CARBON FUNDAMENTALS OF IRON-PENTACARBONYL, Fe(CO)$_5$ (R. S. McDOWELL AND L. H. JONES, PRIVATE COMMUNICATION, DECEMBER 1968)

Designation of mode	Species	Activity	Frequency in cm^{-1}		
			Infrared		Raman
			Vapor	Liquid	Liquid
ν_9	a_2''	IR	not observed		—
ν_{14}	e'	IR, R	$\begin{cases} 104.4 \\ \sim 102^* \end{cases}$	111	~ 113
ν_{15}	e'	IR, R	$\sim 65^*$		68
ν_{18}	e''	R	$\begin{cases} 97.3^\dagger \\ \sim 97^* \end{cases}$		105

* Band position calculated from other combination bands.
† Band position calculated from difference bands in the 5-μ region.
  IR = infrared
  R = Raman

generally valid observation that for low-lying modes the frequency of the vapor band is lower than that of the corresponding mode in the condensed phase.[5] We thus see from Table 5.9 that the frequencies of 104.4 cm^{-1} (vapor) and 111 cm^{-1} (liquid) are now assigned to the e' mode (v_{14}), in agreement with the proposals of Edgell et al.[37] who studied the compound in the liquid phase and in solution in cyclohexane between 65 and 135 cm^{-1} with a Perkin-Elmer Model 301.

Up to now, only one of these low-lying modes has been observed directly in the far-infrared; the positions of the $v_{15}(e')$ and $v_{18}(e'')$ modes have been inferred from indirect observations whereas the mode $v_9(a_2'')$ has not been located as yet.

In Table 5.10 we give a compilation of the recently proposed assignments and frequencies of the fundamentals of Fe(CO)$_5$.

TABLE 5.10

FREQUENCIES AND ASSIGNMENTS OF THE FUNDAMENTALS OF Fe(CO)$_5$ BY W. F. EDGELL, W. E. WILSON, AND R. SUMMITT[38]; L. H. JONES AND R. S. McDOWELL[39]

Species and designation		cm^{-1}	Description*
a_1'	v_1	2117	d(C—O), in-phase
	v_2	2031†	d(C—O), out-of-phase
	v_3	414‡	d(Fe—CO), in-phase
	v_4	377‡	d(Fe—CO), out-of-phase
a_2'	v_5	Inactive	
a_2''	v_6	2034†	d(C—O), out-of-phase, axial bonds
	v_7	620	δ(C—O)
	v_8	474	d(Fe—CO), axial bonds
	v_9	—	δ(Fe—O)
e'	v_{10}	2014†	d(C—O), out-of-phase, equatorial or trigonal bonds
	v_{11}	646	δ(C—O)
	v_{12}	544	δ(C—O)
	v_{13}	431	d(Fe—CO), equatorial or trigonal bonds
	v_{14}	$\begin{cases} 104 \\ 112‡ \end{cases}$	δ(Fe—CO)
	v_{15}	68‡	δ(Fe—CO)
e''	v_{16}	752‡	δ(C—O)
	v_{17}	492‡	δ(C—O)
	v_{18}	95	δ(Fe—CO)

* d(C—O) = C=O stretch, d(Fe—CO) = Fe—CO stretch, δ(C—O) = C—Fe—C and Fe—C=O angle deformation, δ(Fe—CO) = mainly C—Fe—C angle deformation.

† See also H. Haas and R. K. Sheline, J. Chem. Phys. **47**, 2996 (1967).

‡ Liquid phase.

The question on the inequivalence of the three equatorial (or trigonal) metal–carbon bonds on one hand and the two axial metal–carbon bonds on the other hand is of particular interest in $Fe(CO)_5$. In a recent electron diffraction study, Davis and Hanson[40] found that the axial bonds of $Fe(CO)_5$ are 0.045 ± 0.021 Å shorter than the equatorial bonds. The authors point out that the bond shortening correlates with the higher frequency of the axial Fe—CO stretch (v_8, $474 \, cm^{-1}$) over that of the trigonal Fe—CO bond stretch (v_{13}, $431 \, cm^{-1}$). Indeed, if we (admittedly very crudely) set $v_i = (k_i/\mu)^{1/2}$, where k and μ are force constant and reduced mass, we obtain $k_{axial}/k_{trigonal} \sim 1.2$. However, this assumes tacitly that the vibrations v_8 and v_{13} are "pure"; that is, they do not interact with other modes. We point out again that estimates concerning bond strengths and other bond characteristics made on the basis of the positions of observed frequencies or values of force constants are to be considered with caution; it is necessary to do a normal coordinate treatment and to establish the distribution of the potential function among the coordinates in order to accomplish this purpose analytically. In this context we mention the hexacarbonyls of Cr, Mo, and W, where extensive correlations have been drawn between their force constants and the type of bonding proposed for these molecules.[41] Although a general trend seems discernible, we feel that the conclusions need corroboration, particularly with respect to the very small actual differences of a given principal force constant in the series, the approximations necessary for the solution of the normal coordinate analysis, the extraordinary sensitivity of some of the constants on the values of the assigned frequencies, and the lack of recent electron diffraction studies from which a reasonably precise knowledge of the changes of the bond distances in the series could be obtained.

G. HALOGEN DERIVATIVES OF ALKYL COMPOUNDS

In the last few years an increasing amount of observational data on the carbon–halogen stretching and deformation modes have appeared in the literature. For the heavier halogens chlorine, bromine, and iodine, these motions lie close to or within the far-infrared region of the spectrum. What makes a study of carbon–halogen motions particularly rewarding is the fact that their frequencies are generally sensitive to the molecular environment; that is, they indicate the presence of conformers (*intra*molecular interactions) and the degree of rotational freedom in condensed phases or solutions (*inter*molecular interactions). In this chapter we will concentrate on the first of these two subjects.* The effect of conformers on carbon–halogen fundamentals (or on any other susceptible fundamental) is noticed by the appearance of more carbon–halogen (or other) fundamentals than would be expected from the available number of vibrational degrees of freedom. The reason for this is simply that sufficiently different conformations, which means sufficiently different bond angles and interatomic distances between "nonbonded" atoms† in the conformers, lead to different amounts of vibration–vibration interactions. Since the vibrational energies of the carbon–halogen bonds are low, they are also among the most susceptible, and variations in the vibration–vibration interactions involving contributions of the carbon–halogen modes are most likely to show up in the carbon–halogen fundamentals themselves.‡ The vibrational spectrum then exhibits more or less well separated multiple components of a carbon–halogen fundamental with relative intensities which are a function of the respective Boltzmann distribution of the conformers and with a multiplicity equal to the numbers of the conformers of the molecule.§

Figure 5.13 shows the results of some earlier work by Brown and Sheppard[42] on the temperature dependence of the spectra of (liquid) chain alkyl halides. The conformations in Fig. 5.13 are rendered by projection formulas: In the center of each triangle we have to imagine a carbon atom and at each corner a hydrogen atom, unless substitution is indicated by some appropriate symbol. The P,S,T-notation is by Mizushima et al.[43] and denotes whether the carbon atom carrying the halogen is primary, secondary, or tertiary. Each subscript C or H indicates whether a C—C or C—H bond of all those which are next-nearest to the C—halogen bond is in a trans position to this C—halogen bond and, finally, a primed symbol is used if it is necessary to indicate that the skeletal carbon atoms do not lie in a plane (see B' in Fig. 5.13). This nomenclature becomes particularly

* The concept of a conformer will occupy us extensively in Chapters 7 and 8; we define it here for convenience and state that by conformers of a molecule we mean all its stable forms that may be obtained by bending of bond angles or rotating of molecular groups about bonds within the molecule. With regard to the meaning of "stable forms," we would like to consider this an operational concept. For our purposes a stable form should have a lifetime which is sufficiently long to give a corresponding vibrational spectrum.
† The designation "nonbonded" pertains to substituents which are not attached to the same central atom.
‡ It should be realized that this susceptibility of the carbon–halogen fundamentals means, in turn, that the vibrational motion is not a pure carbon–halogen mode. The terminology "carbon–halogen stretch" or "carbon–carbon–halogen bending" is only a very approximate description of the actual displacements of the atoms.
§ Obviously, the fundamentals of a given motion which belong to the various conformers are of the same symmetry species. Conformers which are mirror images do not possess multiple fundamentals at different frequencies.

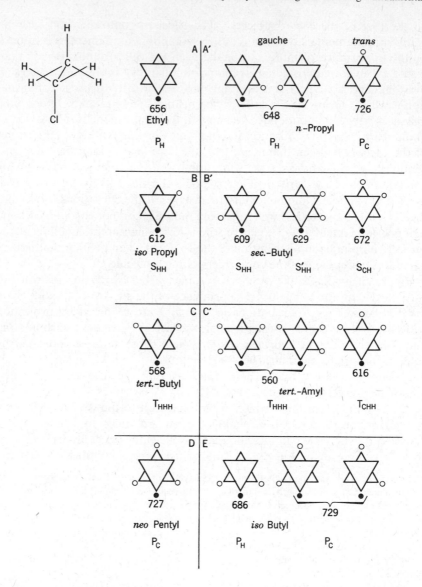

Fig. 5.13   A correlation between C—Cl stretching frequencies (cm^{-1}) and spectroscopically distinguishable conformations of some alkyl chlorides: *A*, ethyl; *A'*, *n*-propyl (two conformers): *B*, isopropyl; *B'*, *sec.*-butyl (three conformers): *C*, *tert.*-butyl; *C'*, *tert.*-amyl (two conformers): *D*, neopentyl; *E*, isobutyl (two conformers); ● chlorine atom, ○ methyl group. *Note.* Each conformation shown in *B'* actually occurs in two optically active forms. The notations of Mizushima *et al.* are given below each conformer (see text). [J. K. Brown and N. Sheppard, *Trans. Faraday Soc.* **50**, 1164 (1954).]

useful for conformations of a more complicated nature and has obvious extensions to symbols such as S_{CC}, S''_{HH}, etc. Table 5.11 gives the frequency range of the carbon–halogen stretches of aliphatic monochlorohydrocarbons of established conformation, reported more recently by Shipman et al.[44] Table 5.11 shows that the carbon–chlorine stretch of the *trans*

TABLE 5.11*

CHARACTERISTIC ABSORPTION FREQUENCIES
FOR ALIPHATIC SATURATED LIQUID
MONOCHLOROHYDROCARBONS

Chemical–geometrical combination	Frequency range, cm^{-1}
P_H	648–657
P'_H (branched)	679–686
P_C	723–730
S_{HH}	608–615
S'_{HH} (bent)	627–637
S_{CH}	655–674
S_{CC}	758†
T_{HHH}	560–581
T_{CHH}	611–632

*J. J. Shipman, V. L. Folt, and S. Krimm, *Spectrochim. Acta* **18**, 1603 (1962). By permission of Pergamon Press.
† Tentative, from 3-chloro-2,2,4,4-tetramethylpentane, which exists only in one conformer. The region may also overlap with that of the methylene-group rocking modes.

conformers (P_C) of *n*-chloroalkanes is at about 10% higher frequencies than the corresponding stretch of the *gauche* (P_H) (see Fig. 5.13) conformers. The reason for this has been demonstrated by Colthup using an analog experiment.[45] It shows that the C_1—Cl stretching frequency is increased by about 10% in the conformer where the C_1—Cl bond is *trans* to the C_2—C_3 bond (planar conformer, P_C) because of a concomitant C_1—C_2—C_3 angle deformation which is not "excited" by the C_1—Cl stretch in the non-planar (*gauche*, P_H) conformer.

The carbon–halogen stretching frequencies of bromo- and iodoalkanes fall into the 700 to 300 cm^{-1} frequency range. An extensive temperature study of their intensities was recently published by Bentley et al.[46] Figure 5.14 gives some of their data. Note particularly the constancy of the positions

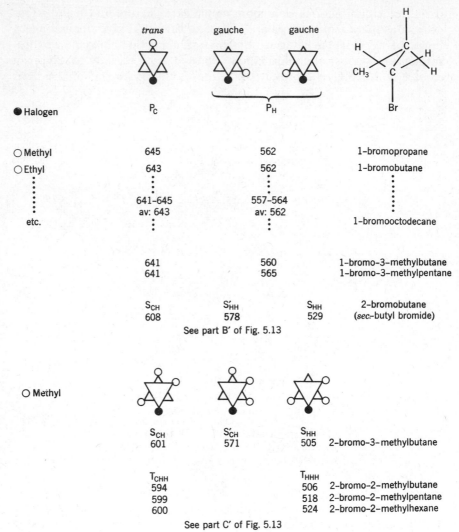

Fig. 5.14   Correlation between carbon–bromine stretching frequencies (cm^{-1}) and spectroscopically distinguishable conformations of some liquid alkyl chain halides. [*After* F. F. Bentley, N. T. McDevitt, and A. L. Rozek, *Spectrochim. Acta* **20**, 105 (1964), by permission of Pergamon Press.]

of the C—Br stretches in the sequence from 1-bromopropane to 1-bromo-octodecane, which clearly indicates that the range of forces which determines the conformations in these compounds does not stretch beyond the third carbon atom group.

The equatorial and planar conformers of the vapors of bromo- and chlorocyclobutane (see Chapter 7, Section E) each possess two widely separated C—halogen stretching fundamentals in the far-infrared and the Raman spectrum; the one at higher frequencies increases with temperature and, on the other hand, vanishes in the spectrum of the low-temperature solid (−175°C). Table 5.12 shows the frequencies and assignments (Rothschild[47]).

TABLE 5.12

FREQUENCIES OF THE CARBON–HALOGEN STRETCHING
FUNDAMENTALS AND CONFORMATIONS OF THE VAPORS
OF THE CYCLOBUTYL MONOHALIDES
(W. G. ROTHSCHILD[47])

cm^{-1} (vapor)		Conformation of
Chloro	Bromo	the carbon ring
631	551	Planar
532.5	487.5	Equatorial

This temperature variation of the C—Br stretching fundamentals of the equatorial and the planar conformers of bromocyclobutane is shown in Figs. 5.15(a) and (b); the evaluation of the temperature dependence of the intensity of the two bands led to an energy difference between planar and equatorial conformers of about 1 kcal/mole.[48]

The carbon–carbon–halogen (abbreviated C—C—X) bending modes of alkyl halides are mainly found in the region from 400 to 100 cm^{-1}. In the alkyl monohalides of point group symmetry C_s the selection rules permit two C—C—X deformation fundamentals with the motion of the halogen atom within the plane of symmetry (species a') or in a direction which has components normal to the plane of symmetry (a'', antisymmetric). It seems, however, that the antisymmetric C—C—X modes are quite weak and that therefore a study of the conformational behavior rests generally on the symmetric deformations in these molecules. Figure 5.16 shows the temperature dependence of the symmetric C—C—X (X = Cl, Br, I) bending fundamentals of 1-X-propane, taken from a study of McDevitt et al.[49] Note that the intensity of the bands towards lower cm^{-1} at 290 (Cl), 273 (Br), and 265 (I) has decreased at lower temperatures (about −170°C); these transitions have therefore been assigned to the less stable conformer of X-propane, the *gauche* form, in analogy with the analyses of the carbon-X stretching modes treated above.

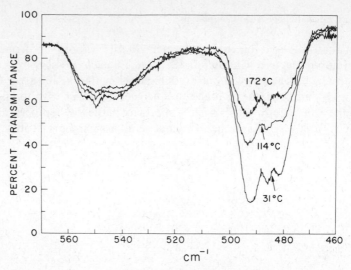

Fig. 5.15*a* Spectral region of 580 to 460 cm^{-1} of bromocyclobutane vapor at three different temperatures. Note how the ratio of the intensities of the carbon-halogen fundamentals of the more stable equatorial conformer (at 487.5 cm^{-1}) and the less stable planar conformer (at 551 cm^{-1}) decreases with increasing temperature. Part of the intensity decrease is due to lower vapor pressures and a small amount of emission from the sample at the higher temperatures. Instrument: Perkin-Elmer 521, glass cell with CsI windows, 11-cm path length, in sample and in reference beam. [W. G. Rothschild, *J. Chem. Phys.* **45**, 1214 (1966).]

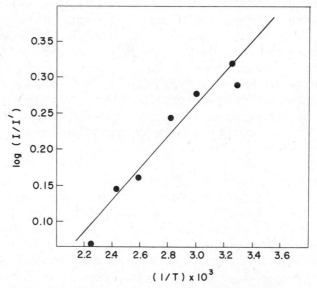

Fig. 5.15*b* Plot of the logarithm of the ratio of the intensities of the 487.5-cm^{-1} band (*I*) and the 551-cm^{-1} band (*I'*) of bromocyclobutane vapor. The slope yields $\Delta E = 1.02$ kcal/mole. [W. G. Rothschild, *J. Chem. Phys.* **45**, 1214 (1966).]

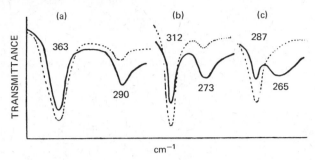

Fig. 5.16 Temperature studies of some 1-halopropanes. (a) 1-chloropropane, (b) 1-bromo-
propane, (c) 1-iodopropane: ——, room temperature: ·····, solid state. The room-temperature
spectra were taken of the liquids in a 0.5-mm thick layer in a polyethylene cell. The spectrometer
was a Perkin-Elmer 301. [N. T. McDevitt, A. L. Rozek, F. F. Bentley, and A. D. Davidson,
J. Chem. Phys. **42**, 1173 (1965).]

Normal propyl halide seems, at present, to be one of the few alkyl halides
for which C—C—X bending fundamentals of different conformers have
been assigned; in all the other alkyl halides only the C—C—X bending
fundamental based on the most stable conformer seems to be observable.
A possible exception is bromocyclobutane. The symmetric C—C—X
deformation $v_{17}{}^{47}$ shows a vapor band contour at 304 cm^{-1} which might
represent an overlap of the two very close C—C—X fundamental bending
motions based on its equatorial and planar conformation, respectively.

In Table 5.13 we have collected observed far-infrared C—C—X bending
fundamentals of some liquid monohaloalkanes. The reversal of the sequence
of the frequencies for the symmetric and antisymmetric fundamentals of
bromocyclopropane with respect to the sequence of these modes in 2-
bromopropane and bromocylobutane is interesting and conceivably
indicates the effects of the large ring strain in the cyclopropyl ring.

We can give a generalization on the different degree of frequency dependence on the conformation exhibited by the carbon–halogen stretching
fundamentals on one hand and the carbon–carbon–halogen deformation
fundamentals on the other hand: As is apparent from the data, and as was
pointed out above, the C—C—X bending modes are less affected by the
conformations than the C—X stretches. In the light of what has been
presented in the above sections, this must mean that the actual atomic
displacements during the C—C—X deformations possess only small contributions of bond-stretching since, loosely speaking, vibrational modes
couple more effectively through a stretching than through a bending coordinate. To be more explicit, since stretching–stretching interaction constants are greater than bending–bending interaction constants by about a
factor of ten, a given stretch probably induces an order of magnitude of
three times greater change in the frequency of an interacting stretch than

TABLE 5.13

FREQUENCIES OF CARBON–CARBON–HALOGEN BENDING FUNDAMENTALS OF SOME
MONOHALOGEN ALKANE LIQUIDS (CM^{-1})

		Chloro		Bromo		Iodo			Confor-
		IR	R	IR	R	IR	R	Species	mation
X-ethane*	(C_s)	335	335	291	292	259	262	a'	
2-X-propane*	(C_s)	338	337	300$^{\parallel}$	295	288$^{\parallel}$	265	a'	
		326	326	290$^{\parallel}$	—	279$^{\parallel}$	—	a''	
2-X-2-methylpropane*	(C_{3v})	365	372	301	303	259	259	e	
1-X-propane*	$(C_s$ or $C_1)$	363	365	312	314	287	285	a'	trans
		290	289	273	271	265	259	a'	gauche
1-X-2-methylpropane*	(C_1)	336	336	301	304	285	284	a	
2-X-butane*	(C_1)	335$^{\#}$	—	317$^{\#}$	317$^{\#}$	297$^{\#}$	303$^{\#}$	a	
		327	335	292	292	265	276	a	
X-cyclopropyl[†,‡]	(C_s)	—	323	272	273	—	—	a'	
		—	323	309	309	—	—	a''	
X-cyclobutyl[§,**]	(C_s)	364	367	298	301	—	—	a'	
		287	287	253	255	—	—	a''	

* See Ref. 49. IR = infrared
[†] See Ref. 50. R = Raman
[‡] See Ref. 51.
[§] See Ref. 47.
$^{\parallel}$ Frequency of the solids.
$^{\#}$ Assignment to C—C—X mode is questionable.
** See also Ref. 52.

an equivalent bending displacement induces in the frequency of an interacting bending mode.

In concluding this section we would like to draw attention to an important
but apparently not sufficiently appreciated observation by Bazhulin and
Osipova[53] on the reversal in the stabilities of two conformers when going
from the vapor to the liquid state. In other words, the conformer which is
the more stable in the vapor phase of the compound becomes the less stable
in the liquid phase, and *vice versa*. El Bermani *et al.*[54] have furnished further
experimental evidence on this phenomenon on the examples of the *gauche*
and *trans* conformers of 1,2-chlorofluoroethane and 1,2-bromofluoroethane.
They investigated the temperature dependence of the *gauche* and *trans*
carbon–chlorine and carbon–bromine, respectively, stretching frequencies
of the liquids and vapors of these two compounds. The $\nu(C—Cl)$ of liquid
CH_2ClCH_2F was at 668 (*gauche*) and 759 (*trans*) cm^{-1}, and the $\nu(C—Br)$
of liquid CH_2BrCH_2F was at 573 (*gauche*) and 631 (*trans*) cm^{-1}.

The final results are collected in Table 5.14, which gives the energy differences between the conformers, the more stable conformer, and the condition of phase of the compounds. As is apparent from these results, the more stable conformer of both compounds reverses from a *trans*-conformation to a *gauche*-transformation when the state of aggregation of both alkyl dihalides changes from the vapor to the liquid. The authors stress that intermolecular forces must play a dominant role in this reversal of stability.

TABLE 5.14*

ENERGY DIFFERENCES BETWEEN THE *gauche*- AND *trans*-CONFORMERS AS A FUNCTION OF AGGREGATION OF 1,2-CHLOROFLUOROETHANE AND 1,2-BROMOFLUOROETHANE

Compound	State of aggregation	More stable conformer	Energy difference in kcal/mole
1,2-Chloro-	Liquid	*gauche*	1.01 ± 0.20
fluoroethane	Vapor	*trans*	0.20 ± 0.08
1,2-Bromo-	Liquid	*gauche*	0.92 ± 0.20
fluoroethane	Vapor	*trans*	$0.30 + 0.08$

* M. F. El Bermani, C. J. Vear, A. J. Woodward, and N. Jonathan, *Spectrochim. Acta* **24A**, 1251 (1968). By permission of Pergamon Press.

We should remember that the measured energy differences displayed in Table 5.14 are, at most, of the order of 1 kcal/mole, which is equivalent to about $1.7\,kT$ or approximately $360\,\mathrm{cm}^{-1}$. It is therefore evident that, generally, the presence of the medium in far-infrared investigations on condensed phases must be properly taken into account. (We shall note various examples of this in this book.) With regard to this section, it is particularly prudent to note carefully the state of aggregation of the various compounds if we wish to draw conclusions from or make comparisons of their conformational behavior.

REFERENCES

1. J. A. A. Ketelaar, F. N. Hooge, and G. Blasse, *Rec. Trav. Chim.* **75**, 220 (1956).
2. F. N. Hooge and J. A. A. Ketelaar, *Rec. Trav. Chim.* **77**, 902 (1958).
3. E. B. Bradley, C. R. Bennett, and E. A. Jones, *Spectrochim. Acta* **21**, 1505 (1965).
4a. E. B. Bradley, M. S. Mathur, and C. A. Frenzel, *J. Chem. Phys.* **47**, 4325 (1967).
4b. E. B. Bradley, C. A. Frenzel, and M. S. Mathur, *J. Chem. Phys.* **49**, 2344 (1968).
5. W. G. Fateley, I. Matsubara, and R. E. Witkowski, *Spectrochim. Acta* **20**, 46 (1964).
6. H. Stammreich and R. Forneris, *Spectrochim. Acta* **8**, 46 (1956).
7. F. A. Miller, D. H. Lemmon, and R. E. Witkowski, *Spectrochim. Acta* **21**, 1709 (1965).
8. Wm. H. Smith and G. E. Leroi, *J. Chem. Phys.* **45**, 1767 (1966).

9. Wm. H. Smith and G. E. Leroi, *J. Chem. Phys.* **45**, 1778 (1966).
10. K. S. Pitzer and S. J. Strickler, *J. Chem. Phys.* **41**, 730 (1964).
11. G. Herzberg, *Infrared and Raman Spectra of Polyatomic Molecules*, Van Nostrand, Princeton, New Jersey, 1962, p. 304.
12. Wm. H. Smith and G. E. Leroi, *J. Chem. Phys.* **45**, 1784 (1966). This reference should be consulted for a more detailed discussion.
13. R. L. Redington, *Spectrochim. Acta* **23A**, 1863 (1967).
14. W. R. Thorson and I. Nakagawa, *J. Chem. Phys.* **33**, 994 (1960).
15. See, for instance, L. Pauling and E. B. Wilson, Jr., *Introduction to Quantum Mechanics*, McGraw-Hill, New York, 1935.
16. E. C. Kemble, *The Fundamental Principles of Quantum Mechanics*, Dover, New York, 1958, pp. 109 and 155.
17. J. E. Connett, J. A. Creighton, J. H. S. Green, and W. Kynaston, *Spectrochim. Acta* **22**, 1859 (1966).
18. J. Y. Beach, *J. Chem. Phys.* **9**, 54 (1941).
19. A collection of publications is found in *Noble Gas Compounds*, H. H. Hyman, Ed., Univ. Chicago Press, Chicago, 1963.
20. P. A. Agron, G. M. Begun, H. A. Levy, A. A. Mason, C. G. Jones, and D. F. Smith, *Science* **139**, 842 (1963).
21. See Ref. 11, pp. 18 and 391.
22. See Ref. 19, p. 297.
23. S. Reichman and F. Schreiner, *J. Chem. Phys.* **51**, 2355 (1969).
24. J. Jortner, S. A. Rice, and E. G. Wilson, *J. Chem. Phys.* **38**, 2302 (1963); *J. Am. Chem. Soc.* **85**, 814 (1963).
25. G. C. Pimentel and R. D. Spratley, *J. Am. Chem. Soc.* **85**, 826 (1963).
26. L. Salem and M. Alexander, *J. Chem. Phys.* **39**, 2994 (1963).
27. H. H. Claassen, C. L. Chernick, and J. G. Malm, *J. Am. Chem. Soc.* **85**, 1927 (1963).
28. W. A. Yeranos, *Mol. Phys.* **9**, 449 (1965).
29. W. G. Rothschild, *J. Opt. Soc. Am.* **54**, 20 (1964); A. A. Mason and A. H. Nielsen, *J. Opt. Soc. Am.* **57**, 1464 (1967); H. H. Hyman, *Science* **145**, 773 (1964).
30. H. H. Claassen, private communication, December 28, 1966.
31. L. C. Hoskins and R. C. Lord, *J. Chem. Phys.* **46**, 2402 (1967).
32. J. A. Pople, W. G. Schneider, and H. J. Berstein, *High-Resolution Nuclear Magnetic Resonance*, McGraw-Hill, New York, 1959.
33. R. S. Berry, *J. Chem. Phys.* **32**, 933 (1960).
34. R. S. McDowell and L. H. Jones, *J. Chem. Phys.* **36**, 3321 (1962).
35. F. A. Cotton, A. Danti, J. S. Waugh, and R. W. Fessenden, *J. Chem. Phys.* **29**, 1427 (1958).
36. H. Stammreich, O. Sala, and Y. Tavares, *J. Chem. Phys.* **30**, 856 (1959).
37. W. F. Edgell, C. C. Helm, and R. E. Anacreon, *J. Chem. Phys.* **38**, 2039 (1963).
38. W. F. Edgell, W. E. Wilson, and R. Summitt, *Spectrochim. Acta* **19**, 863 (1963).
39. L. H. Jones and R. S. McDowell, *Spectrochim. Acta* **20**, 248 (1964).
40. M. I. Davis and H. P. Hanson, *J. Phys. Chem.* **69**, 3405 (1965).
41. L. H. Jones, *Spectrochim. Acta* **19**, 329 (1963).
42. J. K. Brown and N. Sheppard, *Trans. Faraday Soc.* **50**, 1164 (1954).
43. S. Mizushima, T. Shimanouchi, K. Nakamura, M. Hayashi, and S. Tsuchiya, *J. Chem. Phys.* **26**, 970 (1957).
44. J. J. Shipman, V. L. Folt, and S. Krimm, *Spectrochim. Acta* **18**, 1603 (1962).
45. N. B. Colthup, *Spectrochim. Acta* **20**, 1843 (1964).
46. F. F. Bentley, N. T. McDevitt, and A. L. Rozek, *Spectrochim. Acta* **20**, 105 (1964).
47. W. G. Rothschild, *J. Chem. Phys.* **45**, 3599 (1966).

48. W. G. Rothschild, *J. Chem. Phys.* **45**, 1214 (1966).
49. N. T. McDevitt, A. L. Rozek, F. F. Bentley, and A. D. Davidson, *J. Chem. Phys.* **42**, 1173 (1965).
50. M. I. Kay, Ph.D. dissertation, University Microfilm, Inc., Ann Arbor, Michigan, Order No. 63-3377 (1964).
51. W. G. Rothschild, *J. Chem. Phys.* **44**, 3875 (1966).
52. J. R. Durig and W. H. Green, *J. Chem. Phys.* **47**, 673 (1967); J. R. Durig and A. C. Morrissey, *J. Chem. Phys.* **46**, 4854 (1967).
53. P. A. Bazhulin and L. P. Osipova, *Opt. Spectry.* **6**, 406 (1959).
54. M. F. El Bermani, C. J. Vear, A. J. Woodward, and N. Jonathan, *Spectrochim. Acta* **24A**, 1251 (1968). For complete vibrational assignments of the *gauche* and *trans* conformers of 1-chloro-2-fluoroethane, 1-bromo-2-fluoroethane, and 1-fluoro-2-iodoethane, see M. F. El Bermani and N. Jonathan, *J. Chem. Phys.* **49**, 340 (1968).

6 Far-Infrared Spectra of Hydrogen-Bonded Systems

A. INTRODUCTION

In this chapter we discuss the spectroscopic characteristics of a very important bond: the hydrogen bond.* The hydrogen bond is of particular significance in molecular biology but also has aspects of great interest to the physical chemist and the spectroscopist.

The hydrogen bond is an example of a weak bond. It can, in most cases, be partially or completely broken by moderate increase of the kinetic energy or by simple dilution. For our purposes here we consider the hydrogen bond as (part of) the linkage between two molecules, one possessing an end group —A—H, where A is any type of molecular group and A—H is the terminal bond between group A and a hydrogen atom H, and the other molecule having a substituent, B, such that the two molecules can form a complex of the type

$$-A-H\cdots B-.$$

Cases in which the linkage —A—H···B— is formed *intra*molecularly or between more than two molecules also occur. At any rate, we assume that the nature of the linkage —A—H···B— is such that the part A—H is much stronger than the part H···B. Instances in which the characteristics of the A—H bond are completely altered upon complexing, in other words, where the bond H···B would "borrow" so much energy from the A—H bond that the linkage should be written —A—H—B—, will not be considered here since the ensuing bond strength would presumably put the vibrational stretching modes into the 1000 to 2000 cm^{-1} spectral range.

In the past the effects of the linkage H···B— on the linkage —A—H have been extensively studied by the perturbations experienced by the stretching

* See, for instance, a recent review article on past results and future prospects by R. J. Jakobsen, J. W. Brasch, and Y. Mikawa, *Appl. Spectry.* **22**, 641 (1968).

fundamentals of $-A-H$ in the 3000 to 3500 cm^{-1} range. More recently far-infrared spectroscopy has been used to study directly the weak linkage H····B—, the "hydrogen bond."

The limited amount of experimental data obtained up to the present from these far-infrared observations does not permit us to make any closer distinction among the different vibrational modes of $-A-H\cdots B-$ other than a characterization into "stretch" and "deformation." (We shall point out a few instances where a more detailed characterization has been accomplished.) Thus, we apply the currently used nomenclature and we designate in the following—for both intra- and intermolecular hydrogen bonds—a stretch which involves mainly the motion of the weaker bond H····B,

$$-A-\overset{\leftrightarrow}{H}\cdots\overset{\leftrightarrow}{B}-,$$

by the symbol v_σ. A stretch involving predominantly motion of the stronger linkage A—H,

$$-\overset{\leftrightarrow}{A}-\overset{\leftrightarrow}{H}\cdots B-,$$

is designated by v'_s. The unprimed symbol, v_s, is reserved for the corresponding nonhydrogen-bonded mode of A—H. These stretching modes must be expected to interact with each other. However, the great difference in the bond strength between A—H and B····H exhibited by the examples to be discussed makes the above classification meaningful.

The deformation mode or bending mode of the link H····B,

$$-A-H\cdots\overset{\nwarrow}{\underset{\swarrow}{B}}-,$$

is designated by v_β.

The deformation mode

$$-\overset{\curvearrowleft}{\underset{\searrow}{A}}-H\cdots B-,$$

that is, the mode which involves mainly the strong link A—H, is not of great interest here.

The unambiguous assignment of the motions v_σ and v_β is difficult because of the presence of other low-frequency modes in the long-wavelength spectrum. This fact is underlined by the various and often contradictory conclusions found in the literature,[1-4] where transitions previously assigned to v_σ and v_β have been subsequently attributed to other internal modes[1,2] and to spurious effects,[3,4] respectively. Deuteration of the hydrogen atoms is of little value unless an exact calculation of the reduced mass involved in the mode has been performed. Because of the concerted motions of the atoms involved in the hydrogen bond, a simple $1/\sqrt{2}$ dependence of the isotopically shifted frequency of v_σ can by no means be expected to hold. In fact, a $1/\sqrt{2}$ shift would strongly point to a nonhydrogen-bonded, simple A—H mode (v_s).

It is thus necessary to apply some special techniques, techniques other than those of normal coordinate analysis or correlations within a series of very closely related compounds, in order to arrive at the correct assignments and characterizations of the hydrogen bond motions. In the following we shall meet various situations in which we will point out the application of such special techniques for the proper assignments of hydrogen bond fundamentals.

B. SELF-ASSOCIATED PHENOLS

Hurley et al.[5] recently proposed a convincing criterion for the assignment of hydrogen bond modes: A band assigned to the modes v_σ or v_β should disappear completely if the H atom of the hydrogen bond is substituted by a group (or atom) incapable of hydrogen bonding, for instance, the methyl group. Table 6.1 and Fig. 6.1 show some of their data and assignments on low-frequency motions of cresols (methylphenols, $CH_3C_6H_4 \cdot OH$), chlorophenols, and the corresponding anisols ($CH_3C_6H_4 \cdot OCH_3$,

TABLE 6.1*

FAR-INFRARED ABSORPTION IN SOME SUBSTITUTED PHENOLS AND
METHYLATED PHENOLS (NEAT LIQUIDS)

Compound	Position of maximum absorption, cm^{-1}	Assignments[†]
o-Cresol	124 s, b	v_σ, intermol.
	190 s	X[‡]
o-Methyl anisole	108 s, b	—OCH_3 torsion
	178 s	X

m-Cresol	146 s, b	v_σ, intermol.
	214 s	X
	246 s	X
m-Methyl anisole	106 s, b	—OCH$_3$ torsion
	222 vs	X
p-Cresol	124 s, b	v_σ, intermol.
	178 s	X
p-Methyl anisole	98 s, b	—OCH$_3$ torsion
	158 sh	X
o-Chlorophenol	84 s, b	v_σ, intramol.
[structure: benzene ring with OH and Cl (ortho)]	176 s	X
o-Chloro anisole	110 s, b	—OCH$_3$ torsion
[structure: benzene ring with OCH$_3$ and Cl (ortho)]	160 s	X
m-Chlorophenol	130 s, b	v_σ, intermol.
	192 m	X
m-Chloro anisole	108 vs, b	—OCH$_3$ torsion
	188 s	X
p-Chlorophenol	122 m, b	v_σ, intermol.
	160 s	X
p-Chloro anisole	94 s, b	—OCH$_3$ torsion
	144 vs	X

* After W. J. Hurley, I. D. Kuntz, Jr., and G. E. Leroi, J. Am. Chem. Soc. **88**, 3199 (1966). By permission of American Chemical Society.

† The assignments other than those for v_σ are tentative. s = strong, m = medium, b = broad, v = very, sh = shoulder. The spectra were obtained with a RIIC-FS-520 Michelson interferometer with a 0.0005 in. (0.0127 mm) Mylar beam splitter and a Golay detector. High-frequency radiation (>600 cm^{-1}) was removed by a black polyethylene filter.

‡ X denotes a vibrational mode, the frequency of which is sensitive to the particular nature of a substituent X, where X may be halogen, methyl, and so forth.

Cl·C$_6$H$_4$·OCH$_3$). The values (in cm^{-1}) of 124, 146, 124, 84, 130, and 122 are the frequencies at the maximum absorption of v_σ for o-, m-, p-cresol and o-, m-, p-chlorophenol, respectively, since these absorption bands disappear when the OH hydrogen atom is replaced by the methyl group.

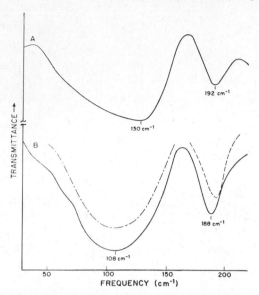

Fig. 6.1   Ratioed spectra (vs cell background) for m-chlorophenol (A) and m-chloroanisole (B). The band at 108 cm^{-1} for m-chloroanisole has a counterpart at 106 cm^{-1} in anisole ($-\cdot-\cdot-\cdot-\cdot$) and is therefore assigned to a OCH$_3$-sensitive mode. The band at 188 cm^{-1} for m-chloroanisole has a counterpart at 194 cm^{-1} in chlorobenzene and is therefore assigned to a Cl-sensitive mode ($-----$); the corresponding band in m-chlorophenol is at 192 cm^{-1} (see curve A). The band at 130 cm^{-1} in m-chlorophenol (curve A) is therefore assigned to the v_σ mode. [W. J. Hurley, I. D. Kuntz, Jr., and G. E. Leroi, *J. Am. Chem. Soc.* **88**, 3199 (1966), by permission of American Chemical Society.]

The value of 84 cm^{-1} for o-chlorophenol, which does not align with the values of the *para*- and *meta*-substituted compounds, is a consequence of *intra*molecular hydrogen bridging.[6,7] The low frequency is probably due to the relatively larger distance H$\cdots$Cl and the smaller electronegativity of the Cl atom relative to the oxygen atom. We show this schematically in Fig. 6.2.

The data of Table 6.1 seem to indicate that the frequency of v_σ is sensitive to the structure of the molecule, as was proposed by Stanevich.[1] For instance, the difference between the hydrogen bond frequencies of *meta*-chlorophenol (130 cm^{-1}) and *para*-chlorophenol (122 cm^{-1}) amounts to 8 cm^{-1}, and that of *meta*-cresol (146 cm^{-1}) and *para*-cresol (124 cm^{-1}) amounts to 22 cm^{-1}. It does not seem likely that such large shifts could be produced by interactions of v_σ with the environment or by a different degree of overlapping with other modes, particularly since the frequency of v_σ for *ortho*- and *para*-cresol coincide. We stress that this is in agreement with the well-known difference in the chemical behavior of *ortho*- and *para*-substituted benzenes as opposed to the *meta*-substituted compound.

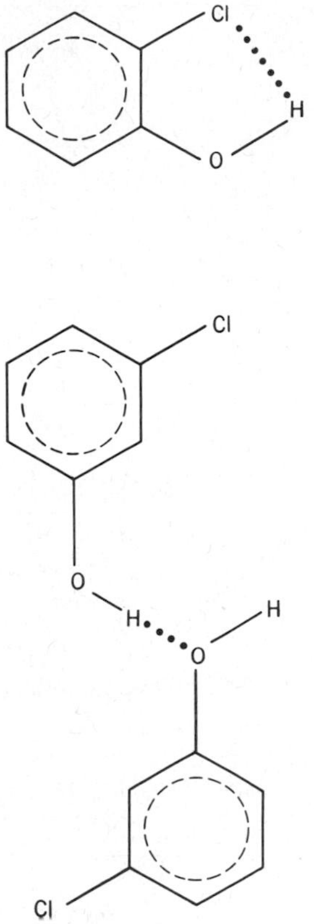

Ortho-chlorophenol

Meta-chlorophenol

Fig. 6.2 *Intra*- and *inter*molecular hydrogen bonds in chlorophenols (schematic).

A different technique for the assignment of hydrogen bond frequencies has been reported by Jakobsen and Brasch, who have utilized a convenient dilution technique using polyethylene as solvent.[8] Either by heating the solution of the compound in polyethylene or by using low concentrations of solute, the hydrogen bridge is broken and the free A—H may be studied; upon recooling, the hydrogen-bonded species A—H····B are reformed. Furthermore, this technique permits us to study low-temperature spectra with minimized interference of lattice modes because of the dispersion of the solute in the polyethylene.[9] Figure 6.3 shows these relations for 2,6-dimethylphenol. The upper curve represents the pure solid and shows a band at 164 cm^{-1} and a typical, hydrogen-bonded OH stretching (ν_s') absorption near 3400 cm^{-1}. The 164 cm^{-1} band disappears in a dilute solution of the phenol in polyethylene at room temperature (middle curve), and the sharp peak due to free OH (ν_s) is now present at 3600 cm^{-1}. The

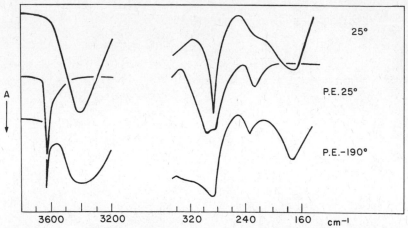

Fig. 6.3   Absorption spectrum of the v_σ mode and of the O—H stretch in 2,6-dimethylphenol. (*Upper curve*): solid film of the compound at 25°C, showing disappearance of free OH and the presence of the broad hydrogen-bridge absorption centered near 160 cm^{-1}. (*Intermediate curve*): dilute solution of 2,6-dimethylphenol in a polyethylene matrix (P.E.) at 25°C. The phenol compound exists mainly as monomeric species. (*Bottom curve*): 2,6-dimethylphenol in polyethylene matrix at −190°C. Some of the hydrogen bridges have reformed at the low temperature, the spectrum representing a mixture of free and hydrogen-bonded OH. The band at 240 cm^{-1}, which does not alter its appearance throughout, does not arise from a hydrogen bond vibration. [R. J. Jakobsen and J. W. Brasch, *Spectrochim. Acta* **21**, 1753 (1965), by permission of Pergamon Press.]

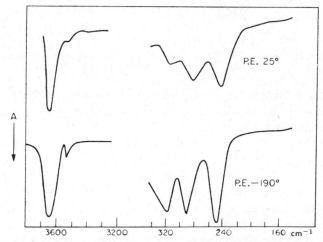

Fig. 6.4   Absorption spectrum of 2,6-di-*tert*-butylphenol in the regions of v_σ and of the O—H stretch. (*Upper curve*): the butyl phenol–polyethylene matrix at 25°C. (*Lower curve*): the same sample at −190°C. There is no formation of hydrogen bonds in the region of v_σ (120–180 cm^{-1}); the only effect of lowering the temperature is a sharpening of the bands. [R. J. Jakobsen and J. W. Brasch, *Spectrochim. Acta* **21**, 1753 (1965), by permission of Pergamon Press.]

lower curve shows that the 164 cm^{-1} band (and v'_s) reappears upon cooling to $-190°$C. The 164 cm^{-1} band is therefore assigned to v_σ. Figure 6.4 shows the same relations for 2,6-*di-tert*-butylphenol, which apparently does not form hydrogen-bonded complexes because of steric hindrance.

TABLE 6.2*

HYDROGEN BOND STRETCHING VIBRATIONS IN
PHENOLS ASSIGNED WITH THE AID OF THE
POLYETHYLENE MATRIX TECHNIQUE

Compound, liquid phase	v_σ $-O \leftrightarrow H \cdots O \leftrightarrow$, cm^{-1}
Phenol	162
o-Cresol	187, 121
m-Cresol	143
p-Cresol	178, 126
o-tert-Butylphenol	127
2,6-Dimethylphenol[†,‡]	150
2,6-Diisopropylphenol	106
2-Methyl-6-*tert*-butylphenol	104
2,6-Di-*tert*-butylphenol	none[§]

* *After* R. J. Jakobsen and J. W. Brasch, *Spectrochim. Acta* **21**, 1753 (1965). By permission of Pergamon Press.
[†] Run at 50°C.
[‡] Counting of substituent positions on the ring begins with OH as position 1.
[§] Sterically hindered.

Table 6.2 presents a partial summary of the frequencies of v_σ of some phenolic compounds.[8]

A comparison of the assignments for those compounds which had been studied by the substitution technique (Table 6.1) and the polyethylene matrix method (Table 6.2) shows that there is not complete agreement in all aspects. Jakobsen and Brasch prefer to assign two frequencies of *p*-cresol at 178 and 126 cm^{-1} (see Table 6.2) to coupled vibrations, both vibrations containing contributions of a v_σ mode and a X-sensitive mode. On the other hand, from the substitution technique used by Hurley *et al.* it would appear that the higher frequency component is predominantly a X-sensitive vibration (see Table 6.1). Hurley *et al.* give, as additional evidence for some of their assignments, the strong intensity of the corresponding Raman shifts of those infrared bands which they claim to represent X-sensitive modes (the polarizability of the hydrogen bond is small). Indeed, for *o*-cresol they found a strong Raman shift for the infrared band at 190 cm^{-1} (Table 6.1). Intense Raman shifts were likewise observed by them in *m*-cresol at 216 and 245 cm^{-1}, corroborating the assignments of the corresponding infrared bands to *X*-sensitive motions. However, no Raman shift was detected in *p*-cresol for the supposedly *X*-sensitive band at 178 cm^{-1}. Although we know that Raman shifts may be very weak (or absent) for a variety of reasons, it seems that only the assignments of $v_\sigma = 121$–124 cm^{-1} for *o*-cresol and $v_\sigma = 143$–146 cm^{-1} for *m*-cresol are definite. Concerning the *para*-compound, it is tempting to prefer Hurley *et al.*'s assignment, namely that $v_\sigma = 124$ cm^{-1}

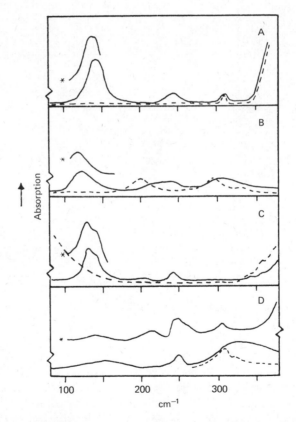

Fig. 6.5   A ——— ~2 M PhOH + ~5 M Me$_3$N in CCl$_4$; * ——— ~2 M PhOD + ~7 M Me$_3$N in CCl$_4$; - - - - - ~5 M Me$_3$N in CCl$_4$; B ———— 2.8 M PhOH in Et$_3$N; * ——— ~3 M PhOD in Et$_3$N; - - - - - liquid Et$_3$N; C ———— 2.8 M PhOH in pyridine; the shoulder at 143 cm^{-1} is spurious (see Ref. 17); * ——— ~3.5 M PhOD in pyridine; - - - - - liquid pyridine; D ———— 2.7 M PhOH in CCl$_4$; * ——— 2.8 M PhOD in CCl$_4$; - - - - - liquid CCl$_4$. [S. G. W. Ginn and J. L. Wood, *Spectrochim. Acta* **23A**, 611 (1967), by permission of Pergamon Press.]

and the band at $178 \, cm^{-1}$ is mainly a X-sensitive mode, by remembering the great similarity of the chemical reactivity of o- and p-substituted benzenes relative to that of the m-substituted isomers.

The reader will have noticed some of the difficulties inherent in the assignments of the hydrogen bond frequencies, even when relatively straightforward methods such as those just described above are applied. As we have indicated above, the difficulties are caused by (1) the richness of the low-frequency modes of the large molecules (dimers, etc.) with the concomitant overlap or proximity of the various fundamentals, by (2) the broad band contours (see, for example, Fig. 6.1) caused, presumably, by the relatively high degree of anharmonicity of the motions (large amplitude vibrations), and by (3) the possibility of coupling between the different modes.

C. Hydrogen Bonds between Phenols and Amines

A study of the hydrogen bond stretching mode of the complexes phenol–trimethylamine, phenol–triethylamine, and phenol–pyridine as well as of their deuterated species has led to an estimate of the corresponding force constants by an abbreviated normal coordinate calculation (Ginn and Wood[10]). Figure 6.5 shows the frequency range of 80–$380 \, cm^{-1}$ of the

TABLE 6.3*

Hydrogen Bond Stretching Mode ν_σ (N$\cdots$H) of Phenol–Amine Complexes and Phenol-d_1–Amine Complexes

Solution[†]	cm^{-1}
Phenol–pyridine, $C_6H_5OH\cdots NC_5H_5$	134
Phenol-d_1–pyridine, $C_6H_5OD\cdots NC_5H_5$	130
Phenol–triethylamine	123
Phenol-d_1–triethylamine	120
Phenol–trimethylamine in CCl_4 solution	143
Phenol-d_1–trimethylamine in CCl_4 solution	141
	ν_σ(O$\cdots$H)
Phenol in CCl_4 solution	150
Phenol-d_1 in CCl_4 solution	143

* *After* S. G. W. Ginn and S. L. Wood, *Spectrochim. Acta* **23A**, 611 (1967). By permission of Pergamon Press.

† For the composition of the solutions, see Fig. 6.5.

spectra of the three complexes and of phenol. The bands centered between 120 and 145 cm^{-1} are assigned in each case to the N$\cdots$H stretch ν_σ since (*1*) they are not present, with comparable intensity, in the pure components or in a solution of the pure components in carbon tetrachloride, respectively, and since (*2*) they occur at frequencies sufficiently far removed from the *intra*molecular vibrations of the components. The respective frequencies at the maximum absorption of the complexes are compiled in Table 6.3. The compositions in moles/l (M) are indicated in the caption of Fig. 6.5.

The computation of the ν_σ force constant for the phenol–amine complexes is based on a 1:1 model structure. In the following we will sketch the course of the computations for the complex phenol–trimethylamine since it may be of interest to see the assumptions which underlie the analysis made for such a complicated system. The assumed configuration and the internal coordinates are given in Fig. 6.6; the transferred force constants are

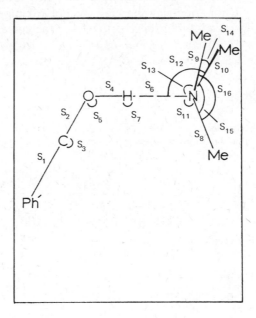

Fig. 6.6   Internal coordinates of 1:1 phenol–trimethylamine complex. [S. G. W. Ginn and J. L. Wood, *Spectrochim. Acta* **23A**, 611 (1967), by permission of Pergamon Press.]

summarized in Table 6.4. The following assumptions were made concerning the masses and distances: (*1*) The hydrogen bridge O—H$\cdots$N is linear and in the direction of the threefold symmetry axis of trimethylamine, (*2*) the phenol moiety is considered as the triatomic molecule $C_5H_5\cdot C\cdot O$,

TABLE 6.4*

STRUCTURAL PARAMETERS AND FORCE CONSTANTS,
PHENOL–TRIMETHYLAMINE COMPLEX

Structural unit	Length, Å or angle, deg	Force const., mdyne/Å
Ph'—C	1.68	1.5
C—O	1.36	6.0
O—H	1.00	6.0
H—N	1.75	—
N—Me	1.47	4.0
Ph'—C—O	180	0.53
C—O—H	120	0.7
O—H—N	180	0.0
H—N—Me	111	0.0
Me—N—Me	108	0.86

* S. G. W. Ginn and J. L. Wood, *Spectrochim,
Acta* 23A, 611 (1967). By permission of Pergamon
Press.

with a C_5H_5 group (Ph') of 65 mass units with the CO group attached at
its center of mass, (3) the methyl groups are considered point masses of
15 mass units, (4) the bond distance of OH and the distance between the
O and N atoms are estimated from an empirical correlation with the value
of the shifted O—H stretch, v_s',* and (5) the C—O distance is taken from the
crystal structure. With regard to the values of the force constants, those of
the trimethylamine moiety are transferred from the free molecule [they
differ little from those of $(CH_3)_3N \cdot HCl$], the C—O stretch constant is
taken from CO, the Ph'—C stretch and Ph'—C—O bending force constants
are estimated from the "triatomic" molecule Ph'CO, the O—H stretch
constant is transferred from free phenol with corrections for its shift due to
hydrogen bonding, and the C—O—H bending force constant is taken from
the corresponding constant of formic acid. The bending constants of the
hydrogen bridge O—H····N and all interaction force constants are set
equal to zero. Similar considerations were applied to the parameters of 1:1
complex structures of phenol–triethylamine and phenol–pyridine. The values
of v_σ for N····H, resulting from the normal coordinate treatment, are summarized in Table 6.5.

The authors have also calculated the force constant for the grouping
O—H····N, that is, all force constants except those of v_σ and v_s' are set equal

* See Ref. 7, p. 85.

TABLE 6.5*

HYDROGEN BOND STRETCHING FORCE CONSTANTS OF SOME
PHENOL–AMINE COMPLEXES

| | | Force constant of v_σ, mdyne/Å | |
| | Simplified normal coordinate | O—H⋯N | PhOH⋯NR |
1:1 Complex			
PhOH–trimethylamine	0.27	0.09[†]	0.44
PhOD–trimethylamine	0.27		0.43
PhOH–triethylamine	0.24	0.07[†]	0.44
PhOD–triethylamine	0.22		0.42
PhOH–pyridine	0.23	0.08[†]	0.45
PhOD–pyridine	0.23		0.43

* *After* S. G. W. Ginn and J. L. Wood, *Spectrochim. Acta* **23A**, 611 (1967). By permission of Pergamon Press.

[†] The values for the complexes with deuterated hydrogen bonds are about 1–3% lower. The O—H bond is assumed to be perfectly stiff ($v'_\sigma \to \infty$).

to zero, and the grouping PhOH⋯N(CH$_3$)$_3$, that is, all force constants except that of v_σ have been set equal to "infinity." The values obtained from these computations are given in columns 3 and 4 of Table 6.5. We note that the value of the force constant based on the complete "diatomic model," for instance, PhOH⋯N(CH$_3$)$_3$, is much closer to the more correct value (normal coordinate treatment) than is the value based on the grouping O—H⋯N, which seems to be far too low. It is not difficult to rationalize why this O—H⋯N model is such a poor approximation: it completely neglects the heavy masses which are attached on either side of the hydrogen bridge. On the other hand, the fact that model structures such as PhOH⋯N(CH$_3$)$_3$ give a value of the hydrogen bond force constant which is at least of the correct order of magnitude indicates that the intramolecular motions of the association partners exert, to a first approximation (zero interaction constants; see above), only a minor influence on the hydrogen bond frequency. Again this facet can be understood since, in general, the energies of hydrogen bonds are much smaller than those of the intramolecular bonds in the molecules. The "diatomic model" is then certainly of some usefulness in assigning and predicting the location of hydrogen bond stretching frequencies, but obviously the model cannot be applied for any detailed characterization of the nature of hydrogen bonds. Some of these properties seem to be indicated by the spectroscopic data; we summarize them as follows. (1) There is a great difference in the frequencies of v_σ for intra and intermolecular hydrogen bonds (see the chlorophenols, Table 6.1). (2) There is an apparent decrease in the frequency of v_σ in the presence of increasingly bulky groups attached near the A—H group (see the 2,6-substituted phenols, Table 6.2). (3) There seems to be a variation of the frequency of v_σ with the relative position of electron-withdrawing or donating substituents on the benzene ring, as evidenced by the o- and p-cresols vs the m-cresol of Tables 6.1 and 6.2 and data on p- and m-nitrophenols.[11]

In the same fashion, Ginn and Wood explored model calculations on some phenolic polymers.[10] Reasonable values for the force constant of v_σ were obtained for an open dimer, an open trimer, and a planar cyclic trimer, whereas the calculations seem to preclude the existence of a cyclic dimer (point group C_{2h}) because of the unreasonably high value of its force constant of v_σ.

D. HYDROGEN BONDING IN CARBOXYLIC ACIDS

The far-infrared spectroscopy of hydrogen bonds in carboxylic acids has received a relatively large amount of attention. The search for the respective absorption bands has been carried to rather long wavelengths, and assignments of hydrogen bond deformation modes and more detailed characterizations of the various modes lumped together under the symbols v_σ and v_β (see Section A) have been proposed. A determination of the frequencies and the characterization of hydrogen bonds of a series of low-molecular weight n-alkyl carboxylic acids has been published by Stanevich in a series of articles.[12] The model structure of the hydrogen-bonded associate was assumed to be a planar dimer in which all intramolecular force constants were set "infinitely" large. As mentioned above, such a model can only be considered as an approximation. However, within the main series of acids investigated (that is, acids differing only by the size of the alkyl group), the approximation of two rigid parts vibrating with respect to each other about the hydrogen bond should at least give accurate relative information. Figure 6.7 shows the model structure and the six possible* hydrogen bond modes, their symmetry species, designation, and activity.

The presence and the position of the lattice modes in the solid acids were established by scanning the spectra of the acids dissolved in carbon tetra-

---

* When two nonlinear molecules combine to form a new species, they change six of their total of twelve translation–rotation degrees of freedom into vibrational degrees of freedom. The motions of the latter are classified in the usual way into stretch, deformation, and torsional fundamentals.

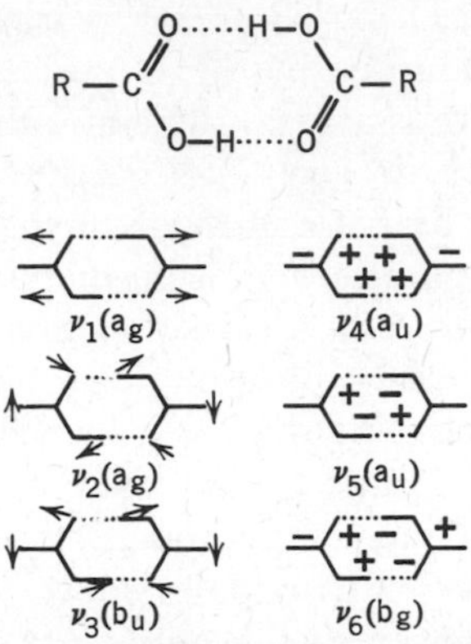

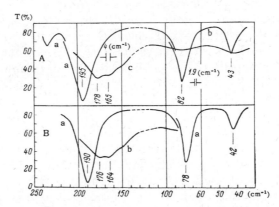

Fig. 6.7   The six vibrational modes of a hydrogen bond in carboxylic acid dimers. Modes of symmetry species b_u and a_u are infrared-active; those of species a_g and b_g are Raman-active. The mode v_σ is identified with v_3, the deformation modes with v_4, v_5. [A. E. Stanevich, *Opt. Spectry.* Suppl. 2, "Molecular Spectroscopy," 104 (1966), by permission of Optical Society of America.]

chloride. The assignments of the hydrogen bond modes was done by ascribing the transition at the highest observed frequency to $v_3(= v_\sigma)$, the lowest lying transition to v_5, and the intermediate frequency to v_4.[13] The spectra of CH_3COOD, CD_3COOD, and CCl_3COOH are shown in Figs. 6.8 and 6.9. Table 6.6 shows the observed frequencies, assignments, and, for v_3, the calculated frequencies. The calculated frequencies were obtained according

Fig. 6.8   Spectra of the region of the hydrogen bond transitions of CH_3COOD and CD_3COOD. CH_3COOD (*curve A*): $a = 10$-μ layer thickness, $-175°C$; $b = 20\,\mu$, $-175°C$; $c = 20\,\mu$, $20°C$. CD_3COOD (*curve B*): $a = 20\,\mu$, $-175°C$; $b = 20\,\mu$, $20°C$. The dashed lines indicate opaque regions due to absorption of the quartz cells. [A. E. Stanevich, *Opt. Spectry.* **16**, 243 (1964), by permission of Optical Society of America.]

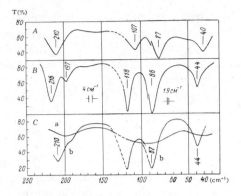

Fig. 6.9   Spectra of the region of the hydrogen bond transitions of CCl_3COOH. *Upper spectrum* (A): 80-μ layer thickness at 20°C; *intermediate spectrum* (B): 80 μ, −175°C; *bottom spectrum* (C): $a = 10\%$ solution by weight in CCl_4, 1-mm layer thickness at 20°C; $b = 25\%$, 0.5 mm, −175°C. The dashed lines indicate strong quartz cell absorption. [A. E. Stanevich, *Opt. Spectry.* **16**, 243 (1964), by permission of Optical Society of America.]

TABLE 6.6*

HYDROGEN BOND FUNDAMENTALS OF SOME CARBOXYLIC ACIDS

Acid	v_3 obs.,[†] cm^{-1}	v_3 calcd.,[‡] cm^{-1}	v_4, obs., cm^{-1}	v_5, obs., cm^{-1}
CH_3COOH, Acetic acid	176	—	~ 75[#]	~ 40[#]
CH_3COOD	178	176	~ 80[#]	~ 40[#]
CD_3COOD	176	171	78[§]	42[§]
CH_3CH_2COOH, Propionic acid	157	159	~ 67,[‖,#] 78[§]	~ 38[#]
$CH_3(CH_2)_2COOH$, Butyric acid	154[§]	161[**]	~ 72[#]	~ 38[#]
CCl_3COOH, Trichloroacetic acid	107	107	77	40
C_6H_5COOH, Benzoic acid	107	124	~ 70[††]	~ 39[††]

* * *After* A. E. Stanevich, *Opt. Spectry.* **16**, 243 (1964). By permission of Optical Society of America.

   [†] At 20°C if not otherwise marked.

   [‡] Force constant $= 0.34 \times 10^5$ dyne/cm.

   [§] At −175°C.

   [‖] In solution of CCl_4.

   [#] Diffuse and broad contour.

   [**] Force constant $= 0.40 \times 10^5$ dyne/cm (for −175°C).

   [††] Broad.

   States of aggregation under normal conditions: CCl_3COOH and C_6H_5COOH are solids; all others are liquids.

to $v = (1/2\pi)(2f/\mu)^{1/2}$, where f is the force constant and μ the reduced mass. The value of f was obtained from the frequency assigned to the v_3 mode of acetic acid. We identify, of course, v_3 with v_σ. Table 6.6 shows that the trend in the frequencies of v_3 of the various acids indeed seems to be mainly a consequence of the different masses of the alkyl radical; in other words, the strength of the hydrogen bonds of these carboxylic dimers is, to a first approximation, constant. Again we state that any effects on the strengths of the hydrogen bonds that might arise from the different nature of the radicals attached to the carboxyl group would not be accounted for by the assumption of a rigid monomer molecule.[13] We might expect, for instance, that the hydrogen bonding between trichloroacetic acid molecules on one hand and between benzoic acid molecules on the other hand should lead to different hydrogen bond strengths in the resulting dimers, simply because of the different strength of the O—H bond in the two monomers. It is indeed shown by the data in the Table 6.6 that one and the same value for the stretching force constant agrees rather poorly with the observations for these two acids.

Jakobsen *et al.* have recently performed normal coordinate analyses on propionic acid[14] and propose, on the basis of their results, that the frequencies of the deformation vibration v_4 and v_5 of propionic and butyric acid (see Table 6.6) be permuted. We thus see that the frequencies of the various hydrogen bond deformation modes of the alkyl acids do not reflect a simple mass relation as do the hydrogen bond stretching fundamentals.

Closer inspection of Fig. 6.8 (curve A.c), which displays Stanevich's work on CH_3COOD, indicates a band at 165 cm^{-1} adjacent to $v_\sigma = 178$ cm^{-1}. This band was also detected in the vapor phase of CH_3COOH (shown to be dimeric) and is now rather convincingly assigned[15] to represent a combination (species B_u) between the (infrared-inactive) deformation $v_6(b_g)$, calculated to lie at 130 cm^{-1}, and the twisting mode $v_5(a_u)$ located at about 40 cm^{-1} (see Fig. 6.7 and Table 6.6). The relatively large intensity of this combination band is believed to arise from Fermi resonance with $v_3(b_u)$.[15]

In conclusion in Table 6.7 we give a compilation of the observed and calculated (normal coordinate analyses) frequencies of the six hydrogen bond fundamentals of the acetic acid dimer.

E. Solvent Effects on Hydrogen-Bond Frequencies

In Section B we mentioned briefly the possibility that an interaction of a hydrogen bond complex with the molecules of the environment in the condensed phase may lead to fairly large solvent shifts of the hydrogen bond frequencies. Hall and Wood[17] have recently shown evidence that such effects are indeed observable. Their observation points again to the

TABLE 6.7

CALCULATED AND OBSERVED FREQUENCIES OF THE
HYDROGEN BOND FUNDAMENTALS OF THE
ACETIC ACID DIMER (C_{2h})

| | | cm^{-1} | |
Mode*	Symmetry species	Calculated	Observed
v_1, $v(OH \cdots O)$	a_g	$210^\dagger$ $193^\ddagger$	IR-inactive
v_3, $v(OH \cdots O)$	b_u	$187^\dagger$ $180^\ddagger$	$176^\S$ $188^{\dagger,\ddagger}$
v_6, $\gamma(OH \cdots O)$	b_g	$130^\dagger$ $128^\ddagger$	IR-inactive
v_2, $\beta(OH \cdots O)$	a_g	$81^\dagger$ $95^\ddagger$	IR-inactive
v_4, $\gamma(OH \cdots O)$	a_u	$79^\dagger$ $80^\ddagger$	$\sim 75^\S$
v_5, $\gamma(twist)$	a_u	$54^\dagger$ $46^\ddagger$	$\sim 40^\S$ $50^{\|}$

* See Fig. 6.7 for a description of the modes. The
designations $v(OH \cdots O)$ for stretching, $\beta(OH \cdots O)$
for in-plane, $\gamma(OH \cdots O)$ for out-of-plane deformation, and γ (twist) are also in use.
† Reference 15 (observed frequencies are those of
the vapor).
‡ Reference 13.
§ Reference 12 (liquid).
$^{\|}$ Reference 16 (vapor).

necessity of considering the dynamical behavior of the entire solute–solvent
system in the far-infrared spectral region since the energy levels of the
vibrations are low and are therefore particularly apt to reflect molecule–
environment interactions (see also Chapter 11).

In particular, Hall and Wood determined the peak frequency of the
v_σ mode of the 1:1 phenol–pyridine complex[18] as a function of the composition of the solutions. In each case the molar ratios of the phenol and pyridine
were adjusted to the ratio 1:2 whereas the mole fraction of the solvent,
carbon tetrachloride, was varied from 0 to 0.85; the excess of pyridine
serves to suppress formation of (phenol)$_n$–pyridine ($n > 1$) complexes. (On
the basis of infrared data,[19] Takahashi et al. have postulated that complex
formation in alcohol–pyridine–CCl$_4$ solutions rich in pyridine leads to
higher than 1:1 entities. On the other hand, Hall and Wood give some

infrared evidence to the effect that only 1 : 1 complexes of phenol and pyridine
are present in their solutions.)

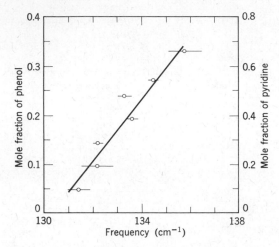

Fig. 6.10   Dependence of the peak frequency of the intermolecular stretching mode (v_σ) of the
1 : 1 phenol-pyridine complex on the composition of the solution. The mole fraction of pyridine
was in all cases twice that of the mole fraction of phenol. The remainder of the solution composition was made up by carbon tetrachloride (mole fraction = 0.85 at 131.5 cm^{-1}, 0 at 135.8
cm^{-1}). The spectrometer was a self-built vacuum grating instrument. The solutions were examined in a cell of 2–3-mm path length, equipped with thick polyethylene windows. The horizontal
bars in the figure on the experimental points denote the estimated uncertainty of the frequency
measurements of the (broad) peaks. [*After* A. Hall and J. L. Wood, *Spectrochim. Acta* **24A**,
1109 (1968), by permission of Pergamon Press.]

Figure 6.10 shows Hall and Wood's far-infrared experimental results.
On the ordinate, the mole fraction of phenol (left-hand side) and of pyridine
(right-hand side) are plotted; on the abscissa the peak frequency of v_σ is
displayed. The plot indicates that the peak frequency of v_σ of the proposed
1 : 1 phenol–pyridine complex increases with increasing concentrations of
phenol/pyridine. Based on their assumptions, Hall and Wood therefore

$$R-N \!:\!\cdots H-O-R'$$

$$R-\overset{+\delta}{N} \!:\!\cdots \overset{-\delta}{H}-O-R'$$

Fig. 6.11   Schematic drawing of charge transfer and the resulting increase in the stability of the
intermolecular bond due to the polarity of the solvent (stabilization of charges).

postulate that this increase of the peak frequency of v_σ is a consequence of the increasing polarity of the solution. In terms of a more detailed picture, the authors draw on the concepts of charge-transfer complexes.[20,21] They view the more polar medium as able to facilitate the charge transfer from the electron donor (the nitrogen atom of the pyridine base) to the electron acceptor of the phenol molecule (the H atom of its OH group). The resulting polar valence bond structure of the complex leads to a greater stability which, in turn, is reflected in the increased frequency of v_σ, the intermolecular bond stretching mode[21] (Fig. 6.11).

F. Concluding Remarks

It is hoped and expected that much more work will be undertaken in the near future on far-infrared spectroscopy of hydrogen-bonded systems, not only in order to corroborate or modify already known data and theories, but mainly to give a more extensive and deeper knowledge and understanding of these exciting and important topics, in particular for materials of biological significance.

REFERENCES

1. A. E. Stanevich, *Opt. Spectry.* **21**, 355 (1966).
2. V. Lorenzelli and A. Alemagna, *Compt. Rend.* **256**, 3626 (1963).
3. G. L. Carlson, R. E. Witkowski, and W. G. Fateley, *Nature* **211**, 1289 (1966).
4. S. G. W. Ginn and J. L. Wood, *Nature* **200**, 467 (1963).
5. W. J. Hurley, I. D. Kuntz, Jr., and G. E. Leroi, *J. Am. Chem. Soc.* **88**, 3199 (1966).
6. J. N. Finch and E. R. Lippincott, *J. Phys. Chem.* **61**, 894 (1957).
7. See p. 173 of G. C. Pimentel and A. L. McClellan, *The Hydrogen Bond*, Freeman, San Francisco, 1960.
8. R. J. Jakobsen and J. W. Brasch, *Spectrochim. Acta* **21**, 1753 (1965).
9. J. W. Brasch and R. J. Jakobsen, *Spectrochim. Acta* **20**, 1644 (1964).
10. S. G. W. Ginn and J. L. Wood, *Spectrochim. Acta* **23A**, 611 (1967).
11. A. E. Stanevich, *Opt. Spectry.* **16**, 425 (1964).
12. A E. Stanevich, *Opt. Spectry. Suppl.* "Molecular Spectroscopy" (Russian original, 1963), 1966, p. 104; *Opt. Spectry.* **16**, 243 (1964); see also Ref. 1.
13. T. Miyazawa and K. S. Pitzer, *J. Am. Chem. Soc.* **81**, 74 (1959).
14. R. J. Jakobsen, Y. Mikawa, and J. W. Brasch, private communication, May 1968.
15. R. J. Jakobsen, Y. Mikawa, and J. W. Brasch, *Spectrochim. Acta* **23A**, 2199 (1967).
16. G. L. Carlson, R. E. Witkowski, and W. G. Fateley, *Spectrochim. Acta* **22**, 1117 (1966).
17. A. Hall and J. L. Wood, *Spectrochim. Acta* **24A**, 1109 (1968). See also Section C of this chapter.
18. See, for instance, A. K. Chandra and S. Banerjee, *J. Phys. Chem.* **66**, 952 (1962).
19. H. Takahashi, K. Mamola, and E. K. Plyler, *J. Mol. Spectry.* **21**, 217 (1966).
20. An intensive review on all aspects of charge-transfer complexes is that of C. Briegleb, *Elektronen-Donator-Acceptor-Komplexe*, Springer, Berlin, 1961.
21. S. G. W. Ginn and J. L. Wood, *Trans. Faraday Soc.* **62**, 777 (1966).

7   Skeletal Modes of Strained-Ring Systems

A. INTRODUCTION

The overall vibrational motion of an isolated molecule is the superposition of its individual $3N$-6 (or $3N$-5) normal modes. It is well known that these normal modes can often be pictured, in an approximate manner, by more or less idealized and separable (that is, from molecule to molecule transferable) motions of individual atoms or groups of atoms. Such motions may involve, for instance, bond stretches or bond angle deformations. Their vibrational frequencies are predominantly influenced by the bond strengths and by the atomic masses although they are affected to a lesser but still appreciable degree by the character of the adjacent bonds and atoms. In short, infrared spectral investigations of such groups yield mainly *structural information*; that is, information about the sequence of the atoms, the type of binding between atoms, and—depending on a resolution of the fine structure—a more or less precise knowledge of bond lengths and bond angles.

With increasing masses and decreasing forces between atoms or groups of atoms, the positions of fundamental modes move towards longer wavelengths until finally, in the far-infrared region of the spectrum, we may encounter fundamental motions which involve the deformation of the entire frame of the molecule. These modes are therefore termed here "skeletal modes." The reason for their special interest lies in the fact that they give information on the *conformational* behavior; that is, on the different and (often) interconvertable spatial arrangements of the molecule[1] rather than structural information. In fact, it is often difficult to elucidate skeletal modes without structural data from near- and medium-infrared, microwave, or electron diffraction studies. In essence, the motions of the skeletal modes arise from twisting and bond deformations about single bonds. This leads in open chains to the "internal rotations" and in ring compounds to "out-of-plane ("ring-puckering") modes." A separation of the skeletal modes

218

into these two groups is convenient and has therefore been applied here (see Chapter 8). In the following, we shall first briefly describe some of the principles of the skeletal modes of flexible ring systems and then discuss a few representative examples from the recent literature.

Skeletal modes in flexible, strained-ring systems are of considerable theoretical interest since they represent and characterize the balance between angle deformation strain and intramolecular repulsive interactions. Substituents on adjacent ring atoms can generally lower their repulsive interactions by twisting the ring. In this way the distance between the repulsive centers is increased, but the angle strain is also increased.

Depending on the number of ring atoms, we can (grossly) classify the behavior of the out-of-plane ring deformation modes (ring-puckering modes):

1. Four-Member Rings. If the planar ring is bent to a dihedral angle of, say, $\gamma = 20°$,* the angle between the ring atoms (say carbon) decreases from 90° to about 88° and the distance between the centers of, say, hydrogen atoms attached to adjacent carbon atoms increases by about 0.05–0.1 Å. These relations may be translated into energy terms with the help of some crude estimates of angle strain and of "nonbonded" repulsions. (Under nonbonded we understand the substituents which are not attached to a common central atom.) The energy required to decrease the C—C—C angle from 90 to 88° amounts to (an order of magnitude of) $E = K_\theta(\theta)^2$, where the constant $K_\theta \sim 17.5$ cal mole^{-1} deg^{-2}.[2] Inserting $\theta = 2°$, we find that the energy required to bend the planar cyclobutane ring is about $4 \times 70 = 280$ cal mole^{-1}. This estimate neglects the concomitant changes of the H—C—H and C—C—H angles, which probably contributes at least the same amount of energy as the C—C—C— angle deformation. Some lower limit of the energy required to bend the planar ring by an appreciable degree is therefore about 0.5 kcal mole^{-1} or 175 cm^{-1}. (As mentioned above, a decrease of the C—C—C angle by 2° corresponds to a dihedral angle of about 20°.) Continuing in the spirit of this rough estimate, it is seen that this ring deformation energy is balanced by the now decreased repulsion energy between the hydrogen atoms, which has been computed to be 0.47 kcal mole^{-1}.[3] Since $kT \sim 0.6$ kcal mole^{-1}, we see that in four-member rings the balance between ring strain and repulsive interactions is delicate. Hence we can expect that the character of the ring-puckering modes is going to depend to a relatively great extent on the nature of the substituents.

---

* The dihedral angle is the angle between two vectors which are normal to the two planes each formed by three (or more) ring atoms.

2. Five-Member Rings. It is apparent that there is an ambiguity in choosing which of the ring atoms undergoes the maximum amplitude during the ring-puckering motion. It is therefore necessary to introduce a phase dependence between the amplitudes of the out-of-plane motion of the ring atoms, in other words, the maximum amplitude travels around the ring (unless the ring flexibility is strongly perturbed by a double bond, such as in cyclopentene). This interesting concept has indeed been established by experiment and is treated in Section F ("Pseudorotation").

3. Six-Member Rings. Their ring skeletal frequencies generally fall toward higher frequencies and we therefore shall not treat them here. The reason for this frequency shift is easy to understand. The ring in cyclohexane compounds (for instance) is essentially strain-free (tetrahedral angles) and the conformation is therefore well "locked in" by the interactions of the substituents. Hence relatively large forces are required to deform the ideal carbon–carbon angles.

Little spectroscopic work has been done on seven or higher-member rings. Likewise, most data have been obtained on ring systems in which the ring contains (often exclusively) carbon atoms.

B. THEORETICAL CONSIDERATIONS

It will be useful to give a short theoretical treatment of the potential functions governing ring-puckering modes. To render the problem tractable, it is assumed that the ring-puckering mode can be treated as a one-dimensional problem in terms of a displacement coordinate z which somehow describes the deviation from planarity. The practical solution of the quantum-mechanical problem is best set up in the matrix formulation. The Hamiltonian

$$\mathscr{H} = -\frac{\hbar^2}{2m}\frac{d^2}{dz^2} + V(z) \tag{7.1}$$

is transformed to a dimensionless form by the substitutions $\alpha z = \xi$, where $\alpha^2 = m\beta/\hbar$, β is a scale factor of a dimension appropriate to the problem, $E = (\frac{1}{4})\hbar\beta E'$, and $V(\xi) = (\frac{1}{4})\hbar\beta V'(\xi)$. The dimensionless Hamiltonian is then[4a,b]

$$\mathscr{H} = -\frac{d^2}{d\xi^2} + (\tfrac{1}{2})V'(\xi) \tag{7.2}$$

and its eigenvalues E' are in units of $(\frac{1}{4})\hbar\beta$. Choosing an (orthonormal) representation $|i\rangle$, the eigenfunctions of Eq. 7.2 are

$$|\psi_j\rangle = \sum_i c_{ij}|i\rangle, \tag{7.3}$$

leading to the Schrödinger equation

$$\mathscr{H} \sum_i c_{ij}|i\rangle = E_j \sum_i c_{ij}|i\rangle. \tag{7.4}$$

Operating with $\mathscr{H}$ on $|i\rangle$ and multiplying from the left by $|h\rangle$, we obtain

$$\sum_i c_{ij}\langle h|\mathscr{H}|i\rangle = E_j \sum_i c_{ij}\langle h|i\rangle, \tag{7.5}$$

$$\sum_i c_{ij}\{H_{hi} - E_j\delta_{hi}\} = 0, \tag{7.6}$$

where $\langle h|\mathscr{H}|i\rangle = H_{hi}$, $\langle h|i\rangle = \delta_{hi}$. Since $c_{ij} \neq 0$,

$$\sum_i \{H_{hi} - E_j\delta_{hi}\} = 0 \text{ for all } h, j. \tag{7.7}$$

The solutions of Eqs. 7.7 and 7.6 yield the eigenvalues E_j and the corresponding eigenvectors c_{ij} which are the desired coefficients of the expansion of $|\psi\rangle$ for each energy level j in terms of the original representation $|i\rangle$. The choice of the $|i\rangle$ depends on the particular form of the potential V and will be discussed below for individual molecules. In general, the expansion of Eq. 7.3 comprises an infinite number of terms. The point at which we wish to truncate the matrix then depends on the desired precision of the numerical values as well as on the characteristics of the available computer. Obviously, we would pick that set of basis vectors which give the most rapid convergence. We also would like to point out that a simple method for the generation of the required matrix elements, which is based on transformation theory, has been published.[5]

C. SYMMETRIC RING-PUCKERING POTENTIAL FUNCTIONS

a. Cyclobutane and Perfluorocyclobutane

Cyclobutane was one of the first molecules for which a low-frequency ring-puckering mode governed by a symmetric double-minimum potential was invoked to account for the observed thermodynamic properties.[6] Since the ring-puckering mode of cyclobutane is infrared-forbidden for a planar as well as a bent conformation, the compound is only of indirect interest here. The same holds for perfluorocyclobutane.[7] Suffice it to say that both compounds assume a nonplanar conformation (D_{2d}) in their first few lowest states which are below a barrier of inversion of (estimated) 400 and 640 cm^{-1}, respectively. As a consequence of the favorable Boltzmann factors an appreciable fraction of molecules—those near and above the top of the barrier—is found in the planar conformation and therefore obeys D_{4h} selection rules.

b. Trimethylene Oxide

The investigation of the ring-puckering mode of trimethylene oxide $(CH_2 \cdot CH_2 \cdot CH_2 \cdot O)$ has culminated in an excellent agreement between the large body of observed and calculated quantities. Trimethylene oxide is a prototype of a perturbed quartic oscillator and we will therefore discuss it in detail. The observed far-infrared Q branches of the transitions and their assignments as given by Chan et al. and by Lafferty and Lord[8] are shown in Table 7.1. The spectral region was measured on three instruments, two of which were grating instruments and the other an interferometer. The deviation of the positions of the observed band maxima from those of the true band origins due to the difference of the rotational constants of the lower and upper vibrational levels were estimated to fall within the experimental errors of the frequency determinations. The assignments in Table 7.1 are based on the frequency of the fundamental $0 \rightarrow 1^*$ at $53.4 \, cm^{-1}$. This agrees with the value obtained from the microwave spectrum of trimethylene oxide,[9] at $60 \pm 20 \, cm^{-1}$, by intensity measurements of the two strongest vibrational satellites. It may be recalled that each pure rotational transition is averaged over a vibrational state of the molecule and, consequently, a series of rotational transitions[10] based on excited vibrational states, the so-called satellites, may appear in the rotational spectrum if the vibrational Boltzmann distribution is favorable. In general, microwave intensity measurements are less precise than wavelength determinations; however, since they measure the fraction of molecules in a certain vibrational state, microwave intensity measurements are independent of vibrational transition probabilities and vibrational absorption coefficients. Thus, they often indicate the position of a far-infrared vibrational mode which may otherwise have been missed on account of its feeble intensity or experimental difficulties, such as occultation by adjacent water lines.

Table 7.1 also lists the observed transitions and vibrational assignments of the ring-puckering mode of trimethylene oxide-d_6. The observed isotope effect, expressed as the ratio of the frequencies of a given transition of protonated and deuterated compound, is also shown. The observed isotope affect agrees reasonably well with the theoretical isotope effect of 1.290, computed on the basis of (1) the potential of the ring-puckering motion is essentially quartic, (2) the dynamical path of the atoms during the ring-puckering involves out-of-plane displacements of the carbon and oxygen atoms such that their magnitudes of displacement from a planar ring are

---

* In two previous publications by Chan et al. [J. Chem. Phys. **33**, 1643 (1960); **34**, 1319 (1961)], the $0 \rightarrow 1$ transition had been assigned to the band at $89.9 \, cm^{-1}$ (see Table 7.1) since the band at $53.4 \, cm^{-1}$ had not been detected. See also A. Danti, W. J. Lafferty, and R. C. Lord, J. Chem. Phys. **33**, 294 (1960).

TABLE 7.1*

OBSERVED AND CALCULATED FAR-INFRARED SPECTRUM OF TRIMETHYLENE OXIDE AND
PERDEUTERO TRIMETHYLENE OXIDE

	Trimethylene oxide			Deuterotrimethylene oxide		
	Observed		Calculated	Observed	Calculated	Observed
Transition	cm^{-1}[†]	cm^{-1}[‡]	cm^{-1}[§]	cm^{-1}[‡]	cm^{-1}[‖]	isotope effect[††]
$0 \to 1$	53.4	53.4	53.4	39	38.7	1.368
$1 \to 2$	89.9	89.8	89.9	69.3	68.6	1.296
$2 \to 3$	104.7	105.2	104.8	80.7	79.9	1.304
$3 \to 4$	118.1	118.3	118.2	91.0	90.5	1.301
$4 \to 5$	128.8	128.9	129.2	99.0	99.0	1.305
$5 \to 6$	138.9	139.0	138.7	106.4	106.4	1.309
$6 \to 7$	147.3	147.6	147.1	112.6	113.0	1.312
$7 \to 8$	155.0	154.9	154.7	118.0	118.9	1.311
$8 \to 9$	161.6	161.8	161.6	124.0	124.4	1.305
$9 \to 10$	168.2		168.0		129.4	
$10 \to 11$	174.9		174.0		134.0	
$11 \to 12$	180.4		179.6		138.4	
$12 \to 13$	185.5		184.9		142.5	
$13 \to 14$	190.1		190.0		146.5	
$14 \to 15$	195.7 (?)		194.7		150.2	
$15 \to 16$	200.1 (??)		200		155.4	
$0 \to 3$	247.5		248.0			
$1 \to 4$	312.8		312.9			
$2 \to 5$	352.0		352.2			
$3 \to 6$	387 (??)		386.1			

* *After* S. I. Chan, T. R. Borgers, J. W. Russel, H. L. Strauss, and W. D. Gwinn, *J. Chem. Phys.*
44, 1103 (1966).

[†] Chan *et al.*[8] The observed frequencies are known within 0.2 to 0.5 cm^{-1}.

[‡] Lafferty and Lord.[8]

[§] $V(\xi) = 11.17(-2.34\xi^2 + \xi^4)\,(\text{cm}^{-1})$. The values are good within 0.1 cm^{-1} except the $15 \to 16$ because of poor convergence.

[‖] $V(\xi) = 8.664(-2.6571\xi^2 + \xi^4)\,(\text{cm}^{-1})$.

[††] cm^{-1} (protonated)/cm^{-1} (perdeutero).

equal, and (3) during the motions of the hydrogen atoms the angles of the methylene groups are always bisected by the plane through the nearest carbon atoms. The motions under (2) and (3) are defined to have no contributions of carbon–carbon or carbon–hydrogen bond stretching.

A further test for the soundness of the assignments is the excellent agreement between the computed and observed frequencies of the first three $\Delta n = 3$ overtones (see Table 7.1).

We will now discuss the potential function of the ring-puckering mode of trimethylene oxide. The monotonic increase of the frequencies of the successive transitions $0 \to 1$, $1 \to 2$, $2 \to 3$, $\cdots$ shown in Table 7.1 indicates that the potential function is steeper than that of the harmonic oscillator; that is, it contains fourth-order (or higher) terms. The question of whether the potential possesses a steep central barrier with respect to the interconversion of the two mirror images of trimethylene oxide, in other words, whether the molecule is permanently bent (the energy levels of the ring-puckering mode are doubly degenerate)* or whether the molecule is planar or nearly so (the energy levels are single), had been decided from the microwave spectrum[9] in favor of an essentially planar ring conformation. The arguments are as follows. If the molecule were planar or nearly planar, the intensities of rotational transitions from the ground ($E = 0$) and successively excited vibrational states ($E_1, E_2, \cdots$) would be in the ratio of $7:9 \exp(-E_1/kT):7 \exp(-E_2/kT) \cdots$ for a-type rotational transitions (ΔK_{-1} even) and $9:7 \exp(-E_1/kT):9 \exp(-E_2/kT) \cdots$ for b-type transitions.† If the molecule were strongly puckered (high barrier) so that successive pairs of vibrational satellites would not be resolved (near-degeneracy), the relative intensities of the rotational transitions based on the ground vibrational level and successive higher excited levels would be

$$(7 + 9):(7 + 9) \exp(-E_1/kT):(7 + 9) \exp(-E_2/kT) \cdots.$$

Furthermore, an indication that the lowest vibrational levels of the ring-puckering mode are perturbed by a small central hump in the potential function was, likewise, first indicated by the rotation–vibration interaction

---

* See M. W. P. Strandberg, *Microwave Spectroscopy*, Wiley, New York, 1954, p. 81, Fig. 6, for a schematic potential correlation diagram.

† Since the nuclear spins of carbon and oxygen are zero, there are $(2 \times (\frac{1}{2}) + 1)^6 = 64$ nuclear spin functions of trimethylene oxide; that is, there are 64 different ways of distributing spins of quantum number $\frac{1}{2}(= \alpha)$ and $-\frac{1}{2}(= \beta)$ among the six hydrogen atoms. Of these 64, there are one of $\alpha\alpha\alpha\alpha\alpha\alpha$, six of $\beta\alpha\alpha\alpha\alpha\alpha$, 15 of $\beta\beta\alpha\alpha\alpha\alpha$, 20 of $\beta\beta\beta\alpha\alpha\alpha$, 15 of $\beta\beta\beta\beta\alpha\alpha$, six of $\beta\beta\beta\beta\beta\alpha$, and one of $\beta\beta\beta\beta\beta\beta(= n!/(n - p)!p!)$. Since the C$\cdots$O diagonal is a twofold axis of symmetry, rotations about it are equivalent to *odd* permutations of three pairs of hydrogen atoms; that is, $C_2 = (\alpha\alpha)(\alpha\beta)(\beta\alpha)$, for instance. Therefore, the relevant number of spin functions to be considered in the nuclear spin statistics amounts to eight; namely, those contributing nonzero character: one each of $\alpha\alpha\alpha\alpha\alpha\alpha$ and $\beta\beta\beta\beta\beta\beta$, and three each of $\alpha\alpha\beta\beta\beta\beta$ and $\beta\beta\alpha\alpha\alpha\alpha$. The representation of the group of permutations of identical atoms which are equivalent to a twofold rotation are therefore n_A(symmetric) $= (\frac{1}{2})(64 \cdot 1 + 8 \cdot 1)$, n_B(antisymmetric) $= (\frac{1}{2})(64 \cdot 1 - 8 \cdot 1)$, $\Gamma = 36A + 28B$, using the characters of the two species A and B of the symmetry group C_2. The statistical weights of the spin–rotation wave functions are then $(36A + 28B) \times A = 36A + 28B$ for a symmetric rotational level (K_{-1} even) and $(36A + 28B) \times B = 36B + 28A$ for an antisymmetric rotational level (K_{-1} odd). Since H atoms are Fermi particles, we count only the number of antisymmetric combinations (species B); hence for a symmetric (antisymmetric) vibrational level, the ratio of the intensities of a symmetric to an antisymmetric rotational level is $\frac{7}{9}$ $(\frac{9}{7})$.

observed in the microwave spectrum. The interaction results in a "zig-zag" variation of the values of the observed rotational constants with increasing vibrational quantum numbers because, in essence, the potential hump perturbs the even vibrational levels to a greater extent than the odd levels (which possess a node at the central position[11]).

Since the potential function has two adjustable parameters (for instance, the height of the potential hump and the distance between the two potential minima), the analytical form

$$V(\xi) = c_2\xi^2 + c_4\xi^4 \tag{7.8}$$

is the most logical to try.* The matrix elements of the mixed quartic oscillator (Eq. 7.8) in the harmonic oscillator representation are well known.[5] The diagonalization of the energy matrix (Eq. 7.2) gave at best ten to eleven accurate levels when based on a linear combination of 40 harmonic oscillator basis functions, this large number being necessary because the potential is essentially quartic. This is unsatisfactory with respect to the great accuracy and wealth of observations on the spectrum of the ring-puckering mode of trimethylene oxide. To obtain the desired agreement between observed and computed values, we may extend the size of the matrix which, as is well known, results in a rapidly increasing amount of computer time for the diagonalization routine. A better approach is to transform the matrix elements of a reasonable number of states of the reduced energy matrix in the harmonic oscillator representation to a representation in which the terms originating from $-(d^2/d\xi^2) + c_4\xi^4$ are on the diagonal; that is, the quartic oscillator representation. There is no need here to go into the details of the computations done on this subject,[12] suffice it to say that an extensive tabulation of various useful (reduced) matrix elements in the quartic oscillator representation has been published.[13]

The agreement obtained between the observed transitions and the values calculated with the mixed quartic oscillator potential in the quartic oscillator representation,

$$V(\xi) = 11.17\,(-2.34\xi^2 + \xi^4)\,(\text{cm}^{-1}), \tag{7.9}$$

where ξ is a dimensionless coordinate, is near perfect.† The computed

---

* The quadratic term must be negative to yield two minima and one maximum. We usually set $V = -c_2\xi^2 + \cdots$, $c_2 \geqslant 0$.

† The Hamiltonian of the perturbed quartic oscillator (see Ref. 13); that is, in our case the quartic oscillator with a small harmonic term, $\mathcal{H} = p^2/2\mu - (\frac{1}{2})a_2z^2 + a_4z^4$ (where p is the conjugate linear momentum of the displacement coordinate z, and $a_4 \gg a_2$), is conveniently transformed in terms of the dimensionless coordinate ξ to the form $\mathcal{H} = (a_4\hbar^4/64\mu^2)^{1/3} \times \{P^2 + \xi^4 - \eta\xi^2\}$, where $\xi = (8\mu a_4/\hbar^2)^{1/6}z$, $P = (8/\mu a_4\hbar^4)^{1/6}p$, and $\eta = (a_2^3\mu/a_4^2\hbar^2)^{1/3}$. Thus $\eta = c_2$, with $c_4 = 1$, in units of the scaling factor (see Eq. 7.8). Since the ratios of successive energy level spacings of the ring-puckering mode, $(E_v - E_{v-1})/(E_{v+1} - E_v)$, are independent of the scaling factor $(a_4\hbar^4/64\mu^2)^{1/3}$, the value of η can be found by comparing the experimentally observed frequencies with the calculated ones using the potential function $V(\xi) = C(\xi^4 - \eta\xi^2)$.

values are given in Table 7.1. It is noteworthy that the energy levels for the deuterated molecule were obtained from the same potential function, merely corrected on the basis of the different reduced mass involved in the vibration.* Figure 7.1 finally shows the ring-puckering potential and the first four energy levels of trimethylene oxide.

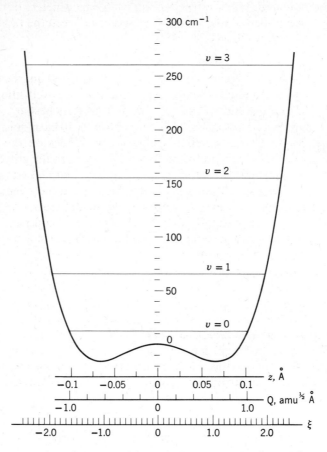

Fig. 7.1   Lower portion of the potential energy function, $V(\xi) = 11.17 \, (\xi^4 - 2.34\xi^2) \, \text{cm}^{-1}$, for the ring-puckering vibration of trimethylene oxide. The barrier height amounts to $15.3 \pm 0.5 \, \text{cm}^{-1}$ and the ground vibrational level lies $12 \, \text{cm}^{-1}$ above the top of the barrier. [S. I. Chan, T. R. Borgers, J. W. Russel, H. L. Strauss, and W. D. Gwinn, *J. Chem. Phys.* **44**, 1103 (1966).]

* The ratio of η (deuterated)/η (protonated) compound (see previous footnote) is given by the ratio of the third power of the respective reduced masses μ_d and μ_p, which in turn had been calculated as described in the text. It follows that $\eta_d/\eta_p = (142.3/97.2)^{1/3} = 1.136$. The ratio of the corresponding scale factors $(a_4 h^4/64\mu^2)^{1/3}$ is given by $(\mu_p/\mu_d)^{2/3} = 0.7756$ (consult the previous footnote). Hence, the potential function for the perdeuterated compound is $V(\xi) = 8.66 \, (-2.66\xi^2 + \xi^4)$.

c. Cyclobutanone

As in trimethylene oxide, the potential function of the ring-puckering mode of cyclobutanone[14,15] is a slightly perturbed quartic oscillator,

$$V(\xi) = 6.655\,(-1.7\xi^2 + \xi^4)\,(\mathrm{cm}^{-1}), \qquad (7.10)$$

or

$$V(z) = -2.88 \times 10^3 z^2 + 4.32 \times 10^5 z^4\,(\mathrm{cm}^{-1})$$

in terms of the cartesian displacement coordinate z (Å). The top of the barrier is below the ground vibrational level and the barrier height is approximately $5\,\mathrm{cm}^{-1}$. The agreement between observed and computed values is not as good as that for trimethylene oxide in so far as the mixed quartic potentials which fit the lower levels deviate at the higher levels. However, a potential function with a Gaussian barrier* and four adjustable parameters fits all levels.[16] The spectrum indicates that the bands become broader with increasing vibrational quantum number and, beyond $\sim 85\,\mathrm{cm}^{-1}$ (the $6 \to 7$ transition), are split by about $1\,\mathrm{cm}^{-1}$ into two components. As a possible explanation Borgers and Strauss suggested[14] that the ring-puckering mode of cyclobutanone couples through its quadratic terms with the C=O out-of-plane deformation $v_{26}(b_2)$ at $395\,\mathrm{cm}^{-1}$;[17] for instance, the matrix element $\langle v|\zeta^2|v'\rangle$ between the levels $v = 4$ and $v' = 8$, $E(8) - E(4) = 350\,\mathrm{cm}^{-1}$, amounts to 0.2.[14] In this respect it may also be significant that the observed isotope effect of $\alpha, \alpha, \alpha', \alpha'$-tetradeuterocyclobutanone,[15] expressed in terms of the ratio of the frequencies for protonated and tetra-deuterated compounds, decreases monotonically from 1.21 for the $1 \to 2$ transition to 1.13 for the $7 \to 8$ transition, thereby probably indicating that admixture of another normal coordinate changes the effective reduced mass.†

d. Perfluorocyclobutanone

The far-infrared spectrum of perfluorocyclobutanone shows no absorption between 35 and $130\,\mathrm{cm}^{-1}$. This indicates that the intensity of the ring-puckering mode is either very weak or that the mode lies below $35\,\mathrm{cm}^{-1}$ (or a combination of both factors).[15] We might conjecture that the reason for the unobservability is mainly a weak intensity since even if the fundamental of the ring-puckering is below $35\,\mathrm{cm}^{-1}$, we should expect to observe some of the transitions between the higher levels at shorter wavelengths. However, this reasoning tacitly assumes that the potential function of per-fluorocyclobutanone is similar to that of cyclobutanone. In other words, that the difference is mainly one of the different reduced masses.

---

* This implies terms of powers higher than four.

† This may be compared with the more random scatter of the values of the isotope effect in trimethylene oxide, see Table 7.1.

e. Trimethylene Sulfide

In this molecule we again encounter a mixed quartic potential function for the ring-puckering mode but here the height of the barrier is relatively high, 274 cm^{-1}, and as a consequence the first four vibrational levels are below the top of the barrier.[14,18] This is shown in Fig. 7.2. The lowest level is nearly degenerate; its splitting of 0.27461 cm^{-1} was obtained from the rotation–vibration interaction in the microwave spectrum of the compound and confirmed by intensity measurements.[18] From the splitting of the ground level we calculate that the coordinate undergoes 8.2 × 10^9 reflections per sec about the position $\xi = 0$ or about two reflections per 10^3 vibrations. The frequencies of the transitions in Fig. 7.2 are those between the corresponding band origins and not between the band maxima of the observed Q branches.*

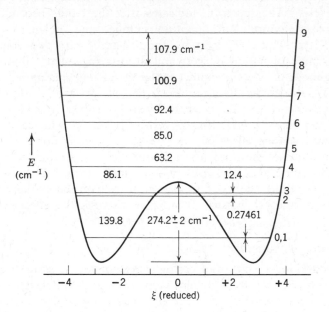

Fig. 7.2   The experimentally determined potential function $V = 7.0207\xi^4 - 87.7581\xi^2$ (cm^{-1}) of the ring-puckering motion of trimethylene sulfide; ξ is the reduced coordinate in the harmonic oscillator representation. The frequencies (cm^{-1}) are those between the calculated band origins (see text). [D. O. Harris, H. W. Harrington, A. C. Luntz, and W. D. Gwinn, *J. Chem. Phys.* **44**, 3467 (1966).]

* The difference Δv between the band maximum (= point of maximum absorption) and the band origin (= point where the pure vibration would be located) is for an A-type band transition (change of dipole moment occurs along axis of the largest rotational constant) $\Delta v = \Delta A J_{max}^2$, where ΔA is the difference of the rotational constant A in the lower and upper vibrational state, and J_{max} is the rotational quantum number of the set of rotational states with the largest relative Boltzmann factor. $J_{max} \sim 32$ and $|\Delta A| \sim 0.001$ cm^{-1}, hence $\Delta v \sim 1$ cm^{-1}.

The differences are small but merit attention, in particular as the $1 \to 2$, $3 \to 4$, and $4 \to 5$ bands are split into two components by about 0.9, 0.5, and 0.8 cm^{-1}, respectively. Attempts to reproduce the magnitudes of these splittings by means of band contour computations of the rotation–vibration interaction and the centrifugal distortion were not entirely successful.[14]

f. Diketene, β-Propiolactone, 1,3-Tetrafluoro-1,3-dithietane, 3-Methylene-oxetane

The formulas of these molecules are, in the above sequence,

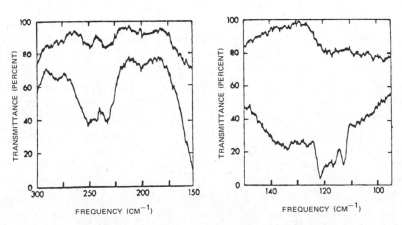

(I)   **(II)**   **(III)**   **(IV)**

Ring-puckering motions were assigned to observed far-infrared bands (1) for diketene (**I**) at 150 cm^{-1}, presumably with a nearly harmonic potential function[19,20] since no "hot" bands were detected at the shorter-wavelength side of the 150 cm^{-1} band; (2) for β-propiolactone (**II**) at 191 cm^{-1}, the potential function possessing probably no barrier to inversion;[20] (3) for 1,3-tetrafluoro-1,3-dithietane (**III**), at 60 cm^{-1}, a harmonic or near-harmonic

Fig. 7.3   Far-infrared spectrum of gaseous 3-methyleneoxetane. The pressure of the sample for the spectrum on the left was 10 mm Hg and a path length of 8.2 m. The pressure of the sample for the spectrum on the right was 1 mm Hg and a path length of 8.2 m. [J. R. Durig and A. C. Morrissey, *J. Chem. Phys.* **45**, 1269 (1966).]

potential,[21] again since no hot bands were observed; and (4) for 3-methylene-oxetane (IV) a near-harmonic potential function, based on three sharp but partly overlapping bands, at 122 cm^{-1} ($0 \rightarrow 1$), 117 cm^{-1} ($1 \rightarrow 2$), 113.5 cm^{-1} ($2 \rightarrow 3$), and an overtone at 241 cm^{-1} ($0 \rightarrow 2$).[22] The corresponding spectrum for the last compound is shown in Fig. 7.3. The three sharp Q branches at 122, 117, and 113.5 cm^{-1} on the background of the R and P branches are clearly discernible; the overtone $0 \rightarrow 2$ at 241 ($122 + 117 = 239$) cm^{-1} has a less distinct band envelope, as we would expect from the relatively larger difference between the rotational constants of the two vibrational levels involved in the transition. The conclusions drawn above for diketene (I) and β-propiolactone (II) are less clear-cut. For instance, there is a rather large discrepancy between the frequencies of the ring-puckering mode of β-propiolactone (II) as obtained from the far-infrared (191 cm^{-1}) and from two independent microwave intensity determinations (average $139 \pm 23 \text{ cm}^{-1}$). We refer the interested reader to the original articles for a more detailed discussion of the experimental conditions and illustrations of the spectra.

D. POTENTIAL FUNCTION AND RING CONFORMATION

We may now relate the shape of the potential functions of the molecules described above to their equilibrium ring conformations. As mentioned in the Introduction, an order of magnitude estimate of the forces which tend to bend the ring can be obtained from the hydrogen–hydrogen repulsion energies between nonbonded hydrogen atoms; that is, hydrogen atoms which are not bonded to the same carbon atom.* In a more concise manner we may designate interactions between hydrogen atoms (or fluorine atoms) on adjacent carbon atoms as "1,2-interactions," and interactions between the atoms attached to carbon atoms along the ring diagonals as "1,3-interactions." In trimethylene oxide, trimethylene sulfide, and cyclobutanone there are eight 1,2-interactions (see Fig. 7.4) and four 1,3-interactions; in β-propiolactone there are four 1,2-interactions; in 3-methyleneoxetane there are essentially eight 1,3-interactions; in diketene there are essentially four 1,3-interactions; and in 1,3-tetrafluoro-1,3-dithietane (III) there are four 1,3-interactions of fluorine. We might therefore expect that these repulsion terms would tend to bend the ring in trimethylene oxide, trimethylene sulfide, cyclobutanone, and perhaps less so in β-propiolactone, since the 1,2-interactions are stronger than the 1,3-interactions (shorter H····H distance).

---

* Repulsion energies due to interactions of the electrons in the bond orbitals are neglected [see, for instance, E. B. Wilson, Jr., *Adv. Chem. Phys.* **2**, 391 (1959)].

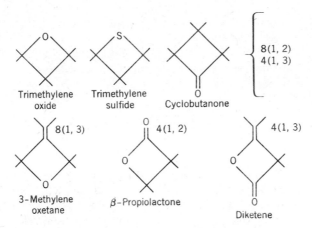

Fig. 7.4 Schematic rendering of the structure of various four-membered ring molecules. The carbon and hydrogen atoms are not further indicated; they are situated at the intersection and ends of the bonds, respectively. The designations (1, 2) and (1, 3) refer to the interaction between nonbonded hydrogen atoms which are attached to carbon atoms on either end of a C–C (1,2-interaction) and C–C–C (1,3-interaction) bond. The numerals give the number of interactions present.

Since all these molecules except trimethylene sulfide have planar rings, the energy required to decrease the angles between any three adjacent ring atoms must exceed the energy which is gained by the decreased hydrogen–hydrogen repulsion in the bent conformation. The reason for a bent ring in trimethylene sulfide, in comparison to trimethylene oxide, is to be sought in the fact that angles with a central sulfur atom are usually a few degrees smaller than those with a central oxygen atom, and that the carbon–sulfur bond is longer than the carbon–oxygen bond.[18] For a given dihedral angle there is less ring strain in trimethylene sulfide than in trimethylene oxide; the non-bonded interactions are therefore readily minimized by twisting the ring into a nonplanar conformation.

As far as cyclobutanone is concerned, its planarity is probably a consequence of the relatively large energy that would be required to change the degree of hybridization on the carbonyl group carbon atom if its angle with the adjacent methylene group carbon atoms were to be changed.

The lack of a fourth-order term in the potential of the ring-puckering mode of 3-methyleneoxetane is interesting. The degree of decrease of the frequencies of successive $\Delta n = 1$ transitions would then indicate that the anharmonicity in the potential is caused by cubic terms. In analogy to what we have discussed above on the example of cyclobutanone, we can expect that in 3-methyleneoxetane the hybridization of the carbon atom carrying the methylene group resists a change in the ring angle. The planarity of the molecule can be rationalized from the absence of the stronger 1,2-hydrogen–hydrogen interactions.

In conclusion, it seems apparent that the interplay of the weak forces which determine the ring conformation of these and similar compounds is far from being perfectly understood.

E. Nonsymmetrical Ring-Puckering Potential Functions

a. Introduction

Unsymmetric potentials (that is, in our case, double minimum potentials with potential wells of unequal depth) have to be employed if the inversion of the coordinate of the motion about the position $\xi = 0$ leads to two different energies, $E(\xi) \neq E(-\xi)$. The characteristics of ring-puckering motions with unsymmetric potentials are quite different from those with symmetric potentials and we will therefore discuss them briefly. To this effect, Fig. 7.5 shows the computed energy levels and amplitudes of the symmetric potential

$$V_s(\xi) = -6\xi^2 + \xi^4 \tag{7.11}$$

and the unsymmetric potential

$$V_u(\xi) = -6\xi^2 + (\tfrac{1}{2})\xi^3 + \xi^4. \tag{7.12}$$

The matrix elements were set up in the harmonic oscillator representation, and the size of the energy matrix was 20×20. The eigenvalues are in units of $(\tfrac{1}{2})\hbar\beta$; the coordinate ξ is dimensionless and indicates the deviation from planarity as described in the section containing the introductory theoretical treatment. We notice that the introduction of the odd power in the displacement coordinate, which destroys the symmetry of the potential, causes a nearly complete decoupling between the two potential wells for all levels below the top of the barrier. That is, only near the top of the barrier (and, of course, above it) is the amplitude of the motion found to be distributed appreciably over the whole range of the coordinate. Below the top of the barrier the amplitude alternates between the left and the right well with increasing vibrational quantum number and, as Fig. 7.5 shows, this alternation is still discernible in the two levels *above* the top of the barrier; more explicitly, the amplitude distribution is greater on the right (left) side of energy level $v = 3$ ($v = 4$) than on its left (right) side. As a consequence, the transitions for the levels below the top of the barrier obey the selection rules "left well–left well" and "right well–right well"; the even–even and

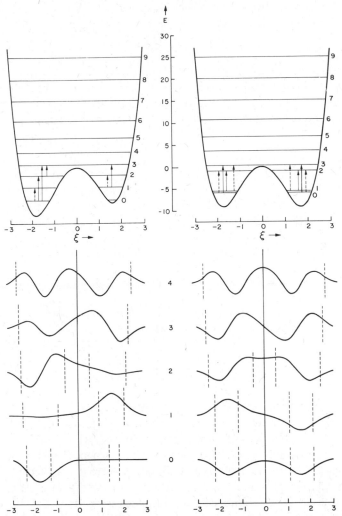

Fig. 7.5 Energies (in units of $(\frac{1}{2})\hbar\beta$) of the first 10 energy levels and amplitudes of the wave equation of the first five energy levels of the potential functions $V(\xi) = -6\xi^2 + (\frac{1}{2})\xi^3 + \xi^4$ (left side) and $V(\xi) = -6\xi^2 + \xi^4$. The dotted lines give the classical turning points (potential energy = total energy). Notice that the effect of the unsymmetric term $(\frac{1}{2})\xi^3$ becomes negligible above the fourth energy level. The solid arrows denote some of the allowed, the dotted arrows some of the forbidden transitions. Since there are no symmetry restrictions on the antisymmetric potential, a mere inspection of the amplitudes of the wave equation (lower left part of the figure) shows all transitions that will be forbidden because of a vanishing amplitude in one (or both) of the energy levels of a transition: in the left well of the potential these are the transitions that involve the level $v = 1$. In the right well of the potential the intensities of transitions that involve the levels 0 and 2 will be forbidden.

odd–odd transitions, which are forbidden for a symmetric potential, may become the most strongly active for the unsymmetric potential.*

b. Bromocyclobutane

Substitution of a hydrogen atom by halogen in cyclobutane makes it possible for the molecule, through the motion of its ring-puckering mode, to assume conformations which are not related to each other by permutations of the same kind of atoms. This is shown in Fig. 7.6. The two conformers

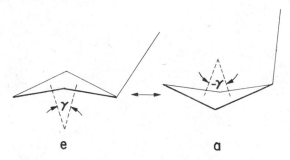

e   a

Fig. 7.6   Equatorial (e) and axial (a) conformations of cyclobutyl-X. The angle γ is the dihedral angle. The molecule belongs to point group symmetry C_s. The ring-puckering motion belongs to the symmetry species a'. [W. G. Rothschild, *J. Chem. Phys.* **45**, 1214 (1966).]

are designated "equatorial" (e) and "axial" (a) in a loose analogy to the nomenclature of the cyclohexyl compounds. The e and a conformers constitute the two limiting cases of all the conformers that the molecule might possess. From microwave spectrum intensity measurements, the two lowest transitions of the ring-puckering mode were obtained at 120 ± 25 and $251 \pm 56 \, \text{cm}^{-1}$. The equilibrium molecular structure of the molecule was found to be strongly bent in the equatorial (e) conformation; rotational transitions based on an axial conformation were not detected.[10] Observation of a relatively intense band at $283 \, \text{cm}^{-1}$ in the far-infrared and failure to find the predicted band of the ring-puckering $0 \to 1$ transition near $120 \, \text{cm}^{-1}$ led then, in accordance with the characteristics of unsymmetric

---

* We may say that, in essence, the intersection of two *equal* potential wells with a range of, say, n levels each, yields a symmetric double minimum potential with 2n levels—which for argument's sake shall all be below the top of the barrier—because of the removal of the twofold degeneracy with respect to inversion of the coordinate by exchange interaction. The intersection of two *unequal* potentials, with n levels each, also yields 2n levels—which shall all be below the top of the barrier—but here there never was any degeneracy with respect to inversion of the coordinate, hence there is no exchange interaction between the left and right side of the combined level and, as a consequence, no delocalization of the amplitude over the whole range of the abscissa. [See also J. Brickmann and H. Zimmermann, *Ber. Bunsenges. Phys. Chem.* **70**, 521 (1966).]

potentials with two minima as discussed in Section E.a, to a tentative assignment of the 283 cm^{-1} band to the lowest even–even $(0 \rightarrow 2)$ transition of the equatorial conformer. The equatorial conformer was believed to be in equilibrium with a less stable axial conformer as shown schematically in Fig. 7.6.[23] This assignment was incorrect since, firstly, the mode at 283 cm^{-1} turned out to possess a depolarized Raman shift and was hence of symmetry species a'' (see Fig. 7.6). Secondly, a weak band of symmetry species a' was subsequently detected in the Raman spectrum of the liquid at 155 cm^{-1} and in the far-infrared spectrum of the vapor at 144 cm^{-1}. Finally, the rotational constants of the first three vibrational levels of the

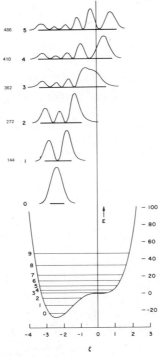

Fig. 7.7 Energy levels and probability distribution of the ring-puckering mode of bromo-cyclobutane. The potential is $V(\xi) = \frac{1}{2}\hbar\beta(-2\xi^2 + 6\xi^3 + 2\xi^4)$, in terms of the dimensionless coordinate ξ. The planar ring conformation is at $\xi = 0$. Negative values of ξ correspond to positive dihedral angles (equatorial conformation). The energy levels are in units of $\frac{1}{2}\hbar\beta$. The heavy horizontal lines denote the classical region. The coefficients of $V(\xi)$ were adjusted to fit the observed and inferred frequencies of the $0 \rightarrow 1$ and $0 \rightarrow 2$ transitions (see Ref. 25 and Fig. 7.9); $\frac{1}{2}\hbar\beta = 15.212$ cm^{-1}. The numbers on the probability distributions of the amplitudes (*upper part* of the figure) denote the quantum number of the level (0 to 5) and its energy (in cm^{-1}) above the ground level. [W. G. Rothschild, *J. Chem. Phys.* **45**, 1214 (1966).]

ring-puckering mode were shown to belong to a ring conformation which monotonically becomes more planar with increasing vibrational excitation rather than to alternate between *e* and *a* conformations.[24] A potential with two distinct wells of unequal depths (that is, the existence of an *axial* conformer) is thereby excluded.*

It was shown in a subsequent study of the temperature dependence of the infrared spectrum of bromocyclobutane (and chlorocyclobutane) that the molecule *does* exist as two conformers. One is the *equatorial* conformer mentioned above and the other, of excess energy of approximately 1 kcal mole^{-1} (or 350 cm^{-1}), is in a *planar* or near-planar conformation.[25] Figure 7.7 shows the proposed potential function and the probability distribution of the amplitude, $\psi_n^2(\xi)$, of the ring-puckering mode of bromocyclobutane. The constants of the potential function were adjusted to reflect the available observations.[10,24,25] The problem was set up in the harmonic oscillator representation. The ring-puckering mode was observed only as a weak and broad band with no pronounced Q branches; the potential function is therefore only a very approximate representation of the actual problem. The concept of conformer is to be understood here in terms of two sets of molecular populations of the ring-puckering mode which are distributed between two different carbon ring conformations such that the three lowest

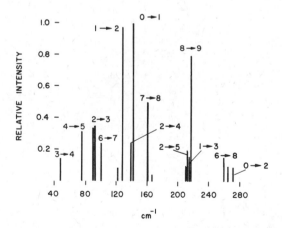

Fig. 7.8 Computed spectrum of the ring-puckering motion of bromocyclobutane between 40 and 280 cm^{-1}. The intensities have been corrected for induced emission. The numbers at the lines indicate the vibrational quantum numbers for the lower and upper energy levels of the transition. Strong transitions between high levels may be expected at frequencies above 280 cm^{-1}. They have not been entered here because of the uncertainties in the assumptions (see text and Figs. 7.9a and b). [W. G. Rothschild, *J. Chem. Phys.* **45**, 1214 (1966).]

* For a more detailed presentation of the pertinent arguments Ref. 24 should be consulted.

levels of the ring-puckering mode belong to an essentially (on the average) equatorial ring conformation whereas the remaining higher levels belong, on the average, to a planar or near-planar conformation of the ring. At room temperature the equatorial set contains a relative population of 0.8, at 170°C it contains 0.7, and at −185°C its content is essentially unity. The two sets of conformations do not overlap and there is, of course, no tunnelling between them.*

The computed spectrum of the ring-puckering mode is shown in Fig. 7.8. Its main absorption agrees, in essence, with the observations[20,26] shown in Fig. 7.9, although the details lack experimental verification because of very weak absorption. In particular, this concerns the position of the weaker long-wavelength wing centered near 80 cm^{-1} (which contains the transitions between levels near the top of the potential hump). Note the predicted weakness of the $n \rightarrow n + 2$ overtones for the lowest values of n. In conclusion, we give the observed and computed dihedral angles for the first six levels of the ring-puckering mode (see Table 7.2).

TABLE 7.2*

COMPARISON BETWEEN OBSERVED AND COMPUTED AVERAGE
DIHEDRAL ANGLES OF BROMOCYCLOBUTANE AS A FUNCTION OF
VIBRATIONAL EXCITATION OF THE RING-PUCKERING MOTION

Vibrational quantum number n	$\langle\psi_n\|\xi\|\psi_n\rangle/\langle\psi_n\|\psi_n\rangle$	$\langle\psi_n\|\gamma\|\psi_n\rangle$	
		Computed	Observed[†]
0	−2.366	(0.511 = 29.3°)	29.3°
1	−2.147	0.464 = 26.6°	27.0°
2	−1.780	0.384 = 22.0°	24.6°
3	−0.753	0.163 = 9.3°	—
4	−0.451	0.0974 = 5.6°	—
5	−0.456	0.0985 = 5.6°	—

* W. G. Rothschild, *J. Chem. Phys.* **45**, 1214 (1966).
[†] See Ref. 24.

c. Chlorocyclobutane

The characteristics of the ring-puckering mode of chlorocyclobutane are quite analogous to those of the bromo compound.[25–27] The fundamental was observed at 158.5 cm^{-1} in the far-infrared spectrum of the vapor. From

---

* The effect of the two different conformations on some higher-frequency transitions in the vibrational spectrum is discussed in Chapter 5.

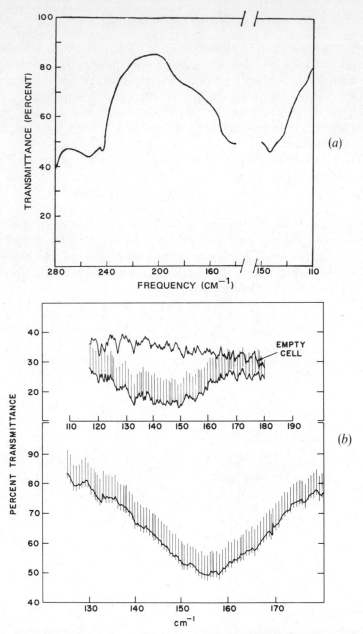

Fig. 7.9 Far-infrared spectrum of bromocyclobutane. (*a*) 16 mm Hg, 8.2-m path length, polypropylene windows, Beckman IR-11. [J. R. Durig, W. H. Green, and N. C. Hammond, *J. Phys. Chem.* **70**, 1989 (1966), by permission of American Chemical Society.] (*b*) (Upper part): 6-m path length, cell empty and filled to 50 mm Hg, slit width ~ 5 cm^{-1}, Perkin-Elmer 301; by courtesy of D. W. Scott, Bureau of Mines, Bartlesville, Oklahoma. (Lower part): 1-mm liquid layer in polyethylene cell, slit width 2 cm^{-1}, Perkin-Elmer 301. [W. G. Rothschild, *J. Chem. Phys.* **45**, 3599 (1966).]

the microwave spectrum the separation of successive transitions $1 \to 2$, $2 \to 3$ was obtained to $140 \pm 20 \, \text{cm}^{-1}$. It is very unlikely, in view of the characteristics of the potential function discussed above, that a weak band at $303 \, \text{cm}^{-1}$ is the $0 \to 2$ overtone of the ring-puckering mode (Durig et al., see Ref. 20), particularly since there is no probability for Fermi resonance with an adjacent transition. The observed effective dihedral angle of the chlorocyclobutane ring in the first three states of the ring-puckering mode is 20, 17, and 14°, respectively.[27] The flattening of the ring with increasing vibrational excitation is therefore more rapid in the chloro- than in the bromo-derivative.

d. Deuterated Bromo- and Chlorocyclobutanes

Durig and Green[28] and Durig and Morrissey[29] have studied the far-infrared spectra of a series of deuterated species of bromo- and chlorocyclo-butanes. Their work confirms the conformational behavior in the cyclobutyl halide series as proposed by Rothschild.[24–26]

e. Fluorocyclobutane

No far-infrared observations of the ring-puckering mode of fluorocyclo-butane have as yet been reported. The spacing between successive energy levels of the ring-puckering mode has been determined from microwave intensity measurements to $140 \pm 20 \, \text{cm}^{-1}$ (up to level $n = 3$). The corres-ponding effective dihedral angle of the first four energy levels amounts to 20, 15.8, 12.3, and 8.6°, respectively.[27]

The existence of a planar conformer rather than an axial conformer as the less stable species of these monohalides of cyclobutane, an unusual phenomenon which is not predicted by computations of nonbonded inter-actions,* has also been deduced by Lambert and Roberts[30] in their nuclear magnetic resonance studies of fluorinated substituted cyclobutanes from the temperature variations of the geminal fluorine–fluorine chemical shifts.

f. Conclusions

The conformational behavior of the cyclobutane derivatives is not only of great interest to the spectroscopist and the theoretical chemist, but also has important implications to reaction kinetics, as has been pointed out by

---

* The computations of the nonbonded interactions (van der Waals and London forces) between all atoms predict that the planar conformer has an excess energy of about 0.3 kcal mole^{-1} ($= 117 \, \text{cm}^{-1}$) with respect to the equatorial conformer. The equatorial and axial conformers are—according to these computations—of approximately equal energy content since 1,2-interactions predominate and these 1,2-interactions are nearly of identical magnitude in the a and e conformers. The contribution of the 1,3-hydrogen–bromine interaction, which is greatest in the a conformer, is relatively small for all dihedral angles on account of the large hydrogen–bromine distance.

Lillien and Doughty.[31] In the past scant attention was paid to this factor, partly because the conformational behavior of cyclobutane derivatives was not understood, partly because it was assumed that the nonbonded interactions could be described in terms of those encountered in the well-known cyclohexane compounds. Lately a considerable amount of work has appeared in the literature of organic chemistry on solvolysis studies* of cyclobutane derivatives which takes into account the effect of conformational behavior on the chemical reactivity in these compounds.[32]

F. PSEUDOROTATION OF FIVE-MEMBERED RINGS

a. Introduction

As we mentioned in the Introduction to this chapter, the ring-puckering vibrations of molecules with five-membered rings, such as cyclopentane, tetrahydrofuran

$$\overline{(CH_2 \cdot CH_2 \cdot CH_2 \cdot CH_2 \cdot O)},$$

pyrrolidine

$$\overline{(CH_2 \cdot CH_2 \cdot CH_2 \cdot CH_2 \cdot NH)},$$

and tetrahydrothiophene

$$\overline{(CH_2 \cdot CH_2 \cdot CH_2 \cdot CH_2 \cdot SH)},$$

can be treated best in terms of a "pseudorotation." For instance, if the equilibrium conformation of the cyclopentane ring were planar (point group D_{5h}), the out-of-plane or ring-puckering mode, which essentially involves a movement of the methylene groups in a direction normal to the plane of the ring, would belong to a doubly degenerate symmetry species (e_2''). Although the general appearance of the infrared spectrum of cyclopentane can be described by a planar model,[33,34] there is a sufficiently great departure from D_{5h} selection rules[33] and, particularly, a divergence between the measured and calculated thermodynamic functions[35,36] to indicate that the equilibrium conformation of the carbon ring is puckered.

To describe the theory of pseudorotation in brief, it is advantageous to discuss, at first, the possible (doubly degenerate) out-of-plane modes of the molecule in the absence of pseudorotation. That is, consider the equilibrium carbon ring structure to be a planar, regular pentagon (D_{5h}). If we designate the two degenerate normal coordinates of the degenerate out-of-plane mode v_i by Q_{ia} and Q_{ib}, we require that the potential energy $V_i = (\frac{1}{2})\lambda_i(Q_{ia}^2 + Q_{ib}^2)$ be invariant under rotations about the fivefold symmetry axis of the molecule.

* Cyclobutane compounds undergo some interesting and important rearrangement reactions.

The transformation equations are the well-known relations[37]

$$Q'_{ia} = Q_{ia} \cos \beta + Q_{ib} \sin \beta$$
$$Q'_{ib} = -Q_{ia} \sin \beta + Q_{ib} \cos \beta,$$

(7.13)

where the angle β can assume the values

$$\beta = \pm \frac{2\pi}{p} l \, ; \, l = 1, 2, 3, \cdots, p - 1$$

(7.14)

with $p = 5$ in our case. Evidently, $l = 1$ is equivalent to $l = 4$ and $l = 2$ is equivalent to $l = 3$. Therefore, we have two species of twofold degenerate out-of-plane vibrations, e''_1 and e''_2.

We first consider the mode $l = 2$; that is, the mode e''_2. According to the above, $\beta = \pm 2(2\pi/5)$. The displacements of successive carbon nuclei $j = 1, 2, \cdots, 5$ of the ring during e''_2 are found by application of the transformation described by Eq. 7.13. Since the displacement vectors of the two mutually degenerate vibrations of e''_2, v_{ia} and v_{ib}, are parallel to the fivefold axis, we must make them orthogonal in order to obtain the correct picture of the normal mode. When this is done,[38] the out-of-plane displacement of successive carbon atoms $j = 1, 2, \cdots, 5$ is given by $z_j^{(a)} = q \cos 2[(2\pi/5)j]$ for v_{ia} and by $z_j^{(b)} = -q \sin 2[(2\pi/5)j]$ for v_{ib}. (The displacement of atom number 5 is arbitrarily set equal to 0 in v_{ib} and equal to q in v_{ia}.) Figure 7.10 shows a picture of the mode e''_2.

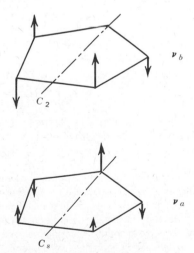

Fig. 7.10 The doubly degenerate out-of-plane fundamental e''_2 of a regular pentagon (D_{5h}). The symbols C_2 and C_s indicate a twofold symmetry axis and a plane of symmetry, respectively. [G. Herzberg, *Infrared and Raman Spectra of Polyatomic Molecules*, Copyright 1962, D. Van Nostrand Company, Inc., Princeton, New Jersey.]

If the same operations are applied to obtain the displacement vectors of e_1'', it turns out[38] that these represent a rotation of the molecule about two mutually perpendicular axes located in the plane of the molecule (non-genuine vibrations). Thus, we are left with only one out-of-plane fundamental of the regular pentagon, e_2''.

The step to the phenomenon of pseudorotation is now easily made: (1) Inspection of Fig. 7.10 shows that we might consider each of the two structures to be a limiting equilibrium conformation in the case of relatively large amplitudes of vibration and a ready conversion between the two conformations (low or no potential barrier). Each of these two limiting conformations retains some symmetry; that is, v_b belongs to point group C_2 and v_a belongs to point group C_s. (2) We now allow all intermediate conformations of the ring by permitting the position of the maximum out-of-plane amplitude of the displacement to move around the ring. This is accomplished by casting the above expressions for z_j into one (normalized) expression containing a phase angle φ, which describes the point of maximum amplitude :[36,39]

$$z_j = (\tfrac{2}{5})^{1/2} q \cos 2\left(\frac{2\pi}{5} j + \varphi\right). \qquad j = 1, 2, \cdots, 5 \qquad (7.15)$$

Figure 7.11 shows this for a phase angle of $\pi/40$. The resulting conformation

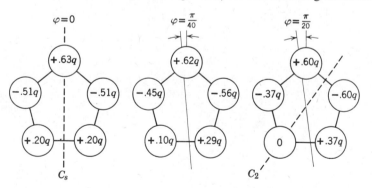

Fig. 7.11 Limiting and intermediate conformations during the ring-puckering (out-of-plane) motion in cyclopentane. Displacements perpendicular to the plane of the paper are given in the circles. The angle φ is the phase angle of the maximum amplitude of the puckering, as given by Eq. 7.15. The dotted line in the left part of the figure indicates the intersection of the symmetry plane with the plane of the paper; this limiting conformation corresponds to a phase angle of $\varphi = 0$. The dotted line in the right part denotes the twofold symmetry axis of the other limiting conformation of the ring (phase angle $\varphi = \pi/20$). The out-of-plane displacements for the example in the middle of the figure are those of an intermediate conformation that possesses no symmetry elements and has $\varphi = \pi/40$ (see also Fig. 7.10). [J. E. Kilpatrick, K. S. Pitzer, and R. Spitzer, *J. Am. Chem. Soc.* **69**, 2483 (1947), by permission of American Chemical Society.]

possesses no symmetry elements. For comparison's sake, it is sandwiched between the two limiting conformations of Fig. 7.10.

Kilpatrick et al.[36] have calculated that the potential energy of the methylene out-of-plane motion in cyclopentane is essentially independent of the phase angle φ; the picture of the displacement is therefore indeed one in which its maximum amplitude moves around the ring. In other words, the normal coordinate of the doubly degenerate mode of the out-of-plane motion of the methylene groups for a planar conformation of point group D_{5h} is now expressed by one nondegenerate normal coordinate (q) describing the oscillatory amplitude of the motion and by a second coordinate (φ) of zero-vibration frequency* describing the phase of the oscillatory amplitude. Since this latter motion can be specified by plane rotorlike solutions and because there is no angular momentum connected with the motion, the phenomenon has been termed pseudorotation. [When the phase moves around the ring, the actual displacement of the methylene groups, $d\mathbf{r}/dt$, is in a direction normal to the direction of the movement of the phase; that is, parallel to the axis of rotation $\mathbf{r}$ of the movement of the phase. Therefore, $\mathbf{L} = m(\mathbf{r} \times d\mathbf{r}/dt) = 0$ (to first order).]

The concept of pseudorotation has met with resistance by some authors. For instance, Miller and Inskeep assumed a nonplanar structure for cyclopentane and assigned two genuine low-frequency modes at 207 and 283 cm^{-1}.[40] The main support for the occurrence of pseudorotation in cyclopentane and related molecules comes from thermodynamic data.[35,36,39,41] A paper on direct far-infrared spectroscopic observations of pseudorotation of five-membered rings has recently appeared in the literature (see below). It supersedes reports of merely indirect evidence, such as the departure of the selection rules from those of a planar molecule and the fuzzy appearance of many pure rotational lines in the Raman spectrum.[41] It should be noted that although both these indirect observations offer strong indications that the ring is puckered, they do not necessarily prove that it undergoes pseudorotation. Strictly speaking, cyclopentane is not a symmetric top since the ring conformation varies between C_s and C_2 (see Figs. 7.10 and 7.11). The effective moments of inertia in the plane of the molecule (that is, I_A and I_B averaged over the vibrational modes) are therefore slightly different and thus cause removal of the symmetric top degeneracy. This may lead to a broadening or even slight splitting of the rotational levels.[†] Because the phase of the puckering motion travels much faster around the ring than the molecule turns end-over-end, the fuzziness of the

---

* This is the coordinate that appears only in exponential expressions in the wave function. See Ref. 11, p. 77, Ref. 42, p. 108, and the Introduction to Ref. 43.

† See, for instance, C. H. Townes and A. L. Schawlow, Microwave Spectroscopy, McGraw-Hill, New York, 1955, Chapters 3 and 4.

pure rotational Raman spectrum definitely indicates that the molecule is puckered, but it does not distinguish between a deviation from D_{5h} symmetry as C_2 or as C_s, or as a time average between C_2 and C_s (pseudorotation).

b. Cyclopentane

The Schrödinger equations of the ring-puckering motion can be written in the polar coordinates q (radial) and φ (angular), which we have introduced above, as follows:[36,39]

$$-\frac{\hbar^2}{2mq}\frac{\partial}{\partial q}\left(q\frac{\partial \psi}{\partial q}\right) + V(q)\psi = E_q\psi, \qquad (7.16)$$

$$-\frac{\hbar^2}{8mq_0{}^2}\frac{\partial^2 \chi}{\partial \varphi^2} = E_\varphi\chi, \qquad (7.17)$$

following the procedure of separation of variables[42] and by noting that, firstly, the angular variable is essentially 2φ (see Eq. 7.15), the differential therefore $2\partial\varphi$, and the boundary condition consequently $\chi(\varphi) = \chi(\varphi + \pi)$. Secondly, the equilibrium value of q, q_0, has replaced the variable q in Eq. 7.17. The meaningful solutions are therefore restricted to small values of $q - q_0$. The general solution of Eq. 7.17 is

$$\chi(\varphi) = \frac{1}{\sqrt{\pi}}\exp(i\sqrt{\alpha}\varphi), \qquad \alpha = (8mq_0{}^2/\hbar^2)E_\varphi. \qquad (7.18)$$

Because of the above boundary conditions on φ, $\sqrt{\alpha}$ must equal $2n$, where n is any integer or zero. It follows then that the angular energy eigenvalues are

$$E_\varphi = n^2\hbar^2/2mq_0{}^2, \qquad n = 0, \pm 1, \pm 2, \cdots \qquad (7.19)$$

If one looks upon $mq_0{}^2$ as a moment of inertia, the similarity of the solutions Eq. 7.19 to those of a free, planar rotor is obvious.

To obtain the solutions of the oscillatory part of the problem (that is, the solutions of Eq. 7.16), we set $\psi(q) = q^{-1/2}f(q)$. Equation 7.16 is then transformed into

$$-\frac{\hbar^2}{2m}\left\{\frac{d^2}{dq^2}f(q) + (\tfrac{1}{4})q^{-2}f(q)\right\} + V(q)f(q) = E_q f(q). \qquad (7.20)$$

If the potential function $V(q)$ is now assumed to be harmonic in the displacement coordinate $q - q_0$ and the factor $(\tfrac{1}{4})q^{-2}$ is approximated by $(\tfrac{1}{4})q_0^{-2}$, the solutions of Eq. 7.20 are those of the harmonic oscillator with the zero-point energy $h\nu/2 - \hbar^2/8mq_0{}^2$:

$$E_q = (v + \tfrac{1}{2})h\nu - \hbar^2/8mq_0{}^2. \qquad (7.21)$$

The solutions of Eqs. 7.19 and 7.21 are only valid for small displacements of q; that is, $q \sim q_0$.

The selection rules for the angular transitions are obtained in the usual fashion from the matrix element of the direction cosine of the dipole moment between the levels of the transition,

$$\langle \chi | \cos 2\varphi | \chi' \rangle = \text{constant} \times \int_0^\pi e^{-2i\varphi n} \cos 2\varphi e^{2in'\varphi} \, d\varphi. \qquad (7.22)$$

The integral is seen to vanish unless $n' = n \pm 1$. Since cyclopentane does not possess a permanent dipole moment, pure pseudorotational transitions cannot be observed. A vibrational transition at 283 cm^{-1} has been assigned to the transition $v = 0 \rightarrow 1$ of the radial ring-puckering oscillation.[36,43]

c. Tetrahydrofuran

The molecule possesses a permanent dipole moment and therefore should exhibit pure pseudorotational transitions. The far-infrared spectrum between 20 and 95 cm^{-1}, obtained with grating spectrometers at two different laboratories,[44] is shown in Fig. 7.12. The observed frequencies

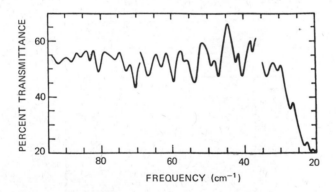

Fig. 7.12 Far-infrared spectrum of tetrahydrofuran redrawn from single-beam sample spectra and instrumental base spectra. The discontinuities occur where gratings and filters were changed. Each continuous span was run with constant slit widths with a resulting decrease in spectral slit width from the high-frequency to the low-frequency limits. A pressure of 40 mm Hg in a cell of 78.4-cm length was used for all spectra. [W. J. Lafferty, D. W. Robinson, R. V. St. Louis, J. W. Russel, and H. L. Strauss, *J. Chem. Phys.* **42**, 2915 (1965).]

of the transitions can be arranged in the sequence $\Delta E = (h^2/8\pi^2 m q_0^2) \times (2n + 1)$, for $n = 3$ to 14 (see Eqs. 7.19 and 7.22). This is shown in Fig. 7.13. A second such sequence of observed transitions, but of less intensity, is shown in Fig. 7.14. The two sets of transitions were assigned to transitions $\Delta n = \pm 1$ between angular levels based on the ground and on the first

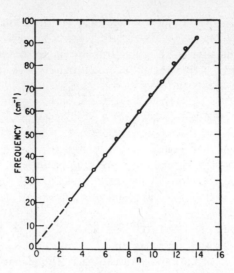

Fig. 7.13   Frequencies in cm^{-1} of far-infrared absorption lines of tetrahydrofuran for the $v = 0$ radial state *vs* the angular quantum number *n*. [W. J. Lafferty, D. W. Robinson, R. V. St. Louis, J. W. Russel, and H. L. Strauss, *J. Chem. Phys.* **42**, 2915 (1965).]

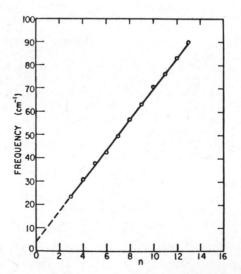

Fig. 7.14   Frequencies in cm^{-1} of far-infrared absorption lines of tetrahydrofuran for the $v = 1$ radial state *vs* the angular quantum number *n*. [W. J. Lafferty, D. W. Robinson, R. V. St. Louis, J. W. Russel, and H. L. Strauss, *J. Chem. Phys.* **42**, 2915 (1965).]

excited radial state, respectively ($v = 0$, Fig. 7.13; $v = 1$, Fig. 7.14). From the slope of the lines a value of the "effective moment of inertia" of the pseudorotation, $mq_0^2 = 8.56 \pm 0.13 \times 10^{-40}$ g cm^2 ($v = 0$) and $8.48 \pm 0.15 \times 10^{-40}$ g cm^2 ($v = 1$) can be calculated by the least squares method. The result agrees well with that from the thermodynamic measurements, 8×10^{-40} g cm^2. The value of mq_0^2 may also be obtained from the intercepts by extrapolation of the straight lines to $n = 0$. The value of mq_0^2 obtained from the intercepts agrees, within experimental error, with that calculated from the slope only for the series based on the ground vibrational state ($v = 0$). According to Lafferty et al.,[44] this may point to an energy dependence of the pseudorotational levels of $an^2 + bn$. However, no convincing explanation why such a linear term in the quantum number n should appear could be construed from various assumptions, such as a barrier to pseudorotation, ellipticity of the potential function, or tunnelling through a central barrier of the planar conformation. There are, indeed, some indications that the assignments of the second series of pseudorotational transitions will have to be modified as higher-resolution data become available.[45] We shall treat this further below, after we have discussed some radial transitions.

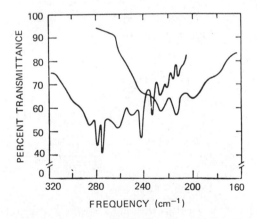

Fig. 7.15 Far-infrared spectrum of tetrahydrofuran and tetrahydrofuran-d$_8$ redrawn from single-beam sample spectra and instrumental base spectra. A pressure of 100 mm Hg in a cell of 30-cm path length was used. [W. J. Lafferty, D. W. Robinson, R. V. St. Louis, J. W. Russel, and H. L. Strauss, *J. Chem. Phys.* **42**, 2915 (1965).]

Figure 7.15 shows the spectrum of the radial transition $v = 0 \to 1$ of tetrahydrofuran and of tetrahydrofuran-d$_8$. The observed frequency of the maximum absorption of the protonated compound agrees well with (*1*) the value of 288 cm^{-1}, which was calculated using 1.3×10^5 dyne/cm for the force constant* and an effective reduced mass appropriately describing the motion of the methylene groups and the oxygen atom,[44] and with (2) the value of $276 \, \text{cm}^{-1}$ observed in the Raman spectrum of the compound.[46] The value for the deuterated compound, $207 \, \text{cm}^{-1}$, calculated using the same radial force constant and the appropriate averaged reduced mass for the CD_2 groups and the oxygen atom, also agrees with observations (about $215 \, \text{cm}^{-1}$, see Fig. 7.15). The fine structure of the radial fundamental, namely the $n = \pm 1$ transitions between the angular states, is seen fairly distinctly for the protonated but less so for the deuterated compound.

In the following we shall give some assignments which have recently been proposed for the pseudorotation of tetrahydrofuran by Greenhouse and Strauss[47] on the basis of new infrared spectra between 15 and $375 \, \text{cm}^{-1}$ with superior resolution ($0.3 \, \text{cm}^{-1}$). A modified Research and Industrial Instruments Company FS-520 Fourier spectrophotometer was used for these experiments. It turned out that the Q branches of the pure pseudo-rotational transitions are doublets split by Coriolis interaction. Both series of lines, represented in Figs. 7.13 and 7.14, thus originate from the $v = 0$ radial state and not, as previously proposed,[44] from the $v = 0$ and $v = 1$ levels, respectively. Transitions belonging to $v = 1$ are also observed in the improved spectra; their relative intensities agree within expectations ($v = 0 \rightarrow 1$ amounts to about $280 \, \text{cm}^{-1}$, see above). The spectrum and assignments for the frequency range 20–100 cm^{-1} are displayed in Fig. 7.16.

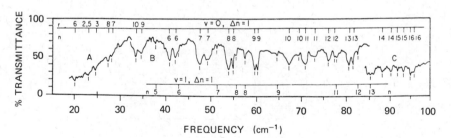

Fig. 7.16 Calculated and observed spectrum of tetrahydrofuran vapor between 20 and 100 cm^{-1}, taken with a modified Research and Industrial Instruments Company FS-520 Fourier spectrophotometer. Cell path length 1 m, high-density polyethylene windows. Vapor pressure 25 mm (A), 40 mm (B), and 101 mm (C). The radial v and angular quantum numbers n are given for the observed transitions $n \rightarrow n + 1$ for the energy regions unperturbed by the barrier to pseudorotation; the doubly degenerate n-levels (see Eq. 7.19) are slightly split by Coriolis interaction. The quantum number r gives the lower angular level of some observed angular transitions $r \rightarrow r + 2$ in the energy region perturbed by the barrier (see also Fig. 7.17). [J. A. Greenhouse and H. L. Strauss, *J. Chem. Phys.* **50**, 124 (1969).]

* This force constant was transferred from cyclopentane, see Ref. 39.

Greenhouse and Strauss furthermore showed that the separation of the radial and angular coordinates does not proceed as outlined in Section F.b since, in contrast to cyclopentane, (1) the ring atoms do not form a regular pentagon, (2) the effective mass of each ring atom is not the same, and (3) the potential energy can no longer be strictly considered independent of coordinate φ. Instead, the potential $V(q, \theta)$ is expanded in a Fourier series in θ and with the running index n (where θ is twice the angle φ defined in Section F.a):

$$V(q, \theta) = \sum_{n=0}^{n} V_{2n}(q) \cos 2n\theta$$

$$= V_0(q) + V_2(q) \cos 2\theta + V_4(q) \cos 4\theta + \cdots.$$

(7.23)

The angular equation is set up similar to Eq. 7.17, except that $V(q, \theta)$ averaged over the radial state v, $\langle v| V(q, \theta)|v \rangle$, is now introduced into the equivalent of Eq. 7.17 as a potential function (barrier to pseudorotation). The terms $\langle v| V_2(q)|v \rangle$ and $\langle v| V_4(q)|v \rangle$ are called the two- and fourfold barrier to pseudorotation for radial state v.

The solution of the angular equation is done as outlined in Section B, using pseudorotator wave equations of the form $\psi_n(\theta) \propto \exp(in\theta)$ (see also Eq. 7.18) as the basis set. Figure 7.17 shows the computed energy levels and

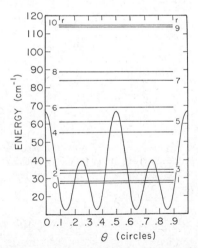

Fig. 7.17 Approximate barrier and energy levels of tetrahydrofuran for radial level $v = 0$. Barrier height is about $50\ cm^{-1}$. The energy levels far above the barrier are doubly degenerate (see Eq. 7.19). They are split by large amounts near the top of the barrier and forced again into nearly degenerate pairs near the bottom of the potential. The individual levels in the barrier region are designated by quantum number r (see also Fig. 7.16). [J. A. Greenhouse and H. L. Strauss, J. Chem. Phys. 50; 124 (1969).]

the approximate potential function; the splitting of the two lowest pairs of pseudorotational levels, 0.68 and 1.5 cm^{-1}, respectively, was obtained from the microwave spectrum of the molecule.[48]

For the sequence of pseudorotational transitions which are not perturbed by the barrier, we again obtain a linear plot of $\Delta E = (h^2/8\pi^2)(2n + 1) \times \langle v|1/mq_0^2|v \rangle = (2n + 1)B_v$. The data yield $B_0 = 3.19$ cm^{-1} (compare this to 8.56×10^{-40} g cm$^2 = 3.27$ cm^{-1} represented by Fig. 7.13) and $B_1 = 3.36$ cm^{-1}. With the help of B_0 and the splittings of the two lowest levels (see above), it is found that $\langle 0|V_2(q)|0 \rangle = 13.5$ cm^{-1} and $\langle 0|V_4(q)|0 \rangle = 20$ cm^{-1}.

d. 2,5-Dihydrofuran

The structure of this molecule is shown in Fig. 7.18. It is an example of a five-membered ring in which the pseudorotation is prevented by the double bond. The low-frequency ring-puckering mode, which mainly involves the out-of-plane motion of the oxygen atom, has been observed as a broad

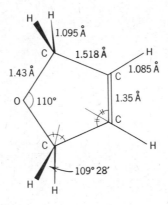

Fig. 7.18 The assumed molecular structure of 2,5-dihydrofuran. The plane of the paper bisects the H–C–H angles of the methylene groups; the =C—H bonds are thus in a staggered conformation to the C—H bonds of the methylene groups in the planar molecule. [T. Ueda and T. Shimanouchi, *J. Chem. Phys.* **47**, 4042 (1967).]

band with distinct Q branches in the region of 100 to 170 cm^{-1}. This is shown in Fig. 7.19, taken from the work done by Ueda and Shimanouchi.[49]

The main interest in 2,5-dihydrofuran arises from the potential function of its puckering motion:

$$V(z) = (1.745z^2 + 82.3z^4) \times 10^4, \tag{7.24}$$

where V is in cm^{-1} and z, the displacement coordinate in the direction of

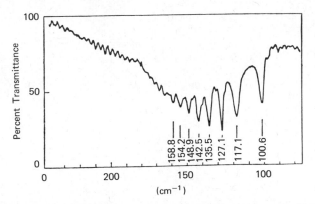

Fig. 7.19   Far-infrared spectrum of the low-frequency ring-puckering mode of 2,5-dihydrofuran. The motion involves essentially the out-of-plane vibration of the oxygen atom. Another ring-puckering mode, which mainly involves a torsional motion about the C=C bond, is at frequencies above $450 \, cm^{-1}$. Instrument: Hitachi FIS-1 vacuum double-beam grating spectrometer. Cell length: 8 cm. Saturated vapor, polyethylene windows. No assignments above $160 \, cm^{-1}$ were attempted because of interference patterns caused by the windows. [T. Ueda and T. Shimanouchi, *J. Chem. Phys.* **47**, 4042 (1967).]

the normal to the plane of the ring, is in Å units. Table 7.3 shows the agreement, on the basis of this potential function, between the computed and measured frequencies of the first eight $\Delta n = 1$ transitions of the low-frequency mode. We see that the potential is represented by a quartic oscillator with a small quadratic term of the same sign; that is, the potential does not possess a central barrier. The planar conformation of 2,5-dihydrofuran is therefore the most stable conformation of this molecule. According to Ueda and Shimanouchi, the reason for the stability of the planar conformation is the absence of the possibility that nonbonded hydrogen atoms could reduce their interactions by assuming a staggered position through ring twisting. We notice from Fig. 7.18 that the nonbonded hydrogen atoms on either end of the single C—C bond are *already* in staggered positions in the planar ring conformation.

Ueda and Shimanouchi have drawn a comparison between the characteristics of the potential functions of the ring-puckering modes of trimethylene oxide (see Section C.b) and 2,5-dihydrofuran; both molecules are essentially planar but trimethylene oxide has a quartic potential with a low barrier. Ueda and Shimanouchi propose that the ring-puckering potential is a composite of a periodic potential, representing the repulsion energies between adjacent substituents which tend to twist the ring bonds if this would lead to a greater distance between their repulsive centers, and the quartic potential representing the ring strain energy which resists further

TABLE 7.3*

CALCULATED AND OBSERVED FREQUENCIES
OF THE LOW-FREQUENCY RING-PUCKERING
MODE OF 2,5-DIHYDROFURAN

Transition	cm^{-1}	
$n \to m$	Observed[†]	Calculated[‡]
0 1	100.6	102.3
1 2	117.1	115.8
2 3	127.1	126.0
3 4	135.5	134.6
4 5	142.5	142.1
5 6	148.9	148.7
6 7	154.2	154.7
7 8	158.8?	160.3

* T. Ueda and T. Shimanouchi, *J. Chem. Phys.* **47**, 4042 (1967).

[†] See Fig. 7.19.

[‡] $V = (1.745z^2 + 82.3z^4) \times 10^4$, $V(cm^{-1})$, z (Å). The matrix elements were set up in the harmonic oscillator representation; size of matrix 40×40 to 60×60 (see Section B).

increases in the dihedral angle of the ring. This is shown in the upper part of Fig. 7.20 for trimethylene oxide, whose potential is an overlap between the quartic potential and the maximum of the repulsive interaction potential (eclipsed positions[†] of adjacent CH_2 groups), and in the lower part of Fig. 7.20 for 2,5-dihydrofuran, in which the minimum of repulsive interactions (staggered positions of adjacent H atoms) overlaps with the quartic ring strain potential.

Ueda and Shimanouchi's ideas can be applied to the corresponding observations in cyclopent*ene*. In this molecule there are three adjacent methylene groups which would be in eclipsed positions with respect to each other if the ring of the molecule were planar. It is therefore not surprising that the potential function of the low-frequency ring-puckering motion of cyclopentene is represented by a quartic oscillator with a central barrier.[50]

In conclusion we mention that Green and Harvey[51] have investigated the ring-puckering vibration of the sulfur analogue of 2,5-dihydrofuran,

---

[†] In an eclipsed position (a designation which obviously has been borrowed from astronomy) the particular substituents are in the line of sight if viewed along the bond which connects the two central atoms to which the substituents are attached.

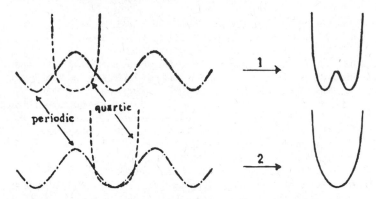

Fig. 7.20 Schematical representation of the origin of the ring-puckering potentials. (*Upper part*): a molecule such as trimethylene oxide in which a planar ring conformation would engender eclipsed positions of nonbonded hydrogen atoms. (*Lower part*): a molecule such as 2,5-dihydrofuran, with staggered positions of nonbonded hydrogen atoms in the planar ring conformation. The periodic potential represents the repulsion energy between adjacent substituents; the quartic potential represents the angle strain of the ring. [T. Ueda and T. Shimanouchi, *J. Chem. Phys.* **47**, 4042 (1962).]

2,5-dihydrothiophene ($\overline{CH_2 \cdot CH = CH \cdot CH_2 \cdot S}$, also called 2,5-dihydrothiofuran). They find that the ring-puckering fundamental can be represented by the potential function

$$V = (1.66z^2 + 46.6z^4) \times 10^4, \qquad (7.25)$$

where V is in cm^{-1} and z, the displacement coordinate normal to the plane of the ring, is in Å units. Comparison of Eqs. 7.24 and 7.25 shows that the potential functions of the ring-puckering mode of 2,5-dihydrofuran and its sulfur analogue, 2,5-dihydrothiophene, are rather similar.

REFERENCES

1. For a definition of the concept of "conformer," see W. G. Dauben and K. S. Pitzer, *Steric Effects in Organic Chemistry*, M. S. Newman, Ed., Wiley, New York, 1956, Chapter 1.
2. See Ref. 1, Chapter 12.
3. H. E. Simmons and J. K. Williams, *J. Am. Chem. Soc.* **86**, 3222 (1964).
4a. E. Heilbronner, Hs. H. Günthard, and R. Gerdil, *Helv. Chim. Acta* **39**, 1171 (1956) (see Ref. 49, footnote 13, with respect to an error in one of the formulas of Ref. 4a).
4b. R. L. Somorjai and D. F. Hornig, *J. Chem. Phys.* **36**, 1980 (1962).
5. D. O. Harris, G. G. Engerholm, and W. D. Gwinn, *J. Chem. Phys.* **43**, 1515 (1965); J. F. Kilpatrick and R. L. Sass, *ibid.*, **42**, 2581 (1965).
6. G. W. Rathjens, Jr., N. K. Freeman, W. D. Gwinn, and K. S. Pitzer, *J. Am. Chem. Soc.* **75**, 5634 (1953). For a normal coordinate treatment see R. C. Lord and I. Nakagawa, *J. Chem. Phys.* **39**, 2951 (1963). A recent structure determination was done by A. Almenningen, O. Bastiansen, and P. Skancke, *Acta Chem. Scand.* **15**, 711 (1961).

7. R. P. Bauman and B. J. Bulkin, *J. Chem. Phys.* **45**, 496 (1966).

8. S. I. Chan, T. R. Borgers, J. W. Russel, H. L. Strauss, and W. D. Gwinn, *J. Chem. Phys.* **44**, 1103 (1966); W. J. Lafferty and R. C. Lord, to be published.

9. S. I. Chan, J. Zinn, J. Fernandez, and W. D. Gwinn, *J. Chem. Phys.* **33**, 1643 (1960).

10. See, for instance, W. G. Rothschild and B. P. Dailey, *J. Chem. Phys.* **36**, 2931 (1962); W. G. Rothschild, *ibid.*, **45**, 1214 (1966), Fig. 7.

11. Compare the shape of the eigenfunctions for even (odd) levels of the harmonic oscillator (for instance, G. Herzberg, *Infrared and Raman Spectra of Polyatomic Molecules*, Van Nostrand, New York, 1951, Fig. 29 on p. 79) with those of the even (odd) levels of the symmetric double-minimum potential (*ibid.*, Fig. 72 on p. 222).

12a. R. P. Bell, *Proc. Roy. Soc. (London)* **A183**, 328 (1945).

12b. R. McWeeny and C. A. Coulson, *Proc. Cambridge Phil. Soc.* **44**, 413 (1948).

12c. A. M. Shorb, R. Schroeder, and E. R. Lippincott, *J. Chem. Phys.* **37**, 1043 (1962).

13. S. I. Chan and D. Stelman, *J. Mol. Spectry.* **10**, 278 (1963); S. I. Chan, D. Stelman, and L. E. Thompson, *J. Chem. Phys.* **41**, 2828 (1964); L. Vescelius and V. D. Neff, *J. Chem. Phys.* **49**, 1740 (1968).

14. T. R. Borgers and H. L. Strauss, *J. Chem. Phys.* **45**, 947 (1966).

15. J. R. Durig and R. C. Lord, *J. Chem. Phys.* **45**, 61 (1966).

16. L. H. Sharpen and V. W. Laurie, *J. Chem. Phys.* **49**, 221 (1968).

17. K. Frei and Hs. H. Günthard, *J. Mol. Spectry.* **5**, 218 (1960).

18. D. O. Harris, H. W. Harrington, A. C. Luntz, and W. D. Gwinn, *J. Chem. Phys.* **44**, 3467 (1966).

19. J. R. Durig and J. N. Willis, Jr., *Spectrochim. Acta* **22**, 1299 (1966).

20. J. R. Durig, W. H. Green, and N. C. Hammond, *J. Phys. Chem.* **70**, 1989 (1966).

21. J. R. Durig and R. C. Lord, *Spectrochim. Acta* **19**, 769 (1963).

22. J. R. Durig and A. C. Morrissey, *J. Chem. Phys.* **45**, 1269 (1966).

23. W. G. Rothschild, *Bull. Am. Phys. Soc.* **9**, 102 (1964).

24. W. G. Rothschild, *J. Chem. Phys.* **44**, 2213 (1966).

25. W. G. Rothschild, *J. Chem. Phys.* **45**, 1214 (1966).

26. W. G. Rothschild, *J. Chem. Phys.* **45**, 3599 (1966).

27. H. Kim and W. D. Gwinn, *J. Chem. Phys.* **44**, 865 (1966). The band at 280 cm^{-1}, assigned by the authors to a $\Delta n = 2$ transition, has a depolarized Raman shift and is properly assigned to be the antisymmetric carbon–chlorine deformation fundamental (see Ref. 26).

28. J. R. Durig and W. H. Green, *J. Chem. Phys.* **47**, 673 (1967).

29. J. R. Durig and A. C. Morrissey, *J. Chem. Phys.* **46**, 4854 (1967).

30. J. B. Lambert and J. D. Roberts, *J. Am. Chem. Soc.* **87**, 3884, 3891 (1965).

31. I. Lillien and R. A. Doughty, *Tetrahedron* **23**, 3321 (1967); *J. Org. Chem.* **32**, 4152 (1967), **33**, 3841 (1968).

32. See, for instance, Refs. 30, 31, and K. B. Wiberg and G. M. Lampman, *J. Am. Chem. Soc.* **88**, 4429 (1966), and References cited therein.

33. B. Curnutte, Jr. and W. H. Shaffer, *J. Mol. Spectry.* **1**, 239 (1957).

34. L. M. Sverdlov and N. I. Prokof'eva, *Opt. Spectry.* **7**, 363 (1959).

35. J. P. McCullough, *J. Chem. Phys.* **29**, 966 (1958).

36. J. E. Kilpatrick, K. S. Pitzer, and R. Spitzer, *J. Am. Chem. Soc.* **69**, 2483 (1947).

37. G. Herzberg, *Infrared and Raman Spectra of Polyatomic Molecules*, Van Nostrand, Princeton, New Jersey, 1962, see p. 94.

38. See Ref. 37, p. 96.

39. K. S. Pitzer and W. E. Donath, *J. Am. Chem. Soc.* **81**, 3213 (1959).

40. F. A. Miller and R. G. Inskeep, *J. Chem. Phys.* **18**, 1519 (1950).

41. K. Tanner and A. Weber, *J. Mol. Spectry.* **10**, 381 (1963).

42. L. Pauling and E. B. Wilson, Jr., *Introduction to Quantum Mechanics*, McGraw-Hill, New York, 1935, pp. 105–107.

43. F. H. Kruse and D. W. Scott, *J. Mol. Spectry.* **20**, 276 (1966). This reference also gives a coordinate analysis based on the puckered conformations C_s and C_2 of cyclopentane as well as a comparison between observed and calculated heat capacities.

44. W. J. Lafferty, D. W. Robinson, R. V. St. Louis, J. W. Russel, and H. L. Strauss, *J. Chem. Phys.* **42**, 2915 (1965).

45. H. L. Strauss, private communication, April 1967.

46. K. W. F. Kohlrausch and A. W. Reitz, *Z. Phys. Chem.* **B45**, 249 (1940).

47. J. A. Greenhouse and H. L. Strauss, *J. Chem. Phys.* **50**, 124 (1969).

48. D. O. Harris, G. G. Engerholm, C. A. Tolman, A. C. Luntz, R. A. Keller, H. Kim, and W. D. Gwinn, *J. Chem. Phys.* **50**, 2438 (1969); G. G. Engerholm, A. C. Luntz, W. D. Gwinn, and D. O. Harris, *ibid.* **50**, 2446 (1969).

49. T. Ueda and T. Shimanouchi, *J. Chem. Phys.* **47**, 4042 (1967).

50. J. Laane and R. C. Lord, *J. Chem. Phys.* **47**, 4941 (1967).

51. W. H. Green and A. B. Harvey, *J. Chem. Phys.* **49**, 177 (1968).

8 Torsional Motions with Periodic Potential Barriers

A. INTRODUCTION

In this chapter we discuss the characteristics of torsional motions in which different parts of a molecule vibrate or rotate against each other with a periodic potential function. We can immediately state two limiting situations which we shall encounter frequently. (*1*) If the height of the potential barrier that separates equivalent relative positions of the pertinent parts of the molecule is much larger than the fundamental frequency of the torsional motion, we can generally approximate this motion by a parabolic potential (at least for the lowest levels). In other words, we can neglect the equivalent positions. (*2*) If, on the other hand, the height of the potential maxima is zero, we speak of free internal rotation. In the intermediate case of a barrier height of the order of magnitude of the $0 \rightarrow 1$ torsional vibration transition, we shall find a rather complex spectrum since tunnelling through the barrier is possible. This is particularly the case for small molecules such as methyl alcohol and hydrogen peroxide.

Far-infrared analysis of the spectra frequently yields values of the parameters of the potential functions—in many cases the results compare satisfactorily with results obtained from other methods, notably microwave spectroscopy. It is clear that detailed knowledge of the low-lying frequencies is important for a complete vibrational analysis and for calculations of the thermodynamic functions from the spectroscopic data.

B. ONE-TOP MOLECULES WITH THREEFOLD BARRIERS

a. Introduction

The torsional motion of a CX_3 group (where X can be a hydrogen, deuterium, or fluorine atom, for instance) with respect to the rest of the molecule is of interest since in many cases it has been found that the three equilibrium positions of the CX_3 group are separated by a potential barrier height of only three to five times the energy difference of the torsional vibration $0 \to 1$ transition; in some cases the barrier height was found to be even lower. Consequently, the degeneracy of the energy levels may be removed by the tunnelling between the equilibrium positions. One of the first such cases extensively treated was the methanol (CH_3OH) molecule. The equivalent positions in this molecule are separated by a potential barrier which is of the order of magnitude of the torsional fundamental frequency. Therefore a very complicated rotation–torsional vibration spectrum results (see below).

Since improved far-infrared instruments have become increasingly available, a considerable number of molecules with CX_3-tops have been investigated, their torsional modes analyzed, and the potential barriers and functions calculated. For such an analysis to be accurate, it is necessary to have reliable structural information; in most cases this information is obtained from microwave spectroscopic investigations which also furnish corroborating evidence in cases of weak far-infrared bands of torsional modes.

There are two different theoretical approaches to treat torsional vibrations (or hindered rotation). One is called the "internal axis method" (IAM). In it the two parts of the molecule are considered to vibrate (or rotate) against each other about an internal axis of the molecule. The other approach, called "principal axis method" (PAM), uses the principal inertial axes of the molecule as the reference system for the description of the internal motion. The different principles of these two methods will become clearer in the next section.

b. Principal Axis Method (PAM) and the Hamiltonian of the Internal
   Rotation for a Symmetric-Top Molecule

Consider a molecule which consists of two parts, (1) and (2), and denote by z a common axis of internal rotation which passes through the respective centers of mass of (1) and (2) (see Fig. 8.1). At present, we do not consider any terms representing the potential energy of hindered rotation. We thus consider first only the kinetic energy, assuming a rigid connection between parts (1) and (2). The total kinetic energy for the pure rotation of this rigid molecule is then obviously

$$T = \tfrac{1}{2}I_x\omega_x^2 + \tfrac{1}{2}I_y\omega_y^2 + \tfrac{1}{2}I_z\omega_z^2, \tag{8.1}$$

where I_x, I_y, and I_z are the principal moments of inertia about the principal axes x, y, and z with the angular velocities ω_x, ω_y, and ω_z about these axes.

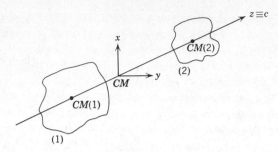

Fig. 8.1   Coordinate system used to describe the internal rotation of the two parts (1) and (2) of the PAM-model. Cartesian coordinates $x, y, z \equiv c$; center of mass: CM.

This system has three degrees of freedom (rotational). We introduce now an internal rotation of part (2) with respect to part (1) as one additional degree of freedom. We call part (1) the "frame" and part (2) the "top"; we further assume cylindrical symmetry for the internal rotation of the top. With part (2) we associate the moment of inertia I_α and the angular velocity $\dot\alpha$, taken about the symmetry axis of the internal top. Since $I_z - I_\alpha$ is now rotating with ω_z and I_α with $\omega_z + \dot\alpha$, for the kinetic energy we obtain the expression:

$$2T = I_x\omega_x{}^2 + I_y\omega_y{}^2 + (I_z - I_\alpha)\omega_z{}^2 + I_\alpha(\omega_z + \dot\alpha)^2$$

or   (8.2)

$$T = \tfrac{1}{2}I_x\omega_x{}^2 + \tfrac{1}{2}I_y\omega_y{}^2 + \tfrac{1}{2}I_z\omega_z{}^2 + \tfrac{1}{2}I_\alpha\dot\alpha^2 + I_\alpha\omega_z\dot\alpha.$$

(In Appendix I of Ref. 1 a rigorous derivation of the kinetic energy of an asymmetric rotor is presented.)

To set up the quantum-mechanical Hamiltonian of the kinetic energy, we must rewrite T as a function of the conjugate momenta. According to well-known principles, the conjugate momenta are given by

$$P_x = \frac{\partial T}{\partial \omega_x} = I_x\omega_x$$

$$P_y = \frac{\partial T}{\partial \omega_y} = I_y\omega_y$$

(8.3)

$$P_z = \frac{\partial T}{\partial \omega_z} = I_z\omega_z + I_\alpha\dot\alpha$$

$$p = \frac{\partial T}{\partial \dot\alpha} = I_\alpha(\omega_z + \dot\alpha).$$

We solve for ω_x, ω_y, ω_z, and $\dot{\alpha}$ as functions of P_x, P_y, P_z, and p, and insert the results into Eq. 8.2:

$$H_{kin} = \frac{P_x^2}{2I_x} + \frac{P_y^2}{2I_y} + \frac{P_z^2}{2(I_z - I_\alpha)} - \frac{pP_z}{I_z - I_\alpha} + \frac{I_z p^2}{2I_\alpha(I_z - I_\alpha)}. \qquad (8.4)$$

We mention here that Eq. 8.4 can be shown to be equivalent to the matrix equation

$$2T = \mathbf{P}^T \mathbf{I}^{-1} \mathbf{P}, \qquad (8.5)$$

where $\mathbf{P}$ is the column vector

$$\begin{pmatrix} P_x \\ P_y \\ P_z \\ p \end{pmatrix}, \qquad (8.6)$$

$\mathbf{P}^T$ its transpose, and $\mathbf{I}$ the matrix

$$\mathbf{I} = \begin{bmatrix} I_x & 0 & 0 & 0 \\ 0 & I_y & 0 & 0 \\ 0 & 0 & I_z & I_\alpha \\ 0 & 0 & I_\alpha & I_\alpha \end{bmatrix}. \qquad (8.7)$$

Equation 8.5, in terms of the angular velocities, is given by

$$2T = \boldsymbol{\omega}^T \mathbf{I} \boldsymbol{\omega} \qquad (8.8)$$

with

$$\boldsymbol{\omega} = \begin{pmatrix} \omega_x \\ \omega_y \\ \omega_z \\ \dot{\alpha} \end{pmatrix}. \qquad (8.9)$$

In order to obtain Eq. 8.4 from Eq. 8.2, we must calculate $\mathbf{I}^{-1}$ of Eq. 8.7.

We note that for a symmetric-top molecule the internal rotation of the CH_3 group does not affect the principal axes.

The operators P_x, P_y, and P_z, which follow the usual commutation rules,[1,2] do not depend on internal coordinates; in other words, they can be expressed in terms of, for instance, the three Eulerian angles θ, ϕ, and χ.

The quantum-mechanical operator for p is

$$p = -i\frac{\partial}{\partial \alpha}\bigg|_{\theta,\phi,\chi}. \qquad (8.10)$$

Since p depends only on the internal coordinate α, it commutes with P_x, P_y, and P_z.

To ascertain whether we have obtained the correct expression of the rigid rotor in the absence of internal rotation, we set $\dot{\alpha} = 0$. Then, from Eq. 8.3

$$p = I_\alpha \omega_z \quad \text{and} \quad P_z = I_z \omega_z. \tag{8.11}$$

It follows that $p = (I_\alpha/I_z)P_z$. If we insert this value into Eq. 8.4, we find indeed the usual expression for the rotational Hamiltonian of a rigid rotor.

We now rewrite the last term of Eq. 8.4 in the form Fp^2 with

$$F = \frac{\hbar^2 I_z}{2I_\alpha(I_z - I_\alpha)}. \tag{8.12}$$

The potential energy hindering the internal rotation of a top with three-fold rotational symmetry, such as CH_3, is periodic with the angle $2\pi/3$. We denote it by $V(\alpha)$. The part of the Hamiltonian which contains the torsional–vibrational part is then written in the form

$$H_T = Fp^2 + V(\alpha), \tag{8.13}$$

where we disregard, at first, the term describing the coupling between the "overall" or "end-over-end" and the internal rotation (see Eq. 8.4). The boundary conditions for the solutions of Eq. 8.13 are periodic.

c. Hamiltonian in the Internal Axis Method (IAM) for a Symmetric-Top Molecule

The Hamiltonian (see Eq. 8.4) contains a coupling term between the end-over-end and the internal rotation; this term is proportional to $P_z p$. We can remove this coupling term with the help of the following transformation:

$$\omega_c' = \omega_z + \frac{I_\alpha}{I_z}\dot{\alpha}$$

$$\dot{\alpha}' = \dot{\alpha}. \tag{8.14}$$

We then obtain from Eq. 8.2

$$T = \tfrac{1}{2}I_x\omega_x{}^2 + \tfrac{1}{2}I_y\omega_y{}^2 + \tfrac{1}{2}I_z\omega_c'^2 + \tfrac{1}{2}I_\alpha\dot{\alpha}'^2\left(1 - \frac{I_\alpha}{I_z}\right), \tag{8.15}$$

and for the corresponding angular momenta

$$P_c' = \left.\frac{\partial T}{\partial \omega_c'}\right|_{\omega_x,\omega_y,\dot{\alpha}'} = I_z\omega_c' = I_z\omega_z + I_\alpha\dot{\alpha}$$

$$p' = \left.\frac{\partial T}{\partial \dot{\alpha}'}\right|_{\omega_x,\omega_y,\omega_c'} = I_\alpha\dot{\alpha}'\left(1 - \frac{I_\alpha}{I_z}\right) = I_\alpha\dot{\alpha}'r \tag{8.16}$$

with

$$r = \left(1 - \frac{I_\alpha}{I_z}\right).$$

P_x and P_y remain unchanged.

   Substitution of the momenta of Eq. 8.16 into Eq. 8.15 gives the Hamiltonian

$$H_{kin} = \frac{P_x^2}{2I_x} + \frac{P_y^2}{2I_y} + \frac{P_c'^2}{2I_z} + \frac{1}{2}\frac{p'^2}{rI_\alpha}. \tag{8.17}$$

We see from Eq. 8.16 that p' is not the angular momentum of the top (which is $I_\alpha \dot{\alpha}$) but the reduced angular momentum $rI_\alpha \dot{\alpha}'$. The transformation Eq. 8.14 with the appropriate momenta (Eq. 8.16), is called the "Nielsen Transformation."[1] It results in a coordinate system of reference, the "internal rotation axes," relative to which both the top and the frame are moving. We can express the transformation (Eq. 8.14) in terms of the angles χ_1 and χ_2 which describe the rotation of $I_z - I_\alpha$ and I_α about the z-direction:

$$\alpha' = \alpha = \chi_2 - \chi_1$$

$$\chi' = \chi_1 + \frac{I_\alpha}{I_z}\alpha = \frac{1}{I_z}[I_\alpha \chi_2 + (I_z - I_\alpha)\chi_1], \tag{8.18}$$

where $\dot{\chi}'$ corresponds to ω_c' and $\dot{\chi}_1$ corresponds to ω_z. If we describe the internal rotation axes in terms of the Eulerian angles θ, ϕ, and χ', we see that this coordinate system rotates about the symmetry axis of the molecule with the angular velocity $(I_\alpha/I_z)\dot{\alpha}$ relative to the frame.

   The angular momenta of the top and the frame cancel each other if taken in the internal rotation axis system:

$$I_\alpha(\dot{\chi}_2 - \dot{\chi}') + (I_z - I_\alpha)(\dot{\chi}_1 - \dot{\chi}') = 0. \tag{8.19}$$

Since P_x and P_y in Eq. 8.17 do not commute with p' (Eq. 8.16), the following transformation is introduced:

$$\begin{pmatrix} P_a' \\ P_b' \end{pmatrix} = \begin{pmatrix} \cos\rho\alpha & \sin\rho\alpha \\ -\sin\rho\alpha & \cos\rho\alpha \end{pmatrix}\begin{pmatrix} P_x \\ P_y \end{pmatrix}$$

$P_c' = P_c$ and $\rho = I_\alpha/I_z$. With this, the Hamiltonian is given by

$$H = \frac{P_a'^2}{2I_x} + \frac{P_b'^2}{2I_y} + \frac{P_c'^2}{2I_z} + \frac{1}{2}\frac{p'^2}{rI_\alpha}, \tag{8.20}$$

where P_a', P_b', P_c' and p' now all commute and where the coupling term is removed.

The "internal rotation part" of the Hamiltonian (Eq. 8.20), together with the potential energy $V(\alpha)$, represents the torsional Hamiltonian

$$H_T = Fp'^2 + V(\alpha), \qquad F = \frac{\hbar^2 I_z}{2I_\alpha(I_z - I_\alpha)}. \tag{8.21}$$

H_T in Eq. 8.21 has formally the same appearance as H_T of Eq. 8.13, but the coupling of overall and internal rotation is now contained in the boundary conditions which are now not periodic in α (see end of Section A.b).

d. Potential Energy

The potential energy $V(\alpha)$ hindering the internal rotation of the top and frame is generally chosen empirically since in most cases its origin is, at present, not clearly understood. For tops of a threefold rotational symmetry, such as CX_3 groups, the potential energy has three equivalent minima and therefore can be represented by a Fourier series of the form

$$V(\alpha) = \frac{V_3}{2}(1 - \cos 3\alpha) + \frac{V_6}{2}(1 - \cos 6\alpha) + \cdots. \tag{8.22}$$

The quantities V_3 and V_6 have to be adjusted to the spectral data. The term V_3 is called the barrier height; V_6 is a correction term. The expression for the potential energy includes a constant term $V_3/2$, chosen for convenience, in order to relate the Schrödinger equation of this problem to the Mathieu equation.

e. Asymmetric-Top Molecules without Structural Symmetry and the Principal Axis Method (PAM)

The kinetic energy of an asymmetric molecule is

$$\begin{aligned} T = {}&\tfrac{1}{2}I_x\omega_x^2 + \tfrac{1}{2}I_y\omega_y^2 + \tfrac{1}{2}I_z\omega_z^2 + I_\alpha\lambda_x\omega_x\dot\alpha \\ &+ I_\alpha\lambda_y\omega_y\dot\alpha + I_\alpha\lambda_z\omega_z\dot\alpha + \tfrac{1}{2}I_\alpha\dot\alpha^2. \end{aligned} \tag{8.23}$$

I_x, I_y, I_z are the moments of inertia around the principal axes x, y, z; ω_x, ω_y, ω_z are the angular velocities about the principal axes; $\lambda_x, \lambda_y, \lambda_z$ are the direction cosines of the symmetry axis of the internal top with respect to the principal axes; I_α is the moment of inertia of the internal top around its symmetry axis; and $\dot\alpha$ is the angular velocity of the internal top about its symmetry axis. (A rigorous derivation of Eq. 8.23 is given in Appendix I of Ref. 1.)

We see that the form of the kinetic energy is similar to that of the symmetric-top molecule discussed above (see Eq. 8.2). The first three terms represent the overall rotation, the last term characterizes the internal rotation, and the fourth, fifth and sixth terms represent coupling between the overall and internal rotations.

Using either a system of equations analogous to Eq. 8.3 or the method mentioned in Eqs. 8.5 to 8.9, we can derive the Hamiltonian in the momentum representation. We include the potential energy function of the type shown in Eq. 8.22:

$$H = A_x P_x{}^2 + B_y P_y{}^2 + C_z P_z{}^2 + \tfrac{1}{2} \sum_{\substack{x,y,z \\ i \neq j}} D_{ij}(P_i P_j + P_j P_i)$$

$$- 2 \sum_{x,y,z} Q_i P_i p + F p^2 + V(\alpha). \tag{8.24}$$

with

$$A_x = \frac{\hbar^2}{2I_x}\left[1 + \frac{\lambda_x{}^2 I_\alpha}{r I_x}\right], \qquad B_y = \frac{\hbar^2}{2I_y}\left[1 + \frac{\lambda_y{}^2 I_\alpha}{r I_y}\right], \qquad C_z = \frac{\hbar^2}{2I_z}\left(1 + \frac{\lambda_z{}^2 I_\alpha}{r I_z}\right)$$

$$D_{ij} = \frac{\hbar^2}{2}\frac{\lambda_i \lambda_j I_\alpha}{r I_i I_j}, \qquad i,j = x,y,z, (i \neq j)$$

$$Q_i = \frac{\hbar^2 \lambda_i}{2 r I_i}, \qquad i = x,y,z, \tag{8.25}$$

$$r = 1 - \sum_{g=x,y,z} \frac{\lambda_g{}^2 I_\alpha}{I_g},$$

$$F = \frac{\hbar^2}{2 r I_\alpha}.$$

We can also express the Hamiltonian of Eq. 8.24 by

$$H = \frac{\hbar^2}{2 r I_\alpha}(p - \Pi)^2 + \frac{\hbar^2}{2}\sum_g \frac{P_g{}^2}{I_g} + V(\alpha)$$

where

$$\Pi = \sum_{g=x,y,z} \frac{\lambda_g P_g I_\alpha}{I_g}.$$

The momenta P_x, P_y, P_z represent the components of the total angular momentum of the entire molecule. (The commutation relations between these components are the usual ones and the P_g commute with p.)

For molecules with a plane of symmetry, either λ_x or λ_y is zero. In the notation of Eq. 8.25, $\hbar^2$ is included in the coefficients A to F, whereas before, in Eq. 8.21, $\hbar^2$ was only included in F.

f. Asymmetric-Top Molecules with a Plane of Symmetry and the Internal Axis Method (IAM)

In this section we will restrict our considerations to asymmetric rotors with one plane of symmetry. As before, one part of the molecule shall be a symmetric top. (The best known example of these molecules is the CH_3OH molecule.)

The classical kinetic energy is:

$$T = \tfrac{1}{2}I_{aa}\omega_a{}^2 + \tfrac{1}{2}I_{bb}\omega_b{}^2 + \tfrac{1}{2}I_{cc}\omega_c{}^2 - I_{bc}\omega_b\omega_c + I_\alpha\omega_c\dot\alpha + \tfrac{1}{2}I_\alpha\dot\alpha^2. \quad (8.26)$$

The c axis passes through the center of mass and is parallel to the symmetry axis of the internal rotor, the b axis is in the plane of symmetry, and the a axis is perpendicular to b and c. I_{aa}, I_{bb}, I_{cc} are the moments of inertia and I_{bc} is the product of inertia. In terms of relations such as Eq. 8.3, we can express the Hamiltonian in momentum operators. Including a potential energy term, we obtain

$$H = A_aP_a{}^2 + B_bP_b{}^2 + C_cP_c{}^2 + D_{bc}(P_bP_c + P_cP_b)$$
$$- 2D_{bc}P_bp - 2C_cP_cp + Fp^2 + V(\alpha), \quad (8.27)$$

with the abbreviations:

$$A_a = \frac{\hbar^2}{2I_{aa}}, \qquad B_b = \frac{(I_{cc} - I_\alpha)\hbar^2}{2d}, \qquad C_c = \frac{I_{bb}\hbar^2}{2d},$$

$$D_{bc} = \frac{I_{bc}\hbar^2}{2d}, \qquad F = \frac{\hbar^2}{2rI_\alpha}, \qquad r = 1 - \sum_{g=y,z}\frac{I_\alpha\lambda_g{}^2}{I_g}, \quad (8.28)$$

$$d = rI_yI_z.$$

The I_x, I_y, I_z are the principal moments of inertia of the whole molecule and λ_y, λ_z are the direction cosines of the internal rotation axis with respect to the principal axes.

In order to remove the coupling terms of overall and internal rotations, $P_b \cdot p$ and $P_c \cdot p$, a transformation is applied similar to the case of the symmetric molecule in Eq. 8.11. The result is a Hamiltonian of the type

$$H = H_R + Fp'^2 + V(\alpha') \quad (8.29)$$

with no coupling terms between overall and internal rotation. The expressions for H_R are given explicitly in Ref. 1 (p. 847) and Ref. 3 (p. 325). The operator p' is, as before, $-i(\partial/\partial\alpha')$, and the coupling of overall and internal rotation is now contained in the boundary conditions. The above considerations have been applied for the interpretation of the far-infrared spectrum of the CH_3OH molecule, as we show in the following section.

g. Far-Infrared Spectrum of Methyl Alcohol

The far-infrared spectrum of methyl alcohol consists of a large number of irregularly spaced absorption lines of varying intensities (see Fig. 8.2). Since CH_3OH is a relatively small molecule, some of the strongest lines of its

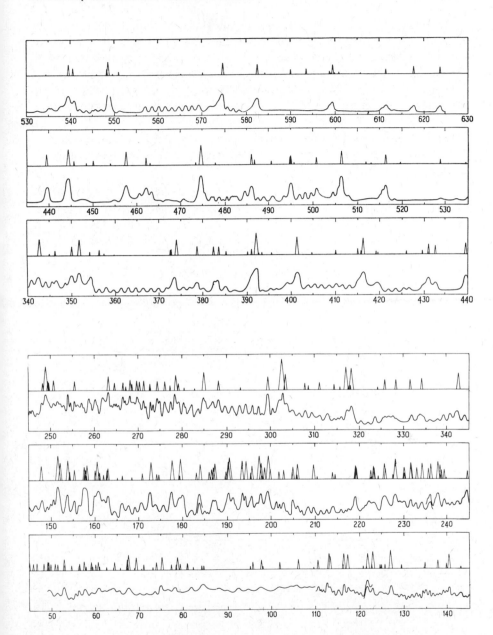

Fig. 8.2   Calculated and observed spectrum of CH_3OH in the 45 to 630 cm^{-1} spectral region. [D. G. Burkhard and D. M. Dennison, *J. Mol. Spectry.* **3**, 299 (1959), by permission of Academic Press, Inc.]

pure rotational spectrum are located in the far-infrared. Although the CH_3OH molecule is an asymmetric top, the asymmetry is small and cannot account for the complexity of the spectrum. The complicated structure of the spectrum is due to a superposition of the pure rotational transitions and transitions arising from the hindered rotational motion. The barrier of hindered rotation of the CH_3- and OH-groups has been determined to be $V = 374.82 \, cm^{-1}$.[4] The barrier is comparable to kT ($\sim 208 \, cm^{-1}$), so that a considerable portion of the molecules are in the higher excited states and transitions are possible from these higher excited states, contributing to the complexity of the spectrum.

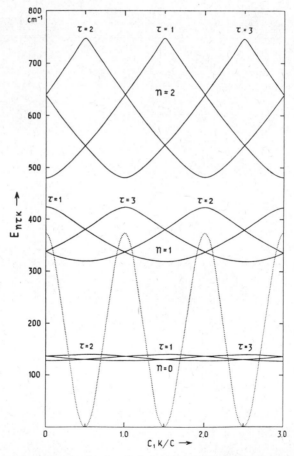

Fig. 8.3 Internal rotation levels of methyl alcohol. The dotted line is the potential function. The torsional vibrational states (internal rotation levels) are given by $n = 0, 1, 2, \ldots$; the torsional vibrational (internal rotational) sublevels are given by $\tau = 1, 2, 3$. [D. G. Burkhard and D. M. Dennison, *J. Mol. Spectry.* **3**, 299 (1959), by permission of Academic Press, Inc.]

In Fig. 8.3 the ground state levels are indicated by $n = 0$ and the first excited state by $n = 1$. We see that the first excited state levels have an energy of about the same value as the barrier height. The case of the CH_3OH molecule is intermediate between a high barrier case, where the barrier height is many times larger than the energy of the first excited state, and which can be described approximately in terms of a parabolic potential function, and the case of free internal rotation, for which we have $V = 0$.

The far-infrared spectrum of methyl alcohol was studied theoretically by Burkhard and Dennison,[4] who used experimental data of the 880 to 50 cm^{-1} spectral region. These data were obtained by different authors with the help of grating spectrometers. Gebbie et al.[5] studied the low-frequency region 80–20 cm^{-1} with a Michelson interferometer and obtained spectra with a resolution of about 0.25 cm^{-1}.

We will restrict our discussion to the far-infrared part and especially to the spectrum from 20 to 80 cm^{-1}. The Hamiltonian for the rotation-torsion motion is given in Ref. 1 (p. 847) and in Ref. 3 (p. 325). The latter reference indicates the perturbation treatment which was used to obtain the solutions of the eigenvalue problem. We will not repeat this here, but we will discuss the origin of the different types of lines appearing in the spectrum. Figure 8.3 shows the energy levels of the internal rotation as function of $(C_1/C)K$. The energy levels are classified according to the vibrational quantum numbers $n = 0, 1, 2, \cdots$, the torsional states $\tau = 1, 2, 3$ (internal rotation sublevels), and the quantum number K of the component of the total angular momentum along the internal rotation axis. This K-dependance of the energy levels arises from the transformation to the "internal rotation axis" system, as discussed above. C_1 is the moment of inertia of the hydroxyl group, C_2 is the moment of inertia of the methyl group, and $C_1 + C_2 = C$, the moment of inertia around the internal rotation axis. The quantum number of the total angular momentum is J, and therefore the energy levels of the overall and internal rotation are characterized by the four quantum numbers J, n, τ, K.

The spectrum may be described as the superimposed pattern of three sets of absorption bands, corresponding to three different sets of selection rules.

1. The first set consists of an overall or end-over-end rotation of the CH_3OH molecule. The selection rules are[4,5] $\Delta J = 0, \pm 1; \Delta K = 0; \Delta \tau = 0;$ $\Delta n = 0$. The slight asymmetry of the methanol molecule from a symmetric rotor is not very important for these transitions. The frequencies of the rotational transitions may be approximated satisfactorily by

$$v_{J-1,J} = 2B_{eff}J - 4D_{JJ}J^3. \tag{8.30}$$

The third-order term represents the influence of the centrifugal stretching of the molecule, with the stretching of the C—O bond the predominant contribution. The intensities of these absorption bands are calculated and represented by lines numbered 13–44 in Fig. 8.4.

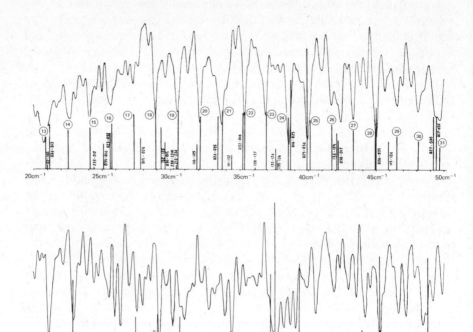

Fig. 8.4   Observed spectrum of CH_3OH in the 20 cm^{-1} to 80 cm^{-1} spectral region. Resolution 0.25 cm^{-1}. Calculated absorption bands are indicated by vertical lines with length proportional to the expected intensities. Set 1 is indicated by numbers 13 to 44. Set 2 is indicated by $(n'', \tau'', K'') - (n', \tau', K')$, and set 3 forms the background. [H. A. Gebbie, G. Topping, R. Illsley, and D. M. Dennison, *J. Mol. Spectry.* **11**, 229 (1963), by permission of Academic Press, Inc.]

2. The second set of transitions contains most of the contributions of the hindered rotation. The selection rules[4] are:

$$\Delta J = 0, \quad \Delta n = \text{any value}, \quad \begin{cases} \Delta K = -1 \quad \text{and} \quad \tau = 1 \to 2, 2 \to 3, 3 \to 1 \\ \Delta K = +1 \quad \text{and} \quad \tau = 2 \to 1, 3 \to 2, 1 \to 3. \end{cases}$$

(8.31)

The corresponding transitions are broadened by the centrifugal stretching effects because lines with different J values do not appear at the same frequency. The intensities of these lines were also calculated and they are indicated in Fig. 8.4 at the lines with the numbers $(n'', \tau'', K'') - (n', \tau', K')$.

3. The third set has the same selection rules as the second set, except that $\Delta J = \pm 1$. Lines resulting from these transitions are very closely spaced and, in general, form the background absorption. Some weak distinct absorptions, not accounted for by the first and second set, may be explained by assuming accidental coincidence of several of the lines of this third set. Although not every observed line is explained, the overall agreement between the theoretical prediction and the observed pattern is satisfactory. There are, at least, no observed strong lines which cannot be accounted for by the theory.

h. Far-Infrared Torsional Vibration Spectra of Various Molecules

In this section we outline the solutions of the wave equations, the eigen functions, energy level schemes, and the characteristics of the potential functions of internal rotors which show a potential barrier several times higher than kT. Many molecules possessing a CX_3-top (X = H, D, F, for example) have been observed to be of this type.[6-13] In such cases the torsional motion is more aptly called torsional vibration instead of hindered internal rotation. We also note that in many of these molecules the rotational fine structure was not observed.

To obtain the Hamiltonian for this strongly hindered torsional motion with the help of the relations discussed in the previous sections, we first note that, formally, we obtain the same Hamiltonian if we (1) replace the non-periodic boundary conditions by periodic boundary conditions in the IAM-representation (see Eq. 8.21) and if (2) we neglect in the PAM-representation the coupling term for the interaction of the end-over-end and the internal rotation (see Eq. 8.13). We use this Hamiltonian for the evaluation of the potential barrier and repeat it here with the same symbols as given above for an asymmetric rotor.

$$H_T = Fp^2 + \frac{V_3}{2}(1 - \cos 3\alpha),$$

where   (8.32)

$$p = -i\frac{\partial}{\partial \alpha}, \qquad F = \frac{\hbar^2}{2I_\alpha r}, \qquad r = 1 - \sum_{g=x,y,z} \frac{\lambda_g^2 I_\alpha}{I_g}.$$

We neglect not only the coupling between overall and internal rotation but also the interaction of the torsional motion with other vibrations: the torsional coordinate is thus considered approximately a normal coordinate.

The Schrödinger equation corresponding to the Hamiltonian of Eq. 8.32 is

$$-F\frac{\partial^2 U}{\partial \alpha^2} + \frac{V_3}{2}(1 - \cos 3\alpha)U = EU.$$ (8.33)

(Correction terms of the form $(V_6/2)(1 - \cos 6\alpha)$ may be included by perturbation methods.)

Equation 8.33 is a Mathieu-type differential equation, and the following substitutions will transform it into a normal form of the Mathieu equation:

$$2x = 3\alpha + \pi, \quad y(x) = U(\alpha)$$

$$V_3 = \tfrac{9}{4}Fs, \qquad E = \tfrac{9}{4}Fb. \tag{8.34}$$

We find

$$y'' + (b - s\cos^2 x)y = 0, \tag{8.35}$$

where b is the eigenvalue and s is a characteristic parameter which determines the solutions as well as eigenvalues. The term $\cos^2 x$ may be rewritten in terms of $\cos 2x$ with a slightly different notation for the eigenvalues and the parameters.

The solutions $U_{v\sigma}$ and eigenvalues $E_{v\sigma}$ of Eq. 8.33 are labeled by the torsional vibration quantum number v and the torsional quantum number σ. The scheme of the energy levels in terms of the quantum numbers v and σ is best presented by considering the two extreme cases of the torsional vibration (Eq. 8.33) in its periodic potential of finite barrier height; namely (1) free rotation of the internal top ($s = 0$) and (2) vibration in a parabolic potential ("infinitely high" barrier, $s \to \infty$).

In the first case, Eq. 8.33 gives

$$\frac{\partial^2 U(\alpha)}{\partial \alpha^2} + \frac{E}{F}U(\alpha) = 0. \tag{8.36}$$

The eigenvalues are $E_m = Fm^2$ with $m = 0, \pm 1, \pm 2, \cdots$. The eigen functions are trigonometric (see Eq. 8.35), as is to be expected for free rotation.

In the second case, that for an infinitely high barrier, we develop the term $(V_3/2)(1 - \cos 3\alpha)$ of Eq. 8.33 into a power series. Neglecting all terms beyond the second, we obtain

$$-F\frac{\partial^2 U(\alpha)}{\partial \alpha^2} + \tfrac{9}{4}V_3\alpha^2 U(\alpha) = EU(\alpha). \tag{8.37}$$

This represents the eigenvalue problem of a harmonic oscillator—as it should—with the eigenvalues

$$E_v = 3\sqrt{FV_3}(v + \tfrac{1}{2}) \qquad v = 0, 1, 2 \cdots. \tag{8.38}$$

The proper linear combinations of the harmonic oscillator eigen functions of Eq. 8.37—in order to reflect the threefold symmetry of the potential function[2]—are given by

$$U_{v,0}(\alpha) = \left(\frac{1}{\sqrt{3}}\right)(u_v^{(1)} + u_v^{(2)} + u_v^{(3)})$$

$$U_{v,1}(\alpha) = \left(\frac{1}{\sqrt{3}}\right)(u_v^{(1)} + \bar{\omega}u_v^{(2)} + \bar{\omega}^2 u_v^{(3)}) \qquad (8.39)$$

$$U_{v,-1}(\alpha) = \left(\frac{1}{\sqrt{3}}\right)(u_v^{(1)} + \bar{\omega}^2 u_v^{(2)} + \bar{\omega}u_v^{(3)}),$$

where $u_v^{(i)}$ are harmonic oscillator eigen functions and the superscript i denotes the ith potential well of the potential function; $\bar{\omega} = \exp(2\pi i/3)$. The two subscripts denote the quantum number v and σ: for nondegenerate levels $\sigma = 0$, for degenerate levels $\sigma = \pm 1$. The interval of periodicity of the angle of internal rotation, α, is 0, $(\frac{2}{3})\pi$ for single and 0, 2π for degenerate levels.

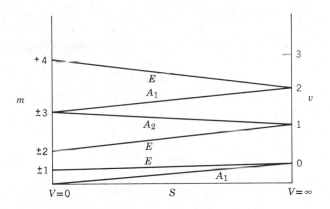

Fig. 8.5   Energy level correlation diagram. (*Left side*): free rotation, $s = V = 0$. (*Right side*): harmonic oscillation, s and $V \to \infty$. A_1, A_2, and E are the species of the corresponding symmetry group which is isomorphous to C_{3v}.

The (approximate) energy level scheme is shown in Fig. 8.5, where the symmetry species of the levels are designated by the symbols A_1, A_2, and E (see Section A.i below). For the extreme $V \to \infty$, we see that all levels are threefold degenerate since no level splittings due to tunnelling are possible. In the other extreme of free internal rotation, $V = 0$, we note that the lowest level is single and that all higher levels are doubly degenerate ($m = 0, \pm 1, \cdots$, see above). (The level scheme is readily established by connecting levels with the proper symmetries, obeying the noncrossing rule.)

The eigenvalues b_{vA} and b_{vE}, where A refers to the nondegenerate and E to the degenerate eigenvalues, have been tabulated. In Herschbach's table,[14] we find them for the range $s \leqslant 100$ (A and E levels), in Hayaski and Pierce's table[15] we find them for the range $100 \leqslant s \leqslant 200$. (The A-levels are also collected in Ref. 16 and E-levels in Ref. 17; see also further tabulations in Refs. 18–20.)

The eigen functions of Eq. 8.33 can be represented by a Fourier series. For the A-functions we obtain

$$U_{v,0}(\alpha) = \sum_{k=0}^{\infty} D_k^v \cos 3k\alpha; \qquad U_{v,0}(\alpha) = \sum_{k=0}^{\infty} D_k^v \sin 3k\alpha, \qquad (8.40)$$

where the first series corresponds to even, the second to odd values of the quantum number v.

The E-eigen functions are given by

$$U_{v,1}(\alpha) = U_{v,-1}^*(\alpha) = \sum_{k=-\infty}^{+\infty} A_k^v \exp\left[i(3k+1)\alpha\right]. \qquad (8.41)$$

The coefficients D_k^v are tabulated in Ref. 16, the A_k^v-coefficients are collected in Ref. 17.

The eigen functions $U_{v,0}$ and $U_{v,\pm 1}$ have the following symmetry properties under a reflection operation

$$\sigma : \alpha \rightarrow -\alpha$$
$$\sigma U_{v,0}(\alpha) = (-1)^v U_{v,0}(\alpha); \qquad \sigma U_{v,1}(\alpha) = U_{v,-1}(\alpha), \qquad (8.42)$$

and under a rotation operation about the symmetry axis of the internal top

$$\delta^n : \alpha \rightarrow \alpha + \frac{2\pi}{3}n, \qquad n = \text{integer}$$

$$\delta^n U_{v,0}(\alpha) = U_{v,0}(\alpha); \qquad \delta U_{v,1}(\alpha) = \overline{\omega} U_{v,1}(\alpha) \qquad (8.43)$$

$$\delta U_{v,-1}(\alpha) = \overline{\omega}^2 U_{v,-1}(\alpha),$$

where

$$\overline{\omega} = \exp(2\pi i/3), \qquad \overline{\omega} + \overline{\omega}^2 = -1. \qquad (8.44)$$

(We note from these formulas that $U_{v,\sigma} = U_{v,-\sigma}^*$, where the starred quantity denotes the complex conjugate.)

Matrix elements in the representation $U_{v,\sigma}$ have been tabulated.[12,14] For instance,

$$(p^n)_{v,v'} = \int_0^{2\pi} U_{v,\sigma}^* p^n U_{v',\sigma} \, d\alpha, \qquad n = 1, 2, 3$$

$$(\cos 6\alpha)_{v,v'} = \int_0^{2\pi} U_{v,\sigma}^* \cos 6\alpha U_{v',\sigma} \, d\alpha. \qquad (8.45)$$

The element $(\cos 6\alpha)_{v,v'}$ is frequently abbreviated by $(6)_{v,v'}$.

i. Selection Rules

The selection rules for the transitions connected with torsional vibrations of the simplified model discussed above (one degree of freedom, coordinate α) are obtained by group-theoretical methods. The symmetry group consists of symmetry elements that transform the molecule into itself. These are symmetry elements of the usual type, such as reflections by a plane of symmetry, but in addition there are included symmetry operations that reflect the equivalent positions of a part of the molecule with respect to internal rotation. Such an operation is the rotation of the CH_3 group around its symmetry axis by multiples of $2\pi/3$. As an example, let us treat the case of a molecule with one plane of symmetry, for instance, CH_3CH_2Cl.

The symmetry operations are:

$$I : \alpha \to \alpha$$

$$\sigma : \alpha \to -\alpha \qquad\qquad (8.46)$$

$$\delta^n : \alpha \to \alpha + \frac{2\pi}{3}n, \qquad n = \text{integer.}$$

The group consists of six elements and has three classes of conjugate elements:

$$(I), \qquad (\delta, \delta^2), \qquad (\sigma, \sigma\delta, \sigma\delta^2).$$

The group is isomorphous to the group of permutations of three elements. Its character table is the same as that of the C_{3v} group (Fig. 8.6). The notation of the energy levels in Fig. 8.5 is taken from Fig. 8.6.

	I	2δ	3σ
A_1	1	1	1
A_2	1	1	-1
E	2	-1	0

Fig. 8.6   Character table for a group isomorphous to group C_{3v}.

The selection rules are usually obtained by investigating whether or not the following integral vanishes:

$$\int U_{v\sigma}(\alpha)\mu U_{v'\sigma'}(\alpha)\,d\alpha \qquad \begin{cases} = 0, \text{ inactive} \\ \neq 0, \text{ active} \end{cases} \qquad (8.47)$$

The $U_{v\sigma}$ are the eigen functions of Eqs. 8.40–8.41 (their symmetry properties are given in Eqs. 8.42–8.44) and μ is the dipole moment. We develop μ into a power series in terms of the coordinate α

$$\mu = \mu_{perm} + \left.\frac{\partial\mu}{\partial\alpha}\right|_0 \alpha + \frac{1}{2}\left.\frac{\partial^2\mu}{\partial\alpha^2}\right|_0 \alpha^2 + \text{higher order terms.} \qquad (8.48)$$

Since we restrict the discussion of the selection rules to the vibrational motion only, we have (1) to substitute for μ in Eq. 8.47 the derivative of the dipole moment with respect to the coordinate (that is, the second term in Eq. 8.48), and (2) we have to require that there is no angular momentum produced by the torsional motion with respect to the rotating system. Therefore, a deflection of the CH_3 group to one side must be compensated by a motion of the rest of the molecule in the opposite direction. In our example of CH_3CH_2Cl, the permanent dipole moment is fixed in the frame of the molecule and is in the plane of symmetry. Consequently, the change of the dipole moment will be in the direction perpendicular to the plane of symmetry. If the potential function of the model (with one degree of freedom) possesses a parabolic potential function (Eq. 8.37), α is the normal coordinate. If all the higher-order terms of α are included in Eq. 8.37, we would arrive finally at the $\cos 3\alpha$ potential function. These higher terms may be treated as perturbations and consequently we would obtain the eigen functions and eigenvalues of Eq. 8.33. As long as this perturbation treatment is meaningful, α is approximately still a normal coordinate. In fact, the perturbation treatment can be used as long as the tunnelling frequency is small compared with the vibration frequency. This is true in almost all observed cases.

We may summarize the above considerations as follows: The top vibrates with a small deflection which is essentially proportional to the normal coordinate α. The energy level scheme is modified by the possibility of tunnelling.

In detail, the selection rules are obtained by group-theoretical arguments as follows: For a molecule with one plane of symmetry, the change of the dipole moment belongs to the symmetry species A_2. We merely have to find out if the direct product of the representations of the initial state Γ_w and final state $\Gamma_{w'}$ contains the symmetry species of μ, that is, Γ_μ, or

$$\Gamma_w \times \Gamma_\mu \times \Gamma_{w'} \in A_1, \qquad (8.49)$$

where A_1 is the totally symmetric representation of the group. The result is that the following transitions are infrared-active:

$$A_1 \leftrightarrow A_2, \qquad E \leftrightarrow E.$$

Figure 8.7 shows the energy level scheme and the selection rules for CH_3CH_2Cl.

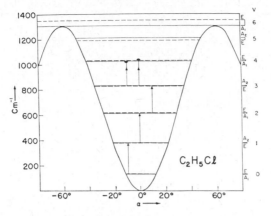

Fig. 8.7 Potential function and energy levels for the torsional motion of CH_3CH_2Cl in terms of the angle of internal rotation α; v is the vibrational quantum number, A_1, A_2, and E are the symmetry species of the vibrational sublevels. The energy is given in cm^{-1}. [W. G. Fateley and F. A. Miller, *Spectrochim. Acta* **19**, 611 (1963), by permission of Pergamon Press.]

For molecules which have no plane of symmetry, the only symmetry elements are I, δ, δ^2. These operations form a group isomorphous to the C_3 group with irreducible representations A and E. The selection rules are

$$A \leftrightarrow A, \qquad E \leftrightarrow E$$

with an energy level schematic as shown in Fig. 8.7 but no distinction between the A_1 and A_2 levels.

Weiss and Leroi[21] have observed the torsional vibration of the CH_3-top in molecules such as CH_3—CH_3 and CH_3—CD_3 under high-pressure and long path-length conditions. These molecules should not have an active torsional vibration according to the discussion presented above. As an explanation of the activity of the torsional vibration in these ethanes, Weiss and Leroi mention the possibility of Coriolis interaction with another vibration. (The influence of the pressure was not considered to be important to induce the activity of these torsional motions.)

j. Spectra and Calculations of the Potential Function Constants

Far-infrared torsional vibration spectra of vapors have been reported by many authors.[6–13,22] Some of the observations are listed in Table 8.1. In most cases the appearance of the spectra is similar to that of ethyl fluoride, shown in Fig. 8.8. There, the sequence of the peaks is interpreted as the Q branches of the $v \rightarrow v'$ transitions, $0 \rightarrow 1$, $1 \rightarrow 2$, and so forth, the R and P branches forming the background absorption. Because of the low energies of the torsional states, higher excited states are considerably populated: Fig. 8.9 shows five peaks, assigned to the $0 \rightarrow 1$, $1 \rightarrow 2$, $\cdots 4 \rightarrow 5$ transitions

TABLE 8.1

TORSIONAL VIBRATION TRANSITIONS (CM^{-1}). THE DISTANCE BETWEEN THE ABSORPTION BANDS IS INDICATED BY Δ

Transition	CH_3CH_2F*	Δ	CH_3CH_2Cl[†]	Δ	CD_3CD_2Br[‡]	Δ	CD_3CD_2Cl[§]	Δ	$CH_3CCl{=}CH_2$‖	Δ	$CD_3CCl{=}CD_2$‖	Δ
$0 \rightarrow 1$	242.7		250.9		180.0		184.1		195.8		145.2	
		17.2		16.2		7.1		8.1		12.7		10.0
$1 \rightarrow 2$	225.5		234.7		172.9		176.0		183.1		135.2	
		17.1		17.6		8.3		7.4		7.6		7.3
$2 \rightarrow 3$	208.4		217.1		164.4		168.6		175.5		127.9	
		31.4		21.3				7.4				5.3
$3 \rightarrow 4$	177.0		195.8				161.2				122.6	
								3.5				4.8
$4 \rightarrow 5$							157.7				117.8	

* Ref. 9.   † Refs. 8 and 12.   ‡ Ref. 12.   § Ref. 13.   ‖ Ref. 22.

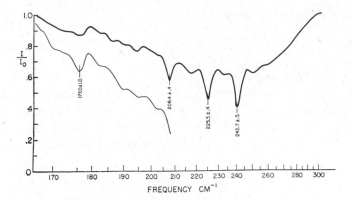

Fig. 8.8 Absorptance spectrum of ethyl fluoride in the 170 to 300 cm^{-1} spectral region. (*Upper trace*): 500 mm Hg pressure in a 30-cm cell. The lower curve was taken with the spectrometer filled with the compound. [G. Sage and W. Klemperer, *J. Chem. Phys.* **39**, 371 (1963).]

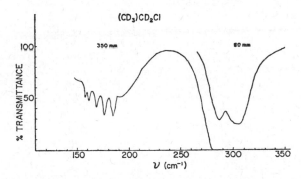

Fig. 8.9 Spectrum of deuterated ethyl chloride in the 150 to 360 cm^{-1} spectral region. Spectral slit width at 180 cm^{-1} is 1.51 cm^{-1}. [K. D. Möller and L. H. London, *J. Chem. Phys.* **47**, 2505 (1967).]

of CD_3CD_2Cl. The model discussed in Sections A.h–i predicts (see Fig. 8.7) that the difference between the $A_1 \rightarrow A_2$ and $E \rightarrow E$ transition for the $v = 3$ to $v = 4$ level is 3–4 cm^{-1} in ethyl chloride. This was observed by Fateley and Miller[8] but not confirmed by Möller *et al.*[12] The corresponding difference of 9 cm^{-1} was not observed in ethyl fluoride either (Sage and Klemperer[9]). Most spectra are complicated by interaction between the internal and external rotation, and therefore the simple level scheme is valid only if this interaction is negligible.

We now show how the potential barrier is estimated from the data, and as an example we use CH_3CH_2Cl. Equation 8.34 yields $V_3 = (\frac{9}{4})Fs$, $E = (\frac{9}{4})Fb$. If a V_6 term is included (see Eq. 8.22), first-order perturbation calculations yield the corrected energy levels

$$E' = E + \frac{V_6}{2}[1 + (6)_{v,v}], \tag{8.50}$$

where $(6)_{v,v}$ is the matrix element given in Eq. 8.45.

From the above equations we see that three transition frequencies are needed to determine V_3, V_6, and F by a best-fit method. For ethyl chloride, it turned out that the "best-fit" was obtained using only V_3 and F (no V_6 term), in good agreement with results from microwave data.[23]

TABLE 8.2

BARRIER OF INTERNAL ROTATION (IN CM^{-1}) FOR ONE-TOP MOLECULES

Molecule	Far-infrared			Microwave		
	V_3	V_6	Ref.	V_3	V_6	Ref.
CH_3CH_2F	1165	-5	9	1159	0	24
CH_3CH_2Cl	1291	0	8	1289	—	23
$CH_3CH=CH_2$	711	-16	8	698	-13	25
$CH_3CF=CH_2$	818	—	8	854	—	26
$CH_3CCl=CH_2$	925	-17	10	934	—	27
$CH_3-\overset{\displaystyle O}{\overset{\diagup\;\diagdown}{CH}}-CH_2$	900	-9	8	895	—	28

In Table 8.2 we have compiled some results on barrier heights obtained by far-infrared and microwave spectroscopic methods. (See also Ref. 29 for additional data on V_3.) The agreement between the V_3 values is rather good, particularly if one considers the approximations included in the theoretical treatment given here. For some molecules, however, this simple theory seems insufficient. This is seen in the variation of Δ, the spacing between the torsional transitions (see Table 8.1): The simple theory, based on Mathieu's equation, predicts that Δ should increase with higher $v \to v'$ numbers—as indeed it does for CH_3CH_2X (X = F, Cl, Br). On the other hand, Δ decreases for CD_3CD_2Cl, $CH_3CCl=CH_2$, and $CD_3CCl=CD_2$. Application of the method described above (best fit of all observed data) yields unreasonable values of the potential constants.[13] (This discrepancy is independent of deuteration.) Probably the coordinate α cannot be considered to be a good normal coordinate here due to the interaction between the torsional and other modes.

Weiss and Leroi[21] have observed the torsional vibrations of C_2H_6, CH_3CD_3, and C_2D_6 and assigned them to 289, 253, and 208 cm^{-1}, respectively. The potential barrier for these three molecules amounts to 1024 cm^{-1}.

C. Two-Top Molecules of C_{2v} and C_s Symmetry with Threefold Barriers

a. Kinetic and Potential Energy

The theory of the torsional vibrations of molecules of C_{2v} symmetry having two internal tops has been investigated by Swalen and Costain[30] and Möller and Andresen.[31] (For molecules of C_s symmetry, see Refs. 12 and 31.)

The Hamiltonian for the torsional part is obtained in a similar way as outlined in Sections A.b,d,e using the PAM representation. The interaction terms between overall and internal rotation and between the torsional vibrations and other vibrations are neglected. The model considers two degrees of freedom. The result is

$$H = H_T + H_I,$$

where

$$H_T = F(p_1{}^2 + p_2{}^2) + \tfrac{1}{2}V(2 - \cos 3\alpha_1 - \cos 3\alpha_2) \tag{8.51}$$

$$H_I = F'(p_1 p_2 + p_2 p_1) + \tfrac{1}{2}V^*(\cos 6\alpha_1 + \cos 6\alpha_2)$$

$$+ V_{12} \cos 3\alpha_1 \cos 3\alpha_2 + V'_{12} \sin 3\alpha_1 \sin 3\alpha_2 .$$

The term H_T describes the independent motion of the two internal tops (see Eq. 8.32) and the term H_I describes the interaction between the torsional motion of the two internal tops. The angles α_1 and α_2 are the angles of rotation for top 1 and 2, respectively, as shown in Fig. 8.10; p_1 and p_2 are the corresponding angular momenta, $p_k = -i(\partial/\partial\alpha_k)$. The potential function of both tops is developed into a Fourier series and V, V^*, V_{12}, and V'_{12} are the coefficients of the expansion.

For C_{2v} molecules, for instance, dimethyl sulfide, we have

$$F = \frac{\hbar^2}{4I_\alpha}\left(\frac{1}{r_z} + \frac{1}{r_x}\right) \qquad r_x = 1 - \frac{2\lambda_x{}^2 I_\alpha}{I_x}$$

$$\tag{8.52}$$

$$F' = \frac{\hbar^2}{4I_\alpha}\left(\frac{1}{r_z} - \frac{1}{r_x}\right) \qquad r_z = 1 - \frac{2\lambda_z{}^2 I_\alpha}{I_z},$$

where I_α is the moment of inertia of one of the tops about its axis and I_x and I_z are the moments of inertia about the principal axes x and z (see Fig. 8.10). The quantities λ_x and λ_z are the direction cosines of the top axes with

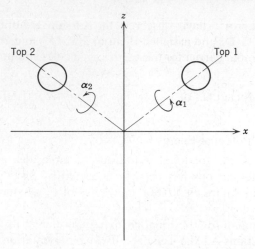

Fig. 8.10   Coordinate system used for the description of the internal rotation in two-top molecules. The coordinates α_1 and α_2 are the angles of internal rotation of the individual tops 1 and 2. The y-axis is perpendicular to the x- and z-axes.

respect to the principal axes x and z. For C_s molecules, for instance dimethyl amine, we have

$$F = \frac{\hbar^2 r'_{xyz}}{2I_\alpha \{ r_{xyz} + 4I_\alpha^2 \rho_{xyz} \}}$$

and

$$F' = \frac{\hbar^2 \left(\dfrac{\lambda_y^2}{I_y} + \dfrac{\lambda_z^2}{I_z} - \dfrac{\lambda_x^2}{I_x} \right)}{2(r_{xyz} + 4I_\alpha^2 \rho_{xyz})}, \tag{8.53}$$

where

$$r'_{xyz} = 1 - I_\alpha \sum_{x,y,z} \frac{\lambda_g^2}{I_g}, \qquad r_{xyz} = 1 - 2I_\alpha \sum_{x,y,z} \frac{\lambda_g^2}{I_g}, \qquad \rho_{xyz} = \frac{\lambda_x^2}{I_x} \left(\frac{\lambda_z^2}{I_z} + \frac{\lambda_y^2}{I_y} \right).$$

I_α is again the moment of inertia of one top about its axis, I_x, I_y, I_z are the moments of inertia about the three principal axes, and λ_x, λ_y, λ_z are the direction cosines of the top axis with reference to the principal axes.

The Hamiltonian in Eq. 8.51 is written in terms of α_1 and α_2 and the corresponding momenta. The coordinates α_1 and α_2 are not normal coordinates (not even approximately). The presentation of H in Eq. 8.51 is chosen for mathematical convenience. The energy level scheme can be obtained by a perturbation treatment using the solutions of H_T as basic functions and H_I as the perturbation term of the Hamiltonian. The basis functions consist of products of the form $U_{v\sigma}(\alpha_1)U_{v'\sigma'}(\alpha_2)$, well known from the foregoing discussion of the one-top problem. The perturbation treatment results in the removal of the degeneracy in the motion of the two equal tops in their excited states. This will become clearer in Section B.b where we will outline a harmonic oscillator approximation, the results of which predict the most important but not all salient features of the spectra. Further below, in Section B.c, we give a more refined perturbation treatment which permits us to explain additional features of the spectra.

b. Harmonic Oscillator Approximation

We shall now outline a simplified consideration which describes the torsion vibration of two equal tops by two normal vibrations (Pitzer[32]). We express this by introducing the normal coordinates $\alpha_+ = (1/\sqrt{2})(\alpha_1 + \alpha_2)$ and $\alpha_- = (1/\sqrt{2})(\alpha_1 - \alpha_2)$ with the corresponding momenta $p_+ = (1/\sqrt{2})(p_1 + p_2)$ and $p_- = (1/\sqrt{2})(p_1 - p_2)$. Inserting these transformed quantities into Eq. 8.51 and expanding the potential function into a power series in terms of α_+ and α_- (neglecting terms higher than second-order), we obtain the following expressions:

$$H = H_+ + H_-$$

$$H_+ = F_+ p_+{}^2 + M\alpha_+{}^2; \qquad H = F_- p_-{}^2 + N\alpha_-{}^2$$

where

$$F_+ = F + F'; \qquad F_- = F - F' \qquad (8.54)$$

and

$$M = (\tfrac{9}{4})\{V - 2(V_{12} - V'_{12})\}; \qquad N = (\tfrac{9}{4})\{V - 2(V_{12} + V'_{12})\}.$$

The eigen functions of the Hamiltonian H are $\psi_{v+}(\alpha_+)\psi_{v-}(\alpha_-)$. The perturbation term to the kinetic energy, $F'(p_1 p_2 + p_2 p_1)$ (see Eq. 8.51), is now implicitly contained in Eq. 8.54. (We note that inclusion of the next higher-order terms of $\alpha_\pm$ would leave us with two anharmonic oscillators).

The energy levels of the harmonic oscillator model are then

$$E = E_+ + E_-, \qquad (8.55)$$

where

$$E_+ = (v_+ + \tfrac{1}{2})\sqrt{4M(F + F')}; \qquad E_- = (v_- + \tfrac{1}{2})\sqrt{4N(F - F')}. \qquad (8.56)$$

In the ground state $v_+ = 0$ and $v_- = 0$; the eigen function is $\psi_{0+}\psi_{0-}$. In the first excited state, we have the combinations $v_+ = 0, v_- = 1$, and $v_+ = 1, v_- = 0$, corresponding to the eigen functions $\psi_{0+}\psi_{1-}$ and ψ_{1+0-}. The second excited state consists of three sublevels, $v_+ = 2, v_- = 0; v_+ = v_- = 1;$

$v_+ = 0$, $v_- = 2$; and so forth. The scheme of the first three states and the allowed transitions (see below) are shown in Fig. 8.11.

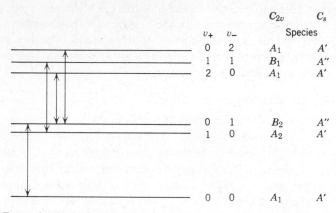

Fig. 8.11   Energy level scheme and selection rules for the harmonic oscillator approximation. v_+ and v_- are the vibrational quantum numbers. The symmetry species of the point group symmetry C_{2v} and C_s are indicated for the energy levels. Transitions are indicated only for C_{2v} symmetry.

For molecules of point group symmetry C_{2v}, the normal coordinate α_+ belongs to the irreducible representation A_2 and α_- belongs to the irreducible representation B_2. For molecules of point group C_s, the corresponding species are A' and A'', respectively. Which of the sublevels of a given state lies higher must be decided by similar arguments as will be discussed in Section C.b;[33] in the case of the first excited state, the asymmetric motion of the tops (B_2 for C_{2v} and A'' for C_s) lies above the corresponding symmetric motion.

The selection rules are as follows. For molecules of point group C_{2v}, the change of the dipole moment is associated with the normal coordinate $\alpha_-(B_2)$; in the fundamental band ($0 \rightarrow 1$), only the $A_1 \rightarrow B_2$ transition is therefore allowed. For molecules of point group C_s, both fundamental bands are allowed.

The allowed transitions for a molecule of point group symmetry C_{2v} between the first three levels are shown in Fig. 8.11. There are no restrictions for C_s molecules.

From Eq. 8.56 we find that the frequencies of the two $0 \rightarrow 1$ transitions are given by

$$v_+ = \sqrt{4M(F + F')},$$
$$v_- = \sqrt{4N(F - F')}. \tag{8.57}$$

Equation 8.57 shows that the first excited state is split into the two sublevels as a consequence of the interaction terms in F', V^*, V_{12}, and V'_{12} (see Eqs. 8.51 and 8.54). These terms are usually small compared to those in V and F. If we neglect the perturbation terms (that is, set $H_I = 0$), we obtain the expression

$$v = 3\sqrt{VF}, \tag{8.58}$$

the frequency for the fundamental transition for two independent internal tops (compare this with Eq. 8.38 in Section A.h).

c. Periodic Potential Model. Symmetry Properties, Energy Level Scheme, Selection Rules

As we mentioned at the end of Section B.a, we now solve Eq. 8.51 with a perturbation treatment using the $U_{v\sigma}(\alpha_1)U_{v'\sigma'}(\alpha_2)$ as basis functions and H_I as perturbation term.[31] Because of the limited resolving power of present infrared equipment, we need not go beyond first-order theory. The evaluation of the matrix elements of the general type

$$\int U_{v\sigma} Z U_{v'\sigma'} \, d\tau, \tag{8.59}$$

where Z is the perturbation operator, is straightforward (see References in Section A.h). The calculations are facilitated if linear combinations of the eigen functions are taken which belong to irreducible representations of the largest symmetry group which leaves the Hamiltonian (Eq. 8.51) invariant. The corresponding symmetry operations are the following (see Fig. 8.10):

Rotation about the z-axis by π: $\qquad C_2(z): \alpha_1 \leftrightarrow \alpha_2$

Reflection on the y, z-plane: $\qquad \sigma_v(yz): \alpha_1 \leftrightarrow -\alpha_2$

$$\delta_1{}^n: \alpha_1 \to \alpha_1 + \frac{2\pi}{3}n$$

Rotation of top (1) and (2) by $2\pi/3$ about their respective symmetry axes $\Bigg\}$: $\qquad\qquad\qquad\qquad$ (8.60)

$$\delta_2{}^m: \alpha_2 \to \alpha_2 + \frac{2\pi}{3}m$$

$$(n, m = 1, 2)$$

(The dimethyl ether molecule, CH_3-O-CH_3, is a good example on which to imagine these operations.)

Looking at Eq. 8.60, we notice that the elements C_2 and σ_v generate the group C_{2v}. However, with the additional operations which describe the rotational symmetry of the internal tops, we end up with a group of order 36—which we shall designate by the symbol G_{36}. (This situation of additional symmetry is similar to the one discussed in the one-top case.)

The group G_{36} can be represented as the direct product

$$G_{36} = C_{3v}^- \times C_{3v}^+,\qquad(8.61)$$

where the elements of C_{3v}^- and C_{3v}^+ are given by

$$C_{3v}^- = \{E, (\delta_1\delta_2{}^2, \delta_1{}^2\delta_2), (C_2, C_2\delta_1\delta_2{}^2, C_2\delta_1{}^2\delta_2)\}$$
$$C_{3v}^+ = \{E, (\delta_1\delta_2, \delta_1{}^2\delta_2{}^2), (\sigma, \sigma\delta_1\delta_2, \sigma\delta_1{}^2\delta_2{}^2)\},\qquad(8.62)$$

with E as unit element. Both subgroups are isomorphous with C_{3v}. The group G_{36} has nine classes of conjugate elements.

As a result, there are now four one-dimensional representations, A_1A_1, A_1A_2, A_2A_1, and A_2A_2; four two-dimensional representations, A_1E, A_2E, EA_1, EA_2; and one four-dimensional representation EE. Thus, for the perturbation treatment the eigen functions are arranged into linear combinations which represent irreducible representations of G_{36}.[34,35] The energy level scheme as obtained with the first-order perturbation treatment[31] is shown in Fig. 8.12. The spacing between the two levels of the first excited state is of the order of 10–$30\,\mathrm{cm}^{-1}$ whereas the spacing between the four sublevels of each level is of the order of $0.01\,\mathrm{cm}^{-1}$.[36] The splitting of these sublevels is too small to be observable in far-infrared spectra; we can thus consider the energy level scheme to be comparable to that obtained from an anharmonic oscillator approximation (see beginning of Section B.b). Of course, this holds only for the first excited states and at barrier heights which are not too low ($s > 30$, see Section A.h).

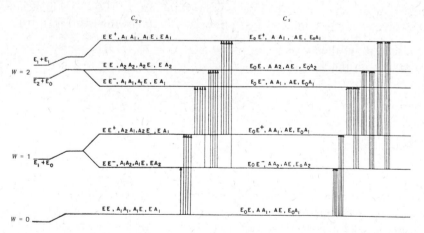

Fig. 8.12   Energy level scheme and selection rules for the 0–1 and 1–2 transitions of the periodic potential model. Symmetry species according to the group G_{36} are indicated for molecules of C_{2v} point group symmetry. For molecules of C_s point group symmetry a similar classification is given. [K. D. Möller and H. G. Andresen, *J. Chem. Phys.* **37**, 1800 (1962).]

We now briefly discuss the selection rules in the group G_{36} (molecule with point group symmetry C_{2v}). In the first approximation, the change of the dipole moment during the torsional motion is associated with the co-ordinate $\alpha_- = (1/\sqrt{2})(\alpha_1 - \alpha_2)$ (see also Section A.i); that is, to the species A_2A_1 of the group G_{36}. Hence, in order for a transition to be active, the direct product

$$\Gamma_w \times A_2A_1 \times \Gamma_{w'} \qquad (8.63)$$

must contain the totally symmetric representation A_1A_1. (As in Section A.i, Γ_w and $\Gamma_{w'}$ are the representations of the lower and upper energy levels, respectively.) The evaluation of Eq. 8.63 leads to the following selection rules (see Fig. 8.12):

$$A_1A_1 \leftrightarrow A_2A_1 \qquad EA_1 \leftrightarrow EA_1$$
$$A_1A_2 \leftrightarrow A_2A_2, \qquad EA_2 \leftrightarrow EA_2 \qquad (8.64)$$
$$A_1E \leftrightarrow A_2E, \qquad EE \leftrightarrow EE.$$

We note that the selection rules permit transitions from the ground level to both sublevels of the first excited state ($w = 1$)—compare Fig. 8.12 with Fig. 8.11. This is interpreted to be a consequence of the tunnelling effect of the CH_3 groups. We therefore expect that this additional transition is of low intensity for high barriers; the corresponding spectrum will then not be different from that computed on the basis of the harmonic oscillator model. However, in cases of low barriers we may be able to observe both sets of $0 \rightarrow 1$ transitions.

The selection rules for molecules whose structure belongs to point group C_s are not different[31] from those established for an harmonic oscillator model (see Section B.b). The expected transitions are also shown in Fig. 8.12.

d. Spectra and the Calculation of the Potential Function Constants

Far-infrared torsional vibration spectra of two-top molecules have been reported by several authors.[6,12,37,38] Figure 8.13 presents the spectra of $(CH_3)_2S$ and $(CH_3)_2C{=}CH_2$.[38] The appearance of the torsional spectrum of both compounds is very similar. According to their structure, both molecules should have strong Q branches. From microwave investigations, the barrier of internal rotation is reported to be $V = 741$ cm^{-1} for dimethyl sulfide[39] and $V = 774$ cm^{-1} for isobutylene.[40] The general appearance of the observed pattern has been discussed by Möller and Andresen,[31] but a detailed interpretation of these bands has not yet been obtained.

Figure 8.14 shows the far-infrared torsional vibration spectrum of acetone. The barrier height, from microwave investigations, is only $V = 274$ cm^{-1}.[30] Following the discussion by Smith et al.[38] the only explanation for the two

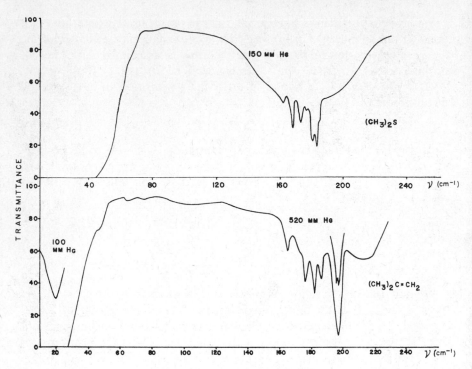

Fig. 8.13 (*Upper curve*): Far-infrared spectrum of dimethyl sulfide. (*Lower curve*): isobutylene. Spectral slit width at 183.5 is 0.75 cm^{-1} in both cases. [D. R. Smith, B. K. McKenna, and K. D. Möller, *J. Chem. Phys.* **45**, 1904 (1966).]

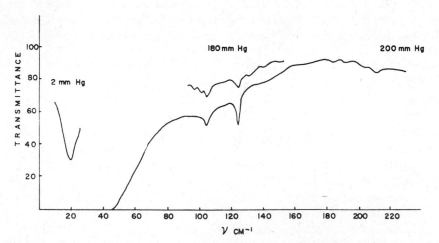

Fig. 8.14   Far-infrared spectrum of acetone. Spectral slit width at 105.0 cm^{-1} is 0.6 cm^{-1}. [D. R. Smith, B. K. McKenna, and K. D. Möller, *J. Chem. Phys.* **45**, 1904 (1966).]

observed bands is that there are two $0 \rightarrow 1$ transitions according to the selection rules of the group G_{36}. Figure 8.15 depicts the torsional vibration band of dimethyl ether, $(CH_3)_2O$, and $(CD_3)_2O$.[12] The spectrum of the fully protonated compound is in excellent agreement with the spectrum previously reported by Fateley and Miller.[37] The striking difference between the band contours of the protonated and deuterated compound is interesting and has not yet been explained. (However, we have to keep in mind that investigations of the vibration-rotation interaction of this torsional band have not yet been performed.) The barrier height (from microwave data) is here $V = 950 \text{ cm}^{-1}$.[41]

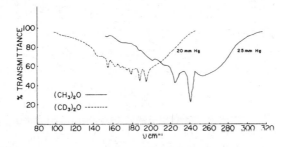

Fig. 8.15   Far-infrared spectrum of dimethyl ether (closed line). Spectral slit width at 241 cm^{-1} is 1.82 cm^{-1}. Dotted line: deuterated dimethyl ether. Spectral slit width at 170 cm^{-1} is 1.17 cm^{-1}. [K. D. Möller, A. R. DeMeo, D. R. Smith, and L. H. London, *J. Chem. Phys.* **47**, 2609 (1967).]

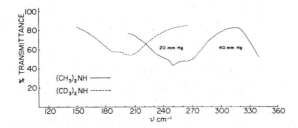

Fig. 8.16   Far-infrared spectrum of dimethyl amine (closed line). Spectral slit width at 257 cm^{-1} is 2.09 cm^{-1}. Dotted line: deuterated dimethyl amine. Spectral slit width at 198 cm^{-1} is 1.60 cm^{-1}. [K. D. Möller, A. R. DeMeo, D. R. Smith, and L. H. London, *J. Chem. Phys.* **47**, 2609 (1967).]

Figure 8.16 shows the torsional bands of two dimethyl amines, $(CH_3)_2NH$ and $(CD_3)_2NH$. The structure of these molecules belongs to point group C_s. The potential barrier can be calculated from the knowledge of $E_1 - E_0$ (see Fig. 8.12 and Eq. 8.58). The influence of the interaction term of the type V_{12} $\cos 3\alpha_1 \cos 3\alpha_2$ and $(V^*/2)$ ($\cos 6\alpha_1 + \cos 6\alpha_2$) is small and will be neglected for all levels. The frequency $v_M = (v_2 - v_1)/2$, where v_1 and v_2 are the transition frequencies from the ground state to the lower and upper level of the first excited state, respectively, corresponds then to $E_1 - E_0$ since the splitting of the first excited states is symmetric. The following values for the potential barrier of dimethyl amines have been obtained with this method: For $(CH_3)_2NH$ we find that $v_M = 257 \, cm^{-1}$. With $F = 6.511 \, cm^{-1}$, we obtain $V = 1266 \, cm^{-1}$. The barrier height from a microwave investigation[42] was estimated to be $V = 1267 \, cm^{-1}$. For $(CD_3)_2NH$ we find $v_M = 198 \, cm^{-1}$ and obtain, with $F = 3.765 \, cm^{-1}$, $V = 1266 \, cm^{-1}$.

D. THREE-TOP MOLECULES (C_{3v}) WITH THREEFOLD BARRIERS

a. Theoretical Considerations

Torsional vibrations of some $(CH_3)_3Y$-type molecules have been investigated in the far-infrared.[12,37] The theoretical analysis, based on the model of the periodic potential function, has been reported by Möller and Andresen.[43] As in the case of the two-top molecules discussed in the previous sections, we shall compare the results of the extended model of three-top molecules with the one obtained from the parabolic potential approximation. As expected, we shall find that the relations are rather similar to those discussed in Section B.c for molecules of C_{2v} symmetry.

The molecules $(CH_3)_3Y$ have structural symmetry C_{3v}. We first take the parabolic potential (see Fig. 8.17): The first excited state of the torsional motion consists of the single sublevel A_2 and the doubly degenerate level E. Noting that the change of the dipole moment belongs to species E, we see that the only infrared-active $0 \to 1$ transition[12,33] is $A_1 - E$ (solid line in Fig. 8.17).

We now take the periodic potential. The Hamiltonian, which is analogous to that described in Eq. 8.51 for two-top molecules, is invariant under a

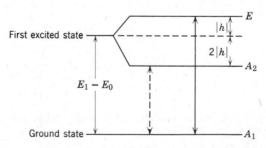

Fig. 8.17   Energy-level scheme and selection rules for the 0–1 transition in the parabolic potential model. An additional transition, which is possible in the case of the periodic potential model, is indicated by a broken line.

group of order 162.[43] The energy schematic is similar to that shown in Fig. 8.17, but the levels A_1, A_2 and E now consist of a number of sublevels. This situation is again similar to the one discussed in detail for two-top molecules with a parabolic and periodic potential function (as shown in Figs. 8.11 and 8.12). Again, the spacing between these sublevels is very small, and the evaluation of the selection rules according to the group of order 162 results in transitions between the sublevels of the ensembles of lines denoted now also by A_1, A_2, and E. Transitions are allowed from sublevels of the ground state A_1 to sublevels of the first excited state denoted by E and indicated in Fig. 8.17 by a solid line. But the selection rules also allow transitions from sublevels of the ground state to sublevels of the first excited state denoted by A_2. These transitions are indicated in Fig. 8.17 by a broken line. The latter transitions are caused by the tunnel effect and, as has been discussed above for the two-top molecules, we expect to observe them only for molecules with a low barrier height.

The energy level scheme as a function of the structural data and potential constants is given in Ref. 43. If interaction terms of the three tops are neglected, the energy difference between the ground state and the first excited state is $E_1 - E_0$, indicated in Fig. 8.17 by a dotted line. The predominant interaction terms shift the ensemble E up by the amount of $|h|$, whereas the ensemble A_2 is shifted down by the amount $2|h|$. The difference $3|h|$ between the A_2 and E levels has been determined by a microwave investigation of trimethyl amine[33] to be about 30 cm^{-1}.

The calculation of the potential barrier is difficult since so far only one far-infrared transition frequency, assigned to the $A_1 - E$ torsional transition, has been reported. Thus, no far-infrared observations are available to estimate the value of $3|h|$, and it is therefore not possible to obtain an experimental value for $E_1 - E_0$ (which could then be used with Eq. 8.58 for the calculation of the potential barrier height). The use of the above-mentioned observed frequency of the torsional vibration of the $A_1 - E$ transition will yield only an upper limit for the potential barrier (Eq. 8.58) because it is too high by the amount $|h|$.

b. Spectra and Potential Barriers

The far-infrared absorption spectrum of trimethyl amine is depicted in Fig. 8.18. The peak frequency at 269 cm^{-1} has been first observed in the far-infrared by Fateley and Miller.[37] A microwave study by Lide and Mann[33] placed the two torsional vibration frequencies at $v_{A_2} = 263 \pm 20$ cm^{-1} and $v_E = 290 \pm 20$ cm^{-1}. From a study of the l-type doubling, they found that $v_E > v_{A_2}$. This is interpreted as follows. In the E-mode one top is moving in the opposite direction with respect to the other two tops and, therefore, more repulsive interaction is present (resulting in a higher tor-

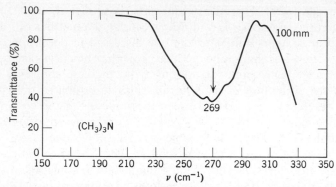

Fig. 8.18   Far-infrared torsional band of trimethyl amine. Pressure, 100 mm Hg. Path length of the absorption cell, about 1 m.

sional frequency) than in the A_2 mode where all the tops move in the same direction. Goldfarb and Khare[44] have reported a torsional vibration frequency of $v = 208\ \mathrm{cm}^{-1}$ for $(CD_3)_3N$.

For tertiary butyl cyanide, $(CH_3)_3CCN$, and tertiary butyl acetylene, $(CH_3)_3CCCH$, small absorption peaks at $v = 271\ \mathrm{cm}^{-1}$ and $v = 284\ \mathrm{cm}^{-1}$, respectively, have been observed.[12] Both absorptions are located in spectral regions where they should occur according to a microwave study by Nugent et al.[45]

We finally add that for trimethyl amine a Raman torsional frequency of $v = 274\ \mathrm{cm}^{-1}$ has been reported by Kohlrausch in an early study.[46]

As mentioned above, the use of these observed torsional frequencies for the calculation of the potential barrier yields only an upper limit of the barrier height (for more details, see Ref. 12). Table 8.3 lists some results for such calculations. We see that the results for the trimethyl amines are satisfactory considering that only an upper limit of the barrier could be obtained.

TABLE 8.3

BARRIERS TO INTERNAL ROTATION (IN CM^{-1}) FOR THREE-TOP MOLECULES

Compound	Upper limit of barrier height V	Barrier height from microwave data
$(CH_3)_3N$	1510 (Ref. 12)	$1530 \pm 10–20\%$ (Ref. 32)
$(CD_3)_3N$	1640 (Ref. 44)	—
$(CH_3)_3CCN$	1640 (Ref. 12)	$1400 \pm 20\%$ (Ref. 45)
$(CH_3)_3CCCH$	1800 (Ref. 12)	$1400 \pm 20\%$ (Ref. 45)

The results for the last two compounds in Table 8.3 are to be considered rather tentative (in both the microwave investigation and in the far-infrared study).

E. TWOFOLD BARRIERS

a. Hydrogen Peroxide and Deuterated Hydrogen Peroxide

One of the smallest molecules with internal rotation is hydrogen peroxide. Its far-infrared spectrum has been studied by Hunt et al.,[47] and a determination of the potential function for internal rotation has been made.

The barrier of internal rotation may be attributed in first approximation[48] to* (1) the interaction between the two nonbonded pairs of p-electrons associated with the oxygen atoms, and (2) to the interaction between the dipole moments of the two OH-groups. The first effect results in a potential energy term represented by a periodic function of $2x$, where x is the angle between

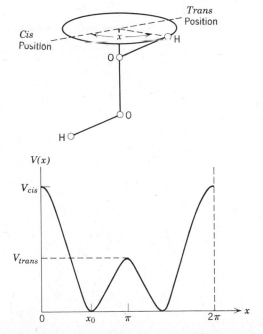

Fig. 8.19   Model of H_2O_2 employed for the discussion of the hindered rotation and the potential function. The angle of internal rotation is designated by x. [R. H. Hunt, R. A. Leacock, C. W. Peters, and K. T. Hecht, J. Chem. Phys. **42**, 1931 (1965).]

* Movement of the oxygen atoms during internal rotation is not taken into account. Therefore, the z-axis of the molecule-fixed coordinate system passes through the oxygen atoms.

the relative positions of the OH-bonds, as indicated in Fig. 8.19. This term can be represented by $V_2 \cos 2x$ with minima between the *cis* $(x = 0)$ and *trans* $(x = \pi)$ positions. The second effect gives a potential energy term periodic in 2π, which can be represented by $V_1 \cos x$. Finer details of the potential energy are taken into account by $V_3 \cos 3x$ or higher-order terms. The potential function used to describe internal rotation in hydrogen peroxide is[47]

$$V = V_1 \cos x + V_2 \cos 2x + V_3 \cos 3x. \qquad (8.65)$$

The potential constants V_1, V_2, and V_3 are related to the potential constants $V(trans)$, $V(cis)$, and to x_0, respectively (see Fig. 8.19 and Ref. 47).

The far-infrared spectrum, which was observed with a grating spectrometer in the 15 to 700 cm^{-1} spectral region under 0.3 cm^{-1} resolution, is depicted in Fig. 8.20.

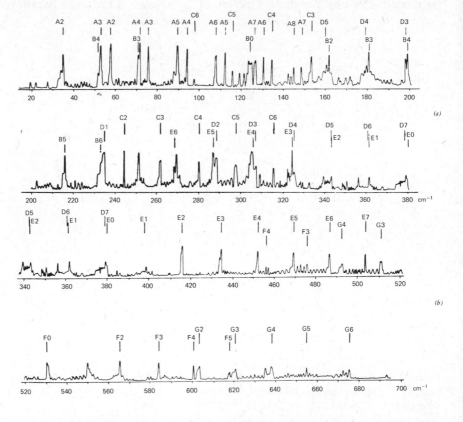

Fig. 8.20   Absorption spectrum of H_2O_2 vapor in the region 700 to 15 cm^{-1}. [R. H. Hunt, R. A. Leacock, C. W. Peters, and K. T. Hecht, *J. Chem. Phys.* **42**, 1931 (1965).]

For the theoretical treatment of the internal rotation of H_2O_2, a semi-rigid model is used in which the OH and OO distances and the HOO angle are considered to remain fixed (see Fig. 8.19). The classical Hamiltonian for the kinetic energy is obtained analogously to the procedure discussed in Section A.c. The coordinate system x, y, z is fixed in the center of mass of the molecule. The Eulerian angles ψ, θ, ϕ are used to describe the orientation of this coordinate system relative to the space-fixed axes X, Y, Z. The angle ϕ is defined by $\phi = \frac{1}{2}(\phi_1 + \phi_2)$, where ϕ_1 and ϕ_2 are the azimuthal angles of the two OH groups, respectively. The angle x^* of internal rotation is given by $x = \phi_1 - \phi_2$; therefore, the semirigid model of three degrees of freedom of overall rotation and one degree of freedom of internal rotation is described by ψ, θ, ϕ, and x.

The classical Hamiltonian for the kinetic energy is then explicitly given by[47]

$$
\begin{aligned}
H = {} & \frac{1}{2}\frac{G}{AG - F^2}P_x^{\,2} + \frac{1}{2}\frac{C}{BC - D^2}P_y^{\,2} + \frac{1}{2}\frac{B}{BC - D^2}P_z^{\,2} \\
& - \frac{D}{BC - D^2}P_yP_z - \frac{2F}{AG - F^2}p_xP_x + \frac{1}{2}\frac{4A}{AG - F^2}p_x^{\,2},
\end{aligned}
\tag{8.66}
$$

where the parameters A to G are functions of the internal rotation angle x;[47] P_x, P_y, and P_z are the components of the total angular momentum along the molecule-fixed axes x, y, z; and p_x is the conjugate momentum to the internal rotation coordinate x. We replace P_x, P_y, P_z, and p_x by their corresponding quantum-mechanical operators which follow the commutation relations

$$
[P_i, P_j] = -i\hbar P_k, \qquad [P_i, p_x] = 0.
\tag{8.67}
$$

The cross term p_xP_x in Eq. 8.66 is removed by transforming P_x, P_y, P_z, and p_x into P'_x, P'_y, P'_z, and p'_x with a transformation matrix analogous to that discussed in Section A.c (but more complicated). The P'_x, P'_y, P'_z, and p'_x follow similar commutation relations as those given in Eq. 8.67 for P_x, P_y, P_z, and p_x. The Hamiltonian for the kinetic energy is then obtained as

$$
\begin{aligned}
H_K = {} & \beta(x)P'^2 + v(x)P'^2_z + \gamma(x)[P'^2_x - P'^2_y] \\
& + \delta(x)[P'_yP'_z + P'_zP'_y] + \alpha(x)p'^2_x,
\end{aligned}
\tag{8.68}
$$

where P'^2 is the square of the total angular momentum. The coefficients of Eq. 8.68 depend on the structural parameters; $\beta(x), v(x), \gamma(x)$, and $\alpha(x)$ are periodic in 2π, $\delta(x)$ is periodic in 4π. These coefficients are expressed in the appropriate Fourier series as a function of the internal rotation angle x.[47]

---

* Following Hunt et al.,[47] we use the symbol x for the angle of internal rotation as well as for one component of the coordinate system of the molecule-fixed axes.

The term $\alpha(x)p_x'^2$ in Eq. 8.68 must be symmetrized before a quantum-mechanical calculation can be performed. The interaction of the overall and internal rotation is contained in the coefficient $\beta(x)$, $\nu(x)$, $\gamma(x)$, and $\delta(x)$ and in the boundary conditions for the internal rotation problem (analogous to what we had discussed in Section A.c). The total Hamiltonian used for the interpretation of the spectrum of hydrogen peroxide is the combination of Eqs. 8.65 and 8.68:

$$H_{tot} = H_K + V(x). \tag{8.69}$$

The hydrogen peroxide molecule turns out to be a nearly symmetric top with internal rotation of the two OH-groups. Therefore, as basis functions for a perturbation treatment of the Hamiltonian in Eq. 8.69, we take product functions of the solutions of a symmetric-top Hamiltonian

$$H_{sym.top} = \beta_0 P'^2 + \nu_0 P_z'^2, \tag{8.70}$$

where $\beta_0 = \beta(0)$ and $\nu_0 = \nu(0)$, and of an internal rotation Hamiltonian,

$$H_{int.rot} = \alpha(x)p_x'^2 + V(x). \tag{8.71}$$

The solutions of Eq. 8.70 are well known. The solutions of Eq. 8.71 [inserting for $V(x)$ the expression given in Eq. 8.65] must be calculated numerically. Subsequently, the influence of the other terms in the Hamiltonian (Eq. 8.69), namely the ones which represent the asymmetric terms of the overall rotation and the interaction terms between the overall and the internal rotation, can be calculated by perturbation theory. As a consequence, the K-degeneracy of the solutions of Eq. 8.70 is removed and the rotational levels are classified by J, K_+, and K_-.

The solutions $M_{n\tau}$ of the internal rotation Hamiltonian (Eq. 8.71) have four basic symmetries. They are symmetric or antisymmetric to the following operations:

$$\sigma_c\,(cis)\!: x \rightarrow -x, \text{ reflection on } x = 0$$

$$\sigma_t\,(trans)\!: x \rightarrow 2\pi - x, \text{ reflection on } x = \pi.$$

In Table 8.4 we have given the Fourier expansion of the $M_{n\tau}$ as well as their correspondence to the four possible symmetry combinations which are indicated and labeled by $\tau = 1$ to 4. The coefficients in the expansion of the $M_{n\tau}$ depend on the vibrational quantum number n.

The transformation which was used to remove the interaction term $p_x P_x$ in the Hamiltonian (Eq. 8.66) introduces a K-dependence into the boundary conditions of the internal rotation problem. Therefore, the boundary condition contains now the coupling between the overall and internal rotation (see Section A.c). As a consequence, the $M_{n\tau}$ depend also on K; their symmetry dependence is also indicated in Table 8.4.

TABLE 8.4*

SERIAL EXPANSION OF THE INTERNAL ROTATION WAVE FUNCTIONS, AND THEIR
SYMMETRY PROPERTIES AND QUANTUM NUMBERS. THE COEFFICIENTS OF THE EXPAN-
SION DEPEND ON THE VIBRATIONAL QUANTUM NUMBER n

K-value (even or odd)	Symmetry under		Quantum number τ	Expansion of internal rotation wave function
	σ_t	σ_c		
e	s	s	1	$M_1(x) = \pi^{-1}\Sigma a(n)\cos nx$
o	s	a	2	$M_2(x) = \pi^{-1}\Sigma b(n)\sin(n+\tfrac{1}{2})x$
o	a	s	3	$M_3(x) = \pi^{-1}\Sigma c(n)\cos(n+\tfrac{1}{2})x$
e	a	a	4	$M_4(x) = \pi^{-1}\Sigma d(n)\sin nx$

* R. H. Hunt, R. A. Leacock, C. W. Peters, and K. T. Hecht, *J. Chem. Phys.* **42**, 1931 (1965).

The total wave function is labeled by J, n, K_+, K_-, τ By taking into account the symmetry properties of the total wave function and of the expression of the dipole moment, $\mu = \mu_0 \cos(x/2)$, the selection rules are derived in the usual manner. They are collected[47] in Table 8.5. (Dreizler[49] has derived more general selection rules, including interaction of the overall and internal rotation, by purely group-theoretical methods.)

TABLE 8.5*

SELECTION RULES FOR H_2O_2

$\Delta J = 0$ transitions	$\Delta J = \pm 1$ transitions
$K+ \rightarrow K'+$	$K+ \rightarrow K'-$
$K- \rightarrow K'-$ with $\Delta K = \pm 1$	$K- \rightarrow K'+$ with $\Delta K = \pm 1$
$n \rightarrow n'$	$n \rightarrow n'$
$\tau = 1 \leftrightarrow 3$ or $\tau = 2 \leftrightarrow 4$	$\tau = 1 \leftrightarrow 3$ or $\tau = 2 \leftrightarrow 4$

* R. H. Hunt, R. A. Leacock, C. W. Peters, and K. T. Hecht, *J. Chem. Phys.* **42**, 1931 (1965).

A schematic of the potential function and the energy levels, labeled according to Table 8.4, is shown in Fig. 8.21. In the limiting case of $V_{cis} \rightarrow \infty$, each of the two sets of levels $\tau = 1, 2$ and $\tau = 3, 4$ for the same n becomes degenerate. (Only reflection by σ_t remains; we have, therefore, alternately symmetric and antisymmetric wave functions in analogy to the corresponding relations in the inversion of the ammonia molecule.) If, in addition, $V_{trans} \to \infty$, all four levels of a given n are degenerate. Thus, the splittings of the levels $\tau = 1, 2$ and $\tau = 3, 4$ are governed by tunneling of the internal rotation motion through the *cis* barrier whereas the separation of the pair $\tilde{\tau} = 1, 2$ and $\tau = 3, 4$ is associated with tunnelling through the *trans* barrier. The corresponding splittings are indicated by Δ *cis* and Δ *trans*, respectively, in Fig. 8.21.

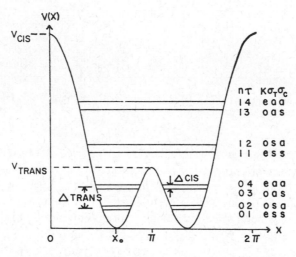

Fig. 8.21 Energy-level scheme describing the hindered rotation of H_2O_2 and the classification of the energy levels by n, τ, K. The symmetry of the levels with respect to σ_c and σ_t is also indicated. [R. H. Hunt, R. A. Leacock, C. W. Peters, and K. T. Hecht, *J. Chem. Phys.* **42**, 1931 (1965).]

The spectrum of H_2O_2, shown in Fig. 8.20, possesses band centers located at 11.43, 198.57, and 370.70 cm^{-1} according to a perpendicular band analysis.[50] These are listed as A, C, E, in Table 8.6. The bands have predominant Q branches. The assignment is indicated in Fig. 8.20. The other band centers, also listed in Table 8.6, are more difficult to obtain because of insufficient observations.

The final energy level scheme and the potential function together with the observed frequencies are presented in Fig. 8.22. The three potential constants are

$$V_{trans} = 386 \pm 4 \text{ cm}^{-1}, \qquad V_{cis} = 2460 \pm 25 \text{ cm}^{-1}, \qquad x_0 = 111.5 \pm 0.5°.$$

$$(8.72)$$

TABLE 8.6*

OBSERVED BAND FREQUENCIES AND BAND ASSIGNMENTS
IN H_2O_2

	Band frequency, cm^{-1}	Internal rotation transition $n\tau \to n'\tau'$
A	11.43	$01 \leftrightarrow 03$ and $02 \leftrightarrow 04$
B	116.51	$11 \to 13$ and $12 \to 14$
C	198.57	$13 \to 21$ and $14 \to 22$
D	242.76	$03 \to 11$ and $04 \to 12$
E	370.70	$01 \to 13$ and $02 \to 14$
F	521.68	$11 \to 23$ and $12 \to 24$
G	557.84	$03 \to 21$ and $04 \to 22$

*R. H. Hunt, R. A. Leacock, C. W. Peters, and
K. T. Hecht, *J. Chem. Phys.* **42**, 1931 (1965).

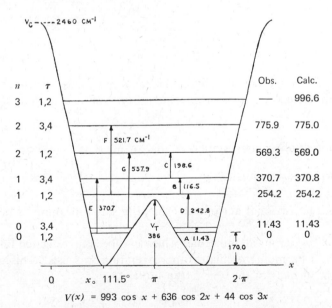

$$V(x) = 993 \cos x + 636 \cos 2x + 44 \cos 3x$$

Fig. 8.22 Potential function, position of the energy levels, and observed frequencies for the internal rotation in H_2O_2. Allowed transitions are indicated by vertical lines (see Table 8.5). [R. H. Hunt, R. A. Leacock, C. W. Peters, *J. Chem. Phys.* **42**, 1931 (1965).]

The structural data which were used are

$$r(\text{O}-\text{H}) = 0.950 \pm 0.005 \text{ Å}, \qquad r(\text{O}-\text{O}) = 1.475 \pm 0.004 \text{ Å},$$

$$\text{angle OOH} = 94.8 \pm 2.0^0. \tag{8.73}$$

A similar investigation for D_2O_2 has been reported by Hunt and Leacock.[51] The potential function found for this molecule is

$$V(x) = 994 \cos x + 641 \cos 2x + 55 \cos 3x, \tag{8.74}$$

with the constants for the barriers given by

$$V_{trans} = 377 \text{ cm}^{-1}, \qquad V_{cis} = 2470 \text{ cm}^{-1}, \qquad x_0 = 110.8^0. \tag{8.75}$$

Comparison with the data for H_2O_2 shows that the difference between the two potential functions is small.

b. Twofold Barriers about a Carbon—Carbon Bond

Investigations of torsional vibrations about carbon–carbon bonds have been reported by Fateley et al.[52] and Miller et al.[53] on molecules such as benzaldehyde, para-substituted benzaldehyde, and substituted ethanes of structure

All results discussed here are obtained from vapor-phase measurements.

The Hamiltonian used for the interpretation is similar to that of Eq. 8.32, namely

$$H_T = Fp^2 + V(\alpha), \tag{8.76}$$

with the potential function similar to that of Eq. 8.65,

$$V(\alpha) = \frac{V_1}{2}(1 - \cos \alpha) + \frac{V_2}{2}(1 - \cos 2\alpha) + \frac{V_3}{2}(1 - \cos 3\alpha), \tag{8.77}$$

where α is the torsional angle and F has the same meaning as defined in Eqs. 8.32.

Since the barrier for all molecules which had been investigated turned out to be high, a harmonic oscillator approximation was used. This gives the following frequency for the $0 \rightarrow 1$ transition:

$$v = \sqrt{F(V_1 + 4V_2 + 9V_3)} = \sqrt{FV^*}. \tag{8.78}$$

The selection rules are those of the harmonic oscillator. The observed frequencies and the calculated V^*-values for some straight-chain molecules are compiled in Table 8.7. Since only one band was observed in each case, little information was obtained. This is similar to some of the cases discussed above (see Section C.b).

TABLE 8.7

Observed Frequencies (in cm^{-1}) and Potential Barriers V^* (in cm^{-1}) for Twofold Barriers about a C—C Bond in Various Molecules

Compound	Frequency, cm^{-1}	V^*, cm^{-1}
Butadiene	163	9820
Isoprene	153	12480
Fluoroprene	147.5	11330
Chloroprene	144	11110
Acrolein	157	7430
Methy vinyl ketone	101	5490
Methacrolein	165	11150
Glyoxal	128	4810
Pyruvaldehyde	105	4340
Biacetyl	48	3520

Table 8.8 shows the results of the studies on benzaldehyde, some of its para-substituted derivatives, and various similar molecules. By reasons of symmetry, the V_1 and V_3 terms (Eqs. 8.76 to 8.78) vanish in these molecules,

TABLE 8.8

Observed Frequencies (in cm^{-1}) and Potential Constants (in cm^{-1}) for Benzaldehyde and Various Para-substituted Compounds and Similar Aldehydes and Ketones

Compound	Frequency, cm^{-1}	V_2, cm^{-1}
Benzaldehyde	111	1630
p-F-benzaldehyde	93.5	1253
p-Cl-benzaldehyde	81.5	983
p-Br-benzaldehyde	73.5	829
p-CH$_3$-benzaldehyde	89.5	1214
Pyridine-4-aldehyde	102	1340
Acetophenone	48	1085
p-F-acetophenone	51	1225

and the calculation of the potential constant V_2 can be based on the single observed frequency (in the harmonic oscillator approximation).

c. Meta- and Ortho-substituted Benzaldehyde and Furan-2-aldehyde

The far-infrared absorption spectra of meta- and ortho-substituted benzaldehyde and furan-2-aldehyde show two absorption lines in the same spectral region where only one line was observed for the para-substituted benzaldehydes. Miller *et al.*[53] have interpreted this, with the help of a relative population study, as evidence for the existence of two rotamers. This interpretation for furan-2-aldehyde is in agreement with a microwave study by Mönnig.[54]

For an interpretation of the data, the potential function is taken to be that of Eq. 8.77; it is assumed that $V_3 = 0$ and that there are two minima at $\alpha = 0$ and $\alpha = \pi$. In the harmonic oscillator approximation, one has then two oscillators centered at these two positions. The potential constants for the two oscillators are

$$V^*_{\alpha=0} = V_1 + 4V_2$$

and   (8.79)

$$V^*_{\alpha=\pi} = -V_1 + 4V_2.$$

TABLE 8.9

OBSERVED FREQUENCIES (IN CM^{-1}) AND POTENTIAL CONSTANTS (CM^{-1}) FOR VARIOUS *cis*- AND *trans*- ROTAMERS

Compound	Frequencies, cm^{-1}		V_1, cm^{-1}	V_2, cm^{-1}	Stable rotamer
m-F-benzaldehyde	95	*O-trans*	479	1449	*O-cis*
	108.5	*O-cis*			
m-Cl-benzaldehyde	94	*O-trans*	231	1414	*O-cis*
	105	*O-cis*			
m-Br-benzaldehyde	90	*O-trans*	504	1407	*O-cis*
	107	*O-cis*			
m-CH$_3$-benzaldehyde	93	*O-trans*	598	1431	*O-cis*
	108.5	*O-cis*			
o-F-benzaldehyde	96	*O-cis*	98	1463	*O-trans*
	106	*O-trans*			
o-Cl-Benzaldehyde	88	*O-cis*	245	1347	*O-trans*
	103	*O-trans*			
Furan-2-aldehyde	134	*OO-cis*	710	2474	*OO-trans*
	145	*OO-trans*			

Two possible F-values are calculated from structural data, for instance,

$$F(O\text{-}cis) \qquad \qquad \text{and} \quad F(O\text{-}trans) \qquad \qquad (8.80)$$

The potential constants can be computed from the observed frequencies with the help of the general formula given in Eq. 8.78. The final assignment is made by comparing the relative Boltzmann intensity of the two observed lines. The observed frequencies, the calculated potential constants, and the stable rotamers are given in Table 8.9.

REFERENCES

1. C. C. Lin and J. D. Swalen, *Rev. Mod. Phys.* **31**, 841 (1959).
2. J. S. Koehler and D. M. Dennison, *Phys. Rev.* **57**, 1006 (1940).
3. C. H. Townes and A. L. Schawlow, Microwave Spectroscopy, McGraw-Hill, New York, 1955.
4. D. G. Burkhard and D. M. Dennison, *J. Mol. Spectry.* **3**, 299 (1959).
5. H. A. Gebbie, G. Topping, R. Illsley, and D. M. Dennison, *J. Mol. Spectry.* **11**, 229 (1963).
6. A. Hadni, Thesis, Université de Paris, 1956.
7. W. G. Fateley and F. A. Miller, *Spectrochim. Acta* **17**, 857 (1961).
8. W. G. Fateley and F. A. Miller, *Spectrochim. Acta* **19**, 611 (1963).
9. G. Sage and W. Klemperer, *J. Chem. Phys.* **39**, 371 (1963).
10. H. Hunziker and H. H. Günthard, *Spectrochim. Acta* **21**, 51 (1965).
11. K. R. Loos and R. C. Lord, *Spectrochim. Acta* **21**, 119 (1965).
12. K. D. Möller, A. R. DeMeo, D. R. Smith and L. H. London, *J. Chem. Phys.* **47**, 2609 (1967).
13. K. D. Möller and L. H. London, *J. Chem. Phys.* **47**, 2505 (1967).
14. D. R. Herschbach, *Tables for the Internal Rotation Problem*, Department of Chemistry, Harvard University, Cambridge, Massachusetts, described in *J. Chem. Phys.* **27**, 975 (1957).
15. M. Hayaski and L. Pierce, *Tables for the Internal Rotation Problem*, Department of Chemistry, University of Notre Dame, Notre Dame, Indiana.
16. National Bureau of Standards, *Tables Relating to Mathieu Functions*, Columbia Univ. Press, New York, 1951.
17. R. W. Kilb, *Tables of Mathieu Eigenvalues and Eigenfunctions for Special Boundary Conditions*, Department of Chemistry, Harvard University, Cambridge, Massachusetts.
18. G. Blanch and I. Rhodes, *J. Wash. Acad. Sci.* **45**, 166 (1955).
19. E. O. Stejskal and H. S. Gutowsky, *J. Chem. Phys.* **28**, 388 (1958).
20. W. G. Fateley, F. A. Miller, and R. E. Witkowski, *Technical Report AFML-TR-66-408*, Air Force Material Laboratory, Wright-Patterson Air Force Base, Ohio.
21. S. Weiss and G. E. Leroi, *J. Chem. Phys.* **48**, 962 (1968).
22. L. H. London and K. D. Möller, *J. Mol. Structure* **2**, 493 (1968).
23. R. H. Schwendeman and G. D. Jacobs, *J. Chem. Phys.* **36**, 1245 (1962).

24. D. R. Herschbach, *J. Chem. Phys.* **25**, 358 (1956).
25. E. Hirota, *J. Chem. Phys.* **45**, 1984 (1966).
26. L. Pierce and J. M. O'Reilly, *J. Mol. Spectry.* **3**, 536 (1959).
27. M. L. Unland, V. Weiss, and W. H. Flygare, *J. Chem. Phys.* **42**, 2138 (1965).
28. D. R. Herschbach and J. D. Swalen, *J. Chem. Phys.* **29**, 761 (1958).
29. J. P. Lowe, "Barrier to Internal Rotation about Single Bonds," in *Progr. Phys. Org. Chem.* **6**, 1 (1968).
30. J. D. Swalen and C. C. Costain, *J. Chem. Phys.* **31**, 1562 (1959).
31. K. D. Möller and H. G. Andresen, *J. Chem. Phys.* **37**, 1800 (1962).
32. K. S. Pitzer, *J. Chem. Phys.* **10**, 605 (1942).
33. D. R. Lide, Jr. and D. E. Mann, *J. Chem. Phys.* **28**, 572 (1958).
34. R. J. Myers and E. B. Wilson, Jr., *J. Chem. Phys.* **33**, 186 (1960).
35. L. Pierce, *J. Chem. Phys.* **34**, 498 (1961).
36. H. Dreizler and G. Dendl, *Z. Naturforsch.* **20a**, 1431 (1965).
37. W. G. Fateley and F. A. Miller, *Spectrochim. Acta* **18**, 977 (1962).
38. D. R. Smith, B. K. McKenna, and K. D. Möller, *J. Chem. Phys.* **45**, 1904 (1966).
39. H. Dreizler and H. D. Rudolph, *Z. Naturforsch.* **17a**, 712 (1962).
40. V. W. Laurie, *J. Chem. Phys.* **34**, 1516 (1961).
41. P. H. Kasai and R. J. Myers, *J. Chem. Phys.* **30**, 1096 (1959).
42. V. W. Laurie and J. Wollrab, *Symposium on Molecular Structure and Spectroscopy*, Columbus, Ohio, 1966, paper AA6.
43. K. D. Möller and H. G. Andresen, *J. Chem. Phys.* **39**, 17 (1963).
44. T. D. Goldfarb and B. N. Khare, *J. Chem. Phys.* **46**, 3379 (1967).
45. L. J. Nugent, D. E. Mann, and D. L. Lide, Jr., *J. Chem. Phys.* **36**, 965 (1962).
46. K. W. F. Kohlrausch, *Ramanspektren*, Becker u. Erler, Leipzig, 1943.
47. R. H. Hunt, R. A. Leacock, C. W. Peters, and K. T. Hecht, *J. Chem. Phys.* **42**, 1931 (1965).
48. W. G. Penney and G. B. B. M. Sutherland, *J. Chem. Phys.* **2**, 492 (1934).
49. H. Dreizler, *Z. Naturforsch.* **21a**, 1628 (1966).
50. G. Herzberg, *Infrared and Raman Spectra of Polyatomic Molecules*, Van Nostrand, New York, 1956.
51. R. H. Hunt and R. A. Leacock, *J. Chem. Phys.* **45**, 3141 (1966).
52. W. G. Fateley, R. K. Harris, F. A. Miller, and R. E. Witkowski, *Spectrochim. Acta* **21**, 231 (1965).
53. F. A. Miller, W. G. Fateley, and R. E. Witkowski, *Spectrochim. Acta* **23A**, 891 (1967).
54. F. Mönnig, *8th European Congress on Molecular Spectroscopy*, Copenhagen, 1965, Abstract 410.

9 Pure Rotational Spectra of Vapors: Water

A. INTRODUCTION

As is well known, evaluation of the pure rotational spectra of vapors of polar molecules primarily yields the structural parameters of the molecule. With the resolution and sensitivity of present day common far-infrared spectrometers, there is little new structural information which could be obtained by studying pure rotational spectra of vapors in the far-infrared. We have mentioned a few simple molecules, whose far-infrared rotational spectra have been well characterized, in Sections H.b,c of Chapter 1.

On the other hand, the rotational transitions which fall into the far-infrared region generally take place between highly excited energy levels; in other words, between states which represent very fast rotations of the molecules. In such a case, then, centrifugal distortion effects become appreciable. These effects can, in turn, be used to obtain the harmonic and anharmonic (cubic and higher-order) force constants of the molecule. The theory of centrifugal distortion is well known,[1] and explicit expressions of the various centrifugal distortion coefficients in terms of these force constants have been published for simple molecules.[2]

Knowledge of the centrifugal distortion effects of the water molecule is of particular importance since intense pure rotational transitions of water vapor stretch throughout the far-infrared region of the spectrum. In Section H.b of Chapter 1 we gave a brief description on the use of the rotational water vapor spectrum for calibration purposes. We shall, in the following, take up this subject again and show some spectra of prominent water lines between 300 and 18 cm^{-1}, observed with different instruments and, in addition, give a tabulation of 278 water vapor frequencies between 12 and 305 cm^{-1}, computed with the help of the hitherto most accurate, published energy levels.

B. Observed Spectra of Atmospheric Water Vapor

In a single-beam vacuum spectrometer, the full length of the optical path of the instrument may be used. The amount of water vapor can be controlled by the pressure if the humidity is known. Since large metal surfaces are present, water outgasses more slowly than nitrogen and oxygen, and this effect may give rise to unexpectedly high water vapor concentrations once the pumps have been turned off at low tank pressures. In a spectrometer flushed with nitrogen or dry air, the water vapor absorption can be controlled by the amount of purging. In addition, gas cells can be used (this, of course, must be done if the spectrometer is used in double-beam operation). The advantage of employing the full optical light path as an "absorption cell" is that the partial vapor pressure can be reduced below what is required in a short absorption cell, and this results in relatively narrow absorption lines due to the reduced amount of pressure broadening.

a. Spectral Region 300 to 150 cm^{-1}

The spectrum of water vapor (15-cm path cell) shown in Fig. 9.1. was taken with a grating monochromator of the Ebert-Fastie type.[3] It shows the first-order spectrum obtained with a grating of grating constant $d = 74.08\,\mu$, having a ruled area of $25 \times 25\,cm^2$ and a blaze angle of 20°. (The grating was used in second-order for the region 320 to 500 cm^{-1}.) The scanning speed amounted to 0.07°/min. Scatter and reststrahlen plates were employed as filtering elements. An aluminum scatter plate, ground with 320-mesh Carborundum, was used for the 500 to 240 cm^{-1} region, and a sooted 220-mesh scatter plate was used for the 240 to 150 cm^{-1} range. The reststrahlen plates were (spectral region in parentheses): $CaCO_3$ (405–320 cm^{-1}), NaF (380–240 cm^{-1}), NaCl (240–190 cm^{-1}), KCl (195–150 cm^{-1}). The maximum resolution appeared to be about 0.5 cm^{-1} throughout the covered region.

b. Spectral Region 150 to 100 cm^{-1}

This portion of the water vapor spectrum is shown in Fig. 9.2[4] and in Fig. 9.3.[5] The spectrum displayed by Fig. 9.2 was obtained with a small grating monochromator (grating size $5.7 \times 7.6\,cm^2$, 320 grooves/in. or about 12.5 grooves/mm) of the Littrow mount. The filter combination was 0.1-mm black polyethylene, 0.5-mm crystal quartz, one KBr reflection plate, and one 600-grooves/in. (about 24 grooves/mm) reflection filter grating. The spectrum was taken with the grating rotated off-blaze.

The spectrum of Fig. 9.3 was obtained with a large grating monochromator (grating size $17.8 \times 22.8\,cm^2$, 160 grooves/in. or 6.3 grooves/mm with blaze angle of 15°) of the Czerny-Turner mount. Filter combination was a KBr

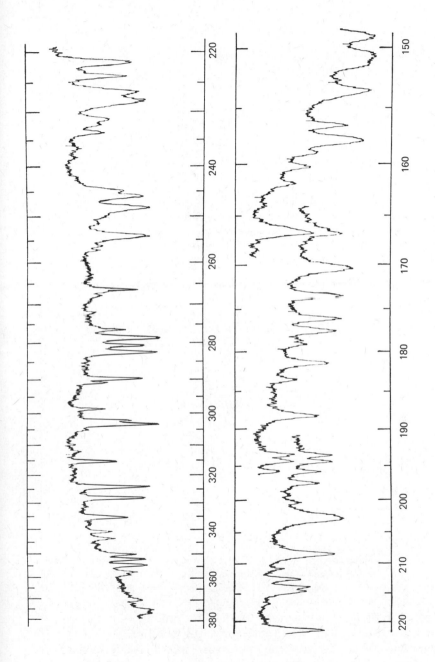

Fig. 9.1 Water vapor absorption in the 300 to 150 cm^{-1} spectral region. Ordinate, transmittance; abscissa, cm^{-1}. [D. W. Robinson, *J. Opt. Soc. Am.* **49**, 966 (1959).]

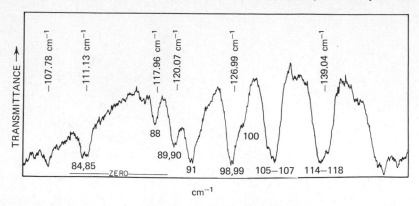

Fig. 9.2 Water vapor absorption in the 150 to 100 cm⁻¹ spectral region. The numbers at the peaks of the absorption lines correspond to the running numbers in Column 1 in the Tabulation of Water Vapor Lines at the end of this chapter. The cm⁻¹-values at the top of the figure give the frequencies calculated by W. S. Benedict. [L. R. Blaine, *J. Res. Natl. Bur. Stand.* **67C**, 207 (1963).]

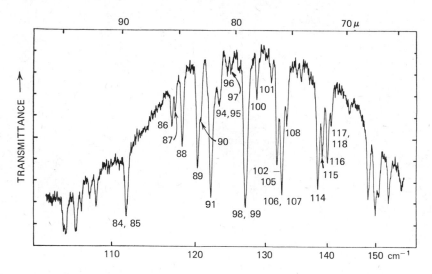

Fig. 9.3 Water vapor absorption in the 150 to 100 cm⁻¹ spectral region. The numbers at the peaks of the absorption lines correspond to the running numbers in Column 1 in the Tabulation of Water Vapor Lines at the end of this chapter (see also Fig. 9.2). [H. Yoshinaga, S. Fujita, S. Minami, A. Mitsuishi, R. A. Oetjen, and Y. Yamada, *J. Opt. Soc. Am.* **48**, 315 (1958).]

chopper, a KBr reststrahlen plate, and a filter grating with $d = 60\mu$. The spectral slit width at 150(100) cm⁻¹ was 0.46(0.22) cm⁻¹ and the time constant was 80 sec. The recording time was 8 min/μ. The radiation source was a mercury lamp. The sample was contained in a long-path gas cell.

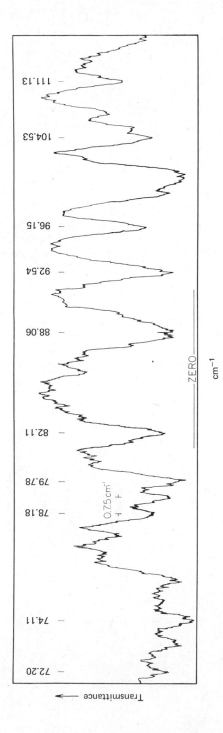

Fig. 9.4 Water vapor absorption in the 111 to 72 cm^{-1} spectral region. The cm^{-1}-values indicated at the top of the figure are frequencies calculated by W. S. Benedict. [L. R. Blaine, *J. Res. Natl. Bur. Stand.* **67C**, 207 (1963).]

c. Spectral Region 111 to 72 cm^{-1}

A spectrum for this region is shown in Fig. 9.4. The spectrum was taken with the same grating and instrument as mentioned under b above.[4] The filter combination was 0.1-mm black polyethylene, 1.0-mm crystal quartz, one CsBr reflection plate, and two 600-grooves/in. (about 24 grooves/mm) reflection filter gratings. The spectrum was taken again with the grating rotated far off-blaze.

d. Spectral Region 80 to 60 cm^{-1}

A spectrogram for this region is shown in Fig. 9.5, using the same instrument as decribed in Ref. 5. An 80-grooves/in. or 3.15-grooves/mm grating with step angle of 15° was used. The filter combination was a KBr chopper, a KRS-5 reststrahlen plate, a reflection filter grating with $d = 60\,\mu$, sooted polyethylene, and a 4-mm thick crystal quartz plate. The spectral slit width at 65 cm^{-1} amounted to 0.28 cm^{-1}, the recording time was 2.4 min/μ, and the time constant was 40 sec.

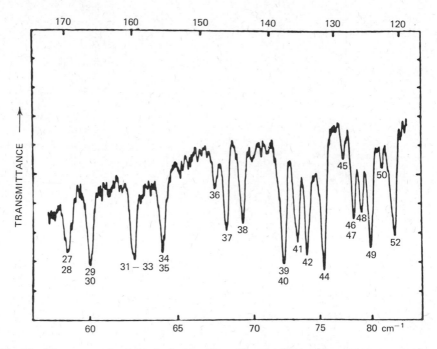

Fig. 9.5   Water vapor absorption in the 85 to 58 cm^{-1} spectral range. The numbers at the peaks of the absorption lines correspond to the running numbers in Column 1 in the Tabulation of Water Vapor Lines at the end of this chapter. [H. Yoshinaga, S. Fujita, S. Minami, A. Mitsuishi, R. A. Oetjen, and Y. Yamada, *J. Opt. Soc. Am.* **48**, 315 (1958).]

e. Spectral Region 64 to 50 cm^{-1}

The part of the water vapor spectrum in this spectral region is shown in Fig. 9.6. The instrument was the same as described in Ref. 4. The grating had 180 grooves/in. (about 7 grooves/mm) and was of the same size as mentioned above in b. The filter combination consisted of 0.2-mm black polyethylene, 1.0-mm crystal quartz, 0.5-mm LiF in transmission, one KRS-5 reflection plate, and two 320-grooves/in. (12.5 grooves/mm) reflection filter gratings.

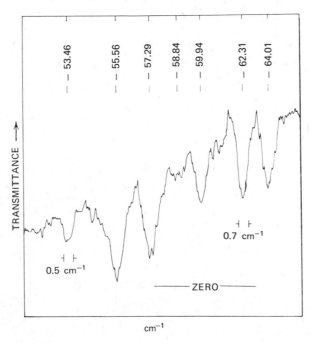

Fig. 9.6   Water vapor absorption in the 64 to 53 cm^{-1} spectral region. The frequencies indicated at the top of the figure are calculated by W. S. Benedict. [L. R. Blaine, *J. Res. Natl. Bur. Stand.* **67C**, 207 (1963).]

f. Spectral Region 53 to 32 cm^{-1}

This region is of particular interest for resolution tests (see Fig. 9.7). Two water lines near 40 cm^{-1} are separated by 0.70 cm^{-1}, a line at 38 cm^{-1} shows a splitting of about 0.32 cm^{-1}, and two lines at 36 cm^{-1} are separated by 0.55 cm^{-1} (For an identification of the respective transitions and the computed frequencies and relative intensities of these lines, refer to the tabulation below). Figure 9.7 shows the spectrum obtained with the instrument described in Ref. 4. A 100-grooves/in. (about 4 grooves/mm) grating was used of size 5.7 × 7.6 cm^2. The filtering combination consisted of 0.2-mm black

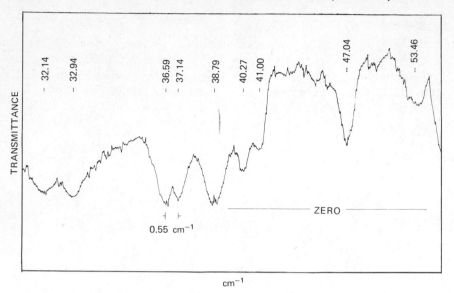

Fig. 9.7   Water vapor absorption in the 53 to 32 cm^{-1} spectral range. [L. R. Blaine, *J. Res. Natl. Bur. Stand.* **67C**, 207 (1963).] Calculated frequencies (W. S. Benedict) are indicated on top of the figure.

polyethylene, 1.0-mm crystal quartz, one 0.2-mm NaF-TlCl crystal powder filter, one KRS-5 reflection plate, and two reflection filter gratings with 240 grooves/in. (about 9.5 grooves/mm).

Figures 9.8(*a*) and 9.8(*b*) show the same portion of the spectrum but were obtained with the help of an interference modulator (see Appendix VI) as the main filtering device.[6] The grating of the monochromator had 90 grooves/in. (about 3.5 grooves/mm) and was approximately 20 × 25 cm^2 in size. Part *a* of the spectrum was obtained at atmospheric pressure (spectral slit width 0.31 cm^{-1}) and Part *b* under the reduced pressure of 30 mmHg (spectral slit width 0.16 cm^{-1}).

g. Spectral Region 30 to 10 cm^{-1}

Figure 9.9(*a*) shows the spectrum of water vapor (instrument filled with ambient atmosphere) in the 33 to 15 cm^{-1} region obtained with a large grating monochromator of the Czerny-Turner mount.[7] The grating constant was $d = 625\,\mu$, its area was 21.5 × 21.5 cm^2, and its step angle was 20°. The filter combination was 1-mm polyurethane (Polyurethan U_0, Farbenfabriken Bayer, Leverkusen, Germany), 2-mm paraffin-soot mixture, one reflection grating with $d = 211\,\mu$, and a crystal quartz plate. The time constant employed in the scan amounted to 200 sec.

WATER VAPOR ABSORPTION AT ATMOSPHERIC PRESSURE
0.31 cm^{-1} SPECTRAL SLIT WIDTH   5.5 mm SLITS

CALCULATED LINE POSITIONS & INTENSITIES:

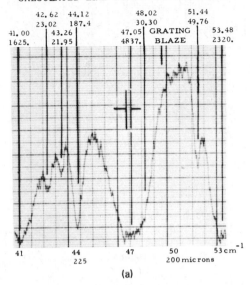

WATER VAPOR ABSORPTION

20 mm Hg PRESSURE

0.16 cm^{-1} SPECTRAL SLIT WIDTH

45 sec TIME CONSTANT   4.5 mm SLITS

35 and 82 minutes RECORDING TIME

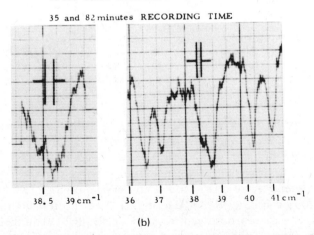

(b)

Fig. 9.8   Water vapor absorption in the 53 to 32 cm^{-1} spectral range. The calculated line positions and intensities are indicated in (a). [E. E. Bell, *Proc. Far-Infrared Symp.* Mission Inn, Calif., 1964.]

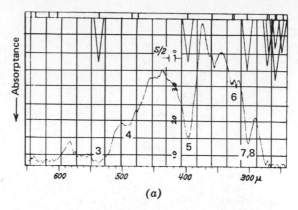

(a)

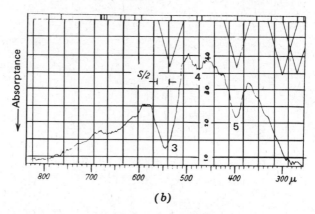

(b)

Fig. 9.9   Water vapor absorption in the 33 to $15\,\text{cm}^{-1}$ (a) and 33 to $14\,\text{cm}^{-1}$ (b) region. The triangles at the top of the figures give the theoretical positions and relative intensities of the strongest transitions. The numbers at the peaks at the absorption lines correspond to those of Column 1 in the Tabulation of Water Vapor Lines at the end of this chapter. [L. Genzel and W. Eckardt, Z. Physik **139**, 578 (1954), by permission of Springer: Berlin-Göttingen-Heidelberg.]

    The spectrum displayed in Fig. 9.9(b) had been scanned with the same instrument and a grating with $d = 833.3\,\mu$. The filter combination in this case consisted of 1-mm polyvinylchloride (Vinidur, Badische Anilin- und Sodafabrik, Ludwigshafen, Germany), 2-mm paraffin-soot mixture, a reflection filter grating of $d = 625\,\mu$, and crystal quartz. The time constant was also 200 sec. Finally, Fig. 9.10 shows the lowest portion of the water vapor spectrum which was observed with the instrument described in Ref. 5. The grating was of the same size as mentioned under b ($17.8 \times 22.8\,\text{cm}^2$)[5] and had 20 grooves/in. (about 0.8 grooves/mm). The filter combination consisted of reflection filter gratings with $d = 500$ and $750\,\mu$, sooted poly-

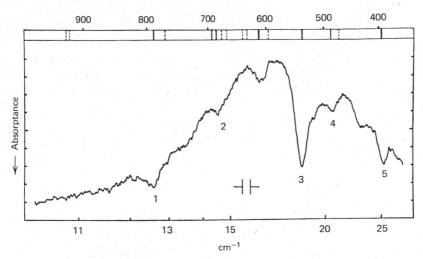

Fig. 9.10   Water vapor absorption in the 25 to 11 cm^{-1} spectral region. The vertical lines at the top of the figure show the locations of the lines calculated with energy levels by W. S. Benedict *et al., J. Res. Natl. Bur. Stand.* **49**, 91 (1952). The numbers at the peaks of the absorption lines correspond to those of Column 1 in the Tabulation of Water Vapor lines at the end of this chapter. The absorption at 620 μ (16.1 cm^{-1}) in the figure coincides approximately with the computed frequency of the $4_{40} \rightarrow 5_{33}$ transition, which, however, is expected to have only $\frac{1}{200}$ of the intensity of line 3. [H. Yoshinaga, S. Fujita, S. Minami, A. Mitsuishi, R. A. Oetjen, and Y. Yamada, *J. Opt. Soc. Am.* **48**, 315 (1958).]

ethylene, and 4-mm thick crystal quartz. At 15 cm^{-1} the spectral slit width amounted to 0.28 cm^{-1}, the time constant was 200 sec, and the recording time was 0.6 min/μ.

C. Computation and Tabulation of Pure Rotational Transitions of Water Vapor

The ideal situation would be one in which the rotational transitions could be computed with the help of the three principal moments of inertia and the centrifugal distortion constants as input parameters. We shall explore this in the following.

The rotational Hamiltonian is given by

$$\mathscr{H}^v = \mathscr{H}^{0,v} + \tfrac{1}{4}\sum {}^v\tau_{\mu\nu\xi\zeta}P_\mu P_\nu P_\xi P_\zeta + O(P^6) + \cdots, \tag{9.1}$$

where $\mathscr{H}^{0,v}$ is the Hamiltonian of the rigid rotor and where it is assumed that all constants are averaged over the particular vibrational state v. In general, the computational problem of evaluating the centrifugal distortion coefficients

τ is not as formidable as might be assumed at first glance. The possible maximum number of $\tau_{\mu\nu\xi\zeta}$ is $3^4 = 81$ because there are three angular momentum components P_α. For water, however, the molecular symmetry (point group C_{2v}) and the inherent symmetry of the τ's reduce this number to four independent τ's (contributing to the first-order of approximation to the energy.[8])

On account of this simplification the solution of the Schrödinger equation is most readily accomplished by rewriting Eq. 9.1 into the equivalent form (we drop, at first, terms higher than P^4):

$$E = \sum_{\substack{\alpha \\ \alpha = a,b,c}} \{U_\alpha \langle P_\alpha^2 \rangle + \tfrac{1}{4}[T_{\alpha\alpha\alpha\alpha} + T_{\alpha\alpha}]\langle P_\alpha^4 \rangle\}, \qquad (9.2)$$

where the U_α are given by $B_\alpha - \tfrac{1}{4}(J + 1)J \cdot T_{\alpha\alpha}$, B_α is the effective rotational constant of the rigid molecule, $\langle P_\alpha^2 \rangle$ and $\langle P_\alpha^4 \rangle$ are the expectation values of the corresponding powers of the P_α's, $T_{\alpha\alpha\alpha\alpha}$ are the centrifugal distortion coefficients $\tau_{\alpha\alpha\alpha\alpha}$ in units of cm^{-1}, and the $T_{\alpha\alpha}$ are given, for instance for $\alpha = a$, by $T_{aa} = T_{bbcc} - T_{ccaa} - T_{aabb}$.[9,10] The letters a, b, c indicate the axes of the rotational constant B_α with $B_a \geqslant B_b \geqslant B_c$. We assume that the molecules are in their vibrational ground state.

With the help of a computer program,[11] which furnishes the expectation values of the angular momenta components and their even powers (in the Wang symmetrized symmetric-rotor representation[12]), we have attempted to reproduce the observed rotational transitions of water[13] using the most accurate molecular constants hitherto published as displayed in Table 9.1.[14]

TABLE 9.1*

ROTATIONAL CONSTANTS AND CENTRIFUGAL DISTORTION COEFFI-
CIENTS OF WATER IN ITS VIBRATIONAL GROUND STATE

$B_a = 27.8761 \pm 0.0034$ cm^{-1}	
$B_b = 14.5074 \pm 0.0090$	
$B_c = 9.2877 \pm 0.0021$	
$T_{aaaa} = -0.1084 \pm 0.0012$ cm^{-1}	$T_{aa} = -0.01602$ cm^{-1}
$T_{bbbb} = -0.0083 \pm 0.0012$	$T_{bb} = -0.02368$
$T_{cccc} = -0.00107 \pm 0.00027$	$T_{cc} = 0.02584$

* After R. T. Hall and J. M. Dowling, J. Chem. Phys. **47**, 2454 (1967).

In Table 9.2 we have collected a sample computation and compared the results with the observed values. It is seen that the improvement over the

TABLE 9.2

COMPARISON BETWEEN COMPUTED AND EXPERIMENTAL ROTATIONAL
LEVELS AND TRANSITION FREQUENCIES (IN CM^{-1})

Level	Energy of level		
	Computed*	Observed[†]	Rigid rotor[‡]
9_{64}	1632.59	1631.27	1662.28
9_{37}	1216.62	1216.27	1224.75
Transition	Transition frequency		
$9_{37} \rightarrow 9_{64}$	415.97	415.00	437.53

* Rotational constants and centrifugal distortion constants given
in Table 9.1. Terms in $\langle P_a^6 \rangle$ are neglected (see text).
[†] Gates et al.[13]
[‡] Rotational constants taken from Table 9.1.

rigid-rotor energy levels is very good but that—for purposes of calibration—
the molecular parameters of Table 9.1 need be known to an even greater
accuracy in order for this approach to be applied usefully. (Addition of a
term in $\langle P_a^6 \rangle$, which has been shown to be significant,[12,14] does not improve
the computational results.) The observed values were not more closely reproduced by the computations due, most likely, to the fact that small uncertainties in the coefficients T are significantly magnified by the (large)
values of the $\langle P_\alpha^2 \rangle$ and $\langle P_\alpha^4 \rangle$.*

The method by which precise pure rotational transition frequencies are
obtained must therefore be based solely on the pure data; in other words,
on many accurately observed frequencies which must be correctly assigned
to the energy levels involved in the transitions. For energy levels of even
parity (see below) the energy levels are easily obtained with the help of the
observed frequency of the allowed transition from the ground state $E = 0$.
The computation of rotational energy levels of odd parity is somewhat more
involved than that of the even-parity levels since transitions to levels of odd
parity from the ground state (which is even) are forbidden.[†] In this case,
certain sum rules which relate the sums of the energies of rotational levels

* For instance, the $\langle P_\alpha^n \rangle$ of the level 9_{64} are, in the sequence $\alpha = a, b, c$: $\langle P_\alpha^2 \rangle = 34.974155$,
33.772138, 21.253700; $\langle P_\alpha^4 \rangle = 1262.1107, 1634.1370, 707.02054$.
[†] See Ref. 15, pp. 92–95 (Section 2 of Chapter 4). The permanent dipole moment vector in
water lies along the inertial axis b of the intermediate moment of inertia.

to the total angular momentum, to the rotational constants, and to the distortion coefficients are employed[14] to obtain the energy of the lowest odd-parity level.

The energy levels thus obtained can then be used to predict transition frequencies (their number depending on the completeness of the sets of known energy levels). In Section D we give a tabulation of 278 pure rotational transitions of water vapor between 12 and 305 cm^{-1} computed in this way from the most recently published energy levels.[13,14] Generally, only absorption lines which are not weaker by about 10^{-4} of the strongest line are tabulated since very weak lines are not of universal usefulness for calibration purposes. In particular, if two very weak lines from accidentally degenerate levels are seen to coincide, usually only the strongest of the two is given. The intensities have been computed using the usual relation

$$I(v) \propto g_N v[1 - \exp(-v/kT)][\exp(-E/kT)]S, \qquad (9.3)$$

where g_N is the nuclear spin factor of the lower level ($= 3$ for odd parity of the wave equation with respect to a rotation by π about the b-axis of inertia, in which case $K_{-1} + K_{+1} =$ odd, and $= 1$ for even parity; that is, $K_{-1} + K_{+1} =$ even),* v is the transition frequency in cm^{-1}, kT assumes the value of ~ 210 cm^{-1}, E is the energy (in cm^{-1}) of the lower level, and S is the line strength of the transition, defined in terms of the dipole moment (μ) matrix elements by $(2J + 1)(J_{K_{-1}K_{+1}}|\mu|J'_{K'_{-1}K'_{+1}})^2$. The allowed transitions obey the rules $J \rightarrow J' = 0, \pm 1$ and $K_{-1}K_{+1} \rightarrow K'_{-1}K'_{+1} =$ even,even $\leftrightarrow$ odd,odd and even,odd $\leftrightarrow$ odd,even. Generally, only those transitions are of considerable line strength in which the K indices change by one unit.[15] Inspection of Eq. 9.3 also shows that the low-frequency transitions are weak because the population difference between lower and upper levels is relatively small $[\exp(-v/kT) \sim 1]$. The values of $I(v)$ in the tabulation are based on the rigid rotor levels of the molecule since the computer program[11] also furnished E, v, and S of a rigid rotor. The deviations with respect to the $I(v)$ based on the actual energy levels and frequencies is immaterial for our purposes; the transition intensities should at any rate not be considered more than rough indications of the actual relations (which depend on pressure broadening, which we have neglected here).

* Since the inertial axis b (see previous footnote) is also the twofold symmetry axis of the water molecule, rotation by π about b permutes the hydrogen atoms of the molecule. Since H atoms are Fermi particles ($\alpha = \frac{1}{2}, \beta = -\frac{1}{2}$), the product of nuclear spin and rotational wave functions must therefore be antisymmetric with respect to such a rotation (assuming the electronic and vibrational ground state). The total number of nuclear spin functions of the H atoms (total nuclear spin $= 1$) is four, namely $\alpha\alpha, \beta\beta, \alpha\beta + \beta\alpha$, and $\alpha\beta - \beta\alpha$. Of these, three ($\alpha\alpha, \beta\beta, \alpha\beta + \beta\alpha$) are symmetric and one ($\alpha\beta - \beta\alpha$) is antisymmetric with respect to a permutation of the H atoms. Thus, an antisymmetric (symmetric) rotational level has a nuclear spin factor of three (one).

D. TABULATION OF PURE ROTATIONAL WATER VAPOR TRANSITIONS BETWEEN 12 AND 305 CM^{-1}

Column 1: Running number of the absorption line.
Column 2: Rotational quantum numbers and indices of lower and upper rotational energy level.
Column 3: Frequency of the transition in cm^{-1}.
Column 4: Relative intensity (the integer indicates powers of ten; for instance, 0.291 − 3 = 0.000291).

The number of significant digits for the values of the transition frequencies indicates the accuracy with which the energy levels are known. For instance, line 26 represents the transition between the energy levels 136.765 and 79.496 cm^{-1};[14] hence the transition frequency is given to three digits after the decimal point and should be considered accurate within about ±0.001 cm^{-1}. Only lines which are in the intensity range of approximately 10^{-4} to 1 are tabulated here; for weaker lines Refs. 13 and 16 should be consulted. All transitions in this tabulation arise from the ground vibrational level of $H_2^{16}O$ between rotational level $J = 1$ to $J = 16$.

No.	Rotational transition $J(K_{-1}, K_{+1}) - J'(K'_{-1}, K'_{+1})$								Frequency (cm^{-1})	Relative intensity
1	3(	2,	1)	−	4(	1,	4)		12.682	.291 −3
2	3(	3,	0)	−	4(	2,	3)		14.946	.247 −3
3	1(	0,	1)	−	1(	1,	0)		18.571	.188 −1
4	4(	4,	1)	−	5(	3,	2)		20.75	.125 −3
5	2(	0,	2)	−	2(	1,	1)		25.086	.125 −1
6	3(	3,	1)	−	4(	2,	2)		30.560	.460 −3
7	4(	3,	1)	−	5(	2,	4)		32.365	.538 −3
8	1(	1,	1)	−	2(	0,	2)		32.951	.900 −2
9	3(	0,	3)	−	3(	1,	2)		36.605	.594 −1
10	0(	0,	0)	−	1(	1,	1)		37.137	.179 −1
11	2(	2,	1)	−	3(	1,	2)		38.469	.892 −2
12	5(	4,	1)	−	6(	3,	4)		38.64	.717 −3
13	3(	1,	2)	−	3(	2,	1)		38.792	.657 −1
14	4(	1,	3)	−	4(	2,	2)		40.283	.207 −1
15	2(	1,	1)	−	2(	2,	0)		40.987	.176 −1
16	6(	5,	2)	−	7(	4,	3)		42.622	.165 −3
17	7(	3,	4)	−	8(	2,	7)		43.25	.224 −3
18	5(	3,	2)	−	6(	2,	5)		44.10	.198 −2

No.	Rotational transition $J(K_{-1}, K_{+1}) - J'(K'_{-1}, K'_{+1})$		Frequency (cm^{-1})	Relative intensity
19	5(1, 4)	– 5(2, 3)	47.05	.534 –1
20	6(3, 3)	– 7(2, 6)	48.061	.322 –3
21	5(4, 2)	– 6(3, 3)	51.433	.476 –3
22	4(0, 4)	– 4(1, 3)	53.444	.254 –1
23	2(1, 2)	– 2(2, 1)	55.406	.622 –1
24	1(0, 1)	– 2(1, 2)	55.705	.155 –0
25	5(0, 5)	– 4(3, 2)	57.12	.650 –3
26	2(1, 2)	– 3(0, 3)	57.269	.150 –0
27	6(2, 4)	– 6(3, 3)	58.772	.118 –1
28	6(4, 2)	– 7(3, 5)	58.89	.452 –3
29	6(1, 5)	– 6(2, 4)	59.871	.143 –1
30	7(2, 5)	– 7(3, 4)	59.94	.190 –1
31	5(2, 3)	– 5(3, 2)	62.30	.598 –1
32	7(0, 7)	– 6(3, 4)	62.73	.130 –3
33	7(5, 2)	– 8(4, 5)	62.84	.309 –3
34	3(1, 3)	– 3(2, 2)	64.029	.332 –1
35	4(3, 2)	– 5(2, 3)	64.04	.102 –1
36	8(2, 6)	– 8(3, 5)	67.23	.326 –2
37	4(2, 2)	– 4(3, 1)	68.060	.278 –1
38	3(2, 2)	– 4(1, 3)	69.191	.180 –1
39	7(5, 3)	– 8(4, 4)	72.13	.143 –3
40	2(0, 2)	– 3(1, 3)	72.186	.966 –1
41	3(2, 1)	– 3(3, 0)	73.259	.796 –1
42	5(0, 5)	– 5(1, 4)	74.11	.802 –1
43	7(4, 3)	– 8(3, 6)	74.88	.116 –2
44	4(1, 4)	– 4(2, 3)	75.523	.110 –0
45	9(3, 6)	– 9(4, 5)	77.36	.327 –2
46	7(1, 6)	– 7(2, 5)	78.19	.308 –1
47	10(3, 7)	– 10(4, 6)	78.28	.375 –3
48	3(2, 2)	– 3(3, 1)	78.914	.295 –1
49	3(1, 3)	– 4(0, 4)	79.776	.108 –0
50	9(2, 7)	– 9(3, 6)	80.97	.481 –2
51	8(3, 5)	– 8(4, 4)	81.62	.297 –2
52	4(2, 3)	– 4(3, 2)	82.153	.108 –0
53	8(4, 4)	– 9(3, 7)	84.51	.180 –3
54	8(5, 3)	– 9(4, 6)	84.97	.108 –3
55	6(4, 3)	– 7(3, 4)	85.63	.360 –2
56	11(3, 8)	– 11(4, 7)	85.80	.374 –3
57	9(4, 5)	– 10(3, 8)	85.88	.144 –3
58	5(2, 4)	– 5(3, 3)	87.760	.313 –1

No.	Rotational transition $J(K_{-1}, K_{+1}) - J'(K'_{-1}, K'_{+1})$		Frequency (cm^{-1})	Relative intensity
59	3(0, 3)	− 4(1, 4)	88.079	.422 −0
60	7(3, 4)	− 7(4, 3)	88.88	.215 −1
61	5(1, 5)	− 5(2, 4)	89.582	.336 −1
62	1(1, 0)	− 2(2, 1)	92.540	.363 −0
63	6(2, 5)	− 6(3, 4)	96.07	.683 −1
64	6(0, 6)	− 6(1, 5)	96.187	.234 −1
65	6(3, 3)	− 6(4, 2)	96.25	.144 −1
66	5(3, 3)	− 6(2, 4)	98.800	.856 −2
67	1(1, 1)	− 2(2, 0)	99.024	.116 −0
68	4(2, 3)	− 5(1, 4)	99.14	.131 −0
69	11(4, 7)	− 11(5, 6)	99.96	.286 −3
70	8(1, 7)	− 8(2, 6)	99.98	.651 −2
71	10(2, 8)	− 10(3, 7)	100.23	.727 −3
72	4(1, 4)	− 5(0, 5)	100.55	.460 −0
73	5(3, 2)	− 5(4, 1)	101.52	.673 −1
74	9(5, 4)	− 10(4, 7)	104.03	.205 −3
75	4(3, 1)	− 4(4, 0)	104.289	.227 −1
76	4(0, 4)	− 5(1, 5)	104.570	.169 −0
77	8(5, 4)	− 9(4, 5)	105.12	.556 −3
78	4(3, 2)	− 4(4, 1)	105.59	.695 −1
79	6(1, 6)	− 6(2, 5)	105.66	.793 −1
80	5(3, 3)	− 5(4, 2)	106.139	.243 −1
81	7(2, 6)	− 7(3, 5)	107.075	.144 −1
82	6(3, 4)	− 6(4, 3)	107.74	.528 −1
83	10(4, 6)	− 10(5, 5)	108.21	.359 −3
84	7(3, 5)	− 7(4, 4)	111.10	.104 −1
85	2(1, 1)	− 3(2, 2)	111.129	.145 −0
86	8(3, 6)	− 8(4, 5)	116.60	.160 −1
87	9(4, 5)	− 9(5, 4)	117.03	.350 −2
88	7(0, 7)	− 7(1, 6)	117.97	.515 −1
89	5(1, 5)	− 6(0, 6)	120.084	.164 −0
90	8(2, 7)	− 8(3, 6)	120.50	.243 −1
91	5(0, 5)	− 6(1, 6)	121.90	.507 −0
92	7(4, 4)	− 8(3, 5)	122.37	.169 −2
93	11(2, 9)	− 11(3, 8)	122.56	.890 −3
94	9(1, 8)	− 9(2, 7)	122.88	.108 −1
95	7(1, 7)	− 7(2, 6)	123.13	.182 −1
96	8(4, 4)	− 8(5, 3)	124.16	.313 −2
97	9(3, 7)	− 9(4, 6)	124.62	.242 −2
98	5(2, 4)	− 6(1, 5)	126.689	.635 −1

No.	Rotational transition $J(K_{-1}, K_{+1}) - J'(K'_{-1}, K'_{+1})$		Frequency (cm^{-1})	Relative intensity	
99	3(1, 2) −	4(2, 3)	126.997	.446	−0
100	7(4, 3) −	7(5, 2)	128.63	.202	−1
101	6(4, 2) −	6(5, 1)	130.81	.110	−1
102	5(4, 1) −	5(5, 0)	131.73	.359	−1
103	7(4, 4) −	7(5, 3)	131.86	.708	−2
104	6(4, 3) −	6(5, 2)	131.89	.336	−1
105	5(4, 2) −	5(5, 1)	132.01	.120	−1
106	8(4, 5) −	8(5, 4)	132.42	.107	−1
107	2(1, 2) −	3(2, 1)	132.666	.353	−0
108	6(3, 4) −	7(2, 5)	133.43	.386	−1
109	9(4, 6) −	9(5, 5)	134.10	.153	−2
110	10(3, 8) −	10(4, 7)	135.18	.295	−2
111	9(2, 8) −	9(3, 7)	135.89	.404	−2
112	10(4, 7) −	10(5, 6)	137.43	.172	−2
113	8(0, 8) −	8(1, 7)	138.84	.107	−1
114	6(1, 6) −	7(0, 7)	138.99	.428	−0
115	6(0, 6) −	7(1, 7)	139.76	.144	−0
116	4(1, 3) −	5(2, 4)	140.708	.135	−0
117	8(1, 8) −	8(2, 7)	141.48	.330	−1
118	9(5, 5) −	10(4, 6)	141.52	.164	−3
119	11(4, 8) −	11(5, 7)	142.81	.191	−3
120	10(1, 9) −	10(2, 8)	144.96	.170	−2
121	11(5, 6) −	11(6, 5)	145.10	.270	−3
122	12(2,10) −	12(3, 9)	145.68	.107	−3
123	11(3, 9) −	11(4, 8)	147.99	.361	−3
124	3(0, 3) −	3(3, 0)	148.656	.262	−2
125	2(2, 0) −	3(3, 1)	149.056	.287	−0
126	12(4, 9) −	12(5, 8)	150.46	.172	−3
127	2(2, 1) −	3(3, 0)	150.519	.870	−0
128	10(5, 5) −	10(6, 4)	150.81	.312	−3
129	6(2, 5) −	7(1, 6)	151.30	.191	−0
130	10(2, 9) −	10(3, 8)	152.50	.541	−2
131	5(1, 4) −	6(2, 5)	153.45	.331	−0
132	9(5, 4) −	9(6, 3)	154.10	.264	−2
133	8(5, 3) −	8(6, 2)	155.76	.202	−2
134	10(5, 6) −	10(6, 5)	156.23	.102	−2
135	9(5, 5) −	9(6, 4)	156.28	.908	−3
136	7(5, 2) −	7(6, 1)	156.36	.108	−1
137	6(5, 1) −	6(6, 0)	156.44	.433	−2
138	6(5, 2) −	6(6, 1)	156.46	.130	−1

No.	Rotational transition $J(K_{-1}, K_{+1}) - J'(K'_{-1}, K'_{+1})$		Frequency (cm^{-1})	Relative intensity
139	8(5, 4) −	8(6, 3)	156.49	.611 −2
140	7(5, 3) −	7(6, 2)	156.53	.362 −2
141	11(5, 7) −	11(6, 6)	156.86	.109 −3
142	7(1, 7) −	8(0, 8)	157.63	.106 −0
143	7(0, 7) −	8(1, 8)	157.87	.320 −0
144	9(0, 9) −	9(1, 8)	158.89	.175 −1
145	9(1, 9) −	9(2, 8)	160.17	.588 −2
146	8(4, 5) −	9(3, 6)	160.18	.507 −3
147	4(0, 4) −	4(3, 1)	161.787	.247 −3
148	12(3,10) −	12(4, 9)	162.43	.358 −3
149	11(1,10) −	11(2, 9)	165.83	.211 −2
150	7(3, 5) −	8(2, 6)	166.24	.135 −1
151	6(1, 5) −	7(2, 6)	166.704	.818 −1
152	11(2,10) −	11(3, 9)	169.90	.720 −3
153	3(2, 1) −	4(3, 2)	170.358	.723 −0
154	7(2, 6) −	8(1, 7)	173.34	.488 −1
155	3(1, 3) −	4(2, 2)	173.503	.899 −1
156	8(1, 8) −	9(0, 9)	176.02	.206 −0
157	8(0, 8) −	9(1, 9)	176.12	.688 −1
158	3(2, 2) −	4(3, 1)	177.534	.248 −0
159	11(6, 5) −	11(7, 4)	177.82	.164 −3
160	10(0,10) −	10(1, 9)	178.49	.280 −2
161	7(6, 1) −	7(7, 0)	178.62	.335 −2
162	7(6, 2) −	7(7, 1)	178.64	.112 −2
163	10(6, 4) −	10(7, 3)	178.87	.167 −3
164	8(6, 2) −	8(7, 1)	179.06	.839 −3
165	8(6, 3) −	8(7, 2)	179.09	.252 −2
166	10(1,10) −	10(2, 9)	179.10	.841 −2
167	9(6, 3) −	9(7, 2)	179.22	.127 −2
168	9(6, 4) −	9(7, 3)	179.36	.423 −3
169	10(6, 5) −	10(7, 4)	179.40	.505 −3
170	10(5, 6) −	11(4, 7)	180.29	.333 −3
171	7(1, 6) −	8(2, 7)	181.38	.163 −0
172	5(0, 5) −	5(3, 2)	183.46	.127 −1
173	12(1,11) −	12(2,10)	185.62	.255 −3
174	12(2,11) −	12(3,10)	187.81	.773 −3
175	4(2, 2) −	5(3, 3)	188.184	.173 −0
176	8(2, 7) −	9(1, 8)	193.45	.917 −1
177	9(1, 9) −	10(0,10)	194.34	.391 −1
178	9(0, 9) −	10(1,10)	194.37	.117 −0

No.	Rotational transition $J(K_{-1}, K_{+1}) - J'(K'_{-1}, K'_{+1})$		Frequency (cm^{-1})	Relative intensity	
179	8(3, 6) –	9(2, 7)	195.80	.312	−1
180	9(4, 6) –	10(3, 7)	197.34	.129	−2
181	8(1, 7) –	9(2, 8)	197.45	.319	−1
182	11(0,11) –	11(1,10)	197.73	.359	−2
183	11(1,11) –	11(2,10)	197.99	.120	−3
184	8(7, 2) –	8(8, 1)	198.36	.632	−3
185	8(7, 1) –	8(8, 0)	198.36	.211	−3
186	9(7, 2) –	9(8, 1)	199.24	.426	−3
187	10(7, 4) –	10(8, 3)	199.94	.192	−3
188	5(2, 3) –	6(3, 4)	202.47	.328	−0
189	3(3, 0) –	4(4, 1)	202.69	1.000	−0
190	3(3, 1) –	4(4, 0)	202.909	.334	−0
191	13(1,12) –	13(2,11)	204.65	.247	−3
192	4(2, 3) –	5(3, 2)	208.49	.522	−0
193	5(1, 4) –	5(4, 1)	210.88	.460	−2
194	10(1,10) –	11(0,11)	212.58	.594	−1
195	10(0,10) –	11(1,11)	212.59	.198	−1
196	4(1, 3) –	4(4, 0)	212.632	.644	−3
197	9(2, 8) –	10(1, 9)	212.66	.164	−2
198	6(2, 4) –	7(3, 5)	213.908	.631	−1
199	9(1, 8) –	10(2, 9)	214.59	.500	−1
200	6(0, 6) –	6(3, 3)	214.830	.248	−2
201	6(1, 5) –	6(4, 2)	214.89	.254	−2
202	2(0, 2) –	3(3, 1)	215.129	.393	−2
203	12(0,12) –	12(1,11)	216.69	.458	−3
204	12(1,12) –	12(2,11)	216.83	.138	−2
205	4(1, 4) –	5(2, 3)	221.72	.170	−0
206	9(3, 7) –	10(2, 8)	221.73	.615	−2
207	7(2, 5) –	8(3, 6)	223.69	.101	−0
208	4(3, 1) –	5(4, 2)	226.264	.238	−0
209	7(1, 6) –	7(4, 3)	227.01	.611	−2
210	4(3, 2) –	5(4, 1)	227.86	.722	−0
211	11(0,11) –	12(1,12)	230.77	.268	−1
212	11(1,11) –	12(0,12)	230.77	.894	−2
213	10(2, 9) –	11(1,10)	231.21	.230	−1
214	10(4, 7) –	11(3, 8)	231.92	.230	−2
215	10(1, 9) –	11(2,10)	232.09	.773	−2
216	8(2, 6) –	9(3, 7)	233.36	.168	−1
217	13(0,13) –	13(1,12)	235.55	.475	−3
218	13(1,13) –	13(2,12)	235.60	.158	−3

No.	Rotational transition $J(K_{-1}, K_{+1}) - J'(K'_{-1}, K'_{+1})$		Frequency (cm^{-1})	Relative intensity
219	9(2, 7)	– 10(3, 8)	244.21	.233 −1
220	10(3, 8)	– 11(2, 9)	244.54	.880 −2
221	5(2, 4)	– 6(3, 3)	245.332	.986 −1
222	3(0, 3)	– 4(3, 2)	245.755	.405 −1
223	5(3, 2)	– 6(4, 3)	247.90	.435 −0
224	8(1, 7)	– 8(4, 4)	248.83	.106 −2
225	12(0,12)	– 13(1,13)	248.87	.362 −2
226	12(1,12)	– 13(0,13)	248.87	.109 −1
227	11(2,10)	– 12(1,11)	249.47	.319 −2
228	11(1,10)	– 12(2,11)	249.87	.960 −2
229	5(3, 3)	– 6(4, 2)	253.82	.150 −0
230	4(4, 0)	– 5(5, 1)	253.98	.233 −0
231	4(4, 1)	– 5(5, 0)	254.01	.699 −0
232	14(1,14)	– 14(2,13)	254.24	.148 −3
233	7(0, 7)	– 7(3, 4)	256.10	.319 −2
234	10(2, 8)	– 11(3, 9)	257.03	.326 −2
235	11(4, 8)	– 12(3, 9)	262.88	.359 −3
236	4(1, 4)	– 4(4, 1)	263.27	.696 −3
237	11(3, 9)	– 12(2,10)	265.19	.118 −2
238	6(3, 3)	– 7(4, 4)	266.24	.749 −1
239	13(0,13)	– 14(1,14)	266.88	.396 −2
240	13(1,13)	– 14(0,14)	266.88	.132 −2
241	12(2,11)	– 13(1,12)	267.59	.356 −2
242	12(1,11)	– 13(2,12)	267.78	.119 −2
243	11(2, 9)	– 12(3,10)	271.85	.371 −2
244	8(2, 6)	– 8(5, 3)	273.01	.498 −3
245	9(2, 7)	– 9(5, 4)	275.36	.954 −3
246	5(1, 5)	– 6(2, 4)	276.142	.321 −1
247	7(2, 5)	– 7(5, 2)	277.45	.158 −2
248	5(4, 1)	– 6(5, 2)	278.27	.432 −0
249	5(4, 2)	– 6(5, 1)	278.49	.144 −0
250	7(3, 4)	– 8(4, 5)	280.35	.100 −0
251	9(1, 8)	– 9(4, 5)	281.21	.118 −2
252	4(0, 4)	– 5(3, 3)	281.912	.209 −1
253	6(3, 4)	– 7(4, 3)	282.25	.237 −0
254	5(1, 5)	– 5(4, 2)	283.481	.464 −3
255	12(3,10)	– 13(2,11)	284.43	.124 −2
256	14(0,14)	– 15(1,15)	284.78	.434 −3
257	14(1,14)	– 15(0,15)	284.78	.130 −2
258	13(2,12)	– 14(1,13)	285.50	.396 −3

No.	Rotational transition $J(K_{-1}, K_{+1}) - J'(K'_{-1}, K'_{+1})$		Frequency (cm^{-1})	Relative intensity	
259	13(1,12)	− 14(2,13)	285.57	.119	−2
260	6(2, 4)	− 6(5, 1)	285.83	.368	−3
261	10(2, 8)	− 10(5, 5)	286.72	.142	−3
262	12(2,10)	− 13(3,11)	287.94	.421	−3
263	6(2, 5)	− 7(3, 4)	289.43	.138	−0
264	12(4, 9)	− 13(3,10)	289.75	.404	−3
265	8(3, 5)	− 9(4, 6)	290.75	.134	−1
266	5(2, 3)	− 5(5, 0)	295.56	.430	−3
267	9(3, 6)	− 10(4, 7)	298.42	.148	−1
268	6(4, 2)	− 7(5, 3)	301.85	.777	−1
269	15(0,15)	− 16(1,16)	302.76	.386	−3
270	15(1,15)	− 16(0,16)	302.76	.129	−3
271	13(3,11)	− 14(2,12)	302.84	.128	−3
272	5(5, 1)	− 6(6, 0)	302.93	.108	−0
273	5(5, 0)	− 6(6, 1)	302.99	.325	−0
274	6(4, 3)	− 7(5, 2)	303.14	.235	−0
275	14(2,13)	− 15(1,14)	303.70	.357	−3
276	14(1,13)	− 15(2,14)	303.72	.119	−3
277	13(2,11)	− 14(3,12)	304.52	.387	−3
278	10(3, 7)	− 11(4, 8)	304.79	.170	−2

REFERENCES

1. J. E. Wollrab, *Rotational Spectra and Molecular Structure*, Academic Press, New York, 1967, Section 3-12, p. 81.
2. K. T. Chung and P. M. Parker, *J. Chem. Phys.* **43**, 3865, 3869 (1965); *ibid.* **38**, 8 (1963); *ibid.* **39**, 240 (1963).
3. D. W. Robinson, *J. Opt. Soc. Am.* **49**, 966 (1959).
4. L. R. Blaine, *J. Res. Natl. Bur. Stand.* **67C**, 207 (1963).
5. H. Yoshinaga, S. Fujita, S. Minami, A. Mitsuishi, R. A. Oetjen, and Y. Yamada, *J. Opt. Soc. Am.* **48**, 315 (1958).
6. E. E. Bell, *Proceedings Far-Infrared Symposium*, Naval Ordnance Laboratory, Corona, California, 1964.
7. L. Genzel and W. Eckardt, *Z. Physik* **139**, 578 (1954).
8. P. M. Parker, *J. Chem. Phys.* **37**, 1596 (1962).
9. Wm. B. Olson and H. C. Allen, Jr., *J. Res. Natl. Bur. Stand.* **67A**, 359 (1963).
10. P. E. Fraley and K. N. Rao, *J. Mol. Spectry.* **19**, 131 (1966).
11. R. A. Beaudet, Doctoral Dissertation, Harvard University, Dept. Chem., 1962. Microfilms Inc., Ann Arbor, Michigan, Order No. 62-3532.
12. See Ref. 1, Section 2-6c, p. 24.
13. By "observed transitions" is meant those based on the most recent energy levels which, in turn, were obtained from many observations and have been tabulated in the literature; see, for instance, D. M. Gates, R. F. Calfee, D. W. Hansen, and W. S. Benedict, *Natl. Bur. Stand. Monograph* **71**, Aug. 3, 1964.
14. R. T. Hall and J. M. Dowling, *J. Chem. Phys.* **47**, 2454 (1967).
15. C. H. Townes and A. L. Schawlow, *Microwave Spectroscopy*, McGraw-Hill, New York 1955, p. 97 and Appendix V.
16. D. E. Burch, *J. Opt. Soc. Am.* **58**, 1383 (1968).

10 Collision-Induced Spectra at Long Wavelengths

A. INTRODUCTION

A homonuclear diatomic molecule, such as the nitrogen molecule N_2, has no vibrational-rotational or pure rotational absorption in the infrared on account of its symmetry. However, appreciable absorption can arise if the molecules form collision clusters, for instance if put under higher pressures, because a dipole moment is induced by the intermolecular forces. Because of the rapid decrease of intermolecular forces with increasing separation of the molecules, the induced dipole moment is a short-range function of the intermolecular distances. Consequently, the induced transition moment persists only during the time of the collision. As is well known, the transition moments of the dipole-allowed motions of an individual molecule do not depend on external coordinates (of the center of mass) of the molecule. This is not the case for collision-induced transition moments since these are dependent on the relative positions of the collision partners during the collision. Consequently, part of the energy of the photon which is absorbed by the collision cluster may appear as an increase in the molecular translational energy after the collision. Obviously, if all of the photon energy is converted into translational motion, we will observe only translational absorption arising from pure translational transitions ($\Delta J = 0, \Delta m = 0$), from translational jumps accompanied by an orientational transition ($\Delta m \neq 0$), or from a "flip-flop" rotational transition in the cluster ($\Delta J_1 = -\Delta J_2$).

In the following we give a somewhat extensive qualitative outline of the various collision-induced phenomena.[1] We begin with the simplest case, the infrared absorption in binary mixtures of rare gases, for instance, in an equimolar mixture of He and Ar. We assume for simplicity's sake that the pressure of the mixture is such that only binary collisions occur. We first note that collisions between two like rare-gas molecules obviously do not lead to the formation of an induced dipole moment because of the $D_{\infty h}$

symmetry of the binary "collision complex." However, if two unlike rare-gas atoms collide, their respective electron clouds are deformed to a different degree and an induced dipole moment persists during the time the two atoms are sufficiently close to each other. This dipole moment can now interact with the electromagnetic radiation. The question then arises as to the frequency at which this absorption will occur, or in other words, which particular motion of the cluster modulates the induced dipole moment. Clearly, this can be no rotational frequency of an individual rare-gas atom since their total angular momenta are zero. The idea comes to mind that the rotational motion of the diatomic cluster could be involved. Indeed, as we shall show below (Section B.c), part of the absorption coefficient may be interpreted to arise in this way. However, its fraction is minor and the rotational motion of the collision complex therefore is essentially not responsible for the optical absorption. There remains only the relative translational motion between the two unlike atoms as a means of modulating the induced dipole moment, that is to say, we shall observe here the pure translational transitions. Since there are no bound translational states, the absorption is not of the resonant type; its frequency stretches continuously from zero on upwards. The absorption band is rather broad. This is related to the time of duration of the collision process (uncertainty principle), as is easily seen by comparing the change of the wave vector Δk of the absorbed photon with the ensueing change of translational energy, ΔE. We have $\Delta E = p\Delta p/m = (\hbar^2 k/m)\Delta k$, where $p = \hbar k$ is the linear momentum. If we now suppose that the change of the wave vector Δk is inversely proportional to the range R_0 of the induced dipole moment, it follows that $\Delta E = (\hbar^2 k/m)\Delta k \approx (\hbar p/m)(1/R_0) = \hbar v/R_0$ (v = velocity) $= \hbar/\Delta t$. Thus we see that the width of the induced translational band is indeed given by the uncertainty in the energy because of the short time, Δt, of the *duration* of the collision process.

We note here that the translational motion of the particles is assumed along prescribed, classical paths with uniform velocity. The assumption $\Delta k \sim 1/R_0$ means that the wavelength of the photon and the range of the induced dipole moment are comparable. We furthermore point out the difference between collision-induced spectra and pressure-broadened dipole-allowed spectra with respect to line broadening. In the latter case the broadening is a function of the rate of collisions; that is, the reciprocal time *between* collisions.

A more complicated phenomenon is the collision-induced absorption in homonuclear diatomic gases, such as nitrogen and hydrogen, which can exhibit collision-induced pure rotational and rotational-vibrational spectra. It should be noted from the outset that the translational motion is inherently coupled to the rotational motion and cannot be separated out, except for very light molecules (such as H_2) with large spacings between rotational levels.

This coupling between translation and rotation in the collision-induced spectra is a consequence of the noncentral intermolecular forces, that is to say, the dependence of the intermolecular potential on the relative orientations of the collision partners. This leads to a wave function of the rotational-translation motion of the form

$$\psi_s(\mathbf{R}^N, \boldsymbol{\omega}^N) = \sum_{r,t} \psi_t(\mathbf{R}^N)\psi_r(\boldsymbol{\omega}^N)(rt|s), \qquad (10.1a)$$

where the $\mathbf{R}^N = R_1, R_2, \cdots, R_N$, $\boldsymbol{\omega}^N = \omega_1, \cdots, \omega_N$ are the position vectors of the centers of gravity and the sets of polar angles of the orientations of the intramolecular axes of the colliding N molecules, respectively. The coefficients $(rt|s)$ are related to certain vector coupling coefficients which describe the coupling of the angular momenta of the colliding molecules with the angular momentum of the intermolecular axis. The reader will remember that in the case of a central intermolecular potential the total wave function is the simple product of the translational and the rotational wave functions. In our case this would mean that

$$\psi_s(\mathbf{R}^N, \boldsymbol{\omega}^N) = \psi_t(\mathbf{R}^N)\psi_r(\boldsymbol{\omega}^N). \qquad (10.1b)$$

We continue our discussion with the understanding that the induced rotational transitions contain the translational transitions as we have explained above. Now, a molecule such as N_2 (point group symmetry $D_{\infty h}$) possesses a molecular quadrupole moment because of its nonspherical charge distribution. The field of this quadrupole moment can induce a dipole moment at the second molecule; a dipole moment the magnitude and direction of which depend on the mutual orientation and intermolecular distance of the two molecules. For simplicity's sake we first consider the second molecule as nonrotating; for instance, we examine a nitrogen-neon mixture.

The quadrupole moment of the N_2 molecule is a symmetric second-order tensor; it rotates twice as fast as the N_2 molecule itself. The induced dipole moment in the Ne atom therefore "vibrates" at twice the rotational frequency of the N_2 molecule. Since it is the N_2–Ne complex which interacts with the electromagnetic radiation, we shall observe transitions at rotational frequencies of N_2 with the selection rule $\Delta J = 2$ (although the dipole moment is induced in the rare gas atom). Clearly, the intensity of these transitions depends on the quadrupole moment of N_2 and the polarizability of Ne.

The complexity of collision-induced phenomena increases if we consider a gas mixture where both collision partners possess a molecular quadrupole moment, for instance $H_2 + N_2$. In this case it is important to distinguish between the two possibilities of neglecting the anisotropy of the polarizability of (one or both of) the molecules or not. In the first case the polarizability is assumed not to depend on the angular coordinates of the molecule, in the second case it does depend on the orientation of the molecule.

We first neglect the anisotropy of the polarizabilities. The following rotational collision-induced transitions may then arise: (*1*) The quadrupole of N_2 induces a dipole moment in H_2. Rotational transitions at a N_2 frequency may be observed with the selection rules $\Delta J(N_2) = 2$, $\Delta J(H_2) = 0$. (*2*) The quadrupole of H_2 induces a dipole moment in N_2. We should expect transitions of $\Delta J(H_2) = 2$, $\Delta J(N_2) = 0$.

We now admit that both molecules possess a nonvanishing anisotropy of polarizability; the induced dipole moment in a molecule is therefore modulated by its own rotational motion. Since the anisotropic polarizability is also a second-order tensor, we shall observe "simultaneous" transitions with the selection rules $\Delta J(N_2) = 2$, $\Delta J(H_2) = 2$. If we were to deal with a homonuclear diatomic gas, for instance pure H_2, we would observe $\Delta J_1(H_2) = 2$, $\Delta J_2(H_2) = 2$ if the anisotropy of polarizability is appreciable ("double rotational transitions"), or $\Delta J_1(H_2) = 2$, $\Delta J_2(H_2) = 0$ if the polarizability is isotropic.

Summarizing, we see that, depending on the degree of anisotropy of the polarizability, more or less intense rotational summation frequencies may appear besides those of the $\Delta J = 2$ frequencies of the individual molecules.

Which collision-induced effects can we expect to observe if the region of the vibrational frequencies is investigated? We remember that in dipole-allowed vibrational spectra it is not the dipole moment but the derivative of the dipole moment (with respect to the normal coordinate) which enters into the pertinent matrix elements. This is, of course, the same in collision-induced vibrational spectra. Furthermore, we understand that in induced vibrational spectra the (much slower) rotational motion of the molecules leads to rotational-vibrational spectra.

Consider a sample of H_2 gas. We neglect at first the anisotropy of the polarizability. We search in the region of the vibrational frequencies of H_2. At room temperature the far greatest number of the H_2 molecules are in their vibrational ground state. We look at an isolated molecule, knowing that interaction with the electromagnetic radiation is not possible since the vibrational motion does not lead to a transition moment. Now we let two molecules collide and thereby obtain a contribution to the induced vibrational absorption due to the following interactions: (*1*) Overlap of the electron clouds of the two molecules leads to an induced dipole moment ("atomic distortion effect") which is modulated by the vibrational motion of the intramolecular bond and thus leads to a pure vibrational frequency, $\Delta n = 1$, $\Delta J = 0$, in one of the colliding molecules (*Q* branch). (*2*) The quadrupolar field $\mathbf{E}$ of molecule j induces a dipole $\boldsymbol{\mu} = \mathbf{a} \cdot \mathbf{E}$ in molecule i. Since the polarizability $\mathbf{a}$ depends on the bond distance and the field $\mathbf{E}$ rotates twice as fast as the molecule (see above), we observe now the transitions $\Delta n = 1$, $\Delta J_j = \pm 2, \Delta J_i = 0$ (S and O branches). Since we are dealing with vibrational modes, the interaction is proportional to the *derivatives* of the quadrupole moment (with respect to the displacement coordinate) and to the polarizability (and vice versa). We also understand that if we do not neglect the anisotropy of the polarizability, double transitions may occur ($\Delta n = 1$, $\Delta J_i = \pm 2, \Delta J_j = \pm 2$).

Summarizing, we see that in the induced vibrational phenomena, the induced dipole moment not only arises from the quadrupole moments of the molecules but also from the overlap of the electron clouds. The relative orientation of the molecules introduces the coupling of the rotation and the vibration. Mostly, only one molecule of a cluster undergoes a vibrational transition but, depending on the degree of anisotropy of polarizability, a rotational-vibrational transition in one molecule may be accompanied by a rotational transition in the other collision partner.

We now ask the question, what effects could take place if (one or both of) the colliding molecules also possess a permanent dipole moment? We stress here that although the absorption due to a permanent dipole moment is generally the largest contribution to the total absorption coefficient, in this book we shall point to various instances which indicate the importance of induced far-infrared absorption in polar molecules. Suppose that we have a mixture of H_2 and HCl. If the polarizability of the hydrogen molecule is considered isotropic, the quadrupole moment of H_2 may induce an enhancement of the intensity of the dipole-allowed transitions in the hydrogen chloride molecule (rotational or vibrational-rotational). If we cannot neglect the anisotropy of the polarizability of the hydrogen, then the dipole induced in it by the HCl molecule also depends on the rotational coordinates of H_2 and we may observe a simultaneous rotational hydrogen/vibrational-rotational hydrogen chloride transition. Such phenomenon has, in fact, been observed by Colpa and Ketelaar in mixtures of carbon monoxide and hydrogen at a frequency of 2730 cm^{-1}, made up by the rotational-vibrational transition $\Delta n = 1$, $\Delta J = \pm 1$ of CO (2143 cm^{-1}) and the rotational transition $J = 1 \rightarrow 3$ of H_2 (587 cm^{-1}).[1] The corresponding simultaneous transition is not observed if CO is replaced by the isoelectronic N_2. Although the permanent dipole moment of CO is rather small, its derivative with respect to the intramolecular axis is large.

The subject of collision-induced spectra is rather complicated but it leads to a good insight into the realm of intermolecular forces. As already mentioned, the effects of collision-induced absorption become more and more apparent—particularly in the far-infrared spectra of solutions and liquids (see Chapter 11)—and a familiarity with their concepts and results therefore seems fruitful.

B. THEORY OF INDUCED ABSORPTION

In the following we treat the theoretical aspects of the collision-induced absorption.[2] Although the pure translational absorption phenomena seem the easiest to understand, in our opinion this is no longer the case if the basic theoretical principles are studied on a quantitative basis or if the translational motion is coupled to the rotational motion (as is the case with all but the smallest molecules).

We therefore begin with the principles of induced vibrational-rotational and pure rotational absorption. In order not to overload the text while still guiding the interested reader, we have attached a relatively large number of explanatory footnotes.

a. Absorption Coefficients, Cluster Functions, and Induced Vibrational-Rotational Transitions

In collision-induced spectra we generally measure the absorption coefficient as a function of pressure, temperature, and admixture of foreign gases. The integrated absorption coefficient $\alpha = \int A(v)\,dv$ per molecule is given by

$$\alpha = (\kappa/V)\sum_{i,f} P_i |\langle s_i|\mu|s_f\rangle|^2 h v_{if}, \tag{10.2a}$$

which is proportional to the absorbed energy, or by

$$\tilde{\alpha} = (hc\kappa/V)\sum_{i,f} P_i |\langle s_i|\mu|s_f\rangle|^2, \tag{10.2b}$$

which is $\alpha \times$ wavelength or the transition probability. The constant κ equals $8\pi^3/3h^2c$, i and f denote the initial and final external states (see Eq. 10.1a) of the molecule, P_i is the Boltzmann factor of the initial state, and V is the total volume. The theoretically more fundamental Eq. 10.2b is not convenient if the spectrum extends to zero frequency.* We note that the transition moment is to be averaged over the rotational-translational (external) states, whether the transition moment is of a vibrational, rotational, or translational nature, because we wish to describe phenomena which depend on the collisions between molecules and therefore on their relative spacial coordinates. (We shall encounter the same principle again in Chapter 11 when we discuss solution spectra.)

Inspection of Eq. 10.2a or 10.2b indicates the degree of mathematical complexity to be expected in the evaluation of the absorption coefficients. If the rotational-translational motion is strongly coupled (as it generally

---

* See the related expression for $I(\omega)$, Chapter 11, Footnote Section C.b. $I(\omega)$ becomes undefined for $\omega = 0$. Rotational absorption does not extend to zero frequency, therefore the energy of the upper state, $E'_{rot} + E'_{transl}$, will always be larger than that of the ground level for all values of E_{transl}.

is the case except for the smallest molecules), we can rewrite Eq. 10.2b in the form (summing over all f)

$$\tilde{\alpha} = (hc\kappa/V) \int\int |\mu(\omega^N, \mathbf{R}^N)|^2 f(\omega^N, \mathbf{R}^N)\, d\omega^N\, d\mathbf{R}^N, \qquad (10.2c)$$

where $\mu(\omega^N, \mathbf{R}^N)$ is the induced transition moment as a function of the relative orientations ω and separations $\mathbf{R}$ of the N colliding molecules, and the distribution function $f(\omega^N, \mathbf{R}^N)$ is defined as $\Sigma_i P_i |\psi_{s_i}(\omega^N, \mathbf{R}^N)|^2$. This can be written more elegantly in the form of an invariant trace with the help of the density operator $\rho = \exp(-H/kT)/\mathrm{Tr}\{\exp(-H/kT)\}$, where H is the external Hamiltonian $H = H_{rot} + H_{transl} + V'$ (V' is the *noncentral* part of the intermolecular potential):

$$\tilde{\alpha} = (hc\kappa/V)\, \mathrm{Tr}\,\{\rho|\mu|^2\} \qquad (10.3)$$

or*

$$\alpha = (\kappa/V)\, \mathrm{Tr}\,\{\rho\mu \cdot [H, \mu]\}. \qquad (10.4)$$

If there is no significant coupling between the rotational and the translational motion ($V' = 0$), the density operator is given by $\rho = \rho_r \cdot \rho_t$, where now $H = H_{rot}$ or H_{transl} in ρ_r or ρ_t, respectively, and the distribution functions simplify to a product of the pure rotational and the pure translational distribution function (see also Eq. 10.1b) $f(\omega^N)$ and $f(\mathbf{R}^N)$. A Lennard-Jones potential is conveniently employed in the expression for the translational distribution function $f(\mathbf{R}^N)$, whereas for the rotational motion the usual spherical harmonic wave functions are used.

On the other hand, if the coupling between rotation and translation is strong ($\rho \neq \rho_r \cdot \rho_t$), complicated coupling coefficients occur (see also Eq. 10.1a) since the motion of the intermolecular axis of the collision partners must now be considered also.

* For instance,

$$\sum_{i,f} |\mu_{if}|^2 h\nu_{if} = \sum_{i,f} \langle i|\mu|f\rangle \langle f|\mu|i\rangle (E_f - E_i).$$

Because of the Hermitian property of μ_{if} and because $H\psi_k = E\psi_k$ ($k = i, f$), this can be rewritten in the form

$$\sum_{i,f} \langle i|\mu|f\rangle \{\langle f|H\mu|i\rangle - \langle f|\mu H|i\rangle\} = \mathrm{Tr}\,\{\mu \cdot [H, \mu]\},$$

where we remember that

$$\exp\left(-\frac{H}{kT}\right)\psi = \left[1 - \frac{H}{kT} + \frac{1}{2!}\left(-\frac{H}{kT}\right)^2 \pm \cdots\right]\psi = \left[1 - \frac{E}{kT} + \frac{1}{2!}\left(-\frac{E}{kT}\right)^2 \pm \cdots\right]\psi$$

$$= \exp\left(-\frac{E}{kT}\right)\psi.$$

For collision-induced rotational and translational spectra the induced transition moment μ_{if} is assumed to be averaged over the ground vibrational state (and, of course, over the ground electronic state). Furthermore, for an induced vibrational spectrum the transition moment $\mu = \mu(r_1, \cdots, r_N, \omega^N, R^N)$, where r_i is the intramolecular distance in molecule i, is averaged over the ground vibrational states $(v_i = 0)$ of all molecules *except* the one that undergoes the vibrational transition. We can write this in the form of the matrix element for a harmonic oscillator (eigen frequency v_0)

$$\mu_{01} = \langle 0|\mu|1 \rangle = \kappa_1 M(\omega^N, R^N). \tag{10.5a}$$

Here, $\kappa_1 = (h/8\pi^2 m v_0)^{1/2}$, with $M = \langle (\partial\mu/\partial r_1)\rangle_{v_2, \cdots, v_N = 0}$. Thus

$$\tilde{\alpha} = (\tilde{\kappa}/V) \, \mathrm{Tr} \, \{\rho|\mathbf{M}|^2\} \tag{10.5b}$$

and $\tilde{\kappa} = 8\pi^3 \kappa_1{}^2/3h$.

During the further computations of α and $\tilde{\alpha}$, use is made of the fact that the induced dipole moment vanishes at intermolecular distances of more than a few molecular diameters. The absorption coefficient per molecule per unit volume is then conveniently developed into a power series of the number density $n = N/V$, where N is the number of molecules and V is the volume:

$$\tilde{\alpha} = \tilde{\alpha}_1 n + \tilde{\alpha}_2 n^2 + \cdots$$
$$\alpha = \alpha_1 n + \alpha_2 n^2 + \cdots \tag{10.6}$$

Here α_1 is the binary, α_2 is the ternary, and so forth, absorption coefficient. Since we carry the reciprocal volume as a factor in the expressions for $\tilde{\alpha}$ and α, the resulting density dependence of the binary (and ternary) absorption coefficients is n^2 (and n^3). This is because the number of molecules per volume as well as the rate at which molecules undergo binary collisions depend on the first power of the pressure.

In the following we will indicate the manner of computing $\tilde{\alpha}_1$ and $\tilde{\alpha}_2$ and point later to the slightly different way of obtaining α_1 and α_2. As we have mentioned above, $\tilde{\alpha}$ does not lead to a useful expression for spectra which extend to zero frequency (such as translational spectra).

We can write the absorption coefficient per molecule and per volume, $\tilde{\alpha}$ of Eq. 10.5b, more explicitly as

$$\tilde{\alpha} = (\tilde{\kappa}/V) \int |M(1 \cdots N)|^2 f(1 \cdots N) \, d1 \cdots dN, \tag{10.7}$$

where $M(1 \cdots N)$ indicates the transition moment induced in a group of molecules $1, 2, \cdots, N$ which are present in a volume V and of which *one* only, say molecule 1, shall undergo the vibrational transition (see above). The function $f(1 \cdots N)$ represents the appropriate distribution function. The expression of $M(1 \cdots N)$ is broken up into a sum of "cluster functions"

$U(1 \cdots N)$ which shall have the property of being equal to zero if the molecules $2, 3, \cdots, N$ are not within the range of the induced transition moment; that is, if they are not near molecule 1.[3] For example, in the case of three molecules 1, 2, 3, the four possible ways of arranging all three molecules in clusters of pairwise interactions are as follows: molecule 1 interacts with 2, and 2 with 3 (12 23); then 23 31, 31 12, and finally 12 23 31. The corresponding threefold cluster function is expressed by $U(123)$ (see Fig. 10.1). The twofold cluster functions of three molecules are $U(12)$ and $U(13)$. We note that the third twofold cluster function, namely $U(23)$, is zero since we have assumed that only molecule 1 undergoes the vibrational transition, and molecule 1 is not contained in the binary cluster 23. The function $U(23)$ must, however, be retained in the corresponding cluster expansion of induced pure rotational and translational transitions since there all molecules are to be treated alike. We shall encounter this differentiation again in the discussion of the effects of the external molecular motion on dipole-allowed spectra in condensed phases (see Chapter 11, Section B).

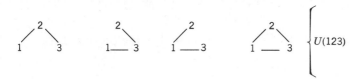

Fig. 10.1   The four different ways of arranging three molecules in clusters of pairwise interactions (schematically).

The transition moment $M(1 \cdots N)$ is now written as the sum of cluster functions of the possible binary, ternary, $\cdots$ clusters among the N molecules of which molecule 1 undergoes the induced vibrational transition.

$$M(12) = U(12)$$

$$M(123) = U(123) + U(12) + U(13)$$

$$M(1234) = U(1234) + U(123) + U(124) + U(134)$$

$$+ U(12) + U(13) + U(14).$$

(10.8)

This expression is inserted into the formula for $\tilde{\alpha}$ (Eq. 10.6), and carried up to and including ternary collisions; that is, up to $\tilde{\alpha}_2$ of Eq. 10.6. Then

$$|M(123)|^2 = |U(123)|^2 + |U(12)|^2 + |U(13)|^2$$

$$+ 2\{U(123) \cdot U(12) + U(123) \cdot U(13)$$

$$+ U(12) \cdot U(13)\},$$

(10.9a)

which is, effectively, equal to

$$|U(123)|^2 + 2|U(12)|^2 + 4U(123) \cdot U(12) + 2U(12) \cdot U(13).$$

Expressing the relevant distribution functions of the two- and three-clusters, $f_2(12)$ and $f_3(123)$, in terms of the general distribution function $f(1 \cdots N)$ by

$$f_3(123) = f(123)$$

$$f_2(12) = \int f(123)\, d3,$$

and setting

$$|U(12)|^2 = A(12)$$

$$U(12) \cdot U(13) + 2U(123) \cdot U(12) + \tfrac{1}{2}|U(123)|^2 = B(123),$$

(10.9b)

the expression for $\tilde{\alpha}$ is given by:

$$\tilde{\alpha} = (\tilde{\kappa}/V)\left\{ 2\int A(12)f_2(12)\, d1\, d2 + 2\int B(123)f_3(123)\, d1\, d2\, d3 \right\}. \quad (10.10a)$$

The absorption coefficient for a total of N molecules is then

$$\tilde{\alpha} = (\tilde{\kappa}/V)\left\{ V^{-1}(N-1)\int A(12)f_2(12)\, d1\, d2 \right.$$

$$\left. + V^{-2}(N-1)(N-2)\int B(123)f_3(123)\, d1\, d2\, d3 \right\},$$

(10.10b)

since molecule 1 (the one which undergoes the induced vibrational transition) can form $N-1$ binary cluster functions, $N-2$ ternary cluster functions, and so forth (see Fig. 10.1).

The distribution function f is also developed into a power series in n:

$$f_2(12) = f_2^{(0)} + f_2^{(1)}n + \cdots$$

$$f_3(123) = f_3^{(0)} + f_3^{(1)}n + \cdots.$$

(10.11)

Collecting terms of the same order, we obtain the following expressions for the binary (n^2) and ternary (n^3) absorption coefficients $\tilde{\alpha}_1$ and $\tilde{\alpha}_2$, respectively:

$$\tilde{\alpha}_1 = (\tilde{\kappa}/V)\int A(12)f_2^{(0)}(12)\, d1\, d2,$$

(10.12)

$$\tilde{\alpha}_2 = (\tilde{\kappa}/V)\left\{ \int A(12)f_2^{(1)}(12)\, d1\, d2 + \int B(123)f_3^{(0)}(123)\, d1\, d2\, d3 \right\}.$$

We see that the binary absorption coefficient, $\tilde{\alpha}_1$, depends on the transition moment induced in a pair of molecules and on the low-density limit of the pair distribution function. The ternary coefficient, $\tilde{\alpha}_2$, consists of three parts: The first, $\tilde{\alpha}_2^{(1)}$, is identical to $\tilde{\alpha}_1$ except that it contains the first-order term of $f_2(12)$ (see Eq. 10.11). The second term of $\tilde{\alpha}_2$, $\tilde{\alpha}_2^{(2)} = (\tilde{\kappa}/V) \int M(12) \cdot M(13) \cdot f_3^{(0)}(123)\, d1\, d2\, d3$ (see Eqs. 10.9b and 10.8), describes the interference effects between the two induced dipole moments $M(12)$ and $M(13)$ because of the correlation of the positions of molecule 2 and molecule 3 by $f_3^{(0)}(123) = f_2^{(0)}(12) f_2^{(0)}(13) f_2^{(0)}(23)$ (where the low-density limit of the triple distribution function has been expressed as a product of the low-density limits of the pair distribution functions). The third term of $\tilde{\alpha}_2$, $\tilde{\alpha}_2^{(3)}$, involves the contribution of $U(123) \cdot U(12)$ and $|U(123)|^2$ of $B(123)$ (see Eq. 10.9b). We shall have no opportunity to use this term and therefore have not given its expression here.

We note that the corresponding coefficients α for induced rotation-vibration, α_1 and α_2, are of the form (see Eqs. 10.3 and 10.4)

$$\alpha_1 = (\kappa/V)\, \mathrm{Tr}\, \{\rho_2^{(0)}(12)\mathbf{\mu}(12) \cdot \hat{\mathbf{\mu}}(12)\},$$

$$\alpha_2 = (\kappa/V)\, \mathrm{Tr}\, \{\rho_2^{(1)}(12)\mathbf{\mu}(12) \cdot \hat{\mathbf{\mu}}(12)\} \tag{10.13}$$

$$+ (\kappa/V)\, \mathrm{Tr}\, \{\rho_3^{(0)}(123)\mathbf{\mu}(12) \cdot \hat{\mathbf{\mu}}(13)\} + \alpha_2^{(3)},$$

where $\hat{\mathbf{\mu}}(ij) = [\tilde{K}, \mathbf{\mu}(ij)]$, $K(i)$ is the kinetic energy of the rotational-translational motion of molecule i, and $\tilde{K} = K(i) + K(j)$.* The ρ_i are the coefficients of the expansion (in N/V) of the density operators.

The remaining calculation, in which the (binary) absorption coefficients are cast into expressions containing molecular parameters of the colliding molecules, are straightforward but rather tedious. Since the computations follow standard methods,[2,4] we shall only outline the general course of the computations, giving more detailed explanations on points which we feel are not frequently encountered in the literature.

We begin by writing down the matrix element of the induced dipole moment between the initial (unprimed) and final (primed) states; as basis functions we use products of vibrational wave functions, ψ_{v_i}, and rotational wave functions, $\psi_{J_i m_i}$. The validity of this assumption is well based on observations: the frequencies of the induced transitions essentially coincide with those calculated from the structure of the individual molecules. The index i assumes the values 1 and 2 since we wish to treat the binary absorption only. Quantum numbers v, J, m are the usual quantum numbers for the vibrational

---

* Since $\mathbf{\mu}(1 \cdots N)$ depends only on the internuclear coordinates, it commutes with the intermolecular potential energy V of the rotational-translational Hamiltonian $H = H_r + H_t + V$. Therefore, H can be replaced by the kinetic energy in the commutator. Any interaction terms, H_{rt}, between H_r and H_t would only influence the statistical factor: if $H_{rt} \neq 0$, then $\rho \neq \rho_r \rho_t$ (see Eqs. 10.3 and 10.4).

level and angular momentum and its components, respectively.[4] Then

$$\langle|\mu'\rangle = \int \psi_{v_1'}^* \psi_{v_2'}^* \psi_{J_1'm_1'}^* \psi_{J_2'm_2'}^* \langle\mu\rangle \psi_{v_1}\psi_{v_2}\psi_{J_1m_1}\psi_{J_2m_2}\, d\tau \qquad (10.14)$$

with

$$\langle\mu\rangle = \int \psi_e^* \sum_i e_i z_i \psi_e\, d\tau, \qquad (10.15)$$

where e_i and z_i are the atomic charges and coordinates, respectively, and ψ_e are the electronic wave functions.

We now develop $\langle\mu\rangle$ into a Taylor series,

$$\mu(R, r_1, r_2, \omega_1, \omega_2) = \mu^0 + \mu_1'(r_1 - r_1{}^0) + \mu_2'(r_2 - r_2{}^0) + \cdots$$
$$\mu_1' = (\partial\mu/\partial r_1)_{r_1{}^0 r_2{}^0} \qquad \mu_2' = (\partial\mu/\partial r_2)_{r_1{}^0 r_2{}^0}. \qquad (10.16)$$

The symbols $r_1{}^0 = r_{ab}^0$, $r_2{}^0 = r_{cd}^0$ denote the equilibrium distances between the nuclei. This is indicated in Fig. 10.2. Since at present we are treating the

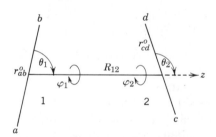

Fig. 10.2 Coordinates of the four-nuclei, two-molecule system (1, 2). r_{ab}^0, r_{cd}^0 = equilibrium intramolecular axes, R_{12} = intermolecular axis.

vibrational-rotational spectra, we are interested only in the terms $(\partial\mu/\partial r)\Delta r$ of the Taylor expansion. The factors $\Delta r_i = r_i - r_i{}^0$, averaged over the ψ_{v_i} and ψ_{v_i} (see Eq. 10.14), give simply $(h/8\pi^2 m v_0)^{1/2}$ for a harmonic intramolecular potential (see Eq. 10.5a). To obtain the part of the transition moment derivative which depends on the orientation of the two molecules 1 and 2, we expand $M_i = \partial\mu/\partial r_i{}^*$ into a linear combination of products of eigen functions which span the subsets of $\omega_1 \equiv \theta_1, \varphi_1$ and $\omega_2 \equiv \theta_2, \varphi_2$, respectively. For our purposes it is convenient to use for the eigen functions the spherical harmonics $Y_{\lambda\mu}(\omega) = P_\lambda^{|\mu|}(\cos\theta) \exp(i\mu\varphi)$, where $P_\lambda^{|\mu|}(\cos\theta)$ is the associated Legendre polynomial of order $|\mu|$, and properly normalized. We then have

$$M_i(R_{12}, \omega_1, \omega_2) = 4\pi \sum_{\substack{\lambda_1\mu_1 \\ \lambda_2\mu_2}} D(\lambda_1\mu_1\lambda_2\mu_2, R_{12}) \cdot Y_{\lambda_1\mu_1}(\omega_1) Y_{\lambda_2\mu_2}(\omega_2). \qquad (10.17)$$

---

* We have tried to adhere to the nomenclature of designating the dipole moment by μ and its change with the coordinate (as in the case of vibrational transitions) by M.

We shall not need coefficients higher than $\mu, \lambda = 2$, as will become clear further below. (We adhere to common usage and designate the order of the associated Legendre polynomial as well as the dipole moment by the symbol μ.) The symmetry of the homonuclear diatomic molecules requires that λ is even.*

Substituting Eqs. 10.16 and 10.17 into Eq. 10.14, expressing the ψ_{Jm} in terms of $Y_{\lambda\mu}(\omega)$, and setting for the vibrational transition $\Delta v_1 = 1$, $\Delta v_2 = 0$, we obtain for the binary absorption coefficient $\tilde{\alpha}_1(B)$ of a vibrational-rotational branch B an expression of the form

$$\tilde{\alpha}_1(B) = \tilde{\kappa} \sum_{\lambda_1 \lambda_2} L_{\lambda_1 \lambda_2}(B) I_0(\lambda_1 \lambda_2), \tag{10.18}$$

and the binary coefficient of the total intensity of the $0 \to 1$ band as

$$\tilde{\alpha}_1 = \sum_B \tilde{\alpha}_1(B) = \tilde{\kappa} \sum_{\lambda_1 \lambda_2} I_0(\lambda_1 \lambda_2). \tag{10.19}$$

Here, the $L_{\lambda_1 \lambda_2}$ are related to certain Racah coefficients,† and $I_0(\lambda_1 \lambda_2)$ is given by the integral

$$I_0 = \int \sum_{\mu_1 \mu_2} |\mathbf{D}(\lambda_1, \mu_1, \lambda_2, \mu_2; R_{12})|^2 g_0(R_{12}) \, dR_{12}$$

with   (10.20)

$$g_0(R_{12}) = \int f_2^{(0)}(12) \, d\omega_1 \, d\omega_2,$$

where $g_0(R_{12})$ is the radial distribution function.

* Inspection of Fig. 10.2 shows that an inversion of the coordinates of the diatomic molecule about its center of mass leads to the same spacial arrangement; that is, M_i is invariant under inversion. If $\mathscr{R}$ is the inversion operator, we have

$$\mathscr{R} Y_{\lambda\mu}(\theta, \varphi) = Y_{\lambda\mu}(\pi - \theta, \pi + \varphi) = (-1)^\lambda Y_{\lambda\mu}(\theta, \varphi).$$

It follows that λ is even. (See, for instance, J. S. Lomont, *Application of Finite Groups*, Academic Press, New York, 1959, p. 151.) For a nonpolar diatomic molecule in which the center of mass and the center of charge distribution do not coincide, for instance HD, odd values of λ will occur.

† The $L_{\lambda_1 \lambda_2}(B)$ are given by

$$\sum_B P_1(J_1) P_2(J_2) L_{\lambda_1}(J_1, J'_1) L_{\lambda_2}(J_2, J'_2)$$

for the rotational transition $J_1, J_2 \to J'_1, J'_2$. The P_i are Boltzmann factors. The quantities $L_\lambda(J, J')$ are of the form

$$L_\lambda(J, J') = \sum_{m,m'} \langle Jm|\lambda\mu J'm'\rangle \langle \lambda\mu J'm'|Jm\rangle = \sum_{m,m'} |\langle Jm|\lambda\mu J'm'\rangle|^2,$$

where $\langle Jm|\lambda\mu J'm'\rangle = \int Y^*_{Jm}(\omega) \cdot Y_{\lambda\mu}(\omega) \cdot Y_{J'm'}(\omega) \, d\omega$; the appearance of a product of six rotational eigen functions as the total integrand of $L(J, J')$ is easily recognized from Eq. 10.14—considering of course only the rotational products—and Eq. 10.17. Frequently used coefficients are $L_2(J, J+2) = 3(J+1)(J+2)/2(2J+3)$, $L_0(J, J) = 2J+1$, $L_1(J, J+1) = J+1$.

The selection rules follow from the evaluation of the integrals in Eq. 10.14:

$$\Delta m_i = 0$$

$$\Delta J_i = 0 \qquad (Q \text{ branch}) \qquad i = 1, 2 \qquad (10.21)$$

$$\Delta J_i = \pm 2 \qquad (S, O \text{ branches}).$$

We now give the coefficients $D(\lambda_1 \mu_1 \lambda_2 \mu_2)$ for homonuclear diatomic molecules (see Eq. 10.17). The main contribution of the atomic distortion effect, that is, the effect which results from the distortion of the overlapping electron clouds of the colliding molecules (see the Introduction), is expressed by the angle-independent term

$$D(0000) = \xi \exp(-R/R_0),$$

with the selection rules   (10.22)

$$\Delta n = 1$$

$$\Delta J_i, \Delta J_j = 0.$$

The quantity ξ denotes the strength, and R_0 gives the range of the induced dipole moment of the collision complex.* Essentially, the form of this coefficient is assumed, although it can be approximated by computing the repulsive overlap between the molecules using Rosen-type hydrogen wave functions.[5]

There are also angle-dependent terms, that is to say, terms which depend on the relative orientation of the two molecules, which contribute to the atomic distortion effect. They may be obtained by arranging the colliding molecules in various fixed positions and repeating the computation of D from Rosen-type wave functions for each orientation. The terms which arise in this way are of the form $D(2000)$, $D(2020)$, and $D(0020)$, since now the angle variables θ_1 and θ_2 are different from zero. (Variations in φ_1 and φ_2 are not considered, hence the index $\mu_1, \mu_2 = 0$.) However, these angle-dependent repulsive terms, arising from the overlap interaction, are significantly smaller than the angle-independent term $D(0000)$. For this reason they are generally neglected and the atomic distortion effect is expressed solely by the coefficient $D(0000)$ as given in Eq. 10.22. [The interested reader will find the details on this in the paper by J. Van Kranendonk and R. B. Bird, *Physica* **17**, 953 (1951).]

Because of the exponential form of $D(0000)$, the repulsive overlap decreases very rapidly with increasing separation of the collision partners ("short-range interaction"). We shall give an example of this below.

We next consider the contribution to the induced absorption arising from the quadrupole-induced dipole interaction, as described qualitatively in

* The reader will find that in the original literature the symbol ρ is used for the range of the induced dipole moment, for the number density, and also for the density operator.

the Introduction to this chapter. The $\mathbf{D}$ coefficients arising from this interaction contribute the significant angle-dependent terms to the sum given by Eq. 10.17. They are readily derived in the following manner. The dipole moment derivative $\mathbf{M}(R_{12}, \omega_1, \omega_2)$ on the left-hand side of Eq. 10.17 is written down explicitly by remembering that the induced dipole $\boldsymbol{\mu}$ is given by $\boldsymbol{\mu} = \mathbf{a} \cdot \mathbf{F}$, where $\mathbf{F}$ is the tensor of the quadrupole field of molecule j at molecule i, and $\mathbf{a}$ is the tensor of the polarizability of molecule i. The components of $\mathbf{a}$ and $\mathbf{F}$ are then expressed in terms of the spherical harmonics $Y_{\lambda_1 \mu_1}(\omega_1)$, $Y_{\lambda_2 \mu_2}(\omega_2)$ of ω_1 and ω_2 (see Fig. 10.2). Equation 10.17 is subsequently solved for $\mathbf{D}$ by performing the necessary averaging over all angular coordinates (the molecules are assumed to rotate freely). We thus obtain the $\mathbf{D}$ coefficients as a function of the intermolecular distance R.

In the spherical harmonic representation the five components of the quadrupole moment tensor assume the indices $\lambda = 2$, $\mu = 0, \pm 1, +2$. The resulting first few nonzero D coefficients are, in terms of the spherical coordinates $0 \leftrightarrow z$, $\pm 1 \leftrightarrow (1/\sqrt{2})(x \pm iy)$:

$$D_0(2000) = (3/\sqrt{5})Q_1' a_2 R^{-4}$$

$$D_0(0020) = -(3/\sqrt{5})Q_2 a_1' R^{-4}$$

$$D_{\pm 1}(2 \pm 100) = (3/\sqrt{15})Q_1' a_2 R^{-4}$$

$$D_{\pm 1}(002 \pm 1) = -(3/\sqrt{15})Q_2 a_1' R^{-4},$$

(10.23)

with the selection rules

$$\Delta n = 1$$

$$\Delta J_1 = \pm 2$$

$$\Delta J_2 = 0.$$

The quantity Q_i is the quadrupole moment of molecule i,* and a_i is its polarizability. The primed quantities in Eq. 10.23 denote the differential quotient with respect to the intramolecular axis (see Fig. 10.2); we remember that we are interested in the derivatives of the induced dipole moment for vibrational spectra.

The polarizability a in the above coefficients is assumed to possess negligible anisotropy. If we cannot neglect the anisotropy of the polarizability,

---

* See, for instance, M. Kauzmann, *Quantum Chemistry*, Academic Press, New York, 1957, pp. 92–97. (The molecular quadrupole moment should not be confused with the nuclear quadrupole moments of nuclei of spin $\geqslant 1$.) The quadrupole moment is defined here as $Q = \frac{1}{2}\int \rho(3z^2 - r^2)\,d\rho$, where ρ is the charge distribution and z is the coordinate in the direction of the intramolecular axis.

coefficients of the form $D(2\mu_1 2\mu_2)$ will appear and double rotational or simultaneous transitions, $\Delta n = 1, \Delta J_i = \pm 2, \Delta J_j = \pm 2$ will become allowed (see the remarks in the Introduction).

TABLE 10.1*

THE FIRST THREE EXPANSION COEFFICIENTS D (IN ESU $\times$ 10^{-10}) OF H_2—H_2 AS A FUNCTION OF THE INTERMOLECULAR SEPARATION R (IN Å UNITS)

R	$D_0(0000)$	$D_0(2000)$	$D_0(0020)$
3	4.1	0.80	-0.89
3.5	1.0	0.43	-0.48
4	0.30	0.25	-0.28
5	0.030	0.10	-0.12
6	0.003	0.050	-0.055

* $D_0(0000) = \xi \exp(-R/R_0) = \delta e \exp[(\sigma - R)/R_0]$. $R_0 = 0.145\sigma$, the range of the overlap dipole moment in terms of $\sigma = 2.93$Å, the molecular diameter of H_2. $\delta = 8.5 \times 10^{-3}$, $e = 4.8 \times 10^{-10}$ esu. The values of δ and R_0 were obtained by fitting the $D_0(0000)$ from Rosen-type wave functions to the data (see text). For $D_0(2000)$ and $D_0(0020)$, see Eq. 10.23. The experimental values for Q_1, Q_1', a_1, and a_1' are 0.61×10^{-26} esu cm^2, 0.76×10^{-18} esu cm, 0.85×10^{-24} cm^3, and 1.2×10^{-16} cm^2, respectively. [See J. van Kranendonk, Physica 24, 347 (1958).]

Table 10.1 shows a comparison between the values of D arising from electronic overlap and those originating from the quadrupolar interaction in compressed hydrogen. The extremely rapid decrease of D(overlap) with increasing separation is seen in contrast to the slower decrease of the D coefficients which stem from the quadrupole-induced dipole interaction.

Before we continue with the computation of the absorption coefficient for induced vibrational-rotational absorption, we summarize briefly the results of this section. (1) An induced vibrational transition moment arises in one molecule of a collision cluster. (2) The induced transition moment matrix elements of binary clusters have as their basis a set of simple rotational-vibrational product wave functions, $\psi_{v_1}\psi_{v_2}\psi_{J_1 m_1}\psi_{J_2 m_2}$, where $\Delta v_1 = 1$, $\Delta v_2 = 0$. (3) The angular variations of the transition moment are developed into a series of spherical harmonics as the most convenient basis set of the rotational wave functions, $Y_{J_1 m_1}(\omega) \cdot Y_{J_2 m_2}(\omega)$. (4) The coefficients of this series are determined by computing the long-range quadrupole-induced dipole interaction between the tensors of the quadrupole field and the polarizability of the two molecules. The resulting transition rules are $\Delta J = \pm 2$ (S, O branches). (5) The short-range interactions by repulsive overlap of the electron clouds are assumed to be governed by an angle-independent exponential term ($\Delta J = 0$, Q branch).

The values of Eqs. 10.22 and 10.23 are now inserted into Eqs. 10.17 and 10.18. Three nonvanishing integrals result,

$$\tilde{\kappa} I_0(00) = (\xi/e)^2 \exp{(-2\sigma/R_0)} \tilde{I} \tilde{\kappa} e^2 \sigma^3$$

$$\tilde{\kappa} I_0(20) = (\eta_1/e\sigma^4)^2 \tilde{J} \tilde{\kappa} e^2 \sigma^3 \qquad \eta_1 = Q_1' a_2 \qquad (10.24)$$

$$\tilde{\kappa} I_0(02) = (\eta_2/e\sigma^4)^2 \tilde{J} \tilde{\kappa} e^2 \sigma^3 \qquad \eta_2 = Q_1 a_2'.$$

The parameter σ is the value of R for which the intermolecular potential, assumed to be a Lennard-Jones potential, V, has its node ($V = 0$). $\tilde{I}$ and $\tilde{J}$ are the integrals over the radial terms, and e is the value of the absolute electronic charge. $\tilde{I}$ and $\tilde{J}$ have the form (setting $x = R/\sigma$)

$$\tilde{I} = 4\pi \int_0^\infty \exp{[\ -2(x-1)\sigma/R_0]} g_0(x) x^2 \, dx,$$

$$(10.25)$$

$$\tilde{J} = 12\pi \int_0^\infty x^{-8} g_0(x) x^2 \, dx.$$

The integrand of $\tilde{I}$ has been expanded by the constant factor $\exp{(2\sigma/R_0)}$. This has been done to permit us to place its inverse factor before the integral since, according to Eq. 10.22, $\xi \exp{(-\sigma/R_0)}$ has the physical meaning of the amplitude of the oscillating overlap dipole at the intermolecular distance $\sigma(V = 0)$. Similarly, the factors η_1/σ^4 and η_2/σ^4 give the corresponding amplitudes for the quadrupole contributions. Finally, the factor $\tilde{\kappa} e^2 \sigma^3 = (8\pi^3 \kappa_1{}^2/3h)e^2\sigma^3$ (see Eq. 10.5a) $= (\pi/3mv_0)e^2\sigma^3$ in g^{-1}cm^{3}sec esu^2 = sec^{-1}cm^6 is seen to be a quantity of an integrated absorption coefficient per (number density)^2 per wavelength; that is, the transition probability.

According to Eq. 10.18 the binary absorption coefficient of a rotational branch is the sum of the three terms of Eq. 10.24, each weighted by some appropriate Boltzmann factor over the various rotational states. The total binary absorption coefficient of the $0 \to 1$ band is simply the sum of the three terms of Eq. 10.24 as it is indicated by Eq. 10.19. Inspection of Eq. 10.24 again shows that the short-range overlap forces induce the Q branch (ΔJ_i, $\Delta J_j = 0$), whereas the long-range quadrupolar forces lead to the S and O branches of the rotation-vibration band ($\Delta J_i = \pm 2$, $\Delta J_j = 0$, assuming an isotropic polarizability, that is to say, $a_\perp \sim a_\parallel$).

b. Induced Pure Rotational Transitions

The binary absorption coefficient of a single induced pure rotational line is given by

$$\tilde{\alpha}_1 = \tfrac{1}{2} hc\kappa \sum (P_r - P_{r'}) \int |\langle r|\mu(\omega_1, \omega_2; R)|r'\rangle|^2 g_0(R)\, dR, \qquad (10.26)$$

where we have written out the dipole moment matrix element between the rotational states r and r'.* The sum extends over all transitions $r(J_1 m_1, J_2 m_2) \to r'(J_1' m_1', J_2' m_2')$ of the molecules 1 and 2, $g_0(R)$ is the low-density limit of the pair distribution function, and the Boltzmann factor P_r' takes care of induced emission. As mentioned in the Introduction, the translational and rotational motions are coupled. For simplicity's sake we assume in this section that both motions are separable and treat only the pure rotational absorption (see also footnote on page 330). The coupled rotational-translational effect will be considered in the next section.

The explicit form of the absorption coefficient for induced pure rotational transitions is, as can be expected, quite similar to that of rotation-vibrational transitions. For the pure rotational spectra, the dipole moment rather than its derivative is now expanded, according to Eq. 10.17, in terms of products of spherical harmonics. The symmetry of the two-molecule system requires that the angle-independent term of the series expansion of μ, $C_0(0000)$, vanishes.[†] Consequently, overlap forces are of much lesser importance here than in the induced vibrational-rotational spectra. Since the computation of the nonisotropic parts of the overlap contributions from Rosen-type wave functions has not led to a good agreement with experiment, the overlap moment is either neglected or approximated empirically. For isotropic polarizabilities the induced binary absorption coefficient for the single rotational line $J_1 \to J_1 + 2, \Delta J_2 = 0$, is given by

$$\tilde{\alpha}_1(J) = \kappa hcL(J)\{I_0(20) + L'I_0(22)\}. \qquad (10.27)$$

The L factors are, as before, related to coupling coefficients (see footnote[†] on page 337 and Ref. 2) and to Boltzmann factors, $I_0(\lambda_1 \lambda_2)$ has the same

---

* The factor $\tfrac{1}{2}$ in Eq. 10.26 and so forth appears in systems of one-component gases since it is the total density of the gas which enters in the expression for the binary absorption coefficient. For induced vibrational-rotational spectra, however, the factor $\tfrac{1}{2}$ is cancelled by the factor 2 which arises from the two terms $\partial\mu/\partial r_i$ (see Eq. 10.16).

[†] The Hamiltonian of the two-molecule system is invariant to inversion about a point midway between the two homonuclear molecules. Hence, applying the inversion operator $\mathscr{R}$, we obtain (see Fig. 10.2) $\mathscr{R} Y_{\lambda_1\mu_1}(\omega_1) \cdot Y_{\lambda_2\mu_2}(\omega_2) = Y_{\lambda_1\mu_1}(\omega_2) \cdot Y_{\lambda_2\mu_2}(\omega_1)$. Consequently, since $\mu(12)$ transforms like a vector, we find that $C_\gamma(\lambda_1\mu_1\lambda_2\mu_2) = -C_\gamma(\lambda_2\mu_2\lambda_1\mu_1)$, thus $C_0(0000) = 0$. The symmetry with respect to rotation about the intermolecular axis requires that $\mu_1 + \mu_2 = \gamma$. It should be noted that even if $C(0000) \neq 0$, the coefficient would be of no interest in pure rotational spectra ($\Delta J_i = 0$).

relation to the $C(\lambda_1\mu_1\lambda_2\mu_2)$ as it had for the D coefficients of the vibrational-rotational absorption; the factor $hc\kappa$ amounts to $8\pi^3/3h$. If we neglect the contributions due to overlap forces, insertion of the proper quantities into Eq. 10.27 yields,

$$\int \frac{A(v)}{v}\,dv = \tilde{\alpha}_1(J) = \tilde{\gamma}[P(J) - P(J+2)]\frac{3(J+1)(J+2)}{2(2J+3)}(aQ/e\sigma^5)^2\tilde{J}, \quad (10.28)$$

where the P's denote the Boltzmann factors for the lower and upper level, $\tilde{\gamma} = 8\pi^3e^2\sigma^5/3h$ (sec^{-1} cm^6), a, Q, e, and σ have their previously defined meaning of average polarizability, quadrupole moment, electronic charge, and molecular diameter appearing in the Lennard-Jones potential, respectively, and $\tilde{J} = 12\pi\int x^{-8}g_0(x)x^2\,dx$ with $x = R/\sigma$. If the anisotropy of the polarizability cannot be neglected, double rotational transitions occur $(\Delta J_i, \Delta J_j = \pm 2)$ because of the presence of coefficients $C(2\mu_1 2\mu_2)$ in the series expansion of the induced dipole moment.

The binary absorption coefficient of the whole rotational band is obtained by the proper summation over the $L(J, J')$ coefficients; we shall not give it here.

c. Induced Pure Translational and Translational-Rotational Absorption

We begin with the simplest case, the induced pure translational absorption in rare-gas mixtures. We have used the description "absorption" rather than "transitions" in order to stress the nonresonant character of the phenomenon (see Introduction to this chapter).

The induced binary absorption coefficient for the gas mixture is that of Eq. 10.13 (see also footnote* on page 342):

$$\alpha_1 = \kappa V^{-1}\,\text{Tr}\{\rho_2^{(0)}(11')\boldsymbol{\mu}(11')\cdot[K(1) + K(1'), \boldsymbol{\mu}(11')]\}, \quad (10.29)$$

where now the trace sums over the motion of the center of mass of the two colliding atoms 1 and 1'. This gives

$$\alpha_1 = \kappa\lambda^3\,\text{Tr}\,\{\exp(-\beta H)\boldsymbol{\mu}\cdot[K, \boldsymbol{\mu}]\}, \quad (10.30)$$

where K and H are the operators for the kinetic and total energy, respectively, $\beta = 1/kT$, and $\lambda = h/(2\pi mkT)^{1/2}$ is equal to $(V/Z)^{1/3}$, with Z the translational partition function of the two-atom system of reduced mass m. We assume, as before, that the induced dipole moment $\boldsymbol{\mu}$ lies along the interatomic axis $\mathbf{R}$ of the two rare-gas atoms. We write the pair dipole moment vector $\boldsymbol{\mu}$ in the form

$$\boldsymbol{\mu} = \mu(R)\cdot\hat{\mathbf{R}}, \quad (10.31)$$

where $\hat{\mathbf{R}}$ is the unit vector along R. We shall permit the magnitude of the pair dipole moment to change as a function of R; that is, we consider the variations of the interatomic distance of the two colliding atoms, and we let the direction of the interatomic axis $\mathbf{R}$ assume all possible orientations in space; that is, we refer $\mathbf{R}$ to a space-fixed coordinate system. The reader will observe that this differs from the procedure used for the induced vibrational-rotational and pure rotational spectra where we had put the coordinate system into the rotating axis $\mathbf{R}$ since we were then not interested in describing explicitly the translational motion of the collision complex.

The rotating coordinate system of μ is obviously most conveniently expressed in terms of the spherical coordinates R, θ, φ. The inner product $\mu \cdot [K, \mu]$ then amounts to

$$\mu(R)\frac{\partial^2}{\partial R^2}\mu(R) + \frac{2}{R}\mu(R)\frac{\partial}{\partial R}\mu(R) + 2\mu(R)\frac{\partial}{\partial R}\mu(R)\frac{\partial}{\partial R} - \frac{2}{R^2}\mu^2(R) \qquad (10.32)$$

in units of $-\hbar^2/2m$.* Multiplying on the left and right by the product wave function $\psi_{nlm} = \chi_{nl}(R)Y_{lm}(\theta, \varphi)$ and integrating over a spherical volume in R-space, we obtain [writing $\mu'(R)$ for $\partial\mu(R)/\partial R$] for the integrated binary translational absorption coefficient of a two-atom system

$$\alpha_1 = (4\pi^2/3mc)\int_0^\infty \{\mu'^2(R) + (2/R^2)\mu^2(R)\}g(R)R^2 \, dR, \qquad (10.33)$$

where $g(R)$ designates the low-density limit of the pair distribution function.[†] Equation 10.33 may be interpreted as a composite of the modulations of the dipole moment by the radial and angular variations of $\mathbf{R}$. As it turns out, the first contribution is by far the more important one; therefore the translational spectrum may not be interpreted as a kind of continuous rotational spectrum due to the turning of the induced dipole moment during a collision (Ref. 2, see also the Introduction.)

The final form of α_1 is obtained by assuming, as we have done previously, that $\mu(R)$ is of the form $\mu(R) = \xi \exp(-R/R_0)$ and that the interaction

---

* The total commutator $[K, \mu] = [\nabla^2, \mu]$ is composed of those of $[\nabla^2, \mu(R)]$ and $[\nabla^2, \hat{\mathbf{R}}]$ by the rule $[B, A_1A_2] = [B, A_1]A_2 + A_1[B, A_2]$. The commutator $[\nabla^2, \mu(R)]$ is obtained, in the usual fashion, as (primes indicate differentiation with respect to R) $\mu'' + (2/R)\mu' + 2\mu'(\partial/\partial R)$. The commutator $[\nabla^2, \hat{\mathbf{R}}]$ equals $-2\hat{\mathbf{R}}/R^2$, which can be derived easily with the help of the unit vectors of $\mathbf{R}$-space, namely $\hat{\mathbf{R}} = \mathbf{i}\sin\theta\cos\varphi + \mathbf{j}\sin\theta\sin\varphi + \mathbf{k}\cos\theta$; $\hat{\theta} = \mathbf{i}\cos\theta\cos\varphi + \mathbf{j}\cos\theta\sin\varphi - \mathbf{k}\sin\theta$; $\hat{\varphi} = -\mathbf{i}\sin\varphi + \mathbf{j}\cos\varphi$. The operator ∇^2 is given by

$$(1/R)^2(\partial/\partial R)\{R^2(\partial/\partial R)\} + (R^2\sin\theta)^{-1}(\partial/\partial\theta)\{\sin\theta(\partial/\partial\theta)\} + (R\sin\theta)^{-2}\partial^2/\partial\varphi^2.$$

[†] The translational wave functions $\chi(R)$ are normalized in a spherical volume of sufficiently large radius R_0. The integral arising from the third term of Eq. 10.32, that is, from the differentiation operator $2\int\chi\mu\mu'(\partial/\partial R)\chi \, d\tau$ (with $d\tau = R^2 \, dR$), is easily calculated by partial integration, $\int u \, dv = vu - \int v \, du$ with $v = \chi$ and $u = \mu\mu'\chi R^2$, keeping in mind that the boundary conditions are $\chi(R) \to 0$ for $R \to R_0$. The result is: $2\int\chi\mu\mu'\chi' \, d\tau = -\int\chi\mu\mu''\chi \, d\tau - \int(2/R)\chi\mu\mu'\chi \, d\tau - \int\chi\mu'\mu'\chi \, d\tau$. (The primes signify differentiation with respect to R.)

potential is of the Lennard-Jones type. We shall have little use for the final expression of the integrated translational absorption coefficient in the form given by Poll and Kranendonk.[2] The reason for this will become apparent when we discuss the experimental data.

There now remains the effect of translational absorption when one or both of the collision partners possess a nonvanishing angular momentum. As we indicated at the beginning of this chapter, the rotational and translational motion are only separable if the intermolecular forces are isotropic. The description of the coupling of the translational and rotational absorption is straightforward. As mentioned just above, we must refer the overall motion to a space-fixed system rather than to a coordinate system which is fixed in the intermolecular axis $\mathbf{R}$, since the motion of the center of mass of the two-molecule system must now be included. To obtain the total wave function, we proceed in the usual way by first coupling the angular momenta of the motion of the axes of the individual collision partners $\mathbf{J}_1$, $\mathbf{J}_2$ to give the intermediate angular momentum $\mathbf{J}_i$, which is then combined with the angular momentum of the motion of the intermolecular axis $\mathbf{L}$ to the total angular momentum $\mathbf{J}$. The induced rotational-translational dipole moment is then, in terms of the eigenfunctions $\psi(J_1 J_2 L, J_i J m ; \mathbf{\Omega}, \mathbf{\Omega}_1, \mathbf{\Omega}_2)$,

$$\mu_\gamma = (64\pi^3/3)^{1/2} \sum_{\lambda_1 \lambda_2 \Lambda L} A(\lambda_1 \lambda_2 \Lambda L; R) \psi(\lambda_1 \lambda_2 L, \Lambda 1 \gamma ; \mathbf{\Omega}, \mathbf{\Omega}_1, \mathbf{\Omega}_2), \quad (10.34)$$

where $\gamma = 0, \pm 1$ gives the spherical component, λ_1 and λ_2 are the values of $\mathbf{J}_1$ and $\mathbf{J}_2$, and $\Lambda = \lambda_1 + \lambda_2, \lambda_1 + \lambda_2 - 1, \cdots, |\lambda_1 - \lambda_2|$; $1 = \Lambda + L, \cdots, |\Lambda - L|$. The coefficients A depend on R; the wave functions ψ depend on the three couples of angles denoted by $\mathbf{\Omega}, \mathbf{\Omega}_1$, and $\mathbf{\Omega}_2$ which give the respective orientations of the intermolecular axis $\mathbf{R}$ and the intramolecular axes $\mathbf{R}_1, \mathbf{R}_2$ of the two collision partners with respect to the space-fixed coordinate system. We note that the symmetry of μ_γ requires that $\lambda_1 + \lambda_2 + L + 1$ is even.

The coefficients $A(\lambda_1 \lambda_1 \Lambda L)$ can be expanded into a series of the coefficients $C(\lambda_1 \mu_1 \lambda_2 \mu_2)$; that is, μ is expressed as a linear combination of products of spherical harmonic basis functions $Y_{\lambda\mu}(\mathbf{\Omega}, \mathbf{\omega}_1, \mathbf{\omega}_2)$ (see Fig. 10.2). The coefficients of this expansion are vector coupling coefficients (see also Eq. 10.1a). The resulting expression for A is

$$A(\lambda_1 \lambda_2 \Lambda L) = \frac{(2L + 1)^{1/2}}{\sqrt{3}} \sum_{\gamma \mu_1 \mu_2} (\lambda_1 \lambda_2 \Lambda ; \mu_1 \mu_2)(\Lambda L 1 ; \gamma 0) C_\gamma (\lambda_1 \mu_1 \lambda_2 \mu_2), \quad (10.35)$$

where we have written the vector coupling coefficients in Rose's notation[6] except that we have left out the letter C to avoid confusion with the C_γ

coefficients.* If the kinetic energy K is written as a sum of the individual contributions of the angle-dependent part (J_1, J_2, L) and the radial part (R), we can write the binary absorption coefficient for a pure diatomic gas in the form

$$\alpha_1 = (2\pi^2/3c) \sum_{\lambda_1 \lambda_2 \Lambda L} \int_0^\infty \left[\left| \frac{1}{m} \frac{\partial}{\partial R} A(\lambda_1 \lambda_2 \Lambda L) \right|^2 \right.$$

(10.36)

$$\left. + \left\{ \frac{\lambda_1(\lambda_1 + 1)}{I} + \frac{\lambda_2(\lambda_2 + 1)}{I} + \frac{L(L + 1)}{mR^2} \right\} |A(\lambda_1 \lambda_2 \Lambda L)|^2 \right] g(R) R^2 \, dR.$$

We arrived at this expression by substituting $K^\dagger$ and Eq. 10.34 into Eq. 10.29 (see footnote* on page 342). As previously, $g(R)$ represents the pair distribution function. The quantity I is the moment of inertia of the molecule. The terms proportional to $(\lambda_1 + 1)\lambda_1/I$ and $(\lambda_2 + 1)\lambda_2/I$ arise from the operators of the angular momentum of the individual molecule and are therefore termed the "pure rotational contributions" to the total intensity (note the close correspondence with Eq. 10.33).

The nonvanishing coefficients $A(\lambda_1 \lambda_2 \Lambda L)$ are easily written down using Eq. 10.35, remembering the allowed values of the λ_i (= even), μ_i (note that $\gamma = \mu_1 + \mu_2$), and $L(= 1, 3, \cdots)$. For a pure diatomic gas we obtain

$$A(2021) = -A(0221) = \xi_1 \exp(-R/R_0),$$

(10.37)

$$A(2023) = -A(0223) = \xi_3 \exp(-R/R_0) + \sqrt{3}(aQ/R^4).$$

* In the coupling scheme $\mathbf{J}_1 + \mathbf{J}_2 = \mathbf{J}_i, \mathbf{J}_i + \mathbf{L} = \mathbf{J}$, the total angular momentum $\mathbf{J}$ is first uncoupled with respect to the intermediate angular momentum $\mathbf{J}_i = \mathbf{J}_1 + \mathbf{J}_2$ and the angular momentum $\mathbf{L}$: $\psi((J_1 J_2)J_i, L, J) = \sum g(J_1 J_2 J_i)h(L)(J_i L|J_i LJ)$, where we have indicated the basis function by a lower case letter and the vector coupling coefficient by (|). The sum stretches over the allowed components of the angular momenta; we have left them out here. In the next step we uncouple $\mathbf{J}_i$: the total wave function is then

$$\psi((J_1 J_2)J_i, L, J) = \sum k(J_1)p(J_2)h(L)(J_1 J_2|J_1 J_2 J_i)(J_i L|J_i LJ).$$

Hence we see that the transformation from a function $\psi(J)$ to a product of wave functions $k(J_1), p(J_2), h(L)$ involves a sum of products of two vector coupling coefficients; coefficients which describe the stepwise uncoupling of the total angular momentum. For a detailed and more exact discussion see M. E. Rose, *Elementary Theory of Angular Momentum*, Wiley, New York, 1957, and A. R. Edmonds, *Angular Momentum in Quantum Mechanics*, Princeton Univ. Press, Princeton, New Jersey, 1957, Chapter 3.

$\dagger$ The computation of the commutator $[K, \mu]$ is simpler than the one discussed for the translational absorption of rare-gas mixtures. The kinetic energy is written in terms of the angular momenta $\mathbf{J}_1, \mathbf{J}_2$, and $\mathbf{L}$: $K = -(\hbar^2/2m)\{\partial^2/\partial R^2 + (2/R)(\partial/\partial R)\} + (\hbar^2/2I)\mathbf{J}_1{}^2 + (\hbar^2/2I)\mathbf{J}_2{}^2 + (\hbar^2/2mR^2)\mathbf{L}^2$. The computations then proceed as described in the footnotes on page 344. The trace over the motion of the center of mass of the pair of molecules here yields $\alpha_1 = \frac{1}{2}(\kappa\lambda^3/Z^2) \text{Tr} \{\exp(-\beta H)\mu \cdot [K, \mu]\}$, where Z is the rotational partition function of a molecule.

For a gas mixture of diatomic and monoatomic molecules the coefficient $A(0001)$ assumes a nonzero value since the restriction $C_\gamma(\lambda_1\mu_1\lambda_2\mu_2) = -C_\gamma(\lambda_2\mu_2\lambda_1\mu_1)$ is lifted. As it turns out, the quadrupolar contribution (long-range) exceeds by far the contribution of the short-range overlap forces in pure compressed hydrogen (compare this with the compilation of vibrational-rotational data in Table 10.1).

If the intermolecular potential is sufficiently isotropic, we can define the binary induced translational band as the intensity arising from all those transitions which leave the rotational energy unchanged; that is, for all rotational transitions for which $\Delta J_1 = -\Delta J_2$, or $\Delta J_i = 0$ for all i. The resulting expression for a pure diatomic gas is

$$\alpha_{1(transl)} = (2\pi^2/3mc) \sum_{\lambda_1\lambda_2\Lambda L} \sum_{\Delta J} L_{\lambda_1}(\Delta J)L_{\lambda_2}(-\Delta J)$$

$$\times \int_0^\infty \left\{ \left|\frac{\partial}{\partial R}A(\lambda_1\lambda_2\Lambda L)\right|^2 + \frac{L(L+1)}{R^2}\left|A(\lambda_1\lambda_2\Lambda L)\right|^2 \right\} g(R)R^2\,dR, \tag{10.38}$$

where the $L_\lambda(\Delta J)$ are sums of Boltzmann factors and vector coupling coefficients.*

To obtain the corresponding expression for a gas mixture of mono- and diatomic molecules, we set $\lambda_2 = 0$ in Eq. 10.36. The formula for $\alpha_{1(transl)}$ is identical to that given by Eq. 10.38 except that we must multiply by a factor of 2 (see also footnote* on page 342) and that the summation is over all λ, L since now $\Lambda = \lambda$, $\Delta J = 0$ (we have dropped the subscript of λ_1).

Substitution of the coefficients of Eq. 10.37 into Eq. 10.38 leads to the binary translational absorption coefficient for diatomic molecules:

$$\alpha_{1(transl)} = \{\beta_1{}^2\tilde{I}_1 + \beta_3{}^2\tilde{I}_3 + \tilde{\mu}^2\tilde{J} + \tilde{\mu}\beta_3\tilde{K}\}L_2(0)\tilde{\gamma}, \tag{10.39}$$

* The L coefficients are of the form

$$L_\lambda(\Delta J) = Z^{-1} \times \sum_J g_J \exp\left(-E_J/kT\right)(J, \lambda, J + \Delta J; 0, 0)^2,$$

where Z is the rotational partition function and $(J, \lambda, J + \Delta J; 0, 0)$ is a vector coupling coefficient (in Rose's notation without the "C"). The other symbols have their usual meaning. If we take the part of Eq. 10.36 describing the "pure rotational contributions," namely

$$(2\pi^2/3c) \sum_{\lambda_1\lambda_2\Lambda L} \{\lambda_1(\lambda_1 + 1)/I + \lambda_2(\lambda_2 + 1)/I\}|A(\lambda_1\lambda_2\Lambda L)|^2 g(R)R^2\,dR,$$

and add it to Eq. 10.38, we find that the resulting sum is not equal to the total absorption coefficient α_1 as given by Eq. 10.36 since $\sum_{\Delta J}L_{\lambda_1}(\Delta J)L_{\lambda_2}(-\Delta J)$ is generally smaller than 1. The translational terms of Eq. 10.36 (that is, the first and the fourth terms) thus give a nonvanishing contribution to the intensity of the rotational lines. This effect often manifests itself experimentally in an asymmetric broadening of the induced rotational lines. The mean frequency of the rotational lines—even after careful separation of the pure translational absorption—is seen to have shifted to higher frequencies with respect to the position calculated for the molecule. We shall remark on an example of this during the discusssion of the experimental data.

where $\beta_L = (\xi_L/e\sigma)\exp(-\sigma/R_0)$ and $\tilde{\mu} = (aQ/e\sigma^5)$ are dimensionless quantites which we have encountered previously (see Eq. 10.24). The quantity $\tilde{\gamma} = \pi e^2\sigma^3/3mc$ has the dimension of a binary absorption coefficient. The coefficient $L_2(\Delta J)$ has been defined in the footnote[†] on page 337. The integrals $\tilde{I}_L$ and $\tilde{J}$ are essentially of the same form as the corresponding quantites in Eq. 10.25 except that $\tilde{I}_L$ contains the additional factor $(\sigma/R_0)^2 + L(L+1)/x^2$ ($x = R/\sigma$) in the integrand. The integral $\tilde{K}$ governs an interference effect between the quadrupolar and overlap moments. A similar term is present in the expression for the pure rotational absorption coefficient but we had not given it there on account of its relatively small magnitude (see Eq. 10.27). For instance, for the binary translational absorption coefficient of equilibrium hydrogen at 300°K the interference effect contributes about 20%, whereas it only contributes about 2% for the pure rotational absorption.[2]

Finally, for a gas mixture consisting of diatomic molecules with a molecular quadrupole moment Q_1 and of atoms with the polarizability a_2 we obtain

$$\alpha_{1(transl)} = \{\beta^2\tilde{I}_1 + \tilde{\mu}^2\tilde{J}L_2(0)\}\tilde{\gamma} \qquad (10.40)$$

with $\beta = (\xi/e\sigma)\exp(-\sigma/R_0)$, $\tilde{\mu} = Q_1a_2/e\sigma^5$. The form of Eq. 10.40 reflects the presence of $A(0001)$ coefficients (see above).

C. Experimental Observations

a. Pure Translational Absorption. Helium-Argon and Neon-Argon Mixtures

In the following sections, in which we discuss some experimental work on collision-induced spectra, we shall see the importance of this far-infrared technique and, in many instances, the unique way by which information on intermolecular parameters can be obtained by it. The experimental work which has appeared in the literature since instrumentation became more readily available has engendered great theoretical interest, and many efforts are being made towards a precise explanation and characterization of the complicated collision phenomena and relevant parameters.

To give a brief outline on what we will mainly discuss here, we first show experiments on the collision-induced absorption in rare-gas mixtures. Because of the relative simplicity of such a system, theoretical studies on line shape functions have progressed to a considerable degree, leading to improved and detailed information on dipole moment functions and interatomic potentials. We then treat collision-induced spectra of diatomic molecules, a more complicated system. Studies of such gases have yielded, for instance, the molecular quadrupole moment; the technique has been extended to find the molecular quadrupole moments even of polar molecules.

Finally, we shall mention induced translation-rotational spectra of homo-nuclear and heteronuclear diatomic molecules dissolved in liquefied rare gases, where we will see that deviations from "ideality" can be treated with theories that had been developed to describe analogous behavior in matrix-isolated molecules.

Much more experimental and theoretical work on collision-induced absorption may be expected to appear in the future, particularly since far-infrared spectroscopy frequently seems to be the only means by which precise values of the relevant parameters can be obtained, and secondly, since the results of the collision-induced spectra have important and close bearings on the elucidation of some characteristics of solution spectra (see Chapter 11).

We begin by discussing the work of Bosomworth and Gush,[7] who have measured the absorption of compressed mixtures of helium–argon and neon–argon within a wide frequency range. Figures 10.3a and 10.3b show their data on Ne–Ar and He–Ar mixtures in terms of $A(v) = (d_A d_B l)^{-1} \times \log(I_0(v)/I(v))$, where d_A and d_B are the partial densities of gases A and B in amagat units,[†] l is the length (cm) of the absorption path, and I_0 and I have their usual meaning of incident and transmitted intensity. Table 10.2 gives the experimental integrated induced absorption coefficient α_1 of the two gas mixtures in units of $\sec^{-1} cm^5$ (Column 3) and in units of $\sec^{-1} cm^{-1}$ (Column 4). Since α_1 was only measured at one temperature, reliable values of the range R_0 and the strength ξ of the induced dipole moment cannot be obtained (see Eq. 10.33).

TABLE 10.2*

INTEGRATED INDUCED BINARY ABSORPTION COEFFICIENTS α_1 OF NEON-ARGON AND HELIUM-ARGON MIXTURES

Gas mixture	Temperature, °K	α_1	
		$\sec^{-1} cm^5$	$\sec^{-1} cm^{-1}$
Ne–Ar	295	4.41×10^{-33}	3.20×10^6
He–Ar	295	8.60×10^{-33}	6.22×10^6

* After D. R. Bosomworth and H. P. Gush, Can. J. Phys. **43**, 751 (1965). By permission of National Research Council of Canada.

† The "amagat" is a unit of density. It is defined as the reciprocal of the molar volume at 0°C and 1 atm pressure.

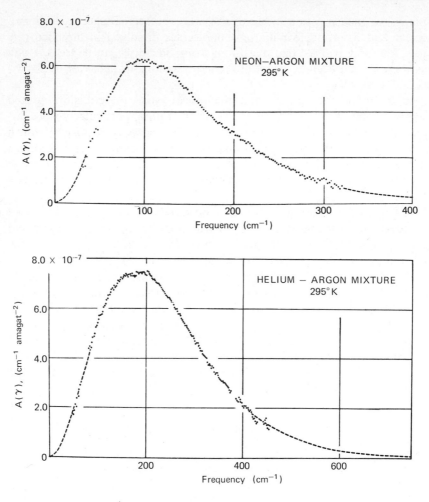

Fig. 10.3   Absorption spectra of (*a*) Ne–Ar (*upper curve*) and (*b*) He–Ar (*lower curve*) mixtures. The dots are experimental points, the dashed lines are extrapolated. Total pressure of the 1 : 1 gas mixtures was about 100 atm. The path length of the absorption cell was 3 m. The polyethylene windows of the absorption cell had strong absorption near 70 cm^{-1}. The experimental procedures and the spectrometers are described in *Can. J. Phys.* **43**, 729 (1965). [D. R. Bosomworth and H. P. Gush, *Can. J. Phys.* **43**, 751 (1965), by permission of National Research Council of Canada.]

However, from the half width of the absorption bands we can roughly estimate the time of the collision process and thereby the range of the dipole moment (see Section A). For instance, in the Ne–Ar mixtures, where the half width is about 200 cm^{-1}, the collision process lasts $\Delta t \sim 1/2\pi \cdot 200 \cdot 3 \cdot 10^{10}$ $\sim 0.3 \cdot 10^{-13}$ sec; the range of the induced dipole moment therefore is of the order of $R_0 = v\Delta t$ $(v = 440 \text{ m sec}^{-1}$ at $300°\text{K}) \sim 0.13 \text{ Å}$. This value is

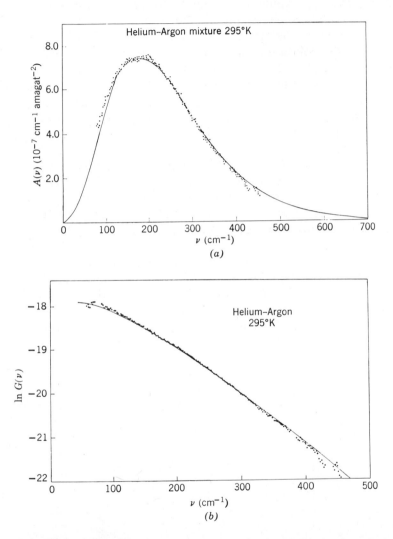

Fig. 10.4 (a) Absorption coefficient (*solid curve*) of He–Ar mixture, computed to the data of Fig. 10.3b with the help of Eq. 10.41e (see text): $A(v) = \alpha(v)(n_A n_B)^{-1}(N_0/V_0)^2$, where $N_0 =$ Avogadro's number, $V_0 =$ molar volume, and n_A and $n_B =$ number of atoms per unit volume. (b) Computed initial-state averaged transition probability for a He–Ar mixture: $G(v) = A(v)v^{-1}\{1 - \exp(-hcv/kT)\}^{-1}$. The dots are the experimental data by D. R. Bosomworth and H. P. Gush, Ref. 7 and Fig. 10.3b. [H. B. Levine and G. Birnbaum, *Phys. Rev.* **154**, 86 (1967).]

only about 1/20 of the Lennard-Jones diameter σ of the mixture (see Eq. 10.24). According to the authors, this signifies that only collisions with impact parameters smaller than σ; that is, molecules in the repulsive region of the potential, contribute significantly to the absorption process.

Next we outline some theoretical studies on line shape functions which were stimulated by the far-infrared data. A calculation of the line shape function of the collision-induced band of compressed rare-gas mixtures, based on classical radiation theory, has been given by Levine and Birnbaum.[8a] The good fit to the experimental data is shown by the solid curves and the dots in Figs. 10.4a and 10.4b, which depict for He–Ar mixtures the absorption coefficient $A(v) = \alpha(v)(n_A n_B)^{-1}(N_0/V_0)^2$ (N_0 = Avogadro's number, V_0 = molar volume, n_A and n_B = number of atoms per unit volume) and the transition probability $A(v)v^{-1}[1 - \exp(-hcv/kT)]^{-1}$ (see, for instance, Section B of Chapter 11) as a function of v (cm^{-1}). [We note here that, in order to agree with the literature, we have made a slight change in notation by designating the absorption coefficient by the symbol $\alpha(v)$, a symbol which in previous sections was reserved to the absorption coefficient integrated over the frequency v. Since we clearly indicate the functional dependence of α on the frequency v, no confusion should result.] Figure 10.5 shows the calculated and experimental absorption band of Ne–Ar; we see that the agreement is not as good here. However, more recent line shape calculations by Birnbaum and Levine[8b] have also led to good agreement with the data in the Ne–Ar system. We shall outline now their calculations, which are based on a few well known principles, in the following. (For the rather involved detailed computations, the reader is referred to the original articles.)

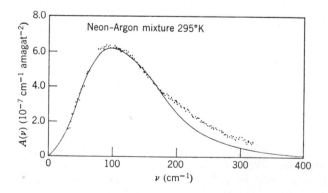

Fig. 10.5   Absorption coefficient (*solid curve*) of Ne–Ar mixture, computed to the data of Fig. 10.3a. (See also Fig. 10.4a). [H. B. Levine and G. Birnbaum, *Phys. Rev.* **154**, 86 (1967).]

The computations first determine the emission spectrum $I(\omega)$ as a function of the (angular) frequency ω. The quantity $I(\omega)$ is the statistical average over the power from various classes of collisions:

$$I(\omega) = \iint N(b, E)\hat{I}(\omega, b, E) \, db \, dE, \qquad (10.41a)$$

where $N(b, E) \, db \, dE$ is the number of collisions (per unit volume per unit time) with impact parameter b in the range $b + db$ and relative energy E in the range $E + dE$ (see, for instance, Fig. 1 of Ref. 8c). The emission intensity $\hat{I}(\omega, b, E)$ is given by the Fourier transform of Larmor's formula of classical electrodynamics (the power radiated by an accelerated charge):

$$\hat{I}(\omega, b, E) = \frac{2\omega^4}{3\pi c^3} \left| \int_{-\infty}^{+\infty} dt e^{i\omega t} \mathbf{\mu}(b, E, t) \right|^2. \qquad (10.41b)$$

The quantity $\mathbf{\mu}(b, E, t)$ is the induced dipole moment in the binary rare-gas collision complex as a function of time for a given set of b and E. The dipole moment is directed along the interatomic axis, as we have previously assumed; that is, it is of the form $\mathbf{\mu}(\mathbf{R}) = (\mathbf{R}/R)\mu(R)$ (see Eq. 10.31). The dependence of $\mathbf{\mu}$ on time, impact parameter, and the interaction energy is, of course, given by the dependence of R on these quantities. The expression for $I(\omega)$ is finally converted into an absorption coefficient $\alpha(\omega)$ by Kirchhoff's law, $\alpha(\omega) = I(\omega)/cu(\omega)$, where $u(\omega)$ is the energy density in the black-body radiation field.

To introduce the detailed physical picture into the above relations, it is necessary to determine or define the specific nature of the quantities $N(\omega, b, E)$ and $\mathbf{\mu}(b, E, t)$. For instance, Levine and Birnbaum have used simple straight-line collision paths $[x = vt \, (v = \text{velocity}), \ y = b, \ z = 0]$ in combination with a dipole function of the form

$$\mu(R) = \mu_0 \gamma R \exp\left(-\gamma^2 R^2\right), \qquad (10.41c)$$

with $R^2 = b^2 + v^2 t^2$. γ is a reciprocal length. This empirical choice of $\mu(R)$, roughly speaking, corrects for the absence of a repulsive potential since the factor γR "shuts off" the dipole moment when $R \lesssim \gamma^{-1}$.[8a] The factor R also guarantees that $\mu(R) \to 0$ for $R \to 0$, a necessary condition under the assumption of straight-line collision paths (see Ref. 8a for more details).

We see that Levine and Birnbaum's form of $\mu(R)$ is a Gaussian-type dipole moment rather than a pure exponential related to overlap interaction (see, for instance, Eq. 10.22). Although it appears to be an "unphysical"

model at first sight,[8d] it is seen[8b] to be analogous to the general approxima-
tion $\mu(t) = \mu(t_0) \exp\left[(\mu''(t_0)(t - t_0)^2/2\mu(t_0)\right]$ (the primes denote differentiation
with respect to time t), a form of dipole moment that roughly covers the
entire time interval $-\infty \leqslant t - t_0 \leqslant \infty$; t_0 is the time at which the minimum
interatomic separation, $r_0 = R(t_0)$, is achieved between the two colliding
atoms with $R = r_0 + \frac{1}{2}a_0(t - t_0)^2 + \cdots$ ($a_0 =$ acceleration). This general
expression for $\mu(t)$ represents a Gaussian function ($\mu''(t_0) < 0$) which reduces
to the Taylor series $\mu(t) = \mu(t_0) + \frac{1}{2}\mu''(t_0)(t - t_0)^2 + \cdots$ for small t and
which decays monotonically to zero for long t (as the true dipole moment
must do). If this Gaussian expression is inserted into the expression for
$\hat{I}(\omega)/cu(\omega)$, the Fourier inversion leads to an energy spectrum for the individual collisions which is very similar to that obtained using $\mu(R) =
\mu_0 \gamma R \exp(-\gamma^2 R^2)$ of Eq. 10.41c.[8a,8b] The averaging over the impact param-
eter b and relative velocities [that is, the steps from $\hat{I}(\omega, b, E)$ to $I(\omega)$] is
very involved, but the results show[8b] that the spectrum takes a rather
similar form to the one using the simple straight-line collision paths model
with the dipole moment "cut-off" (Eq. 10.41c). In fact, the computations on
Ne–Ar, which previously deviated from the data at higher frequencies (see
Fig. 10.5), now fit throughout the whole frequency range. The line shape
function is of the approximate form

$$\alpha(\omega) = (\omega^2 l/c)[\varepsilon'(0) - 1](m/kT)^{1/2}L(x), \qquad (10.41d)$$

where $L(x) = Cx^p K_p(x)$, $x = (\omega l)(m/kT)^{1/2}$, $K_p(x)$ is a modified Bessel
function of the second kind, C is a constant, m is the reduced mass, l is a
parameter of the dimension of length, and $\varepsilon'(0)$ is the zero-frequency dielectric
constant. The line shape function is fitted to the data by choosing the two
parameters p and l and the quantity $\varepsilon'(0) - 1$ such that the moments of the
band, $\int \alpha(\omega)\,d\omega$ and $\int \alpha(\omega)\omega^2\,d\omega$, and the Kramers-Kronig transform[8e]
$\varepsilon'(0) - 1 \propto \int \alpha(\omega)\omega^{-2}\,d\omega$ agree with the experiment.

For the Ne–Ar system, p turns out to be 1.28. For the He–Ar system, the
value of p is 2; this agrees, fortuitously, with the line shape function based
on the dipole moment of Eq. 10.41c, for which $\alpha(\omega)$ has been shown to be of
the form (see Eq. 6.2 of Ref. 8a) of a modified Bessel function of the second
kind of order two:

$$\alpha(\omega) = (\omega^2/3c\gamma)[\varepsilon'(0) - 1](m/kT)^{1/2}x^2 K_2(x)$$

$$= \frac{\mu_0^2 \pi^3 n_A n_B}{12c\gamma^2}\left[\frac{2}{\pi m kT}\right]^{1/2} x^4 K_2(x), \qquad (10.41e)$$

with $x = (\omega/\gamma)(m/kT)^{1/2}$. We therefore understand why the simple dipole
moment function fits so well the data of the He–Ar system but not those of
the Ne–Ar system.

Table 10.3 shows the dipole parameters γ (in Å^{-1}, from Eq. 10.41c) and μ_0 for the rare-gas mixtures Ne–Ar and He–Ar, obtained by a fit of Eq. 10.41e. The corresponding calculated values of the integrated absorption coefficient are also added to Table 10.3 and should be compared with those of Table 10.2. The agreement is very good for He–Ar, as expected.

We note here that the quantity γ of Eq. 10.41c is not equal to $1/R_0$, the distance parameter in the exponential dipole moment $\mu(R) \propto \exp(-R/R_0)$ (see Eq. 10.22), but is related to R_0 through the acceleration a_0 (the acceleration at $R = R_0$) and to t_0, the time for which $R = R_0$ (see above). It turns out that $\gamma \sim 1/\sqrt{5}R_0$.[8b] We have entered the corresponding values of R_0 in Table 10.3 together with a value of l calculated by Birnbaum and Levine for He–Ar using the dipole moment function $\boldsymbol{\mu} = (\mathbf{R}/R)\{\mu_0 \exp(-R/R_0) + \mu_7(\sigma/R)^7\}$. In this type of dipole moment function the long-range attractive interactions, which contribute considerably to the overall dipole moment, are included.[8b,8f]

We conclude with a few remarks on the relative importance of the various parameters on the spectrum. To this effect we present Fig. 10.6 which shows a striking example of the large influence of the "range parameters" R_0 and σ;

TABLE 10.3*

DIPOLE MOMENT PARAMETERS AND BINARY INTEGRATED
ABSORPTION COEFFICIENTS OF NEON–ARGON AND
HELIUM–ARGON MIXTURES AT 295°K

	Gas mixture	
	Ne–Ar	He–Ar
μ_0, Debye	0.223	0.166
γ, Å^{-1}	1.446	1.357
$1/\gamma$, Å	0.693	0.733
l, Å	—	0.625
R_0,[†] Å	0.31	0.33
α_1, $\text{sec}^{-1}\,\text{cm}^{-1}$	2.88×10^6[‡]	6.26×10^6[‡]

* *After* H. B. Levine and G. Birnbaum, *Phys. Rev.* **154**, 86 (1967) and private communication, January 1969.

[†] $\mu(R) \propto e^{-R/R_0}$, $R_0 = 1/\sqrt{5}\gamma$. In the literature, R_0 is frequently designated by ρ. We have chosen the symbol R_0 in order to avoid confusion with the other meanings of the symbol ρ (see Section B).

[‡] Compare with Column 4 of Table 10.2.

parameter in the Lennard-Jones potential function, as defined in Eqs. 10.22 and 10.24, respectively. The curves displayed in the figure are complete model computations on a He–Ar-type system performed by McQuarrie and Bernstein,[8c] based on Levine and Birnbaum's classical formulation (see Eqs. 10.41a and 10.41b) but employing the purely exponential dipole moment model and a standard Lennard-Jones interaction potential. In other words, exact classical collision trajectories were assumed. The line shapes in the figure are given in terms of the reduced quantity $R^* = R_0/\sigma$, with the R^* values of 0.110, 0.117, 0.125, and 0.1275 (from bottom to top). We notice how very sensitive the absorption curves are to the value of R^*. This implies that the experimental studies are a very sensitive method of deducing R^* and that present *ab initio* calculation of R_0 do not seem accurate enough to be useful in calculations of collision-induced absorption which use R_0 as input parameter.[8c]

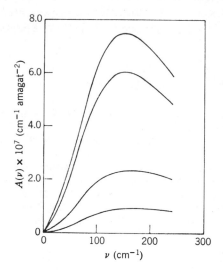

Fig. 10.6 $A(\nu)$ *vs* ν for values of $R^* = 0.110, 0.117, 0.125,$ and 0.1275. The maximum of $A(\nu)$ increases with increasing R^*. [D. A. McQuarrie and R. B. Bernstein, *J. Chem. Phys.* **49**, 1958 (1968).]

We finally show in Fig. 10.7 the temperature dependence of the line shape function from McQuarrie and Bernstein's model calculations. We note that the results are only moderately sensitive to the temperature; they are, in fact, essentially a reflection of the changes in the Boltzmann energy distribution.[8b]

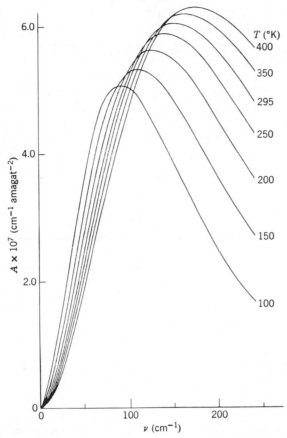

Fig. 10.7 $A(v)$ *vs* v for various temperatures and $R^* = 0.125$. [D. A. McQuarrie and R. B. Bernstein, *J. Chem. Phys.* **49**, 1958 (1968).]

In conclusion we mention that this "dynamics approach" for obtaining line shape functions is not the only one. In fact, a more powerful and general (but more abstract) method is to construct a dipole correlation function and to obtain the power spectrum from it (see Chapter 11, Section B). A discussion of this, we feel, is outside the scope of this chapter; we refer the interested reader to the work of Birnbaum and Levine.[8b]

b. Rotational-Translational Absorption

1. Hydrogen. We have repeatedly mentioned that a separation of the induced far-infrared absorption into the pure translational and pure rotational components generally cannot be made. The hydrogen molecule is an exception. Because of its small size the induced rotational line $S(0)$ is at a sufficiently high frequency to be rather well separable from the translational absorption. This is shown in Fig. 10.8, which represents the data of Bosomworth and Gush[7] and Kiss *et al.*[9] The symbols $S(0)$, $S(1)$, etc., identify in the usual fashion the particular rotational transition ($J = 0 \rightarrow 2, 1 \rightarrow 3$); the translational band can be seen to stretch into the 600 to 900 cm^{-1} region. The agreement of the value of $\alpha_{1(transl)}$ obtained from the data, 5.6×10^{-33} sec^{-1} cm^5,

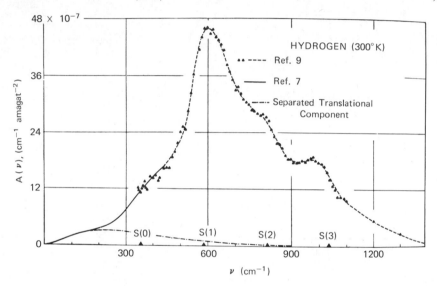

Fig. 10.8   Far-infrared spectrum of compressed hydrogen (37–400 amagat) at 300°K. The computed rotational $\Delta J = 2$ spectrum for the unperturbed molecules is shown by the triangles on the abscissa. [D. R. Bosomworth and H. P. Gush, *Can. J. Phys.* **43**, 751 (1965). By permission of National Research Council of Canada.]

with the theoretical value, 4.0×10^{-33} sec^{-1} cm^5, can be considered to be satisfactory.[10] The agreement improves considerably at lower temperatures (77°K, see Fig. 10.9) because the bands become narrower (see also Fig. 10.7). The respective values for $\alpha_{1(transl)}$ are 2.5×10^{-33} (measured) and 2.7×10^{-33} sec^{-1} cm^5 (calculated).

As pointed out by Bosomworth and Gush in Fig. 10.9, the peak absorption of the hydrogen rotational line is at a higher frequency (by about 17 cm^{-1}) than that of the $S(0)$ line calculated from the rotational constant of the

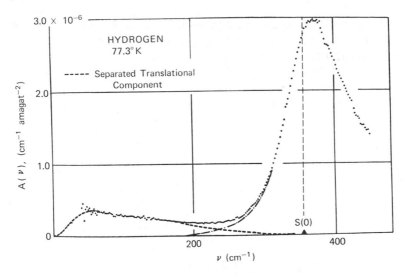

Fig. 10.9   Far-infrared spectrum of compressed hydrogen at 77.3°K, showing the separated translational and rotational components. Notice the difference between the frequency at the peak absorption and that calculated for S(0) (see text). [D. R. Bosomworth and H. P. Gush, *Can. J. Phys.* **43**, 751 (1965). By permission of National Research Council of Canada.]

molecule. This is the nonseparable contribution of the translational motion to the rotational absorption which we have mentioned previously (see Eqs. 10.38 and 10.39 and footnote on page 347).

A knowledge of the pure translational absorption coefficient here is seen to be very useful for a precise determination of the molecular quadrupole moment of the hydrogen molecule. Of course, the value of the quadrupole moment is also accessible through the absorption coefficient of a single induced pure rotational transition (see, for instance, Eq. 10.28) or of the whole rotational band—in fact, this is the way in which the quadrupole moment is frequently obtained. However, the disadvantage is that the rotational and translational components strongly overlap, so that a separation need be done with the help of line shape functions. For the lighter molecules, where the energy spacings of the rotational levels are much larger than those of the translational energies, this can frequently be accomplished.

2. Nitrogen, Oxygen. The far-infrared bands of compressed N_2 and O_2 are of a completely different appearance than that of H_2. Because of the larger size of N_2 and O_2 compared to H_2, the spacings between the rotational transitions are not resolvable. Although the rotational lines are theoretically separated by $8.0 \, \text{cm}^{-1}$ ($= 4B$) and $11.6 \, \text{cm}^{-1}$ ($= 8B$; O_2 has no odd rotational levels on account of its zero nuclear spin), respectively, they are greatly broadened by the uncertainty in the energy. For the half width of N_2

Ketelaar and Rettschnick estimate a value of about $40\,\text{cm}^{-1}$.[11] The induced rotational-translational spectra therefore have the appearance of a broad, continuous absorption.

Figure 10.10 shows a compilation of recent data on compressed N_2 taken from Bosomworth and Gush's paper.[7] Figure 10.11 shows the corresponding spectrum of O_2. The vertical lines give the calculated, unbroadened rota-

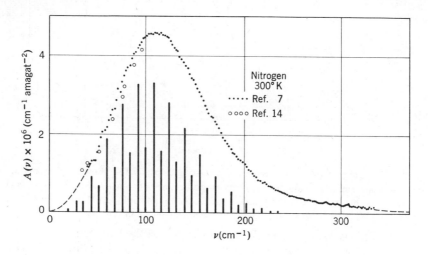

Fig. 10.10 and Fig. 10.11   Absorption profiles of compressed nitrogen (*upper curve*) and compressed oxygen (*lower curve*). The vertical bars indicate the frequencies and intensities of the unbroadened rotational transitions $\Delta J = 2$, computed with Eqs. 10.27 and 10.28. Notice the $2:1$ intensity ratio of adjacent lines in N_2 and the missing transitions of the odd rotational levels for O_2 because of the nuclear spin statistics. [D. R. Bosomworth and H. P. Gush, *Can. J. Phys.* **43**, 751 (1965). By permission of National Research Council of Canada.]

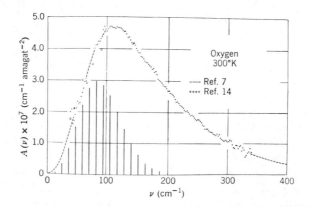

tional $J \to J + 2$ lines of the molecules. The profile of the induced spectrum of N_2 coincides much closer with that of its unperturbed gas Raman spectrum than the profile of compressed O_2 with its Raman spectrum. The much broader absorption in compressed O_2 thus indicates, as discussed in Section A, that the range of the induced dipole moment is shorter in O_2 than in N_2. In fact, if the identical inducing mechanism were effective in both compressed gases, the spectrum of O_2 should be narrower than that of N_2 since the quadrupole moment of O_2 is smaller than that of N_2. The unexpected broadness of the induced O_2 absorption apparently is a consequence of a relatively large contribution due to the short-range overlap forces.[7]

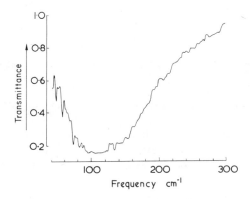

Fig. 10.12   Ratioed spectrum of compressed nitrogen (103 atm) between 20 and 300 cm^{-1}, taken with an interferometer (spectral resolution: 2 cm^{-1}). [H. A. Gebbie, N. W. B. Stone and D. Williams, *Mol. Phys.* **6**, 215 (1963).]

Figure 10.12 shows the ratioed spectrum of compressed N_2 between 20 and 400 cm^{-1}.[12] The molecular quadrupole moment of N_2 was calculated from the integrated absorption coefficient, neglecting overlap contributions—which are small in N_2—and the unknown translational contribution (see Eq. 10.28 and the end of Section B.b). The result $Q = 1.48 \times 10^{-26}$ esu cm^2, is in good agreement with values obtained by other methods.[7]

Stone and Williams[13] have observed a strong far-infrared band in liquid nitrogen with maximum absorption near 75 cm^{-1}. The spectrum, between 33 and 400 cm^{-1}, is shown in Fig. 10.13. The dotted line in the figure indicates regions of spurious absorption due to the formation of ice crystals in the liquid. The authors calculate that compressed N_2 at 77°K, experiencing binary collisions, should have the maximum of intensity of the induced rotational transitions at 52 cm^{-1}, assuming that the lines are narrow (see Eq. 10.28 and Fig. 10.10 which shows compressed N_2 at 300°K). They

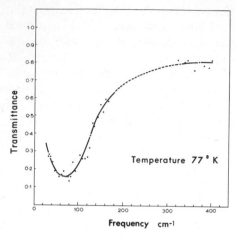

Fig. 10.13   The transmission spectrum of liquid nitrogen as observed at 77°K for an effective
cell length of 10 mm. Instrument: Perkin-Elmer 301. [N. W. B. Stone and D. Williams, *Mol. Phys.*
10, 85 (1965).]

further estimate that broadening at pressures equivalent to the density of
liquid nitrogen would indeed shift the maximum absorption close to the
observed value of 75 cm^{-1}, but that the intensity should be smaller than that
which they observed. They conclude that the difference between the calculated
and the observed profile is made up by a broad absorption which has its
maximum absorption near 50 cm^{-1} and extends to perhaps 300 cm^{-1},
and they ascribe this absorption to translational absorption or to the
effects of ternary collisions. The underlying assumption in these considerations is, of course, that the N_2 molecules are able to rotate freely in the
liquid.

We conclude this section with a reference to remarks by Heastie and
Martin[14] who propose that collision-induced effects may be significant for
the atmospheric energy balance and for "windows" for communication
systems working in the far-infrared. They calculate, for instance, that a path
length of 4 km of dry air at 1 atm would transmit only 38 % of the radiation
at 100 cm^{-1}.

c. Pressure-Induced Quadrupole Spectra. Hydrogen Chloride and Hydrogen Bromide

A paper on the determination of the molecular quadrupole moment of a
polar molecule by observation of its pressure-induced quadrupole rotational
spectrum has been published recently by Weiss and Cole.[15] The method of
separating the intensities of the dipolar and quadrupolar transitions is
demonstrated in Fig. 10.14, which shows the calculated intensities on a

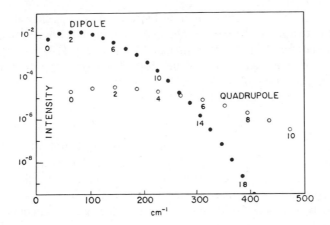

Fig. 10.14   Intensities of dipole transitions ($\Delta J = 1$) and quadrupole-induced dipole transitions ($\Delta J = 2$) in an HCl-SF$_6$ mixture. Numbers opposite points are initial J values for the transition. The intensities are calculated for HCl = 1.07 D, $Q_{HCl} = 5.8 \times 10^{-26}$ esu·cm^2, $d_{HCl} = 0.1$ mole/liter, and $d_{SF} = 0.5$ mole/liter. Note that intensity scale is logarithmic. [S. Weiss and R. H. Cole, *J. Chem. Phys.* **46**, 644 (1967).]

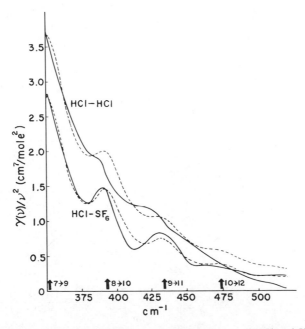

Fig. 10.15   Experimental and calculated (dotted lines) line shapes of HCl and HCl-SF$_6$ mixtures. Halide pressure ~ 1 to 10 atm, SF$_6$ pressure ~ 10 to 25 atm. Temperature 300°K. The half width parameters for the line shape function are 16.5 and 13.5 cm^{-1}, respectively. Instrument: modified Perkin-Elmer 112-6. [S. Weiss and R. H. Cole, *J. Chem. Phys.* **46**, 644 (1967).]

logarithmic scale for HCl broadened by SF_6. It is seen that the pressure-induced quadrupolar absorption becomes prominent above 300 cm^{-1}. The absorption coefficient is given in its usual form, $\alpha(v) = (1/l)\ln(I_0/I)$, where v denotes the frequency in cm^{-1}, l the absorption path length, and I_0 and I the incident and transmitted intensity, respectively.

For the pure hydrogen halide (HX, X = Cl, Br) we have

$$\alpha(v)_{HX} = \gamma(v)_{HX}\, d_{HX}^2 + \delta(v)_{HX}\, d_{HX}, \qquad (10.42a)$$

where d_{HX} is the density of HX and the factors $\gamma(v)_{HX}$ and $\delta(v)_{HX}$ denote the contributions of the pressure-induced quadrupolar and dipole-allowed transitions, respectively. Equation 10.42a therefore describes the total absorption in the self-broadened spectrum of pure HX.

For gas mixtures of hydrogen halide and foreign broadeners (SF_6 and CF_4) we have

$$\alpha(v)_{mixt} = \gamma(v)_{mixt}\, d_{HX} d_m + \alpha(v)_{HX}, \qquad (10.42b)$$

where the term in $\gamma(v)_{mixt}$ gives the contribution of the pair HX–broadener and d_m denotes the density of the broadening gas.

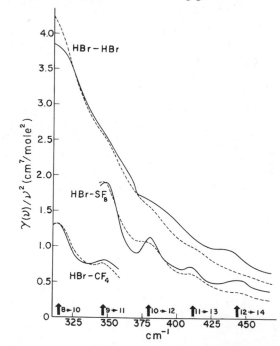

Fig. 10.16 Experimental and calculated (dotted curves) line shapes of HBr, HBr–SF_6, and HBr–CF_4 mixtures. Pressures as in Fig. 10.15; temperature 295°K. The half width parameters used in the line shape functions are 23, 14, and 14 cm^{-1}, respectively. [S. Weiss and R. H. Cole, J. Chem. Phys. **46**, 644 (1967).]

The observed spectra, plotted in terms of $\gamma(v)/v^2$, are shown in Fig. 10.15 for HCl and in Fig. 10.16 for HBr. The thick arrows indicate the frequency of the rotational transitions $J \to J + 2$ calculated with the literature values of the rotational constants of the molecules. We note that the induced rotational transitions are not sufficiently well resolved to permit the evaluation of the data without the help of a line shape function; according to Eq. 10.28 we need to know the induced, integrated absorption coefficient of the individual rotational transition. Fortunately, the values of the quadrupole moments do not depend too critically on the particular form of the line shape function.[15]

The evaluation of the data then follows according to the principles laid out previously (see Section B.c). If we set

$$\int_J \frac{\gamma(v)}{v^2} \, dv = \tilde{A}(J)Q_{HX}^2 a^2 \text{ Int,} \tag{10.43}$$

for the integrated induced absorption coefficient of a single rotational transition,* we find for the transition $J \to J + 2$

$$\tilde{A}(J) = (12\pi^3/hc)\frac{1}{v_{0J}}\left\{\frac{F(J)}{2J+1} - \frac{F(J+2)}{2J+5}\right\}\frac{(J+1)(J+2)}{2J+3}, \tag{10.44}$$

where v_{0J} is the frequency of the transition of the unperturbed molecule, $F(J)$ is the fraction of molecules in rotational state J, and the integral Int has the value

$$\text{Int} = \int_0^\infty R^{-8} \exp\left(-E/kT\right)4\pi R^2 \, dR. \tag{10.45}$$

The term designated by $\tilde{A}(J)$ will be recognized to have come from Eq. 10.28, except that we divide here by the frequency v_{0J} (see footnote below) and

---

* The expression for the absorption coefficient of a perfectly sharp line, $\gamma(v) = (1/l)\ln(I_0/I)$, where l, I_0, and I are the absorption length, incident, and transmitted radiation intensity, respectively, is given by

$$\gamma(v) = (8\pi^3/3hc)(n_i - n_j)v_{i\to j}\sum_{rs}|\mu_{ij}^{rs}|^2,$$

see E. B. Wilson, Jr., J. C. Decius, and P. C. Cross, *Molecular Vibrations*, McGraw-Hill, New York, 1955, Eq. 10, p. 163, and G. Herzberg, *Spectra of Diatomic Molecules*, Van Nostrand, Princeton, New Jersey, 1955, p. 127. Weiss and Cole use a reduced form of the absorption coefficient, $\int [\gamma(v)/v^2] \, dv = (8\pi^3 n/3hc)|\mu^2|l/v_0$, in which they define n to be the number of molecules in the ground state. It would be more precise, however, to set n equal to the difference between the number of molecules in the lower and upper states, since then the factor v_0 in the denominator cancels out for $v_0 \to 0$ (see C. H. Townes and A. L. Schawlow, *Microwave Spectroscopy*, McGraw-Hill, New York, 1955, p. 342). Weiss and Cole's expression, as written above, becomes infinitely large at $v_0 \to 0$.

have inserted the degeneracies of the lower and upper rotational level, $2J + 1$ and $2J + 5$. The factor Int corresponds to the quantity $\tilde{J}$ of Eqs. 10.25 and 10.28. Using the values for Int, the isotropic polarizability a, and the line shape function $f(v, v_{0J})$, we find the value of the molecular quadrupole moment Q_{HX} by comparing the measured absorption at frequency v of any one line with

$$Q_{HX}^2 a^2 \text{ Int } \tilde{A}(J) f(v, v_{0J}),$$

or the experimental total absorption with   (10.46)

$$Q_{HX}^2 a^2 \text{ Int } \sum_J \tilde{A}(J) f(v, v_{0J}).$$

Using literature values for the polarizability a and for the parameters of the Lennard-Jones potential in Int, the molecular quadrupole moments are obtained as

$$Q_{HCl} = 5.8 \times 10^{-26} \text{ esu cm}^2$$
$$Q_{HBr} = 5.5 \times 10^{-26} \text{ esu cm}^2.$$

(10.47)

For the function E of Eq. 10.45 we usually assume a Lennard-Jones potential, $E = 4\varepsilon\{(\sigma/R)^{12} - (\sigma/R)^6\}$. The factor ε gives the depth of the potential well, and σ is the value of R for which E has a node ($E = 0$). We have computed the intermolecular potential E of HCl and SF_6 and of HCl–HCl, using the parameters given by Weiss and Cole (Fig. 10.17). We notice the strong repulsion between the molecules in the region $R < \sigma$; it is usually in this region that the induced overlap dipole moments are formed, as we have already indicated at the beginning of Section C.a.

The experimental uncertainty of the measured quadrupole moments has been estimated by Weiss and Cole to be about 25%. Furthermore, the authors have discussed the possible systematic errors inherent in their technique: (1) Omission of octupole-induced dipole absorption. This is possibly the most important single omission. (2) Omission of the translational absorption. If H_2 serves as a comparison, the neglect of translational absorption does not introduce a significant error (see Fig. 10.8). However, it should be remembered that the angle-independent overlap contributions may be relatively large since nonzero coefficients $A(0001)$ (see Section B.c) will appear for the gas mixtures as well as for the pure halide vapors. (3) Double rotational transitions. (4) Anisotropy of the polarizability. The effects of (3) and (4) are believed to be small. Obviously, we need not worry about dipole-induced dipole absorption since the corresponding bands would appear at the same frequencies as the dipole-allowed transitions.

The values of the quadrupole moments for the two halides (see Eq. 10.47) are higher than those hitherto assumed, be it from line-broadening measure-

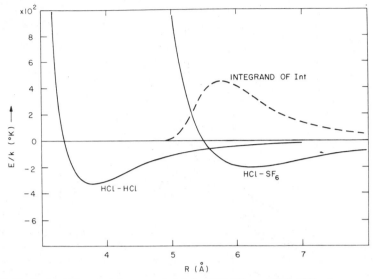

Fig. 10.17   Reduced Lennard-Jones potentials for HCl–SF_6 and HCl–HCl mixtures, computed from the parameters given by Weiss and Cole.[15] The ordinate is in units of the Boltzmann constant k and the abscissa gives the intermolecular distance in Å units. The dotted line represents (in arbitrary units on the ordinate) the integrand of *Int* (see Eqs. 10.25 and 10.45) for HCl–SF_6; notice how the integrand cuts off in the repulsive region of the potential. Notice also that the potential between HCl and SF_6 is "softer" than that between HCl molecules.

ments on the one hand[16] or from molecular orbital calculations on the other hand.[17]

It should be obvious that the method described here is very useful for a determination of molecular quadrupole moments for polar molecules; more work can be expected to be done along these and similar lines. We shall see the importance of the knowledge of the quadrupole moment of polar molecules when we study solution spectra (Chapter 11).

d. Induced Rotation-Translation Spectra in Liquids. Solutions of H_2, D_2, and HD in Liquid Ar

We now discuss some aspects of an interesting phenomenon on the coupling of some of the degrees of freedom of a molecule, a phenomenon which has been extensively studied in the infrared spectra of matrix-isolated molecules.

Holleman and Ewing[18] have recently measured the far-infrared absorption of solutions of H_2, D_2, and HD in liquid Ar at 87°K using about 1 mole-% solutions. The spectra were recorded in terms of a reduced intensity, $\bar{I} = I/v$ (see also footnote on page 365) since they stretch into the medium infrared range. We show the spectra of H_2 and HD in Figs. 10.18 and 10.19, respec-

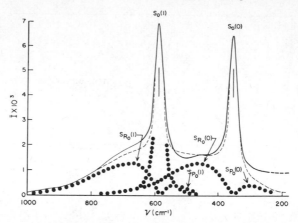

Fig. 10.18   Rotation-translation spectrum of H_2 in liquid argon. Lorentzian and four-parameter equations fitted to the sharp and broad absorption features are indicated by dotted curves. Only the $S_0(1)$ Lorentzian is shown. The dashed curve is the sum of the Lorentzian and four-parameter curves. The experimental data are given by the solid curve. The discrepancy below 300^{-1} is probably due to the translational band (see also Figs. 10.8 and 10.9). Instrument: Perkin-Elmer Model 210 monochromator. The vertical bars give the positions of the gas-phase (Raman) rotational transitions. [G. W. Holleman and G. E. Ewing, *J. Chem. Phys.* **47**, 571 (1967).]

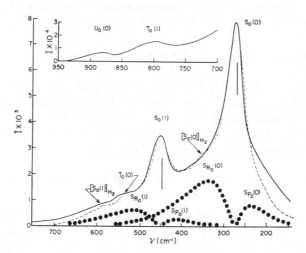

Fig. 10.19   Rotation-translation spectrum of HD in liquid argon. Four-parameter equations fitted to the broad absorption features are indicated by dotted curves. Lorentzian curves fitted to sharp absorption features are not shown. The dashed curve is the sum of the Lorentzian and four-parameter curves. The experimental data are given by the solid curve. The discrepancy below 200 cm^{-1} is probably caused by the continuous translational band. The vertical bars give the positions of the gas-phase (Raman) rotational transitions. [G. W. Holleman and G. E. Ewing, *J. Chem. Phys.* **47**, 571 (1967).]

tively. The sharp intense lines represent the rotational transitions $\Delta J = 2$. Lines are characterized in the usual fashion by the symbols $S, T, U, \cdots$ ($\Delta J = +2, +3, +4, \cdots$); the quantum number of the lower rotational level is given in parentheses, that of the vibrational level as a subscript. The spectra show that the positions of the rotational transitions of HD are shifted to higher frequencies by about 10 cm^{-1} with respect to their computed gas-phase values (unperturbed molecule, Raman lines), whereas the corresponding shifts for the dissolved H_2 (and D_2) are of the order of 1 cm^{-1}.

First we note the distinct rotational lines, indicating that the molecules rotate freely in the liquefied rare gas. The induction mechanism that leads to the optical absorption is already known to us: The molecular quadrupole moment of the hydrogen molecules induces a dipole moment in a neighboring rare-gas atom during a collision. However, it is more precise to talk here in terms of an induced moment located in a "solvent cavity" during the time that the inducing molecule stays within that cavity.[18] It is also clear—from what we have learned about the collision-induced phenomena in compressed gases—that the rotational transitions are accompanied by an increase or decrease in the translational energy. We label the translational levels by a quantum number n, assuming that we can describe the translational motion within the cavity in terms of a convenient intermolecular potential function. Thus, to give an example, the transition $J \rightarrow J + 2$, $\Delta n = \pm 1$ will be designated $S_{R_0}(J)$ and $S_{P_0}(J)$, respectively. Finally, we expect pure translational transitions to occur on account of the angle-independent overlap dipole moment, as we have discussed in the previous sections.

Closer inspection of Figs. 10.18 and 10.19 reveals that the lines of the dissolved HD possess a considerable larger half width than those of the dissolved H_2 (and D_2). The corresponding values, obtained from an empirical profile analysis, amount to 29, 22, and 48 cm^{-1} for the dissolved H_2, D_2, and HD, respectively, on the example of the $S_0(J)$.

It is then necessary to account for the two additional effects found in these solutions; that is, the broadness of the rotational bands of HD and the relatively large shifts in the position of the HD bands.

Holleman and Ewing[18] have discussed these features of HD in terms of a perturbation treatment given by Babloyantz[19] and Friedmann and Kimel.[20] Very briefly, these theories treat the motion of a rotating, trapped diatomic molecule in which the center of mass and the "center of interaction" do not coincide. The center of interaction between the molecule and its surrounding medium is given by the charge distribution of the molecule. The molecule HD is a good example for a species in which the center of interaction and the center of mass are at different positions along the interatomic axis of HD. We now easily see how an interaction term between the rotational motion and the motion of the center of mass arises: use of the inertial system of the molecule (the coordinate system through the center of mass) leads to a simple expression of the rotational energy, whereas the center of interaction coordinate system gives the simplest expression for the intermolecular potential energy. (For a homonuclear diatomic molecule, both systems are the same.) The interaction is obtained in the usual way: The angular momentum operator **L** of the kinetic energy operator in the Hamiltonian is now expressed in terms of an angular momentum operator **P** about the center of interaction,

$$\mathbf{L} = a\mathbf{1} \times \mathbf{p} + \mathbf{P}, \qquad (10.48)$$

where a is the distance between the two centers, $\mathbf{1}$ is the unit vector along the interatomic axis, and $\mathbf{p}$ is the linear momentum operator. To second order in a, the theory leads to a change in the rotational energy levels of diatomic molecules by[20]

$$\frac{\Delta E(J)}{hcB} = \frac{Ma^2}{2I} \left\{ \xi + \frac{\xi}{2J+1} \left[\frac{2J^3}{2J - \xi} - \frac{2(J+1)^3}{2J + 2 + \xi} \right] \right\}. \qquad (10.49)$$

The quantities B, I, and M designate the rotational constant, moment of inertia, and mass of the trapped molecule, respectively; $\xi = v/B$, where v is the frequency of the oscillatory ("rattling") motion of the molecule in its cell, governed by a harmonic oscillator potential function.*

Using this model, Holleman and Ewing have computed the shifts of the rotational transitions of HD due to the rotation-translation coupling.

TABLE 10.4*

ROTATIONAL FREQUENCIES (CM^{-1}) OF HD IN ITS GAS PHASE AND DISSOLVED
IN LIQUID AR

Rotational transition	Experimental		$v_{soln} - v_{gas}$	
	Gas phase[†]	In Ar	Exptl.	Calcd.
$S_0(0)$	267	276	+9	+8
$S_0(1)$	443	453	+10	+18
$T_0(0)$	532	540	+8	+6
$T_0(1)$	794	808	+14	+17
$U_0(0)$	883	890	+7	+5

* After G. W. Holleman and G. E. Ewing, *J. Chem. Phys.* **47**, 571 (1967).
† From the Raman spectrum, B. Stoicheff, *Can. J. Phys.* **35**, 730 (1957).

* See Ref. 20, Eq. 45, for the corresponding expression of a diatomic molecule in a cubic box.

The following values for the parameters of Eq. 10.49 were used: 0.12 Å for the separation of the center of mass and center of interaction a, and $v = 110 \, cm^{-1}$ for the translational energy separation $\Delta n = 1$. A comparison between the measured and calculated values is given in Table 10.4.

REFERENCES

1. J. P. Colpa and J. A. A. Ketelaar, *Mol. Phys.* **1**, 14, 343 (1958).
2. For a detailed discussion on the theory of collision-induced absorption, see J. D. Poll and J. van Kranendonk, *Can. J. Phys.* **39**, 189 (1961) (induced translational spectra); J. van Kranendonk and Z. J. Kiss, *Can. J. Phys.* **37**, 1187 (1959), J. P. Colpa and J. A. A. Ketelaar, *Mol. Phys.* **1**, 343 (1958) (induced rotational spectra); J. van Kranendonk, *Physica* **24**, 347 (1958) (induced vibrational spectra), and references cited therein. Our discussion of the theoretical aspects of collision-induced spectra is taken from these references. A classical derivation of the line shape function and integrated induced absorption coefficients of pure translational absorption in rare-gas mixtures has been published by H. B. Levine and G. Birnbaum, *Phys. Rev.* **154**, 86 (1967).
3. See, for instance, the use of cluster functions in the calculations of the equation of state of real gases, J. de Boer, *Reports Prog. Phys.* **12**, 305 (1949) and the detailed treatment in T. L. Hill, *Statistical Mechanics. Principles and Selected Applications*, McGraw-Hill, New York, 1956, Chapter 5.
4. See, for instance, A. R. Edmonds, *Angular Momentum in Quantum Mechanics*, Princeton Univ. Press, Princeton, New Jersey, 1957.
5. These are of the valence-bond type, $\psi_{ab} = a(1)b(2) + a(2)b(1)$. See C. A. Coulson, *Valence*, Oxford Univ. Press, Oxford, 1956, Chapter V.
6. M. E. Rose, *Elementary Theory of Angular Momentum*, Wiley, New York, 1957.
7. D. R. Bosomworth and H. P. Gush, *Can. J. Phys.* **43**, 751 (1965).
8a. H. B. Levine and G. Birnbaum, *Phys. Rev.* **154**, 86 (1967).
8b. G. Birnbaum and H. B. Levine, private communication by H. B. Levine, January 1969 (to be published).
8c. D. A. McQuarrie and R. B. Bernstein, *J. Chem. Phys.* **49**, 1958 (1968).
8d. J. Van Kranendonk, *Can. J. Phys.* **46**, 1173 (1968).
8e. See, for instance, R. G. Gordon, *J. Chem. Phys.* **38**, 1724 (1963). Note that $n_r = \varepsilon'$, the real part of the complex dielectric constant.
8f. H. B. Levine, *Phys. Rev. Letters* **21**, 1512 (1968).
9. Z. J. Kiss, H. P. Gush, and H. L. Welsh, *Can. J. Phys.* **37**, 362 (1959).
10. J. D. Poll and J. van Kranendonk, *Can. J. Phys.* **39**, 189 (1961). The total far-infrared band at 300°K has an absorption coefficient of $88 \times 10^{-33} \, sec^{-1} \, cm^5$, that is to say, $\alpha_{1(rot)}/\alpha_{1(transl)} \sim 15$. The units of α_1 are here in $sec^{-1} \, cm^5$, obtainable from those of $sec^{-1} \, cm^{-1}$ by multiplication with the square of Loschmidt's number, $(2.687 \times 10^{19} \, molecules/cm^3)^2$. The theoretical value of $\alpha_{1(transl)}$ was computed from Eq. 10.39 by using an experimental value for the polarizability and a theoretical value for the quadrupole moment.
11. J. A. A. Ketelaar, and R. P. H. Rettschnick, *Mol. Phys.* **7**, 191 (1964).
12. H. A. Gebbie, N. W. B. Stone, and D. Williams, *Mol. Phys.* **6**, 215 (1963).
13. N. W. B. Stone and D. Williams, *Mol. Phys.* **10**, 85 (1965).
14. R. Heastie and D. H. Martin, *Can. J. Phys.* **40**, 122 (1962).
15. S. Weiss and R. H. Cole, *J. Chem. Phys.* **46**, 644 (1967).
16. W. S. Benedict and R. Herman, *J. Quant. Spect. Rad. Transfer* **3**, 265 (1963). The values of Q in this paper must be divided by a factor of 2 in order to agree with the definition used here, $Q = \frac{1}{2}\int \rho(3z^2 - r^2)\,d\tau$.

17. A recent compilation of quadrupole moments is given by D. E. Stogryn and A. P. Stogryn, *Mol. Phys.* **11**, 371 (1966).
18. G. W. Holleman and G. E. Ewing, *J. Chem. Phys.* **47**, 571 (1967).
19. A. Babloyantz, *Mol. Phys.* **2**, 39 (1959).
20. H. Friedmann and S. Kimel, *J. Chem. Phys.* **43**, 3925 (1965).

11 Rotational-Translational Motion in Condensed Phases

A. INTRODUCTION

We now come to the treatment of rotational-translational motion in condensed phases, a subject that is of renewed interest and importance. It is studied by a wide variety of experimental techniques, such as dielectric relaxation, spin relaxation, and depolarization of fluorescence as well as by infrared, Raman, and microwave spectroscopy. The difference between spectra in condensed phases and of isolated species in the vapor phase can be attributed to the fact that the magnitude of the intermolecular interactions can no longer be neglected. The total state vector $|\psi\rangle$ of the whole system, that is, the molecule plus its environment, is then dependent on coordinates describing an external state $|e\rangle$ and an internal state $|\beta\rangle$.[1] The separation of $|\psi\rangle$ into a product wave function is done in a manner analogous to the separation of the electronic and nuclear parts of a wave equation by the Born-Oppenheimer approximation; the external state $|e\rangle$, which depends on the center of mass coordinates and the directions of the principal axes of inertia of the molecules, corresponds to the nuclear equation of the Born-Oppenheimer approximation.

The Schrödinger equation of the internal state $|\beta\rangle$ is

$$\mathcal{H}_{internal}|\beta\rangle = (V(\{R\}) + \hbar\omega_0)|\beta\rangle. \tag{11.1}$$

Its eigenvalues $V(\{R\}) + \hbar\omega_0$, which depend parametrically on the set of external coordinates $\{R\}$, yield the intermolecular potential in the Schrödinger equation of the external part of the problem,*

$$\left[(\tfrac{1}{2}) \sum_i \mathbf{P}_i^2 / M_i + (\tfrac{1}{2}) \sum_i \mathbf{L}_i \cdot \mathbf{I}_i \cdot \mathbf{L}_i + V(\{R\}) + \hbar\omega_0 \right] |e\rangle = E|e\rangle. \tag{11.2}$$

---

* This is analogous to the separation of the electronic and nuclear motions of a molecule. The electronic eigenvalues, which depend parametrically on the nuclear separations, yield the potential function for the nuclear wave equation (see, for instance, Chapter X in L. Pauling and E. B. Wilson, Jr., *Introduction to Quantum Mechanics*, McGraw-Hill, New York, 1935).

The first two terms in square brackets denote the kinetic translational-rotational Hamiltonian, $\mathbf{P}_i$ is the total linear momentum operator of molecule i (mass M_i), $\mathbf{I}_i$ is the reciprocal moment of inertia tensor, $\mathbf{L}_i$ is the angular momentum operator for rotation about the principal axes of inertia, E is the rotational-translational energy, and $\hbar\omega_0$ is chosen to cause the inter-molecular potential $V(\{R\})$ to approach zero for large intermolecular distances.

The separation of the total wave function $|\psi\rangle$ into an internal part $|\beta\rangle$ and external part $|e\rangle$ is based on the assumption that the spacings between the internal energy levels are much larger than the spacings between the external energy levels, that is, the energy levels of the overall rotational-translational motions of the center of mass. This assumption is based on the fact that only minor differences between the spectra of a molecule in its vapor and in its condensed phase are generally apparent. Obviously such an assumption would no longer be tenable if the intermolecular forces would amount to a significant fraction of the binding forces within the isolated molecule. Examples of such cases have been treated in the spectra of hydrogen bonds (see Chapter 6).

An example of a simplified general intermolecular potential has been given by Buckingham:[2] The total Hamiltonian is the sum of the Hamiltonian of the vibrational motion of the molecule and a perturbing potential $U(Q_1, Q_2, \cdots, \tau)$ which describes the interaction between the normal coordinates of the isolated molecule, Q_i, and the average configuration τ of the surrounding medium; in other words, the effects of external molecular motions (see Eq. 11.3) are only considered implicitly since the influence of the medium on the pure *vibrational* transitions of the dissolved molecule is treated ("static model"). The potential U is developed into a power series in the normal coordinates Q_i. Second-order perturbation theory then leads to the following result:

$$\langle v_i(\tau)\rangle = v_i^\circ + (\beta_i/hc)\left\{\langle U_{ii}''\rangle - \sum_j (\beta_j/hcv_j)V_{iij}\langle U_j'\rangle\right\}, \qquad (11.3)$$

where $v_i(\tau)$ is the frequency of the vibrational mode i of the molecule in its environment τ, v_i° is the observed frequency of the isolated molecule (vapor phase), U' and U'' are the first and second differential quotients of the power series of U with respect to Q_i, V_{iij} is the constant to the cubic term $Q_iQ_iQ_j$ of the intramolecular potential function of the isolated molecule, and the β_i are scalar factors (not to be confused with vector $|\beta\rangle$ employed above) containing its harmonic frequencies v_i. The angle brackets denote averaging over the configurations of the environment. The quantity $-\{\langle v_i(\tau)\rangle - v_i^\circ\}$ represents what is commonly termed the "solvent shift" of a vibrational transition. Equation 11.3 has been used, for instance, to elucidate quantitatively the solution spectra of HCN and DCN in various polar and non-polar solvents[3] and to characterize the perturbations that are experienced by the infrared-active bending and stretching fundamentals of CS_2 vapor when it is trapped in a polyethylene matrix.[4] These studies showed that the coefficients of the series expansion of the intermolecular potential U generally do not bear a simple relationship to the properties of the molecule and of its environment unless the molecule is highly symmetric. In such a case U_i' and U_{ii}'' may be given explicit meaning as an intermolecular force and force constant, respectively. Unfortunately, because of the high molecular symmetry, the number of observable frequencies are insufficient and the system solute-solvent is underdetermined; that is to say, from solvent shift data alone one cannot obtain in this case numerical values of U' and U''. In general, these two quantities are of the same order of magnitude and it is not, *a priori*, reasonable to neglect one or the other:[4] U_{ii}'' may be negligible only if the motion of v_i is strongly anharmonic (V_{iij} large) and involves heavy atoms (U_{ii}'' small). On the other hand, the term in U_j' may be negligible in the case of strong intermolecular bonds (U_{ii}'' large) to light atoms which undergo nearly harmonic vibrations (V_{iij} small).

From these examples we see that it is generally difficult to interpret the molecular motion in condensed systems on the basis of the shifts of the observed absorption frequencies. The reason for this is not hard to understand. Because of the large density of states of the system, that is, the totality of the wave functions for all the states (molecule plus environment), we would need a large number of off-diagonal matrix elements in order to account for all the possible transitions. These off-diagonal matrix elements would be rather difficult to calculate, let alone to observe and interpret in terms of a useful physical picture. The description of the molecular dynamical motions of an assembly of weakly interacting molecules in terms of observed frequency displacements is therefore generally unsatisfactory.

In the following we shall indicate a different approach for extracting the maximum available information from the experimental observations.[5]

B. THE HEISENBERG PICTURE

The shape of a vibrational absorption band, that is to say, the distribution of the spectral intensity about the band center ω_0, is given by (leaving out unimportant constant factors)

$$I(\omega) = \sum_{if} \rho_i |\langle i|\mathbf{\varepsilon} \cdot \mathbf{m}^v|f\rangle|^2 \delta\{[(E_f - E_i)/\hbar] - \omega\}, \qquad (11.4)$$

where $I(\omega)$ is the transition probability at angular frequency $\omega = 2\pi v$ (v in frequency units), ρ_i is the Boltzmann factor for the initial state i, $\mathbf{\varepsilon}$ is a unit vector along the electric vector of the incident radiation, $\mathbf{m}^v$ is the transition dipole moment for the particular vibrational transition starting from vibrational level v, and E_i and E_f are the eigenvalues of the initial and final rotation-translation quantum state $|i\rangle$ and $|f\rangle$, respectively (external states). The vibrational transition moment $\mathbf{m}^v$ is averaged between the vibrational levels v and v' with $E_{v'} - E_v = \hbar\omega_0$. The frequency ω_0 of this internal transition is "modulated" by the many rotational-translational transitions in the system between the external states $|i\rangle$ and $|f\rangle$ according to $E_f = E_i + \hbar\omega$. The eigen vectors of the total matrix elements (Eq. 11.4) are simple products of the external and internal basis sets (the total Hamiltonian is assumed to have no terms coupling the vibrational and external motions), in analogy to the description of the collision-induced phenomena discussed in Chapter 10 (see Eqs. 10.5, 10.14, 10.16). Hence, $\omega = (E_f - E_i)/\hbar$ is the displacement from the band center ω_0 of the internal transition (see Fig. 11.1).

If we now express the delta function by its Fourier transform,

$$\delta(\alpha) = \frac{1}{2\pi} \int_{-\infty}^{+\infty} \exp(i\alpha t)\, dt, \tag{11.5}$$

$I(\omega)$ is given by

$$I(\omega) = \frac{1}{2\pi} \sum_{if} \rho_i \langle i|\boldsymbol{\varepsilon} \cdot \mathbf{m}^v|f\rangle \langle f|\boldsymbol{\varepsilon} \cdot \mathbf{m}^v|i\rangle$$
$$\times \int_{-\infty}^{+\infty} \exp\{i(E_f - E_i)t/\hbar\} \exp(-i\omega t)\, dt. \tag{11.6}$$

If we express the energy eigenvalues E in terms of their external rotation-translation Hamiltonian H,* it follows that

$$I(\omega) = \frac{1}{2\pi} \int_{-\infty}^{+\infty} \exp(-i\omega t) \sum_{if} \rho_i \langle i|\boldsymbol{\varepsilon} \cdot \mathbf{m}^v|f\rangle$$
$$\times \langle f|e^{iHt/\hbar}\boldsymbol{\varepsilon} \cdot \mathbf{m}^v e^{-iHt/\hbar}|i\rangle\, dt. \tag{11.7}$$

What we have done here is transform $I(\omega)$ from the Schrödinger representation, in which the operators do not explicitly depend on the time but their eigen functions do, to the Heisenberg representation, a representation in which the operators depend on the time but the eigen functions do not. In other words, instead of labeling a multitude of transitions between states $|i\rangle$ and $|f\rangle$ and following the time dependence of the coordinates of all the $|f\rangle$, the system is now described by the (statistical) time dependence of some observable quantity, in our particular case the rotational motion of the

* For the substitution of E by H via the unitary operator $S = \exp\{-(i/h)Ht\}$, see D. I. Blochin-zew, *Grundlagen der Quantenmechanic*, 2nd ed., Deutscher Verlag der Wissenschaften, Berlin, 1957, p. 137.

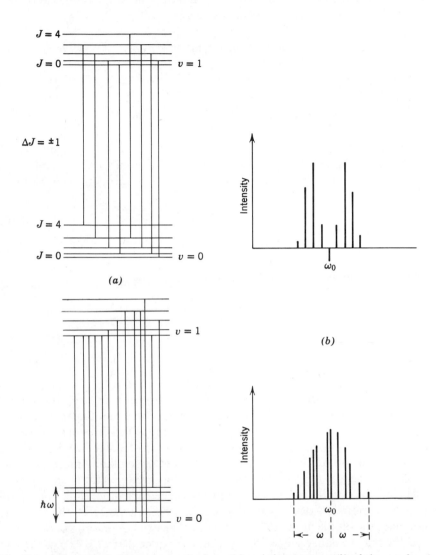

Fig. 11.1   Schematic of the energy level scheme (*a*) and the spectrum (*b*) of the $v = 0 \to 1$ vibrations for an isolated diatomic (*top*) and the dissolved molecule (*bottom*). Five vibrational-rotational levels are shown for the isolated molecule $E = \hbar\omega_0(v + \tfrac{1}{2}) + BJ(J + 1)$ and five energy levels (arbitrarily drawn), for the dissolved molecule. The range of the external energies (rotational-translational states) of the dissolved molecule is indicated by $\hbar\omega$, the corresponding frequency displacement from the band-center ω_0 by ω. Not all of the possible transitions of the dissolved molecule are shown (J may no longer be a good quantum number) and the values of the Boltzmann factors and transition-moment matrix elements (see Eq. 11.4) are chosen arbitrarily.

vector of the vibrational transition moment. This will become apparent after we have summed over the final states $|f\rangle$:

$$I(\omega) = \frac{1}{2\pi} \int_{-\infty}^{+\infty} e^{-i\omega t} \sum_i \rho_i \langle i| \mathbf{\varepsilon} \cdot \mathbf{m}^v (e^{iHt/\hbar} \mathbf{\varepsilon} \cdot \mathbf{m}^v e^{-iHt/\hbar})|i\rangle \, dt, \qquad (11.8)$$

or more concisely,

$$I(\omega) = \frac{1}{2\pi} \int_{-\infty}^{+\infty} e^{-i\omega t} \langle [\mathbf{\varepsilon} \cdot \mathbf{m}^v(0)] [\mathbf{\varepsilon} \cdot \mathbf{m}^v(t)] \rangle \, dt. \qquad (11.9)$$

The square brackets represent an equilibrium statistical average over all initial states.[6] Since we deal here mainly with the liquid phase, we integrate over all directions of polarization. Converting to a normalized spectrum

$$\hat{I}(\omega) = I(\omega) \bigg/ \int_{-\infty}^{+\infty} I(\omega) \, d\omega, \qquad (11.10)$$

we obtain

$$\hat{I}(\omega) = \frac{1}{2\pi} \int_{-\infty}^{+\infty} \langle \mathbf{u}(0) \cdot \mathbf{u}(t) \rangle e^{-i\omega t} \, dt, \qquad (11.11)$$

and its inverse Fourier integral

$$\langle \mathbf{u}(0) \cdot \mathbf{u}(t) \rangle = \int_{-\infty}^{+\infty} \hat{I}(\omega) e^{i\omega t} \, d\omega. \qquad (11.12)$$

Vector $\mathbf{u}(t)$ is now the unit vector along the direction of the (vibrational) transition moment fixed in the molecule. We note here that this requires that we neglect induced absorption; that is, absorption which is a consequence of molecular collisions. As we have shown in Chapter 10, the induced transition moments depend on the relative orientations of the colliding molecules and therefore generally can not be represented by a simple vector such as $\mathbf{u}(t)$.

The product $\langle \mathbf{u}(0) \cdot \mathbf{u}(t) \rangle$, which represents the correlation function of the transition moment, describes the decay of the coherence of the external motion of the dipole moment—and hence of the external motion of the molecule. We may visualize this as follows: Let us assume that we pick a particular molecule among the many molecules which undergo random orientational motions in the liquid. (*Nothing* is assumed here about the particular characteristics of this random motion. That is what we wish to find out.) At an (arbitrary) time $t_0 = 0$, we now note the orientation of the molecule—which we designate by $\mathbf{u}(0)$. We then observe the orientation at successive time intervals $t - t_0$ later, at $t_1, t_2, \ldots$, etc., forming the dot products (direction cosines) $\mathbf{u}(0) \cdot \mathbf{u}(t_1)$, $\mathbf{u}(0) \cdot \mathbf{u}(t_2)$, etc. What can we say about our knowledge of the direction of $\mathbf{u}(0)$ a time t later? For instance, if we observe $\langle \mathbf{u}(0) \cdot \mathbf{u}(t) \rangle$ at *sufficiently* short $t - t_0$, the intermolecular forces will not yet have an observable influence on the orientational motion of the molecule. (For instance, too few or too weak collisions.) Hence, we should observe essentially free rotation within a sufficiently short time interval. On the other hand, at relatively long times $t - t_0$, the many intervening collisions will have influenced the orientational motion of the molecule to such a degree that $\mathbf{u}(t)$ may point into *any* direction: The correlation of the original orientation is lost $[\langle \mathbf{u}(0) \cdot \mathbf{u}(t) \rangle = 0]$. Evidently, the course of $\langle \mathbf{u}(0) \cdot \mathbf{u}(t) \rangle$ during this "time interval of correlation" will give us a quantitative description of the character of the orientational motion of the molecule(s), for instance by indicating the time period of "free rotation" and the time interval after which the molecular orientation is "completely randomized" or in "equilibrium."

It should be noted that the intermolecular potential has been assumed to be the same for the lower and upper vibrational state of the molecule[1]; that is, the solvent shift of the vibrational transition is set equal to zero. If this restriction is lifted, the total Hamiltonians for the two vibrational levels v and v', namely H_v and $H_{v'}$, respectively, are not equal. The relevant part of the integrand of Eq. 11.8 then reads

$$\langle (\boldsymbol{\varepsilon} \cdot \mathbf{m}^v)(e^{iH_{v'}t/\hbar} \boldsymbol{\varepsilon} \cdot \mathbf{m}^v e^{-iH_v t/\hbar}) \rangle. \tag{11.13}$$

This means that in this case the Fourier transform does not lead to a simple dipole correlation function of the form $\langle \mathbf{u}(0) \cdot \mathbf{u}(t) \rangle$.

In a similar manner, the Heisenberg picture for the intensity distribution of a pure rotational transition, taken over all the rotation-translational states, leads to

$$I(\omega) = \frac{1}{2\pi} \int_{-\infty}^{+\infty} e^{-i\omega t} [\langle \boldsymbol{\mu}_1(0) \cdot \boldsymbol{\mu}_1(t) \rangle + \sum_{i \neq 1} \langle \boldsymbol{\mu}_1(0) \cdot \boldsymbol{\mu}_i(t) \rangle] \, dt, \tag{11.14}$$

where $\boldsymbol{\mu}_i$ is the permanent dipole moment of molecule i. The terms $\boldsymbol{\mu}_1(0) \cdot \boldsymbol{\mu}_i(t)$ describe the cross correlation between different dipoles. Obviously, these cross correlation terms vanish with increasing dilution. The Fourier inversion of the band shape gives the correlation function for the pure rotational spectrum:

$$\langle \boldsymbol{\mu}_1(0) \cdot \boldsymbol{\mu}_1(t) \rangle + \sum_{i \neq 1} \langle \boldsymbol{\mu}_1(0) \cdot \boldsymbol{\mu}_i(t) \rangle$$

$$= \int_{-\infty}^{+\infty} \hat{I}(\omega) e^{i\omega t} \, d\omega, \tag{11.15}$$

where $\hat{I}(\omega)$ is the normalized intensity (see Eq. 11.10). Here, ω is not a frequency displacement from a band center—as it was the case for a vibrational transition—but the actual frequency since all rotational transitions have to be treated on a par (see Chapter 10, Section B.a). The quantity $\boldsymbol{\mu}$ (where we have dropped the subscript) is now a unit vector along the direction of the permanent dipole moment.*

C. ROTATIONAL MOTION FROM THE BAND SHAPES OF VIBRATIONAL TRANSITIONS

a. Introduction

Up to the present time, the far-infrared transitions of only a few solution spectra have been evaluated in terms of the molecular motion of the dissolved molecules. The advantage of using the far-infrared region lies in the fact that, firstly, many solvents are transparent at the longer wavelengths or, at least, possess wide "windows," thereby permitting a study of the dependence of the molecular motion of the dissolved molecules on different environments and on varying concentrations. Secondly, the intermolecular potential of the lower and upper vibrational state is essentially the same—as judged by the small solvent shifts usually observed in the range 250 to 450 cm^{-1}. This eliminates one source of uncertainty in the discussion of the characteristics of the molecular motion with the help of the simple dipole correlation function described by Eq. 11.12. Thirdly, the likelihood that the fundamental modes are too close or even overlap is generally smaller in the far-infrared than in the 650 to 1500 cm^{-1} range, at least for the smaller and less flexible molecules. We can then hope to carry out a meaningful integration over the whole band (see Eq. 11.12) without significant truncation errors and without the necessity of having to disentangle the band into the individual contributions of its overlapping components. (The application of empirical line shape functions to the purpose of decomposing an overlapped band contour would, in effect, mean that we try to assume that which we wish to measure.)

The disadvantage of using the far-infrared region lies in the larger probability that the vibrational band shape is distorted by adjacent or overlapping "hot bands"; these are often relatively intense because of the favorable Boltzmann factors at the lower energies. For instance, the band of v_2 of CS_2, the degenerate bending fundamental at 396.7 cm^{-1} (vapor), is overlapped by two hot bands, one hot band at slightly higher and the other at slightly lower frequencies from the sharp Q branch of the fundamental. The two hot bands are hidden beneath the R and P branches in the

---

* It is useful to remember that the random motion of the molecules occurs at all times. The coupling of the given transition moment to the radiation field serves merely as a means for observing the motion.

vapor spectrum. Fortuitously, the peak absorption of the weaker of the two hot bands nearly coincides with the peak absorption of the P branch, whereas the peak absorption of the more intense hot band coincides with the maximum absorption of the R branch. In the solution spectrum, where the R and P branches have collapsed, the hot bands appear and thus give the impression that an "appreciable degree" of quantized rotation of the CS_2 molecule in the polymer matrix persists.[4] We show this in Fig. 11.2a for CS_2 vapor and in Fig. 11.2b for CS_2 dissolved in polyethylene (Marlex). Relative intensity measurements at different temperatures finally established that the side peaks at 403.3 and 387.1 cm^{-1} in the solution spectrum represented hot bands and did not reflect a significant degree of free rotation of the CS_2 molecule in the polymer.[4]

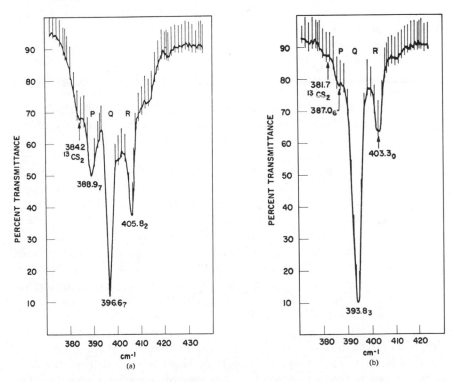

Fig. 11.2   (a) The v_2 mode of gaseous CS_2 (51-mm pressure): 10-cm absorption path, CsI windows. (b) The v_2 mode of CS_2 (30 mg) trapped in Marlex 6002 ($5 \times 5 \, cm^2$): 1.5-mm absorption path. The symbols P, Q, R denote absorption peaks which have undergone an apparent, parallel shift from the P, Q, R branches in (a). The 387.1 cm^{-1} band originates most likely from level $2v_2$; that at 403.3 cm^{-1}, probably from level v_2. [W. G. Rothschild, J. Chem. Phys. **42**, 694 (1965).]

Another disadvantage of the far-infrared region is the intensity due to the continuous absorption (see Section D.d). The continuous absorption band is very broad and its intensity can be subtracted out rather readily from the total intensity (probably with a fair amount of accuracy).

b. Rotational Motion of Methylene Chloride (CH_2Cl_2) in Polystyrene

The characteristics of the motion of a small molecule within a polymer matrix are of great interest. From it we can draw conclusions about the intermolecular forces in plasticized polymeric systems, with obvious relevance to such topics as membrane permeability, diffusion-controlled reaction mechanisms, plasticizer retention, and swelling and solubility properties. It is clear that the Fourier inversion of a suitably chosen infrared band of the dissolved molecule, as was described in Section B, can yield information on its rotational diffusion in the polymer medium without recourse to a model.

An example of such a study is the Fourier inversion of the v_4 fundamental of methylene chloride dissolved in amorphous polystyrene of a very narrow molecular weight distribution (Rothschild[7]). This relatively simple system was chosen in order to avoid complications due to strong specific interactions between like and unlike molecules. The results would therefore be expected to indicate the effect of a viscous medium on the rotational mobility of a small molecule. Although there is ample evidence of a strong correlation between the macroscopic viscosity of a medium and the translatory diffusion of a small molecule dispersed in it, a corresponding correlation between rotatory diffusion and macroscopic viscosity is much less experimentally established for highly viscous media. The fundamental v_4 of methylene chloride was chosen because it lies in a region ($280 \, cm^{-1}$) where the absorption of polystyrene is rather weak and wavelength-independent. The disadvantage of this long wavelength region is that it stretches into the range where the continuous absorption of CH_2Cl_2 is already appreciable (see Section D.d). Thus, the correction for the intensity of the continuous band (that is, the way the base line of the absorption band is drawn) introduces the main uncertainty into the results.

Figure 11.3 shows the v_4 mode of a methylene chloride–polystyrene solution of molar ratio $R = 0.5$ (in terms of the number of unit-chain-segments, $C_6H_5 \cdot CH \cdot CH_2$, per molecule CH_2Cl_2). The transmittance of the continuous band is seen to drop off rapidly towards longer wavelengths. The best base line which could be drawn is shown by the thin curve. The maximum absorption of the vibrational band, at $286 \, cm^{-1}$, has shifted to higher frequencies from its value reported for the vapor phase ($280 \, cm^{-1}$).[8] In view of the simple rotational correlation function for equal intermolecular potentials of the lower and upper vibrational levels (zero solvent shift; see

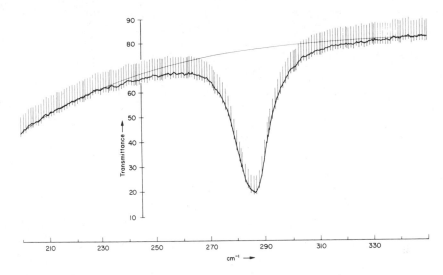

Fig. 11.3 Absorption band of the v_4 fundamental of methylene chloride in polystyrene. Ratio of polystyrene to halide in terms of number of unit-chain-segments of polymer ($C_6H_5CHCH_2$) to number of molecules of halide, $R = 0.5$. The absorption of the polymer was not compensated by a corresponding blank in this run; the polymer absorption is apparent in the wings of the band at 330 and at 235 cm^{-1}. For the sake of convenience the frequency scale was condensed by slowing the speed of the recorder; for a numerical evaluation this scale would be too narrow. Instrument: Perkin-Elmer Model 301, double-beam spectrometer. Temperature: 40°C. [W. G. Rothschild, *Macromolecules* **1**, 43 (1968), by permission of American Chemical Society.]

Eq. 11.13), we have neglected the solvent shift and computed the correlation function about a band center which was obtained from the first moment of the solution band.[7]

The fundamental v_4 (symmetry species a_1) represents a carbon-chlorine deformation; the (internal) transition moment—in this case it is a vibrational transition—lies in a direction parallel to the twofold symmetry axis (C_2) of the methylene chloride molecule (point group symmetry C_{2v}). The dipole correlation functions* computed from v_4 therefore describe the orientational motion of the C_2-axis fixed in the CH_2Cl_2 molecule. The correlation functions

---

* The intensity $I(\omega)$ of Eq. 11.10 is defined as

$$I(\omega) = \frac{\sigma(\omega + \omega_0)}{(\omega + \omega_0)\{1 - \exp(-\hbar[\omega + \omega_0]/kT)\}},$$

which is really an initial-state-averaged transition probability [see H. B. Levine and G. Birnbaum, *Phys. Rev.* **154**, 86 (1967)]. This form of $I(\omega)$ leads to the simple correlation function of Eq. 11.12. The Boltzmann factor in $I(\omega)$ takes care of the induced emission. Note that ω is the displacement from the band center, ω_0, and that $\omega_0 + \omega$ is the actual angular frequency. For pure rotational spectra (see Eq. 11.15), $\omega_0 = 0$. The quantity $\sigma(\omega_0 + \omega)$ represents the absorption coefficient in any convenient unit (since we work with normalized spectra).

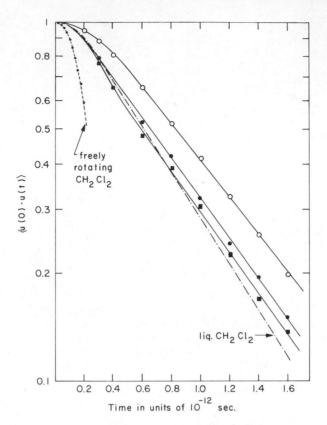

Fig. 11.4 Rotational correlation functions of methylene chloride in polystyrene–methylene chloride solutions obtained from the band contours of the v_4 mode of CH_2Cl_2. The experimental points are the averages of two to four independent determinations; $R = 0.5$ (■), 1.0 (●), and 1.5 (○). The correlation functions of pure liquid CH_2Cl_2 (experimental) and for the freely rotating molecule (computed, see text) are also shown. (The experimental points for the pure liquid CH_2Cl_2, all of which fall close to the drawn curve, have been omitted.) The dipole correlation function of the v_4 mode shows the rotational motion of the direction of the twofold symmetry axis of the molecule in the liquid medium. [W. G. Rothschild, *Macromolecules* **1**, 43 (1968), by permission of American Chemical Society.]

computed from the data for $R = 0$, 0.5, 1.0, and 1.5 are shown in Fig. 11.4 in a semilogarithmic plot. It is seen that the decay of the correlation functions with time is essentially exponential but shows a small initial curved part. The curved portion of $\langle \mathbf{u}(0) \cdot \mathbf{u}(t) \rangle$ indicates that a significant orientation of the molecules occurs by relatively large orientational jumps.[5]

   The dotted curve in Fig. 11.4 shows the initial course of the computed dipole correlation function of (a classical ensemble of) freely rotating CH_2Cl_2

molecules; that is, the correlation function in the absence of intermolecular forces.[9] We notice that this correlation function decays much faster than that of the CH_2Cl_2 molecules in the polymer solutions. This gives an indication of the intermolecular torques that act on the dissolved molecules and therefore slow down their rotational motion.* As is evident from Fig. 11.4, the intermolecular forces begin to affect the rotatory motion of the methylene chloride molecules within a very short time interval after the onset of the observation ($t = 0$). We note here that in Sections C.c and D.b we shall discuss examples in which there is a considerable time lag before the intermolecular forces begin to hinder the motion appreciably.

Inspection of the $\langle \mathbf{u}(0) \cdot \mathbf{u}(t) \rangle$ in Fig. 11.4 shows that the memory of the orientation is more rapidly lost in the more polymer-poor solutions. For instance, $\langle \mathbf{u}(0) \cdot \mathbf{u}(t) \rangle$ has dropped from 1 to 0.8 within about 0.3×10^{-12} sec for the solution with $R = 0.5$ and within $\sim 0.4 \times 10^{-12}$ sec for the polymer-concentrated solution with $R = 1.5$. This variation can be more succinctly expressed with the help of the integral $\int \langle \mathbf{u}(0) \cdot \mathbf{u}(t) \rangle \, dt$, called the "correlation time."[10] This time, designated by the symbol τ_c, indicates how fast the correlation is lost (see Section B); that is to say, how long it takes until $\langle \mathbf{u}(0) \cdot \mathbf{u}(t) \rangle \sim 0$. Because the form of the correlation functions for the CH_2Cl_2 molecules is mainly exponential, the values of τ_c are inversely proportional to the simple, Debye-type diffusion constant of the rotational motion of the C_2-axis; that is, the rotatory diffusion of CH_2Cl_2 in the polymer medium about two axes perpendicular to the C_2-axis. We have computed the τ_c as a function of concentration and collected the results in Table 11.1.

We return now to the question, posed above, of whether there is a significant correlation between the macroscopic viscosity of the medium and the rotatory motion of a small molecule dispersed in it. If we consider that the macroscopic viscosity of the methylene chloride–polystyrene solutions varies over many orders of magnitude when R goes from 0.5 to 1.5, it is seen from the values of τ_c in Table 11.1 that the macroscopic viscosity is essentially unrelated to the characteristics of the rotatory diffusion of the dissolved methylene chloride molecules.

The duration of the curved part of $\langle \mathbf{u}(0) \cdot \mathbf{u}(t) \rangle$, about 0.4×10^{-12} sec, is of the same magnitude as the time of one vibration of a typical skeletal deformation motion of a polymer backbone, $\sim 0.2 \times 10^{-12}$ sec (the fundamental frequency of such a mode is, for polyethylene, about $140 \, cm^{-1}$).[11] Hence, we may be tempted to propose that the period during which a CH_2Cl_2 molecule can orient itself through relatively large angles is—at least in the polymer-rich solutions—given by the closing and opening of

---

* The intermolecular torques can be computed from the fourth moment of the absorption band, $M(4) = \int \hat{I}(\omega)\omega^4 \, d\omega$ (see Ref. 20). This requires very precise absorption measurements.

TABLE 11.1*

CORRELATION TIME OF THE ROTATIONAL MOTION OF THE C_2-
SYMMETRY AXIS OF METHYLENE CHLORIDE IN POLYSTYRENE

Environment	τ_c, 10^{-12} sec
Pure liquid	0.78
$R^{\dagger} = 0.5$	0.81
$R = 1.0$	0.84
$R = 1.5-2.0$	0.94

* W. G. Rothschild, *Macromolecules* **1**, 43 (1968). By permission of American Chemical Society.
† Ratio of numbers of unit-chain-segments of polystyrene to numbers of molecules of methylene chloride.

"solvent cages," in other words, by the (large-amplitude) vibrations of the entangled polymer chains.[7] Of course, we should realize that an *observed* frequency of a skeletal vibration of the polymer chain is not necessarily equal to the maximum of the actual frequency distribution of the chain displacements, in other words, to the maximum of the density distribution of the phonon spectrum of the chain motion (see, for instance, Fig. 12.20). The observed frequency is the only one (of all the many normal skeletal motions of the chain) which is "filtered out" by the infrared selection rules.

We conclude this section by remarking that in order to obtain meaningful results it is necessary to perform accurate (relative) intensity measurements throughout the whole absorption band, including the wings. Although resolution obviously plays no significant role, it would be desirable to keep the slit width small in order to minimize, for instance, the effects of stray radiation. The compensation of the blank must be carefully done; blanks which absorb appreciably will, in most instances, lead to useless data.[7‡]

c. Rotational Motion in Liquid Methyl Iodide

Favelukes *et al.*[12] have published a study on the rotational motion of CH_3I in its liquid phase using attenuated total reflection techniques. Figure 11.5 shows the dipole correlation functions computed by these authors from their data. The lower part (b) of Fig. 11.5 displays a plot of the natural logarithm of the dipole correlation function of the $552\ cm^{-1}$ parallel band of CH_3I. For comparison's sake, the upper part (a) of Fig. 11.5 shows the corresponding function for a perpendicular band (at $884\ cm^{-1}$). In analogy to our discussion in the previous section on the motion of CH_2Cl_2, we learn from the correlation functions of Fig. 11.5 that the CH_3I molecules

‡ It should be kept in mind that we have tacitly assumed that there is insignificant line broadening due to vibrational relaxation.

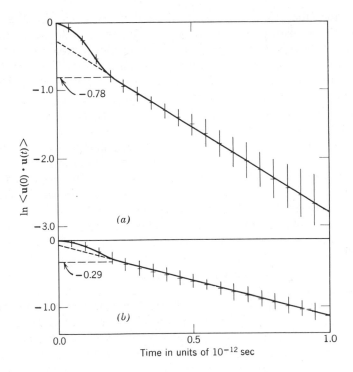

Fig. 11.5   The natural logarithms of the dipole correlation functions of (a) the $884\,\text{cm}^{-1}$ perpendicular band and (b) the $552\,\text{cm}^{-1}$ parallel band of liquid methyl iodide. The ranges of the crosses indicate the experimental uncertainty. [C. E. Favelukes, A. A. Clifford, and B. Crawford, Jr., *J. Phys. Chem.* **72**, 962 (1968), by permission of American Chemical Society.]

rotate relatively unhindered for a period of about 0.2×10^{-12} sec [curved part of $\langle \mathbf{u}(0) \cdot \mathbf{u}(t) \rangle$]. After this time interval, their rotational motion becomes strongly disrupted by collisions and subsequently, $\langle \mathbf{u}(0) \cdot \mathbf{u}(t) \rangle$ is exponential. The particularly interesting aspect of the rotational motion exhibited in Fig. 11.5 is that at the end of this initial time interval of 0.2×10^{-12} see the correlation value for the perpendicular band has dropped to about 0.45 ($\ln 0.45 = -0.78$) but has only fallen to 0.75 ($\ln 0.75 = -0.29$) for the parallel band (lower part of Fig. 11.5). The authors have related this to the magnitudes of the different moments of inertia that are involved in the rotational motion of the CH_3I molecule. We describe this in the following.

In the case of a parallel band, the vibrational transition moment lies along the threefold symmetry axis of the molecule (which is the axis along the carbon–iodine bond). We remember that the dipole correlation function describes the motion of a unit vector that lies along the direction of the particular transition moment and hence, for the parallel band, gives the rotational motion of the *direction* of the threefold symmetry axis of the CH_3I molecule. Since rotation about this symmetry axis does not alter its direction, the velocity of the rotational diffusional motion depends on the other two (equal) moments of inertia which are along axes normal to the symmetry axis. By the same reasoning it is established that in a perpendicular band of CH_3I the dipole correlation function shows a composite of the rotational motion about the threefold symmetry axis and about one axis perpendicular to the symmetry axis. The moment of inertia of CH_3I about its symmetry axis is much smaller than the moments about perpendicular axes; hence the molecules rotate faster about the symmetry axis. Subsequently, $\langle \mathbf{u}(0) \cdot \mathbf{u}(t) \rangle$ of the perpendicular band should decay more rapidly since it involves the faster rotation, as is indeed shown by the data. This is a very satisfying result; in fact, the point can be stretched by estimating the average angle* about which the molecules have turned within the 0.2×10^{-12} sec time interval. From the cosine of 0.75 Favelukes *et al.* found that a molecule of CH_3I rotates by $\sim 41°$ about its two equal perpendicular axes, or by arc cos $(0.75)^{1/2} \sim 30°$ about each perpendicular axis, before disruptive collisions occur. In analogous fashion they found that for the perpendicular band the combined rotation about the symmetry axis and one axis normal to it occurred through an angle of $\sim 63°$ before strong collisions degrade the motion. Rotation about the symmetry axis alone would thus carry the molecule through an angle of $\sim$ arc cos (cos 63°/cos 30°) $\sim 58°$ within the initial time interval of 0.2×10^{-12} sec.

As the authors have stressed, this picture should not be taken too literally. In fact, one would have to show that the CH_3I molecules rotate freely during the initial 0.2×10^{-12} sec, in other words, that the intermolecular forces have not yet become effective, for only then is the use of the simple cosine relations valid in the above estimates.* [We remember that the initial curvature of the correlation function reflects large-angle diffusional jumps but, *per se*, does not necessarily indicate the absence of intermolecular forces (see, for instance, Section C.b)]. We have thus computed the correlation function of the classical ensemble of freely rotating CH_3I molecules[9] for the initial time interval 0.2×10^{-12} sec. Comparison with the results in Fig. 11.5(a) shows that for the perpendicular band (upper part of figure) the molecules indeed rotate freely during this initial time period of 0.2×10^{-12} sec (see Fig. 11.6). For the parallel band the agreement is not quite as good.

---

* The function $\mathbf{u}(0) \cdot \mathbf{u}(t)$ represents the dot product between two vectors. For a single molecule, rotating freely during the time interval t, the value of $\mathbf{u}(0) \cdot \mathbf{u}(t)$ thus gives the cosine of the angle through which the molecule has rotated. See Section B.

It is useful to establish a comparison between the correlation functions computed from attenuated reflection measurements on one hand and from the usual transmission measurements on the other hand. Both techniques have advantages and disadvantages with respect to (*1*) the requirements on the instrumentation to do the experiments properly and to (*2*) the sources of hard-to-avoid errors which influence the accuracy of the computed correlation function. In transmission measurements we generally do not know too accurately the exact zero-absorption line (see, for instance, Section C.b), whereas in attenuated reflection measurements weak bands are difficult to evaluate and very precisely constructed special instrumentation is required.[12a]

The comparison between the results from the two techniques is displayed in Fig. 11.6 which shows the correlation function of the perpendicular band of liquid CH_3I (dotted curve), calculated from transmission measurements (Rothschild[12b]). The solid curve represents the corresponding correlation function from Favelukes *et al.*'s attenuated reflection measurements, transferred from Fig. 11.5(a). The dashed curve shows the corresponding correlation function of freely rotating CH_3I molecules.[12b] We see that the two methods agree on the essentials, namely, that molecular rotation (about the symmetry axis of the molecule and one axis perpendicular to it) in the liquid and in the "infinitely dilute" gas of CH_3I are about equal for 0.1×10^{-12} to 0.2×10^{-12} sec of the motion, and that thereafter the decay to equilibrium in the liquid is exponential.

We believe the significance of these results is the fact that large-angle rotation occurs about *all* axes of inertia: Apparently, rotational motion in liquids is not influenced significantly by the presence or absence of atomic symmetry (Rotation about axes normal to the C—I axis twists the permanent dipole moment, rotation about the C—I axis does not.)

This concludes the section on the evaluation of the molecular motion in solution from band shapes of vibrational transitions. We hardly need to stress the usefulness with which the Heisenberg picture can be applied to many-body problems of this nature.

D. ROTATIONAL MOTION FROM THE BAND SHAPES OF ROTATIONAL SPECTRA

a. Introduction

A considerable amount of work has been done to elucidate the rotational motion of diatomic molecules and small symmetric and spherical tops in solution on the premise that the rotation of such molecules is relatively little impeded by the environment. Most efforts up to the present have been concentrated on a study of the band shapes of vibration-rotation bands of the

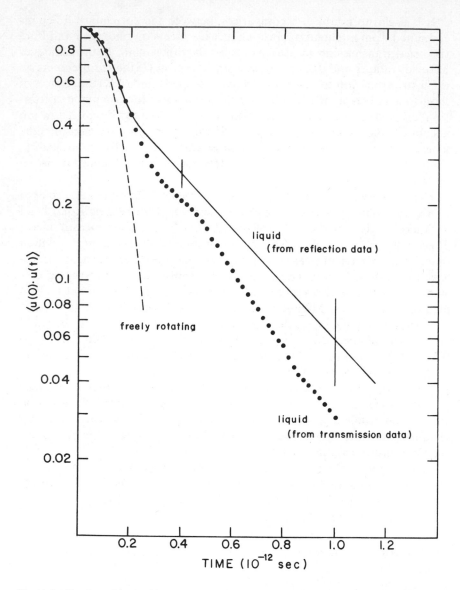

Fig. 11.6   Semilogarithmic plots of the correlation function of the 885-cm^{-1} perpendicular band of CH_3I. (*Dot-curve*): computed, in steps of 2 cm^{-1} from the transmission data (see Ref. 12b). The band center was taken as that obtained from the first moment as described in Section C.b. (*Full curve*): correlation function due to C. E. Favelukes, A. A. Clifford, and B. Crawford, Jr., from their attenuated reflection measurements, transferred from the upper part of Fig. 11.5. (*Dashed curve*): classical ensemble of freely rotating CH_3I molecules, computed according to Ref. 12b. [W. G. Rothschild, *J. Chem. Phys.* **51**, 5187 (1969); **52**, 6453 (1970).]

dissolved molecules in the infrared region of the spectrum, whereas less emphasis has been placed on a study of the band contours of their pure rotational transitions. In the case of vibration-rotation bands of various small dissolved molecules, the band contours exhibited wings which were ascribed to the (remnants of) P and R branches. The presence of these

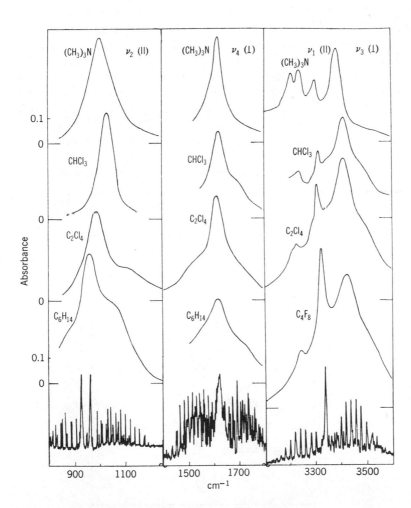

Fig. 11.7 Band contours of the fundamentals v_1, v_2, v_3, and v_4 of NH_3 in various solvents. Perkin-Elmer 421, cell path lengths 0.1 to 0.5 cm. Concentrations: 0.2 to 0.6 moles/l. All solvents were used in their liquid phase (the perfluorocyclobutane was compressed). Note that the v_1 and v_3 fundamentals overlap. [J. Corset, P. V. Huong, and J. Lascombe, *Compt. Rend. Acad. Sci.* (Paris) **262C**, 959 (1966).]

branches in the condensed medium has been taken to indicate that a considerable degree of quantized rotation of the molecule persists in the medium or, expressed differently, that a considerable fraction of the dissolved molecules are in rotational levels sufficiently above a barrier hindering free end-over-end rotation. In the following we shall give some examples taken from mid-infrared and infrared spectra.

Figure 11.7 shows the band envelopes of the fundamentals v_1, v_2, v_3, and v_4 of NH_3 in various solvents.[13] Similar data on v_2 are given in Fig. 11.8.[14] We now compare the band shapes of v_2 and v_4 of the solution spectra with those of the vapor (v_1 and v_3 overlap strongly and are therefore less useful). We see from Fig. 11.7 that the solution spectra of NH_3 in the nonpolar solvents C_6H_{14}, C_2Cl_4, and C_4F_8 exhibit some remnants of the rotation-vibration wings of the corresponding vapor bands. In solution with the polar chloroform ($CHCl_3$), however, only the perpendicular band v_4 shows a wing towards shorter wavelengths; the parallel band v_2 exhibits just the Q branch. It has been proposed that the collapse of the wings of the parallel band in $CHCl_3$ solution is an indication of the complete hindering of the rotation of the NH_3 molecule about axes perpendicular to its threefold symmetry axis due to the formation of a linear, hydrogen-bonded complex

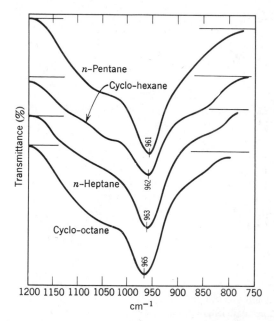

Fig. 11.8 The v_2 band of ammonia dissolved in a variety of hydrocarbons (ammonia pressure 60 psi, cell thickness 0.025 mm). [P. Datta and G. M. Barrow, *J. Am. Chem. Soc.* **87**, 3053 (1965), by permission of American Chemical Society.]

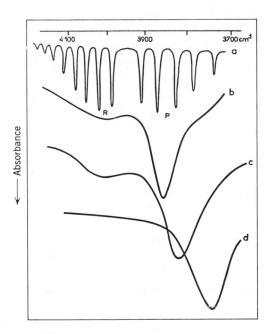

Fig. 11.9   Stretching fundamental of HF in various solvents: (a) vapor phase; (b) CCl_4 solution; (c) CS_2 solution; (d) toluene solution. Instrument: Perkin-Elmer 21, LiF prism. [J. Lascombe, P. V. Huong, and M.-L. Josien, *Bull. Soc. Chim. (France)* **1959**, 1175.]

$CCl_3H\cdots NH_3$.[13] On the contrary, the remnants of the wings of the perpendicular transition of NH_3 would indicate that free rotation about the hydrogen bond axis has not completely ceased.*

Figure 11.9 shows the fundamental stretch of HF dissolved in various solvents.[15] Again, the well pronounced wings are ascribed to the free rotation of the molecule in the solutions and the disappearance of the wings in HF–toluene (curve d) is believed to show prevention of free rotation of

---

* Strictly speaking, the free motion of a symmetric-top molecule such as NH_3 and CH_3Cl does not simply consist of rotations about the figure axis of the molecule (axis of symmetry) and about an axis perpendicular to the figure axis. To see this, inspection of Fig. 7, p. 23, of G. Herzberg, *Infrared and Raman Spectra of Polyatomic Molecules*, Van Nostrand, Princeton, New Jersey, 1962, is helpful. The angular motion of a symmetric top is a composite of (1) the rotation about the molecular figure axis (axis of threefold symmetry, P_z) with an angular frequency proportional to the z-component of the total angular momentum, K, and (2) the nutation of this figure axis about the total angular momentum, J, with angular frequency proportional to $\sqrt{(J+1)J}$. The angle θ between P_z and J is given, approximately, by $\cos\theta \approx K/J$. Thus, in order to obtain $\theta \sim 90°$, the thermal distribution of the energy levels must be peaked around low K and high J.

HF by complex formation with the solvent.[15] Bulanin et al.[16] have estimated barrier heights to free rotation of some diatomic molecules in solution from the intensities of the P, Q, R branches and the P–R spacing relative to its value in the vapor. For instance, a barrier of ~ 0.9 kcal/mole has been computed for HCl in CCl_4, the interpretation being that all molecules in rotational levels with energies above this barrier height, $B(J_m + 1)J_m \geqslant 0.9$ kcal/mole, rotate freely whereas those below this value execute rotational oscillations about some equilibrium position. Using the rotational constant of the HCl vapor molecule, we find that $J_m \geqslant 5$.

b. Individual Molecules. The Hydrogen Halides, Water, and Ammonia

One of the first attempts to observe the pure rotational spectra of dissolved molecules was reported by Silver and Wood,[17] who studied HCl and H_2O in benzene. They did not observe any of the more intense HCl rotational transitions, expected to be at 206.7, 227.3, and 246.3 cm^{-1} ($J = 9 \to 10$, $10 \to 11$, $11 \to 12$). Neither did they succeed in detecting the strong rotational transition of H_2O at 202.8 cm^{-1} ($3_{30} \to 4_{41}$). The authors therefore concluded that quantized rotation of HCl and H_2O, respectively, is absent in both systems. We shall see, however, that the absence of individual pure rotational lines in a solution spectrum, which means the failure to observe resolved transitions of the pure rotational spectrum of the dissolved molecules, does not indicate that its quantized rotational motion has degenerated into Brownian motion.

Figures 11.10 and 11.11 show some recent work by Datta and Barrow[18] on the solution spectra of HF, HCl, DCl, D_2O, and NH_3 in cyclohexane and carbon tetrachloride, scanned in the spectral region of their more intense pure rotational transitions (about 20 to 450 cm^{-1}). The general shape of the absorption profiles of the dissolved molecules is seen to be quite similar to the envelopes of the intensity distribution of the quantized rotational transitions, particularly for cyclohexane as solvent. We also notice that the maximum absorption of the solution bands has shifted to higher frequencies and tails off at higher frequencies than the corresponding vapor spectra. The authors point out that these characteristics are more pronounced the greater the interaction between solute and solvent molecules.

The authors conclude from the data that in these solutions the solute molecules rotate freely to a considerable degree, with a rotational constant that is apparently somewhat larger than that in the vapor molecule.[18] This seems to indicate that the bond distances of the dissolved molecules are smaller and their angular motion faster than in their vapor. In the following we shall present computations by Rothschild[19] of the dipole correlation function of the dissolved hydrogen halide molecules using Datta and Barrow's data of HF in cyclohexane (see Fig. 11.10, *top*). We note that in these examples

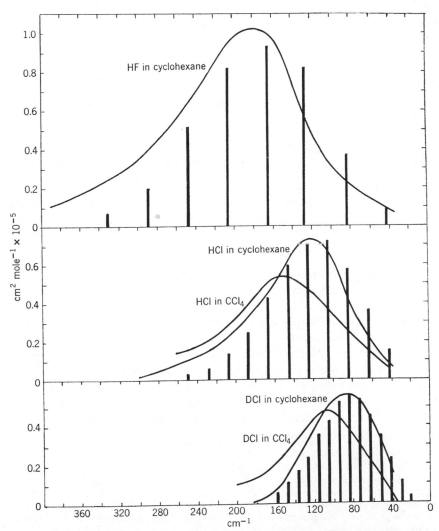

Fig. 11.10 Comparison of the absorption spectra of hydrogen and deuterium halides in cyclohexane and carbon tetrachloride with bar graphs showing the calculated gas-phase rotational spectra. Concentrations of about 0.025M, a cell length of 1 cm, and a temperature of 35°C were used throughout. The ordinate gives the approximate extinction coefficients for the experimental curves. Instrument: Beckman IR-11. [P. Datta and G. M. Barrow, *J. Chem. Phys.* **43**, 2137 (1965).]

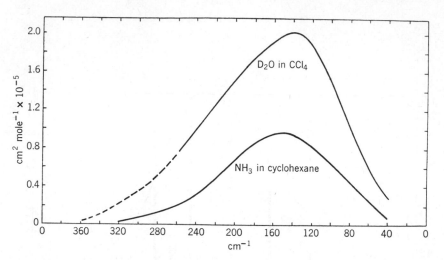

Fig. 11.11   The far-infrared absorption spectra of D_2O in CCl_4 and NH_3 in cyclohexane. Concentrations, cell length, and temperature as for the systems of Fig. 11.10. [P. Datta and G. M. Barrow, *J. Chem. Phys.* **43**, 2137 (1965).]

the (internal) transition is a pure rotational transition (see Sections B and C.b).

The dipole correlation function of hydrogen fluoride dissolved in the hydrocarbon is computed according to Eq. 11.15 (the cross correlation terms are neglected since the solution is dilute). The result is shown in Fig. 11.12 by the dotted curve. Figure 11.12 also contains the correlation function of the classical ensemble of freely rotating HF molecules,[9] shown by the dashed curve.* We notice that for an initial time interval of about 0.08×10^{-12} sec, that is, for about one-fifth of the total decay time of the coherence of the motion, the motion of the dissolved and the unperturbed HF molecules is essentially identical. This is a completely different behavior from that

---

* A few explanatory remarks concerning the correlation function of the freely rotating molecules are useful. The correlation function $\langle \mathbf{u}(t) \cdot \mathbf{u}(0) \rangle$ of a single molecule in state $E(\omega)$, rotating freely with angular velocity ω, is of course periodic and undamped and simply equals $\cos \omega t$, since a molecule in a given state remains in that state if it does not interact with the environment. However, we are interested in an *ensemble* of many noninteracting molecules, and therefore we admit a large number of different states. For the (classical) ensemble, we thus average over all angular frequencies from $\omega = 0$ to $\omega \to \infty$ with a Boltzmann distribution of rotation frequencies $\omega \exp(-E/kT) = \omega \exp(-I\omega^2/2kT)$, where E is the classical rotational energy. Hence $\langle \mathbf{u}(t) \cdot \mathbf{u}(0) \rangle = \langle \cos \omega t \rangle = \int \cos(\omega t) \omega \exp(-I\omega^2/2kT) \, d\omega$, as shown by the dashed curve in Fig. 11.12. For more details, see H. Shimizu, *J. Chem. Phys.* **43**, 2453 (1967). A similar expression was used for CH_2Cl_2, a nonlinear molecule (see dotted curve in Fig. 11.4). However, in such cases this formulation is only a very rough approximation and should be replaced by a more meaningful expression as discussed in Ref. 12b (part 2).

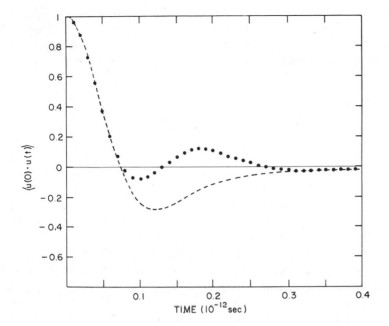

Fig. 11.12 Dipole correlation functions of hydrogen fluoride. (● ● ● ●): 0.025 molar solution of HF in cyclohexane at 35°C. The correlation function was calculated (using Eq. 11.15) from P. Datta and G. M. Barrow's data (corrected for induced absorption, see Fig. 11.13). (– – – –): classical ensemble of freely rotating HF molecules, computed with Eq. III.2 of Ref. 9. [W. G. Rothschild, *J. Chem. Phys.* **49**, 2250 (1968).]

exhibited by methylene chloride (see Fig. 11.4). There is further information contained in the correlation function. The minimum (after about 0.1×10^{-12} sec) indicates that those molecules which are in their most populous thermal state have turned through an angle between $\pi/2$ and π ($\langle \mathbf{u}(0) \cdot \mathbf{u}(t) \rangle$ is negative) within this time interval. The maximum in $\langle \mathbf{u}(0) \cdot \mathbf{u}(t) \rangle$, which occurs after about 0.2×10^{-12} sec following the onset of the observation, indicates that the molecules have regained their original orientation. However, we cannot (necessarily) say that the dissolved HF molecules make, on the average, an end-over-end rotation within 0.2×10^{-12} sec. The statistical picture of the correlation function tells us only that in this system the molecules execute large-angle orientational jumps and, after 0.2×10^{-12} sec, are on the average found to point into the same direction they started with.

From the formulas given by Gordon[20] we can readily compute the average rotational kinetic energy E_{kin} of the dissolved molecules from the second moment of the solution band. The rotational kinetic energy is a useful quantity to obtain since deviations from a value close to kT (after corrections for quantum-mechanical effects) might indicate, besides unsatisfactory intensity measurements, the presence of induced absorption.[19]

Using the expression[20]

$$E_{kin} = (4B)^{-1}\left[\int \hat{I}(v)v^2\, dv - (2B)^2\right],\qquad (11.16)$$

we have computed the average rotational kinetic energy of the dissolved HF and HCl molecules, respectively, in cyclohexane, using the gas-phase rotational constants.[21] The result is $E_{kin} = 228\ \text{cm}^{-1}$ for HF–cyclohexane, which is somewhat higher than kT at 35°C (214 cm^{-1}). For the system HCl–cyclohexane (see Fig. 11.10), the computed value of E_{kin} amounted to 292 cm^{-1}, a value which is far higher than that corresponding to kT.

As it turned out, the high values of E_{kin} computed from the observed bands were indeed a consequence of induced absorption; we shall briefly outline this, in particular for the HF–cyclohexane system.[19] The reader will remember that we pointed out above (Section B) that induced absorption is not considered in the relevant expressions and must thus be subtracted out. The mechanisms which induce absorption have been discussed in detail in Chapter 10 and we merely need repeat the results here. In the system hydrogen halide–cyclohexane the significant inducing mechanism is then as follows: The field of the quadrupole moment of HF induces a dipole moment in a neighboring cyclohexane molecule. This induced dipole moment is modulated with twice the rotational frequency of the HF molecule, hence the collision complex HF–cyclohexane absorbs radiation at $\Delta J = 2$ rotational transition frequencies of the HF molecule. (We know, from the correlation function, that HF executes large-angle rotational jumps in the medium. Hence we assume that J is still a meaningful quantum number.)

We have computed the quadrupole-induced dipole absorption in the HF–cyclohexane system, using Eq. 10.28 and literature values for the quadrupole moment of HF, the polarizability of cyclohexane, and for the parameters of the radial distribution function.[19] Figure 11.13 shows the original absorption band (from Fig. 11.10) together with the true dipolar absorption band, that is, the total absorption band minus the computed quadrupole-induced dipole absorption. From the corrected band we now obtain, with the help of Eq. 11.16, a value of $E_{kin} = 197\ \text{cm}^{-1}$, which is appreciably lower than $E_{kin} = 228\ \text{cm}^{-1}$ obtained from the band uncorrected for induced absorption (see above). These two values are seen to straddle $kT = 214\ \text{cm}^{-1}$, hence the improvement with respect to E_{kin} is not significant in the HF–cyclohexane system. However, in the HCl–cyclohexane system, where we com-

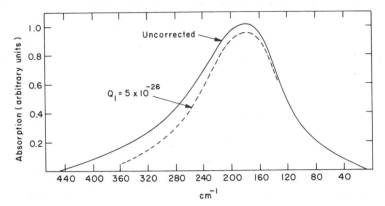

Fig. 11.13 Far-infrared absorption bands of a 0.025 molar solution of HF in cyclohexane at 35°C. (*Solid curve*): Apparent absorption band, taken from Fig. 11.10 (P. Datta and G. M. Barrow). (*Dashed curve*): True dipolar band, corrected for quadrupole-induced dipole absorption by using a molecular quadrupole moment of HF of 5×10^{-26} e.s.u. cm^2 and literature values for the parameters of the intermolecular potential (Lennard-Jones) and the polarizability of cyclohexane (see also Eqs. 10.25, 10.28 and 10.45). [W. G. Rothschild, *J. Chem. Phys.* **49**, 2250 (1968).]

puted an average rotational kinetic energy of the dissolved molecules of 292 cm^{-1} from the uncorrected band (see Fig. 11.10), consideration of the collision-induced absorption must be given if meaningful conclusions are to be drawn with respect to the molecular motion of the molecules.

The coincidence of the decay of the correlation of the motion of the dissolved HF molecules and the freely rotating molecules during the initial 0.08×10^{-12} sec (see Fig. 11.12) indicates that the dissolved molecules essentially rotate unhindered during this time interval. Since it can be shown that the lifetime of a cyclohexane solvent cage exceeds 0.08×10^{-12} sec,[19] it seems likely that the *translational* motion of HF within the solvent cage ("rattling") is a determining factor for this time interval of free rotation. For instance, assuming HF to be in the ideal gas state, we estimate that the most probable displacement of its center of mass is 0.2 Å during a time interval of 0.08×10^{-12} sec. This then would be a (very rough) measure of the "free volume" in the liquid cyclohexane, the volume which is on the average available to the trapped HF molecule—if we assume that the free rotation is strongly perturbed by a collision of the HF molecule with the walls of the surrounding solvent cage.

c. Rotational Motion of a Distortion in Liquid Carbon Tetrachloride

Carbon tetrachloride, a spherical-top molecule, should not possess a permanent dipole moment. However, measurements of the dielectric loss

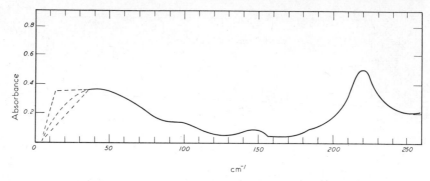

Fig. 11.14   The far-infrared spectrum of about 0.5-cm thickness of liquid carbon tetrachloride taken on a Beckman IR-11 spectrophotometer. The points at 5 cm^{-1} and below were taken from dielectric loss measurements and the dashed lines between 5 and 33 cm^{-1} represent possible extrapolations. Weak vibrational bands appear at 220, 145, and possibly at 97 cm^{-1} and a much stronger vibrational band is off the figure at 315 cm^{-1}. [H. S. Gabelnick and H. L. Strauss, *J. Chem. Phys.* **46**, 396 (1967).]

of liquid CCl_4 have indicated the presence of a small nonzero dipole moment, thereby indicating that the molecule is distorted in the liquid phase from its spherically symmetric symmetry.[22]

Gabelnick and Strauss[23] have investigated the far-infrared spectrum of CCl_4 in the liquid and in the vapor phase. Figure 11.14 displays their data on liquid CCl_4. After accounting for all possible low-frequency internal modes of the molecule, which are found at 97, 145, and 220 cm^{-1}, there remains an additional optical absorption, stretching from the long-wavelength cut-off of the spectrometer towards higher frequencies. This broad absorption band is absent in the vapor spectrum of the compound with comparable amounts of material in the absorption cell of the spectrometer. Because of the great similarity of this additional absorption to the pure rotation spectrum of a polar molecule in solution (see, for instance, Fig. 11.10), the authors have considered this band *as if* it originated from the motion of a CCl_4 molecule possessing a fixed dipole moment. Using Gordon's formulas,[5] they computed from the band contour, among others, the following quantities: (*1*) An effective dipole moment of 0.16 ± 0.02 D from the value of the correlation function[24] at $t = 0$; (*2*) an effective root-mean-square dipole moment[25] of 0.21 ± 0.02 D; (*3*) an effective moment of inertia of CCl_4 which is 24 ± 6 times smaller than the vapor-phase value (with the help of the second moment and assuming that the average value of the rotational kinetic energy is kT. Rotation about an axis parallel to the transition moment is not observed; see Section C.b).

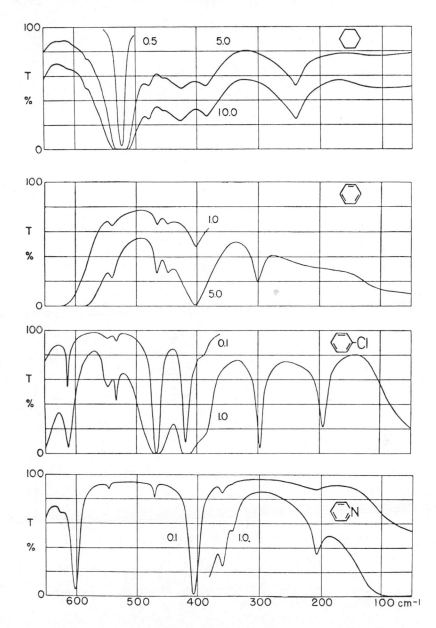

Fig. 11.15 Spectra of eight liquids between 650 and 50 cm^{-1}. The layer thickness of the compounds (in millimeters) is indicated in the figure. The liquids were contained in polyethylene cells. The sharper absorption peaks belong to intramolecular modes (fundamentals and difference bands). Dipole moments (in Debye) C_6H_5Cl, 1.6; CH_2Cl_2, 1.6; $CHCl_3$, 1.2; C_5H_5N,

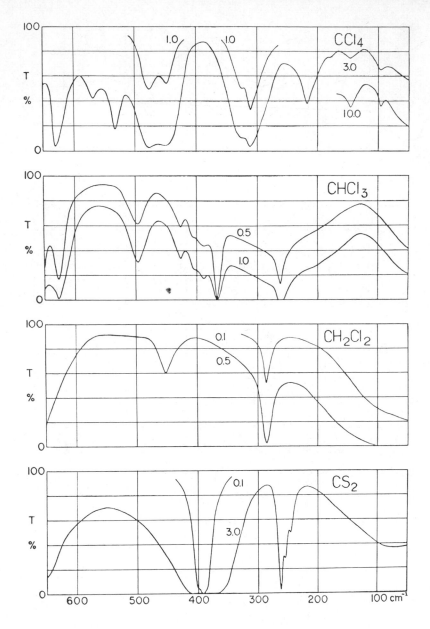

2.3. See, for instance, Landolt-Börnstein, *Zahlenwerte und Funktionen, Atom und Molekular-physik*, Vol. I, Part 3, Springer Verlag, Berlin, 1951. [*From* H. R. Wyss, R. D. Werder, and Hs. H. Günthard, *Spectrochim. Acta* **20**, 573 (1964), by permission of Pergamon Press.]

These results suggest that the additional absorption in liquid CCl_4 is caused by the translational-rotational motion of a distortion in the molecule.[23] A distortion from spherical symmetry which results in a dipole moment of about 0.1 D is obtained by, for instance, bending one carbon-chlorine bond through 6° from its equilibrium position. The energy required to effect this distortion amounts to about 300 cal/mole ($kT \sim 610$ cal/mole), by no means a high value.[22]

d. Molecular Complexes and Clusters in Liquid Phases

The extension of the spectral range to long wavelengths has recently brought to light the fact that many liquids show infrared absorption beyond that allowed by the available vibrational and rotational degrees of freedom of the single molecule. Figure 11.3 gives a striking example, where a broad continuous background at a wavelength as short as 33 μ, which becomes more intense towards longer wavelengths, is apparent. This phenomenon, with the possible exception of individual cases,[23] is not yet fully understood. It offers great promise for a study of intermolecular forces in condensed phases; it is for this reason that we present some of the relevant observations here. The conclusions drawn from them should be regarded as still tentative.

Figure 11.15 shows the spectra of cyclohexane, benzene, chlorobenzene, pyridine, carbon tetrachloride, chloroform, methylene chloride, and carbon disulfide in their liquid phases, scanned between 650 and 50 cm^{-1} with a Perkin-Elmer spectrometer, Model 301.[26] In these examples there is considerable continuous absorption beyond 100 cm^{-1} towards lower frequencies; in methylene chloride and pyridine at wavelengths shorter than 33 μ. A characteristic of this absorption is its extraordinary broadness. Furthermore, it seems that the intensity of the bands is greater the larger the permanent dipole moment of the molecule. In order to observe the effect at a strength comparable to that of polar liquids, rather thick layers of liquid have to be employed for molecules with no permanent dipole moment.

Delorme,[27] Decamps et al.,[28] and Jakobsen and Brasch[29] have proposed that such far-infrared absorption is due to intermolecular modes. Jakobsen and Brasch have specifically ascribed this absorption to vibrational modes between complexes or clusters formed by molecules possessing strong permanent dipole moments. The broadness of the absorption is assigned by these authors to be, at least in part, a consequence of a coupling of the intermolecular motion to intramolecular low-lying vibrational modes, for instance, torsional modes. Some of Jakobsen and Brasch's data are collected in Table 11.2, which gives the respective compound (liquid phase), the value of the dipole moment, and the approximate band center with an indication of its width. The bands are assigned to what Jakobsen and Brasch have termed intermolecular stretching vibrations involving the bond of the respective "dipole–dipole complex."

TABLE 11.2*

FAR-INFRARED ABSORPTION BANDS IN VARIOUS ORGANIC
LIQUIDS

Compound	Dipole Moment[†]	cm^{-1}[‡]
Benzonitrile	3.9	54 vbr
o-Dinitrobenzene	6.0	86 br
p-Dinitrobenzene[§]	0	None observed
Acetonitrile[‖]	3.5	87 vbr
Nitromethane	3.2	60 vbr
Acetone	2.8	< 50 vbr
Bromobutane	2.0	None observed

* *After* R. J. Jakobsen and J. W. Brasch, *J. Am. Chem. Soc.*
86, 3571 (1964). By permission of American Chemical Society.

[†] In Debye (10^{-16} e.s.u. cm).

[‡] vbr = very broad, br = broad. The frequency is that at
maximum absorption.

[§] Solid.

[‖] The vapor has an absorption at 65 cm^{-1}.

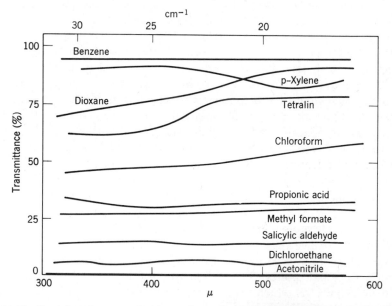

Fig. 11.16   Far-infrared spectra of various polar and nonpolar liquids, taken in a polyethylene
cell of 0.5-mm thickness. [The spectrometers are described in *Rev. univ. des Mines*, Series 9T XV,
No. 5 (1959), A. Hadni and E. Decamps.] [*From* E. Decamps, A. Hadni, and J. M. Munier,
Spectrochim. Acta **20**, 373 (1964), by permission of Pergamon Press.]

It would be desirable to make spectral temperature and dilution studies to confirm these assignments. Furthermore, it does not seem necessary to invoke coupling to intramolecular modes in order to account for the broadness of the absorption of the dipole–dipole complex; in all likelihood, the potential function governing the intermolecular vibration is quite anharmonic. Also, it should be remembered, if two molecules (nonlinear) form a stable complex, six of their translational-rotational degrees of freedom change into vibrational degrees of freedom. In other words, we should expect a few closely spaced or overlapping bands. We might mention that "stable complex" means a complex with a lifetime about an order of magnitude longer than the time of the intermolecular vibrations.

In Fig. 11.16 we show some data by Decamps et al.[28] on the absorption of liquids of varying degrees of polarity at very long wavelengths. As previously observed, we notice that the intensity of the absorption increases with an increasing permanent dipole moment, all other factors considered equal. The extraordinary broadness of, for instance, the absorption of acetonitrile, which stretches out to (at least) $16 \, \text{cm}^{-1}$ without apparent decrease of its

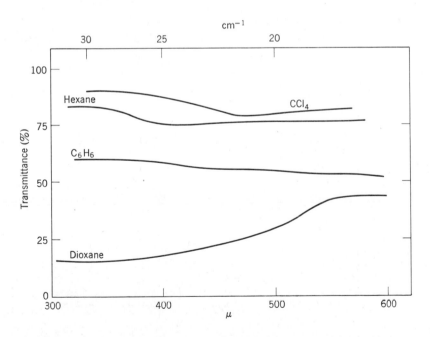

Fig. 11.17   Far-infrared spectra of various nonpolar solvents, taken in a polyethylene cell of 3-mm thickness. [*From* E. Decamps, A. Hadni, and J. M. Munier, *Spectrochim. Acta* **20**, 373 (1964), by permission of Pergamon Press.]

intensity, indicates that the absorption in this frequency range is probably not due to a vibration involving an intermolecular bond of a stable molecular acetonitrile complex (uncertainty broadening). In Fig. 11.17 we show the absorption of some nonpolar liquids between 33 and 16 cm^{-1}, determined by Decamps *et al.* Although the very long wavelength absorption of these nonpolar compounds is relatively weaker than that of polar compounds, it is by no means negligible. In fact, dioxane, a molecule without permanent dipole moment, shows relatively strong absorption (see Figs. 11.16 and 11.17).

We shall return to this peculiar behavior of dioxane later. First, we discuss a publication by Kroon and van der Elsken,[30] who ascribe the far-infrared absorption of dilute solutions of polar molecules in nonpolar solvents to the rotation-translational motion of the molecules, that is, to the collision-induced absorption. As we discussed at length throughout Chapter 10 (see Sections B.b and C.c), the strength of such absorption is proportional to the square of the multipole (dipole, quadrupole, etc.) of the polarizing molecule and to the square of the polarizability of the polarized molecule. We show some of Kroon and van der Elsken's data in Figs. 11.18 to 11.20.

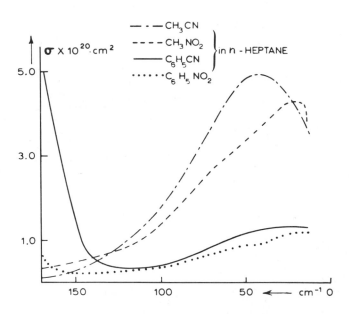

Fig. 11.18   The absorption cross section in the far-infrared region of nitrobenzene, benzonitrile, nitromethane, and acetonitrile in *n*-heptane. [S. G. Kroon and J. van der Elsken, *Chem. Phys. Letters* **1**, 285 (1967), by permission of North-Holland Publishing Company, Amsterdam.]

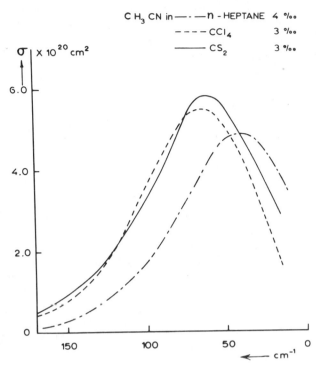

Fig. 11.19 The absorption cross section of acetonitrile in *n*-heptane, carbon tetrachloride, and carbon disulfide. [S. G. Kroon and J. van der Elsken, *Chem. Phys. Letters* **1**, 285 (1967), by permission of North-Holland Publishing Company, Amsterdam.]

Figure 11.18 shows that substitution of a methyl group by phenyl in acetonitrile and nitromethane decreases the absorption of these molecules in their heptane solutions considerably. The authors attribute this to the larger moment of inertia in the phenyl-substituted compounds (see Eq. 10.36, which shows the dependence of the absorption coefficient on the moments of inertia). However, it should be kept in mind that an effect of geometric size may enter. The phenyl group is much larger than the methyl group and, as a consequence, the intermolecular solute-solvent potential is not necessarily the same for the methyl and phenyl derivatives.

Figure 11.19 shows that in the more polarizable solvent the maximum absorption has shifted to shorter wavelengths. (The polarizabilities of *n*-heptane, CCl_4, and CS_2 are in the ratio $1:0.77:0.64$.) The significance of this observation does not seem apparent, since the degree of polarizability influences only the intensity. There is apparently no information available

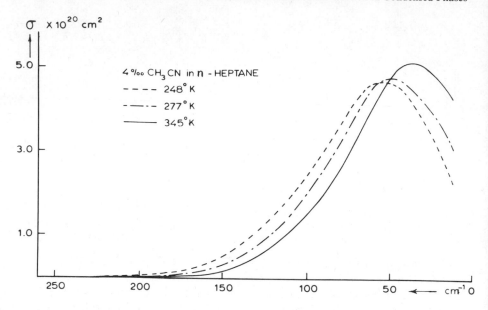

Fig. 11.20   The effect of the temperature on the absorption cross section of acetonitrile in *n*-heptane. [S. G. Kroon and J. van der Elsken, *Chem. Phys. Letters* **1**, 285 (1967), by permission of North-Holland Publishing Company, Amsterdam.]

that would permit a prediction of the frequency of the maximum absorption of the induced far-infrared bands in such complicated systems (see Chapter 10, Section C.a).

Figure 11.20 shows the temperature dependence of the far-infrared absorption band of dilute acetonitrile–*n*-heptane solutions. We notice a shift to lower frequencies of the band and an increase of its peak intensity with increasing temperature. Most important, we see that the total intensity remains essentially unaffected by the temperature variations. This means that stable dimers (or higher complexes) between solute molecules do not exist to any significant degree in these solutions. We remark that the observed shift of the band to lower frequencies with increasing temperature is in the opposite direction to that of the induced pure translational absorption of rare-gas mixtures (see Fig. 10.7).

We return now to the relatively strong absorption of dioxane, displayed in Figs. 11.16 and 11.17. The explanation which comes to mind is simply that although dioxane is a nonpolar compound because of its symmetry, the charge distribution in the molecule gives each "half" considerably polar character, which becomes effective if the molecule aligns itself with its oxygen–

oxygen axis pointing in the direction of a neighboring molecule ("local dipole moment"). This is shown schematically in Fig. 11.21.

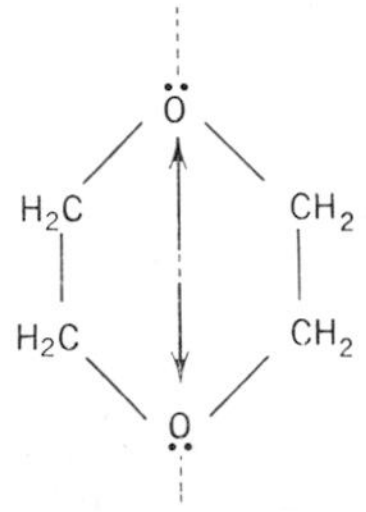

Fig. 11.21   Schematic representation of the structure of dioxane. The dots on the oxygen atoms indicate the lone-pair electrons. The arrow indicates the "local dipole."

Wagner[31] has recently published the far-infrared spectrum of liquid bromine, and Chantry et al.[32] have studied the long-wavelength absorption of various nonpolar liquids using interferometric techniques. This group of authors attribute the long-wavelength absorption to a vibrational motion of the disordered lattice in the liquid, the so-called "liquid lattice absorption." They base their arguments on the coincidence of the lattice absorption of the crystalline material and the maximum of the broad liquid lattice band. We show some of Chantry et al.'s data on carbon disulfide and carbon tetrachloride in Fig. 11.22.

In conclusion, we remark that the low-frequency absorption in solution and liquids exhibits a variety of characteristics which, apparently, are not satisfactorily explainable in terms of one and the same parameter of the system. It is conceivable that most or some of the observed phenomena occur simultaneously, their individual contributions depending on the relative magnitude of molecular parameters such as the dipole-, quadrupole-, and higher moments, the size of the molecules, the strength of intermolecular forces, the distance of nearest approach, the number and duration of collisions, and so forth. It seems, however, fairly well established that absorption at the very long wavelengths has the character of a collision-induced translational-rotational absorption of the molecule in a solvent cage ("rattling").

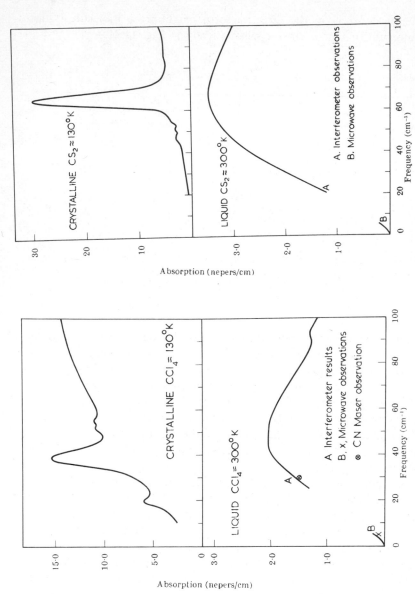

Fig. 11.22 Far-infrared spectra of CCl_4 and CS_2 in their crystalline and liquid phases. The measurements were carried out with inter-ferometric and maser techniques. The points at the lowest frequencies were obtained with the help of dielectric measurements (D. H. Whiffen, *Trans. Faraday Soc.* **46**, 124 (1950). S. K. Gark, H. Kilp, and C. P. Smyth, *J. Chem. Phys.* **43**, 2341 (1965). Compare the spectrum of liquid CCl_4 here with that shown in Fig. 11.14. [G. W. Chantry, H. A. Gebbie, B. Lassier, and G. Wyllie, *Nature* **214**, 163 (1967)].]

REFERENCES

1. We follow here the description given by R. G. Gordon, *J. Chem. Phys.* **39**, 2788 (1963).
2. A. D. Buckingham, *Trans. Faraday Soc.* **56**, 753 (1960).
3. R. H. Mann and W. DeW. Horrocks, Jr., *J. Chem. Phys.* **45**, 1278 (1966).
4. W. G. Rothschild, *J. Chem. Phys.* **42**, 694 (1965).
5. R. G. Gordon, *J. Chem. Phys.* **43**, 1307 (1965); *Advan. Magnetic Resonance* **3**, 1 (1968). H. Shimizu, *J. Chem. Phys.* **43**, 2453 (1965) and references cited therein.
6. U. Fano, *Rev. Mod. Phys.* **29**, 74 (1957).
7. W. G. Rothschild, *Macromolecules* **1**, 43 (1968).
8. J. R. Nielsen and N. E. Ward, *J. Chem. Phys.* **10**, 81 (1942).
9. R. G. Gordon, *J. Chem. Phys.* **44**, 1830 (1966); see especially the Introduction and Eq. III.2.
10. A. Abragam, *The Principles of Nuclear Magnetism*, The Clarendon Press, Oxford, 1961, p. 271.
11. S. Krimm, *Fortschr. Hochpolym. Forschg.* **2**, 51 (1960).
12a. C. E. Favelukes, A. A. Clifford, and B. Crawford, Jr., *J. Phys. Chem.* **72**, 962 (1968). See the references cited therein for a description of the attenuated total reflectance technique and its advantages for band shape determinations.
12b. W. G. Rothschild, *J. Chem. Phys.* **51**, 5187 (1969); **52**, 6453 (1970).
13. J. Corset, P. V. Huong, and J. Lascombe, *Compt. Rend.* **262C**, 959 (1966).
14. P. Datta and G. M. Barrow, *J. Am. Chem. Soc.* **87**, 3053 (1965).
15. J. Lascombe, P. V. Huong, and M.-L. Josien, *Bull. Soc. Chim. Fr.* **1959**, 1175.
16. M. O. Bulanin, N. D. Orlova, and D. N. Shchepkin, *Opt. Spectry.* **19**, 406 (1965); M. O. Bulanin and N. D. Orlova, *ibid.* **15**, 112 (1963).
17. H. G. Silver and J. L. Wood, *Spectrochim. Acta* **19**, 787 (1963).
18. P. Datta and G. M. Barrow, *J. Chem. Phys.* **43**, 2137 (1965); see also *J. Phys. Chem.* **72**, 2259 (1968).
19. W. G. Rothschild, *J. Chem. Phys.* **49**, 2250 (1968).
20. R. G. Gordon, *J. Chem. Phys.* **41**, 1819 (1964); see Eq. 8b (note that the relation between the moment of inertia and the rotational constant is $I = \hbar/4\pi cB$) and Eq. 24 [the quantum-mechanical correction term amounts to about $0.06M(2)$ and has been applied to the final result].
21. W. G. Rothschild, *J. Opt. Soc. Am.* **54**, 20 (1964).
22. D. H. Whiffen, *Trans. Faraday Soc.* **46**, 124 (1950). For a definition and description of dielectric loss, see H. Fröhlich, *Theory of Dielectrics. Dielectric Constant and Dielectric Loss*, Clarendon Press, Oxford, 1949.
23. H. S. Gabelnick and H. L. Strauss, *J. Chem. Phys.* **46**, 396 (1967).
24. See Ref. 5, Eq. III.2.
25. See Ref. 5, Eq. III.3.
26. H. R. Wyss, R. D. Werder, and H. Günthard, *Spectrochim. Acta* **20**, 573 (1964).
27. P. Delorme, *J. Chim. Phys.* **41**, 1439 (1964).
28. E. Decamps, A. Hadni, and J. M. Munier, *Spectrochim. Acta* **20**, 373 (1964).
29. R. J. Jakobsen and J. W. Brasch, *J. Am. Chem. Soc.* **86**, 3571 (1964).
30. S. G. Kroon and J. van der Elsken, *Chem. Phys. Letters* **1**, 285 (1967).
31. V. Wagner, *Phys. Letters* **22**, 58 (1966).
32. G. W. Chantry, H. A. Gebbie, B. Lassier, and G. Wyllie, *Nature* **214**, 163 (1967).

12 Far-Infrared Spectra of Diatomic Cubic Crystals

A. INTRODUCTION

The interaction of long-wavelength electromagnetic radiation with crystals has many of the familiar aspects of molecular spectra. For instance, we shall find that the concept of the harmonic oscillator is also useful and widely applicable, and we shall encounter essentially the same strict selection rules and their relaxation by anharmonic forces and transition moments. However, it is not surprising that we shall encounter some new aspects which are absent in the spectra of vapors and liquids because the long-range periodicity in the atomic or molecular arrangement and the oriented interaction between neighboring particles is lacking in the noncrystalline phases.

Some important phenomena of crystal spectra, for instance the reflection properties of ionic crystals, can be treated to a satisfactory level of approximation without regard to the periodicity of the lattice. This is possible because in such formulation a large number of particles in the crystal is considered rather than only the few which form the basic structural building block, the unit cell. In the more useful and adequate description that takes proper account of the periodicity of the lattice, it is convenient to use the concept of the phonon (quantum of energy of an elastic wave in the crystal). The reason for this is that the excitation which arises from the interaction between the electromagnetic radiation and the crystal vibration is no longer localized within one or a few molecules as in molecular spectra. If we consider that a solid of, say, N diatomic molecules possesses $6N$ degrees of freedom, we see that a macroscopic crystal exhibits a very large number of fundamental modes and therefore has the potential of interacting with a wide range of frequencies of the incident electromagnetic radiation.

In this chapter we shall be mainly concerned with the interaction of electromagnetic radiation with diatomic cubic crystals of the NaCl-, CsCl-, and ZnS-type structures; crystals such as diamond, which have no first-order

412

transition moment, are therefore not considered, the more so since their second-order spectra do not fall into the far-infrared region of the spectrum. The far-infrared spectra of polyatomic crystals, that is, crystals in which more than one atom may occupy a lattice site, will be discussed in Chapter 13.

The information in this chapter, which comes mainly from transmission and reflection measurements, will tell us about such parameters as the effective ionic charge, atomic polarizability, attractive and repulsive potential, crystal anharmonicity effects, dielectric constant, and absorption coefficients.

B. Classical Description of Crystal Spectra

We shall first discuss a series of publications which treat the observed material constants without consideration of the periodicity of the crystal lattices. It will be beneficial to begin with a short discussion of the macroscopic quantities which are either measured directly or calculated from other input data. As we have already indicated, the experimental data are generally obtained through reflection and transmission measurements; these quantities may be characterized by the index of refraction n and the extinction coefficient, κ. The index of refraction and the extinction coefficient are combined to yield the complex dielectric constant.

a. Dielectric Constant, Refractive Index, and Extinction Coefficient

If we eliminate the magnetic field vector from Maxwell's equation, we can deduce the wave equation of the electric field vector $\mathbf{E}$,*

$$\nabla^2 \mathbf{E} = \frac{\varepsilon}{c^2} \frac{\partial^2 \mathbf{E}}{\partial t^2} + \frac{4\pi\sigma}{c^2} \frac{\partial \mathbf{E}}{\partial t}, \tag{12.1}$$

where ε is the dielectric constant, σ the conductivity, and c the velocity of light. A solution of Eq. 12.1 can be written as

$$\mathbf{E} = \mathbf{E}_0 \exp\{i(\mathbf{K} \cdot \mathbf{r} - \omega t)\}, \tag{12.2}$$

where ω is the angular frequency of the radiation, $\mathbf{K}$ is the wave vector of magnitude $|\mathbf{K}| = 2\pi/\lambda$, where λ is the wavelength, and $\mathbf{r}$ is a vector in the x, y, z-coordinate space. Inserting Eq. 12.2 into Eq. 12.1, we obtain for the square of the wave vector

$$\mathbf{K}^2 = \frac{\varepsilon}{c^2}\omega^2 + i\frac{4\pi\sigma\omega}{c^2}. \tag{12.3}$$

* We use the Maxwell equations as given in Ref. 13, p. 219. Since magnetic effects are not considered here, the magnetic susceptibility is set equal to unity.

By using the tilde sign to designate a complex quantity, we may rewrite Eq. 12.3 in the form

$$\tilde{K} = \frac{\omega}{c}\left\{\varepsilon + i\frac{4\pi\sigma}{\omega}\right\}^{1/2}.\tag{12.4}$$

In empty space $\sigma = 0$, $\varepsilon = 1$, and we obtain for the complex wave vector

$$\tilde{K} = K = \frac{\omega}{c}.\tag{12.5}$$

Therefore, we see that in the crystal the phase velocity of the wave is modified by the factor

$$\left\{\varepsilon + i\frac{4\pi\sigma}{\omega}\right\}^{1/2} = \tilde{n} = \tilde{\varepsilon}^{1/2},\tag{12.6}$$

where $\tilde{n}$ is the complex refractive index and $\tilde{\varepsilon}$ is the complex dielectric constant. We write

$$\tilde{n} = n + i\kappa,\tag{12.7}$$

where n is the real refractive index and κ is the extinction coefficient. In the same manner we represent $\tilde{\varepsilon}$ by a real and imaginary part,

$$\tilde{\varepsilon} = \varepsilon' + i\varepsilon''.\tag{12.8}$$

Combining Eqs. 12.6, 12.7, and 12.8, we obtain the following expressions for ε' and ε'':

$$\begin{aligned}\varepsilon' &= n^2 - \kappa^2,\\ \varepsilon'' &= 2n\kappa.\end{aligned}\tag{12.9}$$

From Eqs. 12.7, 12.6, and 12.4 we obtain for the complex wave vector

$$\tilde{K} = \frac{n\omega}{c} + i\frac{\kappa\omega}{c}.\tag{12.10}$$

Inserting Eq. 12.10 into Eq. 12.2 and considering only the z-direction, it follows that

$$\mathbf{E} = \mathbf{E}_0 \exp\left[i\omega\left(\frac{nz}{c} - t\right)\right]\exp\left(-\frac{\kappa\omega z}{c}\right).\tag{12.11}$$

We see that the velocity of the wave in the medium is reduced by the fraction $1/n$, and that the amplitude of the wave is damped by the fraction $\exp(-2\pi\kappa)$ per vacuum wavelength.[†] The damping is, of course, a consequence of the absorption of the electromagnetic radiation in the medium.

In the following we give the expressions, at normal incidence, for the power reflection coefficient R (the ratio of reflected to incident intensity at "infinite" thickness)[‡] and the transmissivity D^*. The expression for R is given by

$$R = \frac{(n-1)^2 + \kappa^2}{(n+1)^2 + \kappa^2}. \tag{12.12}$$

For instance, if n is about 3 and if κ is small, we can expect values of R of the order of ~ 0.25.

The formula for the transmissivity D^* is rather complicated, since it depends on R and on the thickness of the sample (we remember that the energy conservation requires the *sum* of reflection, transmission, and absorption to be constant):

$$D^* = \frac{(1-R)^2 D + 4RD \sin^2\psi}{(1-RD)^2 + 4RD \sin^2(\alpha + \psi)} \tag{12.13}$$

where

$$D = e^{-4\pi\kappa d/\lambda}$$

$$\alpha = 2\pi n d/\lambda$$

$$\tan\psi = 2\kappa/(n^2 + \kappa^2 - 1)$$

with λ = wavelength and d = thickness.

The quantity

$$\frac{4\pi\kappa}{\lambda} = 4\pi\kappa v \quad \left(v = \frac{\omega}{2\pi} \right) \quad \text{(in cm}^{-1}\text{)} \tag{12.14}$$

is called the absorption coefficient. Inspection of Eq. 12.13 shows that D^* can be approximated by simpler expressions under certain conditions; for instance, if R is small, then $D^* \sim D$. Another limiting case is obtained if

---

[†] The term n/c in the exponent of the first exponential (that is, the z-component of the wave vector) gives the reciprocal phase velocity in the z-direction. The relative decrease in the amplitude of the wave vector per unit length is $\exp(-\kappa\omega/c) = \exp(-2\pi\kappa/\lambda) = \exp(-2\pi\kappa)$ per wavelength.

[‡] The power reflection coefficient R corresponds to a real reflection coefficient, namely to the ratio of the squares of the amplitudes of the reflected and incident waves. The "amplitude reflection coefficient" ρ is the ratio of the (complex) amplitudes, $\rho = (\tilde{n} - 1)/(\tilde{n} + 1)$. Therefore, $R = |\rho|^2$ or $\rho = \sqrt{R}\, e^{i\psi}$.

d/λ is very small; in this case, $D^* \sim 1$, independent of the magnitude of R.

We note here that the use of Fourier transform spectroscopy for a determination of n and κ is briefly discussed in Chapter 4, Section J.

b. Dispersion Formulas

In the following we give the necessary formulas to relate the experimentally obtainable macroscopic quantities, for instance n and κ, with molecular parameters of the solid. These relations, which are a function of the frequency (dispersion formulas), are thoroughly discussed in the literature[1a-c] and therefore we need merely give an outline of their derivation, underlying assumptions, and the significance of symbols.

In a phenomenological manner it is assumed that the solid consists of "electric dipole" harmonic oscillators, each atom possessing an effective charge e^* (positive, or negative) and a mass m_+ and m_-, respectively. Under a polarizing field the masses m_+ and m_- are displaced relative to each other about their equilibrium positions—against a restoring force and a friction force—leading to an oscillating induced dipole moment; in other words, to a macroscopic polarization of the medium. The equations of motion,

$$\bar{m}\ddot{\mathbf{u}} + \bar{m}\gamma\dot{\mathbf{u}} + \bar{m}\omega_0{}^2\mathbf{u} = e^*\mathbf{E}_{eff}, \tag{12.15}$$

where $\mathbf{u}$ is the relative displacement and $\bar{m}$ is the reduced mass of m_+ and m_-, $\bar{m}\omega_0{}^2$ is the restoring force constant, $\bar{m}\gamma$ is the damping constant (assumed here to be frequency-independent), and $\mathbf{E}_{eff}$ is the effective polarizing field at the particle, are readily solved. The atomic polarization of one oscillator is given by $p = e^*u$. For N equal oscillators in the volume V, the polarization is accordingly $P = (N/V)e^*u$. Assuming that $\mathbf{E}_{eff}$ equals the external field E, we obtain, with the help of the well-known relation $\varepsilon E = E + 4\pi P$,

$$\tilde{\varepsilon}(\omega) = 1 + \frac{4\pi}{V} \frac{Ne^{*2}}{\bar{m}(\omega_0{}^2 - \omega^2 - i\gamma\omega)}. \tag{12.16}$$

We assume N equal oscillators per volume V; in case there are j different types of oscillators, the second term of Eq. 12.16 is to be summed over all the different oscillators with restoring force $\bar{m}_j\omega_{0j}^2$, damping constant $\bar{m}_j\gamma_j$, and effective charge e_j^*. The real and imaginary parts of Eq. 12.16 are

$$\varepsilon'(\omega) = n^2 - \kappa^2 = \varepsilon_\infty + \sum_j \frac{A_j(1 - (\omega/\omega_{0j})^2)}{[1 - (\omega/\omega_{0j})^2]^2 + (\gamma_j/\omega_{0j})^2(\omega/\omega_{0j})^2},$$

$$\varepsilon''(\omega) = 2n\kappa = \sum_j \frac{A_j(\gamma_j/\omega_{0j})(\omega/\omega_{0j})}{[1 - (\omega/\omega_{0j})^2]^2 + (\gamma_j/\omega_{0j})^2(\omega/\omega_{0j})^2}, \tag{12.17}$$

with

$$A_j = \frac{4\pi}{V} \frac{N_j e_j^{*2}}{m_j \omega_{0j}^2} f_j; \qquad (12.18)$$

f_j is the oscillator strength of oscillator j. In Eqs. 12.17 we have separated out in the usual way the electronic (high-frequency) polarization, $\varepsilon_\infty - 1$, which is a constant in the infrared. Equations 12.17 are frequently found written in terms of the reduced quantities $\Omega_j = \omega/\omega_j$, $\Gamma_j = \gamma_j/\omega_j$. Introducing the static dielectric constant ε_0 [that is, the dielectric constant at zero frequency, $\varepsilon'(0)$] into Eqs. 12.17, we find for a cubic crystal ($j = 1$) $\varepsilon'(0) = \varepsilon_0 = \varepsilon_\infty + A$, hence $A = \varepsilon_0 - \varepsilon_\infty$. This yields the set of equations

$$\tilde{\varepsilon}(\omega) = \varepsilon_\infty + \frac{\varepsilon_0 - \varepsilon_\infty}{1 - (\omega^2/\omega_0^2) - i(\gamma/\omega_0)(\omega/\omega_0)}, \qquad (12.19)$$

$$\varepsilon'(\omega) = n^2 - \kappa^2 = \varepsilon_\infty + \frac{(\varepsilon_0 - \varepsilon_\infty)[1 - (\omega/\omega_0)^2]}{[1 - (\omega/\omega_0)^2]^2 + (\gamma^2/\omega_0^2)(\omega/\omega_0)^2}, \qquad (12.20)$$

$$\varepsilon''(\omega) = 2n\kappa = \frac{(\varepsilon_0 - \varepsilon_\infty)(\gamma/\omega_0)(\omega/\omega_0)}{[1 - (\omega/\omega_0)^2]^2 + (\gamma^2/\omega_0^2)(\omega/\omega_0)^2}. \qquad (12.21)$$

As we shall see, these equations are very frequently used to put a smooth curve through the data points by adjusting ω_0, γ, and f. We notice that when $\omega \sim \omega_0$, κ and n become rather large. Inspection of Eq. 12.12 shows that, as a consequence, the power reflection coefficient R approaches unity. This leads to the phenomenon of the reststrahlen band.

The assumption made above, namely that the field at the particle equals the external polarizing field, is rather crude. As discussed by Szigeti[1b] and Burstein et al.,[1a,1c] the effective field for displacements transverse to the direction of propagation of the radiation field is given by the sum of the external field, the Lorentz field ($\frac{4}{3}\pi \mathbf{P}_T$, where $\mathbf{P}_T$ is the atomic and electronic transverse polarization), and a transverse geometrical depolarization term which, however, vanishes for a slab of large lateral dimensions relative to the wavelength of the radiation.[†] The field $\mathbf{E}_{eff}$ in Eq. 12.15 is thus given by $\mathbf{E} + \frac{4}{3}\pi \mathbf{P}_T$, where $\mathbf{E}$ is the external field. Since $\mathbf{P}_T$ is linear with $\mathbf{u}_T$ (see above), the term due to the Lorentz field effectively introduces an additional harmonic restoring term with a negative sign into Eq. 12.15. Consequently the transverse resonance frequency is no longer ω_0 but less. Remembering the Clausius-Mosotti relation and that the transverse atomic and electronic

† See Ref. 10, Chapter 12.

polarization is $\mathbf{P}_T = (N/V)\{e^*\mathbf{u}_T + \alpha(\mathbf{E} + \frac{4}{3}\pi\mathbf{P}_T)\}$, where α is the electronic polarization per molecule, the transverse resonance frequency is obtained as

$$\omega_T{}^2 = \omega_0{}^2 - \frac{4\pi N e_s^{*2}}{9\overline{m}V}(\varepsilon_\infty + 2)$$

and the dispersion formula reads now

$$\tilde{\varepsilon}_T(\omega) = \varepsilon_\infty + \frac{4\pi N e_s^{*2}(\varepsilon_\infty + 2)^2}{9\overline{m}V(\omega_T{}^2 - \omega^2 - i\gamma_T\omega)},$$

$$\varepsilon_T'(\omega) = n^2 - \kappa^2 = \varepsilon_\infty + \frac{4\pi N e_s^{*2}(\varepsilon_\infty + 2)^2(\omega_T{}^2 - \omega^2)}{9\overline{m}V((\omega_T{}^2 - \omega^2)^2 + \gamma_T{}^2\omega^2)}, \qquad (12.22)$$

$$\varepsilon_T''(\omega) = 2n\kappa = \frac{4\pi N e_s^{*2}(\varepsilon_\infty + 2)^2\gamma_T\omega}{9\overline{m}V((\omega_T{}^2 - \omega^2)^2 + \gamma_T{}^2\omega^2)}.$$

In Eqs. 12.22 we have replaced e^* with e_s^*, the so-called Szigeti charge; we return to this point later.

The longitudinal resonance frequency, that is, the resonance frequency for longitudinal displacements of the atoms (in a direction normal to the wave front of the field $\mathbf{E}$), is readily obtained from Eq. 12.15. Remembering that the longitudinal geometric depolarization term is $-4\pi\mathbf{P}_L$,[†] the longitudinal effective field is $\mathbf{E}_{eff}^L = \mathbf{E}^L - 4\pi\mathbf{P}_L + \frac{4}{3}\pi\mathbf{P}_L$ with $\mathbf{P}_L = (N/V)e^*\mathbf{u}_L + (N/V)\alpha(\mathbf{E}^L - \frac{8}{3}\pi\mathbf{P}_L)$. The resulting longitudinal resonance frequency is $\omega_L{}^2 = \omega_T{}^2 + (4\pi N e_s^{*2}/9\overline{m}V\varepsilon_\infty)(\varepsilon_\infty + 2)^2$.

Inspection of Eqs. 12.22 shows that $\varepsilon_T'(\omega) - \varepsilon_\infty$ becomes negative for $\omega > \omega_T$. In this case $\tilde{K}$ and $\tilde{n}$ are purely imaginary (see Eqs. 12.10 and 12.6) and wavelike solutions of Eq. 12.1 do not exist within the material: The radiation is strongly reflected at the front surface. The angular frequency at the upper limit of this frequency gap,[‡] that is, the frequency for which $\varepsilon_T'(\omega) = 0$, is seen to be near ω_L.

To summarize: Observable transmission of radiation through a thick slab occurs for $\omega \gg \omega_L$, $\omega \ll \omega_T$; strong reflection takes place at frequencies $\omega_T < \omega < \omega_L$. The maximum of $\varepsilon_T''(\omega) = 2n\kappa$ occurs near ω_T, whereas $\varepsilon_T'(\omega) = n^2 - \kappa^2$ vanishes near ω_L.

Solving Eqs. 12.22 for $\varepsilon_0 = \varepsilon_T'(0)$, we obtain $\varepsilon_0 = \varepsilon_\infty + (4\pi N e_s^{*2}/9V\overline{m}) \times (\varepsilon_\infty + 2)^2/\omega_T{}^2$ which, together with $\omega_L{}^2 = \omega_T{}^2 + (4\pi N e_s^{*2}/9V\overline{m}\varepsilon_\infty)(\varepsilon_\infty + 2)^2$, yields the Lyddane-Sachs-Teller relation

$$\frac{\varepsilon_0}{\varepsilon_\infty} = \frac{\varepsilon'(\omega = 0)}{\varepsilon'(\omega \to \infty)} = \frac{\omega_L{}^2}{\omega_T{}^2}. \qquad (12.23)$$

[†] See Ref. 10, Chapter 12.
[‡] See Ref. 10, Fig. 20, p. 153.

The *LST* relation is frequently used to compute ω_L from the data.

In the following we drop the subscript T from $\varepsilon_T(\omega)$ since we are mainly concerned with the transversal displacements. We shall, however, find that longitudinal oscillations can couple to electromagnetic radiation in cases of oblique incidence and thin slabs (see Section C.b).

c. Kramers-Kronig Relations

In what follows we give some of the Kramers-Kronig relations frequently used in the evaluation of the data.

The real and imaginary parts of the dielectric constant, $\varepsilon'(v)$ and $\varepsilon''(v)$ ($v = \omega/2\pi$), are related to one another by

$$\varepsilon'(v) = 1 + \frac{2}{\pi} \int_0^\infty \frac{\varepsilon''(v')v'}{v'^2 - v^2} \, dv', \tag{12.24}$$

$$\varepsilon''(v) = \frac{2v}{\pi} \int_0^\infty \frac{\varepsilon'(v')}{v'^2 - v^2} \, dv'. \tag{12.25}$$

In analogy to the considerations which led to Eqs. 12.17 from Eq. 12.16, we divide the integral of Eq. 12.24 into two parts. One part represents the contribution of the electronic absorption in the high-frequency region, $v > v_i$, the other part gives the contributions due to the lattice vibrations in the infrared, $v \leqslant v_i$. We thus write

$$\varepsilon'(v) = 1 + \frac{2}{\pi} \int_0^{v' \leqslant v_i} \frac{\varepsilon''(v')v'}{v'^2 - v^2} \, dv' + \frac{2}{\pi} \int_{v' > v_i}^\infty \frac{\varepsilon''(v')v'}{v'^2 - v^2} \, dv'. \tag{12.26}$$

For large values of v, the first integral is much smaller than the second integral on account of its smaller range; the value of the second integral, in the limit $v \to \infty$, is simply given by $\varepsilon_\infty - 1$ (see Eq. 12.20). Thus, for infrared frequencies:

$$\varepsilon'(v) = \varepsilon_\infty + \frac{2}{\pi} \int_0^\infty \frac{\varepsilon''(v')v'}{v'^2 - v^2} \, dv'$$

and   (12.27)

$$\varepsilon'(0) = \varepsilon_0 = \varepsilon_\infty + \frac{2}{\pi} \int_0^\infty \frac{\varepsilon''(v')}{v'} \, dv'$$

for the static dielectric constant.

We note that the Kramers-Kronig relations can be formulated in various ways, for instance, between the quantities $n^2 - \kappa^2$ and $2n\kappa$. Another useful formulation employs the power reflection coefficient R and the phase ψ of the amplitude reflection coefficient $\rho = \sqrt{R}e^{i\psi}$ as the interdependent variables, and we remember that ρ gives the ratio of the complex amplitudes of the reflected and incident waves (see footnote[‡] on page 415).

From

$$\rho = \frac{n + i\kappa - 1}{n + i\kappa + 1} = \sqrt{R}e^{i\psi} \tag{12.28}$$

we find that n and κ are related to R and ψ by

$$n = \frac{1 - R}{1 - 2\sqrt{R}\cos\psi + R}, \qquad \kappa = \frac{2\sqrt{R}\sin\psi}{1 - 2\sqrt{R}\cos\psi + R}. \tag{12.29}$$

Writing for $\ln \rho$ of Eq. 12.28

$$\ln \rho = \tfrac{1}{2}\ln R + i\psi, \tag{12.30}$$

the Kramers-Kronig relation between the real and imaginary parts of Eq. 12.28 is (see Eq. 12.25):

$$\psi(v) = \frac{v}{\pi}\int_0^\infty \frac{\ln R(v')}{v'^2 - v^2}\,dv'. \tag{12.31}$$

We are now ready to discuss the experimental data. [For a detailed discussion on the optical constants and their measurements we refer the reader to the article by E. E. Bell, "Optical Constants and Their Measurements," in *Handbuch der Physik*, Vol. 25/2a, Springer-Verlag, Berlin, 1967, pp. 1–58 (in English).]

d. Dispersion Measurements on Alkali Halides

Genzel *et al.*[2] have published a study of transmission measurements on parallel plates of NaCl, KCl, and KBr in the spectral range 0.3 to 3 mm (33 to 3.3 cm^{-1}). A klystron-crystal diode multiplier served as the radiation source of the spectrometer for the very long waves (longer than 1 mm) and gave off various discrete harmonics between 1.09 and 3.20 mm. A Golay cell was employed as detector. Figure 12.1 shows the transmission measurements in terms of the ratio of the transmitted power with and without the sample inserted into the beam, D^*, that is, the transmittance including all reflection losses, multiple interferences, and so forth. The limiting value of this quantity at vanishing absorption ($\kappa = 0$) and under neglect of multiple interferences is designated D_∞^*,

$$D_\infty^* = (1 - R)^2, \tag{12.32}$$

where R is the corresponding reflectivity (see Eqs. 12.12 and 12.13). The data show that the alkali halides become essentially transparent toward wavelengths of about 1 mm (see end of Section B.b).

The extinction coefficient κ is computed from the "transmission at negligible reflection," D (see Eq. 12.13). The connection between D^* and D

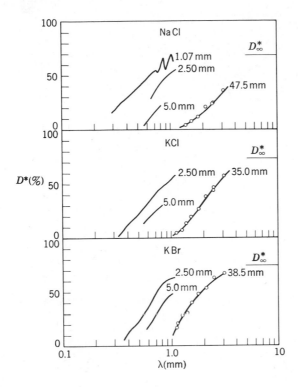

Fig. 12.1 Transmittance of NaCl, KCl, and KBr in the spectral range of 0.3 mm (33 cm^{-1}) to 3 mm (3.3 cm^{-1}). The values on the curves give the thickness of the salt plates in millimeters; D_∞^* is the limiting transmission at vanishing absorption and under neglect of multiple interferences. [L. Genzel, H. Happ, and R. Weber, Z. *Physik* **154**, 13 (1959), by permission of Springer: Berlin-Göttingen-Heidelberg.]

in the region of the eigenfrequency ω_T is rather complicated (see Eq. 12.13), but it can be approximated for the points on the long wavelength side of ω_T (where $n \gg \kappa$) by $(1 - R)^2 D/[(1 - RD)^2 + 4RD \sin^2\alpha]$, and on the short wavelength side by the relation $D^* \approx (1 - R)^2 D$ (since $RD \ll 1$).[†] The dependence of κ on the wavelength is shown in a log-log plot in Fig. 12.2. The extinction coefficient is seen to be essentially proportional to the frequency for NaCl and KCl but to increase somewhat stronger with frequency for KBr. The authors ascribe this to the large experimental errors due to multiple

[†] The corresponding formula has been given incorrectly in Ref. 2—the exponent of two is missing in $(1 - R)^2$ (see Eq. (3) there). The evaluation of the data has been done with the correct formula. The refractive index n is essentially constant in the range of the measurements, $n = \varepsilon_0^{1/2}$ (see Eq. 12.6).

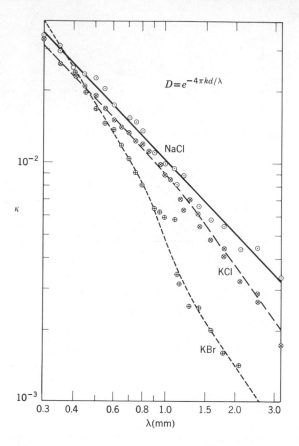

Fig. 12.2   Dependence of the extinction coefficient of NaCl, KCl and KBr on the wavelength
in the range 0.3 mm (33 cm^{-1}) to 3 mm (3.3 cm^{-1}). [L. Genzel, H. Happ, and R. Weber, Z.
Physik **154**, 13 (1959), by permission of Springer: Berlin-Göttingen-Heidelberg.]

interferences at the required large crystal thicknesses (κ small at the longer
wavelengths) and to the possibility that a side maximum may exist near
100 μ in the dispersion curve. We shall return to this point later.

The dispersion curves for the refractive index n and the extinction coefficient κ of NaCl are shown in Fig. 12.3. The curves are the calculated dispersions, see Eq. 12.17, with three infrared terms ($j = 1, 2, 3$). The parameters are collected in Table 12.1.

The good fit of the computed curve to the data, except for an unexplained
deviation near 30 μ, shows the usefulness of the dispersion formulas. We
notice, however, that the parameters of the dispersion formulas are no

TABLE 12.1*

Dispersion Parameters of NaCl[†]

Index of term[‡]	Wave-length, μ	Frequency, $\sec^{-1}$	Damping parameter	Oscillator strength
1[§]	61.1	4.918×10^{12}	1.840×10^{11}	7.740×10^{25}
2[§]	40.5	7.407×10^{12}	10.48×10^{11}	0.4938×10^{25}
3	120[‖]	2.493×10^{12}	41.1×10^{11}	0.2064×10^{25}

* *After* L. Genzel, H. Happ, and R. Weber, *Z. Physik* **154**, 13 (1959). By permission of Springer: Berlin-Göttingen-Heidelberg.
[†] See Eq. 12.17.
[‡] $\varepsilon_\infty = 2.3276$ (three electronic terms).
[§] The parameters for $j = 1, 2$ were taken from Ref. 3.
[‖] Assumed.

longer directly related to a cubic alkali halide crystal structure (which should give rise to only one infrared resonance peak, see Eq. 12.19). The indication is therefore that we apparently deal here with more than one kind of oscillator. It is not immediately apparent how we can relate the dispersion parameters to molecular parameters of the alkali halide structures. The multiple infrared absorption or reflection peaks in the spectrum of a cubic crystal will be explained below.

Geick[4] has measured the transmission and reflection, at nearly vertical incidence (10°), of thin NaCl layers in the very range of ω_T. Instead of employing the dispersion formulas to calculate the optical constants n and κ from the measurements, he obtains n and κ by a graphical analysis (the intersection of the curves of constant reflectivity and the curves of constant transmissivity in the n,κ-plane at wavelength λ). In addition, Geick has attempted to obtain n and κ from the determination of only one optical property, for instance the reflectivity, by computing the phase ψ with the help of the Kramers-Kronig relation (see Eq. 12.31) and, subsequently, calculating n and κ from $\psi(v)$ and $R(v)$ with the help of Eq. 12.29.

Geick carried out his measurements by inserting the crystal into the light beam of the spectrometer for a period of 24 sec and measuring the basis spectrum (without the crystal) for the succeeding 24 sec. The integrator in the electrical circuit of the recorder, a low-pass filter of 32 to 100 sec time constant, was equipped with two capacities, one of which was connected to the signal of the sample cycle and the other to the signal of the background cycle. The essential advantage of this technique lies in the fact that the stability of the optical components of the spectrometer need only be con-

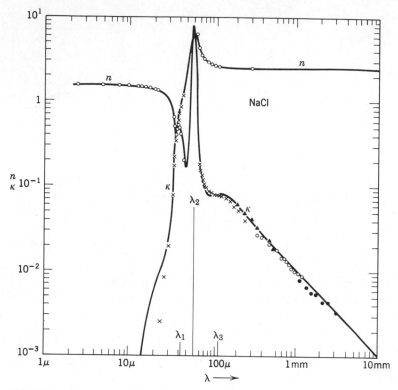

Fig. 12.3   Dispersion curve of NaCl between 3μ $(3300\,\text{cm}^{-1})$ and 3 mm $(3.3\,\text{cm}^{-1})$. The curves were calculated by using three electronic and three infrared terms (see Table 12.1 and Eq. 12.17). The sources of the measured points are L. Genzel, H. Happ, and R. Weber, Z. *Physik* **154**, 13 (1959); L. Kellner, Z. *Physik* **56**, 215 (1929); M. Czerny, Z. *Physik* **65**, 600 (1930); E. F. Nichols and J. D. Tear, *Astrophys. J.* **61**, 36 (1925); H. Rubens and Trowbridge, *Wied. Ann.* **60**, 733 (1897). [L. Genzel, H. Happ, and R. Weber, Z. *Physik* **154**, 13 (1959), by permission of Springer: Berlin-Göttingen-Heidelberg.]

stant over periods of 48 sec.[5] In Fig. 12.4 we show an example of a scan (CsBr reflection).

The transmission and reflection data of NaCl in the region of ω_T are shown in Figs. 12.5 and 12.6; the optical constants computed from the data are collected in Table 12.2.

Fig. 12.7 shows the comparison between the optical constants obtained from reflection measurements and the Kramers-Kronig relations (solid curve), and the optical constants which were computed from the reflection and transmission experiments (single points). We see that the agreement is good for n, but less so for κ in the wavelength range below $30\,\mu$ and above

TABLE 12.2*

OPTICAL CONSTANTS OF NaCl BETWEEN 55 AND 67 μ

Wavelength, μ	Refractive index, n	Extinction coefficient, κ
55	0.54	4.11
56	0.74	4.70
57	1.02	5.28
58	1.45	6.04
59	2.18	6.72
60	3.99	7.51
61	6.67	6.49
62	7.81	3.56
63	7.05	2.02
64	6.31	1.34
65	5.52	0.87
66	5.00	0.64
67	4.61	0.48

* *After* R. Geick, Z. *Physik* **166**, 122 (1962). By permission of Springer: Berlin-Göttingen-Heidelberg.

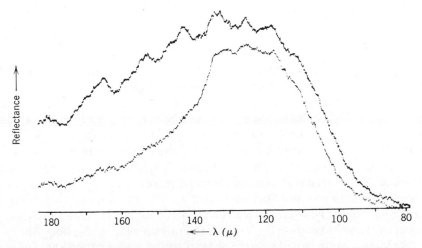

Fig. 12.4   Determination of the reflection of a CsBr reststrahlen plate in the KRS5 frequency region employing the alternating registration method (see text). (*Upper curve*): basis spectrum. (*Lower curve*): additional reflection on CsBr. [R. Geick, Z. *Physik* **163**, 499 (1961), by permission of Springer: Berlin-Göttingen-Heidelberg.]

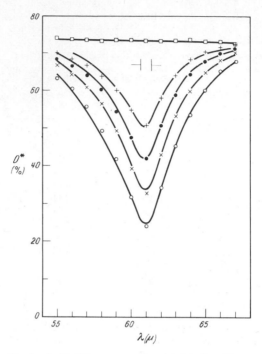

Fig. 12.5 Transmittance of a NaCl layer on a polymer (phthalic acid ester polymer). The curves are calculated (see text). The points indicate the measured values: □, polymer background; +, 0.07-μ thickness; ●, 0.11 μ; ×, 0.17 μ; ○, 0.26 μ. [R. Geick, Z. *Physik* **166**, 122 (1962), by permission of Springer: Berlin-Göttingen-Heidelberg.]

70 μ. In the range of 110 to 170 μ the values of κ which are based solely on reflection measurements become meaningless (negative).

We shall see further below that the common experimental technique used by many authors, namely, the employment of very thin films under oblique incidence of the radiation in reflection measurements (see, for instance, Fig. 12.6), has an interesting consequence: under such conditions, resonance absorption or reflection at the longitudinal frequency ω_L may occur. Since in a thick cubic crystal the electromagnetic waves can only couple to transversal motion (see end of Section B.b), it was tacitly assumed that infrared radiation cannot be used to measure the longitudinal resonance frequency ω_L in *any* sample geometry. We shall take this up again in Section C.b.

We know that in lieu of a direct observation of the longitudinal resonance frequency, it can be computed with the help of the Lyddane-Sachs-Teller relation (Eq. 12.23) from ε_0, ε_∞, and ω_T. Another possibility for an indirect observation of ω_L is to search for the upper value of the angular frequency ω

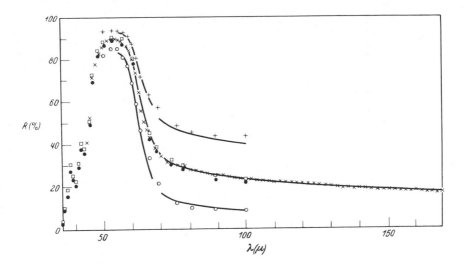

Fig. 12.6. Reflectance of NaCl. The curves are calculated (see text). The points indicate the measured values: ×, at normal incidence (see Ref. 4); □, s-polarized; ●, p-polarized, both at 12° incidence; +, s polarized; ○, p-polarized, both at 52° incidence (from A. Mitsuishi et al., J. Opt. Soc. Am. **50**, 433 (1960). The notation s and p polarization refers to the polarization of the incident light with respect to the plane of incidence (s = normal, p = parallel; see, for instance, F. A. Jenkins and H. E. White, Fundamental of Optics, McGraw-Hill, New York, 1957). [R. Geick, Z. Physik **166**, 122 (1962), by permission of Springer: Berlin-Göttingen-Heidelberg.]

of the incident radiation for which $\varepsilon'(\omega) = 0$. The latter approach serves—depending on the accuracy with which the position of the node of $\varepsilon'(\omega)$ can be located—as a convenient means for a calculation of the static dielectric constant ε_0 and thus $\tilde{\varepsilon}(\omega)$ (see Eq. 12.19). Work along these lines has been reported by Brodsky and Burstein.[6] The authors measured the reflectance (at normal incidence) of lithium hydride, lithium deuteride, and isotopically enriched lithium-6 hydride and lithium-7 hydride over the entire frequency region of the lattice spectrum. With the help of the Kramers-Kronig phase dispersion relation (Eq. 12.31) the refractive index and the extinction coefficient were computed (Eq. 12.29) and therefore $\varepsilon'(\omega)$ and $\varepsilon''(\omega)$. The frequencies ω_T and ω_L were then obtained from the positions of the maximum value of $\varepsilon''(\omega)$ and the node of $\varepsilon'(\omega)$, respectively, as described at the end of Section B.b. Finally, ε_0 is calculated with the help of the Lyddane-Sachs-Teller relation (Eq. 12.23). Then, from Eqs. 12.22, we can solve for the difference $\varepsilon_0 - \varepsilon_\infty = (\varepsilon_\infty + 2)^2 (4\pi N e_s^{*2})/9V\bar{m}\omega_T^2$. Rewriting this, the identity $(e_s^*/e)^2 \equiv 9V\bar{m}\omega_T^2 \times (\varepsilon_0 - \varepsilon_\infty)/4\pi N e^2 (\varepsilon_\infty + 2)^2$, where $\pm Ze$ is the static equilibrium charge of the

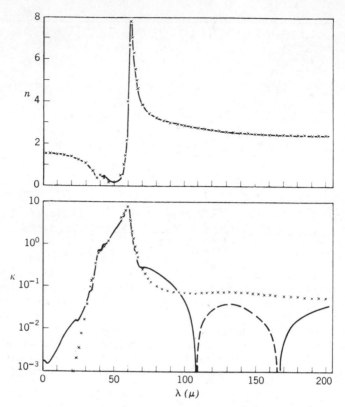

Fig. 12.7   Linear dispersion plot of n and semilogarithmic dispersion plot of κ, computed from reflection data and the Kramers-Kronig relation. The points are the single measurements obtained from the reflection and transmission measurements. The dashed part of the curve indicates that κ is negative (see text). [R. Geick, Z. *Physik* **166**, 122 (1962), by permission of Springer: Berlin-Göttingen-Heidelberg.]

atom (we assume $Z = 1$), is obtained. If this equation holds, e_s^*/e should equal unity. The evaluation of the lithium hydride data (and those of many other salts) indicate, however, that this is not the case.

Before we continue with the discussion of the lithium hydride data, we shall give a short description of the significance of the deviation of e_s^*/e from unity. The Szigeti charge e_s^* is defined by $e_s^* = (\partial \mathbf{P}_T/\partial \mathbf{u}_T)_{\mathbf{E}_{eff}}$, where $\mathbf{P}_T = (N/V)e^*\mathbf{u}_T$ is the electric moment resulting from the relative displacement of the atoms (see Section B.b). The Szigeti charge includes the effects of charge redistribution resulting from short-range interaction between neighboring ions but does not include the long-range effects due to $\mathbf{E}_{eff}$

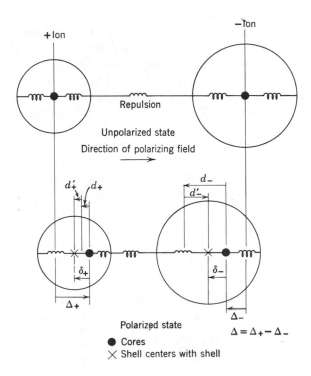

Fig. 12.8 Short-range interaction polarization. Unpolarized state: center of core and shell of each respective ion coincide. Polarized state: $d_{\pm}$ are the displacements of the shells with respect to their cores in the absence of repulsion between the shells; $d'_{\pm}$ are the additional shell displacements resulting from repulsion. The distance between the cores (the masses) has been diminished by Δ, the distance between the shells by $\Delta - \delta_- + \delta_+$. [B. G. Dick, Jr. and A. W. Overhauser, *Phys. Rev.* **112**, 90 (1958).]

(Burstein *et al.*[7a]). To account for deviations from e_s^*/e from unity, Dick and Overhauser[7b] have introduced a model, their so-called "shell model," of short-range repulsive interaction between ions of closed shell configurations. They came to the conclusion that the failure of $e_s^*/e = 1$ to hold is a demonstration of an inadequacy in the dielectric theory and not a consequence of a deviation of the ionic character of the crystals (alkali halides). Dick and Overhauser retain the notion that the ions carry the charge $\pm Ze$, and they seek the deviations in the charge redistribution that results when the electron shells of the moving ions overlap.

The calculations are thus based on a model of an ideal ionic NaCl-type crystal with overlapping ions. In brief, the ions are considered to be made up of an outer spherical shell of electrons (rare-gas atom configuration) and of an inner core (nucleus plus tightly bound inner electrons). The shell retains its spherical shape and moves bodily with respect to the core in an electric field; equilibrium conditions are obtained by restoring forces (spring constants) between core and shell (see Fig. 12.8). In addition, it is assumed that in the crystal repulsive forces act between adjacent shells and not between the cores (or masses). This effect therefore modifies the polarization of the ions.

The shell-shell repulsion and the core-shell restoring forces are expressed in terms of harmonic potentials; the force constants are related to the Born-Mayer compressibility and to the ionic polarizability of the (free) ion. The solution of Lagrange's equation of motion leads to the simple expression[6]

$$e_s^* = e + \frac{A\mu}{1 + A\lambda}$$

$$\mu = \frac{1}{e}\left(\frac{\alpha_+}{n_+} - \frac{\alpha_-}{n_-}\right)$$

$$A = \frac{6R_0}{\beta}$$

$$\lambda = \frac{1}{e^2}\left(\frac{\alpha_+}{n_+^2} + \frac{\alpha_-}{n_-^2}\right) = \frac{1}{k_+} + \frac{1}{k_-},$$

(12.33)

where $\alpha_\pm$ are the polarizabilities, $n_\pm$ is the number of electrons on the shell, R_0 is the lattice constant, β is the compressibility, e is the absolute value of the electronic charge, and $\pm$ refers to cation and anion, respectively. The repulsive spring constant between adjacent shells is given by A; the shell-core spring constants are given by $k_\pm = (n_\pm e)^2/\alpha_\pm$. The anion is, in general, more deformable than the cation. In this case, $\mu < 0$, and therefore e_s^*/e is smaller than unity. We shall see that this is the case in the alkali halides.

In Eq. 12.33 a contribution to e_s^*/e from an effect which Dick and Overhauser designated "exchange-charge polarization" has been omitted since its magnitude is difficult to estimate. Havinga[8] has shown that it is not necessary to invoke this additional effect and that it is more direct to estimate the number of electrons in the shells rather than to assume—as Dick and Overhauser have done—that the numbers in the ion shell and the shell of the isoelectronic rare gas are equal. Havinga's expression is, setting $e_s^* = e \cdot s$,

$$(1 - s)^2 = \frac{\overline{m}\omega_T^2(\varepsilon_0 + 2)}{Z^2 e^2(\varepsilon_\infty + 2)}(\alpha_+^{free} + \alpha_-^{free} - \alpha_\infty),$$

(12.34)

where α_∞ is the polarizability from the Lorenz-Lorentz formula, "free" designates the free ion polarizability, and the other symbols have their previously defined meaning.

We return now to the discussion of the lithium hydride data of Brodsky and Burstein. Figure 12.9 shows the optical path of the fore-optics contained in a vacuum reflection chamber to a Perkin-Elmer Model 112 single-beam, double-pass, prism spectrometer. The three isotopically substituted crystals LiD, ^{6}LiH, and ^{7}LiH were measured with the Model 221-0357 micro-specular reflection attachment of a Perkin-Elmer Model 421. Figures 12.10 to 12.13 show the reflection spectra and the calculated reflectivity, based on one oscillator, for LiH, LiD, and enriched ^{6}LiH and ^{7}LiH. Figures 12.14 and 12.15 show the plots of the real and imaginary parts of $\tilde{\varepsilon}$, from which ω_L and ω_T are read off as explained above. The corresponding curves for the other isotopically substituted compounds are not shown here.

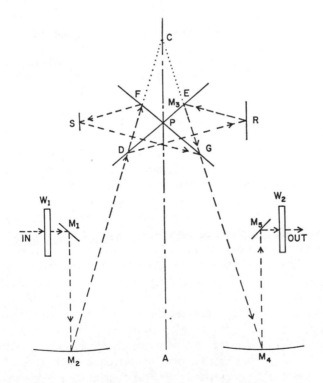

Fig. 12.9 Optical path inside reflection chamber. [M. H. Brodsky and E. Burstein, *J. Phys. Chem. Solids* **28**, 1655 (1967), by permission of Pergamon Press.]

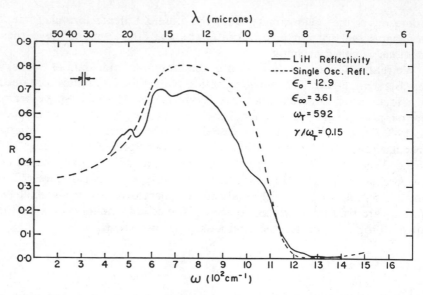

Fig. 12.10   Measured and calculated reflectivity of naturally abundant (92.48 % ^{7}Li) LiH (see Eqs. 12.31, 12.29, 12.9, 12.8, 12.23, and 12.19). The value of the damping constant γ was obtained by adjusting the calculated values to the measured peak reflectivity. The data are shown as taken. [M. H. Brodsky and E. Burstein, *J. Phys. Chem. Solids* **28**, 1655 (1967), by permission of Pergamon Press.]

We note from the figures that measurements and calculations essentially agree. However, again there are side maxima in the spectra (see also above, Genzel *et al.*), in other words, the simple picture of a single oscillator is merely a first approximation. With respect to the experimental technique, we can infer from, for instance, Fig. 12.14 that there may be cases where values of ω_L may contain rather appreciable errors unless the frequency at the upper node of $\varepsilon'(\omega)$ is as well discernible as in this work.

We neglect the appearance of the side maxima here—since they are of no importance within the framework of these studies—and give a summary of the results. The emphasis is on the comparison between the measured and the computed effective charges. The computed values were obtained with the help of the simplified shell model (see Eq. 12.33). The model parameters are collected in Table 12.3 together with those for LiF and NaCl for comparison's sake. We notice that the effective charge of LiH is rather small. According to what we have said above, this is a consequence of the relatively large deformation of the negative ion (H$^-$). The agreement between measured and calculated charges indicates that the shell model gives a meaningful physical picture (see Table 12.4).

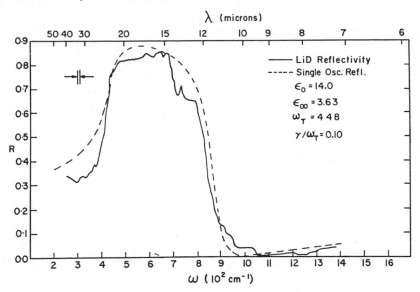

Fig. 12.11   Measured and calculated reflectivity of LiD (92.48% [7]Li). The peak reflectivity is normalized to about 90% and the spectrum is corrected for scattered radiation. The calculations of the one-oscillator model were performed as described in the legend to Fig. 12.10 [M. H. Brodsky and E. Burstein, *J. Phys. Chem. Solids* **28**, 1655 (1967), by permission of Pergamon Press.]

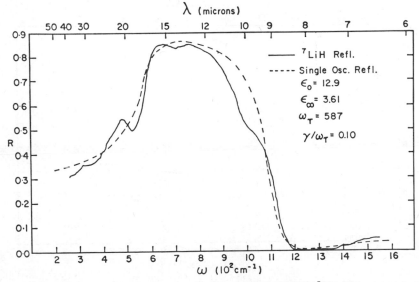

Fig. 12.12   Measured and calculated reflectivity of LiH (99.99% [7]Li). Measurements and calculations as described for LiD. [M. H. Brodsky and E. Burstein, *J. Phys. Chem. Solids* **28**, 1655 (1967), by permission of Pergamon Press.]

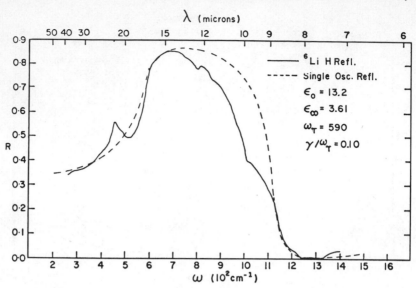

Fig. 12.13   Measured and calculated reflectivity of LiH (95% ^{6}Li). Measurements and calculations as described for LiD. [M. H. Brodsky and E. Burstein, *J. Phys. Chem. Solids* **28**, 1655 (1967), by permission of Pergamon Press.]

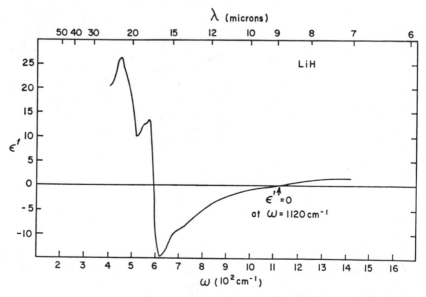

Fig. 12.14   Real part of the dielectric constant of LiH. The longitudinal frequency ω_L is read off at the upper node of ε'. [M. H. Brodsky and E. Burstein, *J. Phys. Chem. Solids* **28**, 1655 (1967), by permission of Pergamon Press.]

TABLE 12.3*

SHELL MODEL PARAMETERS FOR LiH, LiD, LiF, AND NaCl

	LiH	LiD	LiF	NaCl	Reference
R_0 (Å)	2.042	2.034	2.008	2.814	[†,‡]
β (10^{-12} cm^2/dyne)	3.57	—	1.53	3.97	[†]
α_+ (10^{-24} cm^3)	0.029	0.029	0.029	0.179	[§]
α_- (10^{-24} cm^3)	10.1	10.1	1.04	3.66	[§]
n_+	1.12	1.12	1.12	2.4	[§]
n_-	1.12	1.12	2.4	4.35	[§]
k_+ (10^7 dyne/cm)[∥]	0.997	0.997	0.997	0.741	
k_- (10^7 dyne/cm)[∥]	0.00286	0.00286	0.127	0.119	
A (10^4 dyne/cm)[††]	3.43	—	7.85	3.96	
e_s^*/e (theo)	0.4	—	0.87	0.87	
e_s^*/e (exp)	0.53	0.56	0.87	0.74	

* M. H. Brodsky and E. Burstein, *J. Phys. Chem. Solids* **28**, 1655 (1967). By permission of Pergamon Press.

† F. Pretzel, private communication to M. H. Brodsky and E. Burstein.

‡ F. Pretzel *et al.*, *J. Phys. Chem. Solids* **16**, 10 (1960).

§ See Ref. 7b (corrected values).

∥ Force constant (restoring forces).

†† Shell–shell repulsion.

Finally, Table 12.5 gives the ratios of isotopic masses and frequencies. The ratios of the frequencies for both ω_L and ω_T agree very well with the inverse ratios of the square roots of the corresponding reduced masses.

Jones *et al.*[9a] have studied the lattice resonances (reflection and transmission) of a large series of alkali halides at room temperature and 4°K, and

TABLE 12.4*

SUMMARY OF RESULTS

Crystal	ω_T, cm^{-1}	ω_L, cm^{-1}	ε_0	e_s^* Exp	e_s^* Calc	Transmission studies ω_T cm^{-1}
LiH	592 ± 6	1120 ± 10	12.9 ± 0.5	0.53 ± 0.02	0.4	588 ± 7[†]
LiD	444 ± 4	880 ± 9	14.0 ± 0.5	0.56 ± 0.02	—	446 ± 4[†]
^{7}LiH	587 ± 6	1109 ± 10	12.9 ± 0.5	0.52 ± 0.02	(0.4)	588[†]
^{6}LiH	590 ± 6	1129 ± 10	13.2 ± 0.5	0.54 ± 0.02	(0.4)	592[†]

* M. H. Brodsky and E. Burstein, *J. Phys. Chem. Solids* **28**, 1655 (1967). By permission of Pergamon Press.

† W. B. Zimmerman and D. J. Montgomery, *Phys. Rev.* **120**, 405 (1960).

TABLE 12.5*

RATIOS OF ISOTOPIC MASSES AND FREQUENCIES

Crystal	Cation mass Amu	Anion mass Amu	Reduced mass Amu	Inverse ratio of square root of reduced masses	Ratio of ω_T's	Ratio of ω_L's
LiH	6.94	1	0.874			
				1.33	1.32	1.27
LiD	6.94	2	1.55			
^{7}LiH	7	1	0.875			
				0.99	0.995	0.99
^{6}LiH	6	1	0.857			

* M. H. Brodsky and E. Burstein, *J. Phys. Chem. Solids* **28**, 1655 (1967). By permission of Pergamon Press.

TABLE 12.6*

RELATIVE EFFECTIVE CHARGES OF VARIOUS ALKALI HALIDE CRYSTALS

	ε_0	ε_∞	e_s^*/e at room temperature	e_s^*/e at 4°K
NaBr	5.99	2.62	0.70	—
NaI	6.60	2.91	0.71	—
KCl	4.68	2.13	0.81	—
KBr	4.78	2.33	0.78	—
KI	4.94	2.69	0.71	0.76
RbCl	5	2.19	0.84	0.90
RbBr	5	2.33	0.84	0.91
RbI	5	2.63	0.75	0.83
AgCl	12.3	4.04	0.71	0.83
AgBr	13.1	4.62	0.70	0.82
CsCl	7.20	2.60	0.85	0.90
CsBr	6.51	2.78	0.78	0.84
CsI	5.65	3.03	0.67	0.71
TlCl	31.9	5.10	0.80	0.84
TlBr	29.8	5.41	0.82	0.81

* *After* G. O. Jones, D. H. Martin, P. A. Mawer, and C. H. Perry, *Proc. Roy. Soc.* (*London*) **A261**, 10 (1961). By permission of the Royal Society, London.

they have compared the measured values of e_s^*/e with those computed from the shell model (Dick and Overhauser; Havinga). Low-temperature data are useful since disturbing effects, for instance anharmonic contributions due to

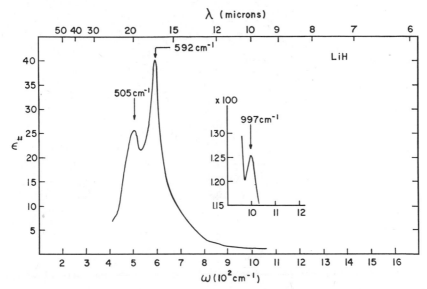

Fig. 12.15 Imaginary part of the dielectric constant of LiH. The transversal frequency ω_T is read off the maximum (592 cm^{-1}). There are two side maxima at 505 and 997 cm^{-1}, respectively. [M. H. Brodsky and E. Burstein, *J. Phys. Chem. Solids* **28**, 1655 (1967), by permission of Pergamon Press.]

large-amplitude motions, are decreased at very low temperatures and the simple theoretical models are therefore more likely to be valid. Table 12.6 gives a summary of the experimental ratios of e_s^*/e. Table 12.7a gives a comparison of the calculated and experimental ratios. We notice that the ratios computed according to the (complete) shell model of Dick and Overhauser are too large and that, in general, the agreement is better using Havinga's modification of the shell model. We remind the reader that in Havinga's approach the number of electrons in the shell is directly calculated and not assumed to equal that of the isoelectronic rare-gas atom (see above). Dick and Overhauser's shell model overestimates the number of electrons.

Of particular interest are the thallium (Tl$^+$) and silver (Ag$^+$) ions—which are not alkali ions—since the outermost electrons are in s and d (and not in p) orbitals. Because of this the ions show a large electronic polarizability as indicated by the high values of ε_∞ in Table 12.6. However, we notice that the corresponding values of e_s^*/e for Tl$^+$ are higher than those of the Na$^+$ salts. Apparently a high electronic polarizability is not simply associated with a high overlap distortion between the ions.

TABLE 12.7a[*][†]

COMPUTED AND MEASURED VALUES OF e_s^*/e

	1	2	3	4
NaBr	0.89	0.71	0.70	
NaI	0.89	0.64	0.71	
KCl	0.93	0.80	0.81	
KBr	0.93	0.81	0.78	
KI	0.91	0.74	0.71	0.76
RbCl	0.95	0.76	0.84	0.90
RbBr	0.92	0.80	0.84	0.91
RbI	0.93	0.76	0.75	0.83
CsCl		0.80	0.85	0.90
CsBr		0.83	0.78	0.84
CsI			0.67	0.71
AgCl			0.71	0.83
AgBr			0.70	0.82
TlCl			0.80	0.84
TlBr			0.81	0.81

* *After* G. O. Jones, D. H. Martin, P. A. Mawer, and C. H. Perry, *Proc. Roy. Soc.* (*London*) **A261**, 10 (1961). By permission of The Royal Society, London.

† Column 1: Dick and Overhauser,[7b] computed. Column 2: Havinga,[8] computed. Column 3: Jones et al.,[9] room temperature (see Table 12.6). Column 4: Jones et al.,[9] 4°K (see Table 12.6).

Table 12.7a shows the expected general trend in the values of e_s^*/e, namely that the ratio decreases with increasing size of the negative ion and that it approaches unity with increasing similarity of the sizes of anion and cation.

The dielectric properties of ZnS-type (zinkblende structure) crystals have been discussed by Burstein et al.[9b] These crystals offer an interesting contrast to the series of alkali halides, since their valence electrons have extended wave functions in both ground and excited electronic states. Therefore the Lorentz field $(4/3)\pi \mathbf{P}$, which is the field from polarization charges inside a spherical cavity cut out of the material with the reference particle at the center, does not couple to electronic and vibrational excitation in the ZnS-type crystals. This means that the effective field $\mathbf{E}_{eff}$ approximately equals the macroscopic field $\mathbf{E}$. We must therefore replace $e_s^* = (\partial \mathbf{P}/\partial \mathbf{u})_{\mathbf{E}_{eff}}$ by the "macroscopic dynamic ionic charge" e_T^*, which is given by $e_T^* = (\partial \mathbf{P}/\partial \mathbf{u})_{\mathbf{E}}$, for a characterization of the polar character of the vibrational mode. Table 12.7b gives a compilation of Burstein et al.'s data,[9b] derived from

TABLE 12.7b

INFRARED (LATTICE VIBRATION) PARAMETERS FOR ZnS-TYPE CRYSTALS

[E. Burstein, M. H. Brodsky, and G. Lucovsky, *Intern. J. Quantum. Chem.* **IS**, 759 (1967).]

Crystal	ε_0	ε_∞	ω_T, cm^{-1}	ω_L, cm^{-1}	e_T^*/e	Ref.
SiC	10	6.7	793	930	2.7*	*
AlSb	12.0	10.2	318	345	2.2[†]	†
GaP	10.2	8.5	366	401	2.0[‡]	‡
GaAs	12.9	10.9	273	297	2.2[§]	§
GaSb	15.7	14.4	231	241	2.0[§]	§
InP	12.6	9.6	307	351	2.7[§]	§
InAs	15.1	12.3	218	243	2.7[§]	§
InSb	17.9	15.7	185	197	2.5[§]	§
ZnS	8.00	5.14	282	352	2.0[‖]	‖, #
ZnSe	8.33	5.90	207	246	1.8[#]	#, **
ZnTe	9.86	7.28	177	206	2.0**	#, **
CuCl	7.3	4.8	155	198	1.2[††]	††

* W. G. Spitzer, D. A. Kleinman, and C. J. Frosch, *Phys. Rev.* **113**, 133 (1959).
† W. J. Turner and W. E. Reese, *Phys. Rev.* **127**, 126 (1962).
‡ D. A. Kleinman and W. G. Spitzer, *Phys. Rev.* **118**, 110 (1960).
§ M. Hass and B. W. Henvis, *J. Phys. Chem. Solids* **23**, 1099 (1962).
‖ S. J. Czyzah, W. M. Baker, R. C. Crane, and J. B. Howe, *J. Opt. Soc. Am.* **47**, 240 (1957).
A. Manabe, A. Mitsuishi, and H. Yoshinaga, *Japan J. Appl. Phys.* **6**, 593 (1967).
** D. T. F. Marple, *J. Appl. Phys.* **35**, 539 (1964).
†† S. Iwasa, Thesis, Physics Department, University of Pennsylvania, 1965.

the infrared lattice spectra of zinkblende structure crystals with the help of the relation $e_T^{*2} = (\varepsilon_0 - \varepsilon_\infty)\bar{m}\omega_T^2/4\pi(N/V)$. Since the *static* ionic charge in these crystals is about zero, the large values of e_T^* are mainly attributed to charge redistribution which possibly also involves some charge transfer from one atom to the other (ZnS-type crystals lack a center of symmetry).

We remember that in molecular spectra the intensities of transitions may be compared with the calculated dipole moment $\langle\psi|\mathbf{M}|\psi\rangle$ as a check on the validity of electronic wave functions $|\psi\rangle$. In the same sense, Burstein *et al.* suggest[9b] that the values of e_T^* of Table 12.7b, which give the vibrational transition moment $|\langle\partial\mathbf{M}/\partial\mathbf{u}\rangle|$, would lend themselves well for a check on the validity of *ab initio* energy band calculations of ZnS-type crystals.

At the end of this section we return briefly to the alkali halide crystals and mention some data on the damping constant of the dispersion formula (see Eq. 12.16). Jones *et al.*[9a] have measured the damping constant of AgBr as a function of temperature. Their data are displayed in Fig. 12.16, from

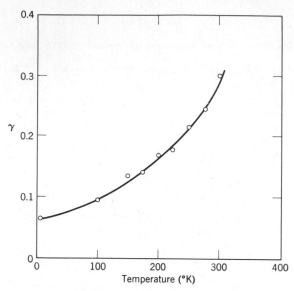

Fig. 12.16   Temperature-dependence of the damping constant of AgBr. [G. O. Jones, D. H. Martin, P. A. Mawer, and C. H. Perry, *Proc. Roy. Soc.* (*London*) **A261**, 10 (1961), by permission of The Royal Society, London.]

which it is apparent that the damping decreases with decreasing temperatures. This is due, in part, to the smaller vibrational amplitudes at the lower temperatures. We shall say more about this when we discuss the quantum-mechanical theory.

In conclusion we can state that the classical dispersion formulas and simple physical models of ionic overlap distortions of the ions explain rather well the experimental data, but that the simple dispersion theory needs modification and extension in order to account for certain observations, such as the occurrence of side maxima in the spectra and the temperature dependence of the damping constant.

C. Effects of Periodicity

a. Lattice Modes of Cubic Crystals

The positive and negative ions (for instance, Na^+ and Cl^-) lie on two interpenetrating face-centered cubic lattices.[10] According to Born-Oppenheimer, the effective potential function U which governs the displacement motions of atoms or ions in a lattice depends only on the nuclear coordinates. We therefore expand the potential function in terms of displacement vectors from nuclear equilibrium positions. Retaining only harmonic terms of the Taylor expansion of U, the equation of motion is given by[11-14]

$$m_k \frac{\partial^2}{\partial t^2} u_\alpha \binom{l}{k} = - \sum_{\beta l' k'} \frac{\partial^2 U}{\partial u_\alpha \binom{l}{k} \partial u_\beta \binom{l'}{k'}} u_\beta \binom{l'}{k'}, \tag{12.35}$$

where $u_\alpha \binom{l}{k}$ is the displacement coordinate of ion k in the lth unit cell in the three orthogonal directions $\alpha([100]$, etc.). The index k pertains to one of the two interpenetrating lattices, for instance, $k = 1$ for the positive, $k = 2$ for the negative ions. We assume the smallest possible cell as the unit cell for the face-centered cubic lattice, that is, a cell containing only one ion of each sign. We introduce the periodicity of the lattice by requiring that the solutions of Eq. 12.35 be of the form of traveling plane waves, of wave vector $\mathbf{q}$ and angular frequency ω:

$$u_\alpha \binom{l}{k} = \frac{Q_\alpha(k)}{m_k^{1/2}} \exp \left[i \left(\mathbf{q} \cdot \mathbf{r} \binom{l}{k} - \omega t \right) \right], \tag{12.36}$$

where $\mathbf{r} \binom{l}{k}$ is the equilibrium position vector of ion $\binom{l}{k}$, and $Q_\alpha(k)$ is the amplitude of the displacement. Both ω and Q are functions of $\mathbf{q}$. Inspection of Eq. 12.36 shows that in each cell l, atom (ion) k undergoes harmonic oscillations with the same amplitude and in the same direction but with a different phase. If the $u_\alpha \binom{l}{k}$ of Eq. 12.36 are inserted into Eq. 12.35, we obtain the secular equation

$$(\mathbf{M} - \omega^2 \mathbf{1})\mathbf{Q} = 0 \tag{12.37}$$

where $\mathbf{M}$ is the "dynamical matrix,"

$$\mathbf{M} = (m_k m_{k'})^{-1/2} \sum_{l'} \frac{\partial^2 U}{\partial u_\alpha \binom{l}{k} \partial u_\beta \binom{l'}{k'}}$$

$$\times \exp \left\{ i\mathbf{q} \cdot \left[\mathbf{r} \binom{l'}{k'} - \mathbf{r} \binom{l}{k} \right] \right\}, \tag{12.38}$$

$\mathbf{1} \equiv \delta_{\alpha\beta}\delta_{kk'}$ is a 6×6 unit matrix (see below), and $\mathbf{Q} \equiv Q_\beta(k')$ is a six-component column vector. The expression for $\mathbf{M}$ can be simplified since the inter-ionic forces in the crystal depend only on the relative positions of the ions in the lattice, $l - l' = h$. We rewrite the force constant tensor and the radius vector in the forms

$$\frac{\partial^2 U}{\partial u_\alpha \begin{pmatrix} l \\ k \end{pmatrix} \partial u_\beta \begin{pmatrix} l' \\ k' \end{pmatrix}} \equiv G^{kk'}_{\alpha\beta}(h), \tag{12.39}$$

$$\mathbf{r}\begin{pmatrix} l' \\ k' \end{pmatrix} - \mathbf{r}\begin{pmatrix} l \\ k \end{pmatrix} \equiv \mathbf{r}_{kk'}(h). \tag{12.40}$$

It follows that

$$\mathbf{M} = (m_k m_{k'})^{-\frac{1}{2}} \sum_h G^{kk'}_{\alpha\beta}(h) \exp\left[i\mathbf{q} \cdot \mathbf{r}_{kk'}(h)\right]. \tag{12.41}$$

An instructive and relatively simple example can be given by considering a linear chain of equidistant masses m_1 and m_2 which are permitted to vibrate in the direction of the chain under a harmonic potential function with nearest-neighbor interactions,

$$U = \tfrac{1}{2} f(u_l - u_{l+a})^2 + \tfrac{1}{2} f(u_l - u_{l-a})^2. \tag{12.42}$$

The force constant is given by f and the distance between m_1 and m_2 by a (repeat distance $= 2a$). The dynamical matrix is of size 2×2. The matrix elements of $\mathbf{M}$ (we are dropping the indices α and β) are

$$M_{11} = m_1^{-1}\{G^{11}(0)e^{iq\cdot 0}\} = m_1^{-1}2f$$

$$M_{12} = M_{21} = (m_1 m_2)^{-\frac{1}{2}}\{G^{12}(a)e^{iqa} + G^{12}(-a)e^{-iqa}\} \tag{12.43}$$

$$= -(m_1 m_2)^{-\frac{1}{2}}2f \cos(qa)$$

$$M_{22} = m_2^{-1}2f.$$

The secular determinant (see Eq. 12.37) is therefore

$$\begin{vmatrix} \dfrac{2f}{m_1} - \omega^2 & -\dfrac{2f}{(m_1 m_2)^{\frac{1}{2}}}\cos(qa) \\[3mm] -\dfrac{2f}{(m_1 m_2)^{\frac{1}{2}}}\cos(qa) & \dfrac{2f}{m_2} - \omega^2 \end{vmatrix} = 0 \tag{12.44}$$

and its roots are given by (after some tedious algebra):

$$\omega^2_\pm = \frac{f}{m} \pm f\left\{\frac{1}{m^2} - \frac{4\sin^2(qa)}{m_1 m_2}\right\}^{\frac{1}{2}} \qquad \left(\frac{1}{m} = \frac{1}{m_1} + \frac{1}{m_2}\right). \tag{12.45}$$

A graphical representation of ω_+ and ω_- is shown in Fig. 12.17. Obviously we obtain all solutions for $-\pi/2a \leqslant q \leqslant \pi/2a$ (Brillouin zone). The solution ω_+ is called the optical phonon branch since the lattice motion may interact with the photons of the incident light (see below). We describe this process

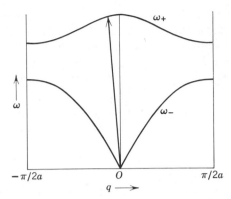

Fig. 12.17 Optical (ω_+) and acoustical (ω_-) phonon branch of the dispersion relation of the diatomic linear chain. The nearly vertical arrow denotes the infrared-active transition at $q \sim 0$. The slope of the arrow corresponds to the velocity of light, the slope of ω_- at $q = 0$ to the velocity of sound. [*After* J. M. Ziman, *Principles of the Theory of Solids*, Copyright 1964, Cambridge University Press.]

in terms of the absorption of the (infrared) photon, of wave vector $\mathbf{K}$ ($|\mathbf{K}| = 2\pi/\lambda$) and energy $\hbar\omega = \hbar c|\mathbf{K}|$, by the crystal and the creation of a phonon of wave vector $\mathbf{q}$ and energy $\hbar\omega'$. In the far-infrared, $|\mathbf{K}|$ is of the order of magnitude of 10 to 10^3 cm^{-1}. The phonon wave vector $|\mathbf{q}|$ may range up to orders of magnitude of 10^8 cm^{-1}. The laws of energy and wave vector conservation require that $\hbar\omega = \hbar\omega'$ and $\mathbf{K} = \mathbf{q}$. Therefore, we need only consider phonons of $\mathbf{q} \sim 0$.* This is presented schematically by the nearly vertical arrow in Fig. 12.17. Indeed, if in Eq. 12.45 we set $q \sim 0$, we notice that $\omega_+ \approx (2f/\bar{m})^{1/2}$. By inserting this value into the secular matrix equation, it easily follows that the two masses m_1 and m_2 move in opposite directions with their center of mass at rest.† We therefore see that the eigen frequency at $q = 0$ of the optical mode of the diatomic linear lattice with force constant f equals that of the fundamental of the corresponding diatomic molecule with force constant $2f$. The motion of the phonon ω_- (acoustical phonon branch), in which the atoms move in the same direction,† is obviously infrared-inactive.

* The condition $\mathbf{q} \sim 0$ holds also for Raman scattering (optical phonons) and Brillouin scattering (acoustical phonons). Because the velocity of sound is much smaller than the velocity of light, the generated phonon $\mathbf{q}$ can carry off only a small fraction of the quantum of the incident photon $\mathbf{K}$ of the ultraviolet or visible electromagnetic radiation. Therefore, the wave vector of the scattered photon, $\mathbf{K}'$, obeys $\mathbf{K}' \sim \mathbf{K}$ and, by the conservation law, $\mathbf{q} \sim 0$. The phonons with $\mathbf{q} = 0$ carry a non-vanishing linear momentum (see Ref. 10, p. 160).

† The relative amplitudes Q_k, which are the components of the eigen vectors, are obtained in the usual way by inserting the solutions of Eq. 12.44 into the matrix equation (Eq. 12.37). We find for $\omega_+^2 = (2f/\bar{m})$ (at $q = 0$) that $m_2^{-1}Q_1 + (m_1 m_2)^{-1/2}Q_2 = 0$; $Q_1/Q_2 = -(m_2/m_1)^{1/2}$ or, in terms of the normal coordinates $u_k = m_k^{-1/2}Q_k$ (see Eq. 12.36), $u_1/u_2 = -m_2/m_1$. For $\omega_-^2 = 0$ (at $q = 0$), one obtains $u_1/u_2 = 1$ (translation of the whole chain). The reader will have noticed the analogy with the (nongenuine) translational and stretching modes of a linear molecule.

It is also of interest to know the maxima of the distribution $q(\omega)$ in the Brillouin zone since the major contribution to the intensity of any absorption band will arise from states of high phonon density, in other words, for large values of $dq/d\omega$ or for $d\omega/dq \sim 0$. For instance, in the linear chain the high phonon density is found at $q = 0$ (for ω_+) and at the zone boundaries (for ω_+ and ω_-), as can be seen from Fig. 12.17.

We note that we have discrete values of q, and that q is bounded. This can be seen, *a priori*, by remembering that (*1*) the meaningful solutions are obviously given by lattice waves which have nodes at the boundaries of the crystal (standing waves), hence q is discrete. (2) Since adjacent nodes of the standing waves cannot be closer than a translational lattice vector, q has a maximum value. This leads to a finite number of q. Assuming k ions per unit cell (with three degrees of freedom per ion) and N unit cells per crystal, we obtain a total number of $3kN$ degrees of freedom. Because of the translational invariance, we need only know the $\mathbf{q}$ of one unit cell, thus there is a total number of N different values of $\mathbf{q}$ per crystal. Inspection of Eq. 12.37 shows that in the diatomic cubic crystal there exist six allowed values of ω for each $\mathbf{q}$. The task of calculating a dispersion relation ($\mathbf{q}(\omega)$ *vs* ω) is therefore rather formidable.

The boundary conditions for the N allowed values of $\mathbf{q}$ in terms of the translation vectors $\mathbf{a}_l$ are (see Chapter 13, Section B.a)

$$\exp{(i\mathbf{q} \cdot \mathbf{a}_l n)} = 1, \qquad \text{thus } \mathbf{q} \cdot \mathbf{a}_l = \frac{2\pi}{n}h_l \qquad \text{with } l = 1, 2, 3, \qquad (12.46)$$

where h_l is an integer and the translation vectors $\mathbf{a}_l$ are those of a volume which has the same shape as the elementary cell[10] and which contains $n^3 = N$ elementary cells. In terms of the reciprocal lattice vectors

$$\mathbf{b}_i = 2\pi\frac{\mathbf{a}_j \times \mathbf{a}_k}{\mathbf{a}_i \cdot \mathbf{a}_j \times \mathbf{a}_k} \qquad \text{with } \mathbf{a}_i \cdot \mathbf{b}_j = 2\pi\delta_{ij}, \qquad (12.47)$$

we see that the boundary conditions require that the components of $\mathbf{q}$ are of the form

$$q_i = \frac{h_i b_i}{n} \qquad \left(\text{since } q_i\mathbf{a}_i \cdot \mathbf{b}_j = \frac{2\pi}{n}h_i b_j = 2\pi q_i\delta_{ij}\right), \qquad (12.48)$$

hence

$$\mathbf{q} = \sum_i \frac{h_i}{n}\mathbf{b}_i = \sum_i k_i\mathbf{b}_i \qquad (h_1, h_2, h_3 = 0, 1, \ldots, n-1). \qquad (12.49)$$

If we abbreviate the dynamical matrix by the symbol $\mathbf{M} \equiv \begin{Bmatrix} k & k' \\ \alpha & \beta \end{Bmatrix}$, the secular determinant for a given $\mathbf{q}$ is of the form ($k, k' = 1, 2; \alpha, \beta = x, y, z$):

$$
\begin{vmatrix}
\begin{Bmatrix} 1 & 1 \\ x & x \end{Bmatrix} - \omega^2 & \begin{Bmatrix} 1 & 2 \\ x & x \end{Bmatrix} & \begin{Bmatrix} 1 & 1 \\ x & y \end{Bmatrix} & \cdot\ \cdot & \begin{Bmatrix} 1 & 2 \\ x & z \end{Bmatrix} \\[2mm]
\begin{Bmatrix} 2 & 1 \\ x & x \end{Bmatrix} & \begin{Bmatrix} 2 & 2 \\ x & x \end{Bmatrix} - \omega^2 & \begin{Bmatrix} 2 & 1 \\ x & y \end{Bmatrix} & \cdot\ \cdot & \begin{Bmatrix} 2 & 2 \\ x & z \end{Bmatrix} \\[2mm]
\begin{Bmatrix} 1 & 1 \\ y & x \end{Bmatrix} & \cdot & \begin{Bmatrix} 1 & 1 \\ y & y \end{Bmatrix} - \omega^2 & \cdot\ \cdot & \begin{Bmatrix} 1 & 2 \\ y & z \end{Bmatrix} \\[2mm]
\cdot & \cdot & \cdot & \cdot\ \cdot & \cdot \\[2mm]
\cdot & \cdot & \cdot & \cdot\ \cdot & \cdot \\[2mm]
\begin{Bmatrix} 2 & 1 \\ z & x \end{Bmatrix} & \cdot & \cdot & \cdot\ \cdot & \begin{Bmatrix} 2 & 2 \\ z & z \end{Bmatrix} - \omega^2
\end{vmatrix} = 0.
$$

$$\tag{12.50}$$

We note that the determinant is symmetric,

$$
\begin{Bmatrix} k & k' \\ x & y \end{Bmatrix} = \begin{Bmatrix} k' & k \\ y & x \end{Bmatrix}.
$$

This follows from the equivalence of the lattice sites of the Na and Cl ions. For certain values of the h_i/n there are further symmetry relations among the coefficients, which split the 6×6 matrix into smaller submatrices.[12]

The computed values of $q(\omega)$ depend strongly on the choice of the particular parameters of the potential function. In the simplest approximation, the polarizability of the ions is neglected and the potential function is a composite of short-range forces between nearest neighbors and long-range Coulomb interactions throughout the crystal (point-dipole model). The Coulomb interactions are computed with the help of the Madelung constant; the charge on each ion is set equal to Ze ($Z = 1$ for NaCl). The short-range potential is related to the compressibility. In the more general treatment, the polarizability of the ions and their repulsive forces are introduced by such models as Dick and Overhauser's shell model (see Section B.d).

As an example, the phonon dispersion relations for the directions [111] and [100] in NaCl, computed with the help of a simple point-charge potential as just described,[12] are shown in Figs. 12.18 and 12.19, respectively. Figure 12.20a, b shows the frequency distribution, that is, the number of lattice modes per frequency interval as a function of the frequency. This distribution is obtained by dividing the frequency scale into a number of small intervals $\Delta\omega$ and counting the number of lattice modes which fall into this range.

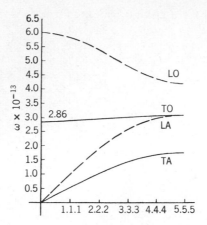

Fig. 12.18   The computed dispersion curves for the direction [111] of NaCl. The ions are considered as point charges. L = longitudinal polarization, T = transversal polarization, O = optical phonon branch, and A = acoustical phonon branch. The TA and TO branches are doubly degenerate. The directions are taken with respect to the plane of the surface of the crystal slab (infinite lateral dimensions). The values on the abscissa give the components of the wave vector (see Eq. 12.49) in convenient units. A set of 48 wave vectors was used. [E. W. Kellermann, *Phil. Trans. Roy. Soc.* (*London*) **A238**, 513 (1940), by permission of The Royal Society, London.]

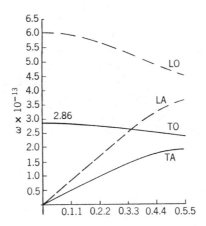

Fig. 12.19   The computed dispersion curves for the direction [100] in NaCl [E. W. Kellermann, *Phil. Trans. Roy. Soc.* (*London*) **A238**, 513 (1940), by permission of The Royal Society, London.]

The frequency marked $2\pi\nu_R$ in Fig. 12.20 represents ω_T, the so-called residual ray or reststrahlen frequency (see Section B.b). The eigen frequency ω_T therefore corresponds to the TO mode at $q \sim 0$.

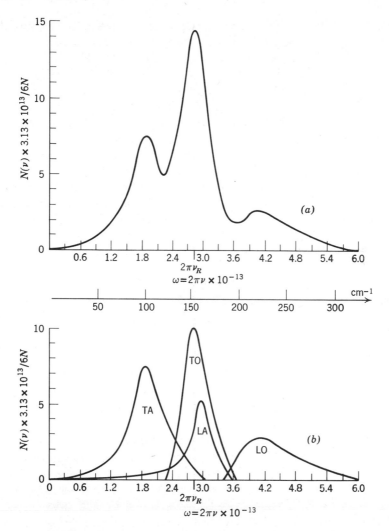

Fig. 12.20 The computed distribution-in-frequency $N(v)$ of the lattice vibrations of NaCl. The angular frequency $2\pi v_R = 2.86 \times 10^{13} \sec^{-1}$ corresponds to the reststrahlen frequency 152 cm^{-1} (observed value: 164 cm^{-1}). N is Avogadro's number. The calculations were based on a point-dipole model. [E. W. Kellermann, *Proc. Roy. Soc.* (*London*) **A178**, 17 (1941), by permission of The Royal Society, London.]

b. Effects of Sample Geometry

In this section we digress briefly from further discussion of the effects of periodicity and instead treat the effects of sample geometry on the optical response of the crystals. As discussed by Burstein *et al.*,[1c] effects in which the geometry of the medium manifests itself on the resonance excitation frequencies may be expected to occur when the effective density of electric-dipole oscillators is high. If radiation (of wave vector $|\mathbf{K}| = 2\pi/\lambda_0 \sim 0$) falls onto a (laterally large) thin crystal slab of thickness less than $\lambda_0/4\pi\kappa$ ("skin depth"), there is appreciable transmission with no change of phase on traversing the slab (see Eq. 12.13). Furthermore, the reflection at the front face is cancelled by the reflection from the back face (phase difference π) at all ω except when $\omega \sim \omega_T$. In this case the resonance absorption attenuates the amount of radiation which reaches the back face and therefore the two reflections do not cancel. In other words, at normal incidence on thin slabs we shall find a reflectance maximum and a transmittance minimum at $\omega \sim \omega_T$. If the electric radiation falls obliquely onto the thin slab, its component normal to the surfaces of the film can couple to the longitudinal excitation mode. We therefore can expect two transmittance minima, at $\omega \sim \omega_T$ and $\omega \sim \omega_L$. Berreman has calculated from Maxwell's equations that the absolute square of the transmitted amplitude (relative to the incident power) for very thin crystal films is given by[15]

$$T_s \approx 1 - \delta\varepsilon''/\cos\theta \qquad (12.51)$$

$$T_p \approx 1 - \delta\left\{\varepsilon''\cos\theta + \frac{\varepsilon''}{|\tilde{\varepsilon}|^2}\frac{\sin^2\theta}{\cos\theta}\right\}, \qquad (12.52)$$

where the subscripts designate the s- (vertical) and p- (parallel) polarized (with respect to the plane of incidence) components of the incident light; δ equals $2\pi d/\lambda_0$ (d = thickness of the film), and θ is the angle of incidence. We note that T_s and T_p both have minima when ε'' is a maximum, and that this is essentially independent of θ near normal incidence ($\theta = 0$). This result is well known to us: the maximum of ε'' is found near ω_T, the zero wave vector ($q \sim 0$) frequency of the TO phonon branch. If we inspect Eq. 12.52 more closely, we notice that under oblique incidence ($\theta \neq 0$) T_p possesses a second minimum wherever $|\tilde{\varepsilon}| \sim 0$, in accordance with what we have discussed just above. We already know that this second minimum occurs near ω_L, the zero wave vector limit ($q \sim 0$) of the LO phonon branch.*

The effect of the sample geometry on the reflectance of s- and p-polarized light is somewhat more involved. We assume that the film is deposited on a nonabsorbing dielectric substrate. The equations for the absolute square of

---

* Designating the effective film thickness $\delta/\cos\theta$ by the symbol Δ, we write the second and third term of Eq. 12.52 in the form $\Delta\varepsilon''\cos^2\theta$ and $\Delta(\varepsilon''/|\tilde{\varepsilon}|^2)\sin^2\theta$. We see that this describes the coupling of the modes ω_T and ω_L with the components of the electric field, $E_T = E_0\cos\theta$ and $E_L = E_0\sin\theta$, respectively (see also Ref. 1c).

the relative reflected amplitude are

$$R_s \approx f(\varepsilon_s, \theta)\left(1 + 4\delta\varepsilon'' \frac{\cos\theta}{\varepsilon_s - 1}\right) \tag{12.53}$$

$$R_p \approx g(\varepsilon_s, \theta)\left[1 + 4\delta \frac{\cos\theta}{\varepsilon_s - 1}\left(\varepsilon'' h(\varepsilon_s, \theta) - \frac{\varepsilon'' \sin^2\theta}{|\tilde{\varepsilon}|^2}\right)\right], \tag{12.54}$$

where $f, g,$ and h are certain functions of the dielectric constant ε_s of the substrate and of θ. Equations 12.53 and 12.54 show that for both s- and p-polarized light there is a reflectance maximum at large ε'' (corresponding to the reststrahlen frequency ω_T, as we already know). Furthermore, there is now—for oblique incidence—a reflectance *minimum* for the p-polarized wave whenever $|\tilde{\varepsilon}|$ is small, in other words, near ω_L. Berreman also points out that deposition of the crystal film on a conductive metal substrate should eliminate the interaction of the incident light with the transverse mode since an electric field cannot exist in a direction parallel and near to a metal surface. However, for the reflectance variations near ω_T to be considerably repressed, it is necessary to use a film thickness much less than the wavelength of the incident radiation, and that $\varepsilon_s \gg |\tilde{\varepsilon}|$.

Figure 12.21 shows Berreman's calculation of the transmittance of s- and p-polarized radiation under 30° incidence on a LiF film of 0.20 μ thickness. The computations were carried out with the help of Eqs. 12.51 and

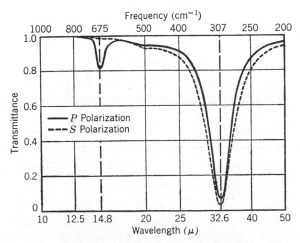

Fig. 12.21 Computed transmittance of s-polarized and p-polarized radiation of a LiF film of 0.20-μ thickness. Angle of incidence = 30°. The dashed lines show the positions of ω_L (14.8 μ or 675 cm^{-1}) and ω_T (32.6 μ or 306 cm^{-1}). [D. W. Berreman, *Phys. Rev.* **130**, 2193 (1963).]

12.52 in combination with a dispersion formula (such as the one given by Eq. 12.19). The dispersion formula was solved for $\Omega (= \omega/\omega_T) \sim 1$ and for $\varepsilon''/|\tilde{\varepsilon}|^2$ near $\Omega_L = (\varepsilon_0/\varepsilon_\infty)^{1/2}$ (see Eq. 12.23), thereby yielding the band shapes at ω_T and ω_L, respectively. The damping function, $\gamma(\omega)$, was obtained from dispersion data on LiF published in the literature. Some of Berreman's data, together with the computed dispersion curves, are shown in Figs. 12.22 and 12.23. Figure 12.23 displays the reflectance of LiF mounted on a silver support. We notice that there is indeed hardly any interaction of the transverse phonon branch with the incident radiation, in accordance with Berreman's prediction (we remember that $\omega_L > \omega_T$).

The data in Fig. 12.23 show that the maximum absorptance corresponding to ω_L is, unexplicably, at slightly different wavelengths for the different deposition temperatures. On the other hand, the film deposited on the heated substrate has a peak absorption which is very close to the frequency predicted for ω_L, but its absorption is greater (throughout the band and at the peak) than predicted. Burstein et al.[1c] have discussed that this discrepancy appears to be a consequence of the fact that Berreman used the damping function $\gamma_T(\omega)$ for the form of the longitudinal dispersion function: $\gamma_L(\omega)$ and $\gamma_T(\omega)$ are not necessarily equivalent for a thin slab.

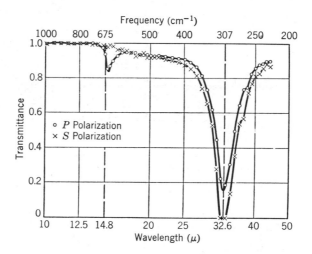

Fig. 12.22   Observed transmittance of s-polarized and p-polarized radiation by a LiF film of 0.20-μ thickness deposited on collodion. Angle of incidence 26–34°. The measurements were performed with a Perkin-Elmer Model 99 double-pass monochromator and CsBr and CsI prisms. The polarizer consisted of pyrolite graphite (8-μ thickness) between CsI plates. [D. W. Berreman, *Phys. Rev.* **130**, 2193 (1963).]

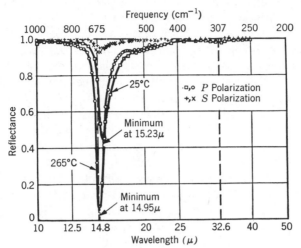

Fig. 12.23   Observed reflectance of s-polarized and p-polarized radiation by a LiF film on silvered glass. Film thickness $\bigcirc$, $\times = 0.325\,\mu$; $\square$, $+ = 0.348\,\mu$. Angle of incidence 26–34°. The minimum reflectance at $15.23\,\mu$ corresponds to a film deposited at 25°C, the minimum reflectance at $14.95\,\mu$ to a film deposited at 265°C. The ratio of the film thickness to the wavelength of the radiation at ω_T is approximately 0.01. [D. W. Berreman, *Phys. Rev.* **130**, 2193 (1963).]

c. Selection Rules and General Appearance of the Spectra

The selection rules for the interaction of electromagnetic radiation with the crystal vibrations are based on the requirements of conservation of energy and momentum of photons and phonons:

$$\hbar\omega^K = \sum_i \pm \hbar\omega^q \tag{12.55}$$

$$\mathbf{K} + n\mathbf{b} = \sum_i \pm \mathbf{q}_i, \tag{12.56}$$

where ω^K and ω_i^q are the angular frequencies of the absorbed photon of wave vector $\mathbf{K}$ and the generated or destroyed phonon of wave vector $\mathbf{q}_i$, respectively, n is an integer, and $\mathbf{b}$ is a translational vector in the reciprocal lattice.[16] Since $\mathbf{K} \sim 0$ in the infrared (see Section C.a), Eq. 12.56 simplifies to

$$n\mathbf{b} = \sum_i \pm \mathbf{q}_i. \tag{12.57}$$

If $n = 0$, we deal with a one- or with a two-phonon process; for a three-phonon process we can have $n = 0, \pm 1$ (we shall not consider higher processes). The one-phonon process is well known to us. This is the absorption of a photon and the generation of a TO (or LO under special circumstances) phonon at the center of the Brillouin zone ($\mathbf{q} \sim 0$), that is, at ω_T (or ω_L). A two-phonon process may take place if there are anharmonicity terms in the potential function or in the series development of the transition moment. We again notice the close correspondence with the phenomena in molecular spectra. There is, however, an important difference: for cubic crystals possessing a center of symmetry (alkali halide and diamond structures), the vectors $\mathbf{q}_i$ and $\mathbf{q}_j$ of the two phonons must belong to different phonon branches. In other words, there is no overtone $2\omega_i$ in such crystals.[17] We shall now describe, in brief, the physical processes which are responsible for the occurrence of two-phonon processes. We note here that it is possible to separate the effects of electric and potential anharmonicity since in a homopolar crystal such as diamond there can be no infrared activity—even in the presence of anharmonic potential terms—without a charge deformation.[18] (We had touched on this in the discussion of the linear diatomic chain in Section C.a).

The effect of the electric anharmonicity can be described in terms of a coupling of the photon to two phonons of quantum numbers n_j and n_i, one of which (compressional mode) produces a shift in the effective charges on the ions and the other simultaneously causes the induced charges to oscillate and to couple to the radiation field.[19] This is shown schematically in Fig. 12.24. If the process creates two phonons, the observed frequency

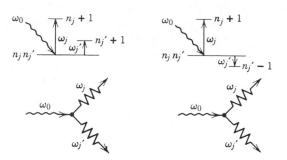

Fig. 12.24 Two-phonon optical processes due to the anharmonicity of the transition moment. The photon is absorbed by coupling directly with two phonons of excitation states n_j, $n_{j'}$ and frequencies ω_j, $\omega_{j'}$, respectively, leading to summation bands $\omega_j + \omega_{j'}$, $q_j = -q_{j'}$, and difference bands $\omega_j - \omega_{j'}$, $q_j = q_{j'}$. Photons are presented by ∿, phonons by ⋀⋀. [E. Burstein, "Interaction of Phonons with Photons: Infrared, Raman, and Brillouin Spectra," in *Phonons and Phonon Interactions*, Th. A. Bak, Ed., Copyright 1964, W. A. Benjamin, Inc., New York.]

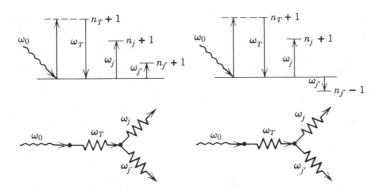

Fig. 12.25 Two-phonon optical processes due to the anharmonicity of the potential function. The photon ω_0 couples first to a $q \sim 0$ phonon n_T, ω_T, which then couples with two phonons n_j, ω_j and $n_{j'}, \omega_{j'}$, leading to summation and difference bands $\omega_j + \omega_{j'}, q_j = -q_{j'}$ and $\omega_j - \omega_{j'}$, $q_j = q_{j'}$, respectively. [E. Burstein, "Interaction of Phonon with Photons: Infrared, Raman, and Brillouin Spectra," in *Phonons and Phonon Interactions*, Th. A. Bak, Ed., Copyright 1964, W. A. Benjamin, Inc., New York.]

would be that of a summation band, $\hbar\omega_0 = \omega_j + \omega_i$, if the irreducible representations of the branches are different, or that of the first overtone, if they belong to the same representation. If the absorption of a photon creates one phonon and destroys another, the resulting frequency is that of a difference band, $\omega_j - \omega_i$.

The process involving anharmonic potential terms is more complicated since a transversal optical phonon serves as an intermediate (virtual) state. This is shown in Fig. 12.25.

The symmetry species of the wave vectors are established by group theoretical arguments similar to those employed in molecular spectra, only the space group of the lattice replaces the point group of the molecule. The computations of character tables can be found in the literature; we shall have no occasion to use them.[20] The selection rules for the infrared spectra of crystals due to higher-order effects have been derived by Szigeti[17] in a lengthy perturbation calculation, the perturbing terms being the anharmonic contributions of the electric transition moment and of the potential function. In Table 12.8 we give some of the infrared-active two-phonon combinations of alkali halides,[21] and in Fig. 12.26 we show a dispersion relation for NaCl in which we have entered some of the allowed transitions. As we mentioned above, for cubic crystals possessing a center of inversion (alkali halide and diamond structure) there are no overtones, 2LO, 2TA, and so forth. We further notice that active combinations are located at the various critical points (or symmetry points) but not at the center of a Brillouin zone. This is a

TABLE 12.8*

INFRARED ACTIVE TWO-PHONON COMBINATIONS OF ALKALI HALIDES

Symmetry point[†]	Active combination
Γ	None
X	None
L	TO + LA, TO + TA, LO + LA, LO + TA
$\left.\begin{array}{l}\Lambda\\\Delta\end{array}\right\}$	TO + LA, TO + TA, LO + LA, LO + TA, TO + LO, TA + LA

* *After* E. Burstein, F. A. Johnson, and R. Loudon, *Phys. Rev.* **139**, A1239 (1965).

[†] See, for instance, Ref. 1a, Fig. 25, p. 332. The symmetry point L is at $(\frac{1}{2}\frac{1}{2}\frac{1}{2})$ in the direction [111], X is at (100) in the direction [100], Λ points are in the direction [111] and Δ points in the direction [100]. (Not all symmetry points and their two-phonon combinations are given here.)

consequence of the selection rule for $n = 0$ (see Eq. 12.57). It is apparent that there is a certain ambiguity with respect to the phonon combination assignments. For instance, along direction [111] (see Fig. 12.26) there is a critical point within the Brillouin zone (Λ) where LO and LA branches have a

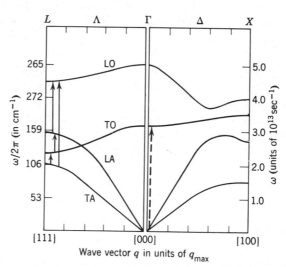

Fig. 12.26   The dispersion relation $q(\omega)$ of NaCl for the [100] and [111] directions, calculated by H. R. Hardy and A. M. Karo, *Phil. Mag.* **5**, 859 (1960), with a model that allows for ionic deformability. The arrows show the allowed difference bands at symmetry point L. The dotted arrow denotes the reststrahlen band. See also Figs. 12.17, 12.18, and 12.19. The location of the symmetry points L, Λ, Γ, X, Δ is shown in Fig. 25 of Ref.1a, p. 332. See also the footnote to Table 12.8. [*After* E. Burstein, F. A. Johnson, and R. Loudon, *Phys. Rev.* **139**, A1239 (1965).]

relative flat course and therefore a high density of states (see Section C.a). We see then that it is *a priori* not obvious whether a two-phonon combination along direction [111] is near symmetry point Λ or L. The unambiguous assignment requires a good knowledge of the dispersion relation between ω and $\mathbf{q}$. We do not wish to go into a description of the procedures that help to distinguish between the different symmetry points, but we merely mention here that empirical correlations between the effective charge and the relative position of the phonon branches are employed.* In spite of these difficulties (it should also be remembered that the transversal phonon branches are doubly degenerate and may split in certain cases) considerable success has been achieved in the empirical assignment of the multiphonon transitions by using only the four branches LO, TO, LA, and TA at the zone boundaries. The apparent success of this approach is not hard to understand since the density of states at the zone boundaries is high. The concept ("density of states maximum") assumes that the set of phonon processes used dominates the other possible phonon processes; we note here a certain analogy to the assignments of resonances in molecular spectra as "group frequencies" rather than as normal modes.

At this point we can understand the general appearance of the infrared spectra, with their reflection or transmission peaks for $\mathbf{q} = 0$ at ω_T and (under certain circumstances) at ω_L, and the (two or higher) phonon combinations at other critical points in the Brillouin zone—the latter assumed to be located mainly at the boundaries. We then see that we can retain the concept of a single fundamental oscillator in these cubic crystals and still satisfactorily explain the occurrence of more than one resonance peak.

As an experimental example of a two-phonon assignment, we here show part of some recent work by Hadni et al.[22] on KBr in reflection and transmission. Figure 12.27 displays the dispersion of the extinction coefficient κ obtained by reflection measurements under 15° incidence on thin films. With the help of the known phonon branches of KBr (from neutron diffraction), Hadni et al. have assigned the side maximum of $\sim 160\,\mathrm{cm}^{-1}$ in Fig. 12.27 to the summation frequency $\mathrm{TO} + \mathrm{TA} = 176\,\mathrm{cm}^{-1}$ at symmetry point L $(\frac{1}{2}\frac{1}{2}\frac{1}{2})$. We show the phonon frequencies of KBr in Table 12.9. Inspection of the table reveals that the $\mathrm{TO}(L) + \mathrm{TA}(L)$ combination approximately coincides with the LO phonon branch at the center of the Brillouin zone (ω_L). Hadni et al. have corroborated this by reflection measurements of a thin KBr film on a brass plate under 60° incidence of the radiation. This is shown in Fig. 12.28.

* See P. J. Gielisse and S. S. Mitra, "Infrared Spectra of Crystals," Air Force Cambridge Research Laboratories, AFCRL Report 65-395, June 1965; see p. 59. A reference to Brout sums, $\sum_i \omega_i(q)$, is also given.

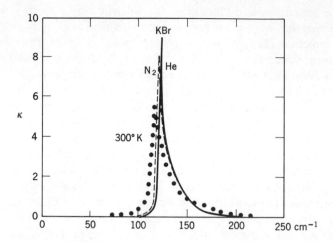

Fig. 12.27   Absorption coefficient κ of KBr computed from a two-oscillator analysis (see Eq. 12.17) of the reflectance measurements at 300°K, liquid-nitrogen, and liquid-helium temperatures. [A. Hadni, J. Claudel, D. Chanal, P. Strimer, and P. Vergnat, *Phys. Rev.* **163**, 836 (1967).]

TABLE 12.9*

CHARACTERISTIC PHONON FREQUENCIES FROM NEUTRON
DIFFRACTION AT 90°K[†]

Phonon	Frequency	
	10^{12} Hz	cm^{-1}
TO $\big\}$(000)	3.60 ± 0.03	120 ± 1
LO	5.00 ± 0.05	162 ± 1.6
LA	2.15 ± 0.03	72 ± 0.8
TA $\Big\}$(001)	1.25 ± 0.02	42 ± 1
LO	4.02 ± 0.07	134 ± 0.02
TO	3.72 ± 0.04	124 ± 1
LA	2.82 ± 0.04	94 ± 1
TA $\Big\}(\frac{1}{2}\frac{1}{2}\frac{1}{2})$	2.20 ± 0.03	74 ± 1
LO	4.34 ± 0.05	145 ± 1.3
TO	3.06 ± 0.05	102 ± 1.3

* A. Hadni, J. Claudel, D. Chanal, P. Strimer, and P. Vergnat, *Phys. Rev.* **163**, 836 (1967).

† A. D. B. Woods, B. N. Brockhouse, R. A. Cowley, and W. Cochran, *Phys. Rev.* **131**, 1025 (1963).

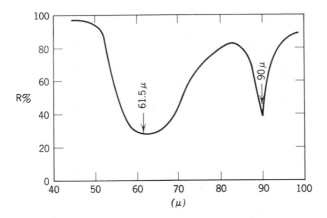

Fig. 12.28   Dispersion of the reflectance of a brass plate covered with a thin layer of KBr under 60° incidence of the radiation (p-component). The frequency of the sharp reflectance minimum at 90 μ corresponds to ω_T, that of the broader and more intense minimum at 61.5 μ (162 cm^{-1}) represents ω_L. The reflectance minimum at ω_T has appeared because the film was apparently not thin enough or ε_s not much larger than $|\tilde{\varepsilon}|$ (see Section Cb). Therefore there was resonance absorption at ω_T and a subsequent decrease in the reflectance of the coated brass. [A. Hadni, J. Claudel, D. Chanal, P. Strimer, and P. Vergnat, *Phys. Rev.* **163**, 836 (1967).]

An example of two-phonon assignments based on the concept of four phonon branches at the zone boundaries (see above) has been published by Stafsudd *et al.* on the spectrum of cadmium telluride, CdTe.[23] The crystal belongs to the cubic zinkblende structure (ZnS), which does not possess inversion symmetry. We therefore expect overtones in the spectrum. The maximum number of different two-phonon combinations amounts to $2^4 = 16$. All these sixteen combinations are considered as allowed. We note here that Birman[24] has given the selection rules for the combination bands in the Brillouin zone of the zinkblende structure. All sixteen possible two-phonon combinations are infrared-active, but not all at all symmetry points. To assign combination bands to the various symmetry points evidently requires a detailed knowledge of the dispersion surfaces of the crystal (such as shown in Fig. 12.26 for NaCl).

Figure 12.29 shows the far-infrared spectrum of a slab ($1 \times 1 \times 0.45$ cm^3) of CdTe at 290, 83, and 18°K. There was no absorption between 330 and 4000 cm^{-1} at all temperatures. The frequencies of the bands observed at 83°K are collected in Table 12.10 in the last two columns. The reason for choosing the spectrum taken at an intermediate temperature as basis for the phonon assignments is—as we can surmise by what we know from molecular spectra

TABLE 12.10*

POSSIBLE TWO-PHONON TRANSITIONS[†]

Designation	v expected in cm^{-1}	λ expected in microns	v observed in cm^{-1}	λ observed in microns	v observed,[‡] cm^{-1}
2LO	302	33.1	302	33.1	304
LO + TO	276	36.2	290	34.5	293
2TO	250	40.0	250	40.0	—
LO + LA	200	50.0	Fundamental optical		253
LO + TA	188	53.2	absorption makes	71.4–52.6	—
TO + LA	175	57.1	sample opaque		247
TO + TA	162	61.7	190–140 cm^{-1}		—
LO − TA	114	87.7	115	87.0	114
LO − LA	100	100 ⎫	Broad band 83–100		—
2LA	100	100 ⎪	cm^{-1} persists at	100–120	208
TO − TA	88	114 ⎬	18°K but reduces		—
TO − LA	75	133 ⎭	in intensity.		—
2TA	74	135	74	135	71

* O. M. Stafsudd, F. A. Haak, and K. Radisavljević, *J. Opt. Soc. Am.* **57**, 1475 (1967).
[†] The total number possible is sixteen; see text.
[‡] G. L. Bottger and A. L. Geddes, *J. Chem. Phys.* **47**, 4858 (1967), studying polycrystalline CdTe films between 40 and 360 cm^{-1}, report essentially the same assignments except those involving LA, which they assign to 103 cm^{-1} (110°K); see also Table 12.11.

—that the difference bands are either already absent on account of their low Boltzmann factors or will rapidly decrease in intensity upon further cooling. In other words, it will be the relative shift of the intensity of the different combinations as a function of temperature which will be useful in phonon assignments. We are now ready to do this on the observed bands in Fig. 12.29.

The broad and intense fundamental (ω_T) of CdTe was totally absorbing even at a 0.2-mm sample thickness. The highest-frequency two-phonon band should be 2LO, since the LO mode is at the highest frequency (see, for instance, Fig. 12.26). Hence it follows that LO = 151 cm^{-1}. This assignment was checked by estimating ω_L with the help of the Lyddane-Sachs-Teller relation (see Eq. 12.23), yielding $\omega_L \sim 172$ cm^{-1}. The bands at 290 and 250 cm^{-1} were identified by their relative positions in the spectrum; they are the next highest, after 2LO, and therefore were assigned to be LO + TO and 2TO. The TA phonon branch is most likely the one at the lowest frequency (see Fig. 12.26); therefore the band at 74 cm^{-1} is assumed to represent 2TA = $2 \cdot 37$ cm^{-1}.

We note that LO − TA would yield $151 - 37 = 114$ cm^{-1}, one of the observable bands in the spectrum. Furthermore, upon cooling the sample

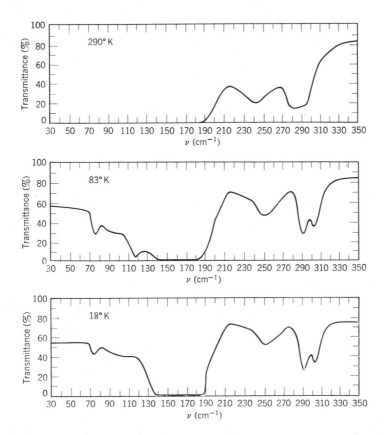

Fig. 12.29   Far-infrared spectrum of cadmium telluride, CdTe, at 290, 83, and 18°K. The spectra were taken with a modified Perkin-Elmer 210 monochromator. Sample size 1 × 1 × 0.45 cm³. The light was transmitted along a [111] axis. [O. M. Stafsudd, F. A. Haak, and K. Radisavljević, *J. Opt. Soc. Am.* **57**, 1475 (1967).]

from 83 to 18°K, we note that this band has disappeared. This behavior, relative to that of the other bands in the spectrum, supports the assignment of a difference frequency.

There remains a broad absorption between 80 and 100 cm⁻¹ which decreases somewhat upon cooling to 18°K but has not completely disappeared at this low temperature. The authors therefore assign the band to a coincidence between the difference frequency LO–LA and the overtone 2LA. We have collected the assignments of the two-phonon frequencies in Table 12.10 and the phonon energies in Table 12.11.

TABLE 12.11*

PHONON ENERGIES FOR CADMIUM TELLURIDE, CM^{-1}

Phonon	Ref. 23 at 88°K	Ref. 25 at 4.2°K	Ref. 26	Ref. 27	At 110°K[†]
LO	151 ± 2	180 ± 3	170		152
TO	125 ± 2	140 ± 3	140	144	142
LA	50 ± 2	105 ± 3			103
TA	37 ± 2	65 ± 3			37
	Density of states maximum		At center of zone		Zone edges

* O. M. Stafsudd, F. A. Haak, and K. Radisavljević, *J. Opt. Soc. Am.* **57**, 1475 (1967).
† G. L. Bottger and A. L. Geddes, see footnote‡ of Table 12.10.

D. TEMPERATURE DEPENDENCE OF PHONON INTERACTIONS

a. Occupation Number

We shall now put the temperature dependence of the absorption spectra, which we have already used for purposes of assignments (see Section C.c), on a more quantitative and fundamental footing. Phonons obey Bose-Einstein statistics; the average excitation (or occupation) number of a phonon at absolute temperature T, $\langle N \rangle$, is given by[28]

$$\langle N \rangle = [\exp(\hbar\omega/kT) - 1]^{-1}. \tag{12.58}$$

The quantum-mechanical treatment of the interaction of the radiation with a system of oscillators yields expressions for the transition probability W for creation and destruction of a phonon of the form[29]

$$\begin{aligned} W_{N,N+1} &\approx |\langle \psi_{N+1,n}|H|\psi_{N,n'}\rangle|^2 \approx N+1 \\ W_{N,N-1} &\approx |\langle \psi_{N-1,n}|H|\psi_{N,n'}\rangle|^2 \approx N. \end{aligned} \tag{12.59}$$

Here, H is the interaction Hamiltonian, N is the occupation state of the phonon (see Eq. 12.58), and n is the quantum number of the radiation field. The net number of phonons which are created in a two-phonon summation process by absorption of a photon is proportional to the difference of the probabilities of creation of two phonons and destruction of two phonons,

$$(1 + N_1)(1 + N_2) - N_1 N_2 = 1 + N_1 + N_2. \tag{12.60}$$

To arrive at this equation we need only remember that the probability of i unrelated events occurring together is given by the product $p_1 \cdot p_2 \cdots p_i$ of the individual probabilities p_i. The net number of phonons which arises if absorption of a photon leads to the creation of one phonon and to the destruction of the other (difference process) is then, analogously,

$$(1 + N_1)N_2 - N_1(1 + N_2) = N_2 - N_1. \tag{12.61}$$

In the same fashion we would write down the probability for a three-phonon process, and so forth. A table summarizing multiphonon events, including three-phonon processes, has been given by Fray et al.[30] Three-phonon processes are frequently found at energies higher than those we wish to consider here; moreover, such multiphonon bands are, as can be imagined, rather difficult to assign unambiguously from the absorption spectra.

Returning to two-phonon processes, we see from Eqs. 12.58, 12.60, and 12.61 that the temperature dependence of a summation band is given by

$$1 + (e^{E_1/kT} - 1)^{-1} + (e^{E_2/kT} - 1)^{-1}, \tag{12.62}$$

and the temperature dependence of a difference band by

$$(e^{E_2/kT} - 1)^{-1} - (e^{E_1/kT} - 1)^{-1}. \tag{12.63}$$

We notice that the expression for the occupation number of a difference band approaches zero as $T \to 0$, whereas that of the summation band stays finite (see also Fig. 12.29).

b. Frequency and Temperature Dependence of the Damping Constant

In Fig. 12.16 we have shown the temperature dependence of the damping constant γ of the dispersion equation (see, for instance, Eq. 12.16); the damping constant was there introduced ad hoc as an adjustable parameter and, in the simplest cases, assumed to be frequency-independent. The damping function can be looked upon to reflect the coupling of the dispersion oscillators with other vibrational modes due to anharmonic terms in the potential function. As discussed by Born and Huang,[14] the resulting broadening of the absorption lines is analogous to the broadening of absorption lines of molecular spectra because of anharmonicity-induced energy exchange with the surrounding medium (see, for instance Eq. 11.3).

We write the frequency-dependent damping function in the form[14]

$$\gamma_{\pm j}(\omega_{\pm}) = \left(\frac{\pi}{\hbar^2}\right) \lim_{\Delta\omega_{\pm} \to 0} \frac{1}{\Delta\omega_{\pm}} \sum_{(s)}^{\omega_{\pm}} |\langle s|\phi_j^A| \pm j\rangle|^2 \tag{12.64a}$$

where ϕ_j^A is the perturbing Hamiltonian (the anharmonic contribution to the potential); the symbol $\sum_{(s)}^{\omega_{\pm}}$ designates summation over the (practically continuous spectrum of) states s which have transition frequencies in the interval $\Delta\omega_{\pm}$, the symbol $\pm j$ defines the states which are obtained when the quantum number n_j of the dispersion oscillator j changes by ± 1, and $\omega_{\pm}$ denotes the transition frequencies $\omega_{\pm j} \pm \omega_s$, that is, in our previous nota-

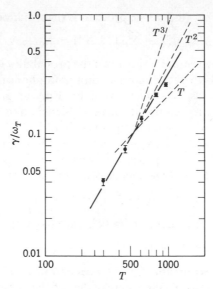

Fig. 12.30   Damping constant of NaCl as a function of absolute temperature T. The limits of error correspond to $\pm 1\%$ in the reflectivity. [M. Hass, *Phys. Rev.* **117**, 1497 (1959).]

tion, $\omega_{\pm} = \pm\omega_T \pm \omega_j(q_j) \pm \omega_{j'}(q_{j'}) = \pm\omega_T \pm \omega$, where ω is the frequency of the involved photon (see Fig. 12.25). Using this expression for $\gamma_{\pm}$, the resulting dispersion formula assumes the form, at $T = 0°K$, of

$$\tilde{\varepsilon}(\omega) \propto \frac{1}{\{\omega_j^2 - \omega^2 - i\omega_j[2\gamma_{+j}(\omega_j - \omega)]\}} \tag{12.64b}$$

in the absorption region. This expression should be compared with, for instance, Eq. 12.19. The constant γ of Eq. 12.19 is here replaced by the frequency-dependent term $2\gamma_{+j}(\omega_j - \omega)$.

If ϕ_j^A is considered to contribute mainly cubic terms, the temperature dependence of $\gamma_{\pm}$ should be proportional to T^3.[14] However, as shown by the data in Fig. 12.30, the experiment suggests a smaller power in T at higher temperatures.

Better agreement with the experimentally found temperature dependence of the damping function is predicted by a theory of Bilz and Genzel.[31a] In addition to the coupling process shown in Fig. 12.25—which is designated a "H_S–H_A process" in which H_S describes the photon–TO ($q \sim 0$) phonon coupling and H_A describes the anharmonicity-induced three-phonon process—the authors take into account the "Umkehr" process, namely H_A–H_S. This H_A–H_S process is shown in Fig. 12.31 on the example of a difference band. We see in this example that the original state of the TO

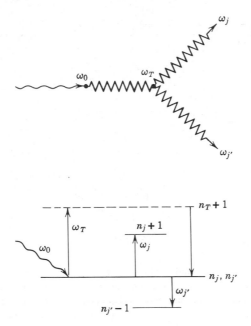

Fig. 12.31   Level scheme of a H_A–H_S process on the example of a difference band. (Compare with Fig. 12.25.)

($q \sim 0$) dispersion oscillator is regained *after* generation of phonon j and destruction of phonon j'.

The scheme of the possible transitions in NaCl-type crystals upon absorption of a quantum of light in the Brillouin zone is shown in Fig. 12.32, taken from Bilz and Genzel's work.[31a] The scheme is to be understood as an approximation for sufficiently long times, that is to say, all intermediate states $|\pm R, \pm s'\rangle$, which do not reach the final state $|0, \pm s\rangle$ determined by $\hbar\omega$, are omitted. Note that the dispersion oscillator is designated by the symbol ω_R and that the damping terms are given in terms of the reduced frequencies. The notation $|\pm R, 0\rangle$ corresponds to Born and Huang's symbol $\pm j$ (see above).

If again we set the temperature to $T = 0°K$, we need only consider absorption processes, which are the upper two processes in Fig. 12.32; in Born and Huang's theory only the lower of these two pathways is considered.

As it turns out, the process H_A–H_S introduces an additional damping term into the dispersion formulas Eqs. 12.64a, b. The effect of this is that the temperature variation of the damping term at high temperatures is now in better agreement with the observed data. We note, in conclusion, that Maradudin *et al.* have performed a numerical calculation on the dispersion

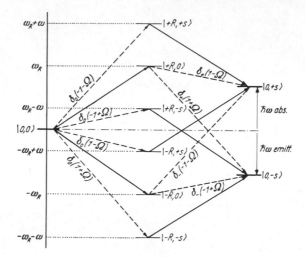

Fig. 12.32   Level scheme of the possible transitions of NaCl-type crystals on interaction with infrared radiation. The scheme shows absorption and induced emission of radiation of energy $\hbar\omega$ in the Brillouin zone. The solid lines represent the dipole transitions of the dispersion oscillator ω_R (usual notation ω_T), and the dotted lines give the anharmonicity-induced three-phonon transitions. The quantities δ designate the damping terms as a function of the reduced frequency $\Omega = \omega/\omega_R$; $|0, 0\rangle$ designates the initial state, $|0, \pm s\rangle$ the final states. The intermediate states are designated by $|\pm R, \pm s\rangle$, where $|\pm R\rangle$ indicates the state of the TO$(q \sim 0)$ phonon, that is, the dispersion oscillator; for instance, $|\pm R, 0\rangle$ means that the dispersion oscillator has gained (lost) one quantum with respect to the ground level $|0, 0\rangle$. [H. Bilz and L. Genzel, Z. Physik **169**, 53 (1962), by permission of Springer: Berlin-Göttingen-Heidelberg.]

of ε'' of NaCl and LiF at room temperature, based on an expression of the damping function which they had derived from first principles. The agreement of their calculation with the data is very good, but discussion of their work[31b] is beyond the scope of this section.

With respect to second-order spectra induced by the anharmonicity in the dipole moment, we mention that work on this has been reported on diamond which, as we discussed in Section C.c, does not possess a first-order infrared spectrum. It has also recently been shown that a first-order spectrum can be induced in diamond by the application of an external static electric field; this field causes the formation of an "effective" ionic charge on the atoms and thus the $q \sim 0$ TO and LO vibrations become infrared-active.

However, these interesting phenomena are generally observable at frequencies which lie in the medium-infrared and infrared; we therefore shall not discuss them in this book. The interested reader is referred to the article by E. Anastassakis, S. Iwasa, and E. Burstein [*Phys. Rev. Letters* **17**, 1051 (1966)], and references cited there.

E. STRUCTURE FAULTS AND PHONON PROCESSES

a. Supertransparency

As early as 1912 it was shown by Rubens and Hertz that crystals become transparent in the far-infrared when cooled to very low temperatures. We have discussed this phenomenon in Section C.c for CdTe, where we showed that the long-wavelength absorption—which occurs mainly by two-phonon differences—disappears at the very low temperatures on account of their Bose-Einstein population (see Eqs. 12.61 and 12.63). The argument can be turned around to state that if there is strong absorption in the far-infrared at room temperature and this absorption disappears on cooling, then the bands are the two-phonon differences, most likely of the acoustic branches since these are at the lowest frequencies in the crystal.

In contrast, glasses mainly absorb by one-phonon processes; their long-wavelength transmission has been shown to vary little upon cooling to the very low temperatures.[32] The reason is that the law of wave vector conservation can no longer be satisfied: the $\mathbf{q} \sim 0$ phonon vector is transformed into a phonon of $\mathbf{q} \neq 0$ at the faults in the structure of the solid because of the local disappearance of the translational symmetry. The formation of two-phonon combination processes is therefore prevented.

The same phenomenon occurs when a crystal possesses faults, either foreign ions (introduced by doping, for instance) or isotopically substituted atoms. The impurity can give rise to localized modes, that is, a phonon which decays exponentially as one moves away from the fault.[33] The absorption frequency of the local phonon is outside the optical and acoustical bands. The local phonon modes are sensitive to changes of the temperature since these affect the local geometry, for instance the interatomic distances, at the fault.

Beside these local phonon modes, the impurity can induce absorption within the lattice by exciting single lattice modes in a one-phonon process, which under certain circumstances can become resonant (large amplitude). This one-phonon absorption, which may occur in the range of the optical *as well as* the acoustical phonon branches of the unperturbed lattice, is broad since the motion affects the whole lattice—in contrast to the local phonon absorption. The effect of the temperature is minor, as we have indicated above for the case of one-phonon absorption in glasses.

In the following, we shall give a brief introductory description of some of the above aspects. For a more thorough treatment we refer the reader to Appendix I.

b. Diatomic Linear Chain with Symmetrically Placed Lattice Faults

Recently, a great deal of theoretical work has been published on the dynamical motion in the linear chain containing impurities which are either

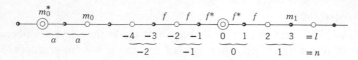

Fig. 12.33 Diatomic linear chain with symmetrically placed impurity mass m_0^*. [*After* L. Genzel, K. F. Renk, and R. Weber, *Phys. Stat. Sol.* **12**, 639 (1965).]

statistically distributed or placed symmetrically.[34-36] Here we treat the second case,[35,36] remembering that the linear chain is much simpler to consider than the cubic lattice and still offers a great deal of insight into the problems (see, for instance, Section C.a).

Figure 12.33 shows the particular linear chain of equidistant masses m_0 and m_1, the fault mass m_0^*, and the force constants f and f^*. The index l gives the position of mass, and n designates the number of the unit cell. The mass m_0^* is put at the symmetry center of each periodicity interval containing N unit cells. Newton's equations of motion in terms of the displacements u of the masses are, for the undistorted region,

$$
\begin{aligned}
- m_1 \ddot{u}_{2l+1} &= f(u_{2l+1} - u_{2l}) + f(u_{2l+1} - u_{2l+2}), & l \neq 0, -1 \\
- m_0 \ddot{u}_{2l} &= f(u_{2l} - u_{2l-1}) + f(u_{2l} - u_{2l+1}), & l \neq 0.
\end{aligned}
\tag{12.65}
$$

For the region of the faults, we have three equations of motion, namely those describing the displacements of m_0^* and the two masses m_1 on either side of the fault:

$$
\begin{aligned}
- m_0^* \ddot{u}_0 &= f^*(u_0 - u_{-1}) + f^*(u_0 - u_1) \\
- m_1 \ddot{u}_1 &= f^*(u_1 - u_0) + f(u_1 - u_2) \\
- m_1 \ddot{u}_{-1} &= f(u_{-1} - u_{-2}) + f^*(u_{-1} - u_0).
\end{aligned}
\tag{12.66}
$$

The set of Eq. 12.65 are those of the unperturbed chain, the solutions of which we gave in Eq. 12.36:

$$
\begin{aligned}
u_{2l} &= A_0 \exp\{i(2lqa - \omega t)\}, \\
u_{2l+1} &= A_1 \exp\{i([2l+1]qa - \omega t)\}.
\end{aligned}
\tag{12.67}
$$

If we substitute Eq. 12.67 into Eq. 12.65, we obtain for the ratio of the amplitudes A_0/A_1 in the unperturbed chain

$$
\frac{A_0}{A_1} = \left(\frac{2f - m_1 \omega^2}{2f - m_0 \omega^2} \right)^{1/2}.
\tag{12.68}
$$

To solve Eq. 12.65 and 12.66 (the whole chain), we take the set of Eq. 12.67 as trial solution except that we must now extend Eq. 12.67 by an additional phase factor, the value of which is obtained from the boundary conditions.[†] The additional phase factor is called the "scattering phase," α. The effects of the impurity mass m_0^* and the corresponding force constant f^* are conveniently expressed in terms of defect parameters defined by

$$\varepsilon = \frac{m_0 - m_0^*}{m_0}, \qquad -\infty < \varepsilon < 1$$

and   (12.69)

$$\gamma = \frac{f - f^*}{f}, \qquad -\infty < \gamma < 1$$

so that the perturbing mass and force constant are given by

$$m_0^* = m_0(1 - \varepsilon), \qquad f^* = f(1 - \gamma), \qquad (12.70)$$

respectively. The correct solutions of Eq. 12.66 must, of course, satisfy the value of A_0/A_1 of Eq. 12.68. We only give the final result here[35,36] in terms of the displacement of masses m_0^* and m_1:

$$u_0^* = \frac{F_0}{\{(\sigma F_0)^2 + (\rho F_1)^2\}^{1/2}} A_0; \qquad u_{\pm 1} = \left[1 - \frac{\Omega^2}{\kappa}\right] u_0^*$$

with   (12.71)

$$\cos \alpha = \frac{\sigma F_0}{\{(\sigma F_0)^2 + (\rho F_1)^2\}^{1/2}},$$

$$F_0 = \left\{\frac{m_1}{\mu}(1 - \Omega^2)\left|1 - \frac{m_0}{\mu}\Omega^2\right|\right\}^{1/2}, \qquad F_1 = \left\{\frac{m_0}{\mu}\Omega^2\left|1 - \frac{m_1}{\mu}\Omega^2\right|\right\}^{1/2},$$

where α is the scattering phase mentioned above, and in terms of the already defined quantities $\mu^{-1} = m_0^{-1} + m_1^{-1}$, $\Omega = \omega/\omega_T$, and $\omega_T = (2f/\mu)^{1/2}$, we have $\kappa = \mu(1 - \gamma)/m_0(1 - \varepsilon)$, $\sigma = 1 - (\gamma/\kappa)\Omega^2$, and $\rho = \varepsilon - (\gamma/\kappa)\Omega^2$.

Inspection of Eq. 12.71 indicates various limiting cases for u_0^* and $u_{\pm 1}$: *(1)* If at a certain $\omega = \omega_\kappa$ we find that Ω^2 equals κ, then $u_{\pm 1} = 0$; in other words, the masses next to m_0^* do not vibrate. *(2)* For $\omega < \omega_\kappa$, the neighboring masses of the impurity mass m_0^* vibrate in phase with m_0^*, whereas for $\omega > \omega_\kappa$ the neighboring masses vibrate with a phase shift of π with the mass at the fault. We therefore see that $\Omega_\kappa = \omega_\kappa/\omega_T$ divides the frequency regions

[†] The set of Eq. 12.67 is not a solution of the set of Eq. 12.65, 12.66; see B. Szigeti, *Proc. Phys. Soc. (London)* **65B**, 19 (1952) and *J. Phys. Chem. Solids* **24**, 225 (1963) for the simpler example of a monoatomic chain containing an impurity mass.

of acoustic (in-phase) and optic (out-of-phase) behavior with respect to the vibrations at the impurity. (3) In case of a weakly bound impurity ($f^* \sim 0$) it follows that $1 - \gamma \ll 1$. Hence $\kappa \approx \kappa/\gamma$ are also small (and nearly equal). We express this by setting $\Omega^2 \ll 1$ in the expression for u_0^* (see Eq. 12.71). It is easily seen then that $u_0^{*2}/A_0^2 \approx 1/[\sigma^2 + (m_0/m_1)\rho^2\Omega^2]$. This approximation leads to a sharp resonance if $\Omega^2 \approx \kappa/\gamma$ because $\sigma \to 0$ and $\rho^2\Omega^2 \to (\varepsilon - 1)^2\kappa/\gamma = (\mu/m_0\gamma)(1 - \varepsilon)(1 - \gamma)$; hence $u_0^{*2}/A_0^2 \approx (m_1\gamma/\mu)/[(1 - \gamma) \times (1 - \varepsilon)]$. The impurity mass m_0^*, driven by the rest of the lattice, performs large-amplitude forced vibrations ("resonant band absorption"). The scattering phase amounts to $\alpha = \pi/2(\sigma = 0)$, indicating the phase shift between the driving force and the damped oscillator.[36] (4) If $\rho = 0$, that is, $\Omega = (\varepsilon\kappa/\gamma)^{1/2}$, the scattering phase is zero (as in the ideal chain); hence $u_0^* = A_0/\sigma = A_0(1 - \varepsilon)^{-1}$.

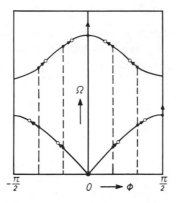

Fig. 12.34   Diatomic linear chain with isotopic fault mass and $2N = 12$ particles per periodicity interval. The abscissa is given in terms of $\phi = q \cdot a$; the ordinate, in reduced frequency units $(\Omega = \omega/\omega_T)$; $\varepsilon > 0$ and $m_0 > m_1$. The perturbed frequencies decrease for $\varepsilon < 0$. [K. F. Renk, Z. Physik 201, 445 (1967), by permission of Springer: Berlin-Göttingen-Heidelberg.]

Finally, we show in Fig. 12.34 a plot[35] which compares the eigen frequencies of an ideal diatomic chain ($m_1 < m_0$) with those of a diatomic chain containing an isotopic fault mass ($f^* = f$) and $2N = 12$ particles per periodicity interval. We remember the dispersion relation of the ideal diatomic chain (see Fig. 12.17) where for each value of $\phi = q \cdot a$ we find an acoustic and an optic frequency. For the perturbed chain, however, one computes (from the scattering phase α and the periodic boundary conditions)

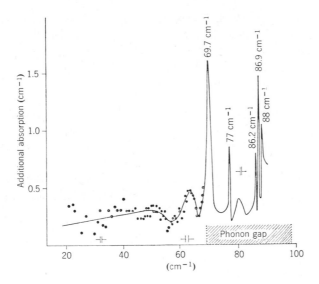

Fig. 12.35 Hydroxyl-ion one-phonon absorption in KI (0.5 mole-% KOH). The reststrahlen band of KI lies at 108 cm^{-1}. The measurements were performed at 5°K with a modulated interferometer. [K. F. Renk, Z. *Physik* **201**, 445 (1967), by permission of Springer: Berlin-Göttingen-Heidelberg.]

that $N\phi$ is given by[35]

$$\tan (N\phi) = \pm \varepsilon \Omega D(\Omega)\left(\frac{m_0}{m_1}\right)^{1/2}\frac{1 - (m_0/\mu)\Omega^2}{1 - 2\Omega^2}$$

(12.72)

with

$$D(\Omega) = |1 - 2\Omega^2|\left\{(1 - \Omega^2)\left(1 - \frac{m_0}{\mu}\Omega^2\right)\left(1 - \frac{m_1}{\mu}\Omega^2\right)\right\}^{-1/2}.$$

We see that for $\phi = 0$, that is $q = 0$, the optical solution ($\Omega = 1$) does not satisfy Eq. 12.72. Likewise, the acoustical solution for $\phi = \pi/2$ (right zone boundary) is lost. [From Eq. 12.45 it follows that (for $m_0 > m_1$) $\omega_- = (2f/m_0)^{1/2}$ at the boundary. Thus, the acoustical solution of the ideal chain is $\Omega = (2f/m_0)^{1/2}/\omega_T = \{m_1/(m_1 + m_0)\}^{1/2}$.] The solutions of Eq. 12.72 are indicated in Fig. 12.34 by open circles and the unperturbed solutions by solid dots. Note that the acoustical and optical frequencies in the perturbed diatomic chain have increased ($\varepsilon > 0$) compared to those of the ideal chain; the degeneracy in ϕ is removed. Furthermore, the optical solution at $q = 0$ has moved into the forbidden range $\Omega > 1$ ("local optical mode") and the acoustical solution at $q = \pi/2a$ has moved into the forbidden gap ("localized gap mode").

A more detailed discussion of the impurity-induced absorption is beyond the scope of this section. We refer the interested reader to Appendix I of this book. In conclusion, in Fig. 12.35 we show some experimental results on the long-wavelength one-phonon absorption in potassium iodide doped with hydroxyl ions.[35] The shaded area in Fig. 12.35 indicates the phonon gap between the acoustical and the optical branches in KI. We notice the (sharp) impurity-induced phonon gap modes between 68 and 88 cm^{-1} and the (broad) impurity-induced band absorption in the acoustical region.

REFERENCES

1a. E. Burstein, "Interactions of Phonons with Photons: Infrared, Raman, and Brillouin Spectra" in *Phonons and Phonon Interactions*, (Aarhus Summer School Lectures, 1963), Th. A. Bak. Ed., Benjamin, New York, 1964.

1b. S. Szigeti, *Trans. Faraday Soc.* **45**, 155 (1949).

1c. E. Burstein, S. Iwasa, and Y. Sawada, *Estratta da Rendiconti della Scuola Internazionale di Fisica "E. Fermi,"* Course XXXIV. See also C. Kittel, *Introduction to Solid State Physics*, 3rd ed., Wiley, New York, 1967.

2. L. Genzel, H. Happ, and R. Weber, *Z. Physik* **154**, 13 (1959).

3. M. Czerny, *Z. Physik* **65**, 600 (1930).

4. R. Geick, *Z. Physik* **166**, 122 (1962).

5. R. Geick, *Z. Physik* **163**, 499 (1961).

6. M. H. Brodsky and E. Burstein, *J. Phys. Chem. Solids* **28**, 1655 (1967).

7a. E. Burstein, S. Perkowitz, and M. H. Brodsky, "The Dielectric Properties of the Cubic IV-VI Compound Semiconductors," in *Proceedings of the 1968 C.N.R.S. International Colloquium on IV–VI Semiconductor Compounds, Gif-sur-Ivette, France.*

7b. B. G. Dick, Jr. and A. W. Overhauser, *Phys. Rev.* **112**, 90 (1958). See also Reference 1a.

8. E. E. Havinga, *Phys. Rev.* **119**, 1193 (1960).

9a. G. O. Jones, D. H. Martin, P. A. Mawer, and C. H. Perry, *Proc. Roy. Soc. (London)* **A261**, 10 (1961).

9b. E. Burstein, M. H. Brodsky, and G. Lucovsky, *Intern. J. Quantum Chem.* **IS**, 759 (1967).

10. C. Kittel, *Introduction to Solid State Physics*, Wiley, New York, 1967, see Chapter 1.

11. J. R. Hardy, *Phil. Mag.* **7**, 315 (1962).

12. E. W. Kellermann, *Phil. Trans. Roy. Soc. (London)* **A238**, 513 (1940).

13. J. M. Ziman, *Principles of the Theory of Solids*, Cambridge Univ. Press, London, 1964, see Chapter 2.

14. M. Born and K. Huang, *Dynamical Theory of Crystal Lattices*, Oxford Univ. Press, London, 1962.

15. D. W. Berreman, *Phys. Rev.* **130**, 2193 (1963).

16. P. J. Gielisse and S. S. Mitra, *Air Force Cambridge Res. Labs. Rept. AFCRL-65-395*, June 1965, 85 pp. In accordance with the literature, in this chapter we have used the same symbol for different quantities. The reader should have no difficulty in keeping the various meanings apart, since they are always clearly defined.

17. B. Szigeti, *Proc. Roy. Soc.* (*London*) **A258**, 377 (1960); *ibid.* **A252**, 217 (1959); see also R. E. Peierls, *Quantum Theory of Solids*, Oxford Univ. Press, London, 1955, Introduction to Chapter III.

18. M. Lax and E. Burstein, *Phys. Rev.* **97**, 39 (1955).

19. See Ref. 1a, pp. 281–283.

20. L. P. Bouckaert, R. Smoluchowski, and E. Wigner, *Phys. Rev.* **50**, 58 (1936).

21. E. Burstein, F. A. Johnson, and R. Loudon, *Phys. Rev.* **139**, A1239 (1965).

22. A. Hadni, J. Claudel, D. Chanal, P. Strimer, and P. Vergnat, *Phys. Rev.* **163**, 836 (1967).

23. O. M. Stafsudd, F. A. Haak, and K. Radisavljević, *J. Opt. Soc. Am.* **57**, 1475 (1967).

24. J. L. Birman, *Phys. Rev.* **131**, 1489 (1963).

25. G. A. Slack, F. S. Ham, and R. M. Chrenko, *Phys. Rev.* **152**, 376 (1966).

26. P. Fisher and H. Y. Fan, *Bull. Am. Phys. Soc.* **4**, 409 (1959).

27. A. Mitsuishi, *Phys. Soc. Japan* **16**, 533 (1961).

28. See Ref. 10, p. 166, and S. Glasstone, *Theoretical Chemistry*, Van Nostrand, New York, 1955, p. 326.

29. J. M. Ziman, *Electrons and Phonons*, Clarendon Press, Oxford, 1960, Chapter III, p. 134.

30. S. J. Fray, F. A. Johnson, and R. H. Jones, *Proc. Phys. Soc.* (*London*) **76**, 939 (1960).

31a. H. Bilz, L. Genzel, and H. Happ, *Z. Physik* **160**, 535 (1960); H. Bilz and L. Genzel, *ibid.* **169**, 53 (1962).

31b. I. P. Ipatova, A. A. Maradudin, and B. F. Wallis, *Phys. Rev.* **155**, 882 (1967).

32. A. Hadni, J. Claudel, X. Gerbaux, G. Morlot, and J.-M. Munier, *Appl. Opt.* **4**, 487 (1965).

33. See, for instance, Ref. 10, pp. 156–158.

34. G. Ernst, *Z. Physik* **203**, 214 (1967), and references cited therein.

35. K. F. Renk, *Z. Physik* **201**, 445 (1967).

36. L. Genzel, K. F. Renk, and R. Weber, *Phys. Stat. Sol.* **12**, 639 (1965).

13 Far-Infrared Spectra
of Polyatomic Crystals

A. INTRODUCTION

In this chapter we discuss the spectra of polyatomic crystals. The wealth and the complexity of far-infrared spectral data which has accumulated in this area makes it rather difficult to select from the many subjects those we wish to discuss in the available space. Fortunately, it is possible to reduce most of the spectral characteristics of polyatomic solids to a combination of simple phenomena, phenomena which we have treated in the previous chapters. We therefore prefer to present here, first, a general description of the characteristics of the spectra of polyatomic crystals and then give, in much greater detail, a few representative examples from the literature. Since the necessary group-theoretical fundamentals have been thoroughly treated in various books[1] and articles,[2] we shall only show their application as required in context and elaborate where we feel it will be beneficial or instructive.

In Chapter 12 we saw that the fundamental motions of a diatomic crystal arise from the oscillations of two interpenetrating lattices. One of the lattices fixes the positions of the positive ions (cations), the other fixes the positions of the negative ions (anions) of the diatomic molecule. We further saw that out of a total of $6f$ fundamental vibrations in a crystal containing f diatomic molecules, it is only necessary to consider those which arise from the six degrees of freedom of the two lattices in the Brillouin zone. We also saw that only fundamental vibrations at the center of the Brillouin zone ($q \sim 0$) can possibly interact with electromagnetic radiation. Finally, we discussed that a fuller understanding of the observed spectra in terms of combination and difference bands ($q \neq 0$) required the consideration of the frequency–wave vector dispersion surfaces of the crystal.

We can discuss the spectra of polyatomic crystals in an analogous manner by dealing only with the primitive unit cell of the crystal. The primitive unit cell is the smallest unit which, by a series of translations of the primitive unit cell, would build up the whole crystal. In other words, the primitive unit cell is the building block of the crystal (for our purposes). Let us assume that there are m atoms per primitive unit cell, yielding a total of $3m$ degrees of freedom. Of this total, three degrees of freedom are required for the translation of the crystal as a whole, the so-called "acoustical modes" (see Chapter 12, Section C.a). They are designated by the symbol T. The remaining $3m - 3$ degrees are distributed among the "optical modes" (rotation of the whole crystal is not permitted as we shall see below). If the forces which bind the single atoms (or monoatomic ions) of the crystal are comparable to each other, that is, if only a single monoatomic ion sits on a lattice point, no further classification of the normal optical modes is possible and they are lumped together under the designation "lattice modes."

If, on the other hand, it is possible to classify the atoms into molecules or groups of atoms in such a way that the interatomic forces are considerably greater than the forces between the molecules or between the particular groups of atoms, respectively, then a division of the optical modes into "internal modes" (symbol n'), "translatory optic modes" (T'), and "rotatory modes" (R') can be performed.

First we mention the internal modes. Their spectrum is frequently very similar to that of the genuine vibrational fundamentals of the molecular spectrum of the particular entity. Thus the fundamental frequencies of the majority of n'-modes lie in the infrared. If there are M (nonlinear) molecules, consisting of m atoms each, in the primitive unit cell, there would be a total of $M(3m - 6)$ internal fundamental modes. If the unit cell, on the other hand, contained $M_1, M_2, \cdots$ separate (nonlinear) groups of atoms containing $m_1, m_2, \cdots$ atoms, respectively, we will find $M_1(3m_1 - 6) + M_2(3m_2 - 6) + \cdots$ internal modes. This is shown in Part d of Fig. 13.1 on the simple example of sodium azide, NaN_3.[3] The rhombohedral crystal contains one molecule per unit cell; the forces binding the nitrogen atoms in the N_3^- anion are much stronger than the Coulomb forces between the N_3^- and Na^+ ions. The (linear) anion has therefore $1(3 \cdot 3 - 5)$ internal modes; obviously, the Na^+ ion has no internal degrees of freedom.

During the rotatory modes (R'), the molecules or atomic groupings perform librations about their center of mass. The total number of these motions is $3M$ for M (nonlinear) molecules per primitive unit cell, or $3(M_1 + M_2 + \cdots)$ for $M_1, M_2, \cdots$ separate (nonlinear) atomic groupings. Figure 13.1, Part c shows these modes for NaN_3. Since the anion is linear and the cation has no rotational degrees of freedom, the total number of rotatory modes is $2M_1 = 2$. These rotatory modes are generally (but not always) the lowest-frequency optical fundamentals and (if active) of rather weak intensity in the infrared and of strong intensity in the Raman spectrum.

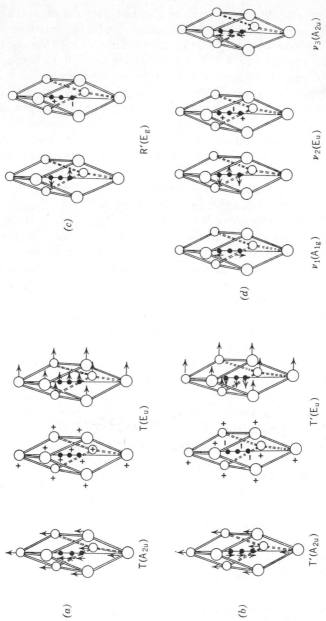

Fig. 13.1 Unit cell modes of sodium azide (rhombohedral, space group D_{3d}^5). The open circles represent Na$^+$ ions, the solid circles the N atoms of the N$_3^-$ anion. The symmetry species are those of the point group D_{3d}. T = translation of the whole unit cell (acoustic mode, nongenuine vibration); T' = translatory optic, R' = rotatory. Series d represents the internal modes, Series b and c the genuine lattice modes. Spectral activity (i = infrared, R = Raman): $\nu_2(i)$, $\nu_3(i)$, $\nu_1(R)$, $\nu_3(i)$, $b(i)$. [J. I. Bryant, *J. Chem. Phys.* **40**, 3195 (1964).]

T(A$_{2u}$)   T(E$_u$)   R'(E$_g$)   (c)

(a)

T'(A$_{2u}$)   T'(E$_u$) ν_1(A$_{1g}$) ν_2(E$_u$) ν_3(A$_{2u}$)   (d)

(b)

We mention now the translatory optic modes (T'), of which there is a total of $3M - 3$ for M molecules or M groups (single atoms included) per primitive unit cell. Hence, for NaN_3 we should find a total of three T' modes ($M = 2$, namely Na^+ and N_3^-); this is shown in Part b of Fig. 13.1. The frequencies of the translatory optic fundamentals are such that they are generally observed in the far-infrared (if active). Furthermore, since their intensity is considerable, we will deal to a large extent with the T' modes in this chapter. If we were inclined to stretch a point and to draw analogies to the spectrum of molecular NaN_3, one might call these modes "$Na-N_3$ stretch" (A_{2u}) and "deformation" (E_u).

Finally, Part a of Fig. 13.1 shows the three acoustic modes of the NaN_3 crystal, that is, the translatory motions of the whole primitive unit cell (see above). Because two directions in the crystal are equivalent, the corresponding translatory mode is doubly degenerate (E_u).

The approach just presented is a (very rough) outline of what is termed the "factor group method." This method is employed for finding all *potentially* infrared- and Raman-active fundamentals—we remember that a crystal consisting of r molecules with, say, s atoms each possesses a total of $3rs$ fundamental modes. We shall discuss the factor group method in the necessary detail in the following sections.

As our concluding remark in this section, we note that if we wish to explain fully the finer details of the spectra, such as combination and difference bands, we have to take into account the periodicity of the lattice in a more complete sense, in other words, consider the acoustical and optical dispersion surfaces ($\mathbf{q} \neq 0$). We saw in Chapter 12 how difficult this is for diatomic cubic crystals, and it is therefore easy to understand that the main efforts in the spectroscopy of polyatomic crystals have, so far, been expended on the study of the fundamentals. Therefore we will restrict our discussion in this chapter almost entirely to the factor group method. We point out, however, that the temperature dependence of the combination and difference bands is used as a means of characterizing them (see Section D of Chapter 12 and Ref. 1, p. 68, this chapter).

B. Lattice Vibration Spectra of NiF_2, CoF_2, and FeF_2

a. Factor Group Analysis. General Principles

Balkanski et al.[4] have investigated the lattice modes of nickel difluoride, cobaltous difluoride, and ferrous difluoride by reflection measurements of single crystals and transmission measurements on sintered powders. The three salts crystallize in a rutile-type lattice (tetragonal) of space group symmetry D_{4h}^{14} (see below). The spectra of these fluorides are of interest because of the similarity of their crystal structure, combined with nearly identical masses of the positive ions (Ni $\sim$ 58, Co $\sim$ 59, Fe $\sim$ 56): the differences in their normal frequencies would thus be essentially due to the specific electronic structures of the different metals. Before we discuss the experimental results we shall establish the nature of the various fundamentals and their spectral activity with the help of the known crystal structure and group-theoretical arguments.

The procedure for finding the number of normal modes (a_l) belonging to a given symmetry species (l), that is, the construction of the representation of the symmetry group which governs this vibrational problem, is carried out in the usual fashion. First, we find the characters $[\chi(R)]$ of all the symmetry operations (R) of the relevant symmetry group by subjecting the molecules to all symmetry operations R. Then we multiply the $\chi(R)$ by the corresponding characters $\chi^{(l)}(R)$ of the irreducible representation l of the symmetry group, and finally sum over all operations R. If the order of the symmetry group is h, a_l is given by the well known relation

$$a_l = (1/h) \sum_R \chi(R)\chi^{(l)}(R). \tag{13.1}$$

The procedure is repeated for all irreducible species l of the symmetry group.

We know that for molecular spectra the relevant symmetry group is the point group of the molecule. For crystal spectra we must consider the space group of the crystal, since we now have translational symmetry of the lattice in addition to any other symmetry elements, such as rotations, reflections, and so forth. As we deal with finite crystals, it is convenient to consider finite rather than infinite space groups since it is then possible to use well known principles of representation theory of finite point groups.* Winston and Halford[2] have shown the relations between the different groups; in the following we give a short summary.

We define the primitive unit cell by the fundamental translation vectors $\mathbf{t}_1, \mathbf{t}_2$, and $\mathbf{t}_3$. Letting N_1, N_2, and N_3 be fixed (positive) integers and n_1, n_2, n_3 be any integer, we assign all elements of the infinite space group which differ in their effect only by a translation

$$\mathbf{T} = n_1 N_1 \mathbf{t}_1 + n_2 N_2 \mathbf{t}_2 + n_3 N_3 \mathbf{t}_3 \tag{13.2}$$

---

* The analysis of the motions of a crystal in terms of finite space groups makes use of Born's cyclic boundary conditions [M. Born, *Proc. Phys. Soc.* **54**, 362 (1942)], which can be interpreted by assuming an infinite crystal made up of periodic repeat units of a block of edges $N_1\mathbf{t}_1, N_2\mathbf{t}_2$, $N_3\mathbf{t}_3$ which consists of primitive unit cells (see Refs. 1 and 2). This imposes the restriction that atoms which are separated by translations $n_1 N_1 \mathbf{t}_1 + n_2 N_2 \mathbf{t}_2 + n_3 N_3 \mathbf{t}_3$ (n_1, n_2, n_3 any integer) have the same displacements. The values of N_1, N_2, and N_3 (the size of the crystal block of unit cells) are immaterial.

to the same element of the finite space group. This group is then a group of symmetry operations of lattices with lattice vectors $N_1\mathbf{t}_1$, $N_2\mathbf{t}_1$, $N_3\mathbf{t}_3$. Arbitrary *translations* by these vectors carry the lattice into itself and are therefore equivalent to the identity. If the infinite space group contains H symmetry elements other than translations, the order of the finite space group just defined will be $N_1N_2N_3H$. The group of $N_1N_2N_3$ translations (Eq. 13.2) is an invariant subgroup (index H) of the finite space group. The H nonequivalent cosets of this invariant subgroup in the finite space group form a factor group of the finite space group. We have already noted that the identity element of the factor group is the invariant translation subgroup of order $N_1N_2N_3$.*

Next, it is necessary to form the representations of the space group, that is, to relate the elements of the group to the (traces of) transformation matrices. Only those representations are derived which contain potentially active fundamental modes (Raman and infrared). They are obtained by assigning to each element of the space group the matrix which corresponds to the coset of the factor group (factor group representation). Since all factor groups of space groups are isomorphous with point groups, we have thus reduced the analysis to a mere but tedious bookkeeping which uses the well-known point group character tables. The contraction is shown schematically in Fig. 13.2 on the example of the monoclinic space group C_{2h}^5.

* Let a group $\mathbf{G}$ consist of the elements $\mathbf{G} = \{E, X_1, X_2, \cdots, X_h\}$. Suppose that $\mathscr{H} = \{E, X_1, X_2, \cdots, X_a\}$ forms a subgroup of $\mathbf{G}$. We now form products, between the elements of $\mathscr{H}$ and the elements of $\mathbf{G}$ (not necessarily contained in $\mathscr{H}$), of the form $X^{-1}\mathscr{H}X$. If $X^{-1}\mathscr{H}X = \mathscr{H}$ for every X in $\mathbf{G}$, then $\mathscr{H}$ is an *invariant subgroup* of $\mathbf{G}$. If every element of the invariant subgroup $\mathscr{H}$ is multiplied by every element of $\mathbf{G}$ not contained in $\mathscr{H}$, we obtain the *cosets* of the invariant subgroup in $\mathbf{G}$. The cosets and the invariant subgroup form a group called the *factor group* (or quotient group), in which the unit element is the invariant subgroup. Example: The group D_2 has the following multiplication table

	E	A	B	C
E	E	A	B	C
A	A	E	C	B
B	B	C	E	A
C	C	B	A	E .

We see that $\mathscr{H} = \{E, A\}$ is an invariant subgroup of D_2 since $B^{-1}AB = A$, $B^{-1}EB = E$, $C^{-1}AC = A$, and $C^{-1}EC = E$ belong to $\mathscr{H}$. The elements of the cosets are obtained by multiplying $\mathscr{H}$ with elements not contained in $\mathscr{H}$. We find $\mathscr{H}C = \{EC, AC\} = \{C, B\}$; $\mathscr{H}B = \{EB, AB\} = \{B, C\}$. Hence, $D_2 = \mathscr{H} + \mathscr{H}B = \mathscr{H} + \mathscr{H}C$. Designating $\mathscr{H}C$ by $\mathscr{F}$, the multiplication table of this factor group of D_2, with the invariant subgroup $\mathscr{H}$ as unit element, is

	$\mathscr{H}$	$\mathscr{F}$
$\mathscr{H}$	$\mathscr{H}$	$\mathscr{F}$
$\mathscr{F}$	$\mathscr{F}$	$\mathscr{H}$.

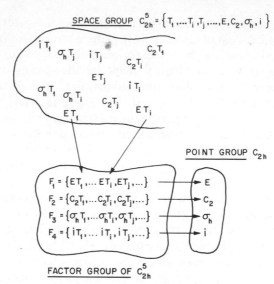

Fig. 13.2   Schematic of the relations between the space group C_{2h}^5 and its factor group. The point group C_{2h} is isomorphous with the factor group of the space group.

We now briefly indicate why only the factor group representations of the space group contain the "activity representations," that is to say, representations which can possibly lead to infrared- and Raman-active fundamentals. To this effect, we consider the fundamental symmetry element of the crystal, namely its translational symmetry, with the following arguments.

The general expression for the representations of the translation group is[5]

$$\chi^{(\kappa)}(T) = \exp\{2\pi i\kappa \cdot \mathbf{t}\}, \tag{13.3}$$

where $\mathbf{t}$ is a translation vector of the group and κ is the vector

$$\kappa = \frac{p_1}{N_1}\mathbf{b}_1 + \frac{p_2}{N_2}\mathbf{b}_2 + \frac{p_3}{N_3}\mathbf{b}_3. \tag{13.4}$$

The p_i are running integers, the N_i are fixed integers (see Eq. 13.2), and the $\mathbf{b}_i$ are the reciprocal lattice vectors ($\mathbf{b}_i \cdot \mathbf{t}_j = \delta_{ij}$). Each fixed value of κ yields a set of exponentials, $\exp(2\pi i\kappa \cdot \mathbf{t})$, which forms a representation $\Gamma(\kappa)$ of the translation group as $\mathbf{t}$ assumes all values. All $N_1 N_2 N_3$ different representations which can be written by allowing the p_i to assume integral values ($p_1 = 0$ to $N_1 - 1$, and so forth) are satisfactory (irreducible) representations of the translation group of order $N_1 N_2 N_3$.[1,2]

Now, it is not too difficult to see that the translation group has only one activity representation, that is to say, an irreducible representation which occurs in the representations of vectors (infrared) or symmetric tensors

(Raman): This is the *totally symmetric* representation. To understand this we simply remember that vectors and symmetric tensors are invariant under a pure translation operation and therefore belong to the totally symmetric representation of the translation group. The characters of the totally symmetric representation are, by definition, equal to unity. Hence, in order to obtain $\exp(2\pi i\mathbf{\kappa}\cdot\mathbf{t}) = 1$ for all $\mathbf{t}$, the wave vector $\mathbf{\kappa}$ must vanish. We are reminded here of the $\mathbf{q} \sim 0$ condition for the optical absorption of, for instance, the TO fundamental of cubic diatomic crystals (see Chapter 12). The physical picture of the restriction $\mathbf{\kappa} = 0$ is as follows: The displacement motion of any atom (ion) is *in phase* with the displacement motion of any other atom (ion) which sits at the same lattice position in any other unit cell.

We need now establish, as the final steps in the symmetry analysis outlined here, how the above selection rules of the translation group are to be translated into the selection rules of the space group. We had decided (see above) to concentrate our attention to fundamentals; in this case it turns out that the analysis under the space group is formally identical with that under the factor group.[1,2] We now give a (very rough) outline on how this comes about.

Since the translation group is a subgroup of the space group, the selection rules of the translation group are the *minimal* set which must be obeyed by the space group. Assuming m atoms per primitive unit cell, there will thus be a total of $3m$ of such potentially active fundamentals in the crystal. [This is easily seen. For our crystal block of edges $N_1 t_1$, $N_2 t_2$, $N_3 t_3$, containing $N_1 N_2 N_3$ primitive unit cells, application of Eq. 13.1 gives the number of motions falling in the representations given by the vector $\mathbf{\kappa}$. Since the only operation of the translation group (of order $N_1 N_2 N_3$) which leaves the positions of the atoms in the crystal invariant is the identity translation T_E, we find that the nonzero contributions to the character are $\chi(T_E) = 3m N_1 N_2 N_3$, $\chi^{(l)}(T_E) = 1$, hence $a_l = 3m$. Thus, *each* representation of the translation group contains $3m$ motions of the crystal.]

We now divide the $3m$ potentially active vibrations among the representations of the factor group of the space group. By the manner in which we have constructed the factor group (see above and Fig. 13.2), we notice that its representations reduce to the totally symmetric representation of the translation group if all nontranslational elements of the space group (rotations, reflections, and so forth) are eliminated. In other words, the factor group representations of the space group are the activity representations of the space group; no other representations of the space group can contain active (infrared or Raman) fundamentals.

In summary, we see that from the total of $3m N_1 N_2 N_3$ fundamental modes of a crystal block consisting of $N_1 N_2 N_3$ primitive unit cells with m atoms each, we need only consider the $3m$ fundamental modes belonging to the factor group representations of the space group of the crystal. Since this factor group is isomorphous with a point group,* the total analysis is thereby reduced to the well known principles of point groups. However, since the factor group may contain elements with translatory components, such as glide reflections and screw rotations, we shall see that the construction of the characters of the various symmetry operations is more complicated than under the point groups used in molecular spectra.

b. Derivation of the Symmetry Species of the Fundamentals

We now apply the considerations discussed in the previous section to the metal difluorides studied by Balkanski et al.[4] The crystal structure (D_{4h}^{14}) is shown schematically in Fig. 13.3; the main symmetry axis has been stretched in order to avoid crowding of the figure. We shall find it more instructive and illustrative to perform the symmetry analysis of these crystals as if we were dealing with well defined molecules (linear) of the type XF_2 rather than with coordinated ions (as is actually the case) and we shall then, at a later point, indicate the physical reality of some of the aspects of the group-theoretical analysis.

The factor group of the space group D_{4h}^{14} is isomorphous with point group D_{4h}. The symmetry operations on the atoms in the unit cell (which contains two molecules) are indicated in Fig. 13.3 in terms of permutations of identical atoms. For instance, the operation

$$(4123)(65)$$

means that among the F atoms we exchange atom 4 with atom 1, atom 1 with atom 2, atom 2 with atom 3, and atom 3 with atom 4; among the metal atoms we exchange atom 6 with atom 5. A designation such as (2) indicates that atom 2 is invariant under the operation. The factor group elements are two fourfold screw axes parallel to the c-axis of the crystal, a twofold rotation axis (C_2) in the same direction as the fourfold axis, four twofold rotation axes in directions normal to the c-crystal axis (two twofold axes are screw axes, C_2'), a plane of symmetry normal to the c-axis (σ_h), two planes of symmetry along the diagonals and parallel to the c-axis (σ_d), two glide symmetry planes (σ_v^g) along the direction of the c-axis and "cutting the crystal in half," two fourfold rotation-reflection axes (S_4), and a center of symmetry, i, at the metal ion. We note here that establishing the different symmetry operations on crystals such as shown in Fig. 13.3 is a difficult part of this analysis. In carrying out the procedure we must remember that the sym-

---

* Note that this isomorphism does not (necessarily) have the consequence that factor groups are point groups. Inspection of Fig. 13.2 shows that the operations of the factor group contain translational components. We later discuss such examples as glide reflections and screw rotations. Since such operations do not leave a point invariant, the factor group cannot be a point group.

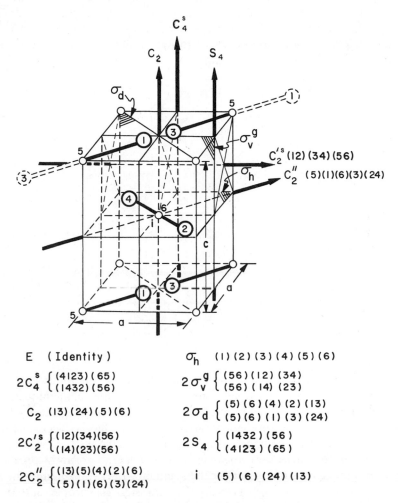

E   (Identity)

$2C_4^s$ $\begin{cases} (4123)(65) \\ (1432)(56) \end{cases}$

C_2 (13)(24)(5)(6)

$2C_2'^s$ $\begin{cases} (12)(34)(56) \\ (14)(23)(56) \end{cases}$

$2C_2''$ $\begin{cases} (13)(5)(4)(2)(6) \\ (5)(1)(6)(3)(24) \end{cases}$

σ_h (1)(2)(3)(4)(5)(6)

$2\sigma_v^g$ $\begin{cases} (56)(12)(34) \\ (56)(14)(23) \end{cases}$

$2\sigma_d$ $\begin{cases} (5)(6)(4)(2)(13) \\ (5)(6)(1)(3)(24) \end{cases}$

$2S_4$ $\begin{cases} (1432)(56) \\ (4123)(65) \end{cases}$

i (5)(6)(24)(13)

Fig. 13.3   Symmetry operations of the factor group of the space group D_{4h}^{14} of NiF$_2$, CoF$_2$, and FeF$_2$ (Rutile structure, tetragonal). The drawing is not to scale: the c-axis has been stretched to avoid crowding the figure. The symmetry elements shown do not necessarily conform to the standard system used in X-ray spectroscopy but are merely intended to show the possible symmetry operations on the two units AB$_2$ in the unit cell. The structure is presented—for simplicity's sake—as if it consisted of molecules AB$_2$ and not of ions (see text). The designations of the various symmetry elements correspond to those of point group D_{4h}, as given by E. B. Wilson, Jr., D. C. Decius, and P. C. Cross, *Molecular Vibrations*, McGraw-Hill, New York, 1955, Appendix X. Note that the symmetry elements C_4^s (fourfold screw axis), $C_2'^s$ (twofold screw axis), and σ_v^g (glide mirror consisting of a reflection and a glide by the translation vector $\{c/2, a/2, a/2\}$) do not represent point group operations.

metry operations are those which exchange all atoms of the same type among each other in the unit cell. For instance, the operation σ_h (see Fig. 13.3) does not only affect the atoms, *2, 4, 6* but also reflects the atoms *1, 3, 5* from "above to below." (Operation σ_h is seen to reflect all atoms of the unit cell into identical positions in this crystal.)

TABLE 13.1

FACTOR GROUP ELEMENTS AND CHARACTERS FOR SOME XF_2 METAL FLUORIDES

Factor group elements	φ	$2 \cos \varphi \pm 1$*	U_R	$\chi(R)$
E	0	3	6	18
C_2	π	-1	2	-2
$2C_2''$	π	-1	4	-4
σ_h	0	1	6	6
$2\sigma_d$	0	1	4	4
$i(=S_2)$	π	-3	2	-6
$2C_2'^s$	π	-1	0	0
$2\sigma_v^{\,g}$	0	1	0	0
$2S_4$	$\pi/2$	-1	0	0
$2C_4^{\,s}$	$\pi/2$	1	0	0

* Plus sign: proper rotation. Minus sign: improper rotation.

We now count the number of those atoms of a unit cell which are left invariant under the symmetry operations R (since all others contribute zero character). The number of invariant atoms under R is simply given by the number of single permutation elements under this R; we have collected them in Table 13.1 under the heading U_R (the description "$u(coset)$" or "$v(coset)$" is also used in the literature), together with the factor group symmetry elements, angles of rotation (φ) under the symmetry operation R, and their character $\chi(R) = U_R(2 \cos \varphi \pm 1)$, with the plus sign for proper and the minus sign for improper rotations.[1,2]

In Table 13.2 we give the characters $\chi^{(l)}(R)$ of the symmetry species of the point group D_{4h} besides some other information which we shall need below.

We now apply Eq. 13.1 to obtain how often each of the ten symmetry species of D_{4h} (order 16) occurs in the representation of the factor group of the crystal. We show this in detail for two cases, A_{1g} and E_u; the others follow accordingly. The first factor in the development is the character for the symmetry operation R, taken from Column 5 of Table 13.1. The second factor is the corresponding character $\chi^{(l)}(R)$ of the symmetry species l of D_{4h} taken from Table 13.2, multiplied by the frequency with which the cor-

TABLE 13.2

Symmetry Species and Activity of the Fundamentals of Some Metal Difluorides. The Characters of the Irreducible Species of Point Group D_{4h} (see Ref. 23) and the Characters under the Operations of the Factor Group of D_{4h}^{14} for the Total, Acoustic, Translatory Optic, and Rotatory Fundamentals are also Given

D_{4h}	E	C_2	$2C_2''$	σ_h	$2\sigma_d$	i	$2C_2'$	$2\sigma_v$	$2S_4$	$2C_4$	n^*	T	T'	R'	n'	Activity†
A_{1g}	1	1	1	1	1	1	1	1	1	1	1	0	0	0	1	R
A_{2g}	1	1	−1	1	−1	1	−1	−1	1	1	1	0	0	1	0	f
B_{1g}	1	1	−1	1	−1	1	1	1	−1	−1	1	0	0	0	1	R
B_{2g}	1	1	1	1	1	1	−1	−1	−1	−1	1	0	0	0	1	R
E_g	2	−2	0	−2	0	2	0	0	0	0	2	0	0	1	1	R
A_{1u}	1	1	1	−1	−1	−1	1	−1	−1	1	0	0	0	0	0	—
A_{2u}	1	1	−1	−1	1	−1	−1	1	−1	1	2	1	0	0	1	IR
B_{1u}	1	1	−1	−1	−1	−1	1	1	1	−1	2	0	1	0	1	f
B_{2u}	1	1	1	−1	1	−1	−1	−1	1	−1	0	0	0	0	0	—
E_u	2	−2	0	2	0	−2	0	0	0	0	4	1	1	0	2	IR
U_R	6	2	4	6	4	2	0	0	0	0						
$U_R(s)$	2	2	2	2	2	2	0	0	0	0						
$\chi(n)$	18	−2	−4	6	4	−6	0	0	0	0						
$\chi(T)$	3	−1	−1	1	1	−3	−1	1	−1	1						
$\chi(T')$	3	−1	−1	1	1	−3	1	−1	1	−1						
$\chi(R')$‡	4	0	−2	0	−2	4	0	0	0	0						

$$\chi(n) = U_R(\pm 1 + 2\cos\varphi)$$

$$\chi(T) = \pm 1 + 2\cos\varphi$$

$$\chi(T') = [U_R(s) - 1][\pm 1 + 2\cos\varphi]$$

$$\chi(R') = U_R(s - v)(1 \pm 2\cos\varphi)^{\ddagger}$$

* n = total fundamentals. T = acoustic. T' = translatory optic. R' = rotatory. n' = internal.
† R = Raman. IR = infrared. f = forbidden.
‡ See text.

responding symmetry element occurs in the group (order of class). Hence:

$$n(A_{1g}) = \frac{1}{16}\{18 \cdot 1 + (-2) \cdot 1 + (-4) \cdot 1 \cdot 2 + 6 \cdot 1 + 4 \cdot 1 \cdot 2 + (-6) \cdot 1\} = 1$$

$$(13.5)$$

$$n(E_u) = \frac{1}{16}\{18 \cdot 2 + (-2) \cdot (-2) + 6 \cdot 2 + (-6) \cdot (-2)\} = 4.$$

Continuing in this way, we find that the resulting factor group representation of the space group D_{4h}^{14} of the crystals of XF_2 (X = Ni, Co, Fe) is

$$\Gamma_n = A_{1g} + A_{2g} + 2A_{2u} + B_{1g} + 2B_{1u} + B_{2g} + E_g + 4E_u, \quad (13.6)$$

a total of 13 normal modes (we count a degenerate mode as one), five of which are doubly degenerate (yielding the expected total of $3 \times 6 = 18$ degrees of freedom).

To obtain the genuine fundamentals, we have to split off the three acoustical modes (the components of the three pure translational motions of a whole unit cell; see Fig. 13.1).* Their representation is easily found by inspection of Fig. 13.3 with regard to the x, y, z translations (there is no need here to go through an analysis as that just described for Γ_n). Since the crystal is tetrahedral, two directions are equivalent, hence one of the symmetry species of the pure translation must be doubly degenerate. A translation vector changes sign with respect to inversion at a center of symmetry. Thus, only the u-species can enter. Furthermore, the z-direction must be symmetric with respect to proper rotations about axes in the z-direction (C_4, C_2) and change sign when rotated by π about an axis perpendicular to the z-direction $(x, y$-plane). This leaves symmetry species A_{2u} as the nondegenerate species of the pure translation (as is easily ascertained from the character table in Table 13.2). The representation of the translation of the whole crystal (acoustic modes) is therefore

$$\Gamma_T = A_{2u} + E_u, \quad (13.7)$$

leaving

$$\Gamma = A_{1g} + A_{2g} + A_{2u} + B_{1g} + 2B_{1u} + B_{2g} + E_g + 3E_u \quad (13.8)$$

as the completely reduced representation of the eleven genuine fundamental vibrations.

We know that the dipole moment operator has the same representation as the translation vector, that is, the representation of the acoustical modes (see above). Inspection of Eqs. 13.7 and 13.8 then shows that we can expect four infrared active modes, namely one single (A_{2u}) and three doubly degenerate (E_u).

---

* We need not concern ourselves about rotational degrees of freedom of the whole crystal since these are not permitted on account of Born's cyclic boundary conditions (see footnote on page 476 and Ref. 2).

In order to establish by first principles how the total thirteen fundamentals break up into lattice and internal modes, we make use of a character table[1,2] appended to Table 13.2 (see bottom of table). It gives explicit expressions for the characters $\chi(R)$ we need in Eq. 13.1 for establishing the number and symmetry species of the translatory and rotatory lattice modes. These characters are designated by $\chi(T')$ and $\chi(R')$, respectively. For completeness' sake we have also added the expressions for $\chi(n)$, the character for finding the representation of the factor group (see Table 13.1, Column 5), and $\chi(T)$, the character for finding the species designation of the acoustic modes, which we had just accomplished above by inspection. The species and number of the internal modes (n') is most readily obtained by difference.

A few explanatory remarks concerning these characters may be helpful. The factors before the trigonometric expression indicate the number of invariant entities (invariant under symmetry operation R) which have to be considered for establishing $\chi(R)$ of the particular vibration (we remember that those atoms, molecules, ions, or other entities which are not invariant under operation R contribute zero character). We now discuss these factors in more detail.

1. U_R is defined as the number of all invariant atoms in the unit cell. Evidently, in order to obtain the character of the acoustic modes, $\chi(T)$, that is the modes where the whole unit cell moves, we must set $U_R = 1$.

2. In the translatory optic modes, units of atoms, molecules and/or single atoms move against each other, as shown in Part *b* of Fig. 13.1 for the example of NaN$_3$. This condition is taken care of by the factor $U_R(s)$, where s is the number of such units among which the unit cell atoms may be distributed and $U_R(s)$ represents the invariant number of these s units.[6]

3. Finally, we consider the rotatory modes for which the number of invariant units is given by $U_R(s - v)$, where v is the number of units consisting of single atoms (these must be subtracted out since they lack rotational degrees of freedom; see also Section A).

We now apply these principles to the fluorides of Co, Ni, and Fe, where we evidently (in agreement with our assumption of considering the crystal to consist of distinct diatomic molecules) can set $s = 2$, $v = 0$. Inspection of the symmetry operations depicted in Fig. 13.3 shows that $U_R(s) = 2$ for the operations E, C_2, C_2'', σ_h, σ_d, and i (remember that we now consider the molecule as a "point" under the symmetry operations), and that $U_R(s) = 0$ for C_4^s, $C_2'^s$, σ_v^g, and S_4. Thus, $[U_R(s) - 1]$ is 1 and -1, respectively, for these two sets of operations. By using these values, we can easily establish the quantities $\chi(T')$ with the help of Table 13.2 and Column 3 of Table 13.1.

It now remains to establish the species of the rotatory lattice modes. For a linear molecule we note that there are only two degrees of freedom of rotation; therefore the character $\chi(R') = U_R(s - v)(1 \pm 2 \cos \varphi)$ of Table

13.2 cannot be used.[7] We will now calculate the character $\chi(R')$ for the rotatory motion of the linear molecules (see Fig. 13.3). The trigonometric factor $1 \pm 2 \cos \varphi$ is the trace of the transformation matrix which describes rotation or reflection of the components of the angular momentum (we note that angular momenta transform as axial vectors). If we rotate or reflect about a symmetry element along which there also lies a linear molecule, the angular momentum about this axis is zero. The transformation matrix therefore lacks the corresponding elements.[8]

The factor $U_R(s - v)$ is equal to zero for the operations $C_2'^s, \sigma_v{}^g, S_4,$ and $C_4{}^s$ since the molecules are not invariant under these operations (for instance, molecule *2 4 6* changes place with *3 5 1*). Thus, $\chi(R') = 0$ for these operations.

For the operations $E, C_2, \sigma_h,$ and i the factor $U_R(s - v)$ amounts to two. The corresponding transformation for the angular momentum under C_2 is

$$\begin{pmatrix} l'_x \\ l'_y \\ l'_z \end{pmatrix} = \begin{pmatrix} 1 & 0 & 0 \\ 0 & \cos \varphi & \sin \varphi \\ 0 & -\sin \varphi & \cos \varphi \end{pmatrix} \begin{pmatrix} l_x \\ l_y \\ l_z \end{pmatrix} \tag{13.9}$$

(the z-axis is chosen to be along molecule *2 4 6*). Since $\varphi = \pi$ and since the l_z-component has to be omitted, the trace of Eq. 13.9 amounts to $1 + \cos \pi = 0$. In an analogous way we obtain for $E(\varphi = 0)$, $\sigma_h(\varphi = 0$, minus sign of improper rotation), and i ($\varphi = \pi$, minus sign of improper rotation) the traces $= 2, 0,$ and 2, respectively, and for $\chi(R') = 4, 0, 4$ in the same sequence.

For the operations C_2'' and σ_d the calculation of $\chi(R')$ is more complicated. We obtain $\chi(R')$ by direct calculation of the traces of the transformation of the components of the angular momentum for *both* molecules of the unit cell (that is, there are two blocks of Eq. 13.9). The six components of the angular momentum are denoted by $l_{x_1}, l_{y_1}, \cdots, l_{z_2}$; the contribution of the l_{z_i} is, of course, zero. For C_2'', l_{x_1} goes into $-l_{x_1}$; $l_{y_1} \to l_{y_1}, l_{x_2} \to -l_{x_2}, l_{y_2} \to -l_{y_2}$. Thus, $\chi(C_2'') = -2$. In a similar way, it follows that $\chi(\sigma_d) = -2$.

Before we summarize the results of the symmetry analysis, it is important to realize that we have performed the classification of the fundamentals of these metal difluorides into rotatory, internal, and translatory optic modes by purely group-theoretical considerations without inquiring into the physical reality of such a picture. According to Bhagavantam and Venkatarayudu, and we quote part of their remarks:[1] "s is the number of groups into which the nonequivalent points may be divided, with due regard to the magnitude of the forces that exist between them".* In a crystal of, for instance Na_2CO_3 (see also our previous example of NaN_3), a separation into the groups Na^+ and CO_3^{2-} is meaningful since the covalent binding

---

* "Group" has here, of course, the meaning of sets, combinations, or units of atoms (or ions); the "nonequivalent points" are the atoms (or ions) within a unit cell.

forces in the CO_3^{2-} group are much stronger than those between the ions Na^+ and CO_3^{2-}. In the metal fluorides studied by Balkanski *et al.*, the forces between the different cations and anions cannot be meaningfully separated into *intra-* and *inter*molecular forces, that is to say, between forces within a species XF_2 and forces between the species XF_2. The crystals are not only ionic (so are NaN_3 and Na_2CO_3), but also the metal ion is surrounded by six nearest neighbor fluorine ions which, besides, are located at approximately the same radius vector distance from the central metal ion. We therefore cannot carry the group-theoretical distinction of Table 13.2, which separates the total number of genuine fundamentals into one translatory optic, three rotatory, and seven internal fundamentals, into the physical reality. We must lump the different aspects of the analysis together and thus predict that the metal fluorides possess a total of $(1 + 3 + 7 =)$ 11 genuine *lattice modes*, of which four are infrared-active ($1A_{2u}, 3E_u$).

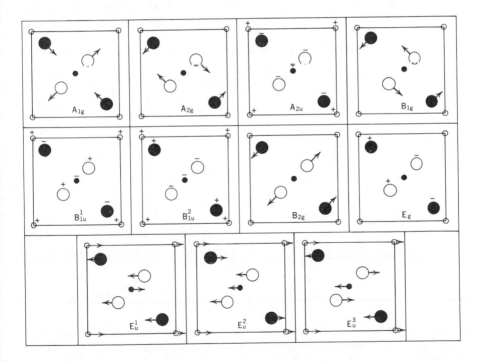

Fig. 13.4 Schematic of the genuine lattice modes of the difluorides of Ni, Fe, and Mn on the similar example of rutile (TiO_2). Large circles show the F ions, smaller circles the metal ions. Shaded circles are at a level $c/2$ above the basal plane. Note that for all even vibrations (symmetric with respect to inversion) the metal ion is at rest. The signs $+$ and $-$ designate motions perpendicular to the plane of the paper. [B. Dayal, *Proc. Indian Acad. Sci.* **32A**, 304 (1950).]

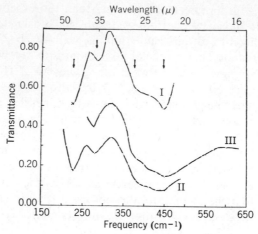

Fig. 13.5   Absorption spectrum of NiF$_2$ between 200 and 500 cm^{-1}. Transmission measurements on polyethylene–NiF$_2$ pellets of varying composition and thickness. [M. Balkanski, P. Moch, and G. Parisot, *J. Chem. Phys.* **44**, 940 (1966).]

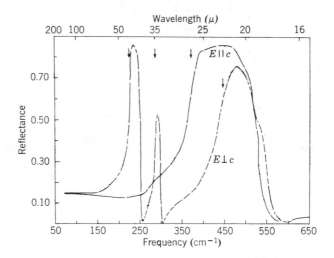

Fig. 13.6   Reflectivity spectrum of NiF$_2$ between 70 and 650 cm^{-1}. Curve $E \perp c$: extraordinary ray; curve $E \| c$: ordinary ray. [M. Balkanski, P. Moch, and G. Parisot, *J. Chem. Phys.* **44**, 940 (1966).]

The ionic displacements of the various genuine fundamentals are shown in Fig. 13.4 on the similar example of rutile (TiO$_2$).

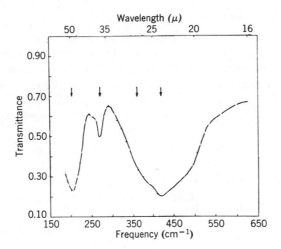

Fig. 13.7 Absorption spectrum of CoF$_2$ between 150 and 650 cm^{-1}. Transmission measurements on a polyethylene–CoF$_2$ pellet. [M. Balkanski, P. Moch, and G. Parisot, *J. Chem. Phys.* **44**, 940 (1966).]

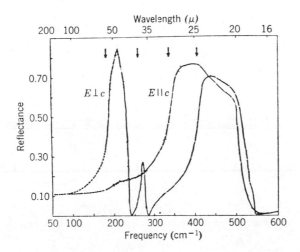

Fig. 13.8 Reflectivity spectrum of CoF$_2$ between 70 and 600 cm^{-1}. Curve $E\|c$: extraordinary ray; curve $E \perp c$: ordinary ray. [M. Balkanski, P. Moch, and G. Parisot, *J. Chem. Phys.* **44**, 940 (1966).]

c. Experimental Results

Figure 13.5 shows the absorption spectrum of NiF_2 powder between 200 and 500 cm^{-1}. Figure 13.6 shows the reflection spectrum of a crystal of NiF_2 between 70 and 650 cm^{-1} (12–17° incidence), with the incident polarized light parallel and perpendicular to the major symmetry axis (c) of the crystal. Figures 13.7 and 13.8 show the corresponding spectra of CoF_2.

The results are internally consistent: The four absorption bands are related to four strong onsets of reflectivity (see Chapter 12, Section B.b). Furthermore, a Kramers-Kronig inversion (see Chapter 12, Section B.c) of the $\perp c$ modes was performed, yielding $\varepsilon'(\omega)$ and $\varepsilon''(\omega)$. The results, for NiF_2, are shown in Fig. 13.9. Table 13.3 and 13.4 show a summary of the results for NiF_2 and CoF_2, respectively. The number of observed strong peaks is in accord with the theoretical predictions presented above (Section B.b). We also see that there are small but definite differences in the fundamental frequencies, in spite of the great similarity of the lattice parameters[4] and the masses of the metal ions (see above).

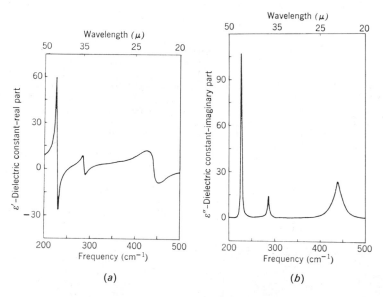

Fig. 13.9 (a) Dielectric constant of NiF_2 for the ordinary ray ($E \perp c$), real part. (b) Dielectric constant of NiF_2 for the ordinary ray ($E \perp c$), imaginary part. [M. Balkanski, P. Moch, and G. Parisot, *J. Chem. Phys.* **44**, 940 (1966).]

TABLE 13.3*

FREQUENCIES OF INFRARED-ACTIVE MODES AS OBSERVED IN NICKEL
FLUORIDE BY TRANSMISSION MEASUREMENTS ON POWDERS, DIRECT
OBSERVATION OF OPTICAL REFLECTION SPECTRA OF SINGLE CRYSTALS,
AND INTERPRETATION OF KRAMERS-KRONIG INVERSION

Lattice mode	Frequency from transmission data of powders, cm^{-1}	Frequency from direct observation of reflection spectra, cm^{-1}	Frequency from KK inversion, cm^{-1}	Polarization
E_u	445	445	440	$\perp c$
A_{2u}	375	370		$\parallel c$
E_u	285	285	286	$\perp c$
E_u	225	225	228	$\perp c$

* M. Balkanski, P. Moch, and G. Parisot, *J. Chem. Phys.* **44**, 940 (1966)

TABLE 13.4*

FREQUENCIES OF INFRARED-ACTIVE MODES AS OBSERVED IN COBALT
FLUORIDE BY TRANSMISSION MEASUREMENTS ON POWDERS, DIRECT
OBSERVATION OF OPTICAL REFLECTION SPECTRA OF SINGLE CRYSTALS,
AND INTERPRETATION OF KRAMERS-KRONIG INVERSION

Lattice mode	Frequency from transmission data of powders, cm^{-1}	Frequency from direct observation of reflection spectra, cm^{-1}	Frequency from KK inversion, cm^{-1}	Polarization
E_u	420	410	412	$\perp c$
A_{2u}	360	340		$\parallel c$
E_u	270	265	268	$\perp c$
E_u	205	190	196	$\perp c$

* M. Balkanski, P. Moch, and G. Parisot, *J. Chem. Phys.* **44**, 940 (1966).

In conclusion, we note that the antiferromagnetism of these salts leads to additional far-infrared absorption at longer wavelengths. This is discussed in Appendix III.

C. Lattice Vibrations of $KNiF_3$

In another publication, Balkanski *et al.* describe far-infrared reflection data (75 to 600 cm^{-1}) on $KNiF_3$.[9] The space group of the crystal is O_h^1 with one molecule per unit cell (perovskite structure). The material is an antiferromagnet below 275°K.

Group-theoretical analysis predicts five threefold degenerate modes. Subtracting the acoustic mode leaves four genuine lattice modes, $\Gamma_n = \Gamma_{25} + 3\Gamma_{15}$, of which three (species Γ_{15}) are predicted to be infrared-active.[9]

Figure 13.10 shows the reflection spectrum of $KNiF_3$ at room temperature (solid line) and at liquid-nitrogen temperature. There are, indeed, three intense high-reflectivity peaks as predicted. The intensity of these peaks

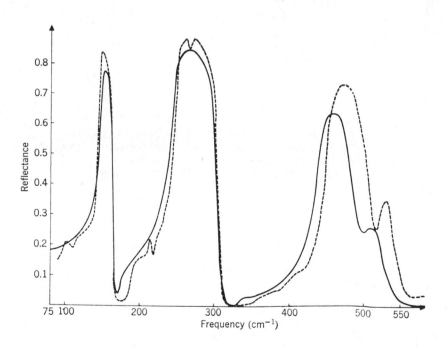

Fig. 13.10   The reflectivity of $KNiF_3$ at room temperature (solid lines) and at liquid-nitrogen temperature (dotted lines). [M. Balkanski, P. Moch, and M. K. Teng, *J. Chem. Phys.* **46**, 1621 (1967).]

remains essentially constant over the temperature variation, again in agreement with the theory (see Chapter 12, Section D.a). There is, however, a fourth peak at 520 cm^{-1} (see Fig. 13.10), the assignment of which is more speculative. According to what we have discussed in detail in Chapter 12, the following explanations for the appearance of the additional peak offer themselves: (1) a two-phonon process and (2) a localized mode due to impurities or crystal defects. A third possibility is that a one-phonon forbidden transition is made active by structural defects at the reflective surface[9]—we remember that one of the fundamental vibrations of the crystal is infrared-forbidden (see above). Balkanski et al. have offered some detailed arguments, based in part on experimental observation of the crystal spectrum in the 5000 to 25000 cm^{-1} region, which favor an explanation in terms of a one-phonon process corresponding to the forbidden mode.[9] Unfortunately, the extensive overlap in the 500 cm^{-1} region of the spectrum did not permit full use of the different characteristics of the temperature variations of one- and two-phonon processes (see Chapter 12, Section D). An explanation of the fourth peak in the spectrum of KNiF$_3$ in terms of a two-phonon process can therefore not be set aside.[9]

TABLE 13.5*

THE VALUES OF OSCILLATOR PARAMETERS DEDUCED FROM KRAMERS-KRONIG ANALYSIS AND THE FREQUENCIES ASSOCIATED WITH THE LONGITUDINAL MODES[†]

Temperature	Mode	v_T, cm^{-1}	$4\pi\rho$	γ	v_L, cm^{-1}
$T = 300°$K	$(\Gamma_{15})_1$	148	1.35	0.026	146
	$(\Gamma_{15})_2$	244	1.40	0.033	306
	$(\Gamma_{15})_2$	441	0.48	0.040	501
	Γ_{25}	501	0.02	0.050	521
$T = 90°$K	$(\Gamma_{15})_1$	144	1.14	0.021	164
	$(\Gamma_{15})_2$	248	1.42	0.016	305
	$(\Gamma_{15})_3$	452	0.50	0.017	521
	Γ_{25}	520	0.02	0.033	540

* M. Balkanski, P. Moch, and M. K. Teng, J. Chem. Phys. 46, 1621 (1967).

† γ = damping constant, ρ = oscillator strength (see Chapter 12, Section B.b). In regard to species Γ_{15} and Γ_{25}, see species F_{1u} and F_{2u}, respectively, in G. Herzberg, Infrared and Raman Spectra of Polyatomic Molecules, Van Nostrand, Princeton, New Jersey, 1962, p. 123, Table 29; L. P. Bouckaert, R. Smoluchowski, and E. Wigner, Phys. Rev. 50, 58 (1936), Table I.

In Table 13.5 we give a summary of the data on the basis that the fourth mode is a one-phonon process as just discussed. The tabulation includes the longitudinal frequencies associated with the fundamentals, measured according to the principles we have explained in detail in Chapter 12. The dispersion parameters are also given.

D. LATTICE VIBRATION IN CRYSTALLINE POLYETHYLENE

The far-infrared spectrum of polyethylene shows a weak absorption band near $71 \, cm^{-1}$,[10] which was later shown to correlate with the degree of crystallinity in the polymer[11,12] and was therefore suggested to arise from a vibration of the crystal lattice.[11]

Krimm and Bank[13] have presented evidence which confirms this assignment. Figure 13.11 shows the pertinent region of the spectrum for commercial high-density polyethylene,* (crystalline) $n\text{-}C_{36}H_{74}$, and a

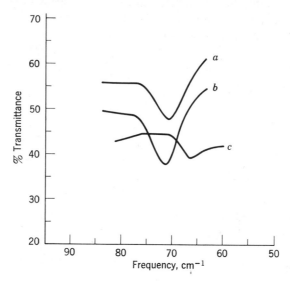

Fig. 13.11 Infrared spectra of (a), commercial high-density polyethylene; (b), a n-paraffin, $C_{36}H_{74}$; and (c), a deuteroparaffin, $C_{100}D_{202}$. Instrument: Perkin-Elmer, Model 301. [S. Krimm and M. I. Bank, *J. Chem. Phys.* **42**, 4059 (1965).]

deuteroparaffin, $C_{100}D_{202}$ (approximate composition). The authors stress that since $n\text{-}C_{36}H_{74}$ also shows the crystalline band at the same frequency as the polymer, the absorption cannot originate from the folded-chain

---

* The greater the degree of crystallinity, the higher the density. Commercial high-density material (Marlex) has a degree of crystallinity of 80–95%. The various aspects of the crystallinity of polyethylene are a rather interesting subject in themselves (see, for instance, P. H. Geil, *Polymer Single Crystals*, Wiley-Interscience, New York, 1963).

crystals in polyethylene but must arise from the vibrations of the linear zigzag chain carbon atoms. The factor group analysis[14] of the unit cell of polyethylene crystals (space group D_{2h}^{16}) predicts one infrared-active translatory lattice mode of symmetry species B_{1u} and one of B_{2u}, polarized along the longest (a) and shorter (b) axis, respectively. Translatory-type modes are indeed to be expected in the 70 to 150 cm^{-1} range of the spectrum; the rotatory-type modes (of which there are four allowed in the infrared) are usually at much lower frequencies (they are expected to be very weak; see Section A). Krimm and Bank assign the 71 cm^{-1} band to the B_{2u} translatory lattice mode; Tasumi and Shimanouchi[15] assign it to the B_{1u} mode. We show, in Fig. 13.12, a cross section of the unit cell of the polymer and, in Fig. 13.13, the displacements of the atoms during the B_{1u} and B_{2u} translatory lattice modes.

The shift from 71 to 66.5 cm^{-1} on deuteration (see Fig. 13.11) is in accord with what we would expect from the frequency dependence of such a translatory lattice mode on the mass m, namely $71/(m_{CD_2}/m_{CH_2})^{1/2} = 66.4$ cm^{-1}.[13] Krimm and Bank point out also that n-$C_{20}H_{42}$, a triclinic crystal with one

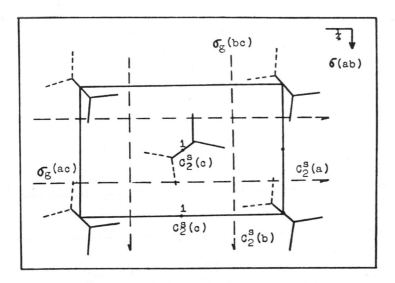

Fig. 13.12   Cross section of unit cell of polyethylene perpendicular to c-axis and bisecting C—C bond. C_2^s—screw diad, σ—mirror plane, σ_g—glide plane (σ_g is a diagonal glide plane, with translation $(b + c)/2$), i—center of inversion. CH$_2$ groups are indicated as above (*solid V*) and below (*dashed V*) the plane. [S. Krimm, C. Y. Liang, and G. B. B. M. Sutherland, *J. Chem. Phys.* **25**, 549 (1956).]

chain-segment per unit cell, does not exhibit absorption in the $70\,\text{cm}^{-1}$ region.

The authors have rationalized an observed decrease of the frequency of the $71\,\text{cm}^{-1}$ band with increasing ambient temperature[11] as follows: With increasing temperature, the separation between chains increases and, therefore, the "effective" force constant, $\{\partial^2 V/\partial q^2\}_{eff}$, decreases. Based on the model of a Lennard-Jones intermolecular potential and assuming that the separation between chains is governed by an average linear expansion coefficient, Krimm and Bank were able to reproduce satisfactorily the observed temperature dependence of the frequency of the translatory absorption band, B_{2u}.

Krimm and Bank predict the B_{1u} translatory mode occurs near $50\,\text{cm}^{-1}$;[13] Tasumi and Shimanouchi, by a normal coordinate analysis, calculate that the translatory mode B_{1u} occurs at $76\,\text{cm}^{-1}$ and that B_{2u} is at $105\,\text{cm}^{-1}$.[15]

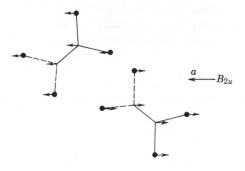

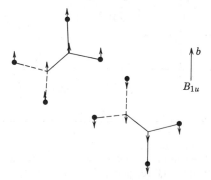

Fig. 13.13   The two infrared-active translatory modes of the unit cell of crystalline polyethylene. The direction of the crystal axes a and b are indicated (see Fig. 13.12). B_{2u} is polarized along the b-, B_{1u} along the a-axis (see also Fig. 9–11 of Ref. 14).

They also discuss models of intermolecular potentials, employing their calculated force constants in combination with various potential functions (describing the interaction between the hydrogen atoms of adjacent chains). They conclude that the interchain interactions are mainly repulsive and that, accordingly, the effective intermolecular force constant decreases rapidly with increasing $H\cdots H$ distances.

In conclusion, we remark that the characteristics of the infrared spectrum of polyethylene can, in its essentials, be readily predicted by considering polyethylene a single linear zigzag chain of CH_2CH_2 units rather than a three-dimensional crystal.[1,14] The success of this approach can be easily understood by remembering that the binding forces within the chain of methylene groups are very much stronger than those between two chains. However, the finer details (mainly the interesting intermolecular effects as discussed here) would be lost by such a treatment. An isolated chain of polyethylene does not possess translatory optic fundamentals (see Fig. 13.13; removing one chain from the unit cell leaves only an acoustic mode).

E. Spectra of Crystalline Nitrogen and Carbon Monoxide

Anderson and Leroi[16] have published a note on the absorption spectra of crystalline films of N_2 and CO at $10°K$ in the frequency range $20–250\ cm^{-1}$. The spectra were obtained by Fourier transformation of the output of a RIIC FS-520 Michelson interferometer; they are shown in Fig. 13.14. The far-infrared spectra of the lattice modes of these isoelectronic molecules are of particular interest with respect to the observed collision-induced absorption band in liquid nitrogen (see Chapter 10, Section C.b), which had been proposed to originate possibly from (induced) translational transitions of the molecule.

A projection of the unit cell of the low-temperature phase of N_2 (space group T^4), taken from an X-ray diffraction study,[17] is shown in Fig. 13.15. The primitive unit cell contains four molecules N_2. As indicated in the caption of the figure, the displacement from a centrosymmetric structure (T_h^6) is slight. The obvious consequence of this is that modes which are forbidden in the strictly centrosymmetric structure are of rather weak intensity in the actual case (see below).

With four linear molecules per unit cell, there is a total of $4\cdot2\cdot3 = 24$ fundamentals of which $4\cdot2 = 8$ are rotatory, $4(6-5)$ are internal, three are acoustic, and the remainder of $9\ (=4\cdot3-3)$ are translatory optic (see Section A). The results of the factor group analysis are shown in Table 13.6. Adding up the dimensions in Column 4 of this table, we obtain a total of twenty-four degrees of freedom, as we should (remembering that species F designates a threefold and species E a doubly degenerate mode).

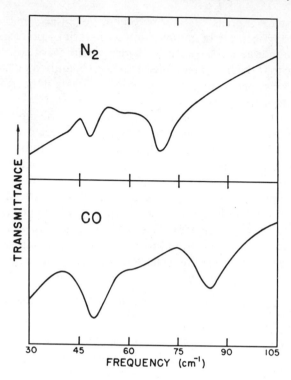

Fig. 13.14   Single-beam far-infrared absorption spectra of nitrogen (*upper*) and carbon mono-xide (*lower*) at 10°K. [A. Anderson and G. E. Leroi, *J. Chem. Phys.* **45**, 4359 (1966).]

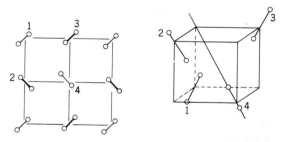

Fig. 13.15   A projection of a unit cell of α-N$_2$: the $P2_13$ (T^4) structure. The displacement of the molecular center away from the origin and along the nonintersecting threefold axes of the unit cell has been exaggerated. The heavy-rendered molecules are in the upper, and the thin-rendered in the lower plane. Rotation by 120° about the threefold axis of the cube turns molecule 1 into molecule 2, 2 into 3, 3 into 1, and 4 into itself (see sketch). [T. H. Jordan, H. W. Smith, W. E. Streib, and W. N. Lipscomb, *J. Chem. Phys.* **41**, 756 (1964).]

TABLE 13.6*

CORRELATION DIAGRAMS FOR THE α PHASES OF CRYSTALLINE
N_2 AND CO

Mode	Molecular symmetry	Site symmetry	Factor-group symmetry [†]
	$N_2 : D_{\infty h}$	C_3	T
ν	Σ_g^+	A	$A + F$
R_x, R_y	π_g	E	$E + 2F$
T_z	Σ_u^+	A	$A + F$
T_x, T_y	π_u	E	$E + 2F$
	$CO : C_{\infty v}$	C_3	T
ν, T_z	Σ^+	A	$A + F$
$T_x, T_y ; R_x, R_y$	π	E	$E + 2F$

* A. Anderson and G. E. Leroi, *J. Chem. Phys.* **45**, 4359 (1966).
[†] In the final column, all species are Raman-active; F species are also infrared-active. One of the F modes is of zero frequency, corresponding to translation of the entire lattice.

Table 13.6 also contains the symmetry species of the particular mode of the isolated molecule (point group $D_{\infty h}$) and of the "site symmetry" (point group C_3). We shall elaborate on this in the following. A "site" is a point which is left invariant by some operations of the space group; it is not difficult to see that these operations belong to some subgroup (which is a point group) of the factor group,[18] called the site group. Inspection of Fig. 13.15 shows that the site group symmetry is C_3. With this we can readily correlate the species designations of the vibrational fundamentals of the isolated molecule of N_2 and the crystal of N_2. For instance, the fundamental stretch, ν, is of species Σ_g^+ in the point group of the molecule, of species A in the site group, and of species A and F in the factor group. The entire correlation diagram is given in Fig. 13.16.

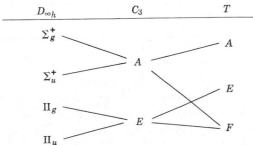

Fig. 13.16 Correlation diagram for crystalline and molecular nitrogen.

We shall not utilize the various other aspects of the "site group analysis" (see Ref. 18) and we will continue with our discussion of the crystal spectra of N_2 and CO. The modes of symmetry species F are infrared active since this is the species of the acoustic mode (see Table 13.6). We therefore expect four infrared-active genuine lattice modes, two of them rotatory (librational) and two translatory. [As the solids of N_2 and CO form molecular crystals with much greater intra- than intermolecular forces, the distinction of the fundamentals into internal and lattice modes is physically meaningful. We note that the internal (stretch) fundamentals lie in the infrared part of the spectrum.] Since the species of the four lattice modes is the same, "mixing" is allowed. However, the rotatory modes are expected to be weak since they are forbidden in a centrosymmetric structure for N_2 (see above). As concerns CO, its dipole moment is—as is well known—rather small; thus the rotatory motion should be weak. (We note in this respect that rotatory lattice modes are rather strong in the Raman spectrum because of the large change of the polarizability with the rotatory motion. In contrast, the translatory lattice modes are weak in the Raman spectrum.) Anderson and Leroi have thus assigned the two observed far-infrared absorption peaks in the spectra of crystals of N_2 and CO to predominantly translational lattice motions.

F. CRYSTAL SPECTRA OF COMPLEXES

One of the most important and interesting classes of compounds are the addition compounds or complexes. The range of their complexity, properties, type, and number of participating atoms is very large, reaching from a simple silver halide-ammonia complex (familiar from classical analytical chemistry) to such complicated structures as chlorophyll and the heteropolyacids of, say, molybdenum and tungsten. Because of this large number of atoms which build up the complex, the latter are generally solids at room temperature. Since the forces between the central atom (or central groups) and the ligands are usually weaker than the forces within the ligands, far-infrared spectroscopy appears to be a very useful and profitable technique for a study of the characteristics of complexes, in particular with regard to their structure and the forces between ligands and the central atom.

a. Silver Halide Complexes of the Types RAg_2X_3, $RAgX_2$, and R_2AgX_3

Bottger and Geddes[19] have investigated the infrared spectra of a large series of complexes of the type indicated in the heading above (R = univalent cation) between 30 and 440 cm^{-1} with a Perkin-Elmer spectrometer, Model 301. The cation R was $N(CH_3)_4^+$, $P(CH_3)_4^+$, $N(C_2H_5)_4^+$, $P(C_2H_5)_4^+$, and Cs$^+$. The symbol X represents Cl, Br, or I. The solid phase of these complexes consists of long chains of tetrahedral AgX_4 units. This is shown in

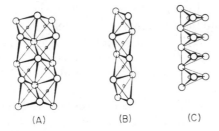

Fig. 13.17   Idealized diagrams showing the arrangement of AgX_4 tetrahedra in compounds of the types RAg_2X_3 (A), $RAgX_2$ (B), and R_2AgX_3 (C). Large circles represent halogen atoms; small circles, silver atoms. Note that silver-halogen bonds have been omitted and that the lines indicate edges of the AgX_4 tetrahedra. [G. L. Bottger and A. L. Geddes, *Spectrochim. Acta* **23A**, 1551 (1967), by permission of Pergamon Press.]

Fig. 13.17. Representative spectra are depicted in Figs. 13.18 through 13.21 (replotted on a contracted abscissa scale). We notice that the spectral pattern divides into two subpatterns. The one towards higher frequencies is seemingly

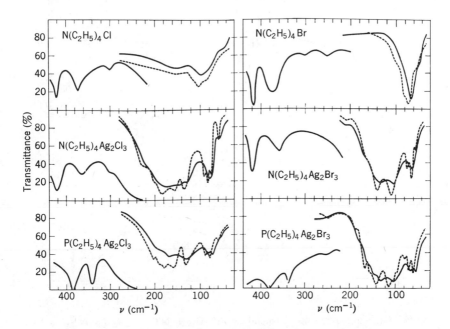

Fig. 13.18   Far-infrared spectra of crystalline $N(C_2H_5)_4Cl$, $N(C_2H_5)_4Ag_2Cl_3$, $P(C_2H_5)_4Ag_2Cl_3$, $N(C_2H_5)_4Br$, $N(C_2H_5)_4Ag_2Br_3$, and $P(C_2H_5)_4Ag_2Br_3$ at 25°C (solid line) and at 143°K (dotted line). [G. L. Bottger and A. L. Geddes, *Spectrochim. Acta* **23A**, 1551 (1967), by permission of Pergamon Press.]

indicative of the internal motions of the group of atoms R. If R is monoatomic, no band should appear in this frequency region. This is indeed the case as seen in the spectra of $Cs^+ \cdots [Ag_2I_3]^-$ and $\cdots [AgI_3]^{--}$ complexes (see Figs. 13.19 and 13.21). The lower frequency range then contains the transitions of the internal motions of the silver-halogen anion and the lattice modes of the complex. Bottger and Geddes apply the following criteria for the respective assignments: The location of the absorption bands of the silver-halogen anion should depend mainly on the mass of the halogen contained in the anion. The ratio of the silver–bromine and the silver–chlorine stretching frequencies should then, for instance for $Ag_2X_3^-$, amount to about 0.79 (square root of the inverse ratio of the reduced masses); that of the silver–iodine and silver–

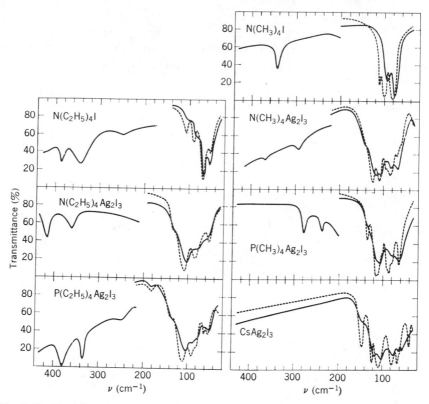

Fig. 13.19 Far-infrared spectra of crystalline $N(C_2H_5)_4I$, $N(C_2H_5)_4Ag_2I_3$, $P(C_2H_5)_4Ag_2I_3$, $N(CH_3)_4I$, $N(CH_3)_4Ag_2I_3$, $P(CH_3)_4Ag_2I_3$, and $CsAg_2I_3$ at 25°C (solid line) and at 143°K (dotted line). Sample preparation: crystalline material dispersed in Nujol or in pressed polyethylene disks. [G. L. Bottger and A. L. Geddes, *Spectrochim. Acta* **23A**, 1551 (1967), by permission of Pergamon Press.]

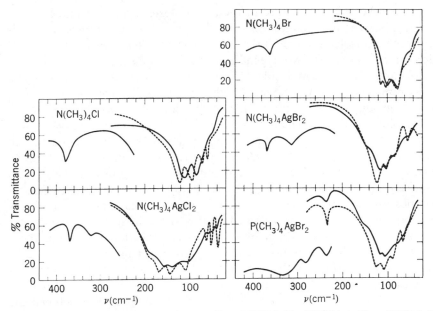

Fig. 13.20   Far-infrared spectra of crystalline $N(CH_3)_4Cl$, $N(CH_3)_4AgCl_2$, $N(CH_3)_4Br$, $N(CH_3)_4AgBr_2$, and $P(CH_3)_4AgBr_2$ at 25°C (solid line) and at 143°K (dotted line). [G. L. Bottger and A. L. Geddes, *Spectrochim. Acta* **23A**, 1551 (1967), by permission of Pergamon Press.]

bromine modes should amount to about 0.86. This tacitly assumes that the silver-halogen anions can be approximated by a diatomic system with a force constant which is the same for X = Cl, Br, and I.

Bands assigned to lattice modes should undergo shifts to lower frequencies if *both* cation and anion are substituted by heavier homologues. (We recall that in the linear diatomic chain the frequency is inversely proportional to the square root of the reduced mass for the TO, $q \sim 0$ mode; see Chapter 12, Section C.a.)

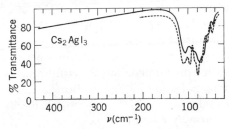

Fig. 13.21   Far-infrared spectra of crystalline Cs_2AgI_3 at 25°C (solid line) and at 143°K (dotted line). [G. L. Bottger and A. L. Geddes, *Spectrochim. Acta* **23A**, 1551 (1967), by permission of Pergamon Press.]

Since the detailed structure of these complexes is not yet known in its entirety, Bottger and Geddes used the "line group theory" approach. In this approach, an isolated chain of the complex is analyzed instead of the three-dimensional crystal. As we have indicated in Section D, this approach is often satisfactory if the intermolecular effects between chains are relatively weak.

We now consider the complexes of the type RAg_2X_3. The factor group of the line group, corresponding to ideal tetrahedra as shown in Fig. 13.17, is isomorphous with D_{2h}. The representation of the ten infrared-active normal modes, obtained from a group-theoretical analysis such as described in Section B.a, is $3B_{1u} + 3B_{2u} + 4B_{3u}$.[20] It turns out to be rather difficult to assign so many modes without ambiguity. Not only does the line group approach introduce a simplification into the predicted spectrum (by neglecting possible splittings due to interactions between chains) but—and this is possibly more serious—the assumed ideal tetrahedral structure results in a classification of modes which imposes stricter selection rules than those actually present on account of the lower symmetry. This may turn out to be particularly disturbing in the assignment of some of the weaker transitions since there might be doubt whether they arise from allowed combinations under the higher symmetry or from the (weak) fundamentals which become allowed when the (slightly) distorted structure is considered.

Bottger and Geddes essentially ascribe the highest observed (and intense) frequencies of the modes which belong to the anionic chain (the low-frequency region in Figs. 13.18 and 13.19) to stretching motions, the lower ones to bending motions, and the lowest observed frequencies to lattice modes. The weaker transitions at the high-frequency tail of the main absorption band are believed to be combinations between stretching and bending modes. It is also assumed that some of the relatively broad bands are unresolved superpositions; for instance, the broad band at $139 \, cm^{-1}$ of the $N(C_2H_5)_4$ and $P(C_2H_5)_4$ complexes of $[Ag_4Br_6]_n^{2n-}$ (Fig. 13.18) is, in this manner, predicted to represent the overlap of two stretching vibrations. Unfortunately, the great stability of complexes makes it frequently difficult to find a suitable solvent; by means of the solution spectra of the complex one can hope to separate out the lattice modes. However, when applying this technique, it is necessary to be sure that the absorbing species is the same for solution and crystal, and that there are no charge transfer reactions between solute and solvent or other complicating phenomena, such as equilibria between different species.[19]

In conclusion, in Table 13.7 we give Bottger and Geddes' tentative assignments for the modes of $[Ag_4X_6]_n^{2n-}$. We notice that the regions of internal and lattice modes strongly overlap.

From our discussion of Bottger and Geddes' data, it is quite apparent that a complete elucidation of the spectra of large complex molecules meets with difficulties. Some of the uncertainties of the assignments of the normal modes

TABLE 13.7*

ABSORPTION FREQUENCIES OF $[Ag_4X_6]_n^{2n-}$ ANIONIC CHAINS AND OF SOME PERTINENT LATTICE MODES[†]

Compound	Combinations $(v''' + \delta'')$	v'	v''	v'''	δ'	δ''	Lattice modes
$N(C_2H_5)_4Cl$							120 w, 108 s, 86 m
$N(C_2H_5)_4Ag_2Cl_3$	228 m, 203 m	184 s	155 s	134 s	89 m	82 s, 75(?)	56 w
$P(C_2H_5)_4Ag_2Cl_3$	200 m	178 s	155 s	129 s	86 m	76 s	55 w
$N(C_2H_5)_4Br$							106 w, 83 m, 73 s, 53 w
$N(C_2H_5)_4Ag_2Br_3$	167 m	139 s		109 s	74 m	64 m	55 w
$P(C_2H_5)_4Ag_2Br_3$	164 m	139 s		112 s	69 m	58 m	—
$N(C_2H_5)_4I$							108 w, 90 w, 68 s, 53 w
$N(C_2H_5)_4Ag_2I_3$	136 w	108 s	86 s	—	53 m		—
$P(C_2H_5)_4Ag_2I_3$	140 w	110 s	90 s	62 m	53 m		<30
$N(CH_3)_4I$							109 m, 98 s, 79 s
$N(CH_3)_4Ag_2I_3$	142 w	123 s	108 s	83 s	67 m	59 w	<30
$P(CH_3)_4Ag_2I_3$	135 w	115 w	110 s	86 s	63 s	—	—
$CsAg_2I_3$	146 m	122 s	107 s	76 s	64 s	58 m	38 m

* G. L. Bottger and A. L. Geddes, *Spectrochim. Acta* **23A**, 1551 (1967). By permission of Pergamon Press.

[†] Frequencies given in cm^{-1}; sample temperature about 143°K; m = medium; s = strong; w = weak. Stretching modes are designated by v', v'', and v'''. See the discussion in the text for their expected amount of isotope shift. Bending modes are designated by δ', δ''.

can, at times, be readily removed by working with single crystals and polarized radiation (infrared dichroism). We shall now discuss an example that employs this technique.

b. Metal-Ligand Force Constants of Acetylacetonates

Various metals form solid complexes with β-diketone, $CH_3 \cdot CO \cdot CH_2 \cdot CO \cdot CH_3$ (acetylacetone, abbreviated in the following by acac). The central atom–ligand bond is established between the metal and the oxygen atom(s). The main interest concerning the application of a spectral analysis on these complexes centers on the characteristics of the metal–oxygen bond, for instance, the variation of its stretching force constant with different central metal ions or with the ligand field splitting.

In what follows we shall give a description of a far-infrared study on the acetylacetonates by Mikami et al.[21] Table 13.8 summarizes the metal atoms used in the study and also gives, for convenience sake, the atomic number and the electronic configuration (neutral atom).

TABLE 13.8

ATOMIC NUMBER AND ELECTRONIC CONFIGURATION OF
SOME TRANSITION METALS

Metal	Atomic number	Configuration*
Cr	24	$KL3s^23p^63d^54s$
Mn	25	$KL3s^23p^63d^54s^2$
Co	27	$KL3s^23p^63d^74s^2$
Cu	29	$KLM4s$
Rh	45	$KLM4s^24p^64d^85s$
Pd	46	$KLM4s^24p^64d^{10}$
Pt	78	$KLMN5s^25p^65d^96s$

* Filled shells: $K = 1s^2$. $L = 2s^22p^6$. $M = 3s^23p^63d^{10}$.
$N = 4s^24p^64d^{10}4f^{14}$.

From previous work on the infrared spectrum of these complexes, it was established that (*1*) the skeletal deformations are localized in the ligands and (*2*) the metal–ligand stretching and bending vibrations lie in the frequency region below 750 cm^{-1}; in particular, it turned out that the position of modes within the 400 to 200 cm^{-1} range was relatively strongly sensitive to the nature of the central metal ion.[22] These latter modes should therefore represent the metal–ligand vibrations. In complexes of molecular composition 1:2 of metal:ligand, the metal is bound to four oxygen atoms in a square-planar metal–oxygen structure; in the 1:3 complexes the metal is bound to six oxygen atoms in an octahedrally coordinated structure. We show the structures schematically in Fig. 13.22.

Figure 13.23 shows the far-infrared spectrum of six complexes on samples mixed with paraffin, Nujol, or polyethylene. Figure 13.24(*a*) shows the spectrum of a single crystal of Cu(acac)$_2$ (monoclinic) with the incident radiation polarized parallel and perpendicular to the *b*-axis of the crystal, respectively. The molecular symmetry of the planar complexes is nearly D_{2h}; the infrared-active vibrations are of the symmetry species B_{1u}, B_{2u}, and B_{3u},[23] with the transition moment parallel to the *z*, *y*, and *x* axes, respectively, of the molecule (see Fig. 13.22).

Figure 13.24(*b*) shows the position of the molecule with respect to the crystal axis *b*. Accordingly, incident light polarized in a direction parallel to the *b*-axis of the crystal can interact with the B_{1u} and B_{3u} modes, whereas light polarized in a direction perpendicular to the *b*-axis of the crystal (see Fig. 13.24(*b*)) interacts only with B_{2u} modes (see Fig. 13.22 for the position of the molecular *x*, *y*, and *z* axes). Thus, we see from Fig. 13.24(*a*) that the

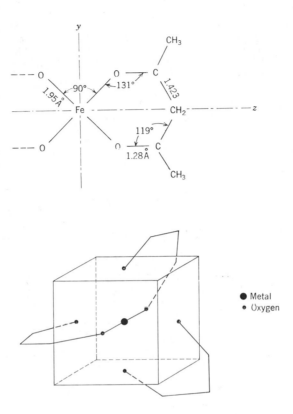

Fig. 13.22 Structures of the 1:2 and 1:3 metalacetylacetone complexes. [*After* M. Mikami, I. Nakagawa, and T. Shimanouchi, *Spectrochim. Acta* **23A**, 1037 (1967), by permission of Pergamon Press.]

451 and 217 cm^{-1} bands of Cu(acac)$_2$ are either B_{1u} or B_{3u}, whereas the mode represented by the band at 431 cm^{-1} belongs to species B_{2u}. If we look again at Fig. 13.23, we note that there is a band between about 450 and 470 cm^{-1} in each 1:2 complex. This band had been previously assigned to a metal–oxygen stretch,[22] in other words, the atomic displacements are in the molecular plane (B_{1u} or B_{2u}; see Fig. 13.24(*b*)). As its dichroism now shows, this would then be a metal–oxygen stretching mode of symmetry species B_{1u} in which the metal atom is displaced along the long axis (z) of the molecule.

There is another in-plane stretching fundamental, namely the one in which the displacement of the metal is in the direction of the y-axis of the molecule (see Fig. 13.22). The symmetry species of this fundamental is B_{2u} (see Fig. 13.24(*b*)) and the assigned frequencies for Cu(acac)$_2$, Pd(acac)$_2$, and Pt(acac)$_2$ are 291, 294, and 280 cm^{-1}, respectively (see Fig. 13.23).

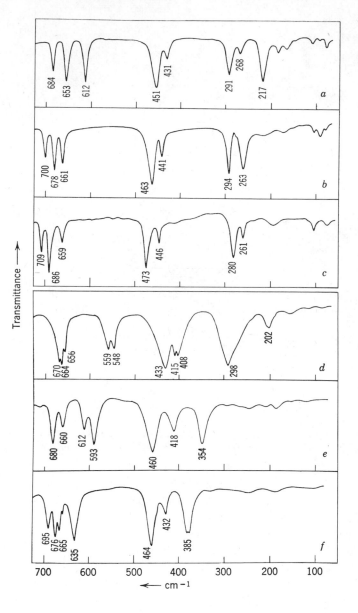

Fig. 13.23 Far-infrared spectra of M(acac)₂ and M(acac)₃: a = Cu(acac)₂, b = Pd(acac)₂, c = Pt(acac)₂, d = Fe(acac)₃, e = Cr(acac)₃, and f = Co(acac)₃. Instrument: Hitachi FIS-1 vacuum double-beam spectrometer. [M. Mikani, I. Nakagawa, and T. Shimanouchi, *Spectrochim. Acta* **23A**, 1037 (1967), by permission of Pergamon Press.]

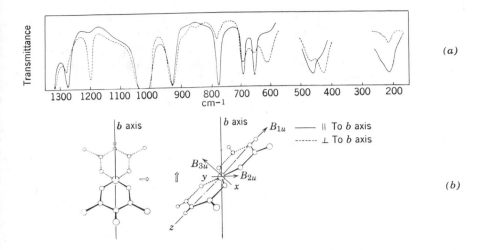

Fig. 13.24   Infrared dichroism of Cu(acac)$_2$. (a) Spectrum of a single crystal, with the incident radiation polarized along the crystal axis b (*solid curve*) and perpendicular to axis b (*dotted curve*). (b) Position of the molecular and crystal axes in Cu(acac)$_2$. The angle between the crystal axis b and the long molecular axis (z) is 36.5°. Instrument: Perkin-Elmer 12C single-beam prism spectrometer with AgCl (or polyethylene) polarizer and Perkin-Elmer Model 85 infrared microscope. [M. Mikami, I. Nakagawa, and T. Shimanouchi, *Spectrochim. Acta* **23A**, 1037 (1967), by permission of Pergamon Press.]

There remains the assignment of the motion in which the metal ion moves in a direction normal to the plane of the molecule (B_{3u}). The authors propose for Cu(acac)$_2$, Pd(acac)$_2$, and Pt(acac)$_2$ the values 217 (see above), 263, and 261 cm^{-1}, respectively.

Mikami *et al.* have also performed a normal coordinate analysis on some of the complexes, using a modified Urey-Bradley force field.[21] In Fig. 13.25 we give the atomic displacements of the atoms, obtained from this analysis, for some low-frequency infrared-active normal modes of the 1:2 complex Cu(acac)$_2$. The calculated frequencies (in cm^{-1}) are also given and may be compared with the observed values presented in Fig. 13.23. Of particular interest are the modes $\nu_{11}(B_{1u})$, $\nu_{25}(B_{2u})$, and $\nu_{32}(B_{3u})$, which describe the translational motions of the central metal atom. From the potential energy distribution, Mikami *et al.* calculate that the ν_{11} and ν_{25} modes are 46 and 67%, respectively, pure copper–oxygen stretch. The mode ν_{32} contains only 23% pure oxygen–copper–oxygen bond deformation. It is clear from these low contributions that a considerable number of the remaining (low-frequency) fundamentals contain contributions of metal–oxygen stretching and oxygen–metal–oxygen deformation motions.

In conclusion, we give the stretching force constants for the metal–oxygen bond of $Cu(acac)_2$, $Pd(acac)_2$, and $Fe(acac)_3$, obtained from the normal coordinate analysis, in mdyne/Å: 1.45, 1.85, and 1.30, respectively. In more general terms, the sequence of the values of the stretching force constants is, for the 1:2 complexes, $Pt > Pd > Cu > Mn$, and for the 1:3 complexes, $Rh > Co > Cr > Fe > Mn$.

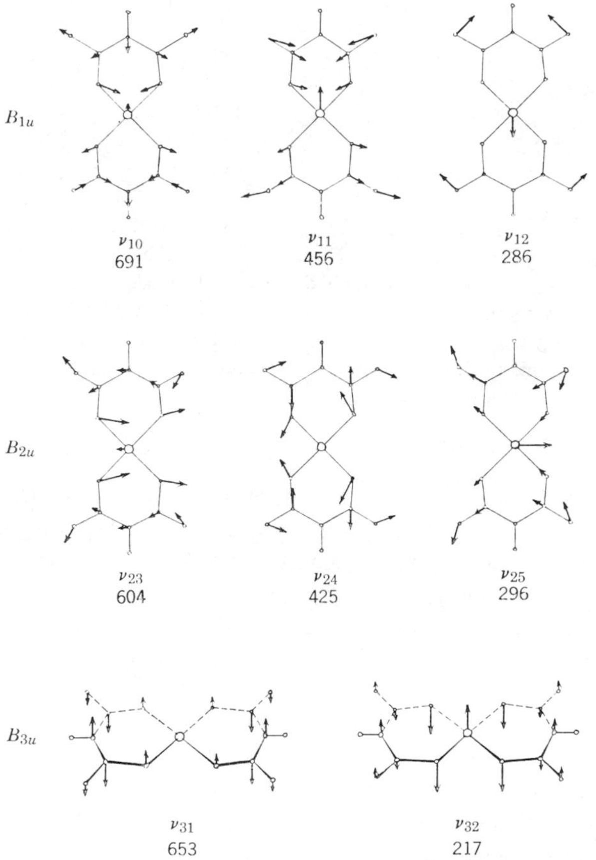

Fig. 13.25 Low-frequency modes and their displacement vectors (from the normal coordinate analysis) of $Cu(acac)_2$. The values below the designations of the fundamentals are the calculated frequencies (in cm^{-1}). Compare with Fig. 13.24a. [M. Mikani, I. Nakagawa, and T. Shimanouchi, *Spectrochim. Acta* **23A**, 1037 (1967), by permission of Pergamon Press.]

G. Temperature and Pressure Dependence of Lattice Modes

In the examples of the "crystalline band" in polyethylene (Section D) and the alkali halides (Chapter 12, Sections B.d and D.b) we have discussed the temperature dependence of the lattice modes and the various conclusions which may be drawn from the results of such experiments—for instance, conclusions about the intermolecular or interionic forces, assignments of two-phonon branches, and so forth. In this section we will discuss the effect of the temperature on the lattice modes of an "order–disorder" type ferroelectric crystal, sodium nitrite, $NaNO_2$. At a temperature of about 163°C the crystal structure goes through a second-order transition in which the long-range order of the unit cell dipoles is lost by a reorientation of the NO_2^- groups[24a]; the spontaneous polarization therefore vanishes and the crystal becomes paraelectric. (This behavior is in contrast to the ferroelectric transition in perovskite-type crystals, which is related to the instability of the lattice to the displacements of certain phonon modes. This is discussed in detail in Appendix II.)

The experimental work we wish to describe has been published by Nakamura *et al.*[24b] The $NaNO_2$ crystals (space group C_{2v}^{20}) were measured, in reflection, along the crystal axes a, b, c with polarized radiation obtained with the help of a polarizer consisting of a pile of polyethylene sheets under 15° incidence. A far-infrared grating spectrometer JASCO-DS501G was used in single-beam operation (spectral purity exceeded 90% for all frequency regions studied). The temperature range studied was 77 to 460°K (Curie point is at 436°K) and the frequency region investigated stretched from 63 to 400 cm^{-1}. The data were evaluated with the help of the Kramers-Kronig relation and the classical oscillator model of the complex dielectric constant as discussed on various occasions in Chapters 12 and 13.

The factor group analysis of ferroelectric $NaNO_2$ predicts (one molecule per unit cell) one each of an infrared-active translational lattice mode of species A_1 (b-polarized), B_1 (a-polarized), B_2 (c-polarized), and two rotatory lattice modes of species B_1 and B_2, respectively. Of the remaining four genuine zero-wave vector modes, one is infrared-inactive (A_2) and three represent the internal motions of the (nonlinear) NO_2^- ion (they are at frequencies above 800 cm^{-1}).

In Fig. 13.26 we show the reflectivity of the A_1-lattice mode, that is, the lattice mode in which the electromagnetic radiation is polarized parallel to the b-axis of the crystal, at 77, 300, 433, and 459°K. This mode should particularly reflect any anomalies when the crystal goes through the transition point since the structural disorder above the Curie point primarily results in equal numbers of NO_2^- groups with their apexes pointing in the $+b$ and $-b$ direction.[24a] Yet, as the data show, no remarkable anomaly is apparent, as

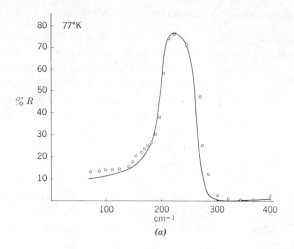

(a)

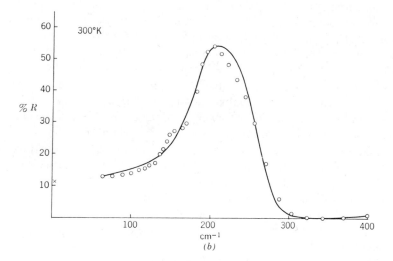

(b)

Fig. 13.26   The far-infrared reflectivity of $NaNO_2$ in *b*-polarization at various temperatures below and above the Curie point (436°K): *(a)* 77° K ; *(b)* 300°K, *(c)* 433°K ; *(d)* 459°K. The crosses on the ordinate denote the low-frequency values (from microwave data) [I. Hatta, *J. Phys. Soc. Japan* **24**, 1043 (1968)]. [K. Suzuki, F. Sugawata, S. Sawada, and T. Nakamura, *Tech. Rep. ISSP (Tokyo)*, Ser. **A**, No. 338, Nov. 1968.]

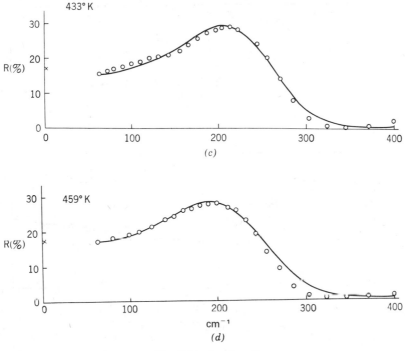

Fig. 13.26   (continued)

can also be ascertained from the normal appearance of the temperature dependence of the frequency, strength ($\varepsilon_\infty + \Sigma_j A_j$, see Eqs. 12.17 and 12.18), and the damping constant of the mode (Fig. 13.27). Summarizing, we see that the results do not indicate any direct evidence for "extra" modes associated with the positional disordering of the crystal.[24a,b]

In the last paragraphs of this section we shall briefly describe the significance of frequency measurements of lattice modes as a function of the external pressure.

Ferraro et al.[25] have published a note describing the effect of pressure on far-infrared crystal spectra. Figure 13.28 shows their data on $NaNO_3$, obtained with a high-pressure cell of the diamond anvil type and a beam condenser which was mounted into a Perkin-Elmer spectrophotometer, Model 301. As can be seen from the figure, the lattice vibration of $NaNO_3$ at 212 cm^{-1} has split into two components at a pressure of 4400 atm; one component lies at 220 cm^{-1}, the other at 250 cm^{-1}. Further increases of pressure cause further separation of the two components. At a pressure of 35000 atm these are observed to have shifted to 224 and 279 cm^{-1}, respectively. Ferraro et al. have also observed that the bands become broader

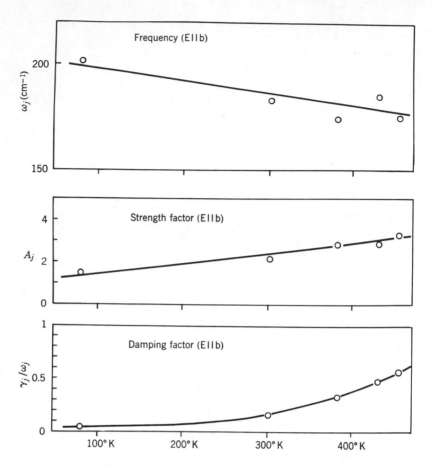

Fig. 13.27   The temperature dependence of the frequency, the strength factor A_j for $\varepsilon_0 = \varepsilon_\infty + \Sigma_j A_j$, and the reduced damping constant of the b-polarized translational lattice vibration of $NaNO_2$ in the classical oscillator approximation. [K. Suzuki, F. Sugawara, S. Sawada, and T. Nakamura, *Tech. Rep. ISSP (Tokyo)*, Ser. A, No. 338, Nov. 1968.]

under pressure and, in some cases, featureless. The authors furthermore note that, generally, the pressure-induced shifts are reversible and that internal modes undergo only minor shifts, if any, under increased external pressures.

In the following we shall briefly discuss some of the principles which are common to the pressure- and temperature-induced effects described above. The temperature dependence of the fundamentals arises from the anharmonicities in the crystal. In Section C.c of Chapter 12 we discussed in detail the interactions of two phonons due to the anharmonicities (mechanical and

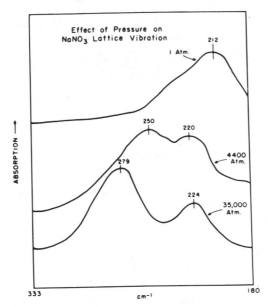

Fig. 13.28 Pressure-induced shifts of lattice vibrations of sodium nitrate. [J. R. Ferraro, S. S. Mitra, and C. Postmus, *Inorg. Nucl. Chem. Letters* **2**, 269 (1966), by permission of Pergamon Press.]

electrical) in the crystal. There is an additional, complicating anharmonicity effect—called the Grüneisen contribution[26]—which arises from the thermal expansion of the nonideal crystal, in other words, from the volume dependence of a one-phonon lattice mode. (We remember that an ideal crystal has zero thermal expansion.) Summarizing, the variations of the frequency of a lattice mode with temperature arise then from the interplay of rather complicated phenomena, especially since the respective frequency shifts due to phonon–phonon interaction and the Grüneisen effect may be of opposite signs.[27]

In contrast to this, the effect of pressure on the fundamental frequencies of lattice modes arises primarily from the Grüneisen contribution, that is, the volume dependence of the one-phonon mode.[25]

In conclusion we mention that Ferraro *et al.* have proposed that pressure- and temperature-induced frequency shift data be combined with regard to a differentiation between the Grüneisen contributions and those caused by phonon–phonon interaction. Furthermore, pressure-shifted spectra could be employed to distinguish between internal and lattice modes.[25]

H. Vibrational Spectra of Orientationally Disordered Crystals. Translational Lattice Vibrations of Ice Ic and Ih

In this last section of this chapter we discuss the far-infrared spectrum of ice Ic and Ih, reported by Whalley and Bertie.[28] The spectra and the property they describe are not only of interest *per se*, but the particular aspects are based on phenomena that are somewhat related with those discussed in Chapter 11.

We begin with a description of the relevant properties of ice (a molecular crystal) and the singular character of its infrared spectrum. The oxygen atoms in ice Ic (cubic) and ice Ih (hexagonal) are in a regular tetrahedral arrangement. There is only one hydrogen atom along each O—O direction; furthermore, a hydrogen atom is considerably closer to one oxygen atom (about 0.95 Å, as in the vapor) than to the other (about 1.8 Å). As a consequence, the hydrogen atoms are in statistical "half-hydrogen" positions,[29] as shown in Fig. 13.29.

Reorientation of a water molecule in the crystal about its oxygen atom (serving as pivot) changes one of the many different configurations of the crystal into another.[30] We express this by stating that there is long-range positional order (since the molecules are on regular positions in the lattice) but that there is long-range orientational disorder.

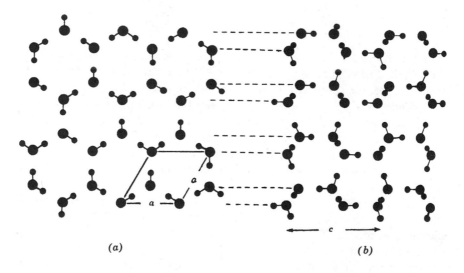

(a)   (b)

Fig. 13.29   Molecular arrangement of the water molecules in ice Ih (hexagonal); only the projections along [001] (part *a*) and [100] (part *b*) are given. [*After* K. Lonsdale, *Proc. Roy. Soc. (London)* **A247**, 424 (1958), by permission of The Royal Society, London.]

In the following we shall briefly enumerate the expected translatory optic fundamentals of ice Ic and Ih based on the factor group analysis of the space group of the crystal (by this we understand the positions of the oxygen atoms as obtained by X-ray diffraction). Thereafter we shall compare the results of the factor group analysis with the experimental infrared spectrum.

The primitive unit cell of ice Ic (diffraction space group O_h^7) contains two molecules. Thus, there is a triply degenerate zero-wave vector translatory optic vibration (see Section A) of species F_{1g}, in which the two interpenetrating cubic close-packed lattices of oxygen atoms vibrate against each other. Based on the oxygen atoms alone (diamond structure), this triply degenerate fundamental is infrared-inactive (and Raman-active). It is expected to be nearly infrared-inactive in the actual ice crystal. Although the motion of the oxygen lattices will cause a change in the dipole moment due to the fact that the hydrogen atoms are not in the center between two oxygen atoms, this very fact also ensures that the amount of change of the dipole moment is very small (the H atom is associated sometimes with one, sometimes with the other oxygen atom). In ice Ih, of diffraction space group D_{6h}^4 with four molecules per primitive unit cell, the relations are quite similar (its nine zero-wave vector translatory optic modes, all infrared-inactive, are A_{1g}, B_{1g}, B_{2u}, E_{1g}, E_{2g}, and E_{2u}).

We now inspect the far-infrared spectrum of ice Ih and Ic (there was no observable difference in the spectra of the two phases), which is shown in Fig. 13.30. According to what we have just said, there should be no ap-

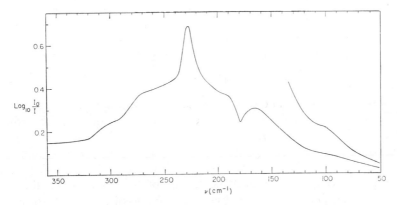

Fig. 13.30   Far-infrared spectrum of ice Ih (100% H_2O) and ice Ic (100% H_2O) at 100°K. The spectra of the two phases were identical. The two curves represent different thicknesses. Instrument: Perkin-Elmer spectrophotometer, Model 301. A conventional low-temperature cell with low-density polyethylene windows was used for most experiments. Stray radiation: 2–3%, 250–220 cm^{-1}; 2%, 100–96 cm^{-1}; less than 1% for other regions between 360 and 80 cm^{-1}. [J. E. Bertie and E. Whalley, *J. Chem. Phys.* **46**, 1271 (1967).]

preciable absorption in the region of the translatory lattice vibrations. Therefore, the zero-wave vector analysis evidently does not predict the data correctly.

However, there is no doubt that the far-infrared absorption band arises from translational motions. Since ice has, strictly speaking, no symmetry, all fundamentals are "impure" and may *a priori* consist of internal, rotational, and translational motion. Yet, the internal vibrations are at much higher frequencies (around 1400 cm^{-1}) and mixing of rotational and translational modes is, most likely, not significant in ice—as shown by comparison with the spectrum of D_2O[28] (the rotational modes, involving mainly motions of the hydrogen atoms, are at higher frequencies than the translatory modes).

We shall now show how this discrepancy between the results of factor group analysis and observation can be resolved by taking into account the orientational disorder of the crystal;[28] for the purposes of our discussion we need not give the detailed formalism of the theory,[31] but shall merely state the important results as they pertain to the interpretation of the observed far-infrared spectrum. (*1*) The irregularity of the positions of the hydrogen atoms causes only a small irregularity in the positions of the oxygen atoms (their root-mean-square displacement due to this cause is about one-tenth of their root-mean-square zero-point amplitude). Hence, the translational lattice vibrations, which involve motions of one water molecule with respect to another, can be described in terms of the *mechanically regular* crystal— in other words, about mean positions of the oxygen atoms as determined by diffraction experiments. (*2*) Because of the statistical distribution of the hydrogen atom positions as discussed above, there are *no rigorous* selection rules. (*3*) Although the most probable value of the transition moment is zero, the most probable value of the squared transition moment is not zero. (*4*) The conclusion is therefore that all lattice vibrations, whether optic, acoustic, transverse, or longitudinal, are in principle *active* in the infrared (and the Raman).

Since the intensity of an disorder-induced translational band is essentially proportional to the square of the frequency,[31] the original absorption spectra of Fig. 13.30 are replotted in terms of optical density divided by v^2 (v in cm^{-1}) as shown in Fig. 13.31 for ice Ih (the plot for ice Ic is essentially the same). Bertie and Whalley assume that the reduced plot of Fig. 13.31 corresponds better to features in the density of vibrational states than the simple intensity factor shown in Fig. 13.30. We shall describe then, in the light of what we have said above, the translational lattice modes (*1*) in terms of a wave vector *vs* frequency (dispersion) plot based on the diffraction space group of the crystal and (*2*) by allowing all lattice fundamentals. For simplicity's sake, we restrict our discussion to the simpler case of cubic ice (*Ic*), with diamond structure ($O_h{}^7$) and two molecules per primitive unit cell (see

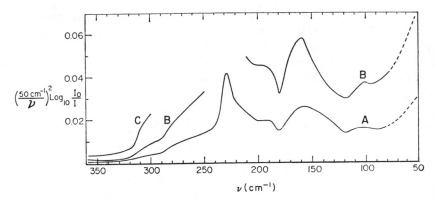

Fig. 13.31   Reduced optical density $(50\,cm^{-1}/v)^2 \log (I_0/I)$ as a function of frequency $v(cm^{-1})$ for ice Ic and Ih. The curves A, B, C denote different thicknesses of material. [J. E. Bertie and E. Whalley, *J. Chem. Phys.* **46**, 1271 (1967).]

above). The chosen dispersion plot is shown in Fig. 13.32. The actual dispersion curves are not known, but Whalley and Bertie use the ones of Fig. 13.32 for a starting point; whether the actual dispersion plot of ice Ic is similar to the one used here needs to be corroborated by experiment.[28]

The six translational degrees of freedom yield three acoustic and three optic branches, labeled in the usual manner (see Chapter 12). We note that the vibrations are only pure in the symmetry directions indicated in the figure and that the twofold degeneracy of the transversal branches is not

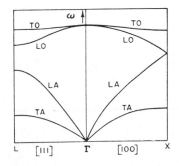

Fig. 13.32   Frequency–wave vector dispersion curves in the [100] and [111] directions of the Brillouin zone for the various branches in a crystal of the diamond structure. All transversal branches are doubly degenerate. The plot is assumed to represent the six translational branches of cubic ice (two molecules per primitive unit cell), as explained in the text. [J. E. Bertie and E. Whalley, *J. Chem. Phys.* **46**, 1271 (1967).]

removed. The assignments of the features of the reduced spectrum (Fig. 13.31) to the dispersion plot are done in the manner we have discussed in Section C.c of Chapter 12 for two-phonon combination and difference bands except that here the *fundamental* spectrum is considered since all selection rules are removed. Quite generally, we know that high densities of vibrational states are found at the center of the Brillouin zone for the TO and LO branches and near where all branches touch the boundary. The dispersion plot of Fig. 13.32 also indicates that there is probably a region of low density of states (gap) between the LO and LA maxima. In detail, we make the following assignments.[28]

(*1*) The strong sharp peak at 229 cm^{-1} is probably due to the TO branch. (*2*) The peak at 164 cm^{-1} probably arises from the LA branch, the feature at 190 cm^{-1} from the LO branch. (*3*) The expected gap lies probably at 180 cm^{-1}. (*4*) The peak due to the TA branch is presumably associated with the feature below 80 cm^{-1} (it is not seen in the direct spectrum (Fig. 13.30)).[†]

We notice then that the gross features of the dispersion plot correspond exactly to the features of the observed spectrum. A more detailed description of the spectrum makes use of the critical points of the dispersion curves. (The advantage is that calculations can be performed with the help of an assumed force field. We shall not go into this here[28].) We give a listing of the more detailed assignments in Table 13.9. In Assignment 1, the strong band at 229 cm^{-1} is assigned to symmetry point Γ; in Assignment 2 it is ascribed to point L.

TABLE 13.9*

FREQUENCY ASSIGNMENTS AT THE ZONE CENTER AND BOUNDARIES FOR ICE *Ic*

Symbol	Description	Branch	1	2
Γ	Center	LO = TO	229.2	240
		LA = TA		
X	Boundary in [100]	TO	200	220
		LO = LA	180.5	180.5
		TA	~ 70	~ 70
L	Boundary in [111]	TO	220	229.2
		LO	190	190
		LA	164	164
		TA	< 50	< 50

Location in Brillouin zone / Assignment, cm^{-1}

* J. E. Bertie and E. Whalley, *J. Chem. Phys.* **46**, 1271 (1967).

† The observed frequencies agree very well with results of recent inelastic neutron scattering experiments [see H. Prask, H. Boutin, and S. Yip, *J. Chem. Phys.* **48**, 3367 (1968)].

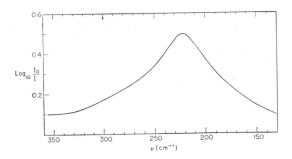

Fig. 13.33 Far-infrared absorption of a film of vitreous ice. [J. E. Bertie and E. Whalley, *J. Chem. Phys.* **46**, 1271 (1967).]

For comparison, we show in Fig. 13.33 the broad and almost featureless spectrum of vitreous ice for the spectral range of 130 to 360 cm^{-1}. As is well known, in a glass there is neither long-range orientational nor translational order; we see that Fig. 13.33 reflects this general lack of long-range symmetry. On the other hand, we show in Fig. 13.34 the infrared lattice spectrum of the ordered crystal ice *II* as reported by Bertie *et al.*[32] We

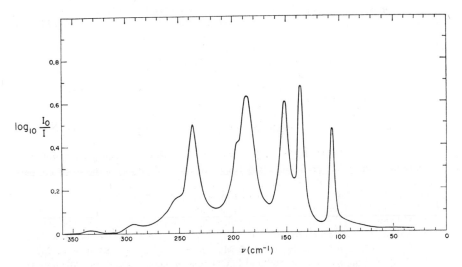

Fig. 13.34 Far-infrared spectrum of ice *II*, an ordered crystal, at 100°K and atmospheric pressure. The space group is C_{3i}^2, with 12 molecules per unit cell. The factor group is isomorphous to point group S_6 and the $3 \times 12 - 3 = 33$ genuine translational lattice modes form the representation $\Gamma = 6A_g + 5A_u + 6E_g + 5E_u$ of S_6. Of these, the 10 modes $5A_u + 5E_u$ are infrared-active (see Section Bb). The far-infrared spectrum shows nine recognizable peaks and shoulders. [J. E. Bertie, H. J. Labbé, and E. Whalley, *J. Chem. Phys.* **49**, 775 (1968).]

notice here the sharp peaks which represent the transitions of the zero-wave vector lattice vibrations of the ordered crystal in contrast to the broad bands of the transitions between the states of maximum phonon density in the orientationally disordered lattices of ice Ic and Ih.

REFERENCES

1. A concise yet rather exhaustive description is found in R. Zbinden, *Infrared Spectroscopy of High Polymers*, Academic Press, New York, 1964, Chapter II. We have based part of our brief treatment on Zbinden's presentation. See, for an extensive treatment, S. Bhagavantam and T. Venkatarayudu, *Theory of Groups and Its Application to Physical Problems*, Bangalore Press, Bangalore City, 1951 (newer editions have been issued). See also S. Bhagavantam, *Crystal Symmetry and Physical Properties*, Academic Press, New York, 1966.
2. H. Winston and R. S. Halford, *J. Chem. Phys.* **17**, 607 (1949) and references cited therein; D. F. Hornig, *J. Chem. Phys.* **16**, 1063 (1948).
3. J. I. Bryant, *J. Chem. Phys.* **40**, 3195 (1964).
4. M. Balkanski, P. Moch, and G. Parisot, *J. Chem. Phys.* **44**, 940 (1966).
5. G. F. Koster, *Solid State Phys.*, Vol. 5, F. Seitz and D. Turnbull, Eds., Academic Press, New York, 1957, p. 173.
6. See S. Bhagavantam and T. Venkatarayudu, Ref. 1, p. 138.
7. J. I. Bryant, *J. Chem. Phys.* **38**, 2845 (1963).
8. J. E. Rosenthal and G. M. Murphy, *Revs. Mod. Phys.* **8**, 317 (1936); see Eq. 51.
9. M. Balkanski, P. Moch, and M. K. Teng, *J. Chem. Phys.* **46**, 1621 (1967).
10. H. A. Willis, R. G. J. Miller, D. M. Adams, and H. A. Gebbie, *Spectrochim. Acta* **19**, 1457 (1963); R. V. McKnight and K. D. Möller, *J. Opt. Soc. Am.* **54**, 132 (1964).
11. J. E. Bertie and E. Whalley, *J. Chem. Phys.* **41**, 575 (1964).
12. A. O. Frenzel and J. P. Butler, *J. Opt. Soc. Am.* **54**, 1059 (1964).
13. S. Krimm and M. I. Bank, *J. Chem. Phys.* **42**, 4059 (1965).
14. S. Krimm, C. Y. Liang, and G. B. B. M. Sutherland, *J. Chem. Phys.* **25**, 549 (1956).
15. M. Tasumi and T. Shimanouchi, *J. Chem. Phys.* **43**, 1245 (1965).
16. A. Anderson and G. E. Leroi, *J. Chem. Phys.* **45**, 4359 (1966).
17. T. H. Jordan, H. W. Smith, W. E. Streib, and W. N. Lipscomb, *J. Chem. Phys.* **41**, 756 (1964).
18. The site group [R. S. Halford, *J. Chem. Phys.* **14**, 8 (1946), and, particularly, Ref. 2] is very useful for an interpretation of the grosser (if not all) features of the spectrum, particularly the selection rules.
19. G. L. Bottger and A. L. Geddes, *Spectrochim. Acta* **23A**, 1551 (1967).
20. M. C. Tobin, *J. Chem. Phys.* **23**, 891 (1955), see Table II.
21. M. Mikami, I. Nakagawa, and T. Shimanouchi, *Spectrochim. Acta* **23A**, 1037 (1967).
22. K. Nakamoto, Y. Morimoto, and A. E. Martell, *J. Phys. Chem.* **66**, 346 (1962); K. Nakamoto, P. J. McCarthy, A. Ruby, and A. E. Martell, *J. Am. Chem. Soc.* **83**, 1066, 1272 (1961).
23. See E. B. Wilson, Jr., J. C. Decius, and P. C. Cross, *Molecular Vibrations*, McGraw-Hill, New York, 1955, Appendix X.
24a. J. D. Axe, *Phys. Rev.* **167**, 573 (1968), see Fig. 1.
24b. K. Suzuki, F. Sugawara, S. Sawada, and T. Nakamura, *Tech. Rep. ISSP (Tokyo)*, Ser. A, No. 338, Nov. 1968; *J. Phys. Soc. Japan* **26**, 1199 (1969); Y. Ishibashi, K. Siratori, and T. Nakamura, *J. Phys. Soc. Japan* **21**, 809 (1966).
25. J. R. Ferraro, S. S. Mitra, and C. Postmus, *Inorg. Nucl. Chem. Letters* **2**, 269 (1966).

26. C. Kittel, *Introduction to Solid State Physics*, Wiley, New York, 1967, Chapter 6.
27. J. R. Jasperse, A. Kahan, J. N. Plendl, and S. S. Mitra, *Phys. Rev.* **146**, 526 (1966).
28. J. E. Bertie and E. Whalley, *J. Chem. Phys.* **46**, 1271 (1967).
29. K. Lonsdale, *Proc. Roy. Soc. (London)* **A247**, 424 (1958).
30. L. Pauling, *J. Am. Chem. Soc.* **57**, 2680 (1935).
31. E. Whalley and J. E. Bertie, *J. Chem. Phys.* **46**, 1264 (1967).
32. J. E. Bertie, H. J. Labbé, and E. Whalley, *J. Chem. Phys.* **49**, 775, 2142 (1968).

APPENDIX I

IMPURITY-INDUCED LATTICE ABSORPTION IN THE FAR-INFRARED

A. J. SIEVERS

Department of Physics and Laboratory of Atomic and Solid State Physics, Cornell University, Ithaca, New York

A. INTRODUCTION

Impurity-induced lattice absorption was first observed in covalent crystals in 1934. Type I diamond[1] was measured in the infrared spectral region but the complex spectrum was not correctly understood until the early fifties when intensive research on the chemical purification of silicon and germanium occurred. The similarity between the infrared spectra for a number of covalent crystals[2] enabled Lax and Burstein[3] to anticipate some impurity absorption processes correctly. For example, they suggested that an impurity which did not strongly modify the vibrational properties of the lattice in the immediate neighborhood of the impurity could induce an absorption spectrum which would roughly map the density of phonon states.

The first quantitative determination of a specific lattice impurity was by Kaiser *et al.*[4] on oxygen in silicon and germanium. The identification of the (oxygen) absorption had been an unresolved puzzle for some time. The carefully controlled experiments by Kaiser *et al.* showed that the oxygen impurities, in fact, originated from the quartz crucibles in which the crystals had been grown! This is a lesson worth remembering as crystal preparation and impurity identification are still major problems for the alkali halide crystals which we consider here.

Before turning to impure crystals, the infrared properties of pure alkali halides must be understood. The infrared absorption processes for ionic crystals have been investigated by many workers,[5,6] and only a summary of the essential results will be given here (see Chapter 12). A cubic lattice with two ions per unit cell has only six branches: three acoustic and three optic. Only one mode in a transverse optic (TO) branch is infrared-active in first order. For this mode the wave vector of the phonon is equal to the wave vector of the incident photon. As the wave vector of the photon is almost zero, the process is usually represented by a "vertical" transition. The longitudinal optic (LO) branch is not degenerate with the TO branch but, depending on the magnitude of the oscillating dipole moment per unit cell, is shifted to higher frequencies, and the crystal is opaque in the intermediate frequency region between the TO and LO branches. This forbidden frequency band is usually referred to as the reststrahlen region. In Fig. AI.1 the absorption coefficient for LiF as a function of temperature is shown. The TO mode

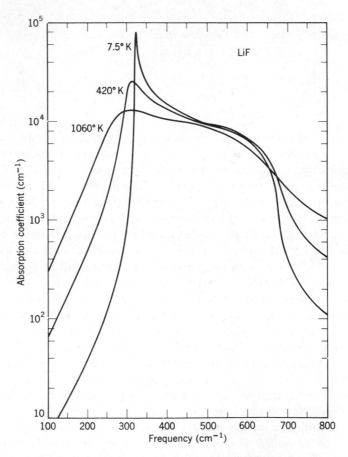

Fig. AI.1   Absorption coefficient of LiF as a function of frequency and temperature. [J. R. Jasperse, A. Kahan, J. N. Plendl, and S. S. Mitra, *Phys. Rev.* **146**, 526 (1966).]

occurs at about $310\,\mathrm{cm}^{-1}$ and the LO mode at about $660\,\mathrm{cm}^{-1}$.[7] The crystal is indeed opaque between the two branches.

At frequencies outside the reststrahlen region, the temperature-dependent absorption coefficient in Fig. AI.1 stems from higher-order, two-phonon processes. The temperature effect is most dramatic below the TO mode, as Rubens and Hertz first showed in 1912.[8] Their crystals were observed to become more transparent as the temperature was decreased. The absorption coefficient in this frequency region varied linearly with temperature. The physical process which describes this temperature dependence has been identified with the anharmonic coupling of the transverse optic phonon ω_T, $\mathbf{k} = 0$, with two other lattice phonons ω_1, $\mathbf{k}_1$ and ω_2, $\mathbf{k}_2$. The coupling originates from the third- and higher-order terms which describe the lattice potential energy. The absorption of a photon ω_p can either create two phonons (summation band) with $\omega_p = \omega_1 + \omega_2$ and $\mathbf{k}_1 = -\mathbf{k}_2$ or create one phonon and destroy a lower energy phonon (difference band) with $\omega_p = \omega_1 - \omega_2$ and $\mathbf{k}_1 = \mathbf{k}_2$. The temperature dependence of the absorption coefficient for the summation band varies as $S(T) = (1 + n_1 + n_2)$ or as the sum of the occupation numbers for the two modes in the high-frequency region. In the far-infrared frequency region the difference band is important and the absorption coefficient varies as $D(T) = (n_2 - n_1)$ or as the difference in occupation numbers of the two modes. Since $n_i = (e^{\hbar\omega_i/kT} - 1)^{-1}$, then at $T \sim 0°\mathrm{K}$, $S(0) = 1$ and $D(0) = 0$ and at very low temperatures the difference band is quenched. At intermediate temperatures the frequency dependence of the absorption is expected to contain structure because the absorption band will mainly involve phonons from regions of $\mathbf{k}$ space with a high density of phonon modes.

For a specific example (which we shall need later) the difference band absorption in potassium iodide is now considered. This particular alkali halide has been chosen because a gap occurs in the spectrum between the optic and acoustic branches. Of the many alkali halide crystals whose frequency spectra have been calculated[9] or measured by inelastic neutron scattering techniques[10,11] only five, LiCl, NaBr, NaI, KBr, and KI, have such a gap. Most experimental investigations of impurity-induced gap modes have been made on KI because it is not hygroscopic as are both NaBr and NaI and also because it has the largest gap of the three remaining crystals.

Recently the normal mode spectrum of KI has been obtained by Dolling et al.[11] by inelastic neutron scattering. From their measured dispersion curves the separation of the branches at the same zone boundary are reproduced in Table AI.1. With a model which takes into account the polarizability of both ions, axially symmetric short-range forces between first and second nearest neighbors, and a variable ionic charge, they have fitted the dis-

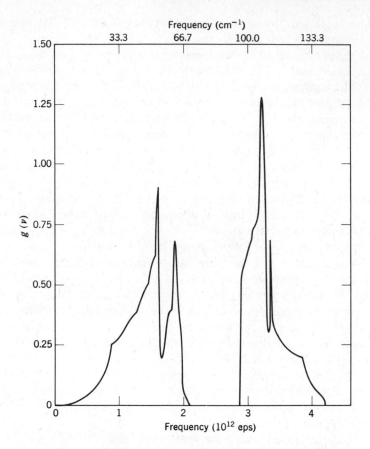

Fig. AI.2   The frequency distribution for the density of phonon modes in KI. A well defined energy gap extends from 69.7 to 95.6 cm^{-1}. [G. Dolling, R. A. Cowley, C. Schittenhelm, and I. M. Thorson, *Phys. Rev.* **147**, 577 (1966).]

persion curve in the principle directions and calculated the frequency distribution function $g(v)$ for KI. Their results are shown in Fig. AI.2. A well defined energy gap extends from 69.7 to 95.6 cm^{-1}. The two peaks in the density of acoustic modes occur at 62.4 and 53.7 cm^{-1}, respectively.

For comparison, the temperature dependence of the far-infrared absorption coefficient in pure KI in the frequency region below the fundamental absorption is shown in Fig. AI.3. A broad maximum in the absorption coefficient at 76 cm^{-1} is clearly visible at 44°K and decreases as the temperature is decreased. With crystals of some centimeters thickness cooled to 2°K, transmission measurements can be readily obtained up to 85 % of the trans-

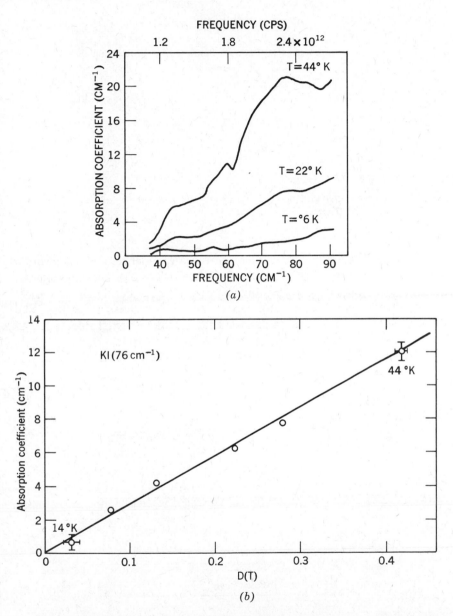

Fig. AI.3   (a) The temperature dependence of the absorption coefficient in pure KI. The broad maximum in the absorption coefficient occurs at 76 cm⁻¹. (b) The temperature dependence of the absorption coefficient at 76 cm⁻¹ vs $D(T)$. The absorption coefficient varies as the difference in the populations of the two phonon modes $n_2 - n_1 = D(T)$, as described in the text. [C. D. Lytle, M.S. Thesis, Cornell University, 1965.]

TABLE AI.1*

<small>SEPARATION OF THE PHONON BRANCHES AT THE ZONE BOUNDARY FOR KI</small>

	[100]	[111]
LA → LO	60.7 cm^{-1}	62.2 cm^{-1}
LA → TO	56.2 cm^{-1}	25.7 cm^{-1}
TA → LO	81.3 cm^{-1}	76.2 cm^{-1}
TA → TO	77.0 cm^{-1}	39.3 cm^{-1}

* *After* G. Dolling, R. A. Cowley, C. Schittenhelm, and I. M. Thorson, *Phys. Rev.* **147**, 577 (1966).

verse optic mode frequency. The temperature dependence of the absorption at 76 cm^{-1} can be readily fitted [12] with $D(T) = (n_2 - n_1)$. The experimental temperature dependence can be obtained either by considering the TA → LO transition for the [111] direction or the TA → TO transition in the [100] direction with the appropriate phonon energies taken from Table AI.1.

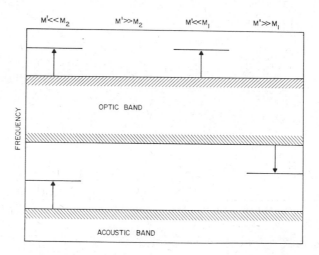

Fig. AI.4   Isotopic impurity in a diatomic linear chain. The light mass is labeled M_1 and the heavy mass, M_2. Four cases are shown: (1) a light impurity mass replaces the heavy mass of the host lattice ($M' \ll M_2$); (2) a heavy impurity mass replaces the heavy mass of the host ($M' \gg M_2$); (3) a light impurity mass replaces the light host atom ($M' \ll M_1$); and (4) a heavy impurity mass replaces the light mass of the host ($M' \gg M_1$).

B. IMPURITY-INDUCED ABSORPTION

In contrast with the one-phonon absorption spectrum of a pure alkali halide crystal which occurs only for the transverse optic branch at $\mathbf{k} = 0$, a number of one-phonon effects are possible in an impure crystal because the impurity destroys the translational symmetry and the lattice wave vector is no longer a good quantum number. Before turning to the spectroscopic investigation of impurity modes, let us see what type of one-phonon effects should be expected for a diatomic lattice (see also Chapter 12, Section E).

Let us begin by considering the results of the theoretical work by Mazur et al.[13] They considered the problem of a diatomic linear chain with a mass defect. An impurity of mass M' replaces either the light mass M_1 or the heavy mass M_2 of the diatomic chain, and the nearest neighbor force constants at the impurity are not altered. The local mode solutions for this problem are shown pictorially in Fig. AI.4. For M' replacing M_2 where $M' \ll M_2$, two impurity modes are generated, a local mode above the top of the optic spectrum and a gap mode between the acoustic and optic branches. Both modes have the same symmetry. In the second case $M' \gg M_2$ and neither local modes or gap modes are predicted. The different behavior for these two cases can readily be understood. At the zone edge of the acoustic branch, the heavy atoms M_2 vibrate by a value of π out-of-phase with each other while the light atoms M_1 are at rest. The converse is true at the zone edge of the optic branch, the light atoms M_1 vibrate by π out-of-phase and the heavy atoms M_2 are at rest.

Let us consider the first case where $M' \ll M_2$. We start by letting $M' = M_2$ and then increase the mass perturbation by reducing M'. Initially the impurity is vibrating out-of-phase with the other nearest M_2 atoms. As M' is decreased the number of atoms participating in this mode decreases and the mode moves up into the gap. The mode at $\mathbf{k} = 0$ moves out of the optic band because of the mutual repulsion between levels of the same symmetry. For the second case, where $M' \gg M_2$, the mode at the zone edge of the acoustic band is depressed. Because the mode is moving away from the optic band, no local mode from the optic band is expected. Two more cases are possible if the impurity M' replaces the lighter atom M_1 in the diatomic chain, and they are also shown in Fig. AI.4. These modes can be interpreted in a similar manner as above.

The first case, where $M' \ll M_2$, is particularly important because this mass perturbation is appropriate (as a first approximation anyway) to describe the hydride ion in alkali halide crystals. Figure AI.4 indicates that not only a local mode should be observed but also another localized lattice mode should occur in the frequency gap between the optic and acoustic phonon branches. This "gap" mode should also be infrared-active. The

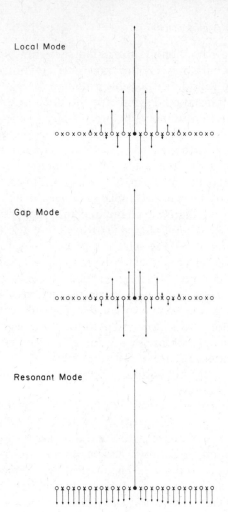

Fig. AI.5   Eigen vector for an infrared-active impurity mode in a diatomic linear chain. Local mode and gap mode by Renk. Resonant mode by Weber. [*After* K. F. Renk, *Z. Physik* **201**, 445 (1967), R. Weber, Ph.D. dissertation, Phys. Inst. University Freiburg i. Br., by permission of Springer: Berlin-Göttingen-Heidelberg.]

amplitude of an impurity ion and its neighbors for a local mode and also for a gap mode have been calculated for a diatomic linear chain by Renk.[14] The atomic amplitudes are represented in Fig. AI.5. For the local mode the negative ions move in one direction while the positive ions move in the opposite sense. This motion is the same as for the $\mathbf{k} = 0$ mode of the transverse optic branch except that the amplitude has a spatial variation. In fact, the maximum amplitude is associated with the impurity ion itself, and the amplitude decays exponentially as the distance from the impurity is increased.

An exponential decay of the amplitude with increasing distance from the impurity ion also occurs for gap modes and is shown in Fig. AI.5. In this mode, like ions vibrate by π out-of-phase with each other, which is a motion characteristic of normal lattice modes at the edge of the Brillouin zone. Because of the opposing motion of like ions, the oscillating dipole moment associated with this gap mode is smaller than for the local mode.

If the nearest neighbor force constants of the impurity ion are changed, then, in addition to the infrared-active modes described above, a number of infrared-inactive local modes can occur in which the impurity ion itself is at rest. Force constant perturbations can also produce interesting changes in the amplitudes of the ions in the in-band frequency region as was first shown by Szigeti[15] for a monatomic linear chain. Genzel et al.[16] have studied the same problem for a diatomic linear chain lattice and find that resonant lattice modes can occur if weak nearest neighbor force constants are assumed for the impurity ion. Such resonant lattice modes were first predicted for a mass defect in a three-dimensional monatomic lattice by Brout and Visscher.[17]

In contrast with the local and gap modes, the resonant mode is not spacially localized but extends far into the lattice. The amplitude of the impurity ion and its neighbors for a low-frequency resonant mode in a linear chain has been calculated recently by Weber.[18] We show his results in Fig. AI.5. Most of the ions move in phase as is expected for a low-frequency acoustic mode. The relative motion of the impurity ion which is out-of-phase gives rise to an oscillating electric dipole moment in an ionic crystal. Thus an infrared-active mode in the acoustic spectrum is possible. Of course, infrared-inactive resonance modes in which the impurity ion does not participate are also possible and some of these should be Raman-active.

A theory which describes the infrared lattice absorption coefficient for crystals with mass defects was first given by Wallis and Maradudin and also Takeno et al.[19] Later, a more complete description of the optical properties associated with the mass defect problem for a homonuclear crystal was given by Dawber and Elliott.[20] Their calculations showed that the induced absorption coefficient associated with a mass defect can be written as

$$\alpha_I(\omega) = \frac{\pi D}{3\sqrt{\varepsilon} c}\left(\frac{\varepsilon + 2}{3}\right)^2 [m_L{}^2(\omega)\delta(\omega) + m_B{}^2(\omega)g(\omega)], \qquad (AI.1)$$

where D is the number of impurities per unit volume, ε is the frequency-dependent dielectric constant of the host crystal, c the velocity of light,

$[(\varepsilon + 2)/3]^2$ is the local field correction factor to account for the presence of the medium, $m_L(\omega)$ is the dipole moment associated with the local mode vibration, $\delta(\omega)$ is the shape function of the local mode to account for the anharmonicity or randomness in the lattice, $m_B(\omega)$ is the dipole moment associated with the band modes, and $g(\omega)$ is the density of modes per unit frequency range per unit volume.

A number of peaks can occur in $\alpha_I(\omega)$. From the first term in Eq. AI.1, the absorption coefficient has a maximum at the local mode or gap mode frequency. From the second term in Eq. AI.1, resonant modes can occur in the band mode region associated with peaking of the functions $m_B(\omega)$ and $g(\omega)$ at frequencies where local modes would like to occur, but are in a sense lifetime-broadened by the finite density of modes. Also, additional structure can appear in the absorption coefficient where maxima occur in the density of "normal" lattice modes.

C. Local Modes

In 1936, a characteristic absorption band in the ultra-violet spectral region (U-band) was identified with an electronic transition of the substitutional hydride ion[21] (U-center). The strength of this band is proportional to the number of hydride centers and represents a transfer of an electron from the hydride ion to a neighboring alkali ion.[22] In 1958, impurity-induced infrared lattice absorption in alkali halide crystals was first observed for the same impurity center by Schäfer.[23] The lattice absorption, which occurred at a frequency far above the maximum phonon frequency of the host lattice, was also found to be proportional to the hydride concentration, and from Smakula's equation, an oscillator strength of 0.5 was estimated. Also, by representing the absorption mechanism as a vibrating mass defect in the form of an Einstein oscillator, a qualitative estimate of the resonant frequency was obtained from the ratio of the host atom mass to the impurity mass.

Some of Schäfer's transmission spectra for the hydride impurity in alkali halide lattices are reproduced in Fig. AI.6. The absorption frequency was observed to depend on the lattice spacing of the host crystal. In particular, the Mollwo-Ivey relation was observed, namely,[24]

$$\omega_0 a^n = \text{const.} \tag{AI.2}$$

This equation, which related (for $n = 1.84$) the absorption frequency ω_0 of an F-center to the lattice constant a, could also be applied to the infrared U-center with $n = 2.22$ for sodium halides, $n = 2.28$ for potassium halides, and $n = 2.53$ for rubidium halides.

Some side band structure on the main line was also found, as shown in Fig. AI.6.

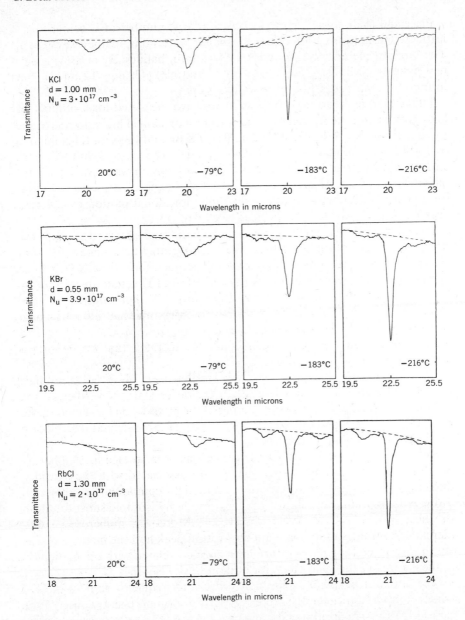

Fig. AI.6   Transmission of alkali halide crystals doped with U-centers as a function of wavelength and temperature. [G. Schäfer, *J. Phys. Chem. Solids* **12**, 233 (1960), by permission of Pergamon Press.]

Price and Wilkinson[25] then measured a large number of U-center host–lattice combinations and observed similar variations in the exponent n. The largest difference was found for the sodium halides. By measuring two new systems NaF:H$^-$ (858.1 $\pm$ 0.1 cm^{-1}) and NaI (430.6 $\pm$ 0.2 cm^{-1}), they found $n = 2.0$ a better fit for the sodium salts.

The deuteride–hydride isotope shift was also measured for a number of host lattices and the frequency ratio was not constant. This ratio was found to be 1.387 for NaCl, 1.398 for KCl, 1.394 for KBr, and 1.378 for KI compared to 1.414 for a simple Einstein oscillator model. The isotope shift was conclusive evidence that the impurity-induced absorption was associated with vibrational motion. Also, the variation in shift frequency demonstrated that the impurity neighbors participated in the vibrational motion by different amounts for the various host lattices.

In addition, the center frequency of the band was observed to shift, the width changed with temperature, and the integrated area under the main absorption line decreased with increasing temperature. Finally Mitsuishi and Yoshinaga[26] were able to destroy the infrared U-center absorption by irradiating a crystal at room temperature with UV light, and Fritz[27] observed an impurity-induced absorption from interstitial hydride ions which were produced by irradiating a U-center crystal with UV light at low temperatures. These initial experiments demonstrated that an important lattice probe was at hand.

To date, no impurity besides the U-center has been observed to produce an infrared-active local mode in alkali halide crystals. Nevertheless, the continuous investigation of the properties of the hydride ion local mode since Schäfer's original work has led to a large body of experimental information on the phonon–phonon interaction.

So far, the local mode from the U-center has been studied in 15 different alkali halide crystals, and the frequency shift associated with the hydride–deuteride substitution has been observed in eight compounds as shown in Table AI.2. A frequency shift of the local mode by isotopic substitution in the host lattice has also been observed.[28] Moreover, numerous experiments[29–32] on the shift of the main absorption peak and the increase in line width with increasing temperature have been correlated with calculations[33] of the lifetime of the local mode for both impurity isotopes.

The different behavior of the half width and frequency shift of the D$^-$ center compared with the H$^-$ center stems from a simple energy conservation consideration. The local mode frequency for the H$^-$ center is usually more than two times the maximum frequency of the host lattice, and three-phonon decay processes are necessary to conserve energy. For the D$^-$ center the frequency is usually less than two times the maximum frequency of the band modes, and two-phonon decay can now occur.

TABLE AI.2

LOCALIZED MODE FREQUENCY FOR U-CENTERS

Host crystal	Impurity ion	Frequency, cm^{-1}	Sample temp, °K	Ref.
^{6}LiF	H$^-$	1030.9	20	*
^{7}LiF	H$^-$	1027.0	20	†
NaF	H$^-$	858.9	20	†
NaF	D$^-$	614.7	20	†
NaCl	H$^-$	562.5	57	‡
NaCl	D$^-$	408.0	90	§
NaBr	H$^-$	496.0	57	‡
NaBr	D$^-$	361.0	90	§
NaI	H$^-$	426.8	10	#
NaI	D$^-$	317.8	10	#
KF	H$^-$	725.5	100	‖
KCl	H$^-$	503.8	100	‖
KCl	D$^-$	360.4	100	‖
KBr	H$^-$	446.7	100	‖
KBr	D$^-$	320.4	100	‖
KI	H$^-$	383.9	100	‖
KI	D$^-$	278.6	100	‖
RbF	H$^-$	703.1	100	‖
RbCl	H$^-$	476.0	90	§
RbCl	D$^-$	339.0	90	§
RbBr	H$^-$	425.0	90	§
RbI	H$^-$	360.0	57	‡
CsCl	H$^-$	423.9	100	‖
CsBr	H$^-$	366.6	100	‖

* A. J. Sievers and R. L. Pompi, *Solid State Commun.* **5**, 963 (1967).
† H. Dötsch, W. Gebhardt, and C. Martius, *Solid State Commun.* **3**, 297 (1965).
‡ G. Schäfer, *Phys. Chem. Solids* **12**, 233 (1960).
§ B. Fritz, U. Gross, and D. Bäuerle, *Phys. Stat. Sol.* **11**, 231 (1965).
‖ W. C. Price and G. R. Wilkinson, University of London King's College, Molecular and Solid State Spectroscopy Technical Report No. 2 (1960); unpublished.
D. Bäuerle and B. Fritz, *Phys. Stat. Sol.* **24**, 207 (1967).

Recently the threefold degeneracy of the excited state of the local mode has been lifted by the application of uniaxial stress.[34] In this manner the coupling of the local mode excited state to the band modes with spherical, tetragonal, and trigonal symmetries has been measured. As an example, for KCl:H$^-$

the shift of the U-center mode with lattice constant change is

$$-\frac{\Delta\omega}{\omega} = 3.6\frac{\Delta a}{a},\qquad\qquad (AI.3)$$

where ω is the U-center frequency and a the lattice constant. Thus a large lattice constant change of 0.1 % produces a small local mode shift of -1.8 cm^{-1}. (This result will be contrasted with the giant shifts for resonant modes described in Section E.)

The absorption spectrum from U-centers consists not only of the resonance peak but also of side bands displaced from the main peak by the band mode frequencies. This side band structure had been observed in Schäfer's original work (see Fig. AI.6), but its significance was not recognized at that time. Somewhat later Fritz[30] properly identified the side bands. The resultant structure occurs because of the anharmonic crystal potential. Transitions are allowed in which one lattice phonon is created or destroyed in addition to the creation of a local mode phonon. These one-phonon side bands are related to the one-phonon density of states of the pure crystal. However, because the coefficients which couple the local mode to the different band modes are frequency-dependent, only the most obvious features of the vibration spectrum of the pure crystal are easily identified in the side band. An example of such a side band absorption spectrum of KBr:KH is shown in Fig. AI.7. The absorption coefficient is given as a function of the frequency from the local mode absorption line. The frequency gap between the optic and acoustic phonon bands is clearly observed in the side band. Some of the known critical points in the phonon spectrum for pure KBr are indicated

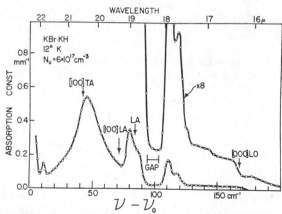

Fig. AI.7   The absorption spectrum of KBr:KH for wavelengths between 22.4 and 16 μ plotted as a function of frequency separation from the local mode. Some of the known critical points in the phonon spectrum are indicated by arrows. [T. Timusk and M. V. Klein, *Phys. Rev.* **141**, 664 (1966).]

by arrows. The coupling of the local mode to the acoustic modes is much larger than the coupling to the optic modes. A theory that explains qualitatively the shape of the side band spectrum in terms of the anharmonic coupling of the H^- ion to its nearest neighbors has been developed by Timusk and Klein.[35]

The presence of foreign alkali or halogen ions in the crystal has been observed to perturb the local mode frequency.[23] In mixed crystals new lines appear which are shifted to both higher and lower frequencies with respect to the main U-center absorption line. Groups of these new lines have been identified with perturbed U-centers which have foreign ions in one of the next three shells of neighboring ions.[34,36] In addition to these secondary absorption lines, the entire absorption spectrum undergoes a concentration-dependent frequency shift and broadening which is related to the change in the average lattice constant of the mixed crystal.

D. GAP MODES

a. Monatomic Impurities

Although the U-center investigations have yielded a wealth of detailed information on the mechanics of the impurity–lattice coupling, only the U-center impurity has been observed to produce infrared-active localized lattice modes at frequencies that are higher than the maximum phonon frequency of the host lattice. In contrast with the experimental situation for high-frequency localized modes, lower-frequency localized modes have been observed for a variety of impurities at frequencies in the gap region between the optic and acoustic phonon branches of potassium iodide.[37] Let us look at some of the successes and failures of the far-infrared studies for this particular alkali halide crystal.

To date, the only metal impurity ion which induces an infrared-active gap mode in KI is Cs^+ (see Table AI.3). A sharp absorption line is observed at 83.5 cm^{-1} with an additional broad absorption band occurring near the top of the acoustic spectrum. This experimental observation clearly demonstrates that a force constant change and not the impurity charge determines the gap mode frequency.

The most complete experimental investigation of gap modes has been on the system $KI : Cl^-$.[38,39] The Cl^- gap mode for a nominally pure KI sample is shown in Fig. AI.8. The doublet arises from the presence of the two stable chlorine isotopes. As the sample temperature is increased, the sharp doublet does not change appreciably, except to appear on a broad background absorption. The background absorption becomes strong enough by a temperature of $15°K$ to control the absorption in this frequency region. This broad absorption is just the difference band absorption which has been

TABLE AI.3

GAP MODES IN POTASSIUM IODIDE

Impurity in KI	Gap mode frequency, cm^{-1}	Band mode frequency, cm^{-1}
H^-	—	62
D^-	—	59.4
$^{35}Cl^-$	77.10	58, 61, 66
$^{37}Cl^-$	76.79	58, 61, 66
Br^-	73.8	56, 61, 67
Na^+	—	53, 63.5
Cs^+	83.5	60
Tl^+	—	55, 64.5
CN^-	81.2	65
NO_2^-	71.2, 79.5	55, 63

described in Section A for the pure crystal. In addition to the sharp lines in the gap, a band composed of at least three broad lines occurs for frequencies near the top of the acoustic branch (see Table AI.3). Below 30 cm^{-1}, the impurity-induced absorption is very small. The total integrated absorption strength has been measured for a number of concentrations and varies linearly with the chemically determined impurity concentration. This linear dependence leads us to define an oscillator strength, f, for the absorption line in analogy with the strength defined for the electronic transitions. The oscillator strength f is defined by[20]

$$\int \alpha(\omega)\, d\omega = \frac{\pi e^2}{\sqrt{\varepsilon_0} c^2 M^*}\left[\frac{\varepsilon_0 + 2}{3}\right]^2 N f\, [cm^{-2}], \qquad (AI.4)$$

where ε_0 is the dielectric constant of the host crystal at the local mode frequency, e the electronic charge, M^* the effective mass, and N the number of impurities/cm^3. For the sharp absorption line at 77 cm^{-1}, we find $f = 0.016$. (Later we shall find that $f \sim 0.1$ for resonant modes.)

Because the phonon gap between the acoustic and optic branches in KI extends from 69.7 to 96.5 cm^{-1}, the three broad absorptions centered at 58, 61, and 66 cm^{-1} occur in the acoustic phonon spectrum. We shall see that similar peaks occur at 56, 61, and 67 cm^{-1} in KI:Br$^-$. In neither case do the absorption peaks seem to identify critical points in the KI density of modes spectrum.[11] However, frequency shifts in the absorption spectrum are to be expected for the two impurities if the absorption lines are to be identified with resonant or incipient resonant modes which occur in the

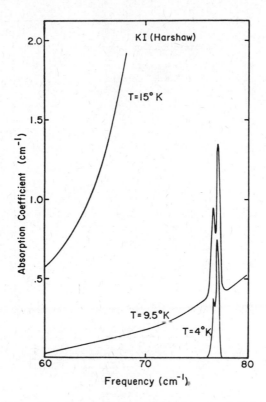

Fig. AI.8   Infrared absorption in a nominally pure Harshaw sample between 60 and 80 cm^{-1} as a function of temperature. The sharp absorption at 77 cm^{-1} is due to an intrinsic Cl^{-} impurity content of about 0.5×10^{16} ions/cm^3. [I. G. Nolt, R. A. Westwig, R. W. Alexander, Jr., and A. J. Sievers, *Phys. Rev.* **157**, 730 (1967).]

frequency region of the large density of modes. A satisfactory method for investigating the nature of these broad resonant modes has not yet been found, and we now turn to the study of the gap mode absorption.

At a higher resolving power of 380 the gap mode at 77 cm^{-1} is shown in Fig. AI.9 to consist of two lines with center frequencies of 77.10 and 76.79 $\pm$ 0.05 cm^{-1}. The frequency separation is estimated to be 0.31 $\pm$ 0.05 cm^{-1}. The ratio of the line strengths for the doublet components is about 3 to 1 which corresponds nicely to the natural abundance ratio of the two stable chlorine isotopes.

The activation of gap modes by bromide impurities in KI has also been studied and these measurements are shown in Fig. AI.10. An absorption spectrum very similar to that found previously for the chloride doping is observed, namely, three fairly broad absorption bands below the acoustic

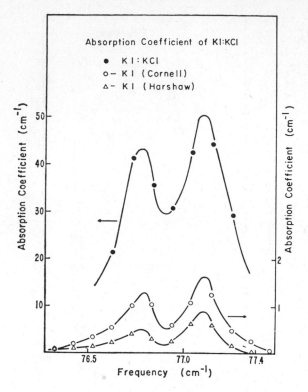

Fig. AI.9   Gap mode absorption in KI:Cl⁻ between 76.5 and 77.4 cm⁻¹. Note the different left and right ordinate scales. This figure shows the isotope frequency shift due to chloride impurities of mass-35 and mass-37 in their natural abundance ratio of approximately 3:1. [I. G. Nolt, R. A. Westwig, R. W. Alexander, Jr., and A. J. Sievers, *Phys. Rev.* **157**, 730 (1967).]

edge at 56, 61, and 67 cm⁻¹, plus a sharp absorption line at 73.8 cm⁻¹. An additional absorption line at 77 cm⁻¹ is attributed to an unwanted Cl⁻ impurity, which a chemical analysis showed to be present in a concentration of 3×10^{18} Cl⁻/cm³ in these crystals. The gap mode at 73.8 cm⁻¹ should also show an isotope shift; however, a resolving power of 380 is insufficient to measure the frequency difference.

Let us see what can be concluded from these experimental results. Band modes are observed for all impurities whereas gap modes appear for about one-half of the crystal–dopant combinations. At least some of the broad absorption bands in the acoustic spectrum must occur because of the large density of states in this frequency region. Also, additional absorptions could arise from resonant modes as has been discussed by Maradudin *et al.*[38]

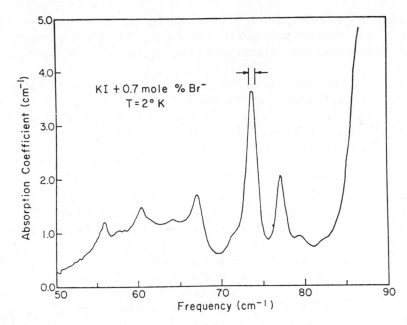

Fig. AI.10   Impurity-induced absorption in $KI:Br^-$. The Br^--induced gap mode absorption
has a center frequency of 73.8 cm^{-1}. The line at 77 cm^{-1} is due to the chloride impurity. The
instrumental resolution is indicated by the arrows. [I. G. Nolt, R. A. Westwig, R. W. Alexander,
Jr., and A. J. Sievers, *Phys. Rev.* **157**, 730 (1967).]

A comparison of the gap mode frequency with the fractional mass change
demonstrates that the experimental results cannot be interpreted with a
simple mass defect model. Some success in fitting the gap mode frequencies
has been obtained by introducing an additional parameter, the nearest
neighbor coupling parameter to the impurity. Nolt *et al.*[39] have fitted the
frequencies by using the simple lattice model of Mitani and Takeno.[40] A
comparison of the impurity ion diameter[41] with the force constant change
gives consistent results in that a large mismatch in diameters corresponds to a
large change in coupling constant.[42] Also, if the impurity diameter is larger
than the host diameter such as Cs^+, then the coupling constant is larger than
the host value, while if the impurity diameter is less than the host diameter,
the coupling constant is less than the host value.

The internal consistency of these fits can be checked with the isotope
shift data. For Cl^- with $K'/K = 0.38$, the shift in frequency associated with
the Cl^- mass change is calculated to be 1.74 cm^{-1} compared to the measured
value of 0.31 cm^{-1}. With $K'/K = 0.81$ for the Br^- impurity, the model
predicts an isotope shift of 0.26 cm^{-1} while the experimental shift is less than

$0.15 \, \text{cm}^{-1}$. For both impurities the model fails to account for the small isotope shift observed although the calculated shifts are, indeed, much less than the maximum possible shift given by the square root of the isotope mass ratio.

Apparently the observed shift cannot be obtained within the harmonic approximation. Benedek and Maradudin[43] recently calculated the isotope shift for Cl^- using Hardy's deformation dipole model and also Cochran's shell model for the host crystal. Both models of the host crystal fail to predict the small isotope frequency shift.

b. Molecular Impurities

Because of the cubic symmetry (O_h) at the monatomic impurity site, the first excited state of infrared-active gap modes is threefold degenerate. A molecular impurity, on the other hand, usually is described by a lower symmetry point group and hence reduces the local symmetry of the lattice. This more complex defect must be represented by not only a mass change but also by an anisotropic coupling to the surrounding lattice. The reduced symmetry of the local lattice then influences the degree of degeneracy of the infrared-active gap mode.

Two molecular impurities have been studied in KI. They are $KI:KCN$[12] and $KI:KNO_2$.[12,44,45] In Fig. AI.11 low-resolution impurity-induced spectra for the $CN^-(C_{\infty v})$ center and the $NO_2^-(C_{2v})$ center are compared with the Cl^- spectrum. Absorption lines are found in the gap region and broad bands are located at the edge of the acoustic spectrum. For the CN^- center only one infrared-active mode has been observed in the gap although two are to be expected from the impurity symmetry. The center frequency is given in Table AI.3. For the NO_2^- center two modes are found in the gap region and two broad bands in the acoustic spectrum as shown in Fig. AI.11. The center frequencies are given in Table AI.3. A number of attempts have been made to explain the dynamics of the nitrite ion in KI but as yet no model has been truly successful.[44-48]

E. RESONANT MODES

a. Model

Impurity-induced one-phonon absorption which is associated with large peaks in the phonon density of modes has been observed by Angress *et al.*[49] and also by Balkanski and Nusimovici.[50] At this time we shall not deal with this band mode absorption but focus our attention on some additional sharp absorption lines that have been observed in a number of doped alkali halide crystals at still lower frequencies by far-infrared optical techniques.[51-53] For these crystals substitutional defects can give rise to discrete

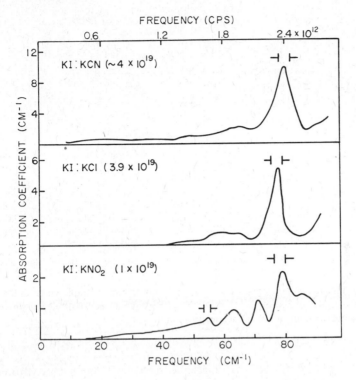

Fig. AI.11 Impurity-induced absorption coefficient for two molecular impurities in KI at 2°K. The absorption spectra for KI:KCN (4×10^{19} impurities/cm³) and for KI:KNO₂ (1×10^{19} impurities/cm³) are compared with the absorption spectrum for a monatomic impurity KI:KCl (4×10^{19} impurities/cm³). The instrumental resolution is shown. [C. D. Lytle, M.S. Thesis, Cornell University, 1965.]

absorptions which are electric dipole in nature. The observed center frequencies occur in a phonon frequency region where the host density of lattice modes is expected to increase monotonically with increasing frequency. The absorption process has been identified with the excitation of a quasi-localized mode or resonant lattice mode. Brout and Visscher[17] first calculated the frequency and width of such a mode for the mass defect. Later Visscher[54] considered the possibility of lattice resonant modes arising from both mass and coupling constant changes. With the aid of a computer he calculated some relevant properties in a simple cubic lattice with impurity mass and nearest neighbor force constants different from those of the host lattice. The calculations indicated that either a large impurity mass or a weak coupling constant could give rise to a lattice resonant mode. The optical

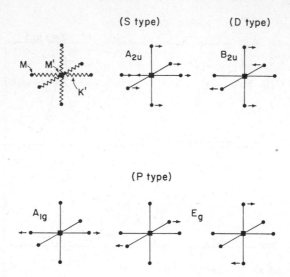

Fig. AI.12   Vibrational modes associated with an impurity in a simple cubic lattice. The impurity has mass M' and is coupled to the lattice with a coupling constant K'. The three types of vibrational mode occur for the model described in the text.

properties of resonant modes associated with a mass defect perturbation for a homonuclear crystal were developed later by Dawber and Elliott.[20]

In order to understand the far-infrared experimental results in some detail, the following lattice model is outlined. A point defect of mass M' is placed in a simple cubic lattice which contains host ions of mass M. For the impurity, the nearest neighbor coupling constants K' are assumed to be different from the coupling constants of the unperturbed lattice K. The impurity and nearest neighbors are shown schematically in Fig. AI.12. The position of the substitutional impurity atom is taken as the origin and long-range interaction forces which would introduce coupling between each component of the motion are not included. Also, the distortion of the lattice around the impurity is neglected and the electronic polarizability of the ions is not explicitly included. Finally, to make the solutions tractable, the central and noncentral force constants are set equal to each other. Incorporating these restrictions into the $3N$ equations of motion, three types of vibrational modes can be found which are associated with the impurity. They are S-type modes which transform as the B_{2u} representation and P-type modes which transform as the A_{1g} and E_g representations. A typical motion for each of these types is displayed in Fig. AI.12. With our simple model the x, y, and z components of the motion are completely uncoupled from each other. For the ionic crystals only the S-type modes give rise to an oscillating electric dipole moment.

Let us look at the solution for the S-type mode more closely. Experimentally, we know the absorption frequencies are far down in the acoustic spectrum. To determine the eigen frequency of the S-type mode, a secular equation[53] is solved for the special case where $\omega_r/\omega_D \ll 1$. In this case the eigen frequency for the resonant mode reduces to

$$\left(\frac{\omega_r}{\omega_D}\right)^2 = \frac{1 + \mu}{3\lambda + \mu\lambda - 2\mu} \tag{AI.5}$$

where

$$\mu = \frac{K'}{K} - 1 \quad \text{and} \quad \lambda = \frac{M'}{M} - 1.$$

For the harmonic approximation, the line width of this mode is given by

$$\Gamma = \frac{3\pi}{2}|\lambda|\left(\frac{\omega_r}{\omega_D}\right)^3 \omega_r. \tag{AI.6}$$

Some of the physical characteristics of the resonant mode can be determined by investigating two specific limiting cases. Consider the case of a weakly coupled impurity where $(K'/K) \to 0$. Then Eq. AI.5 reduces to

$$\omega_r^2 = 6\frac{K'}{M'},$$

which is the expression for an Einstein oscillator. On the other hand when $(K'/K) = 1$, that is, only a mass change is allowed, Eq. AI.6 simplifies to

$$\Gamma = \frac{\pi}{2}\left(\frac{\omega_r}{\omega_D}\right)\omega_r. \tag{AI.7}$$

Comparing Eqs. AI.6 and AI.7 we find, for a given resonant mode frequency, the width is much smaller when both mass and coupling constant changes occur than for a mass change alone. The physical significance of the difference in width can be readily understood. For a simple mass defect, the line width is determined by the density of lattice modes at the resonant frequency. If, in addition to the mass change, the impurity–lattice force constant K' is much less than K, the impurity is uncoupled from the lattice by the additional factor

$$\frac{K'}{K} \approx \left(\frac{\omega_r}{\omega_D}\right)^2.$$

Thus Eq. AI.7, the line width expression for the isotope approximation, times the uncoupling factor K'/K gives essentially Eq. AI.6.

All lattice resonant modes that have been observed in the low-frequency region where the host density of lattice modes increases monotonically with increasing frequency satisfy the condition that $K'/K \ll 1$. By Eq. AI.6 the resonant modes will always appear as sharp absorption lines in the far-infrared spectral region.

b. Observations

The first optically active resonant modes were identified with the substitutional silver ion impurity in alkali halide crystals.[51,52] The far-infrared absorption spectra found for silver-activated potassium chloride and bromide were quite similar. The spectra are characterized by an absorption line superimposed on the low-frequency wing of a broad absorption band. The center frequencies are recorded in Table AI.4.

TABLE AI.4

RESONANT MODES IN ALKALI HALIDE CRYSTALS*

Resonant mode system	Experimental resonant frequency, cm^{-1}	$\frac{\omega_l}{\omega_h}$, experimental	$\frac{\omega_l}{\omega_h}$, theoretical
NaCl : Li$^+$	44.0		
NaCl : ^{63}Cu$^+$	23.57 ⎫		
NaCl : ^{65}Cu$^+$	23.20 ⎬	1.016 ± 0.002	1.017
NaCl : Ag$^+$	52.5		
NaBr : Ag$^+$	48.0		
NaI : Ag$^+$	36.7		
NaI : Cl$^-$	5.44		
KCl : Li$^+$	40.0		
KBr : ^{6}Li$^+$	17.71 ⎫		
KBr : ^{7}Li$^+$	16.07 ⎬	1.105 ± 0.004	1.085
KBr : Ag$^+$	33.5		
KI : ^{107}Ag$^+$	17.37 ⎫		
KI : ^{109}Ag$^+$	17.23 ⎬	1.008 ± 0.002	1.0099
RbCl : Ag$^+$	21.4; 26.4; 36.1		

* Columns 3 and 4 give a comparison of the experimental resonant mode frequency ratios with the ratios predicted by Eq. AI.5. Here ω_l is the resonant mode frequency for the light isotope and ω_h the resonant mode frequency for the heavy isotope.

Silver concentrations of 10^{18} ions/cm^3 in potassium iodide yield a much sharper absorption line centered at 17.3 cm^{-1} with a full width at one-half maximum absorption of 0.45 cm^{-1}. Using isotopically pure silver in the form of Ag metal as a dopant, it has been possible to detect the isotopic frequency shift for the two silver isotopes 107 and 109.[55] These center frequencies are also recorded in Table AI.4. This result shows first that the resonant mode is due to silver and second that the frequency ratio varies approximately as the square root of the mass ratio.

Another sharp absorption line associated with a low-frequency resonant mode has been observed in NaCl:Cu$^+$ by Weber and Nette.[56] By using isotopically pure ^{63}Cu$^+$ and ^{65}Cu$^+$ the isotope shift has been resolved.[55] The impurity-induced absorption spectrum is shown in Fig. AI.13. Again the frequency ratio is given quite accurately by a simple Einstein oscillator dependence as shown in Table AI.4.

Far-infrared absorption lines have been observed to occur far down in the acoustic continuum for the lithium impurity in NaCl, KCl, and KBr host lattices. The measured frequencies are given in Table AI.4.

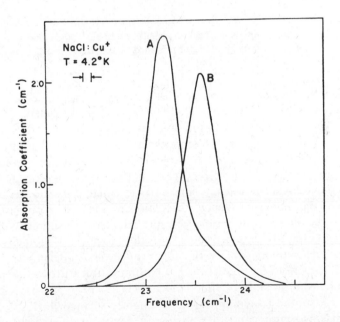

Fig. AI.13 Impurity-induced absorption coefficient *vs* frequency for ^{63}Cu$^+$ and ^{65}Cu$^+$ impurities in NaCl. Curve A has 1.5×10^{17} ^{65}Cu$^+$ ions/cm^3 and curve B has 1.06×10^{17} ^{63}Cu$^+$ ions/cm^3. The resolution is indicated by the separation of the arrows. [R. D. Kirby, I. G. Nolt, R. W. Alexander, Jr., and A. J. Sievers, *Phys. Rev.* **168**, 1057 (1968).]

The most complete spectroscopic measurements have been carried out on the system $KBr:Li^+$. In this case a sharp low-frequency lattice mode is found with additional structure occurring at higher frequencies.[53] The measured width for the resonant mode absorption is about a factor two larger than given by Eq. AI.6 (see Table AI.5). The sharp absorption line

TABLE AI.5

LINE WIDTHS FOR ISOTOPIC IMPURITIES*

Resonant mode system	Γ, cm^{-1} experimental	Γ, cm^{-1} theoretical
$NaCl:^{63}Cu^+$	0.38 ± 0.05	0.57
$NaCl:^{65}Cu^+$	0.40 ± 0.05	0.56
$KI:^{107}Ag^+$	0.47 ± 0.05	0.72
$KI:^{109}Ag^+$	0.45 ± 0.05	0.72
$KBr:^6Li^+$	0.60 ± 0.05	0.21
$KBr:^7Li^+$	0.43 ± 0.05	0.15

* A comparison of the experimental resonant mode line widths with the line widths predicted by Eq. AI.6.

does allow the lithium 6, 7 isotope shift to be easily measured. The shift for the low-frequency absorption is slightly larger than predicted by Eq. AI.5. The absorption coefficients found for both $^6Li^+$ and $^7Li^+$ are shown in Fig. AI.14. With the higher $^6Li^+$ concentration (curve A), in addition to the sharp absorption line recorded in Table AI.4, prominent bands are observed at 45.5 and 83 cm^{-1}. For $^7Li^+$, again in the more concentrated sample (curve C), two higher-frequency bands are observed at 43 and 83 cm^{-1}.

Most of the higher-frequency absorption in Fig. AI.14 probably corresponds to the one-phonon absorption spectrum from the impurity-activated phonons of the host lattice. The absorption coefficient should vary as a weighted one-phonon density of states; however, sufficient detail is not observed to verify this prediction.[57] One unusual feature is the small but observable isotope shift of the band at 43 cm^{-1} for $^7Li^+$ to 45.5 cm^{-1} for $^6Li^+$. Also, the bands appear to be slightly stronger for $^6Li^+$ than for $^7Li^+$. This absorption can either be identified with another slightly infrared-active impurity mode or as the second overtone of the low-frequency absorption. If anharmonic forces are important for this system, the second overtone can become infrared-active in an octahedral environment.[53,58,59]

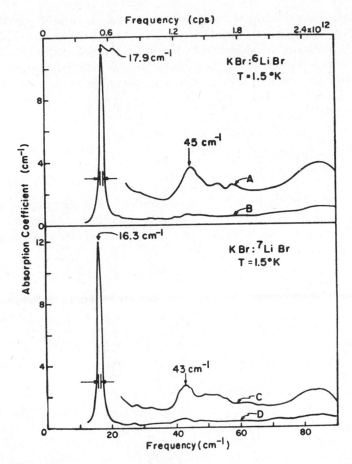

Fig. AI.14   Impurity-induced absorption coefficient vs frequency for the two lithium isotopes in KBr. For curves A and B the concentrations of KBr:^{6}LiBr are 1×10^{19} and 1.5×10^{18} ^{6}Li$^+$ ions/cm^3, respectively. For curves C and D the concentrations of KBr:^{7}LiBr are 1.2×10^{19} and 1.3×10^{18} ^{7}Li$^+$ ions/cm^3, respectively. [A. J. Sievers and S. Takeno, *Phys. Rev.* **140**, A1030 (1965).]

c. Stress Effects

These low-lying lattice modes have the interesting property of being very sensitive to external perturbations.[60,61] With the application of uniaxial stress along different crystallographic directions, the coupling of the resonant mode excited state to lattice distortions of different symmetries can be distinguished. Consider the electric dipole transition between the two vibrational states ψ which transform according to the A_{1g} and T_{1u} irreducible representations of the O_h group appropriate to the defect site. The local symmetry of the lattice is lowered by the uniaxial elastic deformation and a splitting of the degenerate T_{1u} state can occur. The lattice distortions which will perturb the resonant mode excited state can be obtained as follows. The matrix elements associated with the stress perturbation H' are for the ground state:

$$\Delta E_g = \langle \psi(A_{1g})|H'|\psi(A_{1g})\rangle \tag{AI.8}$$

and for the excited state

$$\Delta E_e = \langle \psi(T_{1u})|H'|\psi(T_{1u})\rangle. \tag{AI.9}$$

For these matrix elements to be invariant under all symmetry operations of the octahedral group O_h, they must transform like the A_{1g} representation. Thus, the stress perturbation must transform according to the representations contained in the direct product of

$$T_{1u} \times T_{1u} = A_{1g} + E_g + T_{2g}, \tag{AI.10}$$

and the only modes which interact with the excited state of the resonant mode are the long-wavelength acoustic distortions of A_{1g} (spherical), E_g (tetragonal and orthorhombic), and T_{2g} (trigonal) symmetry. Three coupling coefficients A, B, and C then determine the dependence of the resonant mode frequency upon the fully symmetric, the tetragonal, and the trigonal strain components, respectively.

These coupling coefficients can be estimated from the slopes of the absorption frequency vs applied stress data if the stiffness constants of the crystal are known. Unfortunately, an unsettled question is whether or not the stiffness constants of the host crystal need be modified to account for a softening of the lattice in the neighborhood of the impurity. For the coefficients which we give in Table AI.6 the stiffness constants of the unperturbed crystal have been used.

From Table AI.6 it is seen that the resonant mode for $KBr:Li^+$ is most strongly coupled to long-wavelength spherically symmetric modes. Also,

TABLE AI.6*

STRESS COUPLING COEFFICIENTS FOR $KBr:Li^+$ AND $KI:Ag^+$

	$KBr:Li^+$	$KI:Ag^+$
$A(A_{1g})$	830 ± 100	$390 \pm 90 \ cm^{-1}$/unit strain
$B(E_g)$	360 ± 40	$510 \pm 70 \ cm^{-1}$/unit strain
$C(T_{2g})$	170 ± 30	$-15 \pm 10 \ cm^{-1}$/unit strain

* *After* I. G. Nolt, Ph.D. Thesis, Cornell University, Ithaka, New York, 1967.

these coupling coefficients are about the same size as has been found for the U-center mode.[34] Hence, for both the U-center local mode and the Li^+ resonant mode similar stress shifts on the order of a few cm^{-1} can be observed. Of course, the shift is much more dramatic for the resonant mode because of the low frequency involved.

Also recorded in Table AI.6 are the coefficients for $KI:Ag^+$. The resonant mode is most strongly coupled to tetragonal distortions of the surrounding lattice and progressively less to those of spherical and trigonal symmetry. The presence of appreciable noncentral force components is indicated by this result. Such components could arise from some covalent bonding associated with the Ag^+ impurity. Preliminary experimental results indicate that for $NaCl:Cu^+$ the tetragonal coupling coefficient is again largest and all three coefficients are similar to those found for Ag^+. Both of these impurity ions are no doubt bound in their respective alkali halide lattice in a similar manner.

For all of these defect systems the resonant mode is very sensitive to the lattice constant. Because the A_{1g} hydrostatic coefficient gives the resonant mode frequency shift per unit volume change, the dependence of the frequency upon the lattice constant can be written as

$$-\frac{\partial \ln \omega}{\partial \ln a} = \frac{3A}{\omega}, \qquad (AI.11)$$

where a is the lattice constant, ω is the resonant frequency, and A is the hydrostatic coupling coefficient. For $KBr:Li^+$ this dimensionless coefficient is $3A/\omega = 155$ whereas for the U-center in KCl the coefficient is 3.6.

d. Temperature Effects

In contrast with temperature-dependent studies on gap modes which, as we have seen, are hindered by the strong temperature dependence of the host lattice absorption,[38] temperature-dependent studies on resonant modes at moderately low temperatures are relatively straightforward. One of the most striking features which has been observed with some of these low-frequency absorptions is the rapid temperature dependence of the line strength.[62] For both $KBr:Li^+$ and $KI:Ag^+$ the strength decreases rapidly as the temperature is increased and disappears by about 40°K. For $NaCl:Cu^+$ the strength decreases much more slowly and the absorption line can still be measured at liquid-nitrogen temperatures.[56] This different behavior, which has been a puzzle for some time, can now be understood from the stress results. Because of the large linear coupling of the resonant mode to long-wavelength phonons, the modulation of the resonant mode energy levels by the lattice modes will introduce characteristic properties into the spectrum. The sharp zero-phonon lines which have been studied in the optical spectral region are a well known example of this motional effect of the lattice.[63-66]

A characteristic property of a zero-phonon line is that the line strength varies with temperature similar as with a Debye-Waller factor. The magnitude of this exponent is controlled by the square of the coupling constant of the defect mode to long-wavelength phonons—a quantity which is readily obtained from stress measurements.[67] For the three resonant modes we have considered here the coupling decreases in the following order: $Li^+ > Ag^+ > Cu^+$. It is now clear that the strength of the lithium resonant mode should vary with temperature much more rapidly than the strength of the copper resonant mode.

F. Summary

Whereas only the U-center modes have been observed at frequencies above the phonon spectrum of the pure crystal, a large number of gap modes have been found. These gap modes produce extremely sharp absorption lines with widths less than $0.2 \, cm^{-1}$. The magnitude of the isotope frequency shift observed for the Cl^- gap mode in KI has not yet been accounted for satisfactorily. With low-symmetry centers such as polyatomic molecules, the complexity of the spectra in the gap region is offset to some degree by the additional spectroscopic measurements which are now possible in the near-infrared and optical region.

Several resonant modes have been studied in alkali halide crystals. Sharp absorption lines have been identified with weak coupling of the impurity ion to the host lattice. Because the simple Debye spectrum of modes can be used for the crystal in this low-frequency region, some of the simplicity characteristic of the U-center mode is regained for resonant modes. Hence, the qualitative features of these resonant modes are fairly well understood, but to predict which impurity–lattice system will lead to a lattice resonant mode is not yet possible.

REFERENCES

1. R. Robertson, J. J. Fox, and A. E. Martin, *Phil. Trans. Roy. Soc.* **A232**, 463 (1934).
2. R. J. Collins and H. Y. Fan, *Phys. Rev.* **93**, 674 (1954).
3. M. Lax and E. Burstein, *Phys. Rev.* **97**, 39 (1955).
4. W. Kaiser, P. H. Keck, and C. F. Lange, *Phys. Rev.* **101**, 1264 (1956).
5. H. Bilz and L. Genzel, *Z. Physik* **169**, 53 (1962).
6. J. R. Jasperse, A. Kahan, J. N. Plendl, and S. S. Mitra, *Phys. Rev.* **146**, 526 (1966) and references contained therein.
7. E. Burstein, *Phonons and Phonon Interactions*, Th. A. Bak, Ed., Benjamin, New York, 1964, p. 276.

8. H. Rubens and G. Hertz, *Berl. Ber.* **14**, 256 (1912).

9. A. M. Karo and J. R. Hardy, *Phys. Rev.* **129**, 2024 (1963).

10. B. N. Brockhouse, *Phonons and Phonon Interactions*, Th. A. Bak, Ed., Benjamin, New York, 1964, p. 221.

11. G. Dolling, R. A. Cowley, C. Schittenhelm, and I. M. Thorson, *Phys. Rev.* **147**, 577 (1966).

12. C. D. Lytle, M.S. Thesis, Cornell University (1965). Materials Science Center Report #390, unpublished.

13. P. Mazur, E. W. Montroll, and R. B. Potts, *J. Wash. Acad. Sci.* **46**, 2 (1956).

14. K. F. Renk, *Z. Physik* **201**, 445 (1967).

15. B. Szigeti, *J. Phys. Chem. Solids* **24**, 225 (1963).

16. L. Genzel, K. F. Renk, and R. Weber, *Phys. Stat. Sol.* **12**, 639 (1965).

17. R. Brout and W. M. Visscher, *Phys. Rev. Letters* **9**, 54 (1962).

18. R. Weber, Ph.D. Thesis, Physikalisches Institut der Universität, Freiburg im Breisgau, Germany (1967).

19. For a list of references on the mass defect problem see: A. A. Maradudin, E. W. Montroll, and G. H. Weiss, *Theory of Lattice Dynamics in the Harmonic Approximation, Solid State Physics Suppl. 3*, F. Seitz and D. Turnbull, Eds., Academic Press, New York, 1963.

20. P. G. Dawber and R. J. Elliott, *Proc. Roy. Soc. (London)* **273**, 222 (1963); *Proc. Phys. Soc. (London)* **81**, 453 (1963).

21. R. Hilsch and R. W. Pohl, *Gött. Nachr. Math. Phys. Neue Folge* **2**, 139 (1936); R. Hilsch and R. W. Pohl, *Trans. Faraday Soc.* **34**, 883 (1938).

22. J. H. Shulman and W. D. Compton, *Color Centers in Solids*, Macmillan, New York, 1963.

23. G. Schäfer, *J. Phys. Chem. Solids* **12**, 233 (1960).

24. E. Mollwo, *Z. Physik* **56**, 85 (1933); and H. Ivey, *Phys. Rev.* **72**, 341 (1947).

25. W. C. Price and G. R. Wilkinson, *Molecular and Solid State Spectroscopy Technical Report No. 2*, University of London King's College, (1960), unpublished.

26. A. Mitsuishi and H. Yoshinaga, *Prog. Theoret. Phys. (Kyoto) Suppl.* **23**, 241 (1962).

27. B. Fritz, *J. Phys. Chem. Solids* **23**, 375 (1962).

28. A. J. Sievers and R. L. Pompi, *Solid State Commun.* **5**, 963 (1967).

29. A detailed review of this topic has been given by A. A. Maradudin, *Solid State Physics*, Vol. 19, F. Seitz and D. Turnbull, Eds., Academic Press, New York, 1966, p. 1.

30. B. Fritz, *Proc. Inter. Conf. Lattice Dynamics, Copenhagen 1963*, R. F. Wallis, Ed., Pergamon Press, New York, 1965, p. 485.

31. D. N. Mirlin and I. I. Reshina, *Soviet Phys.-Solid State* (English Transl.) **5**, 2458 (1964).

32. B. Fritz, U. Gross, and D. Bäuerle, *Phys. Stat. Sol.* **11**, 231 (1965).

33. H. Bilz, D. Strauch, and B. Fritz, *J. Phys. Rad. Suppl.* **27**, (1966).

34. W. Barth and B. Fritz, *Phys. Stat. Sol.* **19**, 515 (1967).

35. T. Timusk and M. V. Klein, *Phys. Rev.* **141**, 664 (1966).

36. D. N. Mirlin and I. I. Reshina, *Soviet Phys.-Solid State* (English Transl.) **8**, 116 (1966).

37. A. J. Sievers, *Low Temperature Physics*, J. Daunt, D. Edwards, F. Milford, and M. Yaqub, Eds., Plenum, New York, LT9, Part B, p. 1170.

38. A. J. Sievers, A. A. Maradudin, and S. S. Jaswal, *Phys. Rev.* **138**, A272 (1965).

39. I. G. Nolt, R. A. Westwig, R. W. Alexander, Jr., and A. J. Sievers, *Phys. Rev.* **157**, 730 (1967).

40. Y. Mitani and S. Takeno, *Prog. Theoret. Phys. (Kyoto)* **33**, 779 (1965).

41. L. Pauling, *Nature of the Chemical Bond*, Cornell Univ. Press, Ithaca, New York, 1945.

42. A. J. Sievers, NATO Advanced Study Institute, Cortina d'Ampezzo, Italy, 1966, p. VI.1 (unpublished).

43. G. Benedek and A. A. Maradudin, *J. Phys. Chem. Solids* **29**, 423 (1968).

44. A. J. Sievers and C. D. Lytle, *Phys. Letters* **14**, 271 (1965).
45. K. F. Renk, *Phys. Letters* **14**, 281 (1965).
46. T. Timusk and W. Staude, *Phys. Rev. Letters* **13**, 373 (1964).
47. H. S. Sack and M. C. Moriarty, *Solid State Commun.* **3**, 93 (1965).
48. V. Narayanamurti, W. D. Seward, and R. O. Pohl, *Phys. Rev.* **148**, 481 (1966).
49. J. F. Angress, S. D. Smith, and K. F. Renk, *Proc. Inter. Conf. Lattice Dynamics, Copenhagen 1963*, R. F. Wallis, Ed., Pergamon Press, New York, 1965, p. 467.
50. M. Balkanski and M. Nusimovici, *Phys. Stat. Sol.* **5**, 635 (1964).
51. A. J. Sievers, *Phys. Rev. Letters* **13**, 310 (1964).
52. R. Weber, *Phys. Letters* **12**, 311 (1964).
53. A. J. Sievers and S. Takeno, *Phys. Rev.* **140**, A1030 (1965).
54. W. M. Visscher, *Phys. Rev.* **129**, 28 (1963).
55. R. D. Kirby, I. G. Nolt, R. W. Alexander, Jr., and A. J. Sievers, *Phys. Rev.* **168**, 1057 (1968).
56. R. Weber and P. Nette, *Phys. Letters* **20**, 493 (1966).
57. A. D. B. Woods, B. N. Brockhouse, R. A. Cowley, and W. Cochran, *Phys. Rev.* **131**, 1025 (1963).
58. G. Benedek and G. F. Nardelli, *Phys. Rev.* **155**, 1004 (1967).
59. M. V. Klein, *Physics of Color Centers*, W. Beall Fowler, Ed., Academic Press, New York, 1968, Chapter 7.
60. I. G. Nolt and A. J. Sievers, *Phys. Rev. Letters* **16**, 1103 (1966).
61. I. G. Nolt, Ph.D. Thesis, Cornell University, Ithaca, New York (1967).
62. S. Takeno and A. J. Sievers, *Phys. Rev. Letters* **15**, 1020 (1965).
63. S. S. Mitra and R. S. Singh, *Phys. Rev. Letters* **16**, 694 (1966).
64. G. F. Imbusch, W. M. Yen, A. L. Schawlow, D. E. McCumber, and M. D. Sturge, *Phys. Rev.* **133**, A1029 (1964).
65. R. J. Elliott, W. Hayes, G. D. Jones, H. F. Macdonald, and C. T. Sennett, *Proc. Roy. Soc. (London)* **A289**, 1 (1965).
66. R. H. Silsbee, *Phys. Rev.* **128**, 1726 (1962); R. H. Silsbee and D. B. Fitchen, *Rev. Mod. Phys.* **36**, 423 (1964).
67. R. W. Alexander, Jr. and A. J. Sievers, to be published. See also *Diss. Abstr. B* **1968**, 29(6), 2168.

APPENDIX II

DIELECTRIC PROPERTIES AND OPTICAL PHONONS IN PARA- AND FERROELECTRIC PEROVSKITES*

C. H. PERRY

Department of Physics, Northeastern University, Boston, Massachusetts

A. INTRODUCTION

Many compounds possessing the perovskite ABO_3 crystal structure (see Fig. AII.1) or some distorted modification exhibit unusual properties such as ferroelectric and antiferroelectric behavior.[1] Extremely large values of the static dielectric constant, $\varepsilon_0 = \varepsilon'(0)$, are observed at microwave and lower frequencies, especially in the region of the Curie temperature

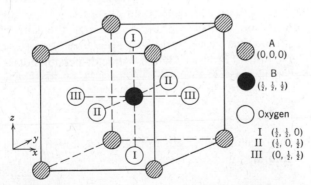

Fig. AII.1 The cubic perovskite ABO_3 crystal structure. [R. A. Cowley, *Phys. Rev.* **134**, A981 (1964).]

* This contribution was made when the author was at the Massachusetts Institute of Technology, Cambridge, Massachusetts.

where ε_0 may vary from 100 to 10000 or higher depending on the material.[2] These materials are often characterized by Curie-Weiss law behavior[1,3] of the static dielectric constant such that

$$\varepsilon'(0, T) = \frac{C}{T - T_c} \qquad \text{(AII.1)}$$

in their paraelectric phase. The transition to a ferroelectric, antiferroelectric, or some other structure can be shown to be closely connected with the phonon modes of the crystal. In 1960 Cochran[4,5] and Anderson[6] independently proposed a theory of ferroelectricity in which the ferroelectric transition in crystals such as $BaTiO_3$ is related to an instability of one or more transverse optical phonon branches of the lattice. In the long-wavelength limit the frequency of the mode drops[7,8] and tends to zero (finally becoming unstable) as the temperature is lowered towards the Curie temperature T_c such that

$$\omega_T \propto (T - T_c)^{1/2}. \qquad \text{(AII.2)}$$

These extremely temperature-dependent low-frequency unstable modes, ω_T, are called "Cochran" modes or "soft" modes and have large oscillator strengths. They can account for more than 90% of the value of the static dielectric constant, as indicated by the Kramers-Kronig relation connecting the low-frequency real part of the dielectric constant and the integral over the loss spectrum (the imaginary part)

$$\varepsilon'(0, T) - \varepsilon_\infty = \frac{2}{\pi} \int_0^\infty \frac{\varepsilon''(\omega', T)\,d\omega'}{\omega'}, \qquad \text{(AII.3)}$$

where $\varepsilon_\infty = n_\infty^2$ is the limiting dielectric constant at frequencies above about 4000 cm^{-1} (where phonon effects are unimportant).

From the continuum analysis derived by Huang,[9] the isotropic dielectric constant is a summation over a series of Lorentz oscillators corresponding to the transverse optical branches in the lattice dispersion curves of angular frequency, ω_j, such that:

$$\varepsilon(\omega, T) = \varepsilon_\infty + \sum_j \frac{s_j(T)}{\omega_j^2(T) - \omega^2 - i\omega\gamma_j(T)}, \qquad \text{(AII.4)}$$

where $s_j(T)$ is the oscillator strength and $\gamma_j(T)$ is assumed to be a frequency-independent damping function. The static dielectric constant can be written

$$\varepsilon'(0, T) = \varepsilon_\infty + \sum_j \frac{s_j(T)}{\omega_j^2(T)}. \qquad \text{(AII.5)}$$

Noting that for paraelectric crystals the dielectric constant obeys the Curie-Weiss law, it can easily be seen that if any of the ω_j's should decrease markedly on cooling of the crystal, the static dielectric constant would increase strongly and a dielectric transition would occur near any temperature at which an ω_j tends to zero (assuming that s_j is almost temperature-independent). This is precisely the case for $SrTiO_3$[7] and $KTaO_3$,[8] which illustrate the lattice dynamics which are important to the instability in other perovskite ferroelectrics (although neither of these materials become ferroelectric). For the alkali halides, experiments show only an increase of ε_0 of the order of 10–20 % on cooling to 4°K ; the phenomena leading to ferroelectricity are not found in these materials,[10] although the thallium halides may begin to exhibit this tendency.[11]

Analysis of the temperature-dependent infrared reflectance measurements as discussed by Barker and Tinkham for $SrTiO_3$[7] and by Perry and McNelly[8] for $KTaO_3$ provides the most direct observations of the onset of the ferroelectric instability from a lattice-dynamical viewpoint. The results are in good agreement with the theories proposed independently by Cochran[4,5] and Anderson.[6] (Both materials were also investigated by Spitzer et al.[12] but only at room temperature.) The frequency vs wave vector dispersion curves have also been obtained for $SrTiO_3$[13] and $KTaO_3$[14] from inelastic neutron spectroscopy. The results for $SrTiO_3$ were successfully fitted by Cowley[15] using a shell model approach. This, perhaps, gives the most complete confirmation of these theories. Nevertheless, the infrared measurements provide the $k \sim 0$ normal mode frequencies and associated polarizations in a considerably easier and more straightforward way (experimentally at least).

The dielectric dispersions of other perovskites have been obtained by Perry et al.[16–18] These include $CaTiO_3$, $PbTiO_3$, $CaZrO_3$, $SrZrO_3$, $BaZrO_3$, $PbZrO_3$, and $PbHfO_3$. Most of the samples were in the form of ceramics. $CaTiO_3$, $PbTiO_3$, $PbZrO_3$, and $PbHfO_3$ show Curie-Weiss law dependence of the dielectric constant in the paraelectric phase. All the materials exhibited similar frequency values of the infrared modes. In the case of $PbZrO_3$,[16] which is antiferroelectric at room temperature, about 90 % of the value of the low-frequency dielectric constant was found to be the result of a low-frequency optically active mode.

$BaTiO_3$ has been studied by a number of workers at both room temperature[12,19] and elevated temperatures.[20] Ballentyne's measurements[20] do not conclusively show Cochran-type behavior if only the frequency of the loss peak (ε'') is considered. However, Barker[21] has shown that ε'' contains a large and frequency-dependent damping function which modifies the true "mode frequency" in $BaTiO_3$. Reference should be made to this work for further discussion. The Raman spectrum of $BaTiO_3$, reported by Perry and Hall,[22] shows the temperature dependence in the various ferroelectric phases.

Barker has also measured the optical modes in a mixed $KTa_{0.65}Nb_{0.35}O_3$ crystal[23]; this material shows two well defined modes in positions close to those observed in $KTaO_3$. The low-frequency mode was predicted from the Lyddane-Sachs-Teller relationship[24] and is comparable with the lowest mode found in $BaTiO_3$.[20] The mixed crystal system $(Na, K)TaO_3$ has been investigated by Tornberg, Davis, and Perry,[25] and soft mode behavior is observed with Na concentrations up to 40%. The measured infrared spectra are more complicated than those of the $K(Ta, Nb)O_3$ or $KTaO_3$ systems as various modes split with increasing Na content even at room temperature.

Barker and Loudon[26] have extensively studied $LiNbO_3$, a distorted perovskite, in its ferroelectric state. Their results indicate that an A_1 phonon mode becomes unstable and contributes the largest part to the temperature dependence of ε_0. Further details on mode stability and possible ion motions are discussed in their paper.

Bosomworth[27] has accurately measured the indices of refraction of $LiNbO_3$ at 300 and 80°K; large far-infrared birefringence and dispersion is observed. These properties may suggest the use of $LiNbO_3$ for optical mixing, but residual absorption even at low temperatures indicates that difference frequency generation would be possible only at frequencies below about 50 cm^{-1}.

B. LATTICE MODES BASED ON THE SHELL MODEL

Consider an ionic diatomic lattice such that the anions are assumed to consist of a positive core coupled by an isotropic force constant to a negative shell representing the electrons.[28] This allows the equations of motion to be written as follows:

$$m_+\ddot{u}_+ = K_+(u_s - u_+) + Z_+eE_l$$
$$m_c\ddot{u}_c = K_c(u_s - u_c) + Z_ceE_l \qquad \text{(AII.6)}$$
$$0 = K_+(u_+ - u_s) + K_c(u_c - u_s) + Z_seE_l,$$

where m_+, m_c, and m_s and u_+, u_c, and u_s are the masses and displacements of the cation, the core, and the shell of the anion, respectively. The first two equations describe the cation-to-shell interaction (force constant K_+) and the positive core-to-shell interaction (force constant K_c), and the electrostatic action of the local field, E_l, on cation and core charge. The third equation shows that the electronic shell is always in an equilibrium position relative to the instantaneous position of the nuclei. Z_+, Z_c, and Z_s are the corresponding charges, and e is the electronic charge. The locally acting or effective field consists of the uniform macroscopic field E and the Lorentz field $4\pi P/3$, where P is the macroscopic polarization field.

For transverse waves

$$E_l(T) = E + \frac{4\pi P}{3}. \tag{AII.7}$$

For longitudinal waves the depolarization field $-4\pi P$ must also be included and

$$E_l(L) = E - \frac{8\pi P}{3}. \tag{AII.8}$$

Rearranging the Eqs. AII.6 with reference to the displacement of the two nuclei against one another gives

$$m_+ \ddot{u}_+ = -K_+(u_+ - u_c) - K_+(u_c - u_s) + Z_+ e E_l$$

$$m_c \ddot{u}_c = K_c(u_+ - u_c) - K_c(u_+ - u_s) + Z_c e E_l \tag{AII.9}$$

$$0 = K_+(u_+ - u_c) - (K_c + K_+)(u_s - u_c) + Z_s e E_l,$$

which reduces to

$$m_+ \ddot{u}_+ = -K'(u_+ - u_c) + Z' e E_l$$

$$m_c \ddot{u}_c = K'(u_+ - u_c) - Z' e E_l, \tag{AII.10}$$

where

$$K' = \frac{K_c K_+}{K_c + K_+} < K_+ \quad \text{and} \quad Z' = Z_+ + \frac{K_+ Z_s}{K_c + K_+} < Z_+ \tag{AII.11}$$

and

$$Z_+ + Z_c + Z_s = 0. \tag{AII.12}$$

Equations AII.10 can be combined into one equation for the reduced mass:

$$m_\mu(\ddot{u}_+ - \ddot{u}_c) = -K'(u_+ - u_c) + Z' e E_l, \tag{AII.13}$$

where

$$m_\mu = \frac{m_+ m_c}{m_+ + m_c}. \tag{AII.14}$$

An optical phonon mode, of a wavelength which is long in comparison with atomic distances, causes a displacement of the three Bravais arrays and thus an atomic polarization proportional to the relative displacement of the undistorted sublattices. It follows that

$$P = N[Z'e(u_+ - u_c) + \alpha E_l], \tag{AII.15}$$

where α is the electronic polarizability and N is the number of positive ion–negative ion pairs per unit volume. The electronic polarizability can be determined from the dielectric constant at optical frequencies (n_∞^2) as

$$\frac{n_\infty^2 - 1}{n_\infty^2 + 2} = \frac{4\pi N\alpha}{3} = \frac{4\pi\alpha}{3v}, \tag{AII.16}$$

where v is the unit cell volume. The local field from Eqs. AII.7 and AII.8 can be written

$$E_l(T) = \frac{E + (4\pi/3)NZ'e(u_+ - u_c)}{1 - (4\pi/3)N\alpha} \quad \text{for transverse waves,} \tag{AII.17}$$

$$E_l(L) = \frac{E - (8\pi/3)NZ'e(u_+ - u_c)}{1 + (8\pi/3)N\alpha} \quad \text{for longitudinal waves.} \tag{AII.18}$$

Hence for transverse waves

$$m_\mu(\ddot{u}_+ - \ddot{u}_c) = \left[-K' + \frac{\frac{4}{3}\pi N(Z'e)^2}{1 - 4\pi N\alpha/3}\right](u_+ - u_c) + \frac{Z'eE}{1 - 4\pi N\alpha/3}, \tag{AII.19}$$

and for longitudinal waves

$$m_\mu(\ddot{u}_+ - \ddot{u}_c) = \left[-K' - \frac{\frac{8}{3}\pi N(Z'e)^2}{1 + 8\pi N\alpha/3}\right](u_+ - u_c) + \frac{Z'eE}{1 + 8\pi N\alpha/3}. \tag{AII.20}$$

Introducing the parameter

$$u = (m_\mu/v)^{1/2}(u_+ - u_c),$$

we obtain

$$\ddot{u}(T) + \left[\frac{K'}{m_\mu} - \frac{(4\pi/3)N[(Z'e)^2/m_\mu]}{1 - 4\pi N\alpha/3}\right]u = \frac{Z'eE(N/m_\mu)^{1/2}}{1 - 4\pi N\alpha/3}, \tag{AII.21}$$

$$\ddot{u}(L) + \left[\frac{K'}{m_\mu} + \frac{(8\pi/3)[N(Z'e)^2/m_\mu]}{1 + 8\pi N\alpha/3}\right]u = \frac{Z'eE(N/m_\mu)^{1/2}}{1 + 8\pi N\alpha/3}. \tag{AII.22}$$

Substituting Eq. AII.16 into Eqs. AII.21 and AII.22, neglecting the externally applied field E, and assuming the displacements (and fields) have the time dependence $e^{i\omega t}$, it follows that

$$\omega_T = \left[\frac{K}{m_\mu} - \frac{4\pi(n_\infty^2 + 2)(Z'e)^2}{9vm_\mu}\right]^{1/2} < \left(\frac{K}{m_\mu}\right)^{1/2} \tag{AII.23}$$

and

$$\omega_L = \left[\frac{K}{m_\mu} + \frac{8\pi(n_\infty^2 + 2)(Z'e)^2}{9vn_\infty^2 m_\mu}\right]^{1/2} > \left(\frac{K}{m_\mu}\right)^{1/2}. \tag{AII.24}$$

The relation between the frequencies (ω_L) and (ω_T) and the static dielectric constant (ε_0) and high-frequency dielectric constant ($\varepsilon_\infty = n_\infty^2$) was first derived by Lyddane *et al.*[24] and is given by

$$\frac{\omega_L{}^2}{\omega_T{}^2} = \frac{\varepsilon_0}{\varepsilon_\infty}. \tag{AII.25}$$

For any diagonally cubic crystal, Cochran[5] has shown that with n atoms per primitive unit cell this relation (for the $\mathbf{k} \sim 0$ modes) can be written

$$\prod_{j=2}^{n} \frac{(\omega_j{}^2)_L}{(\omega_j{}^2)_T} = \frac{\varepsilon_0}{\varepsilon_\infty}, \tag{AII.26}$$

where there are n doubly degenerate transverse modes and n longitudinal modes. (One of each is an acoustical mode with $\omega_1 = 0$.)

From Eq. AII.23 it follows that the transverse lattice vibration (for $\mathbf{k} \sim 0$) has a smaller resonance frequency (due to the polarization) than the mechanical resonance frequency $\omega_0 = (K'/m_\mu)^{1/2}$. A lowering of about 30% is found in the alkali halides.[10] However, should the "catastrophe" denominator[29] $1 - 4\pi\alpha/3v$ in Eq. AII.21 be sufficient to cause the equality

$$\frac{K'}{m_\mu} = \frac{4\pi(Z'e)^2/3vm_\mu}{1 - 4\pi\alpha/3v} = \frac{4\pi(n_\infty^2 + 2)(Z'e)^2}{9vm_\mu}, \tag{AII.27}$$

the transverse mode becomes unstable; if the crystal is not to disrupt exponentially with time, a transition to another structure must take place.

From Eq. AII.25 and the Curie-Weiss law ($\varepsilon_0 \propto 1/(T - T_c)$) behavior of displacive ferroelectric materials, it can easily be seen that as T approaches T_c, ε_0 approaches a large value whenever there is a drastic lowering of the transverse optical mode frequency. It should also be noted that whereas the α/v contribution causes a lowering of the transverse mode, it actually raises the longitudinal mode frequency ω_L (see Eqs. AII.22 and AII.24) and makes it more stable. (For further discussion see, for example, the work of Nakamura.[30])

Cochran[5] has shown that if one considers a short-range effective potential between two Bravais lattices and restricts the discussion to the "Z" direction only, the restoring force can be written in the form

$$\tilde{R} = Ku_Z + Bu_Z{}^3 + Cu_Z{}^5, \tag{AII.28}$$

and the displacive Coulomb force

$$\tilde{D} = \frac{4\pi(n_\infty^2 + 2)(Z'e)^2}{9v} u_Z. \tag{AII.29}$$

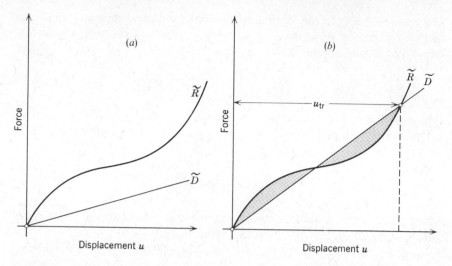

Fig. AII.2   Onset of ferroelectricity resulting from a near cancellation of short-range (restoring force $\tilde{R}$) and Coulomb (displacive force $\tilde{D}$) interactions that cause an abnormally low transverse optic mode at $\mathbf{k} \sim 0$. (a) Above the transition temperature T_{tr} stable but anharmonic oscillations are possible. (b) At the transition (e.g., cubic-to-tetragonal phase) temperature the free energies of the two phases are equal, as shown by the shaded areas. Between T_{tr} and T_c the crystal is metastable. [W. Cochran, *Adv. Phys.* **9**, 387 (1960).]

B is taken to be negative. The two forces are plotted in Fig. AII.2 for reasonable values of B and C.

From Eq. AII.21 or AII.23 it can be seen that the value of $\omega_T{}^2$ is proportional to the difference in slope of the two lines at the origin, whereas ε_0 is inversely proportional to the same quantity.

As the temperature is lowered, the slope of the displacive force increases until it intercepts the upper curve. At this point stable yet extremely anharmonic vibrations are possible but the transition to a stable tetragonal phase will not take place until the free energies of the two phases are the same. The temperature when this occurs is T_{tr}, as given by Cochran.[5] Above T_{tr} the crystal is stable. Between T_c and T_{tr} the crystal is metastable. At T_c the crystal is unstable; it undergoes a first-order phase change and exhibits ferroelectric behavior (that is to say, it has a spontaneous polarizability for $T < T_c$). However, there is some experimental evidence that certain crystals first undergo a second-order phase change to a ferroelectric state and then undergo a structural change.[31] Consequently, this metastable temperature region appears to be quite extensive for these crystals.

For the perovskite lattice (see Fig. AII.1) the motions of the B ions against the oxygen ions which lie in a linear chain (Z-direction say) have been shown to possess Lorentz factors larger than $4\pi/3$. This type of motion favors the ferroelectric polarization.[3,26]

Barker has derived the equation of motion and the transverse frequency of this mode for $SrTiO_3$.[32] These are written as

$$\ddot{u}(T) + \left\{ \frac{K'}{m_\mu} + \frac{[(Z'e)^2/m_\mu v](4\pi/3 + 30.08)}{1 - (4\pi/3 + 15.04)(2\alpha/v)} \right\} u$$

$$= \frac{Z'eE}{1 - (4\pi/3 + 15.04)(2\alpha/v)} \qquad \text{(AII.30)}$$

$$\omega_T = \left[\frac{K'}{m_\mu} - \frac{(Z'e)^2(4\pi/3 + 30.08)/m_\mu v}{1 - (4\pi/3 + 15.04)(2\alpha/v)} \right]^{1/2} \qquad \text{(AII.31)}$$

as compared with

$$\omega_T = \left[\frac{K'}{m_\mu} - \frac{(Z'e)^2(4\pi/3)/m_\mu v}{1 - 4\pi\alpha/3v} \right]^{1/2} \qquad \text{for an alkali halide, say.} \qquad \text{(AII.32)}$$

The terms 30.08 and 15.04 on the right-hand side of Eq. AII.31 considerably increase the chance for instability and tend to lower the restoring force to zero. Hence, due to the unique structure of the perovskite lattice, this type of transverse mode has a very large "feedback effect" favoring the ferroelectric transition. Figures AII.3, AII.4 and AII.5 show the mode shift in $KTaO_3$[8]

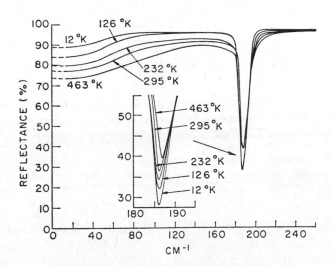

Fig. AII.3   Far-infrared reflectivity of $KTaO_3$ as a function of temperature over the frequency range 10–250 cm^{-1}. The error in the measurement at the low-frequency end is about $\pm 2\%$. [C. H. Perry and T. F. McNelly, *Phys. Rev.* **154**, 456 (1967).]

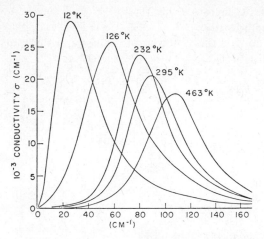

Fig. AII.4   Conductivity calculated from the reflectivity curves shown in Fig. AII.3, for the five temperatures. [C. H. Perry and T. F. McNelly, *Phys. Rev.* **154**. 456 (1967).]

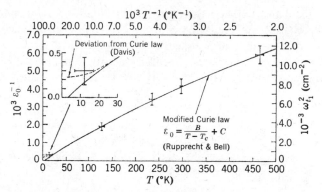

Fig. AII.5   Square of the frequency (in cm^{-1}) of the unstable soft mode plotted as a function of temperature. Vertical lines indicate the $\omega_{TO_1}^2$-error and the horizontal lines the temperature variation. The solid curve shows the reciprocal of the dielectric constant from the results of Rupprecht and Bell.[2] The extrapolated curve gives a Curie temperature, $T_c = 2.9°$K. The dashed curve shows the low-temperature deviation (below 30°K) from the modified Curie law (Rupprecht and Bell) obtained by Davis[31] and Wemple.[33] [C. H. Perry and T. F. McNelly, *Phys. Rev.* **154**, 456 (1967).]

in the temperature range 463–10°K. The results indicate a Cochran-type relationship[5] between the lowest-frequency transverse optical mode ω_{T_1} for $\mathbf{k} \sim 0$ and the static dielectric constant,[2,31,33] ε_0, such that

$$\omega_{T_1}^2(T) = \frac{A}{\varepsilon_0} = \left[\frac{B}{T - T_c} + C \right]^{-1}, \qquad \text{(AII.33)}$$

where $A = 1.9 \times 10^6\,\mathrm{cm}^{-2}$, $B = 5.99 \times 10^4\,^\circ\mathrm{K}$, $C = 39$, and T_c (extrapolated) $\sim 2.8^\circ\mathrm{K}$.

C. CRYSTAL SYMMETRY AND SPECIES OF THE VIBRATIONAL MODES

Many ABO_3 compounds possess the cubic perovskite or slightly distorted crystal structure, as is mentioned in the Introduction. Those materials, which have the cubic structure, belong to the space group[34] $O_h^1(Pm3m)$ and contain one molecule per unit cell. The A ions (see Fig. AII.1) are located at the corner of a cube, the B ions at the body center position, and the oxygen ions at the face center positions of the six cube faces. To a certain approximation there are two types of oxygen ions when motion parallel to the cube edges is considered. Four oxygen ions move in the plane of their surface (O_I and O_{II}) and two move normal to their surface plane (O_{III}).

Procedures for determining selection rules for optical transitions in crystals have been devised by Bhagavantam and Venkatarayudu[35] and by Winston and Halford[36] among others; they are discussed in Chapter 13. For convenience's sake we repeat here that the number of normal modes of a particular symmetry species is given by n_i, the number of times the irreducible representation Γ_i corresponding to that species is contained in the completely reduced representation Γ of the space group of the crystal. The group-theoretical expression for n_i is

$$n_i = (1/N) \sum_\rho h_\rho \chi'_\rho(\rho) \chi_i(\rho), \tag{AII.34}$$

where N is the order of the group, h_ρ is the number of group operations falling in the class ρ, $\chi'_\rho(\rho)$ and $\chi_i(\rho)$ are, respectively, the characters of the group operation ρ in the representation Γ and Γ_i, and $\chi'_\rho(\rho)$ is given by

$$\chi'_\rho(\rho) = U_\rho(\pm 1 + 2\cos\phi_\rho). \tag{AII.35}$$

Proper rotations by ϕ_ρ take the positive sign and improper rotations take the negative sign. The symbol U_ρ designates the number of atoms (ions) in the primitive unit cell which under the operation ρ remain invariant. When applied to ABO_3 cubic perovskites,[37] the above considerations yield $\Gamma = 4F_{1u} + F_{2u}$ as the representation of the space group of the crystal; these are the five (triply degenerate) normal modes. The effect of the macroscopic electric field in the crystal results, for $\mathbf{k} \sim 0$, in (1) a doubly degenerate transverse optic mode (TO) and an associated longitudinal optic mode (LO) for each of the three F_{1u} infrared-active species, (2) one doubly degenerate TO mode and one LO mode (which are assumed degenerate at the zone center) for the F_{2u} species and which are not first-order infrared- or Raman-active, and (3) the acoustic modes, TA (doubly degenerate) and LA, that correspond to the F_{1u} species and represent the translation of the whole unit cell (see Chapter 13, Section C).

Such a conclusion is in agreement with experimental data for $SrTiO_3$, $KTaO_3$,[12] and other perovskite crystals.[16,17] Three peaks in the imaginary part of the dielectric constant are observed in a Kramers-Kronig analysis of the infrared reflectance measurements. The method described in Section F of this chapter provides both the TO and LO mode frequencies. However, in the case of $CaTiO_3$ and $PbTiO_3$, which possess slightly distorted perovskite structures and become orthorhombic and tetragonal, respectively, additional peaks are observed. These distortions from ideal cubic symmetry can split degenerate modes and additional frequencies may be observed (see Table AII.1). The various phonon dispersion curves at small wave vector $\mathbf{k}$ and their expected infrared or Raman activity are shown in Fig. AII.6 for some of the structures found in perovskite materials.

For paraelectric $SrTiO_3$ at 90°K Cowley has fitted a shell model.[15] Table AII.2 shows the shell model ion displacements and charges for his "Model IV." Axe[38] has also reported the apparent ionic charges and vibrational eigenmodes for $SrTiO_3$, $BaTiO_3$, and $KTaO_3$. These results are summarized in several tables and figures in his paper.

Raman spectra of $SrTiO_3$[39-42] and $KTaO_3$[39,43] have been reported by several researchers. From group-theoretical arguments neither of these crystals should exhibit a first-order spectrum. Nevertheless, several authors[40-42] have claimed their results to be first-order, whereas others[39,43] have suggested that the spectra are essentially combined density of states curves of phonon pairs at critical points in the Brillouin zone. The available neutron scattering data[14] substantiate the latter viewpoint. More conclusive evidence, however, has been presented by Fleury and Worlock, who used an electric field to produce first-order spectra in $KTaO_3$[44] and $SrTiO_3$.[45] Recent Raman work on $(Na_xK_{1-x})TaO_3$[25] mixed crystals indicates that for low concentrations a second-order spectrum is obtained. At higher sodium concentrations where the structure is distorted, and even for low concentrations below the Curie temperature, first-order spectra are superimposed. Moreover, these results are in agreement with far-infrared measurements (see Table AII.3).

D. INFRARED REFLECTION MEASUREMENTS

To obtain the infrared dielectric dispersion of these perovskite crystals it is necessary to measure the reflectance spectrum over as wide a frequency range as possible. This is essential for a Kramers-Kronig or classical dispersion analysis, as will be discussed later. With the advent of new low-temperature detectors[47] and increased use of interferometric techniques[48]

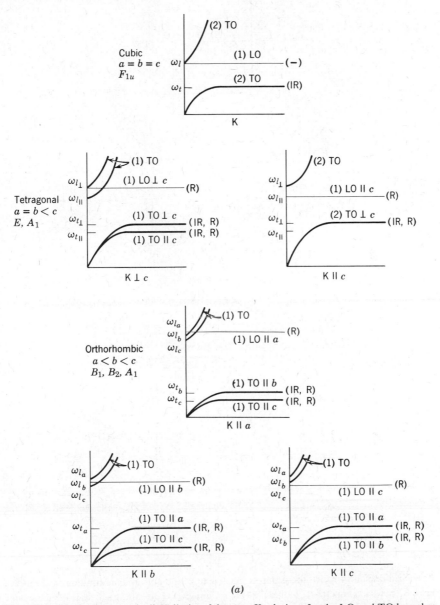

Fig. AII.6 The long-wave (small-K) limits of the ω vs. K relations for the LO and TO branches of cubic, uniaxial, and biaxial crystals for which electrostatic forces predominate over anisotropy in the interatomic forces. (a) represents a crystal in which only one group (or a well separated group) of three lattice vibrational branches is infrared-active (in the cubic phase). Shown also are the associated activities if such a crystal were to undergo structural changes to tetragonal and orthorhombic phases (e.g., the phase changes that occur in $BaTiO_3$, $KNbO_3$ and $PbTiO_3$).

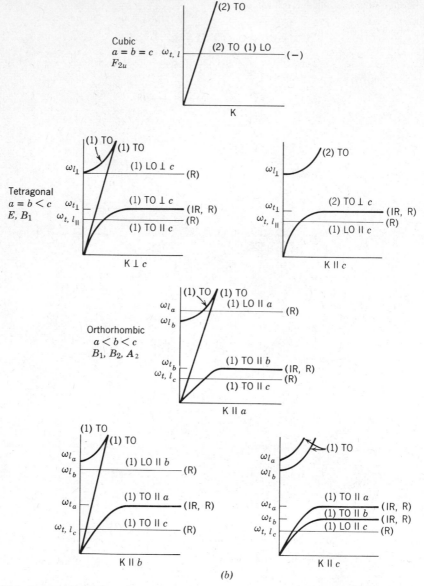

(b)

(b) is for a crystal in which the group is inactive (e.g., the out-of-plane bending mode in the perovskites with symmetry F_{2u} in the cubic phase) and gives the resulting activities when the same phase transitions take place. The sloping line is the dispersion of an electromagnetic wave in the crystal and its presence indicates that there is no interaction between the lattice vibration and the infrared electromagnetic wave. It should be noted that if polarized light is used in conjunction with an oriented sample, the frequencies of the various modes of different symmetries can be obtained but their detection will depend on the allowed activity (due to the anisotropy of the crystal and the associated dielectric constants), the alignment of the phonon wave vector with the axes of the crystal, and the instrumental resolution. [D. B. Hall and C. H. Perry, *Res. Lab. Electr. Quat. Rept. No. 78*, Massachusetts Institute of Technology, 1966.]

TABLE AII.1

Frequencies (in cm⁻¹) and Approximate Symmetry of Infrared-Active Modes Obtained from a Kramers-Kronig Analysis of the Reflectance Data and Raman Data in Perovskite (ABO₃) Titanates and Zirconates*

Sample	Crystal symmetry	TO_1 AB-O_6 mixed mode (soft mode)†	TO_2 A-BO_6 lattice mode†	TO_3 $\oplus \ominus (B) \oplus \ominus \oplus$ out-of-plane bend†	TO_4 O-B-O in-plane stretch†
		(IR)	(IR)		(IR)
Titanates					
$CaTiO_3$ [16]	Orthorhombic (multiple cell)	148 (B_1, B_2, A_1)	179 (B_1, B_2, A_1)	(—) 443 (B_1, B_2, A_2)	549 (B_1, B_2, A_1)
$SrTiO_3$ [16]	Cubic	100 (F_{1u})	185 (F_{1u})	265 (F_{2u})(neutron)[15]	555 (F_{1u})
$BaTiO_3$ [17,20]	Tetragonal	12 (E, A_1)	175 (E, A_1); 184	— (B_1); 305 (E)	491 (E, A_1)
$PbTiO_3$ [16]	Tetragonal	83 (A_1); 155 (E)	220 (A_1); 350 (E)	— (B_1); 290 (E)	510 (A_1)
Zirconates					
$CaZrO_3$ [17]	Orthombic (multiple cell)	153 (B_1, B_2, A_1); 96	228 (B_1, B_2, A_1); 186	340 (B_1, B_2, A_2)	515 (B_1, B_2, A_1)
$SrZrO_3$ [17]	Cubic (multiple cell)	143 (B_1, B_2, A_1)	240 (B_1, B_2, A_1)	325 (B_1, B_2, A_2)	552 (B_1, B_2, A_1)
$BaZrO_3$ [17]	Cubic	115 (F_{1u})	210 (F_{1u})	— (F_{2u})	505 (F_{1u})
$PbZrO_3$ [17]	Tetragonal (multiple cell)	80 (E, A_1); 34	221 (E, A_1)	290 (B_1, E)	508 (E, A_1)

* The Refs. are also given. TO_1, TO_2, ···, gives the numbering of the transverse modes.   † See Fig. AII.1.

TABLE AII.2*

FREQUENCIES (CM^{-1}), MODE DISPLACEMENTS [IN Å ALONG (001)], AND IONIC CHARGE FOR $SrTiO_3$ AT 90°K

Frequency (cm^{-1})		Ti	Sr	O_{111}	O_1	O_{11}
TO_1	40	0.077	0.022	−0.093	−0.129	−0.129
TO_2	173	0.097	−0.074	0.035	0.040	0.040
TO_3	265	0	0	0	−0.177	0.177
TO_4	544	0.007	0.004	−0.217	0.087	0.087
LO_1	178	0.080	−0.077	0.050	0.065	0.065
LO_2	265	0	0	0	−0.177	0.177
LO_3	473	0.057	0.001	0.107	−0.143	−0.143
LO_4	815	0.075	0.008	−0.207	−0.032	−0.032
Charge (in units of e)		1.26	4.64	−1.97	−1.97	−1.97

* R. A. Cowley, *Phys. Rev.* **134**, A981 (1964).

it is now possible to extend the reliable range of reflectance measurements as far as $1000\,\mu$. The usual practice as adopted in the Spectroscopy Laboratory at M.I.T. is to use conventional grating spectrometers from 2.5–$200\,\mu$ and a Michelson-type interferometer from 25–$1000\,\mu$. Consequently, considerable overlap in the various spectral regions can be obtained so that the influence of various filter, grating, and beam-splitter changes in the different instruments can be minimized. Due to the considerable absorption by atmospheric water vapor throughout the far-infrared, it is a decided advantage to use evacuable instruments, as this provides for more accurate intensity measurements. It is necessary that the observed area of the sample should be "flat" to the shortest wavelength under investigation and not rounded at the edges, as a slight convex surface may cause a severe reduction in signal. The sample intensity measurements are usually compared by using a freshly aluminized mirror, and it is imperative that the sample and mirror be exactly interchangeable without disturbing the optical system. Reflectance measurements are conveniently made as close as possible to normal incidence (for ease in calculating the optical constants). If it is possible to orient the sample, polarization measurements should be performed. For the majority of the perovskite materials it is usually either unnecessary (because of their cubic structure in the paraelectric phase) or impracticable to obtain polarized infrared spectra in the various ferroelectric phases since only relatively small single crystals (often multi-domain) or ceramics are available. The temperature-dependent measurements are achieved by attaching the sample either to a heat sink in an

TABLE AII.3

Comparison of $KTaO_3$,[8,39] $(Na, K)TaO_3$,[25] and $K(Ta, Nb)O_3$,[23] at Various Temperatures

Sample	$KTaO_3$					$(Na_{0.12}K_{0.88})TaO_3$			$(Na_{0.40}K_{0.60})TaO_3$				$(Na_{0.55}K_{0.45})TaO_3$								$K(Ta_{0.65}Nb_{0.35})O_3$
°K	463	295	232	126	12	295	80	10	295	80	10	30	295	80	109	70	10	57	24	65	300
Activity	IR	IR	IR	IR	IR	IR	IR	IR	IR	IR	IR	R	R	IR	R	IR	IR	IR	R	R	IR
Raman	No first-order Raman bands observed down to 5°K					No first-order Raman bands observed down to 30°K			No first-order Raman bands observed to 37°K												Predicted from the LST relation
Structure	Cubic					Cubic			Cubic-tetragonal				Tetragonal[31]								Cubic
TO_1*	106	88	79	58	25	75	50	48	102 / 138	81 / 132	39 / 129	42 / 128	75 / 132 / 144	132 / 147	130	132 / 147	132 / 147	130	130	130 / 150	17
LO_1	184	184	184	183		186	183	183	186	186	186										185
TO_2	199	198	198	196		198	198	198	204	201	198	200	196	196	200	205	246				199
LO_2, TO_3											255	255	255 / 270	258 / 270	261	270					
LO_3	421	421	421			420	420	420	414	414	414		414	414					450	450	421
TO_4	547	547	547			540	540	540	576	573	570	572	584	584					580	580	537
LO_4	838	838	838			819	819	819	849	849	849	850	864	864							825

*TO_1, LO_1, etc. refer to the transverse and longitudinal modes described in Table AII.1 and used by Cowley for $SrTiO_3$ (see Table AII.2 and Ref. 15). The modes LO_2 and TO_3 are of species F_{2u} and are inactive in the cubic phase.

appropriate cryostat or to a heated holder for high-temperature investigations. In the regions of the various phase transitions it is necessary to have some type of feedback control mechanism for adjusting the temperature and keeping it stable during the time of observation. The measurement of low temperatures is best obtained by firmly attaching a calibrated germanium resistance thermometer (2–100°K) and a calibrated thermocouple (above 20°K) to the back of the sample and a similar element to the heat sink in close proximity to the sample for purposes of comparison.

E. Analysis of Reflection Spectra by Means of a Kramers-Kronig Relation or a Fit with a Classical Dispersion Formula

In the infrared region polar crystals exhibit a reflection spectrum with reststrahlen bands because of the infrared-active lattice modes. These modes, which have a linear dipole moment with respect to their normal coordinate, cause resonances in the infrared dispersion. When the reflectance of such a crystal has been measured over a sufficiently wide frequency range, two kinds of analyses are generally used to extract the available information, such as eigen frequencies, oscillator strengths, and so forth. One method is the Kramers-Kronig analysis,[49] and the other is an optimum fit of the data by means of a classical dispersion formula. The latter method will be discussed in Section E.b with respect to some theoretical aspects and its practical application.

a. Kramers-Kronig Analysis

In the interaction of nonmagnetic materials with electric fields the polarization of the material, as expressed in terms of the real and imaginary parts of the dielectric constant, is a significant quantity. At low (microwave or radiowave) frequencies the complex dielectric constant can be measured directly; at optical and infrared frequencies only the fractional power reflected from or transmitted through a sample can be measured.

The necessary equations for this analysis have been given in Chapter 12, (see Eqs. 12.29 and 12.31). Equations 12.29 are combined to give $\varepsilon' = n^2 - \kappa^2$ and $\varepsilon'' = 2n\kappa$; the desired dielectric constants may now be computed if the reflectivity R is known over a sufficiently large frequency range. For convenience' sake, we change notation here and replace the symbols κ (extinction coefficient) and ψ (phase) by k and θ, respectively. These equations are evaluated using the values of reflectance measured over the 0 to 4000 cm^{-1} region. The reflectance is assumed constant at the 4000 cm^{-1} value to "infinite" frequency. This approximation is valid, since there is no structure in the reflectivity above 4000 cm^{-1} caused by the physical processes in question, and values of $\theta(\omega)$ are usually only computed at frequencies

$< 1000 \text{ cm}^{-1}$ (the localized nature of Eq. 12.31 permits negligible contributions from the spectrum above 1000 cm^{-1}). The solution of Eqs. 12.29 and 12.31 is provided by a high-speed computer; the integral in Eq. 12.31 is evaluated by representing $\ln R(\omega')$ by straight-line segments between data points. The limiting values of R obtained from the infrared measurements can be compared with the values of R calculated from known values of the high-frequency and low-frequency dielectric constants where these values are known. It is assumed that $k \ll n$ at these limiting frequencies so that $R = (\varepsilon^{\frac{1}{2}} - 1)^2/(\varepsilon^{\frac{1}{2}} + 1)^2$.

The accuracy of the reflectivity analysis depends on (1) the validity of the assumptions of a linear material response to the measuring field, (2) normal incidence for the radiation, (3) accuracy of the measured reflectivity data, and (4) the precision with which the measured data is fed to the computer. The assumption of a linear material response is justified on the basis of the extremely small field strengths present in the infrared spectrometer or interferometer beams.

The validity of the assumption of normal incidence was discussed by Barker and Tinkham,[7] who measured the reflectivity of $SrTiO_3$ at an angle of incidence of 45° and calculated approximate values of ε' and ε'' using the Fresnel equation for normal incidence. The approximate values of ε' and ε'' so determined were used to compute the expected values of reflectivity at normal incidence and at 45°; the expected reflectivities at the two angles of incidence were found to differ by less than the experimental error of their measurement (about $\pm 2\%$). The assumption of normal incidence is therefore valid for these high-dielectric materials as the usual average angle of incidence is only 10–15° with an $f/4$ or $f/5$ beam. However, if we know the angle of incidence and use infrared radiation polarized either perpendicular or parallel to the plane of incidence, the "true" values of ε' and ε'' can be calculated exactly from the appropriate values determined above with the help of the Fresnel equation.[50] Although this extra step is not usually necessary for materials like $SrTiO_3$, it must be done for materials which have a relatively low index of refraction. For example, the apparent frequency of the longitudinal optic mode in NaCl is approximately 20 cm^{-1} higher at an average of incidence of 75° than the value calculated from the Fresnel equation. The apparent frequency of the transverse optic mode, however, is changed negligibly.

Errors introduced by the straight-line segment approximation to the reflectance curves can be minimized by sampling them at frequency intervals corresponding to 2% changes in the values of the reflectance. The main factor which determines the accuracy of the computed ε' and ε'' values is therefore the accuracy of the measured reflectance, especially where the reflectance is small and its negative logarithm is large.

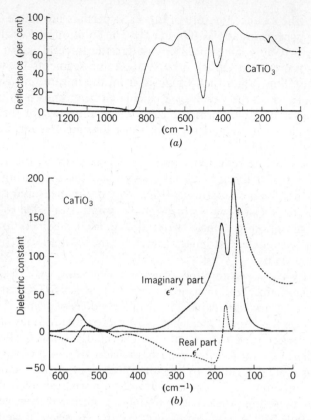

Fig. AII.7 Infrared reflectance spectra and the real and imaginary parts of the dielectric constant obtained from a $CaTiO_3$ ceramic sample at room temperature. [C. H. Perry, B. N. Khana, and G. Rupprecht, *Phys. Rev.* **135**, A408 (1964).]

   Values of the permittivity should be calculated for curves representing the upper and lower limits of the experimental error. The shapes of the ε' and ε'' curves will be similar but the position of the lowest frequency peak in ε'' is quite sensitive to reflectance values at low frequencies; the absolute magnitudes of ε' and ε'' can be appreciably effected. The real and imaginary parts of the dielectric function for a $CaTiO_3$ sample are shown in Fig. AII.7 together with the measured reflection data.[16]
   The frequencies of the normally optically active transverse modes are derived from peaks in the electrical conductivity $\sigma(\omega) = \varepsilon_j''\omega/2$ and the associated magnetic loss parameter ε_j''/ω, where ε_j'' is the contribution of the jth resonance to the imaginary part of the dielectric constant (see Figs. AII.3 and AII.4 and Section F of this appendix). The longitudinal frequencies

are similarly derived from peaks in the electrical resistivity $\rho(\omega) = 2\eta_j''/\omega$ and the associated magnetic loss parameter $\eta_j''\omega$, where $\eta'' = \varepsilon''/(\varepsilon'^2 + \varepsilon''^2)$.

b. Classical Dispersion Analysis

Polar and cubic crystals with two particles per primitive unit cell have only one infrared-active lattice fundamental. The classical treatment of the interaction of electromagnetic waves and this lattice mode yields the classical dispersion formula[9] for the complex dielectric constant of the crystal as a function of the frequency as given in Eq. AII.4, with j now equal to unity (see also Chapter 12).

If a more complicated crystal is considered with more than one infrared-active mode, the dispersion formula contains a sum (or product) of dispersion terms, one for each active mode. In certain cases it may be necessary to take the interaction of at least two modes into acount, and Barker and Hopfield[51] have derived a dispersion formula for the case in which two modes strongly interact. In the equations of motion an interaction term was added:

$$\ddot{y}_1 + \rho_{11}\dot{y}_1 + a_{11}y_1 + a_{12}y_2 = a_{13}E$$
$$\ddot{y}_2 + \rho_{22}\dot{y}_2 + a_{21}y_1 + a_{22}y_2 = a_{23}E. \tag{AII.35}$$

The transformation diagonalizing the force constant matrix

$$\begin{pmatrix} a_{11} & a_{12} \\ a_{21} & a_{22} \end{pmatrix}$$

yields off-diagonal terms for the damping constants:

$$\ddot{x}_1 + \gamma_{11}\dot{x}_1 + \gamma_{12}\dot{x}_2 + \omega_1^2 x_1 = b_{13}E$$
$$\ddot{x}_2 + \gamma_{21}\dot{x}_1 + \gamma_{22}\dot{x}_2 + \omega_2^2 x_2 = b_{23}E. \tag{AII.36}$$

From these two equations the following dispersion formula, in which a third dispersion term has been added for a third infrared-active eigen vibration which does not interact with the other two, is obtained*:

$$\varepsilon = \varepsilon' - i\varepsilon'' = \varepsilon_\infty$$

$$+ \frac{s_1(\omega_2^2 - \omega^2 - i\omega\gamma_2) + 2i(s_1 s_2)^{1/2}\omega\gamma_{12} + s_2(\omega_1^2 - \omega^2 - i\omega\gamma_1)}{(\omega_1^2 - \omega^2 - i\omega\gamma_1)(\omega_2^2 - \omega^2 - i\omega\gamma_2) + \omega^2\gamma_{12}^2}$$

$$+ \frac{s_3}{\omega_3^2 - \omega^2 - i\omega\gamma_3}, \tag{AII.37}$$

* The complex dielectric constant is written here in the form $\varepsilon = \varepsilon' - i\varepsilon''$ (See Eq. 12.8).

where the subscripts 1, 2, and 3 refer to the three active modes, and γ_{12} is the damping term arising from the interaction of modes 1 and 2.

Since the damping constants γ_v and $\gamma_{v\mu}$ are introduced *ad hoc* in the classical treatment, this derivation of a dispersion formula cannot show the physical meaning of the damping terms. The results of a proper quantum-mechanical treatment of the infrared dispersion, however, will show the damping related to the dissipation of energy from the active lattice modes (dispersion oscillators) to other lattice modes by means of anharmonic terms in the lattice potential.[52] If a third-order anharmonic potential only is taken into account and if all nonlinear terms in the dipole moment are neglected, the following expression holds for the dielectric constant of a cubic crystal with two infrared-active lattice modes:[53]

$$\varepsilon = \varepsilon_\infty + \frac{s_1(\omega_2{}^2 - \omega^2 - i\delta_2) + 2i(s_1 s_2)^{1/2}\delta_{12} + s_2(\omega_1{}^2 - \omega^2 - i\delta_1)}{(\omega_1{}^2 - \omega^2 - i\delta_1)(\omega_2{}^2 - \omega^2 - i\delta_2) + \delta_{12}^2}.$$

(AII.38)

This is the same formula obtained by the classical treatment, but instead of damping constants γ there are frequency-dependent damping functions δ:

$$\delta_j = \text{const} \sum_{j'j''} \int \left| \Phi \begin{pmatrix} 0 & k & -k \\ j & j' & j'' \end{pmatrix} \right|^2 \frac{[n' + \tfrac{1}{2}] \pm [n'' + \tfrac{1}{2}]}{\omega'\omega''}$$

$$\times \; \Delta(\omega \pm \omega' - \omega')\,dk^3,$$

(AII.39)

$$\delta_{jl} = \delta_{lj} = \text{const} \sum_{j'j''} \int \Phi \begin{pmatrix} 0 & k & -k \\ j & j' & j'' \end{pmatrix} \Phi \begin{pmatrix} 0 & k & -k \\ l & j' & j'' \end{pmatrix}$$

$$\times \; \frac{[n' + \tfrac{1}{2}] \pm [n'' + \tfrac{1}{2}]}{\omega'\omega''} \Delta(\omega \pm \omega' - \omega')\,dk^3,$$

where $\Phi \begin{pmatrix} 0 & k & -k \\ j & j' & j'' \end{pmatrix}$ is the third-order potential coefficient coupling modes $\begin{pmatrix} 0 \\ j \end{pmatrix}$ [wave vector 0, branch j], $\begin{pmatrix} k \\ j' \end{pmatrix}$, and $\begin{pmatrix} -k \\ j'' \end{pmatrix}$; ω', ω'', n', and n'' are frequencies and the thermal averages of phonon occupation numbers of modes $\begin{pmatrix} k \\ j' \end{pmatrix}$ and $\begin{pmatrix} -k \\ j'' \end{pmatrix}$, respectively; Δ refers to the Dirac delta-function; and the integrals are to be taken over the volume of the first Brillouin zone in reciprocal space.

Essentially, the damping functions δ_1 and δ_2 are the probabilities for all the two-phonon summation processes (+ sign) or two-phonon difference processes (− sign), via the dispersion oscillators 1 and 2, which are consistent with wave vector and energy conservation at a given photon frequency ω.

The quantity δ_{12} is the interference of the processes contained in δ_1 and δ_2. (See also Chapter 12, Section D.b).

For our purpose, the analysis of a reflectance spectrum is carried out by means of the classical formula with constant damping terms. The reflectance spectrum exhibits only the main features of the infrared dispersion and not the details of the absorption spectrum, especially outside the reststrahlen bands where the reflectance is given mainly by the refractive index; even in the reststrahlen bands it is not possible to evaluate more than one damping function per mode from the experimental data. Consequently, the use of the classical formula may be justified but the damping constants should be understood to be an average of damping functions, mainly in the neighborhood of the eigen frequencies. In the actual analysis, the constants in Eq. AII.37 have to be chosen by trial and error, and then ε is evaluated for the frequency range under consideration. From ε' and ε'' the optical constants n and k are obtained by means of the equations

(a) $\varepsilon' > 0$ $n = [\tfrac{1}{2}(\varepsilon' + \sqrt{\varepsilon'^2 + \varepsilon''^2})]^{1/2}, \qquad k = \dfrac{\varepsilon''}{2n}$

(b) $\varepsilon' = 0$ $n = k = \left[\dfrac{\varepsilon''}{2}\right]^{1/2}$ (AII.40)

(c) $\varepsilon' < 0$ $k = [\tfrac{1}{2}(-\varepsilon' + \sqrt{\varepsilon'^2 + \varepsilon''^2})]^{1/2}, \qquad n = \dfrac{\varepsilon''}{2k}.$

The reflectivity for normal incidence,

$$R = \frac{(n-1)^2 + k^2}{(n+1)^2 + k^2},$$ (AII.41)

can be evaluated and compared with experimental data. By variation of the constants in Eq. AII.37, an optimum fit to the experimental data is obtained. This analysis yields the eigen frequencies, the oscillator strengths, and the damping constants of the infrared-active modes.

Note that Eq. AII.37 can also be used in the case of a crystal having two interacting eigen vibrations and additional dispersion and absorption caused by free carriers (for example, semiconductors).[54] By setting $\omega_3 = 0$, the last dispersion term in Eq. AII.37 is converted into a Drude-term, s_3 and γ_3 are determined by the carrier concentration N, the effective mass m^*, and the mobility μ of the free carriers,

$$s_3 = \frac{N}{m^* \varepsilon_{vac}} \left(\frac{e}{2\pi c}\right)^2 \ [\text{cm}^{-2}]; \qquad \gamma_3 = \frac{e}{2\pi c m^* \mu} \ [\text{cm}^{-1}],$$

where e is the electron charge, c is the velocity of light in vacuum, and ε_{vac} is the dielectric constant in vacuum. The units for ω_v, γ_v, and $\gamma_{v\mu}$ are in cm^{-1}, s_v in cm^{-2}, and ε_∞ is the dielectric constant relative to that of vacuum. The reflectivity computed with the initial sets of constants has to be compared with the experimental data. The agreement, at first, will probably be rather poor for most of the reststrahlen bands. The computed values of ε' and ε'' can be compared with the Kramers-Kronig data. In order to improve the fit, it is often advisable to make use of the following rules.

1. The static dielectric constant which determines the reflectivity at the low-frequency end of the spectrum outside the reststrahlen bands is given by

$$\varepsilon_0 = \varepsilon_\infty + \frac{s_1}{\omega_1{}^2} + \frac{s_2}{\omega_2{}^2} + \frac{s_3}{\omega_3{}^2} + \cdots$$

2. Where the real part ε' changes from positive to negative values, the reflectivity exhibits a pronounced minimum and a sharp rise to high values with decreasing frequencies, provided ε'' is sufficiently small.

3. The height of a reststrahlen band is determined by ε'', that is, mainly by the damping γ_v, while the interaction damping $\gamma_{v\mu}$ becomes significant in the region between two eigen vibrations.

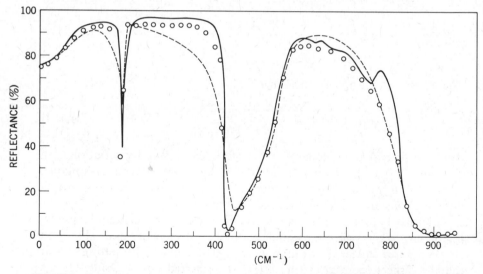

Fig. AII.8   Reflectivity spectrum of pure KTaO$_3$ at room temperature (*solid curve*) from 10–950 cm^{-1}. Dashed curve represents a best fit with a classical dispersion model using three oscillators. The open circles show the improved fit with the introduction of two interacting damping terms, γ_{12} and γ_{13} (see Eq. AII.37). (Similar curves were first obtained by Barker and Hopfield, *Phys. Rev.* **135**, A.1732 (1964)). [T. F. McNelly, B.S. Thesis, Dept. Physics, Massachusetts Institute of Technology, 1967.]

The comparison of the measured reflectivity for $KTaO_3$ at room temperature and the best fits with and without interaction damping are shown in Fig. AII.8.

A similar treatment by Barker and Hopfield for $SrTiO_3$, $BaTiO_3$, and $KTaO_3$ is illustrated in their paper.[51] Barker[23] has also expressed the frequency-dependent damping in terms of the Lyddane-Sachs-Teller relationship; its importance is stressed in the analysis of the ferroelectric mode in $BaTiO_3$. In treating the frequency-dependent damping function $\gamma(\omega)$ analytically, Barker[23] chooses a simple form which satisfies the causality relations:

$$\varepsilon(\omega) = \varepsilon_\infty + \frac{s_1(1 - s_2/\omega_2{}^2)}{\omega_1{}^2 - \omega^2 - s_2\omega_1{}^2/(\omega_2{}^2 - \omega^2 - i\omega\gamma_2)} \qquad \text{(AII.42)}$$

$$\varepsilon(\omega) = \varepsilon_\infty + \frac{s_1(1 - s_2/\omega_2{}^2)}{\omega_1{}^2 - \omega^2 - i\omega\Gamma(\omega)}. \qquad \text{(AII.43)}$$

The main contribution to the dielectric function is the resonance near ω_1 of strength $s_1/\omega_1{}^2$ and a damping function $\Gamma = \Gamma' - i\Gamma''$ which is peaked near ω_2. The value of $s_2/\omega_2{}^2$ controls the coupling of the optical mode to other phonons centered near ω_2 with half width γ_2; $1 - s_2/\omega_2{}^2$ is included to simplify the form of $\varepsilon_0 = \varepsilon_\infty + s_1/\omega_1{}^2$. Equation AII.42 has two complementary pairs of poles and its form is quite different from adding two constant-damping oscillators. The damping function Γ has the effect of causing the main mode ω_1 to have a frequency-dependent force constant $\omega_1{}^2(\omega)$. If $\omega_2 \gg \omega_1$, which is approximately true for the 184 cm^{-1} or the 491 cm^{-1} modes in $BaTiO_3$ compared with the main oscillator ω_1 at ~ 12 cm^{-1} (from Ballentyne),[20] then for ω near ω_1, the denominator in Eq. AII.42 is approximately $\omega_1{}^2(1 - s_2/\omega_2{}^2) - \omega^2$, that is to say, the frequency occurs at a frequency lower than ω_1. Barker has also shown that the LST relation can be rewritten as

$$\frac{\varepsilon_0}{\varepsilon_\infty} = \frac{|\omega_{L_1}|^2|\omega_{L_2}|^2}{\omega_1{}^2(1 - s_2/\omega_2{}^2)\omega_2{}^2}; \qquad \text{(AII.44)}$$

ω_{L_1} and ω_{L_2} are the complex frequencies defined when $\varepsilon' = 0$. The transverse mode parameter which is connected with the LST relation is not $\omega_1{}^2$ but $\omega_1{}^2(0) = \omega_1{}^2(1 - s_2/\omega_2{}^2)$.

From Eqs. AII.1 and AII.42, Cochran's relation for mode instabilities reads $\omega_1{}^2(1 - s_2/\omega_2{}^2) \approx \omega_1{}^2$ (low-frequency fit) $\propto 1/\varepsilon_0(T)$ for stable high-frequency modes. Figure AII.9 shows several Curie law and mode frequency plots of $BaTiO_3$ taken from the work of various authors and Barker's mode frequency fits.[55] Barker notes that in $BaTiO_3$ the peak in ε'' occurs at a

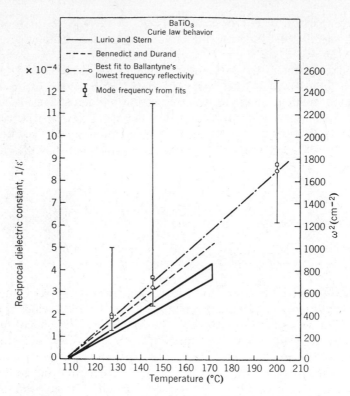

Fig. AII.9   Curie law and mode frequencies for $BaTiO_3$.[23,55] The circles and the solid, dashed, and dot–dash curves are read against the left-hand scale. The dot–dash curve is the best fit to the circles and represents the Curie law derived from Ballantyne's data. The squares (read against the right-hand scale) are the mode frequencies in $BaTiO_3$. Uncertainty bars are included with the mode frequencies. The dot–dash curve can also be read against the right-hand scale and then corresponds to Cochran's theory for the mode frequencies. [A. S. Barker, Jr., *Ferroelectricity*, E. F. Weller, Ed., Copyright 1967, Elsevier Publishing Co., Amsterdam.]

frequency at least three times lower than the frequency ω_1 and he points out that the peak of ε'' may be quite misleading in testing for instability using the Lyddane-Sachs-Teller relation.

F. THE DIELECTRIC RESPONSE FUNCTION

Some general properties of the dielectric response function have been derived by Parrish.[56] From Maxwell's equations it follows that

$$\mathbf{D}(\mathbf{r}, t) = \mathbf{E}(\mathbf{r}, t) + 4\pi\mathbf{P}(\mathbf{r}, t). \qquad (AII.45)$$

Since the applied electric field is small, $\mathbf{P}$ and $\mathbf{D}$ may be approximated by linear functions of $\mathbf{E}$.

Taking the complex Fourier transform of $\mathbf{D}(\mathbf{r}, t)$ assuming that it is some linear function of $\mathbf{E}(\mathbf{r}, t)$ defined at $\mathbf{r}$:

$$\mathbf{D}(\mathbf{r}, t) = \sum_{s=1}^{s} s \int_{-\infty}^{\infty} D_s(\mathbf{r}, \omega) \frac{e^{-i\omega t}}{(2\pi)^{1/2}} \, d\omega$$

$$D_s(\mathbf{r}, \omega) = \sum_q \varepsilon_{sq}(\omega) \varepsilon_q(\mathbf{r}, \omega). \tag{AII.46}$$

In order that $\mathbf{D}$ and $\mathbf{E}$ be real vectors,

$$D_s(\mathbf{r}, -\omega) = D_s^*(\mathbf{r}, \omega)$$

and $\tag{AII.47}$

$$\varepsilon_s(\mathbf{r}, -\omega) = \varepsilon_s^*(\mathbf{r}, \omega),$$

which implies that $\varepsilon_{sq}(-\omega) = \varepsilon_{sq}^*(\omega)$.

Since $D_s(\mathbf{r}, \omega)$ and $\varepsilon_s(\mathbf{r}, \omega)$ can have no essential singularities, $\varepsilon_{sq}(\omega)$ must have no essential singularities. All of the poles and zeros of $\varepsilon_{sq}(\omega)$ must be isolated and in order that $\varepsilon_{sq}(-\omega) = \varepsilon_{sq}^*(\omega)$, they must be symmetrically distributed about the imaginary axis of the ω-plane.

If there is a pole or zero of $\varepsilon_{sq}(\omega)$ at $\omega_1 = \omega_r + i\omega_i$, there must be a complementary pole or zero at $\omega_2 = -\omega_r + i\omega_i$.

In the investigation of the dielectric properties of a material with orthorhombic or higher symmetry, the dielectric response function can be diagonalized with its principle axes parallel to the axes of a Cartesian coordinate system.

$$D_x(\mathbf{r}, \omega) = \varepsilon_{xx}(\omega)\varepsilon_x(\mathbf{r}, \omega) \tag{AII.48}$$

and

$$\varepsilon_x(\mathbf{r}, \omega) = \frac{D_x(\mathbf{r}, \omega)}{\varepsilon_{xx}(\omega)} = \eta_{xx}(\omega)D_x(\mathbf{r}, \omega). \tag{AII.49}$$

A zero of ε_{xx} and a pole of $\eta_{xx}(\omega)$ occurs whenever a frequency ω_l exists such that $D_x(\mathbf{r}, \omega_l) \equiv 0$ and $\varepsilon_x(\mathbf{r}, \omega_l) \neq 0$. Likewise a pole of $\varepsilon_{xx}(\omega)$ and a zero of $\eta_{xx}(\omega)$ occurs whenever a frequency ω_t exists such that $\varepsilon_x(\mathbf{r}, \omega_t) \equiv 0$ and $D_x(\mathbf{r}, \omega_t) \neq 0$. For single-mode behavior, each $\varepsilon_{xx}(\omega)$ may be expressed by:

$$\varepsilon_{xx}(\omega) = \varepsilon_{xx}(\omega \to \infty) + \frac{s}{\omega_0^2 - \omega^2 - i\omega\gamma_0}, \tag{AII.50}$$

where

$$\omega_0{}^2 = |\omega_1|^2 = \omega_r{}^2 + \omega_i{}^2$$

and

$$\gamma_0 = -2\omega_i.$$

In this form, it is similar to Eq. AII.4 for a single classical oscillator, except that in this formulation s may be complex and include a term proportional to ω.

Equation AII.50 also can be written as a product of the poles and zeros such that

$$\varepsilon_{xx}(\omega) = \varepsilon_{xx}(\omega \to \infty)\frac{(\omega - \omega_l)[\omega - (-\omega_l^*)]}{(\omega - \omega_t)[\omega - (-\omega_t^*)]} \qquad \text{(AII.51)}$$

or

$$\varepsilon_{xx}(\omega) = \varepsilon_{xx}(\omega \to \infty)\frac{\omega_L{}^2 - \omega^2 - i\omega\Gamma_L}{\omega_T{}^2 - \omega^2 - i\omega\Gamma_T}, \qquad \text{(AII.52)}$$

where

$$\omega_L{}^2 = |\omega_l|^2$$

$$\Gamma_L = -2\,\text{Im}\,[\omega_l]$$

$$\omega_T{}^2 = |\omega_t|^2$$

$$\Gamma_T = -2\,\text{Im}\,[\omega_t].$$

We define $\varepsilon_{xx}(\infty) = \varepsilon_{xx}(\omega \to \infty)$ and $\varepsilon_{xx}(0) = \varepsilon_{xx}(\omega \to 0)$. Then

$$\frac{\varepsilon_{xx}(0)}{\varepsilon_{xx}(\infty)} = \frac{\omega_L{}^2}{\omega_T{}^2}, \qquad \text{(AII.53)}$$

where $\omega_0{}^2 = \omega_T{}^2$ and $\Gamma_T = \gamma_0$,

$$\varepsilon_{xx}(0) = \varepsilon_{xx}(\infty) + \frac{\text{Re}[s]}{\omega_0{}^2}, \qquad \text{(AII.54)}$$

and

$$s = \varepsilon_{xx}(\infty)[\omega_L{}^2 - \omega_T{}^2 - i\omega(\Gamma_L - \Gamma_T)]$$

are the equations relating the two formulas for $\varepsilon(\omega)$. These can be generalized to give

$$\varepsilon_{xx}(\omega) = \varepsilon_{xx}(\infty) + \sum_{j=1}^{N}\frac{s_j}{\omega_j{}^2 - \omega^2 - i\omega\gamma_j}, \qquad \text{(AII.55)}$$

where s_j is complex and ω_j and γ_j are real, or

$$\varepsilon_{xx}(\omega) = \varepsilon_{xx}(\infty) \prod_{j=1}^{N} \frac{\omega_{L_j}^2 - \omega^2 - i\omega\Gamma_{L_j}}{\omega_{T_j}^2 - \omega^2 - i\omega\Gamma_{T_j}} \qquad \text{(AII.56)}$$

where ω_v and Γ_v are real.

Equation AII.56 has the advantage that both the poles and the zeros are explicitly stated in terms of real numbers, and the limits on the parameters ω_v and Γ_v can be conveniently derived. Moreover, the results of the classical oscillator models discussed earlier, including the model with a frequency-dependent damping term (see Eq. AII.42), can be expressed in Eq. AII.56 with the same number of poles.

For the condition that the magnitude of a plane wave propagating through a dielectric either remain constant or be exponentially damped with time, we find

$$\text{Im}\,[\varepsilon_{xx}(\omega)] \geqslant 0 \qquad \text{for all } \omega \geqslant 0. \qquad \text{(AII.57)}$$

This implies that

$$\Gamma_{T_j} \geqslant 0, \qquad \Gamma_{L_j} \geqslant 0,$$
$$\sum_j \Gamma_{L_j} \geqslant \sum_j \Gamma_{T_j}, \qquad \text{(AII.58)}$$

and

$$\sum_j \frac{\Gamma_{T_j}}{\omega_{T_j}^2} > \sum \frac{\Gamma_{L_j}}{\omega_{L_j}^2}$$

are the necessary conditions for a physically possible component of the dielectric response function. Also, if one has a dielectric response function with single-mode behavior (or if there is a pole and a zero much closer to each other than to any other poles and zeros which may be incorporated into $\varepsilon_{xx}(\infty)\Pi(\omega_{L_j}^2/\omega_{T_j}^2)$ for all $\omega_{T_j}^2$ and $\omega_{L_j}^2$ higher than the mode of interest), then the value of ω_T^2 can be estimated from the behavior of $\varepsilon'' = \text{Im}\,[\varepsilon_{xx}(\omega)]$ and ω_L^2 can be estimated from the behavior of $\eta'' = \text{Im}\,[\eta_{xx}(\omega)]$.

From $\Gamma_T \leqslant \Gamma_L \leqslant \Gamma_T\omega_L^2/\omega_T^2$ it can be shown that

$$\frac{\partial(\varepsilon''/\omega)}{\partial\omega}\bigg|_{\omega=\omega_T} \leqslant 0 \qquad \text{and} \qquad \frac{\partial(\omega\varepsilon'')}{\partial\omega}\bigg|_{\omega=\omega_T} \geqslant 0.$$

Hence the peak in ε''/ω, at ω_1 say, occurs such that $\omega_1/\chi_T \leqslant \omega_1 \leqslant \omega_T$. For $\omega_1 = \omega_T$, $\Gamma_L = \Gamma_T\omega_L^2/\omega_T^2$, where $\chi_T^2 = 1 + \Gamma_T^2/4\omega_T^2$. The peak in $\omega\varepsilon''$, at ω_2 say, occurs such that $\omega_T \leqslant \omega_2 \leqslant \omega_T\chi_T$. For $\omega_2 = \omega_T$, $\Gamma_L = \Gamma_T$.

Similarly, the minimum in η''/ω, at ω_3 say, occurs such that $\omega_L/\chi_L \leqslant \omega_3 \leqslant \omega_L$. For $\omega_3 = \omega_L$, $\Gamma_L = \Gamma_T(\omega_L^2/\omega_T^2)$, where $\chi_L^2 = 1 + \Gamma_L/4\omega_L^2$ and the minimum in $\omega\eta''$, at ω_4 say, occurs such that $\omega_L \leqslant \omega_4 \leqslant \omega_L\chi_L$. For $\omega_4 = \omega_L$, $\Gamma_L = \Gamma_T$. The above relations hold approximately for nearly isolated modes in a multi-mode dielectric response function. In general, the transverse optic mode frequency, ω_T, occurs between ω_1 and ω_2 whereas the longitudinal optic mode, ω_L, occurs between ω_3 and ω_4.

It is interesting to note that electric conductivity is defined as $\varepsilon''\omega/2$ and electric resistivity is defined as $2\eta''/\omega$. The quantities $\eta''\omega$ and ε''/ω are associated with the corresponding magnetic loss parameters. Thus for well separated modes, the dielectric parameters ω_L^2 and ω_T^2 are bracketted by the peaks in the electric and magnetic loss parameters, which may be obtained from Kramers-Kronig analyses of experimental data.

G. Paraelectric Crystals

Both $SrTiO_3$ and $KTaO_3$ crystals have been extensively studied as a function of temperature[7,8,13–15] (see Figs. AII.3–AII.5) and exhibit classic examples of soft mode behavior. Both materials show a deviation in the static dielectric constant at low temperature and neither material actually undergoes a phase transition to the ferroelectric state.

Various other perovskites such as $CaTiO_3$,[2] some zirconates, and hafnates[3] exhibit weak Curie-Weiss law behavior[2] and also remain permanently in the paraelectric phase. As shown in Table AII.1, there is considerable similarity in the frequencies of the modes[16–18] and the major deviation occurs in the lowest-frequency transverse optic mode. Attempts to follow this mode in some of these ceramic materials have not been successful.

H. Ferroelectric Crystals

Of all the ferroelectric materials known, $BaTiO_3$ has probably been the most extensively investigated for its electric properties. The discussion toward the latter part of Section E, and the more detailed analysis given by Barker[55] in his original paper, probably offer the most reasonable explanation for the mode behavior in $BaTiO_3$. This material has never been investigated (except in a preliminary work by Last[19]) in the infrared through the tetragonal-to-orthorhombic-to-rhombohedral transitions. Raman measurements by Hall and Perry[22] are the only low-temperature measurements in these phases.

The mixed-crystal systems $K(Ta,Nb)O_3$ and $(Na,K)TaO_3$ both undergo transitions to the ferroelectric phase. The former material, having a concentration $K(Ta_{0.35}Nb_{0.65})O_3$, has been investigated by Barker[23] at room temperature just above the cubic-to-tetragonal transition. The spectrum is shown in Fig. AII.10 together with the room temperature spectrum of cubic

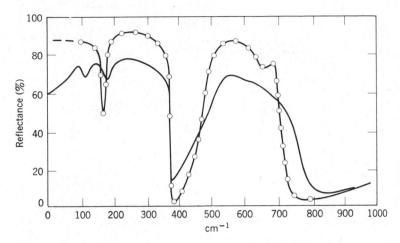

Fig. AII.10   Reflectivity of $K(Ta_{0.65}Nb_{0.35})O_3$ (open circles)[23] and $(Na_{0.40}K_{0.60})TaO_3$ (*solid curve*).

$(Na_{0.40}K_{0.60})TaO_3$ for comparison. The spectrum of $K(Ta,Nb)O_3$ looks very similar to that of $KTaO_3$, whereas $(Na,K)TaO_3$ has a well defined splitting of the center mode. Figure AII.11 shows the low-frequency reflectance behavior for the $(Na_{0.40}K_{0.60})TaO_3$ sample as a function of temperature, and Table AII.3 compares the infrared and Raman frequencies for the $k \sim 0$ modes of various mixtures of $(Na,K)TaO_3$ and of Barker's results for $K(Ta_{0.35}Nb_{0.65})O_3$. The lowest frequency in the $K(Ta,Nb)O_3$ sample was estimated from the LST-relationship as Barker's measurements did not extend below 70 cm^{-1}. The $(Na,K)TaO_3$ system is particularly interesting as the Curie temperature rises with sodium concentration to a maximum of 65°K for a sample with 48% $NaTaO_3$ and falls with higher $NaTaO_3$ concentration and, finally, becomes negative at about 72%. A further increase in sodium content causes the structure to become tetragonal even at room temperature and it finally shears to a pseudo-monoclinic form as the concentration approaches pure $NaTaO_3$.[31] Consequently, as can be seen by the spectra in Figs. AII.10 and AII.11, the structure is considerably more complex for the $(Na_{0.85}K_{0.15})TaO_3$ sample.

However, there is reasonably good comparison with the infrared and Raman frequencies where the crystal symmetry is such that there is no mutual exclusion of the infrared and Raman activities. The measurements facilitated interpretation of the strong second-order Raman spectra, which is extremely complex.[25] (Soft mode behavior has again been observed in samples below 40% $NaTaO_3$ as measured by Davis[57] using electric-field-induced Raman spectra.)

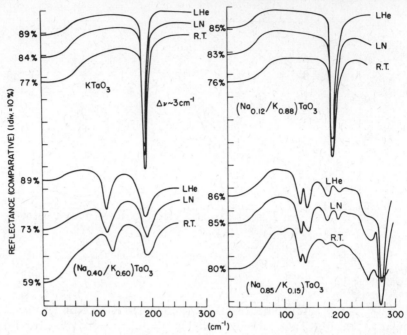

Fig. AII.11 Low-frequency infrared reflectivity of various samples of $(Na, K)TaO_3$ as a function of temperature. The curves are displaced for optimum display. The high-frequency reflectivity is similar to that in Fig. AII.8 and Fig. AII.10 and shows very little temperature-dependence apart from a rise in the reflectivity around 800 cm^{-1} when the temperature is lowered. [N. E. Tornberg, T. G. Davis, C. H. Perry, and K. Knable, *Natl. Bur. Stand. Special Publication* 301, June, 1969, R. S. Carter and J. J. Rush, Eds.]

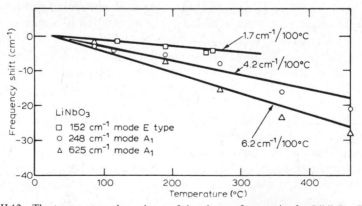

Fig. AII.12 The temperature dependence of the phonon frequencies for $LiNbO_3$. The shifts are measured in the Raman spectra and are used to compare the stability of the modes shown. [A. S. Barker, Jr., *Ferroelectricity*, E. F. Weller, Ed., Copyright 1967, Elsevier Publishing Co., Amsterdam.]

$LiNbO_3$ is a distorted perovskite, and Cochran's theory applied to this material predicts an A_{2u} phonon mode becoming unstable so that the change in mode frequency produces the temperature dependence of ε_0. Barker and Loudon[26] observed both the polarized infrared and Raman spectra. Figure AII.12 shows the temperature dependence of the phonon frequencies as measured in the Raman spectra. With the help of these data it was seen that the 248 cm^{-1} mode is the softest mode (and therefore contributes the largest part to the temperature dependence of ε_0). Barker and Loudon also list the ion displacements, charges, and Lorentz local field factors as well as make comparisons between $LiNbO_3$ and $SrTiO_3$.

I. CONCLUSIONS

The dielectric behavior and the optical phonon modes in a large number of perovskite and related materials have been measured using infrared reflectance techniques. The results have established that the expected number of modes from group-theoretical predictions agree in most cases, especially where good single crystals are available. The perovskites possess the simplest structure which undergoes a ferroelectric transition, and the infrared investigation of these materials in the paraelectric phase has to a large degree confirmed the theory of ferroelectricity as proposed by Cochran.[5] Here, the ferroelectric transition in certain of these crystals is associated with an instability in one of the phonon modes such that the frequency of the relevant optic mode is lowered and tends to zero as $T \rightarrow T_c$. This behavior is most clearly observed in $SrTiO_3$ and $KTaO_3$, which do not themselves actually make the transition to the ferroelectric state but do, in fact, experience the same restoring force–displacive force interaction. For the mixed-crystal system $(Na,K)TaO_3$ and $K(Ta,Nb)O_3$, the pure $KTaO_3$ is very valuable for a clearer understanding of these more complicated materials. $BaTiO_3$ will continue to be of interest, and structurally more complicated materials possessing these unusual properties will no doubt be investigated in the future. The infrared techniques remain a powerful tool for these investigations, but Raman spectroscopy offers complementary results. A combination of both techniques will provide extensive information on the $\mathbf{k} \sim 0$ modes, especially if oriented single crystals can be obtained. Neutron spectroscopy has the advantage of being able to obtain the complete dispersion curves directly, but the accessability to this type of investigation remains limited and other experimental considerations often preclude these measurements. Electric-field-induced Raman spectra have provided more precise measurements of the soft mode frequency and temperature dependence of the phonon life-times. Tuning of the soft mode in $KTaO_3$ from 25–40 cm^{-1} with electric fields from 0 to 20000 V/cm has been observed and offers the possibility of a stimulated Raman scatterer and a far-infrared laser, as well as electro-optic applications. The possibility of instability of modes other than a $k \sim 0$ infrared phonon has been discussed by Cochran[58] (an acoustic mode in certain piezoelectric crystals may become unstable). Antiferroelectric crystals remain to be investigated further.[17,18] Cochran[58] also discusses the possible instability in a zone-boundary phonon.

The dielectric properties of other ferroelectric materials, such as hydrogen-bonded ferroelectrics,[59,60] certain nitrates, and other complex oxides, still await investigation. This type of infrared measurement and data analysis, together with Raman spectroscopy, will be necessary in order to understand the modes of vibration and their effect on the dielectric properties of these structurally more complicated materials.

REFERENCES

1. F. Jona and G. Shirane, *Ferroelectric Crystals*, Pergamon Press, New York, 1963.
2. G. Rupprecht and R. O. Bell, *Phys. Rev.* **135**, A748 (1964).
3. H. D. Megaw, *Ferroelectricity in Crystals*, Methuen, London, 1957.
4. W. Cochran, *Phys. Rev. Letters* **3**, 412 (1959).
5. W. Cochran, *Adv. Phys.* **9**, 387 (1960).
6. P. Anderson, in *Fizika Dielektrikov*, G. I. Skanani, Ed., Akad. Nauk SSSR, Fiz. Inst. P. N. Lebedeva, Moscow, 1960.
7. A. S. Barker, Jr. and M. Tinkham, *Phys. Rev.* **125**, 1527 (1962).
8. C. H. Perry and T. F. McNelly, *Phys. Rev.* **154**, 456 (1967).
9. M. Born and K. Huang, *Dynamical Theory of Crystal Lattices*, Clarendon Press, Oxford, 1954.
10. D. H. Martin, *Adv. Phys.* **14**, 39 (1965).
11. R. P. Lowndes, *Phys. Letters* **21**, 26 (1966).
12. W. G. Spitzer, R. C. Miller, D. A. Kleinman, and L. E. Howarth, *Phys. Rev.* **126**, 1710 (1962).
13. R. A. Cowley, *Phys. Rev. Letters* **9**, 159 (1962).
14. G. Shirane, R. Nathans, and V. J. Minkiewicz, *Phys. Rev.* **157**, 396 (1967).
15. R. A. Cowley, *Phys. Rev.* **134**, A981 (1964).
16. C. H. Perry, B. N. Khana, and G. Rupprecht, *Phys. Rev.* **135**, A408 (1964).
17. C. H. Perry, D. J. McCarthy, and G. Rupprecht, *Phys. Rev.* **138**, A1537 (1965).
18. C. H. Perry, *Japan J. Appl. Phys.* **4**, Suppl. 1, 564 (1965).
19. J. T. Last, *Phys. Rev.* **105**, 1740 (1957).
20. J. M. Ballentyne, *Phys. Rev.* **136**, A429 (1964). Also T. Nakamura and Y. Ishibashi, *J. Phys. Soc. Japan* **21**, 1467 (1966).
21. A. S. Barker, Jr., *Phys. Rev.* **145**, 391 (1966).
22. C. H. Perry and D. B. Hall, *Phys. Rev. Letters* **15**, 521 (1965).
23. A. S. Barker, Jr., in *Ferroelectricity*, E. F. Weller, Ed., Elsevier, Amsterdam, 1967, p. 236.
24. R. H. Lyddane, R. G. Sachs, and E. Teller, *Phys. Rev.* **59**, 673 (1941).
25. N. E. Tornberg, T. G. Davis, C. H. Perry, and N. Knable, *Molecular Dynamics and Structure of Solids*, R. S. Carter and J. J. Rush, Eds., National Bureau of Standards Special Publication 301, June 1969.

26. A. S. Barker, Jr. and R. Loudon, *Phys. Rev.* **158**, 433 (1967).
27. D. R. Bosomworth, *Appl. Phys. Letters* **9**, 330 (1966).
28. B. G. Dick, Jr. and A. W. Overhauser, *Phys. Rev.* **112**, 90 (1958).
29. C. Kittel, *Introduction to Solid State Physics*, 2nd ed., Wiley, New York, 1960, Chapters 7 and 8.
30. T. Nakamura, *J. Phys. Soc. Japan* **21**, 491 (1961).
31. T. G. Davis, M.S. Thesis, Dept. of Electrical Engineering, Massachusetts Institute of Technology, 1965 (unpublished).
32. Ref. 23, p. 218.
33. S. H. Wemple, *Phys. Rev.* **137**, A1575 (1964).
34. R. W. G. Wyckoff, *Crystal Structures*, Wiley-Interscience, New York, 1948, VII,A5.
35. S. Bhagavantam and T. Venkatarayudu, *Proc. Indian Acad. Sci.* **9A**, 224 (1939).
36. H. Winston and R. S. Halford, *J. Chem. Phys.* **17**, 607 (1949).
37. G. R. Hunt, C. H. Perry, and J. Ferguson, *Phys. Rev.* **134**, A689 (1964).
38. J. D. Axe, *Phys. Rev.* **157**, 429 (1967).
39. C. H. Perry, J. H. Fertel, and T. F. McNelly, *J. Chem. Phys.* **47**, 1619 (1967).
40. L. Rimai, J. L. Parsons, and A. L. Cedarquist, *Bull. Am. Phys. Soc.* **12**, 60 (1967).
41. D. C. O'Shea, R. V. Kolluri, and H. Z. Cummins, *Solid State Commun.* **5**, 241 (1967).
42. R. F. Schaufele and M. J. Weber, *J. Chem. Phys.* **46**, 2859 (1967).
43. W. G. Nilson and J. G. Skinner, *J. Chem. Phys.* **47**, 1413 (1967).
44. P. A. Fleury and J. M. Worlock, *Phys. Rev. Letters* **18**, 665 (1967).
45. P. A. Fleury and J. M. Worlock, *Phys. Rev. Letters* **19**, 1176 (1967).
46. I. Lefkowitz, *Proc. Phys. Soc.* (*London*) **80**, 868 (1962)
47. F. J. Low, *J. Opt. Soc. Am.* **51**, 1300 (1961).
48. C. H. Perry, R. Geick, and E. F. Young, *Appl. Opt.* **5**, 1171 (1966).
49. T. S. Robinson, *Proc. Phys. Soc.* (*London*) **B65**, 910 (1952).
50. D. M. Roessler, *Brit. J. Appl. Phys.* **16**, 1359 (1965).
51. A. S. Barker, Jr. and J. J. Hopfield, *Phys. Rev.* **135A**, 1732 (1964).
52. H. Bilz, L. Genzel, and H. Happ, *Z. Physik* **160**, 555 (1960); H. Bilz and L. Genzel, *Z. Physik* **169**, 53 (1962).
53. R. Wehner, Ph.D. Thesis, University of Freiburg, Germany, 1964 (unpublished).
54. R. Geick, W. J. Hakel, and C. H. Perry, *Phys. Rev.* **148**, 824 (1966).
55. A. S. Barker, Jr., *Phys. Rev.* **145**, 381 (1966).
56. J. F. Parrish, Ph.D. Thesis, Dept. of Physics, Massachusetts Institute of Technology (1969).
57. T. G. Davis, Ph.D. Thesis, Dept. of Elec. Eng., Massachusetts Institute of Technology (1968).
58. W. Cochran, *Adv. Phys.* **10**, 401 (1961).
59. R. M. Hill and S. K. Ichiki, *Phys. Rev.* **128**, 1140 (1962); **130**, 150 (1963).
60. R. Blinc and S. Svetina, *Phys. Rev.* **147**, 430 (1966).

APPENDIX III

MAGNETIC PHENOMENA
IN THE FAR-INFRARED

ISAAC F. SILVERA

North American Aviation Science Center, Thousand Oaks, California

A. Introduction

A review of the applications of far-infrared techniques to the study of the magnetic properties of solids is given here. The magnetic properties and related phenomena in matter can in general be described in terms of the low-lying energy states which involve spin and orbital angular momentum reorientations. This statement is manifested by the Zeeman Hamiltonian

$$\mathcal{H} = -\mathbf{H} \cdot \mathbf{M} = -\beta \mathbf{H} \cdot (\mathbf{L} + 2\mathbf{S}), \tag{AIII.1}$$

where β is the Bohr magneton, $\mathbf{H}$ the magnetic field strength, and $\mathbf{L}$ and $\mathbf{S}$ the orbital and spin angular momenta, respectively. This Hamiltonian can often be generalized to the form

$$\mathcal{H} = -g\beta \mathbf{H}_{eff} \cdot \mathbf{s}'. \tag{AIII.2}$$

Here $\mathbf{s}'$ is called the fictitious spin of the system and represents the number of angular momentum energy levels which are needed to adequately describe the system (for $s' = 1$, the number of levels is $2s' + 1 = 3$). $\mathbf{H}_{eff}$ is either an external or internal field that splits the levels by an amount proportional to the g-factor. (The g-factor can, in general, be a tensor.) We can define the effective field by

$$\beta g \mathbf{H}_{eff} = -\mathbf{\nabla}_{s'} \mathcal{H}. \tag{AIII.3}$$

For a g-value of two, the order of magnitude of an effective field that places the excitation spectrum in the far-infrared spectral region of 10 to 100 cm^{-1}

is 10^5 to 10^6 G. Hence the effective fields that we consider are quite large compared to normal laboratory fields.

The study of magnetic phenomena in the far-infrared usually requires cryogenic techniques. Again referring to the region 10 to 100 cm^{-1} in terms of the energy equivalent temperature kT, this region covers 14.4 to 144°K. For a system with a ground state splitting in this region, we could observe electromagnetic transitions from the populated ground state to the excited states at liquid-helium temperatures, whereas the absorption strength would approach zero at room temperature due to thermal equalization of the populations. Likewise, in ordered magnetic materials the most interesting effects are observed below the critical or ordering temperature which in insulators is generally lower than room temperature.

One final introductory point to consider is the absorption strength of a magnetic dipole transition as compared to an electric dipole transition. The ratio

$$\frac{|\langle \mathbf{M} \cdot \mathbf{H} \rangle|^2}{|\langle \mathbf{P} \cdot \mathbf{E} \rangle|^2},$$

which is proportional to the ratio of the absorption intensities, is approximately 10^{-4} for allowed transitions; that is, magnetic dipole transitions are much weaker than electric dipole transitions. This relative weakness of the magnetic interaction is in some ways very fortunate, for it allows one to study magnetic properties of solids by the technique of transmission spectroscopy and easily extract information about the magnetic susceptibility. For strong electric dipole transitions such as the lattice absorption lines in concentrated solids, typically ranging from 100 to 2000 cm^{-1}, we must use the technique of reflection spectroscopy which yields spectra requiring substantially more complicated analyses. In the transmission technique, if we neglect multiple reflections, the normal incidence transmission through a plate of thickness d is

$$T = \left[\frac{4n}{(n+1)^2 + \kappa^2} \right] e^{-\alpha d}, \tag{AIII.4}$$

where n is the index of refraction, κ the extinction coefficient and $\alpha = 2\omega\kappa$ is the absorption coefficient at circular frequency ω (see Eq. 12.14). The magnetic permeability

$$\mu = \mu' - i\mu'' = 1 + 4\pi(\chi' - i\chi'') \tag{AIII.5}$$

and the dielectric constant *

$$\varepsilon = \varepsilon' - i\varepsilon'' \tag{AIII.6}$$

* In Chapter 12, the complex quantities are written in the form $a + ib$.

are related to the complex index of refraction by $(n + i\kappa)^2 = \varepsilon\mu$ giving

$$\mu'\varepsilon' - \mu''\varepsilon'' = n^2 - \kappa^2$$
$$\mu'\varepsilon'' + \mu''\varepsilon' = 2n\kappa.$$

(AIII.7)

In the magnetic case, because of the weakness of the interaction, a fairly accurate approximation can be made[1] which allows the absorption coefficient to be divided into a lattice part α_l and a magnetic part α_m so that $\alpha = \alpha_l + \alpha_m$, with $\alpha_m = 8\pi^2 n_0 v \chi''(v)$. Here n_0 is the index of refraction in the absence of the magnetic interaction, the magnetic interaction giving it a negligible change, and v is the frequency in cm^{-1}. To separate α_l from α_m, we need only measure the transmission spectrum below and above some critical temperature above which the magnetic absorption disappears, or compare the spectra of the magnetic crystal to that of a similar nonmagnetic crystal. A third technique is to shift the magnetic absorption in frequency with an externally applied magnetic field.

B. Cryostats

A very useful double-Dewar cryogenic assembly that can be used for doing zero-magnetic field (or low-field work in large gap magnets) transmission experiments at variable temperatures is shown in Fig. AIII.1. Light-pipe optics[2] are used to carry chopped radiation from an infrared source through one of four positions on a rotatable sample platform and onto a bolometer detector in the same assembly at ~1°K. The samples are shielded from dc

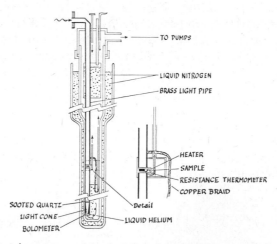

Fig. AIII.1   Cryogenic system used for far-infrared transmission experiments on samples at temperatures between ~1.5 and 150°K. The four-position rotatable sample holder is not shown.

room temperature radiation by cooled quartz, sapphire, or alkali halide crystals; they are cooled to $\sim 1.5°$K by a copper braid that rests in the liquid helium and is in good thermal contact with the sample platform. A carbon resistor heater and a thermometer glued to the sample platform with their electrical leads running up the hollow stainless steel rotating shaft are used for sample temperature control. The light-pipe below the sample enclosure is made of thin-wall stainless steel to thermally isolate the bolometer detector from the sample temperature. The copper braid can be externally raised out of the liquid helium by means of a sliding shaft through a rubber cork or vacuum fitting. With this arrangement, sample temperatures of the order of 150°K can be attained with little effect on the detector temperature or sensitivity.

A cryostat for high-magnetic field work is shown in Fig. AIII.2. A high-field superconductivity magnet (75 kG) with a $1\frac{1}{2}''$ diameter (~ 38 mm) bore sits in helium at 4.2°K. A single-walled evacuated isolation Dewar has its smaller lower section in the magnet bore. The isolation Dewar contains the transmission cryostat which is centered by insulated spacers. The upper can contains nitrogen for thermal isolation from room temperature; the lower can contains pumped-on helium. A copper rod from the helium can to the bolometer mount cools the bolometer which is shielded from the magnetic field by a superconducting shield. The horizontal sample platform has three positions; the sample temperature can be varied as in the previous cryostat. A three-position vertical platform for studies with a field parallel to the propagation vector of the radiation incorporates a "zigzag" assembly shown in the detail. A somewhat different design has been used by Richards.[3]

C. ANTIFERROMAGNETIC INSULATORS

The antiferromagnetic state in paramagnetic solid insulators is that state in which the atomic magnetic moments of the crystal order spontaneously into two or more subsets, or sublattices, of equal and parallel magnetic moments with the distinguishing property that the net moment of a magnetic unit cell is zero. As the temperature of such a paramagnetic crystal is lowered through a certain critical temperature called the Néel temperature (T_N), the crystal generally undergoes a second-order phase transition associated with the onset of long-range order of the antiferromagnetic state.

The Hamiltonian or potential that best describes this state in insulators is

$$\mathcal{H} = \mathcal{H}_{ex} + \mathcal{H}_{anisotropy}. \tag{AIII.8}$$

Here, $\mathcal{H}_{ex}$ has the form of the Heisenberg-Dirac exchange Hamiltonian which couples together the spins of different atoms and, in the simplest case, takes

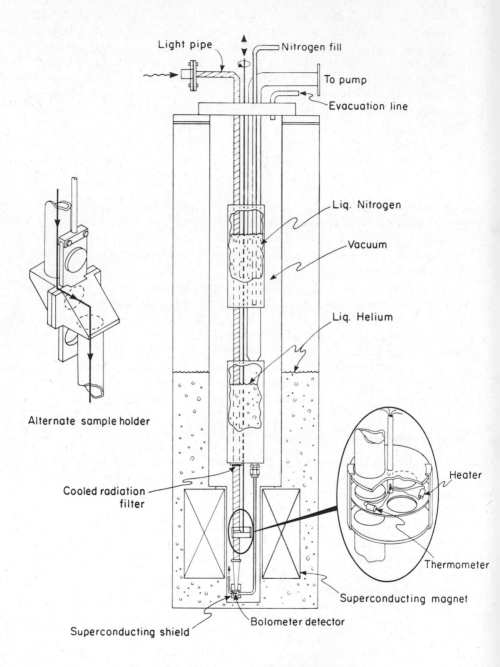

Light pipe

Nitrogen fill

To pump

Evacuation line

Liq. Nitrogen

Vacuum

Liq. Helium

Heater

Thermometer

Alternate sample holder

Cooled radiation filter

Superconducting magnet

Superconducting shield

Bolometer detector

Fig. AIII.2   High-magnetic field transmission cryostat.

on the isotropic form

$$\mathcal{H} = \sum_{i>j} \mathcal{H}_{ex}^{ij} = \sum_{i>j} 2J_{ij}\mathbf{S}_i \cdot \mathbf{S}_j, \qquad (AIII.9)$$

where J_{ij} is the exchange integral and $\mathbf{S}_i$ and $\mathbf{S}_j$ are spin operators for the respective atoms. The principal exchange effects are actually due to super-exchange, that is, the magnetic ions are coupled by electron transfer and covalency through one or more intermediate diamagnetic ions (ligands). This interaction, which allows the exchange to be transmitted over relatively large distances, was first put forward by Kramers[4] and later developed by Anderson and others.[5]

For positive exchange integrals in Eq. AIII.9, the magnetic energy is minimized by an antiparallel alignment of the interacting spins. It is necessary to include the second part, the anisotropy energy, to stabilize the spins with respect to some crystalline direction as is observed. The sources of anisotropy include (1) magnetic dipole-dipole interaction,[6] (2) single-ion crystalline field effects,[6] (3) anisotropic superexchange,[6] and (4) g-tensor anisotropy.[7]

D. SPIN WAVE EXCITATION SPECTRA

We shall consider a model of the simple two-sublattice antiferromagnet with antiparallel alignment of the sublattices. For simplicity, assume that the atoms are in S states so that the magnetic moment arises only from the electron spin. At $T = 0$ the sublattice magnetization takes on its maximum value along the crystalline direction of quantization; at T_N it goes to zero. The decrease in magnetization with increase in temperature is due to the excitation of spin waves or magnons. These elementary excitations represent quantized spin deviations away from the axis of quantization.

In the ordered state, we consider the quantization of the spin of the entire system rather than that of a single atom. Due to the cosine form of the exchange interaction, it costs the system less energy to spread a quantum of spin deviation over all the spins by giving each one a small tilt rather than putting all the deviation on one site. To understand these cooperative excitations more clearly, we shall treat the case in which the spins have only nearest neighbor exchange interaction and a uniaxial single-ion anisotropy along the crystalline z-direction of the form $-g\beta(H_a/2)S^{z^2}$ which causes the spin sublattices to align in the z-direction.

The Hamiltonian for two adjacent spins (on opposing sublattices) in the presence of an applied magnetic field H_0 in the z-direction is

$$\mathcal{H}_{2n} = 2J\mathbf{S}_{2n} \cdot (\mathbf{S}_{2n+1} + \mathbf{S}_{2n-1}) - \frac{g\beta H_a}{2}S_{2n}^{z^2} - g\beta H_0 S_{2n}^z \qquad (AIII.10)$$

$$\mathscr{H}_{2n+1} = 2J\mathbf{S}_{2n+1} \cdot (\mathbf{S}_{2n} + \mathbf{S}_{2n+2}) - \frac{g\beta H_a}{2} S_{2n+1}^{z2} - g\beta H_0 S_{2n+1}^z. \quad \text{(AIII.11)}$$

The object is now to find the dispersion relation and the modes of motion of the spins. One approach is to solve the equations that result from setting the time derivative of the spin equal to $i/\hbar$ times the commutator of the Hamiltonian with the spin. A more lucid approach that gives the same result is to use the coupled Bloch equations which state that the time derivative of the magnetic moment is proportional to the torque on the moment due to the effective field. Or since $\mathbf{M} = g\beta\mathbf{S}$, we have

$$d\mathbf{S}/dt = \gamma\mathbf{S} \times \mathbf{H}_{eff}, \qquad \gamma = g\beta/\hbar. \quad \text{(AIII.12)}$$

Using Eq. AIII.3, we find the effective fields on the $2n$th and the $(2n + 1)$th spins are

$$\mathbf{H}_{2n} = -\frac{J}{g\beta}(\mathbf{S}_{2n+1} + \mathbf{S}_{2n-1}) + H_a S_{2n}^z \mathbf{z} + H_0 \mathbf{z} \quad \text{(AIII.13)}$$

$$\mathbf{H}_{2n+1} = -\frac{J}{g\beta}(\mathbf{S}_{2n} + \mathbf{S}_{2n+2}) + H_a S_{2n+1}^z \mathbf{z} + H_0 \mathbf{z}. \quad \text{(AIII.14)}$$

Here $\mathbf{z}$ is a unit vector in the z-direction. For a system with N spins, using Eq. AIII.12, we find there exist $2N$ coupled simultaneous equations of the form

$$\frac{1}{\gamma}\frac{dS_{2n}^x}{dt} = -\frac{2J}{g\beta}[S_{2n}^y(S_{2n+1}^z + S_{2n-1}^z) - S_{2n}^z(S_{2n+1}^y + S_{2n-1}^y)]$$

$$+ (H_a S_{2n}^z + H_0)S_{2n}^y \quad \text{(AIII.15)}$$

$$\frac{1}{\gamma}\frac{dS_{2n}^y}{dt} = -\frac{2J}{g\beta}[S_{2n}^z(S_{2n+1}^x + S_{2n-1}^x) - S_{2n}^x(S_{2n+1}^z + S_{2n-1}^z)]$$

$$- (H_a S_{2n}^z + H_0)S_{2n}^x. \quad \text{(AIII.16)}$$

To find the dispersion relation we need only consider the four equations for two adjacent spins, say the $2n$th and $(2n + 1)$th. For small spin deviations about the z-axis we can take $S_{2n}^z = -S_{2n+1}^z = S$, a constant. Due to the translational invariance under a displacement of lattice parameter $2a$ and infinitesimal translation in time, we seek a solution of the form

$$S_{2n}^{x,y} = S_A^{x,y} e^{i(\omega t - 2nak)} \quad \text{(AIII.17)}$$

$$S_{2n+1}^{x,y} = S_B^{x,y} e^{i[\omega t - (2n+1)ak]}. \quad \text{(AIII.18)}$$

Substituting these solutions into the four equations of the form of Eqs. AIII.15 or AIII.16 and ignoring quadratic terms in transverse (x or y) components of the spin since they are small, we find four simultaneous linear equations. Because of the uniaxial form of the anisotropy, we expect circularly polarized motion of the spins. The solution is simplified if we substitute for S^x and S^y the independent variables $S^{\pm} = S^x \pm iS^y$ for both A and B sublattices, which gives the four equations

$$(\omega + \omega_0 + \omega_e + \omega_a)S_A^+ + (\omega_e \cos ka)S_B^+ = 0 \qquad \text{(AIII.19a)}$$

$$(-\omega_e \cos ka)S_A^+ + (\omega + \omega_0 - \omega_e - \omega_a)S_B^+ = 0 \qquad \text{(AIII.19b)}$$

$$(\omega - \omega_0 - \omega_e - \omega_a)S_A^- - (\omega_e \cos ka)S_B^- = 0 \qquad \text{(AIII.19c)}$$

$$(-\omega_e \cos ka)S_A^- + (\omega - \omega_0 + \omega_e + \omega_a)S_B^- = 0, \qquad \text{(AIII.19d)}$$

where $\omega_e = 4\gamma JS/g\beta$, $\omega_a = \gamma H_a S$, $\omega_0 = \gamma H_0$. There are four solutions:

$$\omega_1^+(k) = -\omega_1^-(k) = [(\omega_e + \omega_a)^2 - \omega_e^2 \cos^2 ka]^{1/2} + \omega_0 \qquad \text{(AIII.20)}$$

$$\omega_2^+(k) = -\omega_2^-(k) = [(\omega_e + \omega_a)^2 - \omega_e^2 \cos^2 ka]^{1/2} - \omega_0. \qquad \text{(AIII.21)}$$

Of the four, only two are physically different solutions since ω_1^+ and ω_1^- represent the same mode of motion and ω_2^+ and ω_2^- another mode (we shall refer to these solutions as ω^+ and ω^-). Solving Eqs. AIII.19 for the ratio of the amplitudes of the motion of the spins, we find

$$\frac{S_A^+}{S_B^+} = \frac{S_B^-}{S_A^-} = \frac{-\omega_e \cos ka}{\omega_e + \omega_a + [(\omega_e + \omega_a)^2 - \omega_e^2 \cos^2 ka]^{1/2}}. \qquad \text{(AIII.22)}$$

Since the $S^{\pm}$ refer to circular motion, this means that the ends of adjacent spins trace out circles whose relative radii and phases are given by Eq. AIII.22 and shown in Fig. AIII.3 for $ka = 0$ and $\pi/2$. Figure AIII.4 is a plot of the dispersion relations, Eqs. AIII.20 and AIII.21.

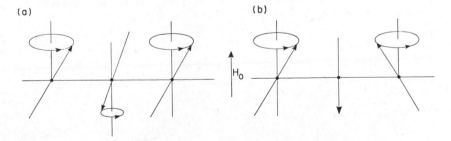

(a)   (b)

H_0

Fig. AIII.3   One of the two spin wave modes in an applied magnetic field for (a) $k = 0$ and (b) the Brillouin zone edge $ka = \pi/2$. The $k = 0$ mode has a net oscillating moment per unit cell and can absorb radiation. The mode in (b) has no net oscillating moment and is inactive.

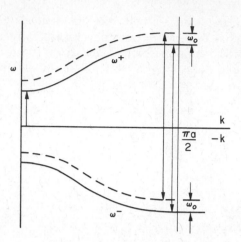

Fig. AIII.4   The dispersion relations for the two branches of the antiferromagnetic spin wave modes. The solid lines are the zero-field curves; the dotted lines are for applied field parallel to the spin quantization axis. The single arrow represents the excitation of the uniform mode; the double-headed arrows represent two-magnon excitations in zero and applied magnetic field.

E. Single-Magnon Absorption

If we let $k = 0$ in Eqs. AIII.20 and AIII.21, we find the antiferromagnetic resonance (AFMR) frequencies

$$\omega^{\pm} = [(2\omega_e + \omega_a)\omega_a]^{1/2} \pm \omega_0. \qquad \text{(AIII.23)}$$

If there is no applied magnetic field, $\omega_0 = 0$ and the modes are degenerate. These modes are active and will absorb the magnetic component of electromagnetic radiation at the given frequency. From Fig. AIII.3a we see that the resultant of the motion of the A and B sublattices is a circularly polarized magnetic moment rotating transverse to the z-direction. These modes will thus absorb radiation of the same polarization in the transverse plane. The strength of this absorption can be obtained by introducing a circularly polarized oscillating magnetic field of the form $(H_x \pm iH_y)e^{i\omega t}$ into the equations of motion and solving for the rf susceptibility

$$\chi'^{\pm} = \frac{M^{\pm}}{H^{\pm}} = \frac{2g\beta S\omega_a}{[\omega - \omega_1^{\pm}(0)][\omega - \omega_2^{\pm}(0)]} \qquad \text{(AIII.24)}$$

with $M^{\pm} = M_A^{\pm} + M_B^{\pm}$. This analysis also yields the result that the only mode excited by the rf magnetic field is the $k = 0$ or uniform precession mode.

Tinkham[8] has shown, using the Kramers-Kronig relationships, that

$$\chi'^{\pm}(\omega) = \frac{\chi''(\omega')\Delta\omega}{\pi(\omega' - \omega)},$$ (AIII.25)

where ω' is the resonance frequency and $\Delta\omega$ the linewidth. By use of Eqs. AIII.24 and AIII.25 one can calculate the absorption coefficient $\alpha_m = 4\pi n\omega\chi''(\omega)$.

Antiferromagnetic resonance was first considered theoretically by Kittel[9] and Nagamiya[10] in 1951. It was first observed by using far-infrared techniques by Ohlmann and Tinkham[11] at 52.7 cm^{-1} in the antiferromagnetic crystal FeF$_2$, which is similar to the model we have used. This resonance line was found to be degenerate in zero applied magnetic field, splitting linearly in field as expected from the theory. The AFMR frequency which is proportional to the magnitude of the spin was also observed to decrease with increasing temperature since the mean value of the spin behaves in a similar manner. AFMR has since been observed in many other antiferromagnets. A partial list is given in Table AIII.1.

In an antiferromagnet with rhombic anisotropy, the zero-field degeneracy of the AFMR is generally lifted because there are different effective anisotropy fields along the crystalline axes. In this case the modes will be plane polarized;

TABLE AIII.1

ANTIFERROMAGNETIC RESONANCES OBSERVED BY FAR-INFRARED TECHNIQUES

Compound	T_N, °K	Resonant frequency, cm^{-1}	Remarks	Ref.
FeF$_2$	78.4	52.7	Two-magnon line also observed	11, 17
MnF$_2$	67.7	8.7	Two-magnon line also observed	24, 21
CoF$_2$	38	28.5, 36	Anomalous	23
NiF$_2$		3.33, 31.14	Weakly canted antiferro-magnet	23
MnO	120	27.5	Only one mode observed	16
NiO	523	36.6	Only one mode observed	16
FeCl$_2$		16.3	Metamagnetic transition	14
CoCl$_2$		19.2		14
α-CoSO$_4$	12	20.6, 25.4, 35.8	Four-sublattice canted antiferromagnet	7
β-CoSO$_4$	~15	20.2, 22.4, 24.5, 40.1	Four-sublattice canted antiferromagnet	25
UO$_2$	30.8	79, 99	Anomalous	26

the zero-field frequency is given by Eq. AIII.23 with ω_a taking on a different value for each mode. These frequencies depend quadratically on the magnetic field rather than linearly as in the degenerate case. Examples of this type of antiferromagnet are MnO and NiO.[12]

So far we have considered the simple two-sublattice case. In general, the number of modes that can exist (at $k = 0$) is equal to the number of sublattices. This is because a three-component spin is constrained (its length and z-component are constant to a good approximation), leaving but one degree of freedom for each sublattice. (This argument excludes the possibility of spin wave excitation from the ground spin state to several molecular field excited spin states as could possibly be the case in CoF_2 which has a complex set of low-lying spin levels.[13]) An example is the antiferromagnet[14] $FeCl_2$ which has an AFMR frequency at $16.3 \, cm^{-1}$. $FeCl_2$ is a metamagnetic crystal: in an applied field greater than 10.5 kOe, it undergoes a transition from a two-sublattice antiferromagnet to a ferromagnet. The AFMR line is split into a doublet for fields less than 10.5 kOe; however, for fields above this value, the AFMR lines disappear in favor of a new line appearing at $16.7 \, cm^{-1}$. The new line is due to the ferromagnetic resonance of the single sublattice; it lies at relatively high frequencies for ferromagnetic resonance because of large anisotropy fields.

A more complicated case is that of α-$CoSO_4$,[7] a four-sublattice canted antiferromagnet. In this crystal the magnetic moments are canted away from the crystalline axes by a large angle of 25° in such a manner that the magnetic unit cell has zero magnetic moment. Theoretically, we predict four modes of which only three of the $k = 0$ modes have net oscillating magnetic moments and can absorb radiation. These have been observed at 20.6, 25.4, and $35.8 \, cm^{-1}$.

The rare-earth ion garnets are another class of ordered magnetic materials with a rich spectrum in the far-infrared due to magnetic excitations. The garnets have the formula $5Fe_2O_3 \cdot 3R_2O_3$ where R is a trivalent rare-earth ion. The ferric ions are located at two different symmetry sites which order into opposing sublattices with a net ferromagnetic moment below the Curie point ($\sim 550°K$). The rare-earth ions are relatively weakly coupled to the iron and at low temperatures order with their spins opposed to its net moment. The problem is complicated by the many inequivalent rare-earth sites in a unit cell and by anisotropic interactions. The spectra of several of these compounds have been observed and interpreted via a simplified two- and three-sublattice model.[15,16] Two types of excitation are observed: collective or spin wave excitation and single-ion transitions. In the former, the $k = 0$ spin wave is one wherein the iron and rare-earth sublattices mutually precess around one another with a net oscillating moment. These are distinguished by their characteristic temperature dependence due to the temperature

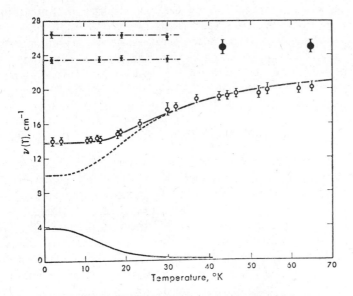

Fig. AIII.5 Temperature dependence of the resonance frequencies in ytterbium iron garnet. The two highest lines are temperature-independent and represent single-ion transitions. The intermediate mode is a $k = 0$ spin wave excitation fitted with a theoretical solid curve that includes anisotropy energy; the dashed curve represents an isotropic model. The low-frequency curve is a predicted ferromagnetic resonance frequency. [A. J. Sievers, III and M. Tinkham, *Phys. Rev.* **124**, 321 (1961).]

dependence of the rare-earth magnetization. The latter type of excitations are single-ion transitions of the rare earth within energy levels determined by the crystalline electric field and the iron exchange field. These, on the other hand, show little temperature dependence at low temperatures because the iron sublattice does not disorder. The temperature dependence of the modes observed in ytterbium iron garnet is shown in Fig. AIII.5.

F. TWO-MAGNON ABSORPTION

Recently a new type of line due to two-magnon absorption has been observed in the higher-frequency range of the far-infrared.[17] Measurement of the absorption frequency permits a precise determination of magnon energies at certain points on the Brillouin zone boundary. Although AFMR had been observed in FeF_2 ($T_N = 78°K$) much earlier at $52.7 cm^{-1}$, it was not until a specific search in the region $\sim 2kT_N$ (roughly twice the Brillouin edge magnon energy) was performed that an absorption comparable in strength to the AFMR line was revealed at 154.4 cm^{-1}. The interpretation of the absorption which disappeared at temperatures above T_N was complicated by the fact that it was an electric dipole transition as determined with polarized infrared light and that a large applied magnetic field has no visible effect on the spectrum.

The phenomenological Hamiltonian that describes the two-magnon absorption is

$$\mathscr{H} = \mathbf{E} \cdot \Sigma \pi_{ij} \mathbf{S}_i \cdot \mathbf{S}_j, \tag{AIII.26}$$

where $\mathbf{E}$ is the rf electric field and π is an effective electric dipole moment which couples to the spins. Equation AIII.26 can be treated as a perturbation on the Hamiltonian, Eq. AIII.8, giving rise to absorption by two magnons with equal and opposite $k \neq 0$. This Hamiltonian can be calculated on a microscopic quantum-mechanical basis. In the transition metal fluorides parity is a rigorous quantum number for the single-ion magnetic states and as a consequence, electric dipole transitions are not permitted. Three mechanisms have been proposed to calculate π: a dipole-quadrupole interaction between ions which couples to the spin via the spin-orbit coupling,[17] an off-diagonal exchange coupling,[18] and an electric dipole coupling to magnons via infrared-active phonons which modulate the exchange.[19] All three mechanisms give rise to the general form of Eq. AIII.26. Although it is difficult to accurately calculate the strength of the couplings, the first has been proposed to explain the two-magnon line in FeF_2, while the latter two have been applied to MnF_2. The crystalline symmetry can be used to eliminate many of the components of π in Eq. AIII.26; in the presence of a center of inversion symmetry such as in $RbMnF_3$, π disappears and a two-magnon line is not expected.

The absorption strength is proportional to a wave vector-dependent transition rate weighted by the magnon density of states. Selection rules can be calculated within the Brillouin zone.[20] The result is peaked at values of absorption at the selected points of the Brillouin zone boundary. The experimental observation in both FeF_2[17] and MnF_2,[21] namely that the absorption is independent of the applied magnetic field, is due to the two magnons being excited on opposite sublattices (in Fig. AIII.3 we note that at the zone boundary all of the spin deviation is on one sublattice) or opposite spin wave branches (Fig. AIII.4). In this case the frequency for the model of the last section is

$$\omega_{2mag}\left(ka = \frac{\pi}{2}\right) = [(\omega_e + \omega_a) + \omega_0] + [(\omega_e + \omega_a) - \omega_0]$$

$$= 2(\omega_e + \omega_a), \tag{AIII.27}$$

which is independent of magnetic field. The density of states in both MnF_2 and FeF_2 is similar since they have the same magnetic structure : this function is shown for MnF_2 in Fig. AIII.6. In Fig. AIII.7 we show the absorption spectra for both MnF_2 and FeF_2. In MnF_2 an extended range interaction of the form $\pi = \pi_0 \exp(-|r_i - r_j|/r_0)$ between ions i and j was used to fit the line shape.[21] In both cases, a much weaker absorption was observed with $E \perp c$, peaked at $100 \, cm^{-1}$ for MnF_2 and $\sim 154 \, cm^{-1}$ for FeF_2. These arise from different points in the Brillouin zone. Two-magnon lines have also been observed in CoF_2 at $\sim 120 \, cm^{-1}$ and $MnCO_3$ at $\sim 50 \, cm^{-1}$.[22]

G. CRYSTALLINE ELECTRIC FIELD EFFECTS

The angular momentum states of a free magnetic ion are greatly modified by placing the atom in a crystal lattice in interaction with its environment. In an ionic crystal the atom sits in a crystalline electric field due to its environment which has the symmetry of the atomic site. The low-lying energy states are determined by a consideration of the effects of the combined Stark effect and the spin-orbit coupling on the appropriate free ion wave functions and energy levels.[27]

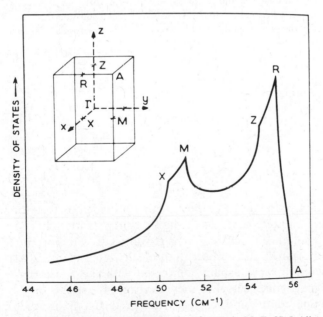

Fig. AIII.6   Relevant portions of the magnon density of states in MnF_2 [S. J. Allen, R. Loudon, and P. L. Richards, *Phys. Rev. Letters* **16**, 463 (1966).]

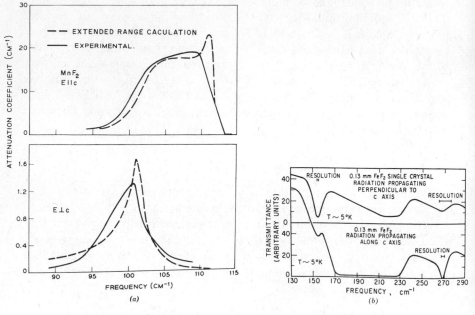

Fig. AIII.7 (a) The observed and calculated line shape in MnF_2. [S. J. Allen, R. Loudon, and P. L. Richards, *Phys. Rev. Letters* **16**, 463 (1966).] (b) The transmission spectra of FeF_2 showing the proximity of the two-magnon lines at 154 cm^{-1} to strong lattice absorptions. [I. F. Silvera and J. W. Halley, *Phys. Rev.* **149**, 415 (1966).]

The full Hamiltonian is

$$\mathcal{H} = \mathcal{H}_F + V \qquad (AIII.28)$$

where

$$\mathcal{H}_F = \frac{-\hbar^2}{2m}\sum_i \nabla_i^2 - \sum_i \frac{Ze^2}{r_i} + \frac{1}{2}\sum_{i \neq j}\frac{e^2}{r_{ij}} + \sum_i \xi_i \mathbf{l}_i \cdot \mathbf{S}_i \qquad (AIII.29)$$

is due to the free ion and V is the potential provided by the atomic environment (sometimes referred to as the ligand field potential). In order to treat V as a perturbation on the free ion eigenvalues it is important to know its relative importance with respect to the last two terms in Eq. AIII.29. In the rare-earth ions we are usually concerned with the f-electrons which give rise to the magnetic effects. For these electrons, the spin orbit term $\sum_i \xi_i \mathbf{l}_i \cdot \mathbf{S}_i$ is large compared to V and we treat V as a perturbation on $\mathcal{H}_F$. On the other hand, in the first transition group we are concerned with d-electrons which usually satisfy the inequality $\xi_i \mathbf{I}_i \cdot \mathbf{S}_i < V < e^2/r_{ij}$, in which case V and the spin-orbit term must be treated together as perturbations on the free ion. The third case is called the strong crystalline field case and applies for $e^2/r_{ij} < V$, a condition which usually arises in covalent complexes.

The low-lying spectrum of angular momentum orientational states which arise from the crystalline field effects determine the magnetic properties of a crystal. A simple example of the use of far-infrared and its interplay with other techniques is in the investigation of cerium magnesium double nitrate (CMN) and cerium zinc double nitrate (CZN). The former is well known to low-temperature physicists for its useful magnetic properties, especially in adiabatic demagnetization for achieving low temperatures. The magnetic properties arise from the $4f^1$-configuration which is split into Kramers doublets. Magnetic relaxation experiments gave estimates for the splitting Δ_1 between the two lowest Kramers doublets to be $\sim 23.6 \, \mathrm{cm}^{-1}$, whereas magnetic susceptibility measurements give values in the range of 23 to $30 \, \mathrm{cm}^{-1}$. A direct measurement was made of its far-infrared spectrum to give a value $25.2 \, \mathrm{cm}^{-1}$.[28] A similar spectrum was found for CZN with a splitting $\Delta_1 = 21 \, \mathrm{cm}^{-1}$ as shown in Fig. AIII.8. The electronic transition was distinguished from the phonon absorption whose onset is indicated by Δ_L by its temperature dependence: as T is raised the Δ_1 transition weakens due to depopulation of the ground state, while the phonon spectrum increases in intensity.

Examples of more complex spectra are erbium, dysprosium, and samarium ethyl sulfate[29], which have many far-infrared absorption lines due to both

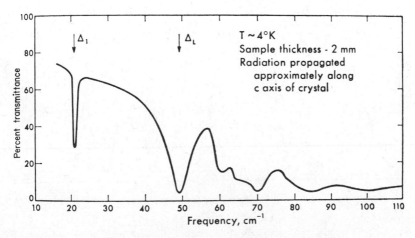

Fig. AIII.8 The far-infrared spectrum of cerium zinc double nitrate (CZN); Λ_1 marks the crystal-field transition, whereas Δ_L indicates the onset of the phonon spectrum which marks higher-frequency electronic transitions. [J. H. M. Thornley, *Phys. Rev.* **132**, 1492 (1963).]

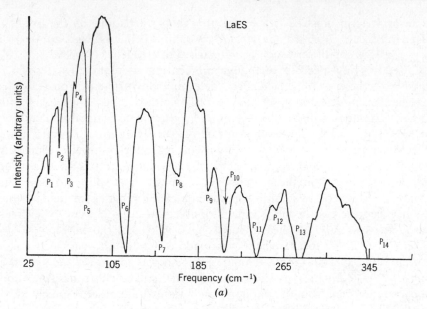

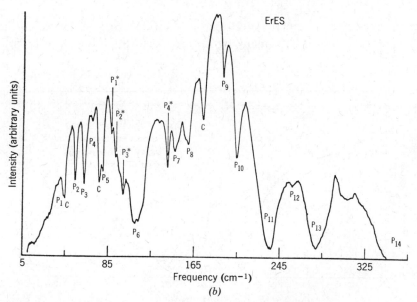

Fig. AIII.9 (a) Unnormalized transmission spectrum of lanthanum ethyl sulfate (LaES) absorption lines. P_1–P_{14} are due to phonon excitation. (b) The spectrum of erbium ethyl sulfate (ErES), showing the phonon lines P_1–P_{14}, P_1–P_4^*, and the crystal-field lines marked C. [J. C. Hill and R. G. Wheeler, *Phys. Rev.* **152**, 482 (1966)]

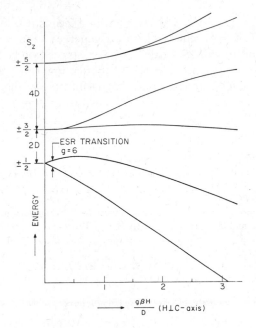

Fig. AIII.10   The splitting of the $S = \frac{5}{2}$ ground state spin manifold of Fe^{3+} in protoheme chloride due to the perturbation $DS_z^2 + g\beta S_{z'}$. The axis of the applied field z' is perpendicular to z, the uniaxial axis of symmetry. [G. Feher and P. L. Richards, *Intern. Conf. Magnetic Resonance Biol. Systems*, Stockholm 1966.]

narrow-line phonon transitions and crystal-field electronic transitions. The separation and identification of the two types of absorptions was accomplished by two techniques. In Fig. AIII.9(a) we show an unnormalized spectrum of lanthanum ethyl sulfate (LaES) taken with an interferometer. Since lanthanum is diamagnetic there are no crystal-field transitions and we see a spectrum due only to phonons. In the spectrum of erbium ethyl sulfate, Fig. AIII.9(b), additional crystal-field lines are observed. Since the phonon transitions are not affected by a magnetic field, the crystal-field transitions were separated by observing their frequency shift in an applied field.

Recently far-infrared techniques have been applied to the study of certain biological systems.[30] The ferric ion Fe^{3+} in protoheme chloride is in a $S = \frac{5}{2}$ spin state. In the absence of spin-orbit coupling the ground state would have a sixfold degeneracy. The interplay of the spin-orbit coupling gives rise to a perturbation of the form DS_z^2. (This is the same form of perturbation that would be responsible for the anisotropy energy on our two-sublattice antiferromagnet model.) This term, when diagonalized within the six S_z states for the spin $\frac{5}{2}$ manifold, partially lifts the degeneracy, giving doublets for $S_z = \pm\frac{1}{2}$, $\pm\frac{3}{2}$, and $\pm\frac{5}{2}$. The addition of a Zeeman term $g\beta\mathbf{H}\cdot\mathbf{S}$ for an applied magnetic field $\mathbf{H}$ removes all the degeneracy as shown in Fig. AIII.10. The zero-field splitting of $2D$ for the $\pm\frac{1}{2} \to \pm\frac{3}{2}$ transition in a powder sample was observed to be 13.9 cm^{-1} at 4.2°K. By heating the sample to 20°K the $\frac{3}{2}$ level was thermally populated and the $4D$ splitting of the $\frac{3}{2} \to \frac{5}{2}$ transition was observed.

REFERENCES

1. A. J. Sievers, Thesis, Univ. of Calif., Berkeley, 1962.
2. R. C. Ohlmann, P. L. Richards, and M. Tinkham, *J. Opt. Soc. Am.* **48**, 531 (1958).
3. P. L. Richards, *Phys. Rev.* **138**, A1769 (1965).
4. H. A. Kramers, *Physica* **1**, 182 (1934).
5. For a comprehensive review and further references see P. W. Anderson, *Solid State Physics*, Vol. 14, F. Seitz and D. Turnbull, Eds., Academic Press, New York, 1963.
6. For a general review of antiferromagnetism see T. Nagamiya, K. Yosida, and R. Kubo, *Adv. Phys.* **4**, 2 (1955).
7. I. F. Silvera, J. H. M. Thornley, and M. Tinkham, *Phys. Rev.* **136**, A695 (1964).
8. M. Tinkham, *Phys. Rev.* **124**, 311 (1961).
9. C. Kittel, *Phys. Rev.* **82**, 565 (1951).
10. T. Nagamiya, *Prog. Theoret. Phys.* (*Kyoto*) **4**, 324 (1951).
11. R. C. Ohlmann and M. Tinkham, *Phys. Rev.* **123**, 425 (1961).
12. A. J. Sievers and M. Tinkham, *Phys. Rev.* **129**, 1566 (1963).
13. M. E. Lines, *Phys. Rev.* **137**, A982 (1965).
14. I. S. Jacobs, S. Roberts, and P. E. Lawrence, *J. Appl. Phys.* **36**, 1197 (1965).
15. A. J. Sievers, III and M. Tinkham, *Phys. Rev.* **124**, 321 (1961).
16. A. J. Sievers, III and M. Tinkham, *Phys. Rev.* **129**, 1995 (1963).
17. J. W. Halley and I. F. Silvera, *Phys. Rev. Letters* **15**, 654 (1965); I. F. Silvera and J. W. Halley, *Phys. Rev.* **149**, 415 (1966); J. W. Halley, *Phys. Rev.* **149**, 423 (1966).
18. Y. Tanabe, T. Moriya, and S. Sugano, *Phys. Rev. Letters* **15**, 1023 (1965); T. Moriya, *J. Phys. Soc. Japan* **21**, 926 (1966).
19. J. W. Halley, *Phys. Rev.* **154**, 458 (1967).
20. R. Loudon, *Adv. Phys.* **17**, 243 (1968).
21. S. J. Allen, R. Loudon, and P. L. Richards, *Phys. Rev. Letters* **16**, 463 (1966).
22. P. L. Richards, *J. Appl. Phys.* **38**, 1500 (1967).
23. P. L. Richards, *J. Appl. Phys.* **35**, 850 (1964).
24. D. Bloor and D. H. Martin, *Proc. Phys. Soc.* (*London*) **78**, 774 (1961).
25. I. F. Silvera, Thesis, Univ. of Calif., Berkeley, 1965.
26. K. Aring and A. J. Sievers, *J. Phys.* **38**, 1496 (1967); S. J. Allen, *J. Appl. Phys.* **38**, 1478 (1967); S. J. Allen, *Phys. Rev.* **166**, 530 (1968).
27. Two excellent sources are C. J. Ballhausen, *Introduction to Ligand Field Theory*, McGraw-Hill, New York, 1962, and M. T. Hutchings, *Solid State Physics*, Vol. 16, F. Seitz and D. Turnbull, Eds., Academic Press, New York, 1964; a reference on more advanced techniques is B. R. Judd, *Operator Techniques in Atomic Spectroscopy*, McGraw-Hill, New York, 1963.
28. J. H. M. Thornley, *Phys. Rev.* **132**, 1492 (1963).
29. J. C. Hill and R. G. Wheeler, *Phys. Rev.* **152**, 482 (1966).
30. G. Feher and P. L. Richards, *International Conference on Magnetic Resonance in Biological Systems*, Sweden, 1966.

APPENDIX IV

FAR-INFRARED SPECTROSCOPIC STUDIES OF SEMICONDUCTORS

R. KAPLAN

Naval Research Laboratory, Washington, D.C. 20390

Recent years have witnessed an increasing interest in the optical properties of semiconductors in the far-infrared region of the electro magnetic spectrum. To an extent this interest has been spurred by the need for measurements of phenomena that occur at far-infrared wavelengths and by the increasing availability of high-quality material for samples. The greatest impetus, however, has come from the technological advances that have made spectroscopy in the far-infrared an almost routine procedure. Fourier transform interferometers, cryogenic detectors, laser sources, interference filters, to name but a few of these advances, are now in common use. Their application to spectroscopic studies of semiconductors is widespread.

This appendix presents a discussion of some recent and current areas of interest in the far-infrared spectroscopy of semiconductors. Since this is a very large and varied subject, the completeness of a review article has not been attempted. Rather, several general areas of research have been outlined, and within each area a number of specific investigations have been described in some detail. The topics have been selected to demonstrate both the application of standard techniques, such as transmission and reflection spectroscopy, and the use of specialized methods for particular problems.

The first section is concerned with far-infrared studies of optically active lattice vibrational modes in semiconductors, and the second with some effects due to the presence of impurities. In these research areas the spectral regions of interest are determined by the natural frequencies of the particular phenomena in the chosen materials. The situation is somewhat different for cyclotron resonance investigations, which are treated in the third section.

There, the frequency or wavelength of resonance depends on the magnitude, and sometimes also the direction, of the applied magnetic field, and so may be tuned over various spectral regions. However, the quality or even the nature of the information obtained may depend on the particular spectral region in which the resonance is observed. Finally, the fourth section deals with interactions that can occur between lattice, free carrier, and cyclotron resonance modes in polar semiconductors. These effects are observed in the vicinity of the normal lattice modes, but they may also be studied in other regions of the spectrum.

A. OPTICAL LATTICE VIBRATIONS AND FREE CARRIER EFFECTS

The optical lattice vibrational frequencies of a number of semiconductors fall in the far-infrared. With the advent of modern techniques, far-infrared measurements of lattice transmission and reflection have become fairly routine. However, optical effects due to free carriers become more pronounced at longer wavelengths, and for high-temperature measurements or impure samples these effects must be taken into account. In this section some recent, and in many respects typical, studies based on transmission and reflection measurements are first described. Following this a description is given of the technique of emission spectroscopy and its application to lattice vibration studies.

The development of high-intensity, monochromatic laser sources of visible and near-infrared radiation has led to increased interest in Raman spectroscopy. This technique can provide information on a variety of effects, including far-infrared lattice modes. However, since it is basically concerned with the observation of fine structure associated with the inelastic scattering of visible or near-infrared photons, Raman spectroscopy will not be considered further here.

a. Transmission and Reflection Measurements

In a typical study of lattice vibrations in a semiconductor, monochromator techniques were used to study the reflection from and transmission through bulk samples of indium antimonide in the wavelength range 20–200 microns.[1] The observed transmission and reflection spectra are shown in Figs. AIV.1 and AIV.2. The reflection spectra show the usual reststrahlen band, and in addition an onset of high reflection at longer wavelengths that is associated with the presence of free carriers. From the transmission spectra it is evident that the sample is much too thick for accurate measurement of the lattice absorption. However, an overtone of this absorption is visible and allows an accurate determination of the transverse optical phonon frequency. In the wavelength region between the reststrahlen band and about 120 microns,

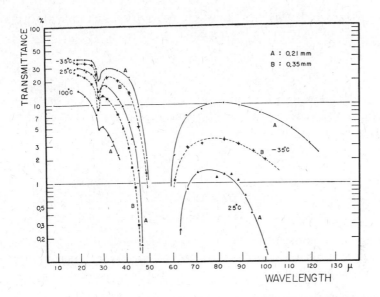

Fig. AIV.1   Transmission of InSb crystals at −35, 25, and 100°C. [H. Yoshinaga and R. A. Oetjen, *Phys. Rev.* **101**, 526 (1956).]

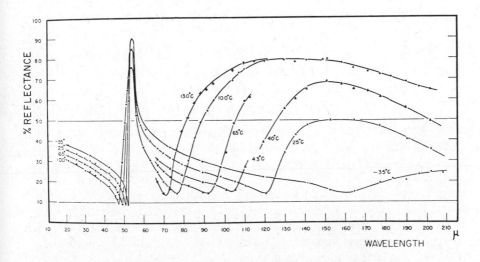

Fig. AIV.2   Reflectivity of InSb crystal between 20- and 200-μ wavelength at various temperatures. [H. Yoshinaga and R. A. Oetjen, *Phys. Rev.* **101**, 526 (1956).]

the transmission and reflection data were used to estimate the absorption constant α according to the following equation:

$$T = \frac{(1 - R)^2 \exp(-\alpha t)}{1 - R^2 \exp(-2\alpha t)}. \tag{AIV.1}$$

This equation takes into account multiple reflections inside the sample; t is the sample thickness. In the same spectral region, the optical constants n and K were then calculated from[†]

$$K = \frac{\alpha \lambda}{4\pi}; \tag{AIV.2}$$

$$R = \frac{(n - 1)^2 + K^2}{(n + 1)^2 + K^2}. \tag{AIV.3}$$

Since the far-infrared spectral behavior was thought to be due to free carriers, the optical constants were used to calculate the free carrier concentration N and mobility μ, as follows:

$$nKv = \frac{Ne\mu}{1 + (v/\gamma)^2}; \tag{AIV.4}$$

$$n^2 - K^2 = \varepsilon_0 - \frac{Ne^2}{\pi m^*(v^2 + \gamma^2)}. \tag{AIV.5}$$

The quantities e and m^* are the magnitude of the electron charge and the free carrier effective mass, respectively, v is the frequency of the infrared radiation, and ε_0 the static dielectric constant in the absence of free carriers. The damping constant or inverse relaxation time $\gamma = 1/\tau$ is given by

$$\gamma = \frac{e}{2\pi m^* \mu}.$$

Calculation of N and μ by Eqs. AIV.4 and AIV.5 yielded values which varied with wavelength just beyond the reststrahlen band, but became independent of wavelength for $\lambda \gtrsim 90$ microns. In the latter range, 90–120 microns, the optical constants were therefore assumed to be determined principally by the free carriers. With N and μ obtained in this manner for each sample temperature, the free carrier reflection was calculated over the entire wavelength range beyond the reststrahlen band. The degree to which the calculations were consistent with the observed reflectivity is shown in Fig. AIV.3. The onset of strong free carrier reflection was accounted for accurately, although the magnitudes were in error. This entire procedure constituted

---

[†] See Eqs. 12.14 and 12.12 (In Chapter 12, K is replaced by the symbol κ.)

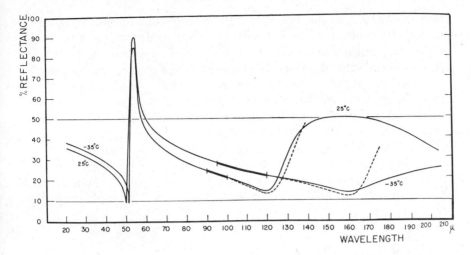

Fig. AIV.3   Comparison of calculated (*broken lines*) and observed (*solid lines*) reflectivity of InSb crystal at -35 and $25°C$. Heavy parts of the solid lines indicate the values used in the calculation. [H. Yoshinaga and R. A. Oetjen, *Phys. Rev.* **101**, 526 (1956).]

a method of fitting the data to a free carrier model, starting from transmission and reflection results in the wavelength range 90–120 microns. If transmission data had been obtainable at longer wavelengths, a more direct procedure could have been used.

Examination of the $25°C$ reflectivity data of Fig. AIV.2 has shown[2] that the assumption of a single classical oscillator to represent the reststrahlen band is valid. Values deduced for the oscillator constants were $\gamma = 5.4\,\text{cm}^{-1}$ and $v_0 = 177.6\,\text{cm}^{-1}$. The maximum absorption constant at 56.4 microns wavelength was estimated from the theory as $1.7 \times 10^4\,\text{cm}^{-1}$. From this estimate it is evident that in order to resolve the lattice absorption peak in a transmission experiment, a sample thickness of about 1 micron would be required.

More recently, the far-infrared optical properties of the same samples used in the work described above have been remeasured[3] by newer techniques, which eliminated second- and higher-order radiation from the monochromator gratings. These measurements have removed the discrepancies between experiment and theory which appeared in the earlier work at long wavelengths.

Another example of the application of far-infrared spectroscopy to the study of semiconductor lattice properties is provided by recent transmission measurements[4] on lead telluride. This material, as well as the related salts lead sulfide and lead selenide, are expected to have large values of the static dielectric constant ε_0 due to their large ionicity. Measurements of ε_0 by electrical and neutron scattering techniques have yielded values in the range 200–500. An optical measurement of ε_0 is of interest, since electrical measurements may be affected by lattice–free carrier interactions that are not entirely understood at present. In the optical measurements, the far-infrared transmission by epitaxial lead telluride films approximately 0.2 microns thick was investigated at sample temperatures of about 4.2°K. For this purpose a Fourier transform interferometer was used in the frequency range 7–70 cm^{-1}, or about 140–1400 microns wavelength. A sharp minimum in the transmission was observed at a frequency of 31.5 cm^{-1}, with a half width of 0.8 cm^{-1}. This minimum was identified with the transverse or fundamental lattice frequency v_{TO} of lead telluride. Application of the Lyddane-Sachs-Teller relation

$$\frac{v_{LO}^2}{v_{TO}^2} = \frac{\varepsilon_0}{\varepsilon_\infty},$$

using previously determined values for the optical dielectric constant ε_∞ and the longitudinal lattice frequency v_{LO}, yielded the result $\varepsilon_0 = 397 \pm 30$. It is interesting to note that the lattice frequencies are in the range of the plasma and cyclotron resonance frequencies for reasonable values of carrier concentration and magnetic field. Under the proper experimental conditions, interactions of these modes may occur. Interactions of this type will be discussed in Section D.

b. Measurement of Emission

A typical semiconductor sample will absorb a fraction A_λ of radiation of wavelength λ incident on it, while reflecting a fraction R_λ and transmitting a fraction T_λ. The sum of these quantities, respectively the spectral absorptance, reflectance, and transmittance, is unity:

$$A_\lambda + R_\lambda + T_\lambda = 1. \tag{AIV.6}$$

The thermal radiation emitted by a sample at temperature T is given by Kirchhoff's law as follows:

$$W'_\lambda(T) = E_\lambda W_\lambda(T) = A_\lambda W_\lambda(T), \tag{AIV.7}$$

where E_λ is the spectral emittance of the sample, and W_λ the spectral emissive power of a blackbody, that is, the emitted power per unit area per unit wavelength. At normal incidence, the emittance or absorptance is given approximately by

$$E_\lambda = A_\lambda = \frac{(1 - r^2)[1 - \exp(-\alpha t)]}{1 - r^2 \exp(-\alpha t)}, \tag{AIV.8}$$

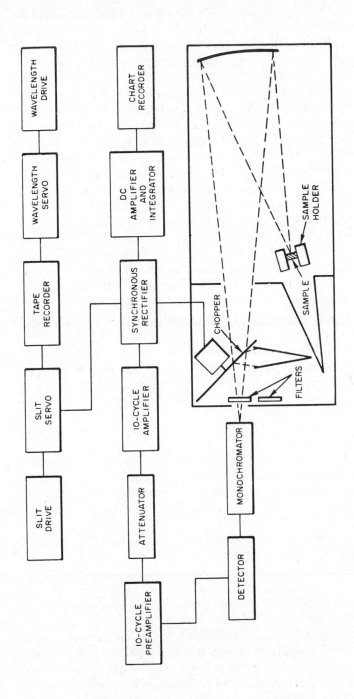

Fig. AIV.4 Block diagram of system used for measuring emittance. The sample chamber, monochromator, and detector were maintained at $25 \pm 0.05°C$ in an evacuated isothermal chamber. [D. L. Stierwalt and R. F. Potter, Rept. No. 630, Naval Ordnance Laboratory, Corona, 1965.]

where α is the absorption constant, t the sample thickness, and r the Fresnel reflection coefficient. (R_λ and T_λ may similarly be written in terms of α and r.)

In a recent series of experiments,[5,6] the spectral emittance of a number of semiconductors has been studied over a wide temperature range to wavelengths as long as 125 microns. A simple analysis of the experimental technique can be given as follows[5] (see Fig. AIV.4). The radiation emitted by a sample S, or a reference blackbody at the position of the sample, is analyzed by a monochromator and sensed by a detector. This radiation is chopped by an opaque chopper wheel, C. The signal S_s, with the sample in place, is proportional to the difference in thermal radiation sensed by the detector for the open and shut positions of the chopper wheel. For normal operation the entire system is kept at a constant temperature, with the exception of the sample which may be either warmer or cooler than its surroundings. With the chopper open, S_s receives the contributions $E_s W_s$ from the sample directly, and $(T_s + R_s)W_r$ from background radiation transmitted through the sample and reflected from its surface. (W_s and W_r are the emissive power of a blackbody at the temperature of the sample and of the reference background, respectively.) With the chopper closed, contributions $E_c W_c$ due to emission from the chopper, and $R_c W_r$ from background emission reflected by the chopper, are received. The net signal is thus

$$S_s = K\{E_s W_s + (T_s + R_s)W_r - E_c W_c - R_c W_r\}, \qquad \text{(AIV.9)}$$

where K is a constant. Using Eq. AIV.6 and $T_c = 0$ for the opaque chopper blade, Eq. AIV.9 may be rewritten

$$S_s = K\{E_s(W_s - W_r) - E_c(W_c - W_r)\}. \qquad \text{(AIV.10)}$$

The last term in Eq. AIV.10 may be eliminated experimentally by using a highly reflecting chopper blade for which $E_c \to 0$, and maintaining the chopper at the background temperature to minimize $(W_c - W_r)$. Thus

$$S_s = KE_s(W_s - W_r). \qquad \text{(AIV.11)}$$

If the sample is now replaced by a blackbody at the same temperature, the observed signal is

$$S_B = K(W_s - W_r). \qquad \text{(AIV.12)}$$

In practice the signals S_s and S_B can be compared automatically to yield directly the spectral emittance E_s of the sample.

A wide variety of effects can be studied by this technique with a single sample and experimental setup. Reference to Eq. AIV.8 shows that in regions of high transmission $\alpha t \ll 1$, the observed quantity $E_s \to \alpha t$. Thus extremely weak absorption processes, that is, multiple-phonon absorption, can be observed. For opaque samples, $\alpha t \gg 1$, $E_s \to (1 - r^2) = (1 - R_s)$. In this

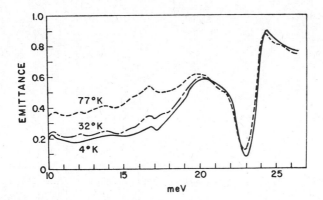

Fig. AIV.5   Spectral emittance of InSb at 4.2, 32, and 77°K. Sample was *n*-type, 1.23 mm thick, with a carrier concentration of 7×10^{13} cm^{-3}. [D. L. Stierwalt, *J. Phys. Soc. Japan* **21**, Suppl., 58 (1966).]

limit, which is pertinent to regions of strong lattice absorption, the signal readily yields the reflectance of the sample. In the intermediate region, $\alpha \sim 1$, structure in α is observable and quantitative results can be obtained through straightforward calculation.

Typical emittance spectra[6] for an indium antimonide sample are shown in Fig. AIV.5 for three different temperatures. The reststrahlen band at 23 meV (1 meV $\sim$ 8.07 cm^{-1}) appears in the data as $(1 - R)$, and as such is seen to have the usual shape of a lattice reflection band. At lower energy, emittance peaks are observed which are due to absorption by multiple phonon summation and difference bands. (One of these bands is also visible in the data at 25.4 meV.) In Fig. AIV.6 the spectral emittance in the reststrahlen region for seven compound semiconductors at 77°K is shown. The information obtained by experiments such as these complements that obtained by the more usual optical techniques described previously. The full potential of the emittance measurement technique has not as yet been explored, although attempts to apply it to magnetooptical experiments are in progress.[7]

B. IMPURITY ABSORPTION AND PHOTOCONDUCTIVITY

The presence of impurity atoms in a semiconductor produces a change in the electronic structure of the host crystal and is thereby responsible for many of its electrical and optical properties. For this reason, much effort has been directed toward gaining an understanding of the energy states associated with impurity atoms in semiconductors, in both the high-concentration region, where the interaction between impurities must be con-

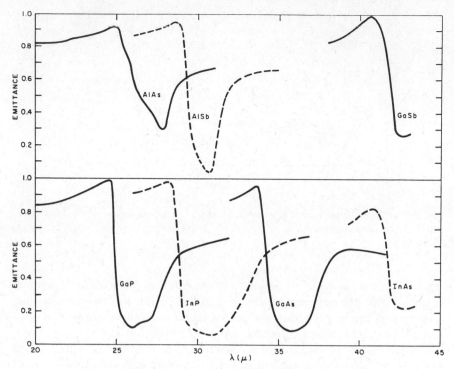

Fig. AIV.6   Spectral emittance at 77°K for seven compound semiconductors in their reststrah-len regions. In these regions the emittance has the form $(1 - R)$. [D. L. Stierwalt and R. F. Potter, Rept. No. 630, Naval Ordnance Laboratory, Corona, 1965.]

sidered, and the low-concentration limit, where they may be treated as non-interacting. In this section, some recent research in far-infrared impurity absorption and photoconductivity in both concentration domains is described.

a. Energy States of Impurity Atoms in Germanium and Silicon

Individual, widely separated impurities in semiconductors give rise to series of discrete energy levels in the otherwise forbidden energy gap between the conduction and valence bands. The impurity ground states have been of particular technological interest in view of their role in the performance of extrinsic photoconductive infrared detectors. Also of interest in this connection are the photon cross sections of impurity atoms, which are central to the efficiency of incident radiation in producing a photocurrent. Theoretical attempts have been made to calculate the ground and excited state energies and impurity wave functions for various systems, while infrared transmission and photoconductivity measurements have sought to obtain these quantities directly. Most recently, magnetic fields and uniaxial stress have been used to study fine details of the "excitation spectra" of impurities, that is, the sum of transitions from ground to excited states observed in transmission or photoconductivity experiments.

The usual theoretical[8] approach to the calculation of impurity wave functions and energy levels is made within the framework of the effective mass approximation. In this approximation the charge carrier (hole or electron) is pictured as moving in the periodic potential of the pure host crystal under the influence of the perturbing potential of magnitude $e^2/\varepsilon_0 r$ due to the impurity ion. (The quantities ε_0 and r are the static dielectric constant of the host crystal and the separation of the charge carrier and impurity ion, respectively.) Recent effective mass calculations[9,10] have yielded excited state energies in good agreement with observed values, but the agreement is much poorer for the ground state energies. This is to be expected, since the ground state wave functions have large amplitude in the immediate neighborhood of the impurity, where the effective mass approximation is not valid. In particular, this means that the theories cannot account for the observed "chemical shifts," that is the difference in ionization energy observed in a given host crystal for various impurities from the same group in the periodic table. Thus an entirely satisfactory theory of impurity states, particularly with respect to acceptor impurities, has not yet been achieved. Recently an attempt has been made to calculate photoionization cross sections using the quantum defect method. In this approach[11] the effective mass theory is assumed to be valid at large distances from the impurity ions with the perturbing potential asymptotically approaching $-e^2/\varepsilon_0 r$. Rather than solving for the energy eigenvalues, experimental values were used to determine the asymptotic form of the wave functions for particular impurities. Knowledge of the asymptotic form is sufficient for obtaining quantitative results for the spectral dependence and magnitude of the photoionization cross section. Results were in good agreement with experiment for a number of different impurities in silicon, germanium, and gallium arsenide.

In a recent experimental study[12] of impurity energy levels, photoconductive, optical, and electrical measurements were made on germanium crystals doped with beryllium. A typical absorption spectrum is shown in Fig. AIV.7. The absorption peaks are due to transitions from the ground state to various excited states, with the principle transitions identified by letters. It is found that the energy spacings between the observed peaks and the D-peaks are almost equal to the corresponding spacings for other acceptor impurities in germanium. This result is not unexpected, since the energy spacings in question represent the energy differences between various

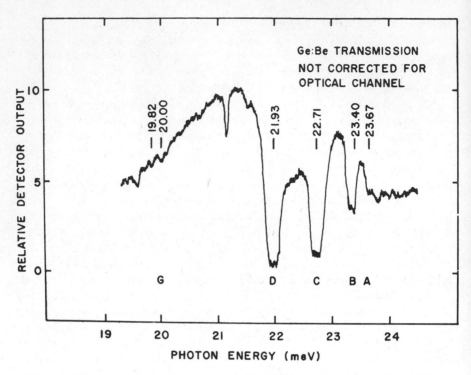

Fig. AIV.7 The absorption spectrum of beryllium-doped germanium in the vicinity of the impurity ionization energy in zero magnetic field at 4.2°K. [H. Shenker, E. M. Swiggard, and W. J. Moore, *Trans. Met. Soc. AIME* **239**, 347 (1967), by permission of American Institute of Mining, Metallurgical, and Petroleum Engineers.]

excited states. The excited state wave functions, having small amplitudes in the immediate vicinity of an impurity ion, are not sensitive to the chemical identity of the ion. Thus the theoretical result[10] that the neutral impurity ionization energy is 2.53 meV greater than the energy of the D-peak, is reliable. From data similar to that of Fig. AIV.7, the ionization energy of neutral beryllium in germanium determined in this way was 24.3 ± 0.1 meV. Application of a magnetic field to the sample was observed to split the absorption peaks in a manner that will be discussed presently. Photoconductivity measurements showed that beryllium-doped germanium is a sensitive detector of infrared radiation in the wavelength range 15–52 microns.

Impurity energy levels in germanium and silicon have now been studied rather extensively by the techniques of transmission and photoconductive spectroscopy. The excitation spectra show a close correlation with the results of effective mass theories for the excited state energies. However, the theories

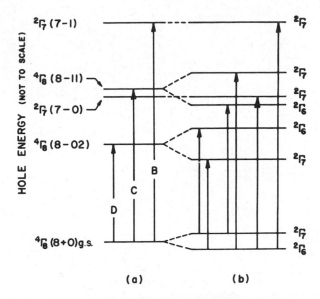

Fig. AIV.8 (a) Energy states of a Group III impurity in germanium in zero stress. (b) Behavior of acceptor states under uniaxial stress along a ⟨100⟩ direction. Also shown are the observed transitions B, C, and D for zero stress and the theoretically allowed transitions for light polarized with its electric field vector parallel to the applied stress. [R. L. Jones and P. Fisher, *Solid State Commun.* **2**, 369 (1964), by permission of Pergamon Press.]

can be further tested by experiments that reveal the degeneracy of the various energy levels, and level assignments can be made more convincingly if they can be shown to yield the expected degeneracy. Application of a magnetic field or of a uniaxial stress has been used to study the degeneracy of the various impurity levels. In a recent investigation[13] of the excitation spectrum of the Group III acceptor impurity thallium in germanium, a uniaxial stress was applied along various crystallographic directions and the infrared transmission observed with polarized light. The predicted energy levels and transitions in the absence and presence of stress along a ⟨100⟩ direction are shown in Fig. AIV.8 for light polarized along ⟨100⟩. According to theory[8] for Group III impurities in germanium, a fourfold degenerate state splits into two twofold degenerate states under stress along ⟨100⟩, while a twofold degenerate state remains unchanged. The predicted degeneracy is indicated by the first numeral (superscript) in the label of a given level in Fig. AIV.8. It is evident that transitions between fourfold degenerate states such as that indicated by arrow-D can yield as many as four components under stress, while transitions to a twofold degenerate final state can yield only two. In order to observe all possible components, infrared radiation of

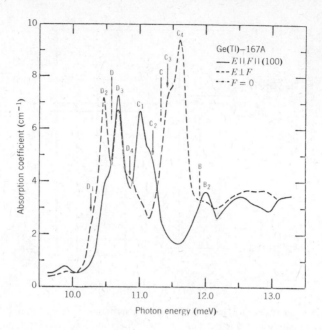

Fig. AIV.9   The stress-induced excitation spectrum of thallium impurities in germanium for uniaxial stress along a ⟨100⟩ direction. Sample temperature 8°K; acceptor concentration $4 \times 10^{14}\,\mathrm{cm}^{-3}$. [R. L. Jones and P. Fisher, *Solid State Commun.* **2**, 369 (1964), by permission of Pergamon Press.]

various polarizations must be used. A typical experimental result is shown in Fig. AIV.9. It may be seen that the C-line splits into a quartet, thus showing that the final state for this transition is fourfold degenerate in the absence of stress. This indicates that the final state in question must be $^4\Gamma_8$ (8–11) and not the nearby $^2\Gamma_7(7-0)$. (A discrepancy was noted in the polarization of one of the components, however, which requires further elucidation; see below.) The interpretation of the data of Fig. AIV.9 is consistent with the assignment of twice the energy separation of the B- and B_2-peaks to the splitting of the ground state, since the final state of the B-transition is only twofold degenerate in the absence of stress. These conclusions provide a sampling of the kind of detailed information which may be obtained from observation of the effects of uniaxial stress on impurity excitation spectra.

Quite recently, this type of investigation has been expanded to include gallium and indium as well as thallium impurities in germanium. These experiments[14] were performed with an apparatus that allowed the application of calibrated variable stress on the samples. It was thus possible to

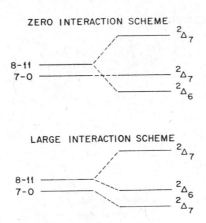

Fig. AIV.10   Schematic diagram indicating two possible models for the $\langle 100 \rangle$ strain splitting of the excited states associated with the C-transition. [D. H. Dickey and J. O. Dimmock, *J. Phys. Chem. Solids* **28**, 529 (1967), by permission of Pergamon Press.]

follow individual stress-induced components back to their zero-stress energies, thereby facilitating the identification of the many components. Furthermore, use was made of the measured energy splitting of the impurity ground state under calibrated uniaxial stress to obtain values of the valence band deformation potential parameters of the host germanium crystal. Finally, the C-transition discussed in the previous work was reexamined, in particular with respect to disagreement between theory and experiment regarding the polarization of one of the components. It was shown that the data are consistent with the inclusion in the theory of a strain-induced interaction between impurity states. This interaction becomes important if its strength is comparable to the energy separation between the states at zero strain. The interacting components must have the same symmetry in the presence of strain. In the case of the C-transition (see Fig. AIV.8), the large-interaction model which gives agreement with experiment is shown in Fig. AIV.10.

b. Hydrogenic Impurity States in Indium Antimonide

An interesting impurity system that has been a recent subject for study occurs in *n*-type indium antimonide. Due to the large dielectric constant and small effective mass in this material, the impurity ionization energy or "effective Rydberg" Ry^* is extremely small, and the impurity ground state radius r is correspondingly large. According to the simple hydrogenic atom model, Ry^* and r are given by

$$Ry^* = \frac{m^* e^4}{2\hbar^2 \varepsilon_0{}^2} = 0.7 \text{ meV}, \qquad (\text{AIV.13})$$

$$r = \frac{\hbar^2 \varepsilon_0}{m^* e^2} = 5.4 \text{ microns}, \qquad (\text{AIV.14})$$

where the values $m^* = 0.015\, m_0$ and $\varepsilon_0 = 17$ have been used. In fact, r is so large that overlap of neighboring impurity ground states occurs even in the purest material available, and the donor electrons are not "frozen out" at low temperatures but remain in the conduction band. However, it is known[15] that application of a magnetic field **H** causes the impurity state to shrink, particularly in the plane transverse to **H**. In moderate fields, the overlap can be reduced sufficiently to allow freezeout to occur at 4.2°K. Thus the magnetic field changes the situation from one of strongly interacting impurities to one of largely independent impurity atoms. But the field also produces an orbital quantization, and in the transverse plane the conduction band continuum is replaced by a series of discrete energy states or Landau levels, of energy $(N + \frac{1}{2})\hbar\omega_c$. Here N takes the values $0, 1, 2, \cdots$; ω_c is the cyclotron frequency

$$\omega_c = \frac{eH}{m^* c}. \qquad (\text{AIV.15})$$

For sufficiently large magnetic fields, the ratio γ of the zero-point energy to the ionization energy can be made larger than unity. Thus

$$\gamma \equiv \frac{\frac{1}{2}\hbar\omega_c}{Ry^*} = 4.66H \times 10^{-4} \, \text{G}^{-1}, \qquad (\text{AIV.16})$$

with $m^* = 0.015 m_0$ and $\varepsilon_0 = 17$ for indium antimonide. This requires fields greater than about 2000 G. (To satisfy this condition for hydrogen atoms, a field of about 2.4×10^9 G would be necessary.) Detailed calculations[16] of the impurity energy levels with $\gamma \to \infty$ yield the result

$$E_{00} = \tfrac{1}{2}\hbar\omega_c - I(H) \qquad (\text{AIV.17})$$

for the ground state, with $I(H)$, the ionization energy in the field H, given by the theory, and

$$E_{N,n} = (N + \tfrac{1}{2})\hbar\omega_c - \frac{Ry^*}{n^2}, \qquad n = 1, 2, \cdots \qquad (\text{AIV.18})$$

for the excited states. This energy level scheme is shown in Fig. AIV.11. It is evident that the excited states associated with each Landau level form a

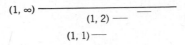

Fig. AIV.11   Energy level scheme for the hydrogenic impurity atom in InSb. The levels are identified by the quantum numbers (N, n) as in Eqs. AIV.17 and AIV.18.

Balmerlike series of levels, and that the first excitation spectrum, that is, transitions from the ground state to the first such series of levels, occur at the energies

$$hv = I(H) - \frac{Ry^*}{n^2}, \qquad n = 1, 2, \cdots. \qquad \text{(AIV.19)}$$

The required wavelengths for observation of these states are in the submillimeter spectral region. Some of the transitions have been observed[17] by means of Fourier transform spectroscopy using a photoconductivity technique; typical spectra are shown in Fig. AIV.12. It is found that the quantity Ry^* in Eqs. AIV.18 and AIV.19 actually varies with magnetic field for finite values of γ. However, for a given field, the observed excited state energies do in fact form a Balmerlike series below the Landau level. A somewhat similar impurity system should occur in other low-mass, high-dielectric constant semiconductors which can be obtained in a sufficiently pure state. It should be pointed out that the experiment just described could not have been performed in a different spectral region, because the magnetic field dependence of the transitions is quite weak.

Several years ago it was suggested that indium antimonide in a magnetic field at low temperature should act as a far-infrared photoconductive

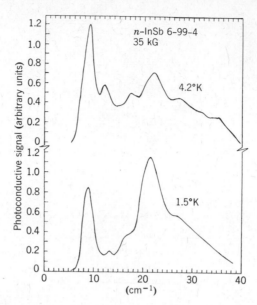

Fig. AIV.12   Photoconductivity spectra at long wavelengths for an InSb sample in a field of 35 kG. [R. Kaplan, *J. Phys. Soc. Japan* **21**, Suppl., 249 (1966).]

detector. In large fields the electrons would be frozen out, and they could be ionized by incident radiation. Detectors[18,19] built along these lines turned out to have their peak sensitivity in fields smaller than 10 kG, despite the fact that larger fields would produce considerably more freezeout. Furthermore, they did not exhibit the low-energy cutoff characteristic of photoconductive detectors. It has since been shown[20] that this so-called "Putley detector" operates as an electron bolometer; that is, the absorbed radiation causes a rise in the average energy of the conduction electrons, affecting their mobility and thus modulating the detector resistance. The role played by the magnetic field is basically one of impedance matching the detector to a preamplifier; a combination of magnetoresistance and partial freezeout raises the resistance of the detector element to a few thousand ohms from a few ohms in the absence of a field. The Putley detector is extremely useful in the far-infrared. The photoconductive response shown in Fig. AIV.12 more nearly approximates that expected from the photoionization detector proposed at the beginning of this paragraph. However, the large magnetic field leads to high noise, and the transitions responsible for the photoconductivity are only weakly allowed. Thus the response shown in Fig. AIV.12 is not particularly suited for detector applications.

c. Far-Infrared Resonant Absorption Due to Donor Impurity Pairs in
   Silicon

The research described thus far in this section has been concerned with
effects due to impurities acting independently of each other. With increasingly
heavy doping of a host semiconductor, interactions between impurity atoms
become important. For example, dc resistivity measurements on heavily
doped samples at low temperatures reveal two general mechanisms for
electrical conduction. In the "intermediate" region of moderate doping,
the impurities are situated close to one another, but not so close that ground
state overlap would occur. The dc conductivity is then due to a phonon-
assisted hopping of charge carriers between neighboring impurities. In
more heavily doped samples overlap does occur, resulting in the formation
of an impurity conduction band, and metallic conduction results. Recent
theoretical and experimental work on *photon*-assisted hopping in the intermediate concentration region has shown that this process can result in
resonant far-infrared absorption under the proper conditions.

The research on photon-assisted hopping has rested on the localized donor
pair model. In this model a neutral donor and an ionized (and therefore
positively charged) donor are both located near the site of an ionized nega-
tively charged acceptor. The two donors are at points of different potential
energy since their distances from the acceptor are generally different. In
the theoretical treatment[21,22] the two lowest energy states, corresponding
to localization at one or the other of the donors, were calculated. The energy
difference between these states corresponds to radiation in the far-infrared
spectral region. Absorption constants for photon-induced transitions
between these states were obtained for particular acceptor and donor
concentrations in silicon and germanium. It was found that the total absorption constant has two major contributions and may be written

$$\alpha_{total} = \alpha_1 + \alpha_2. \qquad \text{(AIV.20)}$$

The first contribution comes from direct photon absorption without phonon
participation, the second from direct photon absorption with emission of
one longitudinal acoustic phonon. (In an indirect process the photon
absorption is phonon-assisted; that is, the photon energy is smaller than
the energy difference between the impurity states.)

The experiments[23] were performed at 2.5°K in the wavelength range
100 microns–1 mm on partially compensated silicon samples having donor
concentrations in the range 1.4–$3.5 \times 10^{17}\,\text{cm}^{-3}$. An interferometric
spectrometer modified for low-temperature solid state studies was employed.
By measuring the reflection loss from the samples, it was possible to determine the absorption coefficient from transmission measurements. Experi-

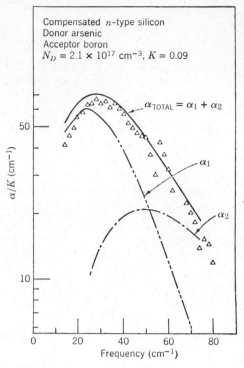

Fig. AIV.13   Theoretical and experimental far-infrared absorption by localized donor pairs in a silicon sample. The separate contributions to the total absorption constant are indicated. [L. J. Neuringer, R. C. Milward, and R. L. Aggarwal, *J. Phys. Soc. Japan* **21**, Suppl., 582 (1966).]

mental results obtained for a typical sample are shown in Fig. AIV.13, together with theoretical predictions of α_1, α_2, and α_{total} computed for this sample. The agreement between theory and experiment is excellent considering that no adjustable parameters are involved.

C. Cyclotron Resonance Investigations

Cyclotron resonance measurements are of great importance in the study of semiconductors, because they yield directly the parameters which characterize the motion and scattering of free charge carriers and determine many features of the band structure. Since the resonance angular frequency ω_c varies with the magnitude H of the applied magnetic field as

$$\omega_c = \frac{eH}{m^*c},$$

(AIV.21)

the resonance may be studied in various spectral regions, depending on the effective mass of the free carriers and the choice of magnetic field. The early experiments were performed at wavelengths in the centimeter range using microwave cavity spectrometers. This technique is severely limited, however, by the need for satisfying the condition $\omega_c\tau > 1$. (The quantity $\omega_c\tau$ is approximately equal to $H_0/\Delta H$, where H_0 is the field at resonance and ΔH the absorption line width.) Except for a few cases such as germanium and silicon, the scattering time τ of the carriers in the samples of interest is too small to yield $\omega_c\tau > 1$ for centimeter wavelengths.

In order to extend the usefulness of cyclotron resonance measurements to semiconductors with relatively low carrier mobility or scattering time, it is necessary to perform the experiments at shorter wavelengths and higher fields. Two different approaches have been used to achieve this end. With the latest developments in microwave equipment it has been possible to extend waveguide and cavity techniques to wavelengths as short as 2 mm. However, this appears to be the limit of usefulness for cavity microwave spectrometers. The second approach has entailed the use of optical infrared methods combined with pulsed or very high sustained magnetic fields. Most recently, the use of far-infrared techniques has filled in the remaining wavelength gap, making it possible to observe cyclotron resonance over a continuous range of nearly four decades of the electromagnetic spectrum.

Aside from the necessity of satisfying the $\omega_c\tau$ criterion, there are various reasons for selecting a particular spectral region in which to observe cyclotron resonance in a given sample. Some of these reasons will be considered in this section in the course of descriptions of a number of experimental investigations. A very basic reason may be simply the need to avoid spectral regions where there are competing processes. For example, in materials which cannot be obtained with low carrier concentrations, free carrier reflection and absorption may dictate avoidance of the long wavelength region.

The cyclotron resonance investigations described below have been grouped according to the general nature of the techniques used. Experiments employing a mixture of microwave and far-infrared optical components will be discussed first. This will be followed by a discussion of purely optical investigations. Studies using microwave techniques exclusively will not be considered.

a. Cyclotron Resonance Studies Using Mixed Microwave-Optical Techniques

The Group VI elemental semiconductor tellurium has recently been a subject of theoretical[24] and experimental[25] interest. Tellurium atoms crystallize along circular helices or "chains;" the axes of these chains are parallel and form a hexagonal array. Attempts have been made to calculate the energy bands corresponding to this structure, but the lack of experimental data has been a hampering factor. Early cyclotron resonance results[26] obtained at a wavelength of 4 mm have recently been refined by experiments in the submillimeter spectral region.[25] In these experiments, CSF carcinotron sources were used at wavelengths between 500–1000 microns. The microwave output of these tubes was carried to cooled samples via oversize circular waveguide or light-pipes. In one set of experiments, radiation transmitted by the samples was detected by a cooled germanium bolometer. Another type of measurement involved monitoring the sample absorptivity by means of a bolometer in thermal contact with the sample. Although the carcinotron is a wavelength-tunable source, its power output varies rapidly and erratically with tuning. Thus it was necessary to perform the measurements at fixed wavelength by varying the magnetic field.

Since the tellurium samples used were p-type, the single absorption line observed was associated with the cyclotron resonance of holes. From the magnetic field strength at resonance, the experimental effective mass m^* was determined. It was found that when the magnetic field direction was varied in a plane transverse to the c-axis of the sample, the observed mass $m_\perp^*$ was unvarying: $m_\perp^* = 0.170m_0$. However, when the field direction was varied in a plane containing the c-axis and a binary axis, the observed mass varied between $m_\perp^* = 0.170m_0$ for the field perpendicular to the c-axis, $\mathbf{H}_\perp c$, and $m_\parallel^* = 0.109$ for $\mathbf{H}_\parallel c$. These observations are consistent with the representation of the valence band maximum by an ellipsoid of revolution about the c-axis. The mass parameters are then given by[27]

$$m_\parallel^* = m_t; \qquad m_\perp^* = \sqrt{m_t m_l}$$

which yield the result $m_t = 0.109m_0$, $m_l = 0.264m_0$. These values, together with the derived form of the valence band maximum, provide a valuable basis for the calculation and evaluation of a theoretical valence band structure for tellurium.

The experimental technique used in the cyclotron resonance studies of tellurium is typical of the marriage of microwave and optical methods for far-infrared applications. Generally the microwave portion of a combined system constitutes the source and waveguide components for matching the radiation to the optical or quasi-optical spectroscopic apparatus. There are two basic reasons for this procedure. First, microwave sources are strong and monochromatic, making high-resolution work straightforward, whereas traditional sources such as mercury arc lamps are weak at long wavelengths and thus unsuited for high-resolution spectroscopy. Second, microwave components, even straight sections of waveguide, are exceedingly difficult to manufacture and extremely lossy for wavelengths shorter than about

2 mm. The combined systems for far-infrared spectroscopy thus acquire the strengths of the microwave and optical methods and avoid their weaknesses. It must be recognized, however, that these systems are most useful for magnetooptical applications where an external magnetic field can be varied. The microwave sources are generally not tunable over a wide range.

Microwave sources other than carcinotrons, typical klystrons, and harmonic generators have been used in combined systems. An example is provided by a recent study[28] of the cyclotron resonance of holes and electrons in germanium at wavelengths of 2 and 4 mm. The sources used were a 4 mm klystron and a second-harmonic generator. Samples were exposed simultaneously to the microwave energy and to room temperature blackbody radiation that served to ionize some of the impurities, producing holes in p-type and electrons in n-type material. The resistance of the samples was monitored by an ac technique while an external magnetic field was swept through the resonance condition. Signal peaks were observed at known resonance field values, as shown in Fig. AIV.14. This technique is a variation of one used earlier,[29] in which high-energy background radiation was used to excite hole-electron pairs. The resulting spectra were considerably more

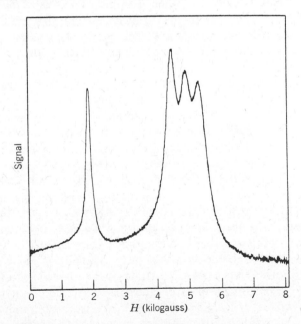

Fig. AIV.14 Cyclotron resonance of electrons at 4-mm wavelength and 4.2°K in arsenic-doped germanium, as observed by a photoconductivity technique. The applied magnetic field direction was nearly parallel to a ⟨111⟩ crystal axis. [R. Kaplan, *Solid State Commun.* **3**, 35 (1965), by permission of Pergamon Press.]

complicated, containing peaks due to both holes and electrons, with the signal phase and amplitude varying strongly with small changes in sample orientation or background light intensity.

The peaks in the spectrum of Fig. AIV.14 occurred when microwave energy was absorbed during cyclotron resonance. Since the sample conductivity is proportional to both the carrier concentration N and mobility μ, changes in either quantity could have produced the signal peaks. Subsequent measurements showed that the mechanism responsible was the following. Radiation absorbed at resonance raised the average free carrier energy slightly above its off-resonance value, thus increasing the carrier recombination time. The subsequent increase in the steady-state concentration N caused a drop in sample resistance. Thus the signal peaks actually reflect resonant decreases in the dc voltage across the sample. The four peaks in the spectrum represent the four effective mass ellipsoids in the conduction band of germanium.

A typical spectrum obtained from p-type germanium by this technique is shown in Fig. AIV.15. The peak at 2 kG is due to light holes, while the remaining spectrum exhibits the partially resolved quantum transitions of the heavy holes.[30] This provides another example of the need for extending cyclotron resonance measurements to wavelengths shorter than those obtainable using normal microwave techniques. The valence band of germanium at $\mathbf{k} = 0$ is fourfold degenerate, and in the presence of a magnetic field four series of energy levels, or "Landau ladders," result.[31] Within each ladder the lower energy levels are unequally spaced. At higher energies their spacings become approximately equal, being heH/m_h^*c for the two heavy hole ladders and heH/m_l^*c for the light hole ladders. If the thermal energy kT of the carriers is smaller than the smallest level spacings, only the lowest levels are occupied and a complicated cyclotron resonance spectrum such as that shown in Fig. AIV.15 results. However, at higher temperatures the higher levels begin to be populated, and when the condition $kT \gg heH/m^*c$ is satisfied, only two cyclotron resonance peaks corresponding to the masses m_h^* and m_l^* are observed. The lower energy levels no longer contribute to the resonance absorption since the population differences between adjacent low-lying levels tend toward zero as the temperature rises. At a wavelength of 2 mm the quantum condition is not fully satisfied for easily realizable sample temperatures. Thus a complete study of the quantum effects requires the use of one of the submillimeter or far-infrared techniques.

The spectra shown in Figs. AIV.14 and AIV.15 illustrate another technique that can be very useful in particular circumstances. This technique entails the use of the sample as its own detector by means of the photoconductivity signal that occurs at resonance. It is necessary that either N or μ change sufficiently due to the absorption of radiation at resonance. In the experi-

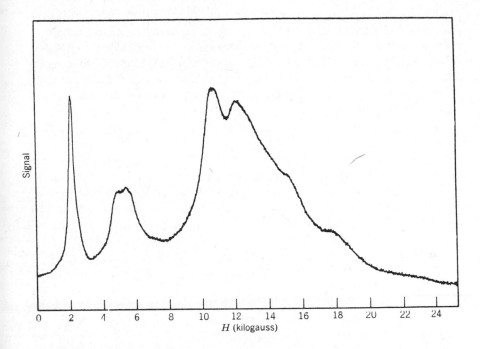

Fig. AIV.15   Cyclotron resonance of holes at 2-mm wavelength and 4.2°K in gallium-doped germanium, as observed by a photoconductivity technique. [R. Kaplan, *Solid State Commun.* **3**, 35 (1965), by permission of Pergamon Press.]

ments just described, N changed due to a resonance radiation-induced increase in the lifetimes of the carriers excited by steady background illumination. Increase in the average carrier energy can also lead to a change in μ, if the scattering mechanism which determines μ is energy-dependent. This principle has been used to study cyclotron resonance and related effects in indium antimonide, as will be described shortly. It has also been used[32] to observe the spin resonance of conduction electrons in indium antimonide as well as the impurity transitions described earlier.

b. Cyclotron Resonance Studies Using Purely Optical Techniques

It was mentioned earlier that in order to observe cyclotron resonance in low-mobility materials, relatively short wavelength radiation must be used. Early experiments on the III–V compound semiconductors were therefore performed in high magnetic fields, using high-temperature radiation sources, monochromators, and focusing optics. In a typical series of experiments,[33,34] the cyclotron resonance of electrons in indium arsenide and indium phosphide was observed at wavelengths in the range 23–160 microns. It was pointed out[33] that in calculating m^* from the minima in the transmission spectra, corrections due to magnetoplasma effects might be necessary. These effects were expected to become pronounced at the longer wavelengths where ω_c was no longer much larger than the plasma frequency $\omega_p = (4\pi Ne^2/m^*\varepsilon_0)^{1/2}$. For $\omega \sim \omega_p$ and $\omega\tau \gg 1$, the resonance condition is no longer $\omega = \omega_c$ but becomes

$$\omega_c = \omega(1 - L\omega_p^2/\omega^2)^{1/2}, \tag{AIV.22}$$

where L is a depolarization factor which depends on the geometry of the experiment. For a thin slab sample, $L = 0$ for the "Faraday orientation" in which propagation is along the field direction and normal to the sample. For the "Voigt orientation," with the field in the plane of the sample and normal to the direction of propagation, $L = 1$ and a resonance shift will occur. Differences in the peak absorption positions were indeed observed for the Faraday and Voigt orientation spectra at the longest wavelengths used. Magnetoplasma reflection may also cause an apparent shift in the resonance peak when ω is only slightly larger than ω_p. This effect, also observed at the longer wavelengths, occurs because the transmission can be altered by the rapidly varying reflectivity in the region near ω_p when the magnetic field is varied. Unless experiments can be performed at frequencies well beyond the plasma edge, the influence of magnetoplasma effects on the data must be taken into account.

The ability to observe cyclotron resonance at very high magnetic fields enhances the possibility of resolving structure in the absorption lines. Two sources of structure may be deduced from Fig. AIV.16,[34] which shows the conduction Landau levels in indium arsenide at $k_z = 0$ as a function of magnetic field. Each Landau level is split by the interaction of the electron spin with the magnetic field. The magnitude of the splitting, $|g|\beta H$, decreases in successively higher Landau levels because $|g|$ is a decreasing function of energy. In addition, since m^* increases with energy, the spacing between successive Landau levels decreases. Thus the four transition energies indicated by dashed lines in Fig. AIV.16 are not equal, as they would be in the absence of energy dependence of g and m^*. These transitions were all observed[34] in the high-field work described above. By varying the sample temperature, the initial state populations for various transitions could be changed, thus facilitating the observation and identification of the absorption peaks.

It will be noticed that the energy levels in Fig. AIV.16 are not linear functions of magnetic field. This reflects the "nonparabolicity" of the energy bands, that is, their departure from pure k^2 dependence. The nonparabolicity manifests itself in many ways, the most direct being the magnetic field dependence of the effective mass. Figure AIV.17 shows the observed[35]

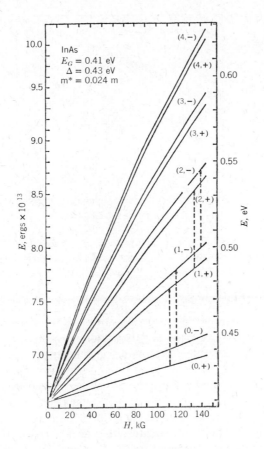

Fig. AIV.16   Magnetic field dependence of the energy levels in the conduction band of InAs. Four transitions that were observed are indicated by the dashed lines. [E. D. Palik and J. R. Stevenson, *Phys. Rev.* **130**, 1344 (1963).]

variation of mass with field for electrons in indium antimonide from about 50–340 kG. These fields were obtained with a pulsed magnet. In other experiments the mass has been measured from 50 kG down to about 120 G. Thus electron cyclotron resonance has been observed in this material over the entire wavelength range from 8.5–12500 microns.

Magnetooptical investigations in the far-infrared can now be performed conveniently by means of Fourier transform spectrometers. In a typical adaptation of such an instrument for this purpose,[36] the radiation is led

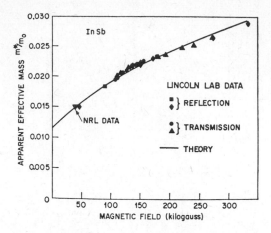

Fig. AIV.17   Variation of the measured effective mass with magnetic field in InSb at room temperature. [B. Lax, J. G. Mavroides, H. J. Zeiger, and R. J. Keyes, *Phys. Rev.* **122**, 31 (1961).]

from the spectrometer sample chamber, via metal light-pipe, through a superconducting magnet and sample holder immersed in liquid helium. Cryogenic detectors mounted below the sample holder are used for transmission experiments, while leads may be attached to the samples for photoconductivity measurements. Adaptation for reflection experiments would also be possible. Either the Faraday or Voigt orientation can be achieved by means of condensing cones and mirrors at the sample position. This system has been used in measurements of the cyclotron resonance of electrons in indium antimonide in the wavelength range 35–350 microns. In pure samples the situation is somewhat complicated by the fact that some of the free carriers become localized at impurities. The Coulomb interaction responsible for localization is weak compared to the interaction with the magnetic field for fields greater than a few thousand gauss. As a result the localized electrons also undergo cyclotron resonance, but at a slightly higher energy. A doublet absorption line results with the relative intensities of the components dependent on the fraction of carriers which have become localized.

In an earlier section a description was given of the mechanism by which indium antimonide acts as a far-infrared detector. In pure material the mobility is a function of the average carrier energy, which can be increased by the absorption of radiation. At wavelengths between a few hundred microns and the fundamental absorption edge near 7 microns, high-purity material is transparent except for the reststrahlen band. If a suitable magnetic field is applied, the absorption constant in the spectral region is very small (outside the lattice band), except for a strong peak at the position of cyclotron resonance. A measurement[36] of the sample photoconductivity as a function

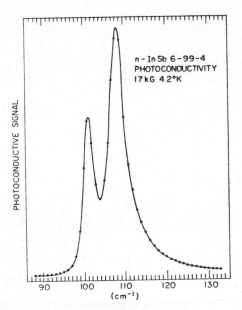

n - In Sb 6-99-4
PHOTOCONDUCTIVITY
17 kG 4.2°K

PHOTOCONDUCTIVE SIGNAL

90   100   110   120   130
(cm⁻¹)

Fig. AIV.18   Free carrier and impurity cyclotron resonance in *n*-type InSb as measured by a photoconductivity technique. [R. Kaplan, *Appl. Opt.* **6**, 685 (1967).]

of wavelength reproduces the spectral dependence of the absorption constant, as shown in Fig. AIV.18. The situation is similar to that described earlier for germanium, but the effects differ in the origin of the photoconductivity. This illustrates the principle of the magnetically tuned detector:[37] only radiation within the spectral bandwidth of the cyclotron resonance absorption is detected. The photoconductivity spectrum shown was obtained from an interferogram similar to that of Fig. AIV.19(*a*). The smaller period of oscillation in the recorder trace determines the centroid of the resonance doublet, while the period of the envelope oscillation determines the splitting of the components. When the sample temperature was lowered, carrier freezeout became almost complete, the lower energy component was greatly weakened, and the beating effect in the interferogram was reduced (Fig. AIV.19(*b*)). It is possible to obtain much information by direct inspection of these interferograms without the need for computerized Fourier transformation. Furthermore, no detector other than the sample itself is needed over a wide range of wavelengths. This provides another example of the usefulness of the photoconductivity technique in particular situations.

The use of Fourier transform spectroscopy is limited to situations where spectra are obtained as a function of wavelength. In many magnetooptical experiments, however, it is desirable to use a fixed wavelength and vary the

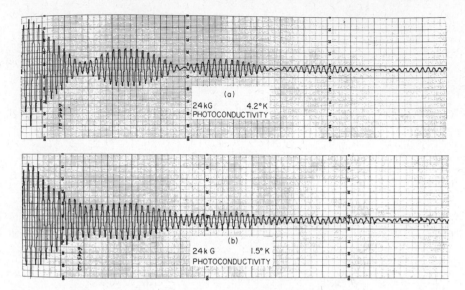

Fig. AIV.19 Interferograms obtained in measurements of free carrier and impurity cyclotron resonance in InSb, using a photoconductivity technique. At the lower temperature of 1.5°K most of the carriers were frozen out. [R. Kaplan, *Appl. Opt.* **6**, 685 (1967).]

magnetic field strength. The carcinotrons described earlier provide radiation at fixed wavelengths as short as 500 microns. For shorter wavelengths in the far-infrared, a variety of laser sources are being developed. The incorporation of a cyanide radical laser, with an output at 337 microns, in a simple magnetic resonance spectrometer has recently been described.[38] With the increasing refinement of laser sources now taking place, this technique will prove very useful.

D. COUPLED PHONON–FREE CARRIER MODES

The subjects of far-infrared investigation that have been discussed in the preceding sections may be classified into two general groups. In the first group, comprising lattice and impurity effects, the resonant frequencies are characteristic of the material and not appreciably affected by external fields (aside from splittings due to the lifting of degeneracy). In the second, in which the example discussed was cyclotron resonance, the resonant frequencies can be shifted through the spectrum by varying an experimental parameter. Also belonging to the second group are plasma effects, since the plasma frequency ω_p varies as $N^{1/2}$ and can be "swept" by changing the carrier concentration. A third group arises due to the possibility of bringing the resonant frequencies of different phenomena into some mutual relationship. This can be accomplished by applying a suitable magnetic field, or by choosing the carrier concentration properly. In this way, interactions between different types of motion, that is, that of lattice ions and free carriers, can be studied.

In the investigation of free carrier effects in semiconductors, the carriers can generally be considered as acting in a "single particle" rather than a "collective" manner if their concentration is sufficiently small. For most purposes the latter requirement is satisfied if the plasma frequency ω_p is small compared to any other frequency that appears in the description of the interaction of interest. In this final section a brief account will be given of some recent investigations of free carrier–lattice interactions. One of these is concerned with single particle, the other with collective, free carrier effects in polar semiconductors.

a. Polaron Modes

An interaction of considerable interest occurs in low-concentration, polar semiconductors between individual conduction electrons and the longitudinal optical lattice mode. This interaction is responsible for the polaron, a quasi-particle which in most situations behaves almost like a conduction electron. A polaron theory has been proposed for weak coupling and small wave vector $\mathbf{k}$, according to which the polaron energy is given by[39]

$$E \approx \frac{\hbar^2 k^2}{2m^*}\left(1 - \frac{\alpha}{6}\right) - \alpha\hbar\omega_{LO}, \tag{AIV.23}$$

where the coupling constant α determines the strength of the interaction. The polaron energy is then found to be different from that of a conduction electron in two respects. First, there is an increase in the effective mass by the factor $(1 - \alpha/6)^{-1}$, and second, there is a further decrease in the energy by the amount $\alpha\hbar\omega_{LO}$. For the polar semiconductor indium antimonide, for which $\alpha \sim 0.02$, the effects are too small to be detected experimentally. When a magnetic field is applied, the polaron energies at low field are given by[40]

$$E(n) = \left(n + \frac{1}{2}\right)\hbar\omega_c\left(1 - \frac{\alpha}{6}\right) - \alpha\hbar\omega_{LO} \tag{AIV.24}$$

and the cyclotron transition energy is

$$\hbar\omega_{cR} = \hbar\omega_c\left(1 - \frac{\alpha}{6}\right). \tag{AIV.25}$$

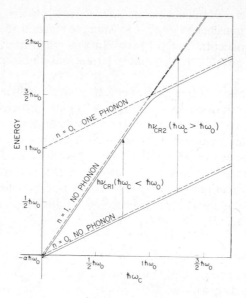

Fig. AIV.20   Theoretical behavior of Landau levels in the presence of electron–LO phonon interaction in InSb, assuming $\alpha = 0.02$. The dashed lines are the unperturbed energy levels. [D. H. Dickey, E. J. Johnson, and D. M. Larsen, *Phys. Rev. Letters* **18**, 599 (1967).]

As the magnetic field is increased, $\hbar\omega_c$ approaches $\hbar\omega_{LO}$, and detailed calculations[40,41] predict that at a critical field value the $n = 1$ polaron level becomes double-valued. These theoretical results are illustrated in Fig. AIV.20. They predict the following behavior of the observable resonance frequency ω_{CR}: (*1*) ω_{CR} initially increases linearly with field, but saturates as the critical field is approached, and never exceeds ω_{LO}. (*2*) Above the critical field there is a second resonant frequency which is slightly higher than ω_c. Transmission experiments[42] designed to test these predictions have been performed on thin samples of indium antimonide. The predictions of the theory were verified, but the interesting spectral region between ω_{TO} and ω_{LO} could not be investigated due to the reststrahlen band. Related polaron effects have been observed in interband magnetooptical studies[41] of indium antimonide. These studies are of interest because the reststrahlen band does not interfere with investigations at magnetic field values for which $\omega_c \sim \omega_{LO}$.

Intraband studies of the polaron, such as that described, must evidently be performed in a spectral region containing the longitudinal optical frequency. In indium antimonide this region is at the edge of the far-infrared, while in the highly polar semiconductors lead sulfide, lead selenide, and lead telluride, it is at still longer wavelengths. The lead salts should be favorable subjects for polaron measurements, in part because of their relatively large values of the separation of ω_{LO} from ω_{TO}. Other intraband methods of investigating polaron effects have been suggested and will no doubt be applied in the near future.

b. Coupled Collective Plasma Cyclotron–Longitudinal Optical Phonon Modes

In high-concentration materials, polaron effects do not occur because the interaction responsible for the polaron is shielded out. Instead, strong collective interactions may be observed when at least two of the frequencies ω_p, ω_c, and ω_{LO} are of the same order of magnitude. One of these interactions makes it possible to couple electromagnetic radiation to the longitudinal optical mode of a polar crystal at normal incidence, as will now be described.

In the absence of a magnetic field, electromagnetic radiation propagating at normal incidence to a sample will couple only to the transverse lattice modes since the electric field of the radiation is purely transverse. This situation is not changed if a magnetic field is applied along the direction of propagation since the Lorentz force on the carriers is then purely transverse. With the field in the plane of the sample, however, the Lorentz force on the carriers has both longitudinal and transverse components. Thus, the coupled modes involving the collective plasma cyclotron excitations and longitudinal optical phonons are transverse as well as longitudinal in character. These modes may then interact with radiation at normal incidence.

The frequencies of the normal modes which arise from the coupling of the collective plasma cyclotron excitations with the longitudinal optical phonons via the macroscopic longitudinal electric field are given by[43]

$$2\omega_\pm^2 = \omega_p{}^2 + \omega_c{}^2 + \omega_{\text{LO}}^2 \pm [(\omega_p{}^2 + \omega_c{}^2 + \omega_{\text{LO}}^2)^2 - 4(\omega_p{}^2\omega_{\text{TO}}^2 + \omega_c{}^2\omega_{\text{LO}}^2)]^{1/2}.$$

$$(\text{AIV.26})$$

Analysis of Eq. AIV.26 reveals a similarity, especially for relatively low ω_p, with the magnetic field dependence of the polaron modes discussed earlier. This dependence for the coupled modes is shown in Fig. AIV.21. The solid lines were calculated from Eq. AIV.26, while the data show the observed[44] absorption peak positions for three samples of indium antimonide having different carrier concentrations. From the three sets of curves it is evident that the strength of the interaction, indicated by the separation of the ω_- and ω_+ modes for a given sample, increases rapidly with carrier concentration. Further information could be obtained from the relative widths and strengths of the various absorption peaks. As with the polaron experiments, these coupled mode effects should be generally observable in polar semiconductors in which the conditions on ω_p, ω_c, and ω_{LO} are satisfied.

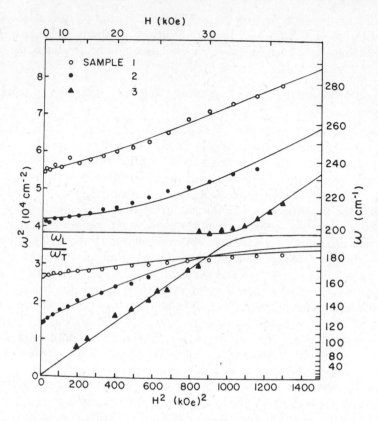

Fig. AIV.21   Frequency positions of the coupled mode absorption lines in InSb as a function of magnetic field. The data points refer to samples with the following free electron concentrations: No. 1, $1.4 \times 10^{17}\,\mathrm{cm}^{-3}$; No. 2, $5.5 \times 10^{16}\,\mathrm{cm}^{-3}$; No. 3, $1 \times 10^{15}\,\mathrm{cm}^{-3}$. The solid lines were calculated by using Eq. AIV.26. [R. Kaplan, E. D. Palik, R. F. Wallis, S. Iwasa, E. Burstein, and Y. Sawada, *Phys. Rev. Letters* **18**, 159 (1967).]

The required spectral regions are those which include the longitudinal frequencies of the materials of interest. For certain materials, these will lie in the far-infrared.

The approach used in this appendix has been to describe specific experiments which serve to illustrate general areas of application of far-infrared spectroscopy to semiconductor research. It may have been noticed that the wide variety of experiments described were performed on a relatively small number of different materials. The reason for this is not that the physical effects of interest were specific to these materials, but rather that the initial far-infrared studies have concentrated on those semiconductors which can readily be obtained as single crystals of high purity. It is to be expected that with the increasing availability both of high-quality sample material and convenient spectroscopic techniques, the far-infrared study of semiconductors will become widespread.

REFERENCES

1. H. Yoshinaga and R. A. Oetjen, *Phys. Rev.* **101**, 526 (1956).
2. T. S. Moss, *Optical Properties of Semiconductors*, Butterworths, London, 1961; p. 235.
3. R. B. Sanderson, *J. Phys. Chem. Solids* **26**, 803 (1965).
4. E. G. Bylander and M. Hass, *Solid State Commun.* **4**, 51 (1966).
5. D. L. Stierwalt and R. F. Potter, *Report No. 630*, Naval Ordnance Lab., Corona (1965).
6. D. L. Stierwalt, *J. Phys. Soc. Japan* **21**, Suppl., 58 (1966).
7. E. D. Palik, Private Communication.
8. W. Kohn, *Solid State Physics*, Vol. 5, F. Seitz and D. Turnbull, Eds., Academic Press, New York, 1957, p. 257.
9. D. Schechter, *J. Phys. Chem. Solids* **23**, 237 (1962).
10. K. S. Mendelson and H. M. James, *J. Phys. Chem. Solids* **25**, 729 (1964).
11. H. B. Bebb and R. A. Chapman, *J. Phys. Chem. Solids* **28**, 2087 (1967).
12. H. Shenker, E. M. Swiggard, and W. J. Moore, *Trans. Met. Soc. AIME* **239**, 347 (1967).
13. R. L. Jones and P. Fisher, *Solid State Commun.* **2**, 369 (1964).
14. D. H. Dickey and J. O. Dimmock, *J. Phys. Chem. Solids* **28**, 529 (1967).
15. Y. Yafet, R. W. Keyes, and E. N. Adams, *J. Phys. Chem. Solids* **1**, 137 (1956).
16. H. Hasegawa and R. E. Howard, *J. Phys. Chem. Solids* **21**, 174 (1961).
17. R. Kaplan, *J. Phys. Soc. Japan* **21**, Suppl., 249 (1966).
18. E. H. Putley, *Proc. Phys. Soc. (London)* **73**, 280 (1959).
19. E. H. Putley, *Proc. IEEE* **51**, 1412 (1963).
20. R. F. Wallis and H. Shenker, *NRL Memorandum Report 1493* (1964).
21. J. Blinowski and J. Mycielski, *Phys. Rev.* **134**, A246 (1964).
22. J. Blinowski and J. Mycielski, *Phys. Rev.* **136**, A266 (1964).
23. L. J. Neuringer, R. C. Milward, and R. L. Aggarwal, *J. Phys. Soc. Japan* **21**, Suppl., 582 (1966).
24. M. Hulin, *J. Phys. Chem. Solids* **27**, 441 (1966).
25. J. C. Picard and D. L. Carter, *J. Phys. Soc. Japan* **21**, Suppl., 202 (1966).
26. J. H. Mendum and R. N. Dexter, *Bull. Am. Phys. Soc.* **9**, 632 (1964).
27. B. Lax, *Proceedings of the International School of Physics, Course XXII, Semiconductors*, Academic Press, New York, 1963, p. 248.
28. R. Kaplan, *Solid State Commun.* **3**, 35 (1965).
29. H. J. Zeiger, C. J. Ranch, and M. E. Behrndt, *J. Phys. Chem. Solids* **8**, 496 (1959).
30. J. J. Stickler, H. J. Zeiger, and G. S. Heller, *Phys. Rev.* **127**, 1077 (1962).
31. R. F. Wallis and H. J. Bowlden, *Phys. Rev.* **118**, 456 (1960).
32. M. Guéron and I. Solomon, *Phys. Rev. Letters* **15**, 667 (1965).
33. E. D. Palik and R. F. Wallis, *Phys. Rev.* **123**, 131 (1961).
34. E. D. Palik and J. R. Stevenson, *Phys. Rev.* **130**, 1344 (1963).
35. B. Lax, J. G. Mavroides, H. J. Zeiger, and R. J. Keyes, *Phys. Rev.* **122**, 31 (1961).
36. R. Kaplan, *Appl. Opt.* **6**, 685 (1967).
37. M. A. C. S. Brown and M. F. Kimmet, *Brit. Comm. Elec.* **10**, 608 (1963).
38. K. J. Button, H. A. Gebbie, and B. Lax, *IEEE J. Quant. Elec.* **QE-2**, 202 (1966).

39. H. Fröhlich, *Polarons and Excitons*, C. G. Kuper and G. D. Whitfield, Eds., Plenum Press, New York, 1963, p. 1.
40. D. M. Larsen, *Phys. Rev.* **135**, A419 (1964).
41. E. J. Johnson and D. M. Larsen, *Phys. Rev. Letters* **16**, 655 (1966).
42. D. H. Dickey, E. J. Johnson, and D. M. Larsen, *Phys. Rev. Letters* **18**, 599 (1967).
43. S. Iwasa, Y. Sawada, E. Burstein, and E. D. Palik, *J. Phys. Soc. Japan* **21**, Suppl., 742 (1966).
44. R. Kaplan, E. D. Palik, R. F. Wallis, S. Iwasa, E. Burstein, and Y. Sawada, *Phys. Rev. Letters* **18**, 159 (1967).

APPENDIX V

SUPERCONDUCTIVITY

GERHART K. GAULÉ

Electronic Components Laboratory, United States Army Electronics Command, Fort Monmouth, New Jersey

A. Introduction

a. Far-Infrared Radiation and Superconductivity as Low-Energy Phenomena

Far-infrared photons may have energies of the order of 10 meV (1 meV = 10^{-3} electron volt = 1.6×10^{-16} joule $\sim 8.07\,\text{cm}^{-1}$); the "condensation energy" for one electron in a superconductor is of the order of 1 meV. Thus, both phenomena are investigated with low-energy experimental techniques, as evidenced by radiation shields, cryostats, low-level amplifiers, and the like. The past decade has brought an increasing interest in the study of superconductors at higher and higher photon (and phonon) frequencies, with emphasis on the range corresponding to the condensation energies. At the same time, scientists studying infrared phenomena placed increasing efforts on the low-energy range of their field; they developed greatly refined instrumentation for the submillimeter wavelength range. Most of the experiments which combined ideas and techniques from both fields of study were then performed in this range. We shall first discuss the general advantages which arise using such low temperatures in optical and electronic systems and then proceed to the special phenomena and benefits related to superconductivity.

b. General Features of Low-Temperature Optical and Electronic Systems. Noise

Any optical detector generates an output unrelated to the signal input, or noise, because of the radiation of the surfaces surrounding the detector. A black surface of temperature T (°K) emits into a frequency interval Δf incoherent electromagnetic radiation with a power[1]

$$S_f = (2\pi h f^3/c^2)[\exp(hf/kT) - 1]^{-1}\Delta f \quad (W/m^2), \quad (AV.1a)$$

where f is the frequency (in Hz = sec^{-1}), h is Planck's constant, c the velocity of light, $k = 1.38 \times 10^{-23}$ $J/°K$ is Boltzmann's constant in joule/°K, $2\pi h/c^2 = 4.63 \times 10^{-50}$ in units of $W \cdot \sec^4/m^2$, and $k/h = 2.08 \times 10^{10}$ in units of $(°K \cdot \sec)^{-1}$. For high frequencies, that is, when photon energies greatly exceed thermal energies, $hf \gg kT$, Eq. AV.1a becomes[2]

$$S_f = (2\pi h/c^2)f^3 \exp(-hf/kT)\Delta f, \quad (AV.1b)$$

while for low energies, $hf \ll kT$, Eq. AV.1a reduces to

$$S_f = (2\pi k/c^2)f^2 T\Delta f, \quad (AV.1c)$$

with $2\pi k/c^2 = 0.965 \times 10^{-39}$ $(W \cdot \sec^3/°K \cdot m^2)$.

For surfaces that are not black, S_f must be multiplied by the dimensionless quantity $\varepsilon(f, T)$, the emissivity, which is less than unity and which decreases with T.[1] The radiative power that enters an optical detector from the surrounding surfaces at temperature T thus always decreases in proportion to T or faster. The radiation tends to raise the temperature of the detector or the sample. It will lead to a dc background signal for a detector responsive to power (or temperature). However, even a detector with a limited bandwidth Δf_{mod} centered about a modulation frequency f_{mod}—which thus responds only to modulated signals—will have a noise output due to the radiation from the surrounding walls because this radiation is not constant in time; the Eqs. AV.1 merely give the time average. Fluctuations about this average in effect modulate the radiation. While the spectrum of the black- or gray-body radiation vanishes rapidly towards zero frequency (see Eqs. AV.1), the spectrum for the fluctuations is essentially flat (modulation or fluctuation frequencies are assumed to be much smaller than the dominant radiation frequencies). Putley, who has discussed these problems extensively,[3] showed that the root-mean-square of the fluctuations of the radiation power from a surrounding black hemisphere at temperature T reaching a detector of area A and emissivity ε is given by

$$P_{rms} = (\Delta P^2)^{1/2} = 2(2\varepsilon C_s kT^5 A\Delta f_{mod})^{1/2} \quad (W), \quad (AV.2a)$$

where the constant

$$C_s = 2\pi^5 k^4/15c^2 h^3 \quad (AV.2b)$$

is Stefan's constant. If the detector is at a temperature comparable to that of the surrounding hemisphere, a similar contribution arises from fluctuations in the emission of radiation from the detector surface. Not much is gained then by cooling the detector while the surrounding housing is not cooled accordingly. The noise input into the detector is further reduced by choosing a detector surface with a small emissivity ε. This is not feasible for infrared detectors because it would decouple the detector from the signal source as well.

It is instructive to discuss briefly the analogous problems of noise reduction in electronic circuits. If all the losses in the circuit are represented by a resistor R, this resistor is the only noise source of the circuit. It is capable of delivering the available noise power P_{av}

$$P_{av} = kT\Delta f \quad \text{(W)}. \qquad \text{(AV.3)}$$

One way of reducing the noise power entering the subsequent amplifier stages is accomplished by reducing the circuit temperature T; another possibility is reducing the amplifier bandwidth. Still more is gained, however, by observing that P_{av} is the *available* noise power, available to the subsequent circuit only when that circuit is impedance-matched to R. A mismatch thus hinders the transfer of noise power. With reactive elements, inductors, and capacitors, it is possible to build circuits with impedances which may be orders of magnitude higher or lower than R.

Superconductive inductors or capacitors or resonant cavities permit us to pursue this approach to the extreme. Such resonators have Q-values up to 10^9 and higher; that is, after 10^9 oscillations the voltage and current amplitudes have decayed only by a factor $1/e$. Still, there is decay and thus losses, and the noise power as given by Eq. AV.3 is available from the superconducting circuit as well. However, the decoupling by impedance mismatch with respect to the noise source is better by a factor of 10^4 to 10^6 for superconducting resonators than for normal resonators.

For the low-level stages of an electronic or an optical system which may or may not contain superconducting components, there is thus one common strategy to prevent noise from entering the subsequent stages, namely (*1*) reduce the temperature, (*2*) reduce the bandwidth, (*3*) eliminate lossy elements (resistances, black or gray surfaces, absorbing filters), and (*4*) decouple from lossy elements by using reactive circuit elements, mirrors, selective non-absorbing filters, and so forth.

Thus far we have dealt with Johnson noise and its optical analogue, and we have shown how it can be reduced by lowering the operating temperature. Other types of noise are, however, basically temperature-independent. They arise from the uncorrelated electron transits in normal metals and in semiconductors, from recombination events in semiconductors, and the like. Noise resulting from transit effects is called shot noise; it prevails at low frequencies and its power is proportional to the current. Indirect gains from cooling may thus be realized when it leads to a reduction in operating current.

For example, cooling reduces the dark current of a photoconductor and the leakage currents in a transistor.[4] Supercurrents are altogether free of shot noise because the electrons move as a superfluid whereby they perform transits not independently but coherently. Superconductor devices containing interfaces, such as superconductor-insulators, may exhibit flicker noise which resembles the recombination noise in semiconductors. A superconductor containing trapped magnetic flux and carrying a current may generate flicker noise.[5]

The noise power from flicker or shot noise decreases with frequency f as $1/f$ or faster. It is thus not directly sensed by a high-frequency or infrared detector; it can, however, still influence the detector output via modulation of certain detector parameters. If the detector is temperature-sensitive, temperature fluctuations will be sensed in a similar fashion, as was shown by Newhouse.[6]

c. Cooling Requirements for Superconductors

As shown above, any long-wavelength detector or experimental apparatus will require some sort of cooling of its sensitive parts. The change with temperature of the electronic and optical properties of a semiconductor, a normal metal, or a dielectric, is usually only a gradual one[7]; therefore, an apparatus not containing superconductors may, at some sacrifice to sensitivity and noise figure, be cooled by one of the moderately cold cryogenic liquids. These are liquid hydrogen (normal boiling point 20.4°K), liquid neon (27.2°K), or even liquid nitrogen (77.37°K). Against the much colder (4.2°K) liquid helium, ^{4}He, these liquids give the advantage of a much higher heat of evaporation and of easier liquefaction. Also, the design and operation of a cryostat is simpler where only moderate cooling is required. Thus far, only one superconductor, the pseudo-binary compound $Nb_3Al_{1-x}Ge_x$ ($0.25 \leqslant x \leqslant 0.3$) has a critical temperature T_c above the normal boiling temperature of hydrogen.[8] Progress in developing materials with higher T_c values has been slow so far.[9]

The term critical temperature indicates an abrupt change of the thermal and the low-frequency electric and magnetic properties of the material (in weak magnetic fields). However, we shall show below that the change in the high-frequency properties and, in particular the optical properties near T_c, is only gradual. Except for cases where the superconductor serves as a thermal sensor (bolometer) or where its near-critical properties are investigated, temperatures much below T_c ($T < 0.5T_c$) are usually required. For the present superconductors this implies T below 10°K. Of course, many sensitive semiconductor applications or studies require the same temperature range.

Only ^{4}He is available as cryogenic liquid for simple open systems below 10°K. Besides its low heat of evaporation, boiling helium has another disadvantage : the boiling takes place in the bulk of the liquid where large bubbles which generate undesirable vibrations are formed. Furthermore, the bubbles make the liquid a poor optical medium.

Temperatures below 4.2°K are easily obtained in a closed system by reducing the external pressure above the boiling ^{4}He below atmospheric pressure. In practice, this is accomplished with the aid of large-capacity rotating pumps. Bubbling stops when a temperature of 2.17°K is reached; at this point the ^{4}He has become a superfluid and evaporation occurs only at the surface. This, and the very high heat conductivity of superfluid He, make it the ideal heat transfer medium for very sensitive technical applications or experiments.

Temperatures as low as 1.3°K can be obtained with pumped ^{4}He under reasonable technical effort. Temperatures as low as 0.3°K are obtained with pumped ^{3}He, which has a normal boiling point of 3.2°K and does not become a superfluid. ^{3}He must be precooled with ^{4}He and is always recaptured and recirculated because of its high price. For still lower temperatures (down to 0.05°K) solutions of ^{3}He in ^{4}He provide the most convenient means of cooling.[10,11] The range of millidegrees and below can be reached by the adiabatic demagnetization of a feasible material after precooling it with the aid of one of the methods just described.

Reliable closed-cycle refrigerators for semiautomatic operation for non-specialist personnel are now commercially available even for the millidegree temperature range.[12] Removal of heat from the sample is effected in a cryostat permanently connected to the refrigerator by a liquid, a gas, or by solid-state heat conduction. The latter means is often preferred for optical work because the sample can then be positioned in a vacuum.

In planning experiments or new devices, it must be borne in mind that even in an ideal system the removal of heat from the cryostat becomes more and more difficult as one goes to lower and lower temperatures (this is a simple consequence of the second law of thermodynamics). This is shown in Fig. AV.1, where "Carnot" describes the performance of an ideal, that is, perfectly reversible thermodynamic process. It takes 300 W input power to remove a heat load of 1 W from a 10°K cryostat and to transfer it into a heat reservoir at ambient temperature of 300°K—cooler ambient reservoirs are seldom conveniently available. To transport the same amount from a cryostat at 1°K requires ten times as much input, and so forth. The performance of actual refrigerators, also shown in Fig. AV.1, is of course much less efficient. The gap between the ideal (Carnot) and the actual performance widens from a factor of three near 100°K to a factor of more than hundred for 1°K. The latter factor should improve in the foreseeable future. Since heat capacity and heat conductivity of most materials is greatly reduced below 20°K, continuous cooling—as made possible by the new semiautomatic refrigerators— is often more economical than frequent warm-ups and cool-downs.

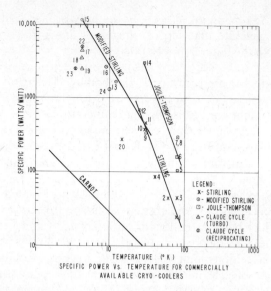

Fig. AV.1   Specific power (input power/cooling power) as function of the sample temperature for the ideal (Carnot) case and for commercially available closed-cycle refrigerators. [Ch. A. Stochl and E. R. Nalan, "Current Status and Future Trends of Cryogenic Coolers for Electronic Applications," Rept. AD610015, July 1964. Modifications by R. A. Ackermann.]

Cooled optical filters which reject unwanted parts of the spectrum have already been mentioned as a means of reducing noise. Such filters obviously also reduce the sample heat load. In an electrooptical application or experiment, a cooled electric bandpass or, better still, a cooled preamplifier will prevent the influx of power and of interference signals from outside the cryostat via the electrical connections. Direct coupling between a low-temperature detector and a conventional amplifier usually curtails the detector performance.[3] Mechanical shock or vibrations can produce an interference signal, and for very low-temperature samples, cause a significant acoustical heat inflow. Many closed-cycle refrigerators are still unsatisfactory in regard to the problem of vibrations.

B. Homogeneous Superconductors in Weak Fields

We limit the scope of this section to weak magnetic (not exceeding tens of Oersteds) and electric fields and to superconductors without nonuniformities which might be significant either electrically or optically. These restrictions allow us to treat the physical effects in terms of simple but instructive models.

No.	Manufacturer	Sample temperature and cooling power of unit
1	Malaker Labs	10 W at 77°K
2	Hughes (Culver City)	10 W at 60°K
3	Norelco	10 W at 80°K
4	Malaker Labs	4 W at 40°K
5	Hughes (Santa Barbara)	5 W at 80°K
6	Hughes (Santa Barbara)	2 W at 79°K
7	Air Products	2 W at 80°K
8	Hughes (Santa Barbara)	0.5 W at 82°K
9	Hughes (Culver City)	2 W at 30°K
10	Malaker Labs	1 W at 29°K
11	Norelco	1 W at 30°K
12	Hughes (Culver City)	1 W at 25°K
13	Hughes (Culver City)	0.4 W at 12°K
14	Air Products	0.35 W at 30°K
15	Arthur D. Little	0.5 W at 4.2°K
16	Cryomech (Syracuse)	1 W at 9.5°K
17	General Electric	1 W at 4.4°K
18	General Electric	2 W at 4.4°K
19	General Electric	10 W at 4.4°K
20	Norelco	40 W at 15 and 100 W at 60°K
21	Norelco	9 Liters/hr liquid He
22	Air Products	1 W at 4.4°K
23	Arthur D. Little	2 W at 4.0°K
24	Arthur D. Little	1.5 W at 10°K

a. The Two-Fluid Model. Resistive Bolometers

The "two-fluid" model of superconductivity visualizes two inter-penetrating fluids inside a superconductor; one "normal," the other a "superfluid." The normal fluid comprises individually moving electrons which interact with the metal lattice in a fashion analogous to that which causes room temperature resistivity. The normal fluid thus incurs friction or ohmic resistance. The superfluid is composed of coherently moving electrons. The coherence interdicts individual interactions with the metal lattice: the superfluid moves without friction. The concomitant electrical current is called a supercurrent.[13,14] The fraction of the total number of free electrons of the metal which form the superfluid is in the following designated by the symbol W. The fraction W depends on the temperature T as[14]

$$W = 1 - t^4, \qquad 0 < t = \frac{T}{T_c} \leqslant 1. \tag{AV.4}$$

For $T \geqslant T_c$, $W = 0$: the metal is normal. Reducing T to $0.5T_c$ brings 93% of the free electrons into the superfluid. At $T = 0.1T_c$, the normal fraction is only $1 - W = 10^{-4}$.

If a small constant current is run through the superconductor and the temperature is raised, no potential drop is observed until a temperature very near T_c is reached. At this point the superfluid is too depleted to carry the given current. Transition from zero (or unmeasurably small) resistance to the full normal resistance occurs in a temperature span of 10^{-8} degrees or less in the most pure and homogeneous samples.[15] Most superconductor specimens, because of imperfections, have a broader transition range from a millidegree to a fraction of a degree. Superconductors with a well defined, linear transition range can thus be used for measurements of temperatures and, in particular, for temperature changes within the transition range. Any kind of radiation supplying heat to the superconductor can thus be sensed.

The most prominent application of such superconducting bolometers is in long-wavelength infrared systems. An advanced detector for the 70 to 1000 μ wavelength range has been constructed and described in detail by Bloor and collaborators.[16] Their bolometer has a very small heat capacity. It is formed by a thin film of tin evaporated on a 1-μ thick mica strip (E in Fig. AV.2) which is suspended in vacuum. The pressure within a helium reservoir (H, bottom of Fig. AV.2) is adjusted for a temperature of 3.7°K, the center of the 0.2°K wide transition region of the superconductor. The slitted brass cylinder S serves as heat sink for the bolometer. Together with the thermal resistance of the nylon cylinder B, it forms a thermal filter which prevents short-lived temperature fluctuations of the helium reservoir H from influencing the bolometer. The mechanical support and the electrical connection is provided by thin nylon threads coated with lead. Because of the higher transition temperature (7.2°K) of the latter, it is superconducting throughout the operation. The leads T thus have zero electrical resistance but a high thermal resistance (the thermal resistance of a superconductor is usually much higher than that of a comparable normal conductor). A typical bolometer resistance R is 50 Ω, and its change with temperature is $dR/dT = 5000$ ohm/°K at the transition. This figure exceeds that of fully normal metals by several orders of magnitude. For a detector area of 3 mm^2 the minimum signal detectable with a conventional amplifier was less than 10^{-12} W and the response time was 1.25 sec. Such long response times can be tolerated in many spectroscopical and astronomical investigations, but the application of long-wavelength infrared for the purpose of communications, for the sensing of fast-moving objects, and similar applications requires extremely short response times.

A superconducting resistive bolometer with a very short response time was obtained by Bertin and Rose,[17] who inserted a thin evaporated tin film

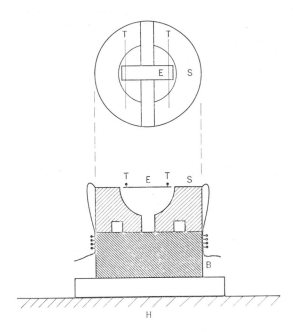

Fig. AV.2   Superconducting bolometer: E, mica strip, 1 μ thick, carrying the tin film; H, liquid-helium reservoir; S, slitted brass cylinder (intermediate heat sink); B, nylon cylinder; and T, lead-coated nylon filaments for the electrical and mechanical connections. [D. Bloor, T. J. Dean, G. O. Jones, D. H. Martin, P. A. Mawer; and C. H. Perry, *Proc. Roy. Soc.* (*London*) **A260**, 510 (1961), by permission of The Royal Society, London.]

into a helium bath of appropriate temperature. In this case, the liquid helium represents the heat sink. The "Kapitza-resistance," which inhibits the flow of heat from a solid into liquid helium, is the thermal resistance analogous to that of the nylon threads of the arrangement shown in Fig. AV.2, but is much smaller in magnitude. The response times are thus decreased to 20 nsec (20×10^{-9}) or even less.[18] The sensitivity seems to increase for longer wavelengths, especially for wavelengths beyond 3 mm. Figure AV.3 compares several of the best fast and slow long-wavelength infrared detectors; it shows clearly the sacrifice in sensitivity or noise-equivalent-power (NEP, expressed in units of $W \cdot Hz^{-\frac{1}{2}}$) required to attain short response times τ.

The basic noise limitations of the resistive (only partly superconducting) bolometers will become evident when they are connected to more advanced amplifiers. The superconductor, acting as a resistor, is usually made part of a resistance bridge and the Johnson noise generated in this network is

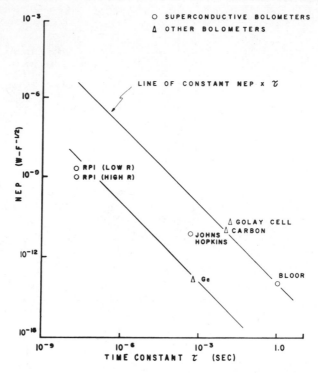

Fig. AV.3   Comparison of several far-infrared bolometers. "RPI" refers to Ref. 17 and "Bloor" refers to Ref. 16. Note the sacrifice in time constant which is necessary to achieve high sensitivity. [E. H. Putley, *J. Sci. Instr.* (*London*) **43**, 857 (1966).]

effectively coupled into the subsequent stages. A typical circuit is shown in Fig. AV.4; its performance and its noise characteristics are discussed by Newhouse.[6]

During a reduction of radiation influx and, thus, a lowering of the temperature, the restoration of superconductivity may be delayed in some regions by supercooling. Therefore, in addition to the Johnson noise from its normal regions, the superconductor exhibits flicker noise generated by the jumps of the normal–superconducting phase boundaries. The resulting resistance jumps produce an erroneous output. Because of these noise limitations, the resistive bolometer will probably be superseded by one of the "parametric" temperature sensors which are fully superconducting at all times. Examples of this are discussed further below in Section E.

b. The Energy Gap and Its Optical Significance

As shown above, a constant current driven through a superconductor will not produce any potential drop as long as this current stays below a

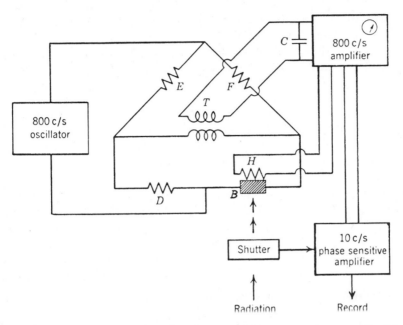

Fig. AV.4 Typical bridge circuit for reading the resistance of a bolometer operated in the superconducting to normal temperature range. (L. Newhouse, "Applied Superconductivity," Copyright 1964, John Wiley and Sons Inc., New York.)

certain critical value. The total current is carried by the resistance-free superfluid; the normal fluid, not experiencing any electric field, remains at rest. If, however, the driving current is alternating, an electric field results from the inertia of the superfluid, stemming from the masses of the electrons which make up the superfluid. Where only the superfluid is present ($T \leqslant T_c$), the electric field will lead the driving current by a phase angle of 90°. The superfluid has a kinetic inductance, which augments the familiar magnetic inductance. In the general case the presence of the normal fluid is not insignificant; it will carry a current in-phase with the field. Actual current distributions are computed by setting the physical current density equal to the real part of the complex quantity $\mathbf{j}$, defined by

$$\mathbf{j} = \sigma \cdot \mathbf{E}, \tag{AV.5a}$$

where

$$\sigma = \sigma_1 + i\sigma_2, \qquad \mathbf{E} = \mathbf{E}_0 \exp\left(-i\omega t\right), \tag{AV.5b}$$

and $\mathbf{E}_0$ is independent of time t. (The plus and minus signs of Eq. AV.5b are sometimes reversed in the literature; a positive σ_2, however, indicates inductive behavior regardless of the sign convention.)

Losses will be noticeable whenever σ_1 is comparable to or larger than σ_2. On the other hand, losses should be insignificant for $\sigma_1 \ll \sigma_2$. Since σ_1 is proportional to the density of the normal fluid, the latter case should occur for $T \ll T_c$ (see Eq. AV.4). For low temperatures, therefore, a superconductor should act as a nearly loss-free inductor.

The simplest way to test this assertion is through constructing the familiar resonant circuits comprising superconducting elements such as inductors, capacitors, tank circuits, microwave cavities, and so on, and to measure losses as function of temperature.[2] From studies of this kind, Hartwig and his collaborators[19,20] concluded that up to 10 MHz (MHz $= 10^6$ Hz) the losses caused by the superconductor are negligible in comparison to the losses caused by the dielectric components in the same circuit. The nearly loss-free operation of the circuits is expressed by the high Q-values, which ranged up to 10^9 (see Section A.b). Even better values ($Q = 1.8 \times 10^{10}$) have been reported by McAshan[21] for microwave cavities operating at 11 GHz (GHz $= 10^9$ Hz). These cavities use the vacuum as a dielectric. Pure materials, appropriate surface treatment, and a protection from external magnetic fields are important in obtaining high Q-values.

Our assertion that under the stated conditions a superconductor will carry an alternating current with only negligible loss is justified for frequencies reaching into the tens of GHz (cm-waves). On the other hand, optical experiments, including some in the intermediate infrared, show no drastic changes in the reflectance or transmittance of superconducting samples upon cooling through T_c and below. We must therefore conclude that for optical frequencies σ retains its normal value as the superconducting state is entered. A significant change in the response of the superconductor to electromagnetic radiation must thus occur in some range between the intermediate infrared on one side and the microwave frequency region on the other.

Because of the well known difficulties in extending either microwave or optical experimental techniques into the millimeter and sub-millimeter regions of the electromagnetic spectrum, it was not until 1956 when measurements of σ_1 and of σ_2 in this region were begun.[22] This work and subsequent further developments have been reviewed by Tinkham.[23,24] The results of Tinkham and his collaborators lead to the general relationships, shown in Fig. AV.5, which are apparently valid for all superconductors.[25] It is seen that σ_2 vanishes beyond a characteristic frequency ω_g, while σ_1 approaches the normal conductivity σ_N. For $\omega = 2\pi f \gg \omega_g$ we thus have the optical region discussed above. For nonzero frequencies below ω_g, σ_1 becomes insignificant (sufficiently low temperatures are assumed) while σ_2 increases as $1/\omega$, in agreement with the asserted loss-free inductive behavior at microwave and lower frequencies. The dc properties are represented by a delta function for σ_1 at $\omega = 0$.[25]

Microwave or optical measurements do not directly determine σ_1 or σ_2 but rather the surface impedance, a complex quantity defined as the ratio of the tangential electric and magnetic field intensities at the surface (in MKS units). The currents linked to the fields have a shielding effect which prevents the radiation from penetrating into the bulk of the metal. A superconductor will carry such a shielding current even at zero frequency in the presence of a constant weak magnetic field. Supercurrents and magnetic field are in effect limited to a depth λ under the surface. This penetration depth λ depends on the temperature as[14]

$$\lambda(t) = \lambda(0)\{1 - t^4\}^{-\frac{1}{2}}, \tag{AV.6}$$

where $t = T/T_c$ is the reduced temperature as previously defined. For low temperatures λ is thus only slightly larger than $\lambda(0)$ (which is usually of the order of 10^{-5} cm) but it becomes large near $t = 1$. A skin depth δ can be computed for the normal fluid itself by the conventional formula[26]

$$\delta = (2/\mu\sigma_1\omega)^{\frac{1}{2}}, \tag{AV.7}$$

where μ is the permeability of the superconductor in its normal state. In contrast to λ, δ is frequency-dependent and becomes small for large values of

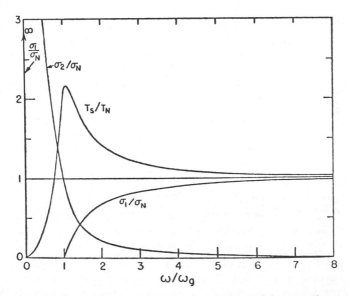

Fig. AV.5   Ratio of the transmittance T_s of a superconductor and the transmittance T_N of the same sample in the normal state. Note the transmission "hump" near the "gap" frequency ω_g. Also shown are the ratios of σ against their normal counterparts; $T \ll T_c$ is assumed. [D. M. Ginsberg and M. Tinkham, *Phys. Rev.* **118**, 990 (1960).]

ω and σ_1. The actual field penetration into the superconductor is always smaller than either of the two quantities λ and δ. At low temperatures it is usually near $\lambda(0)$ for $\omega < \omega_g$.

The onset of the absorption of electromagnetic energy by a super-conducting sample for photon energies exceeding $\hbar\omega_g$, as shown in Fig. AV.5, resembles the onset of the absorption by an intrinsic semiconductor as the photon energy exceeds the width of the energy gap. This, together with observations of the specific heat and of other properties, lead to the stipulation of an energy gap for superconductors. In 1957 Bardeen, Cooper, and Schrieffer[27] published a theory (now generally called the BCS-theory) which is based on a detailed microscopic model of the superfluid: it predicts an energy gap of

$$E_g = 2\Delta = \hbar\omega_g \approx 3.5kT_c \text{ (eV)}, \qquad (AV.8a)$$

where

$$k = 0.86 \times 10^{-4} \text{ (eV/°K)} \qquad (AV.8b)$$

is Boltzmann's constant. The T_c-values of superconductors in technical applications and in most experiments range from, roughly, 1 to 20°K; the corresponding frequencies are 73 GHz and 1.46 THz (1 THz $= 10^{12}$ Hz) and the vacuum wavelengths 4.2 and 0.21 mm, respectively. The energy gaps of superconductors are thus much narrower than those of semiconductors, with the latter typically in the 1 eV range. The conduction electrons of a metal occupy energy levels within the conduction band, which is generally several eV wide. Occupation extends up to the Fermi energy, usually near the center of the conduction band. As the metal becomes superconducting, the narrow superconducting energy gap is established symmetrically about the Fermi energy. The positions of the conduction band and the Fermi energy are not influenced by the normal-to-superconducting transition. Equation AV.8a gives a gap value which has been extrapolated to zero temperature. Generally, the gap value changes with the temperature as does the superfluid density (see Eq. AV.4).

Experimental results confirm the energy gap values predicted by the BCS-theory. Moreover, an extension of the theory by Mattis and Bardeen[28] yielded σ_1 and σ_2 as functions of ω beyond the gap value ω_g in excellent agreement with the observations (see Fig. AV.6).

An intrinsic *semi*conductor will be transparent for photons lacking the energy to excite electrons across the energy gap. However, a *super*conductor will respond with an essentially loss-free shielding current in the analogous case. It will therefore act as a loss-free reflector. If the superconductor is thin, that is, thinner than the penetration depth λ, the shielding will be incomplete; fields at the far surface of the film will be significant and will emit radiation.

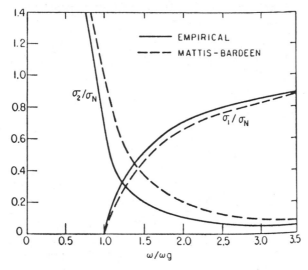

Fig. AV.6   General relationships of theoretical (Ref. 28) and empirical (Ref. 25) conductivities for frequencies exceeding the gap frequencies. The abscissa gives the reduced frequency, ω/ω_g (see Fig. AV.5). [P. L. Richards and M. Tinkham, *Phys. Rev.* **119**, 575 (1960).]

In this case the film acts as a nearly loss-free beam splitter. Since the reflectance of most metal surfaces at low temperatures is already very near unity in the normal state, the small increment due to the transition into the superconducting state is difficult to observe. It is easier to measure the transmittance of a thin film. According to Fig. AV.5, the transmittance should have a characteristic hump near ω_g. This is qualitatively understood by observing that, at ω_g, σ_1 is too far below its normal value to give much attenuation, while σ_2 is yet too small to give much shielding.

The first transmittance measurements with frequencies near the gap frequency ω_g were made in 1956 on 20 Å thick Pb and Sn films by Glover and Tinkham.[22,29,30] In 1960, Ginsberg and Tinkham extended the transmittance measurements to In and Hg films.[25] Their results with Pb films, reproduced here in Fig. AV.7, clearly show the transmittance maximum near $\hbar\omega = 5kT_c$ in accordance with the Mattis-Bardeen theory, which had appeared in 1958 (see Fig. AV.5). The experimental apparatus is similar to the one shown in Fig. AV.8, except that a second light-cone and a bolometer, rather than a thermometer, were placed behind the film.

Tinkham and his group also explored the temperature dependence of the transmittance.[22] The curves shown in Figs. AV.5, AV.6 and AV.7 represent the case $T \ll T_c$. As the temperature is increased beyond $T_c/2$, the transmittance at the hump near $\hbar\omega = 5kT_c$ begins to decrease markedly while the

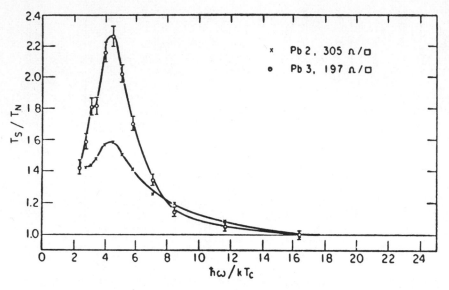

Fig. AV.7   Transmissions of very thin films (20 Å) of lead. The resistance per unit area, $\Omega/\square$ (*upper right corner*), was measured in the normal state to give an indication of thickness and continuity. [D. M. Ginsberg and M. Tinkham, *Phys. Rev.* **118**, 990 (1960).]

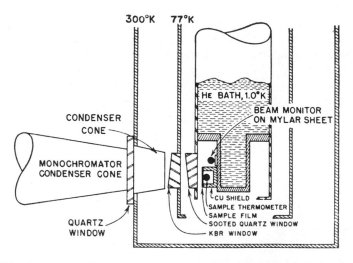

Fig. AV.8   Apparatus for the calorimetric measurement of the absorption of thick films. The windows are wedge-shaped to avoid interference fringes. Either carbon or germanium resistors can be used for the temperature measurements (see text). [S. L. Norman, *Phys. Rev.* **167**, 393 (1968).]

transmittance below ω_g increases. This is consistent with the qualitative picture of an increase of the normal fluid (or σ_1) and a decrease of the super-fluid (or σ_2).

Superconductors in bulk form, particularly single crystals, usually have better defined electrical and thermal properties than films, especially thin ones. Therefore, most of the more recent infrared experiments with super-conductors are absorption experiments. With sensitive thermometers it is possible to detect the absorbed energy calorimetrically. This was shown by Norman,[31] whose apparatus for the study of thick samples (thicker than the penetration depth λ), is shown in Fig. AV.8. Norman gives results for Pb, Sn, In, V, Ta, and Nb and compares them with those of other authors. Large samples can be shaped as nonresonant cavities, within which the radiation undergoes multiple reflections. Absorption by the superconductor's inner surface hereby competes with the absorption by the bolometer in the cavity. Thus, for constant radiation input into the cavity, a reduction in bolometer temperature indicates an increase of absorption by the super-conductor. Figure AV.9 outlines one of the multiple-reflection cavities which were prepared by the casting of a number of elemental or alloy supercon-ductors in the early absorption studies by Richards and Tinkham.[32,33] In later work, Richards[34] also used well oriented, electro-polished single crystal surfaces for some of the cavity walls. In this fashion he was able to demonstrate the anisotropy of the energy gap in pure tin.

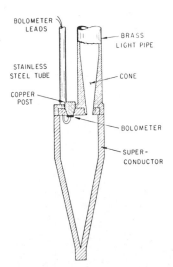

Fig. AV.9   Superconductor shaped (case) as a nonresonant multireflection cavity. [P. L. Richards and M. Tinkham, *Phys. Rev.* **119**, 575 (1960).]

Tinkham and his collaborators used a mercury arc and a grating mono-chromator in most of their work, although a klystron with subsequent semi-conductor frequency-multiplier also served to generate low-frequency radiation. Biondi and Garfunkel[35] made single crystals of zinc or aluminum as the endplates of a microwave system. By measuring the temperature increase of the plate, they could establish the absorption as a function of frequency. The energy gaps and their anisotropy were thus determined.

A microwave setup permitting the measurement of the transmittance as well as the reflectance of a thin superconducting film has been described by Rugheimer et al.[36] The shortest vacuum wavelength obtainable with these microwave systems was 3 mm. Therefore, the energy gaps of only those superconductors with T_c near or below 1°K could be studied (see Eq. AV.8). In a recent development, Palmer and Tinkham[37] used Cassegrain optics to obtain near-normal incidence of the infrared radiation on a superconducting film deposited *in situ*. Simultaneous transmittance and reflectance measure-ments were performed with this apparatus. Drew and Sievers[38] wound a superconducting foil, backed by an insulating foil, into a spiral. The spiral was then inserted into a light-pipe or waveguide structure as shown in Fig. AV.10. The transmission of light along the axis of the spiral is an indication of the absorption by the foil. Highly reflecting materials, such as lead, can thus be studied without building the multiple-reflection cavities needed in earlier work (see Fig. AV.9).

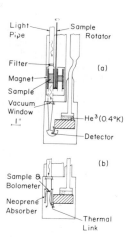

Fig. AV.10   Apparatus for the study of thick foils. The foil, with an insulating, transparent backing, is wound into a spiral. The sample rotator places the spiral samples into a coaxial position in the light beam of the light-pipe. The superconducting magnet coil is also shown. [H. D. Drew and A. J. Sievers, *Phys. Rev. Letters* **19**, 697 (1967).]

C. HOMOGENEOUS SUPERCONDUCTORS IN STRONG FIELDS

Some of the experiments discussed in the previous section have been repeated in high magnetic fields. Theoretically, a thin (100 Å thickness or less) film in a parallel field should experience a uniform reduction in the superfluid density W as the field increases; W should vanish as the critical field, H_c, is attained. This behavior would be analogous to that caused by a temperature increase (see Eq. AV.4). In optical experiments, the reduction in W is seen as a closing of the energy gap and as an increase in absorption losses. Experimental absorption curves of Pb films, performed by Martin and Tinkham,[39] are reproduced in Fig. AV.11. They show the expected normal behavior for the critical field of 39.4 kOe (critical fields for thin films are generally much higher than those for the corresponding bulk materials; the latter are only a few hundred oersteds for nontransition elements such as Pb). However, the low-frequency absorption for subcritical fields is unexpectedly large and does not fully disappear as the applied field is returned to zero (the lowest dashed curve in Fig. AV.11). The authors attribute these anomalies to an unintended perpendicular component of the applied field. A perpendicular field will penetrate a superconducting film of the kind discussed here via a regular array of spots, each spot containing an integer number of (usually one) magnetic flux quanta surrounded by the appropriate eddy or vortex supercurrent. The flux quantum is given by

$$\phi_0 = h/2e = 2.1 \times 10^{-15} (\text{V} \cdot \text{sec}). \tag{AV.9}$$

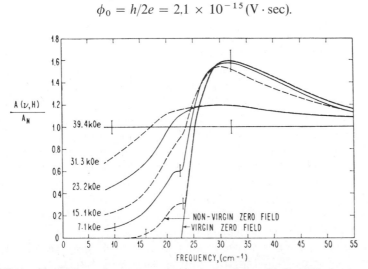

Fig. AV.11   Absorption ratios (superconducting/normal) for a Pb film with the magnetic field as parameter. The applied field was apparently not quite parallel to the film (see text). Note the flattening of the absorptions hump for high fields. [W. S. Martin and M. Tinkham, *Phys. Rev.* **167**, 421 (1968).]

Some of the flux will remain frozen as the external field is turned off. Due to the spotwise concentration of the penetrating magnetic flux, the field inside a vortex (typical diameter 10^{-4} cm) is high (near critical) and the superfluid concentration is low. Vortices are thus regions of high optical absorption. It should be noted that because of the flux concentration, such regions can be created by cooling the film in even very weak external fields (the earth's field, for instance). External fields should thus be compensated or shielded off in critical experiments.

Norman,[31] using the method discussed in the previous section, observed the disappearance of the energy gap of a thick (10^4 Å) Pb film at a parallel field slightly higher than the critical field of the bulk material. A perpendicular high field will penetrate a type I superconductor (nontransition elements, many of their alloys and compounds[9]) via microscopic normal lamellae or a similar intermediate structure. On the other hand, a type II superconductor (transition metals and their compounds, many alloys[9]) will form a more or less regular vortex array. Microwave studies have shown that additional losses are caused by normal currents through the high-field regions and also by the lossy motion of the normal to superconducting boundaries of the vortices.[2,20,21,40] Analogous infrared studies, which appear difficult but challenging, are still to be made.

D. Composite and Weakly Coupled Superconductors

Superconductivity is based on a long-range interaction. The coherent behavior of the electrons will extend from a superconductor into an adjacent material even though the latter may itself not be superconducting. Superconductivity induced by this proximity effect will extend no farther than one coherence length (up to 10^{-4} cm in pure materials). The induced energy gap is relatively small. Fanelli and Meissner[41] studied the microwave absorption of (basically nonsuperconducting) Au or Cu films on superconducting Sn and confirmed the expected proximity effect. From other experiments, we would expect the converse effect, namely the reduction of the basic energy gap of a superconducting film due to the proximity of a normal, and even more so, of a ferromagnetic material. Finally, an enhancement of superconductivity is expected from the proximity of certain dielectric layers. Optical studies of these composite structures with the aid of far-infrared methods discussed above appear promising.

Coherence also extends across a thin (~ 20 Å) insulating layer separating two superconductors. The thin insulator forms a weak but superconductive link between the two superconductors. Narrow metallic bridges, too, may form weak links. The most striking feature of a weak link, first predicted by Josephson,[42] is its response to an applied bias voltage V_0 with a sinusoidal

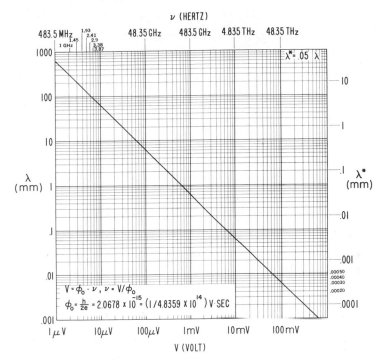

Fig. AV.12 Relation between bias voltage V, frequency ν, and vacuum wavelength λ for a Josephson diode. The symbol λ^* gives the wavelength of the electromagnetic radiation inside a typical coupling dielectric.

current

$$I = I_0 \sin(2\pi f t) \qquad f = V_0/\phi_0, \qquad \text{(AV.10)}$$

where ϕ_0 is the flux quantum defined by Eq. AV.9. A Josephson diode, which can also be made of oxidized pressure contacts,[2,43] will thus oscillate at a frequency of $f = 45.35$ GHz under a bias $V_0 = 100\,\mu$V. Figure AV.12 graphically presents this relationship. It also gives the concomitant wavelength in a vacuum (designated here by λ) and in a typical weak-coupling dielectric (this wavelength is given the symbol λ^*). The Josephson diode undoubtedly constitutes the simplest oscillator with respect to tuning and to power requirements. It is difficult to extract microwave power from a single diode of this type because currents and voltages are small and because the diode impedance amounts to only milliohms in the GHz frequency range (waveguides have impedances of, at least, several ohms in this range). Nevertheless, Dayem and Grimes,[44] using a niobium junction, were able to transmit 10^{-10} W at 9.2 GHz. The power conversion efficiency was 0.1%.

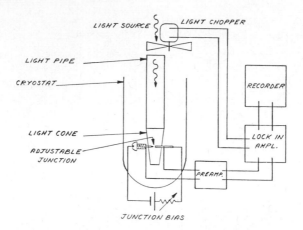

Fig. AV.13   Josephson infrared detector. The diode is made of an adjustable Nb "point" contact. Oxide layers on the Nb provide the weak coupling'dielectric. [S. Shapiro, "Josephson Effect Far-Infrared Detectors," Rept. AD661848, 1967.]

The bias voltage V_0 and thus the frequency is limited by the equivalent gap energy. Still, by using wide gap (high T_c) materials and appropriate dielectrics, it should be possible to develop a coherent emitter for far-infrared photons based on the Josephson effect.

A Josephson diode can also be used as a coherent detector. Equation AV.10 indicates that an unbiased diode carrying an induced high-frequency current $I_0 \sin(2\pi ft)$ will build up a dc voltage V_0. A detailed analysis[45] shows that the current-voltage characteristics will display a number of voltage steps of equal width (but not of equal height), the width given by Eq. AV.10. This was first demonstrated by Shapiro.[46] More recently, Shapiro showed[47] that light quanta of the far-infrared range (wavelength 0.3 mm and longer) are detected in a similar fashion. His apparatus is outlined in Fig. AV.13. Since the quanta had energies exceeding the energy gap, the effect is unexpected and not explainable in terms of Josephson's theory. Nevertheless, the observed NEP of only 5×10^{-13} W $\cdot$ Hz$^{-\frac{1}{2}}$ and the observed response time of better than 10 nsec offer the promise for a useful new far-infrared detector.

A detector with a sensitivity and selectivity much higher than available from the "Shapiro effect" itself becomes feasible through mixing techniques. It can be shown[48] that the ac part of the weak supercurrent crossing the junction behaves inductively according to

$$V = IL, \quad L = L_0\{I - (I/I_0)^2\}^{-\frac{1}{2}},$$
$$L_0 = \phi_0/I_0. \tag{AV.11}$$

The inductance L becomes highly nonlinear as I approaches I_0, the critical current of the junction. Another nonlinearity enters as I_0 is exceeded. Current is then carried by "single particle tunnelling" and other resistive processes; a resistive voltage develops across the junction. According to Matisoo,[49] the transition from the superconductive to the resistive mode occurs in less than 1 nsec. Mixing experiments were performed by Gaulé and collaborators[48] at 12 GHz. More recently, the same authors eliminated the external local oscillator and used the mixer diode itself to oscillate at the local frequency.

A number of technical difficulties must be overcome in order to make these novel coherent detector principles available for the far-infrared. One difficulty to be met is the construction of diodes with wide gap materials, another is a more efficient coupling of the radiation energy into the thin dielectric which carries the weak supercurrents.

E. Superconductivity in Conjunction with Other Solid State Effects. Parametric Thermometer.

Giaever[50] recently replaced the usual dielectric in a Josephson junction with a photoconductor (CdS) and obtained a response of the critical current I_0 to visible light. In view of the very low operating temperature of this new device, it appears possible to use an infrared-sensitive photoconductor as the weak-coupling link, thus creating a novel infrared detector.

Hartwig[50] used n-type silicon as a photodielectric at 4°K and 290–810 MHz to change the capacity and thus the resonance frequency of a high Q re-entrant superconducting cavity. Use of the system as a wide-bandwidth optical detector is predicted. Hartwig expects that infrared-sensitive photodielectrics for analogous applications will be found.[52]

The magnetic fields needed in many solid-state applications are now provided by superconducting magnet coils, which are small and yield fields of a constancy, uniformity, and intensity not available from conventional magnets.

As shown in Eq. AV.11, a Josephson diode has an essentially lossless inductance (except for radiation losses which can be, in principle, avoided) that depends strongly on the ratio of the bias current I and the critical current I_0. Since I_0 is a function of the temperature in the same way as the superfluid density W, an inductance measurement can give a sensitive indication of the temperature. Nonlinear inductors, made by Little[53] of superconducting filaments carrying a high (but still subcritical) bias current, show similar potential characteristics.

In addition to the nominal bias voltage V_0, a Josephson diode will experience a voltage resulting from the analogue of Johnson noise. This will cause a modulation of the frequency about the value given by Eq. AV.10. Since Johnson noise is temperature-dependent according to Eq. AV.3, the frequency modulation is an indication of temperature. This effect was used by Kamper to sense changes in the temperature.[54]

One or several of these parametric fully superconducting temperature sensors should eventually form a superior substitute for the resistive and fundamentally more noisy bolometer discussed above in Section B.b.

REFERENCES

1. *American Institute of Physics Handbook*, 2nd edition, D. E. Gray, Ed., McGraw-Hill, New York, 1963.
2. G. K. Gaulé, "High-Frequency and Detector Applications of Superconductors," 4th Annual Cryogenic Technology Symposium, Chicago, Ill., June, 1968. Published in *Applications of Cryogenic Technology*, R. W. Vance, Ed., Tinnon-Brown, Los Angeles, 1969.
3. E. H. Putley, *J. Sci. Instr. (London)* **43**, 857 (1966).
4. R. L. Petritz, "Cooled Low-Noise, High-Frequency Transistor," U.S. Patent 3,009,085 (1961).
5. G. J. Van Gurp, *Phys. Rev.* **166**, 436 (1968).
6. L. Newhouse, *Applied Superconductivity*, Wiley, New York, 1964.
7. H. M. Rosenberg, *Low-Temperature Solid State Physics*, Clarendon Press, Oxford, 1963.
8. S. Foner, E. J. McNiff, B. T. Matthias, and E. Corenzwit, "11th International Conference on Low-Temperature Physics," St. Andrews, Scotland, August, 1968, Paper IB11b.
9. B. W. Roberts, *Superconductive Materials and Some of Their Properties*, NBS Technical Note 408, U.S. Govt. Printing Office, Washington, D.C., 1966. See also *Progress in Cryogenics*, Academic Press, New York, 1964, vol. 4, p. 159.
10. J. C. Wheatley, *Am. J. Phys.* **36**, 181 (1968).
11. M. F. Wood and D. Phillips, "The ³He Refrigerator," *Proceedings of the International Cryogenic Engineering Conference*, Kyoto, Japan, 1967.
12. Ch. A. Stochl and E. R. Nolan, *Current Status and Future Trends of Cryogenic Coolers for Electronic Applications*, Report AD 610015, July 1964. Federal Clearing House, Department of Commerce, Springfield, Va. 22151.
13. F. London, *Superfluids*, Vol. I, Dover, New York, 1960.
14. E. A. Lynton, *Superconductivity*, Wiley, New York, 1962.
15. M. Strongin, O. F. Kammerer, J. Crow, R. S. Thompson, and H. L. Fine, *Phys. Rev. Letters* **20**, 922 (1968). See also R. E. Glover, *Phys. Letters* **25A**, 542 (1967).
16. D. Bloor, T. J. Dean, G. O. Jones, D. H. Martin, P. A. Mawer, and C. H. Perry, *Proc. Roy. Soc. (London)* **A260**, 510 (1961).
17. G. L. Bertin and K. Rose, *J. Appl. Phys.* **39**, 2561 (1968).
18. G. L. Bertin, Private Communication.
19. J. L. Stone, and W. H. Hartwig, *J. Appl. Phys.* **39**, 2665 (1968).
20. J. M. Victor and W. H. Hartwig, *J. Appl. Phys.* **39**, 2539 (1968).
21. M. S. McAshan, "The Application of Superconductors in the Constructions of High-Q Microwave Cavities", in *Proceedings of the Symposium Physical Superconducting Devices*, Charlottesville, Va., 1967. Report AD 661848, p. C-1. Federal Clearing House, Department of Commerce, Springfield, Va. 22151.
22. R. E. Glover, III and M. Tinkham, *Phys. Rev.* **108**, 243 (1957).
23. M. Tinkham, "Spectroscopy of Solids in the Far-Infrared," in *Science* **145**, 240 (1964) (Review Article).

24. M. Tinkham, *Superconductivity*, Gordon and Breach, New York, 1965.

25. D. M. Ginsberg and M. Tinkham, *Phys. Rev.* **118**, 990 (1960).

26. W. K. Panofsky and M. Phillips, *Classical Electricity and Magnetism*, Addison-Wesley, Reading, Mass., 1962.

27. J. Bardeen, L. N. Cooper, and J. R. Schrieffer, *Phys. Rev.* **108**, 1175 (1957).

28. D. C. Mattis and J. Bardeen, *Phys. Rev.* **111**, 412 (1958).

29. R. E. Glover, III and M. Tinkham, *Phys. Rev.* **104**, 844 (1956).

30. M. Tinkham, *Phys. Rev.* **104**, 845 (1956).

31. S. L. Norman, *Phys. Rev.* **167**, 393 (1968).

32. P. L. Richards and M. Tinkham, *Phys. Rev. Letters* **1**, 318 (1958).

33. P. L. Richards and M. Tinkham, *Phys. Rev.* **119**, 575 (1960).

34. P. L. Richards, *Phys. Rev. Letters* **7**, 412 (1961).

35. M. A. Biondi, M. P. Garfunkel, and W. A. Thompson, *Phys. Rev.* **136**, A1471 (1964).

36. N. M. Rugheimer, A. Lehoczky, and C. V. Briscoe, *Phys. Rev.* **154**, 414 (1967).

37. L. H. Palmer and M. Tinkham, *Phys. Rev.* **165**, 588 (1968).

38. H. D. Drew and A. J. Sievers, *Phys. Rev. Letters* **19**, 697 (1967).

39. W. S. Martin and M. Tinkham, *Phys. Rev.* **167**, 421 (1968).

40. C. R. Haden and W. H. Hartwig, *Phys. Rev.* **148**, 313 (1966).

41. R. Fanelli and H. Meissner, *Phys. Rev.* **147**, 227 (1966).

42. B. D. Josephson, *Rev. Mod. Phys.* **36**, 216 (1964); *Adv. Physics* (*London*) **14**, 419 (1965).

43. G. K. Gaulé, K. Schwidtal, J. T. Breslin, R. L. Ross, and J. J. Winter, "Correlation Between Observed Currents and Coupling Mechanisms in Coupled Superconductors," in *Proceedings of the 10th International Conference on Low-Temperature Physics*, M. P. Malkov, Ed., Viniti, Vol. IIB, Moscow, 1967, pp. 357–361.

44. A. H. Dayem and C. G. Grimes, *Appl. Phys. Letters* **9**, 47 (1966).

45. B. N. Taylor, "The ac Josephson Effect," in *Proceedings of the 10th International Conference on Low-Temperature Physics*, M. P. Malkov, Ed., Viniti, Vol. IIA, Moscow, 1967, p. 59.

46. S. Shapiro, *Phys. Rev. Letters* **11**, 80 (1963).

47. S. Shapiro, "Josephson Effect Far-Infrared Detectors," in *Proceedings of the Symposium on Physical Superconducting Devices*, University of Virginia, Charlottesville, Va., 1967. Report AD 661848, p. I-1. Federal Clearing House Department of Commerce, Springfield, Va. 22151. See also *J. Appl. Phys.* **39**, 3905 (1968).

48. G. K. Gaulé, R. L. Ross, and K. Schwidtal, "Microwave Mixing with Weakly Coupled Superconductors", in *Proceedings of the Symposium on Physical Superconducting Devices*, University of Virginia, Charlottesville, Va., 1967. Report AD 661484, p. P-1, Department of Commerce, Clearing House, Springfield, Va. 22151.

49. J. Matisoo, *J. Appl. Phys.* **39**, 2587 (1968).

50. I. Giaever, *Phys. Rev. Letters* **20**, 1286 (1968).

51. G. D. Arndt, W. H. Hartwig, and J. L. Stone, *J. Appl. Phys.* **39**, 2653 (1968).

52. W. H. Hartwig, Private Communication.

53. W. A. Little, "Device Application of Superinductors", in *Proceedings of the Symposium on Physical Superconducting Devices*, University of Virginia, Charlottesville, Va., 1967. Report AD 661848, p. S-1. Federal Clearing House, Department of Commerce, Springfield, Va. 22151; R. Meservey and P. M. Tedrow, *J. Appl. Phys.* **40**, 2028 (1969).

54. R. A. Kamper, "Millidegree Noise Thermometer," in *Proceedings of the Symposium on Physical Superconducting Devices*, University of Virginia, Charlottesville, Va., 1967. Report AD 661848, p. M-1. Federal Clearing House, Department of Commerce, Springfield, Va. 22151.

APPENDIX VI

RAPID-SCAN FOURIER SPECTROSCOPY

ERNEST V. LOEWENSTEIN

Air Force Cambridge Research Labs., Bedford, Massachusetts

A. INTRODUCTION

Fourier spectroscopy was considered at some length in Chapter 4. In this appendix we discuss the special technique of rapid scan.

Ordinarily, an interferometer for Fourier spectroscopy is driven once through the desired range of path difference and the digitized record then Fourier-transformed to give the spectrum. In the rapid-scan method the interferometer is driven rapidly through the predetermined range of path difference many times. The records are then added, either before or after the Fourier transform is made. In one application of the rapid-scan interferometer—the case of the order sorter and interference modulator—the record need not even be transformed.

Since the scanning function is the fundamental quantity of the interferometer that governs the appearance of the spectrum, we will recall a few points from Chapter 4. Every interferogram is truncated at some finite length L, and the Fourier transform of the "boxcar" function of that length is

$$\text{sinc } 2\pi v L = \frac{\sin 2\pi v L}{2\pi v L}. \qquad (\text{AVI.1})$$

The spectrum produced from the interferogram will be the "true" spectrum convolved with the scanning function, where the true spectrum is defined as that which would be obtained if the interferometer were driven to an infinite path difference. As was shown in Chapter 4, the scanning function can be altered by apodizing, but no apodizing function can give lower side lobes without widening the narrow central peak given by the sinc function. The usual choice is to reduce the side lobes and accept a wider central peak, which means somewhat reduced resolution.

It is apparent that the scanning function is independent of the speed at which the interferometer is driven because time appears explicitly only in the noise and not in the signal analysis of the interferometer. Thus, two spectra corresponding to the same path difference will look the same (disregarding noise) regardless of the time required to record the interferograms. Figure AVI.1(a) schematically illustrates the variation in path difference as a function of time for the aperiodic interferometer, and Fig. AVI.1(b)

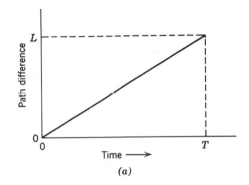

(a)

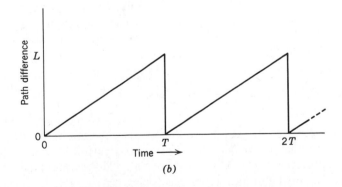

(b)

Fig. AVI.1 (a) The variation of path difference as a function of time for the aperiodic (single sweep) interferometer. (b) The implicit repetetiveness that develops from the calculation of the finite Fourier series with the drive scheme of (a).

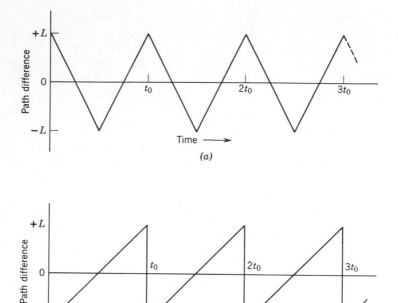

Fig. AVI.2   (a) The time variation of path difference for Genzel's interference modulator and for the order sorter. (b) The ideal variation of path difference in the rapid-scan technique of Mertz.

shows the implicit assumption of periodicity. Figure AVI.2(a) shows the periodic drive mode used by Genzel in his "interference modulation" technique, and Fig. AVI.2(b) shows the idealized drive scheme for the rapid-scan technique of Mertz (zero return time).

The equation that governs the fringe frequency is

$$f = 2vs, \tag{AVI.2}$$

where f is the frequency of the fringes generated from a spectral element of frequency v (cm^{-1}) when the movable mirror of the interferometer is driven at speed s. The factor of 2 applies because the path difference is twice the mirror displacement for Michelson or lamellar-grating interferometers, which are the most common types used.

We will consider three applications of the periodically driven interferometer: the order sorter, Genzel's interference modulator, and the rapid-scan coadding technique of Mertz. The first two will be discussed together, as they amount to essentially the same thing even though the experimental realization is quite distinct.

B. ORDER-SORTING AND INTERFERENCE MODULATION

The earliest work on interferometric modulation as an order-sorting technique was by Strong and Madden,[1] who used an interference filter of continuously varying thickness evaporated on a cylinder. In this method, radiation from a grating spectrometer is reflected from the cylinder, which is placed at a focus and rotated at constant speed. The radiation at each frequency is modulated in accordance with Eq. AVI.2, and overlapping grating orders can then be separated by electronic filtering since the resulting electrical frequencies are well separated.

By using very narrow band electrical filters it becomes possible to dispense with the grating spectrometer altogether, the role of the spectrometer being merely to produce a series of narrow lines. This is illustrated in Fig. AVI.3

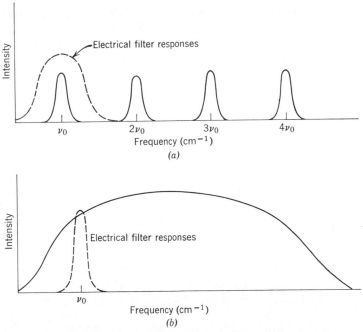

Fig. AVI.3 Two ways of selecting a narrow frequency interval in the spectrum. (a) The narrow interval is provided by a grating and all orders except the desired one are eliminated by a relatively broad electrical filter. (b) The broad background is provided by a filtered source and the narrow band is supplied by a sharply tuned amplifier-filter combination following the interferometer which modulates the entire spectral band.

which shows (*1*) the narrow grating orders used with a relatively broad electrical filter, and (*2*) a continuous spectrum used with a narrow electrical filter. The parameters can be adjusted to make the product of the optical band and electrical band equal in the two cases.

Genzel and Weber,[2-4] recognizing this principle, built a lamellar-grating interference modulator for the far-infrared and used a lock-in or phase-sensitive amplifier to obtain a narrow electrical filter. The system performed satisfactorily in practice, and the mathematical details were worked out by Genzel.[4] It must be recognized that the system operates as a monochromator and thus the multiplex advantage of Fourier spectroscopy is lost. The tuned amplifier is set to a single (electric) frequency, and only the wavelength corresponding to that frequency is read out at one time; information from all other wavelengths is discarded.

The physical realization of an interference modulator depends on fulfilling a number of mechanical conditions. First, since the synchronous demodulation of the fringes must be referred to the shaft of the motor driving the interferometer, the fringe frequency has to be the same for all wavelengths. This requires a drive system whose speed can be adjusted inversely proportional to the wavelength. Second, an integral number of cycles of the interference fringes must be used, demanding precise control of the distance of drive, which must also be varied with wavelength. Third, in order to have an integral number of wavelengths of travel on both sides of zero, the point at which the interferometer crosses zero path difference must be precisely at the center of the travel. These requirements add up to a need for great mechanical rigidity, thereby implying massive components. Since the periodic reversal of the direction of drive must be accomplished in the shortest possible time, high accelerations are incurred at turn-around, with consequent wearing of parts. Vance[5] has made an analysis of the tolerances required under the conditions mentioned here and finds that the drive must be linear to within 1% with the maximum path difference accurate to $\lambda/40$. To avoid significant distortion of the scanning function, the phase error must be less than 4%. Clearly, these tolerances can be met only for wavelengths in the far-infrared.

Besides Genzel's interference modulator, the only other rapid-scanned lamellar-grating in existence is Bell's at Ohio State University. Used purely as an order sorter in connection with a large grating spectrometer, it has consistently been producing very high spectral purity in the far-infrared.

C. RAPID-SCAN FOURIER SPECTROSCOPY

A more important technique, which not only utilizes rapid scanning but also retains the multiplex advantage, is the rapid-scan coadding technique described by Mertz.[6] The path difference variation as a function of time is a sawtooth as illustrated in Fig. AVI.2(*b*). If the available observation time is T, then the number of interferograms recorded is $n = T/t_0$. These n interferograms are added together before being Fourier-transformed by means of a special digital memory unit called a "coadder." The result is one interferogram which has effectively been recorded over a total time T, but each point of which has in fact been sampled at n different times during the interval of observation.

The application of this technique offers several advantages that may be significant. First, the audiofrequency fringes allow dispensing with the chopper, thus doubling the energy obtained by the more conventional method. Second, the fringe frequency is at the experimenter's disposal and may be chosen to lie outside the range of important noise frequencies. This is significant, for example, in astronomical interference spectroscopy where the scintillation is known to be restricted to a definite frequency band. Third, rapid-scan interferometry can be used for investigating a source whose intensity varies in time but whose spectrum remains the same, whereas in a conventional slow-scan the interferogram would acquire level variations that would ruin the spectrum. Hence, a hand-held Fourier spectrometer is practical when using rapid-scan. Any interferogram taken when the source is not in the field of view of the interferometer neither contributes to nor detracts from the result, the only penalty being a certain loss of observation time "on target." This can be an important consideration in a field program or in astronomy where the improbabilities of being able to repeat an observation make it necessary to ensure that at least some useful data will be obtained during the available observation period.

It is not of course necessary to coadd the interferograms at the time the data are taken. Each interferogram may be recorded separately and examined individually if the signal-to-noise ratio is high enough, and those to be added are then chosen according to any criteria the experimenter chooses. Moreover, the fast digital Fourier transform algorithm (due to Cooley and Tukey[7]), now makes it unnecessary to add the interferograms. They may be transformed separately and the spectra inspected and added.

D. SUMMARY

We have discussed the technique of rapid-scan Fourier spectroscopy which has two important manifestations: (*1*) interference modulation or order-sorting and (*2*) rapid-scan coadding. The first pair have been shown to be different manifestations of the same basic approach. The second pair, discussed from the point of view of extending the usefulness of Fourier spectroscopy, is made possible by the multiplex advantage of Fourier spectroscopy. In principle, rapid-scan grating spectroscopy would also be possible, but to the best of our knowledge there is no case yet of a grating spectrometer that could furnish sufficient energy for its application.

REFERENCES

1. R. Madden and J. Strong, *J. Opt. Soc. Am.* **44**, 352 (1954).
2. L. Genzel and R. Weber, *Z. Angew. Phys.* **10**, 127 (1958).
3. L. Genzel and R. Weber, *Z. Angew. Phys.* **10**, 195 (1958).
4. L. Genzel, *J. Mol. Spectry.* **4**, 241 (1960).
5. M. Vance, Ph.D. Dissertation, Ohio State University, Columbus, Ohio (1962). Available from University of Microfilms, Ann Arbor, Mich.
6. L. Mertz, *Transformations in Optics*, Wiley, New York, 1965, p. 44.
7. See, for example, W. T. Cochran *et al.*, *IEEE Tr. Audio and Electroacoustics*, **AU 15**, 45 (1967).

APPENDIX VII

A FAR-INFRARED BIBLIOGRAPHY

E. D. PALIK

Naval Research Laboratory, Washington, D.C.

This bibliography brings up to date, through 1969, the original *A Far Infrared Bibliography, J. Opt. Soc. Am.* **50**, 1329 (1960) and NRL Bibliography No. 21, April 1963 (available from Office of Technical Services, Department of Commerce, $1.50) and the *First Supplement to A Far Infrared Bibliography*, April 1965 (available from the author, Semiconductors Branch, Solid State Division, Naval Research Laboratory, Washington, D.C. 20390). While the second and third bibliographies were cross-indexed by author and subject matter, the present one contains only the author index.

This bibliography is arranged in chronological order. For a given year, the articles are alphabetized by the first author's last name. The sequence of information for each entry is author(s), title of paper, and reference. Journal abbreviations usually follow the form given in the *American Institute of Physics Style Manual*.

The bibliography covers roughly the region 25–1000 μ from 1892 to 1963. In 1964 the far-infrared was redefined as 50–1000 μ with liberal extensions to longer wavelengths by optical techniques. In the years 1964 to 1969, many vibrational analyses of various molecules were done. Since these data usually involved only a few bands at wavelengths longer than 50 μ, with the bulk of the work done at wavelengths shorter than 50 μ, only some representative papers are included. We have also omitted the Raman effect as a means for studying low-frequency phenomena.

Several bibliographies in books and articles have proved helpful. Among these are *Infrared Spectroscopy*, by R. B. Barnes, R. C. Gore, U. Liddel, and V. Z. Williams (Reinhold Publishing Corp., New York, 1944); *Infrared, a Bibliography*, Part I by C. R. Brown, M. W. Ayton, T. C. Goodwin, and

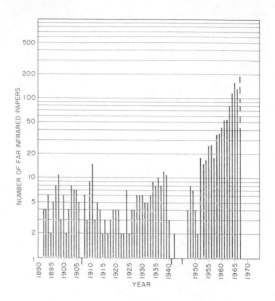

Fig. AVII.1   Far-infrared studies since 1890.

T. J. Derby (The Library of Congress, Technical Information Division, Washington, D.C., 1957); *Analytical Applications of Far Infrared Spectra, I. Historical Review, Apparatus and Techniques*, by F. F. Bentley, E. F. Wolfarth, N. Srp, and W. R. Powell, in *Spectrochim. Acta* **13**, 1 (1958); and *Chemical Far Infrared Spectroscopy*, J. W. Brasch, R. J. Jakobsen, and Y. Mikawa, *Appl. Spectry.* **22**, 641 (1968).

A graph illustrating the growth of far-infrared studies precedes the main bibliography. In previous issues of this bibliography the ordinate was linear. It now is appropriate to plot the number of papers in each year on a logarithmic scale.

1892

1. H. du Bois and H. Rubens, ÜBER EIN BRECHUNGSGESETZ FÜR DEN EINTRITT DES LICHTES IN ABSORBIERENDE MEDIEN, *Wied. Ann.* **47**, 203.
2. H. du Bois and H. Rubens, ÜBER POLARISATION ULTRAROTER STRAHLEN BEIM DURCHGANG DURCH METALLDRAHTGITTER, *Berl. Ber.* **1892**, 1129.
3. H. Rubens, ÜBER DISPERSION ULTRAROTER STRAHLEN, *Wied. Ann.* **45**, 238.

4. H. Rubens and B. W. Snow, ÜBER DIE BRECHUNG DER STRAHLEN VON GROSSER WELLENLÄNGE IN STEINSALZ, SYLVIN UND FLUORIT, *Wied. Ann.* **46**, 529.

1893

5. H. du Bois and H. Rubens, ÜBER POLARISATION VON WÄRMESTRAHLEN DURCH METALLDRAHTGITTER, *Naturw. Rdsch.* **8**, 453.
6. H. du Bois and H. Rubens, POLARISATION UNGEBEUGTER ULTRAROTER STRAHLUNG DURCH METALLDRAHTGITTER, *Wied. Ann.* **49**, 593.
7. H. du Bois and H. Rubens, MODIFIZIERTES ASTATISCHES GALVANOMETER, *Wied. Ann.* **48**, 236.
8. H. Rubens and B. W. Snow, ON THE REFRACTION OF RAYS OF GREAT WAVE-LENGTH IN ROCK-SALT, SYLVINE, AND FLUORITE, *Phil. Mag.* **35**, 35.

1894

9. H. du Bois and H. Rubens, PANZERGALVANOMETER, *Electrotech. Z.* **13**, 321.
10. H. Rubens, EINIGE NEUERE GALVANOMETERFORMEN, *Naturw. Rund.* **9**, 47.
11. H. Rubens, ZUR DISPERSION DER ULTRAROTEN STRAHLEN IM FLUORIT, *Wied. Ann.* **51**, 381.
12. H. Rubens, PRÜFUNG DER DETTELER-HELMHOLTZ'SCHEN DISPERSIONS-FORMEL, *Wied. Ann.* **53**, 267.
13. H. Rubens, PRÜFUNG DER HELMHOLTZ'SCHEN DISPERSIONSTHEORIE, *Naturw. Rdsch.* **9**, 389.
14. H. Rubens, ZUR HELMHOLTZ'SCHEN DISPERSIONSLEHRE, *Naturw. Rdsch.* **9**, 606.

1895

15. H. Rubens, DIE KETTELER-HELMHOLTZ'SCHE DISPERSIONSFORMEL, *Wied. Ann.* **54**, 476.
16. H. Rubens, VIBRATIONSGALVANOMETER, *Wied. Ann.* **56**, 27.

1896

17. H. Rubens, AUFSTELLUNG UND ASTASIERUNG EINES EMPFINDLICHEN SPIEGELGALVANOMETERS, *Verhandl. Deut. Phys. Ges.* **15**, 11.
18. H. Rubens, ÜBER DAS ULTRAROTE ABSORPTIONSSPEKTRUM VON STEIN-SALZ UND SYLVIN, *Verhandl. Deut. Phys. Ges.* **15**, 108.
19. H. Rubens, DEMONSTRATIONSVERSUCHE MIT ELEKTRISCHEN WELLEN, *Z. Phys. Chem. Unterr.* **9**, 241.
20. H. Rubens and E. F. Nichols, ÜBER WÄRMESTRAHLEN VON GROSSER WELLENLÄNGE, *Naturw. Rdsch.* **11**, 545.
21. H. Rubens and E. F. Nichols, BEOBACHTUNG ELEKTRISCHER RESONANZ AN WÄRMESTRAHLEN VON GROSSER WELLENLÄNGE, *Berl. Ber.* **1896**, 1393.

1897

22. E. F. Nichols, ÜBER DAS VERHALTEN DES QUARZES GEGEN STRAHLEN GROSSER WELLENLÄNGE, UNTERSUCHT NACH DER RADIOMETRIS-CHEN METHODE, *Wied. Ann.* **60**, 401.

23. E. F. Nichols, A METHOD FOR ENERGY MEASUREMENTS IN THE INFRARED SPECTRUM AND THE PROPERTIES OF THE ORDINARY RAY IN QUARTZ FOR WAVES OF GREAT WAVELENGTH, *Phys. Rev.* **4**, 297.

24. H. Rubens, VERSUCHE MIT KURZEN ELEKTRISCHEN WELLEN, *Z. Phys. Chem. Unterr.* **10**, 239.

25. H. Rubens and E. F. Nichols, VERSUCHE MIT WÄRMESTRAHLEN VON GROSSER WELLENLÄNGE, *Wied. Ann.* **60**, 418.

26. H. Rubens and E. F. Nichols, HEAT RAYS OF GREAT WAVELENGTH, *Phys. Rev.* **4**, 314.

27. H. Rubens and E. F. Nichols, CERTAIN OPTICAL AND ELECTRO-MAGNETIC PROPERTIES OF HEAT WAVES OF GREAT WAVELENGTH, I., *Phys. Rev.* **5**, 98.

28. H. Rubens and E. F. Nichols, CERTAIN OPTICAL AND ELECTRO-MAGNETIC PROPERTIES OF HEAT WAVES OF GREAT WAVELENGTH, II., *Phys. Rev.* **5**, 152.

29. H. Rubens and E. F. Nichols, RECHERCHES SUR LES RADIATION DE GRANDE LONGUEUR D'ONDE, *Séances Soc. Franc. Phys.* **1897**, 40.

30. H. Rubens and A. Trowbridge, BEITRAG ZUR KENNTNISS DER DISPERSION UND ABSORPTION DER ULTRAROTEN STRAHLEN IN STEINSALZ UND SYLVIN, *Wied. Ann.* **60**, 724.

1898

31. C. E. Guillaume, THE EXTREME INFRARED RADIATIONS, *Smith. Rept.* **1898**, 161.

32. E. Hagen and H. Rubens, ÜBER DAS REFLEXIONSVERMÖGEN VON METALLEN, *Verhandl. Deut. Phys. Ges.* **17**, 143.

33. H. Rubens, ÜBER EINE NEUE THERMOSÄULE, *Z. Instr.* **18**, 65.

34. H. Rubens, EINE NEUE THERMOSÄULE, *Z. Phys. Chem. Unterr.* **11**, 126.

35. H. Rubens and E. Aschkinass, DIE RESTSTRAHLEN VON STEINSALZ UND SYLVIN, *Wied. Ann.* **65**, 241.

36. H. Rubens and E. Aschkinass, BEOBACHTUNGEN ÜBER ABSORPTION UND EMISSION VON WASSERDAMPF UND KOHLENSÄURE IM ULTRAROTEN SPEKTRUM, *Wied. Ann.* **64**, 584.

37. H. Rubens and E. Aschkinass, ÜBER DIE EIGENSCHAFTEN DER REST-STRAHLEN DES STEINSALZES, *Verhandl. Deut. Phys. Ges.* **17**, 42.

38. H. Rubens and E. Aschkinass, ÜBER DIE DURCHLÄSSIGKEIT EINIGER FLÜS-SIGKEITEN FÜR WÄRMESTRAHLEN VON GROSSER WELLENLÄNGE, *Wied. Ann.* **64**, 602.

39. H. Rubens and E. Aschkinass, ÜBER DIE RESTSTRAHLEN VON STEINSALZ UND SYLVIN, *Naturw. Rdsch.* **13**, 185.

40. H. Rubens and E. Aschkinass, OBSERVATIONS ON THE ABSORPTION AND EMISSION OF AQUEOUS VAPOR AND CARBON DIOXIDE IN THE INFRA-RED SPECTRUM, *Astrophys. J.* **8**, 176.

41. H. Rubens and A. Trowbridge, ON THE DISPERSION AND ABSORPTION OF INFRA-RED RAYS IN ROCK SALT AND SYLVINE, *Am. J. Sci.* **155**, 33.

1899

42. E. Hagen and H. Rubens, DAS REFLEXIONSVERMÖGEN VON METALLEN UND BELEGTEN GLASSPIEGELN, Z. Instr. **19**, 293.
43. H. Rubens, ÜBER DIE RESTSTRAHLEN DES FLUSSPATES, Wied. Ann. **69**, 576.
44. H. Rubens and E. Aschkinass, ISOLIERUNG LANGWELLIGER WÄRMESTRAHLEN DURCH QUARZPRISMEN, Wied. Ann. **67**, 459.

1900

45. H. du Bois and H. Rubens, PANZERGALVANOMETER, Ann. Physik **2**, 84.
46. H. du Bois and H. Rubens, PANZERGALVANOMETER, Z. Instr. **20**, 65.
47. E. Hagen and H. Rubens, DAS REFLEXIONSVERMÖGEN VON METALLEN UND BELEGTEN GLASSPIEGELN, Ann. Physik **1**, 352.
48. H. Rubens, LE SPECTRE INFRAROUGE, Rapport du Congrès International de Physique **2**, 141.
49. H. Rubens and F. Kurlbaum, ÜBER DIE EMISSION LANGWELLIGER WÄRMESTRAHLEN DURCH DEN SCHWARZEN KÖRPER BEI VERSCHIEDENEN TEMPERATUREN, Berl. Ber. **1900**, 929.
50. H. Rubens and E. F. Nichols, RECHERCHES SUR LE SPECTRE INFRAROUGE, LA RESONANCE ELECTRIQUE DES RAYONS DE CHALEUR. Rev. Gen. Sci. Pur. Appl. **11**, 7.

1901

51. H. Rubens and F. Kurlbaum, ANWENDUNG DER METHODE DER RESTSTRAHLEN ZUR PRÜFUNG DES STRAHLUNGSGESETZES, Ann. Physik **4**, 649.
52. H. Rubens and F. Kurlbaum, ON THE HEAT-RADIATION OF LONG WAVELENGTH EMITTED BY BLACK BODIES OF DIFFERENT TEMPERATURES, Astrophys. J. **14**, 335.

1902

53. E. Hagen and H. Rubens, DAS REFLEXIONSVERMÖGEN EINIGER METALLE FÜR ULTRAVIOLETTE UND ULTRAROTE STRAHLEN, Ann. Physik **8**, 1.
54. E. Hagen and H. Rubens, DAS REFLEXIONSVERMÖGEN EINIGER METALLE FÜR ULTRAVIOLETTE UND ULTRAROTE STRAHLEN, Z. Instr. **22**, 42.
55. E. Hagen and H. Rubens, DIE ABSORPTION ULTRAVIOLETTER, SICHTBARER UND ULTRAROTER STRAHLEN IN DÜNNEN METALLSCHICHTEN, Verhandl. Deut. Phys. Ges. **4**, 55.
56. E. Hagen and H. Rubens, DIE ABSORPTION ULTRAVIOLETTER, SICHTBARER UND ULTRAROTER STRAHLEN IN DÜNNEN METALLSCHICHTEN, Ann. Physik **8**, 432.

1903

57. E. Hagen and H. Rubens, ÜBER BEZIEHUNGEN ZWISCHEN DEM REFLEXIONSVERMÖGEN DER METALLE UND IHREM ELEKTRISCHEN LEITVERMÖGEN, Verhandl. Deut. Phys. Ges. **5**, 113.

58. E. Hagen and H. Rubens, ÜBER BEZEIHUNGEN ZWISCHEN DEM REFLEXIONS-VERMÖGEN DER METALLE UND IHREM ELEKTRISCHEN LEITVERMÖGEN, *Berl. Ber.* 1903, 269.

59. E. Hagen and H. Rubens, DAS EMISSIONSVERMÖGEN DER METALLE FÜR LANGE WELLEN, *Berl. Ber.* 1903, 410.

60. E. Hagen and H. Rubens, DAS EMISSIONSVERMÖGEN DER METALLE FÜR STRAHLEN GROSSER WELLENLÄNGE, *Verhandl. Deut. Phys. Ges.* **5**, 145.

61. E. Hagen and H. Rubens, ÜBER BEZIEHUNGEN DES REFLEXIONS UND EMIS-SIONSVERMÖGEN DER METALLE ZU IHREM ELEKTRISCHEN LEITVER-MÖGEN, *Ann. Physik* **11**, 873.

62. E. Hagen and H. Rubens, ÜBER BEZIEHUNGEN ZWISCHEN DEM REFLEXIONS-VERMÖGEN DER METALLE UND IHREM ELEKTRISCHEN LEITVERMÖGEN, *Naturw. Rdsch.* **18**, 185.

63. H. Rubens and E. Hagen, DIE OPTISCHEN UND ELEKTRISCHEN EIGEN-SCHAFTEN DER METALLE, *Phys. Z.* **4**, 727.

64. H. Rubens, VERSUCHE MIT RESTSTRAHLEN VON QUARZ UND FLUSSPAT, *Phys. Z.* **4**, 726.

1904

65. H. du Bois and H. Rubens, ÜBER POLARISATION LANGWELLIGER WÄRME-STRAHLUNG DURCH DRAHTGITTER, *Verhandl. Deut. Phys. Ges.* **6**, 77.

66. E. Hagen and H. Rubens, EMISSIONSVERMÖGEN UND ELEKTRISCHE LEIT-FÄHIGKEIT DER METALLEGIERUNGEN, *Verhandl. Deut. Phys. Ges.* **6**, 1.

67. E. Hagen and H. Rubens, EMISSIONSVERMÖGEN UND ELEKTRISCHE LEIT-FÄHIGKEIT DER METALLEGIERUNGEN, *Verhandl. Deut. Phys. Ges.* **6**, 128.

68. E. Hagen and H. Rubens, SUR LES RAPPORTS ENTRE LES QUALITÉS OPTIQUES ET ÉLECTRIQUES DES MÉTAUX, *Ann. Chim. Phys.* **1**, 185.

69. E. Hagen and H. Rubens, SUR LE POUVOIR ÉMISSIF ET LA CONDUCTIBILITÉ ÉLECTRIQUE DES ALLIAGES, *Ann. Chim. Phys.* **2**, 441.

70. E. Hagen and H. Rubens, ON SOME RELATIONS BETWEEN THE OPTICAL AND THE ELECTRICAL QUALITIES OF METALS, *Phil. Mag.* **7**, 157.

71. E. Hagen and H. Rubens, L'OPTIQUE DES MÉTAUX POUR LES ONDES DE GRANDE LONGUEUR, *Rev. Gen. Sci. Pur. Appl.* **15**, 928.

1905

72. E. Hagen and H. Rubens, SUR QUELQUES RELATIONS ENTRE LES PROPRIÉTÉS OPTIQUES ET ÉLECTRIQUES DES MÉTAUX, *J. Phys. Theor. Appl.* **4**, 264.

73. H. Rubens, DAS EMISSIONSSPEKTRUM DES AUERSTRUMPFES, *Verhandl. Deut. Phys. Ges.* **7**, 346.

74. H. Rubens, DAS EMISSIONSSPEKTRUM DES AUERSTRUMPFES, *Phys. Z.* **6**, 790.

75. H. Rubens, ÜBER DAS EMISSIONSSPEKTRUM DES AUERBRENNERS, *Ann. Physik* **18**, 725.

76. H. Rubens, SPECTRE D'ÉMISSION DES MANCHONS AUER, *Le Radium* **2**, 397.

77. H. Rubens, OPTISCHE KONSTANTEN VON METALLEN, Landolt-Börnstein, Berlin, 1905.

78. H. Rubens and E. Ladenburg, ÜBER DAS LANGWELLIGE ABSORPTIONSSPEK-TRUM DER KOHLENSÄURE, *Verhandl. Deut. Phys. Ges.* **7**, 171.

1906

79. H. Rubens, LE RAYONNEMENT DES MANCHONS À INCADESCENCE, *J. Phys. Theor. Appl.* **5**, 306.
80. H. Rubens, ÜBER DIE TEMPERATUR DES AUERSTRUMPFES, *Phys. Z.* **7**, 186.
81. H. Rubens, EMISSIONSVERMÖGEN UND TEMPERATUR DES AUER-STRUMPFES BEI VERSCHIEDENEM CERGEHALT, *Ann. Physik* **20**, 593.
82. H. Rubens, ÜBER DIE TEMPERATUR DES AUERSTRUMPFES, *Verhandl. Deut. Phys. Ges.* **8**, 41.
83. H. Rubens, DAS EMISSIONSSPEKTRUM DES AUERSTRUMPFES, *Verhandl. Ges. d. Naturf. Ärzte* **2**, 44.

1907

84. H. Rubens and E. Ladenburg, ÜBER LICHTELEKTRISCHE ERSCHEINUNGEN AN DÜNNEN GOLDBLÄTTCHEN, *Verhandl. Deut. Phys. Ges.* **9**, 749.

1908

85. E. Hagen and H. Rubens, ÄNDERUNG DES EMISSIONSVERMÖGENS DER METALLE MIT DER TEMPERATUR, *Verhandl. Deut. Phys. Ges.* **10**, 710.
86. E. F. Nichols and W. S. Day, NEW GROUPS OF RESIDUAL RAYS IN THE LONG-WAVE SPECTRUM, *Phys. Rev.* **27**, 225.
87. H. Rubens, ÜBER DIE DISPERSION VON STEINSALZ UND SYLVIN FÜR LANGE WELLEN, *Ann. Physik* **26**, 615.
88. H. Rubens and E. Hagen, ÄNDERUNG DES EMISSIONSVERMÖGENS DER METALLE MIT DER TEMPERATUR, *Phys. Z.* **9**, 874.
89. H. Rubens and E. Ladenburg, DAS REFLEXIONSVERMÖGEN DES WASSERS, *Berl. Ber.* **1908**, 274; *Verhandl. Deut. Phys. Ges.* **10**, 226.
90. H. Rubens and E. Ladenburg, DAS REFLEXIONSVERMÖGEN DES ÄTHYL-ALKOHOLS, *Berl. Ber.* **1908**, 114.

1909

91. E. Hagen and H. Rubens, ÜBER DIE ABHÄNGIGKEIT DES EMISSIONS-VERMÖGENS DER METALLE VON DER TEMPERATUR, *Berl. Ber.* **1909**, 478.
92. H. Rubens and E. Ladenburg, LA PROPRIÉTÉ OPTIQUE DE L'EAU DANS LE SPECTRE INFRA-ROUGE, *Le Radium* **6**, 33.
93. H. Rubens and E. Ladenburg, ÜBER DIE DISPERSION DES WASSERS IM ULTRAROTEN SPECTRUM, *Verhandl. Deut. Phys. Ges.* **11**, 16.

1910

94. E. Hagen and H. Rubens, ÜBER DIE ÄNDERUNG DES EMISSIONSVERMÖGENS DER METALLE MIT DER TEMPERATUR IM KURZWELLIGEN ULTRA-ROTEN SPECTRUM, *Berl. Ber.* **1910**, 467.

95. H. Rubens, ÜBER DIE ÄNDERUNG DES EMISSIONSVERMÖGENS DER METALLE MIT DER TEMPERATUR IM KURZWELLIGEN TEIL DES ULTRAROT, *Phys. Z.* **11**, 139.

96. H. Rubens and E. Hagen, ÜBER DIE ÄNDERUNG DES EMISSIONSVERMÖGENS DER METALLE MIT DER TEMPERATUR IM KURZWELLIGEN TEIL DES ULTRAROTS, *Verhandl. Deut. Phys. Ges.* **12**, 172.

97. H. Rubens and H. Hollnagel, MESSUNGEN IM LANGWELLIGEN SPEKTRUM, *Berl. Ber.* **1910**, 26.

98. H. Rubens and H. Hollnagel, VERSUCHE MIT LANGWELLIGEN WÄRMESTRAHLEN, *Verhandl. Deut. Phs. Ges.* **12**, 83.

99. H. Rubens and H. Hollnagel, MEASUREMENTS IN THE EXTREME INFRA-RED SPECTRUM, *Phil. Mag.* **19**, 761.

100. H. Rubens and R. W. Wood, ISOLIERUNG LANGWELLIGER WÄRMESTRAHLUNG DURCH QUARZLINSEN, *Berl. Ber.* **1910**, 1122.

101. A. Trowbridge and B. J. Spence, REFLECTING POWER OF ICE IN THE EXTREME INFRARED, *Phys. Rev.* **31**, 61.

102. R. W. Wood, THE ECHELETTE GRATING FOR THE INFRARED, *Phil. Mag.* **20**, 770.

1911

103. H. du Bois and H. Rubens, POLARISATION UNGEBEUGTER LANGWELLIGER WÄRMESTRAHLEN DURCH DRAHTGITTER, *Ann. Physik* **35**, 243.

104. H. du Bois and H. Rubens, POLARISATION LANGWELLIGER WÄRMSTRAHLUNG DURCH HERTZSCHE DRAHTGITTER, *Verhandl. Deut. Phys. Ges.* **13**, 431.

105. H. du Bois and H. Rubens, ON POLARIZATION OF UNDIFFRACTED LONGWAVE HEAT RAYS BY WIRE GRATINGS, *Phil. Mag.* **22**, 322.

106. D. Owen, SHORT ELECTRIC WAVES AND LONG HEAT WAVES, *The Electrician* **58**, 504.

107. H. Rubens, BEMERKUNG ZU DER ARBEIT VON RUBENS UND WOOD: EINFACHE METHODE ZUR ISOLIERUNG SEHR LANGWELLIGER WÄRMESTRAHLUNG, *Verhandl. Deut. Phys. Ges.* **13**, 179.

108. H. Rubens, ÜBER LANGWELLIGE RESTSTRAHLEN DES KALKSPATS, *Verhandl. Deut. Phys. Ges.* **13,** 102.

109. H. Rubens and O. V. Baeyer, ÜBER EINE ÄUSSERST LANGWELLIGE STRAHLUNG DES QUECKSILBERDAMPFS, *Berl. Ber.* **1911**, 339.

110. H. Rubens and O. V. Baeyer, ÜBER DIE ENERGIEVERTEILUNG DER VON DER QUARZQUECKSILBERLAMPE AUSGESSANDTEN LANGWELLIGEN STRAHLUNG, *Berl. Ber.* **1911**, 666.

111. H. Rubens and O. V. Baeyer, ON EXTREMELY LONG WAVES EMITTED BY THE QUARTZ MERCURY LAMP, *Phil. Mag.* **21**, 689.

112. H. Rubens and O. V. Baeyer, SUR LES RAYONS DE LONGEUERS D'ONDE EXTRÈMEMENT GRANDES ÉMIS PAR LA LAMPE À MERCURE EN QUARTZ, *Le Radium* **8**, 139.

113. H. Rubens and H. Wartenberg, ABSORPTION LANGWELLIGER WÄRMESTRAHLEN IN EINIGEN GASEN, *Phys. Z.* **12**, 1080.

114. H. Rubens and H. Wartenberg, ABSORPTION LANGWELLIGER WÄRMESTRAHLEN IN EINIGEN GASEN, *Verhandl. Deut. Phys. Ges.* **13**, 796.

115. H. Rubens and R. W. Wood, FOCAL ISOLATION OF LONG HEAT WAVES, *Phil. Mag.* **21**, 249.
116. H. Rubens and R. W. Wood, EINFACHE ANORDNUNG ZUR ISOLIERUNG SEHR LANGWELLIGER WÄRMESTRAHLUNG, *Verhandl. Deut. Phys. Ges.* **13**, 88.
117. H. Rubens and R. W. Wood, ISOLEMENT DES RAYONS CALORIFIQUES DE GRANDE LONGUEUR D'ONDE A L'AIDE DE LENTILLES DE QUARTZ, *Le Radium* **8**, 44.

1912

118. N. Bjerrum, DIE ULTRAROTEN ABORPTIONSSPEKTREN DER GASE; DIREKTE MESSUNG DER GRÖSSE VON ENERGIEQUANTEN, *Nernst Festschrift*, p. 90.
119. H. Rubens, VÉRIFICATION DE LA FORMULE DU RAYONNEMENT DE PLANCK DANS LE DOMAINE DES GRANDES LONGUEURS D'ONDE, *Rapports et Discussions de la Reunion Tenue à Bruxelles*, Paris.
120. H. Rubens and G. Hertz, ÜBER DEN EINFLUSS DER TEMPERATUR AUF DIE ABSORPTION LANGWELLIGER WÄRMESTRAHLEN IN EINIGEN FESTEN ISOLATOREN, *Berl. Ber.* **1912**, 256.

1913

121. E. V. Bahr, ÜBER DEN EINFLUSS DES DRUCKES AUF DIE ABSORPTION SEHR LANGWELLIGER STRAHLEN IN GASEN, *Verhandl. Deut. Phys. Ges.* **15**, 673.
122. H. Rubens, ÜBER DIE ABSORPTION DES WASSERDAMPFES UND ÜBER NEUE RESTSTRAHLENGRUPPEN IM GEBIET DER GROSSEN WELLENLÄNGEN, *Berl. Ber.* **1913**, 513.
123. H. Rubens and O. V. Baeyer, ÜBER DEN EINFLUSS DER SELEKTIVEN ABSORPTION DES WASSERDAMPFES AUF DIE ENERGIEVERTEILUNG DER LANGWELLIGEN QUECKSILBERDAMPFSTRAHLUNG, *Berl. Ber.* **1913**, 802.
124. R. W. Wood, RESONANZVERSUCHE MIT DEN LÄNGSTEN WÄRMEWELLEN, *Phys. Z.* **14**, 189.
125. R. W. Wood, RESONANCE EXPERIMENTS WITH THE LONGEST HEAT-WAVES, *Phil. Mag.* **25**, 440.

1914

126. W. V. Ignatowski, THEORIE DER GITTER, PRÜFUNG AN DEN BEOBACHTUNGEN VON DUBOIS-RUBENS, *Ann. Physik* **44**, 369.
127. T. J. Meyer, REFLEXION LANGWELLIGER WÄRMESTRAHLEN AN RAUHEN FLÄCHEN UND GITTERN, *Verhandl. Deut. Phys. Ges.* **16**, 126.
128. H. Rubens and K. Schwarzschild, SIND IM SONNENSPEKTRUM WÄRMESTRAHLEN VON GROSSER WELLENLÄNGE VORHANDEN? *Berl. Ber.* **1914**, 702.
129. H. Rubens and H. Wartenberg, BEITRAG ZUR KENNTNIS DER LANGWELLIGEN RESTSTRAHLEN, *Berl. Ber.* **1914**, 169.

1915

130. H. Rubens, ÜBER REFLEXIONSVERMÖGEN UND DIELEKTRIZITÄTSKON-
STANTE ISOLIERENDER FESTER KÖRPER UND EINIGER FLÜSSIGKEITEN,
Berl. Ber. **1915**, 4.
131. H. Rubens, ÜBER NORMALE UND ANOMALE DISPERSION IM LANGWELLI-
GEN SPEKTRUM UND ÜBER HERR DEBYES THEORIES DER MOLEKULA-
REN DIPOLE, *Verhandl. Deut. Phys. Ges.* **17**, 315.

1916

132. H. Rubens, ÜBER REFLEXIONSVERMÖGEN UND DIELEKTRIZITÄTSKON-
STANTE EINIGER AMORPHER KÖRPER, *Berl. Ber.* **1916**, 1280.
133. H. Rubens and G. Hettner, DAS LANGWELLIGE WASSERDAMPFSPEKTRUM
UND SEINE DEUTUNG DURCH DIE QUANTENTHEORIE, *Berl. Ber.* **1916**, 167.
134. H. Rubens and G. Hettner, DAS ROTATIONSSPEKTRUM DES WASSERDAMP-
FES, *Verhandl. Deut. Phys. Ges.* **18**, 154.

1917

135. H. Rubens, DAS ULTRAROT SPEKTRUM UND SEINE BEDEUTUNG FÜR DIE
BESTÄTIGUNG DER ELEKTROMAGNETISCHEN LICHTTHEORIE, *Berl. Ber.*
1917, 47.
136. H. Rubens, ÜBER DIE BRECHUNGSEXPONENTEN EINIGER FESTER KÖRPER
FÜR KURZE HERTZSCHE WELLEN, *Berl. Ber.* **1917**, 556.

1918

137. H. Hollnagel, TRANSPARENCY OF CERTAIN CARBON COMPOUNDS TO
WAVES OF GREAT LENGTH, *Phys. Rev.* **11**, 505 (Abstract).
138. H. Hollnagel, ON THE RESIDUAL RAYS OF ROCK SALT, *Phys. Rev.* **11**, 135
(Abstract).
139. H. Rubens, DIE ENERGIEQUELLEN DER ERDE, *Berl. Ber.* **1918**, 941.

1919

140. A. Eucken, ÜBER DIE ANWENDUNG DER QUANTENTHEORIE AUF DIE
ROTATIONSBEWEGUNG DER GASMOLEKÜLE, *Radioak. Elekt.* **16**, 361.
141. T. Leibisch and H. Rubens, ÜBER DIE OPTISCHEN EIGENSCHAFTEN EINIGER
KRISTALLE IM LANGWELLIGEN ULTRAROT SPEKTRUM, *Berl. Ber.* **1919**,
198.
142. T. Leibisch and H. Rubens, ÜBER DIE OPTISCHEN EIGENSCHAFTEN EINIGER
KRISTALLE IM LANGWELLIGEN ULTRAROTEN SPEKTRUM, *Berl. Ber.* **1919**,
876.
143. H. Rubens, ÜBER DIE DREHUNG DER OPTISCHEN SYMMETRIEACHSEN
VON ADULAR UND GIPS IM LANGWELLIGEN SPEKTRUM, *Berl. Ber.* **1919**,
976.

1920

144. A. Eucken, ROTATIONSBEWEGUNG UND ABSOLUTE DIMENSIONEN DER MOLEKÜLE, *Z. Elektrochem.* **26**, 377.
145. H. Rubens, ÜBER DIE OPTISCHEN UND ELEKTRISCHEN SYMMETRIEACHSEN MONOKLINER KRISTALLE, *Z. Physik* **1**, 11.
146. E. C. Wente, THE SELECTIVE REFLECTION OF HEAT WAVES BY LINEAR RESONATORS, *Phys. Rev.* **16**, 133.
147. H. Witt, ÜBER NEUE APPARATE UND MESSUNGEN IM LANGWELLIGEN SPEKTRUM, *Phys. Z.* **21**, 374.

1921

148. T. Leibisch and H. Rubens, ÜBER DIE OPTISCHEN EIGENSCHAFTEN EINIGER KRISTALLE IM LANGWELLIGEN ULTRAROTEN SPEKTRUM, *Berl. Ber.* **1921**, 211.
149. H. Rubens, GITTERMESSUNGEN IM LANGWELLIGEN SPEKTRUM, *Berl. Ber.* **1921**, 8.
150. H. Rubens and G. Michel, BEITRAG ZUR PRÜFUNG DER PLANCKSCHEN STRAHLUNGSFORMEL, *Berl. Ber.* **1921**, 590.
151. H. Rubens and G. Michel, PRÜFUNG DER PLANCKSCHEN STRAHLUNGSFORMEL, *Phys. Z.* **22**, 569.

1922

152. G. Laski, DIE LANGWELLIGE STRAHLUNG DER QUARZQUECKSILBERLAMPE BEI VERSCHIEDENER BELASTUNG, *Z. Physik* **10**, 353.
153. H. Rubens and K. Hoffman, ÜBER DIE STRAHLUNG GESCHWÄRZTER FLÄCHEN, *Berl. Ber.* **1922**, 424.

1923

154. M. Czerny, ÜBER EINE NEUE FORM DER RUBENSSCHEN RESTSTRAHLENMETHODE, *Z. Physik* **16**, 321.
155. W. Weniger, SUMMARY OF INVESTIGATIONS IN THE INFRARED SPECTRUM OF LONG WAVELENGTHS, *Rev. Sci. Instr.* **7**, 517.

1924

156. A. Glagolewa-Arkadiewa, EINE NEUE STRAHLUNGSQUELLE DER KURZEN ELEKTROMAGNETISCHEN WELLEN VON ULTRAHERTZSCHER FREQUENZ, *Z. Physik* **24**, 153.
157. A. Glagolewa-Arkadiewa, SHORT ELECTROMAGNETIC WAVES OF WAVELENGTH UP TO 92 μ, *Nature* **113**, 640.
158. T. H. Havelock, OPTICAL DISPERSION AND SELECTIVE REFLECTION WITH APPLICATION TO INFRA-RED NATURAL FREQUENCIES, *Proc. Roy. Soc. (London)* **105A**, 488.

159. G. Laski, ULTRAROTFORSCHUNG, *Ergeb. Exakt. Naturwiss.* **3**, 86.
160. M. Lewitski, EIN VERSUCH VON DEN KURZEN ELEKTRISCHEN ZU DEN LANGEN WÄRMEWELLEN ÜBER ZU GEHEN, *Phys. Z.* **25**, 107.
161. H. Witt, PRÜFÜNG EINER SPEKTROMETRISCHEN METHODE IM LANG-WELLIGEN SPEKTRUM, *Z. Physik* **28**, 236.
162. H. Witt, ÜBER SERIEN IM ABSORPTIONSSPEKTRUM DES WASSER-DAMPFES, *Z. Physik* **28**, 249.

1925

163. M. Czerny, MESSUNGEN IM ROTATIONSSPEKTRUM DES HCl IM LANG-WELLIGEN ULTRAROT, *Z. Physik* **34**, 227.
164. E. F. Nichols and J. D. Tear, JOINING THE INFRA-RED AND ELECTRIC-WAVE SPECTRA, *Astrophys. J.* **61**, 17.

1926

165. C. Leiss, ÜBER EIN NEUES GROSSES SPIEGELSPEKTROMETER FÜR GIT-TERMESSUNGEN IM LANGWELLIGEN SPEKTRUM, *Z. Physik* **37**, 681.
166. M. Lewitski, ELEKTRISCHE WELLEN IM GEBIETE DES ÄUSSEREN ULTRA-ROT, *Phys. Z.* **27**, 177.
167. O. Reinkober, NEUE RESTSTRAHLEN UND OBERSCHWINGUNGEN VON RESTSTRAHLEN, *Z. Physik* **39**, 437.
168. R. C. Tolman and R. M. Badger, A NEW KIND OF TEST OF THE CORRESPON-DENCE PRINCIPLE BASED ON THE PREDICTION OF THE ABSOLUTE IN-TENSITIES OF SPECTRAL LINES, *Phys. Rev.* **27**, 383.

1927

169. R. M. Badger, ABSOLUTE INTENSITIES IN THE HYDROGEN CHLORIDE ROTATION SPECTRUM, *Proc. Natl. Acad. Sci. U.S.* **13**, 408.
170. R. M. Badger, TWO DEVICES FACILITATING SPECTROMETRY IN THE FAR-INFRARED, *J. Opt. Soc. Am.* **15**, 370.
171. M. Czerny, DIE ROTATIONSSPEKTREN DER HALOGENWASSERSTOFFE, *Z. Physik* **44**, 235.
172. M. Czerny, DIE DARSTELLING DER ULTRAROTEN ABSORPTIONSSPEK-TREN DER HALOGENWASSERSTOFFE NACH DER SCHRÖDINGERSCHEN THEORIE, *Z. Physik* **45**, 476.

1928

173. R. M. Badger, THE PURE ROTATION SPECTRUM OF AMMONIA, *Nature* **121**, 942.
174. O. Fuchs and K. L. Wolf, RESTSTRAHLFREQUENZEN, EIGENFREQUENZEN, UND DISPERSION IM ULTRAROTEN, *Z. Physik* **46**, 506.

175. F. Jentzsch and G. Laski, BESONDERE METHODE DER SPEKTROSKOPIE. SPEZIELLE MESSMETHODIK IM ULTRAROT, *Geiger-Scheels Handbuch der Physik* **19**, 802.
176. F. Kruger, O. Reinkober, and E. Koch-Holm, RESTSTRAHLEN VON MISCHKRISTALLEN, *Ann. Physik* **85**, 110.
177. J. Lecomte, LE SPECTRE INFRAROUGE, Les Presses Universitaires de France, Paris.
178. C. V. Raman and K. S. Krishnan, MOLECULAR SPECTRA IN THE EXTREME INFRARED, *Nature* **122**, 278.

1929

179. R. M. Badger and C. H. Cartwright, THE PURE ROTATIONAL SPECTRUM OF AMMONIA, *Phys. Rev.* **33**, 692.
180. M. Czerny, ZUM RAMAN-EFFEKT DES QUARZES, *Z. Physik* **53**, 317.
181. L. Kellner, UNTERSUCHUNGEN IM SPEKTRALGEBIET ZWISCHEN 20 UND 40 μ, *Z. Physik* **56**, 215.
182. W. Kroebel, ÜBER DIE ENTSTEHUNG DER LANGWELLIGEN ULTRAROTEN STRAHLUNG DES QUECKSILBERS, *Z. Physik* **56**, 114.
183. M. Murmann, ÜBER DIE OPTISCHEN KONSTANTEN DÜNNER METALLSCHICHTEN IM LANGWELLIGEN ULTRAROT, *Z. Physik* **54**, 741.
184. F. I. G. Rawlins and A. M. Taylor, INFRA-RED ANALYSIS OF MOLECULAR STRUCTURE, Cambridge Univ. Press, Cambridge.

1930

185. C. R. Bailey, THE INFRARED SPECTRUM OF WATER VAPOR, *Trans. Faraday Soc.* **26**, 203.
186. C. H. Cartwright, BLACK BODIES IN THE EXTREME INFRA-RED, *Phys. Rev.* **35**, 415.
187. M. Czerny, MESSUNGEN AM STEINSALZ IM ULTRAROTEN ZUR PRÜFUNG DER DISPERSIONSTHEORIE, *Z. Physik* **65**, 600.
188. H. G. Heisekorn, FILTERUNTERSUCHUNGEN IM ULTRAROTEN SPEKTRUM, *Ann. Physik* **6**, 985.
189. O. Reinkober and M. Bluht, RESTSTRAHLEN VON EINWERTIGEN UND ZWEIWERTIGEN FLUORIDEN, *Ann. Physik* **6**, 785.
190. C. Schaeffer and F. Matossi, DAS ULTRAROTSPEKTRUM, Springer, Berlin.

1931

191. R. B. Barnes and M. Czerny, MESSUNGEN AM NaCl UND KCl IM SPEKTRALBEREICH IHRER ULTRAROTEN EIGENSCHWINGUNGEN, *Z. Physik* **72**, 447.
192. R. B. Barnes and M. Czerny, CONCERNING THE REFLECTION POWER OF METALS IN THIN LAYERS FOR THE INFRARED, *Phys. Rev.* **38**, 338.
193. C. H. Cartwright, LAMINARY REFLECTION GRATINGS FOR INFRARED INVESTIGATION, *J. Opt. Soc. Am.* **21**, 785.
194. J. Strong, INVESTIGATIONS IN THE SPECTRAL REGION BETWEEN 20 AND 40 μ, *Phys. Rev.* **37**, 1565.
195. J. Strong, INVESTIGATIONS IN THE FAR INFRARED, *Phys. Rev.* **38**, 1818.

1932

196. R. B. Barnes, MEASUREMENT IN THE LONG WAVELENGTH INFRARED FROM 20 to 135 μ, *Phys. Rev.* **39**, 562.
197. R. B. Barnes, DIE ULTRAROTEN EIGENFREQUENZEN DER ALKALIHALO-GENIDKRISTALLE, *Z. Physik* **75**, 723.
198. H. M. Randall, INFRARED SPECTROMETER OF LARGE APERTURE, *Rev. Sci. Instr.* **3**, 196.
199. J. Strong, APPARATUS FOR SPECTROSCOPIC STUDIES IN THE INTER-MEDIATE INFRARED REGION 20–40 μ, *Rev. Sci. Instr.* **3**, 810.
200. J. Strong and S. C. Woo, FAR INFRARED SPECTRA OF GASES, *Phys. Rev.* **42**, 267.

1933

201. M. Blackman, DIE FEINSTRUKTUR DER RESTSTRAHLEN, *Z. Physik* **86**, 421.
202. M. Born and M. Blackman, ÜBER DIE FEINSTRUKTUR DER RESTSTRAHLEN, *Z. Physik* **82**, 551.
203. C. H. Cartwright, DISPERSIONSMESSUNGEN AM NaCl IM LANGWELLIGEN ULTRAROT, *Z. Physik* **85**, 269.
204. K. Korth, DISPERSIONSMESSUNGEN AM KALIUMBROMID UND KALIUM-JODID IM ULTRAROTEN, *Z. Physik* **84**, 677.
205. J. Kuhne, MESSUNGEN IM ROTATIONSSPEKTRUM DES WASSERDAMPFES, *Z. Physik* **84**, 722.
206. N. Wright and H. M. Randall, THE FAR INFRARED ABSORPTION SPECTRA OF AMMONIA AND PHOSPHINE GASES UNDER HIGH RESOLVING POWER, *Phys. Rev.* **44**, 391.

1934

207. R. B. Barnes AN IMPROVED WIRE GRATING SPECTROMETER FOR THE FAR INFRARED, *Rev. Sci. Instr.* **5**, 237.
208. R. B. Barnes, W. S. Benedict, and C. M. Lewis, ROTATION SPECTRA OF NH_3 AND ND_3, *Phys. Rev.* **45**, 347.
209. C. H. Cartwright, DURCHLÄSSIGKEITSMESSUNGEN IM SPEKTRALBEREICH VON 50 BIS 240 μ, *Z. Physik.* **90**, 480.
210. C. H. Cartwright and M. Czerny, DISPERSIONSMESSUNGEN AM NaCl UND KCl IM LANGWELLIGEN ULTRAROT, *Z. Physik* **90**, 457.
211. M. El'yashevich, AN ANALYSIS OF THE PURE ROTATIONAL SPECTRUM OF THE WATER MOLECULE, *Compt. Rend. Acad. Sci. URSS* **3**, 248.
212. J. Fock, DAS ULTRAROTE SPEKTRUM VON MAGNESIUMOXYD, *Z. Physik* **90**, 44.
213. A. Mentzel, UNTERSUCHUNG DES ABSORPTIONSVERLAUFES VON KCl UND KBr AUF DER KURZWELLIGEN SEITE IHRER ULTRAROTEN EIGEN-SCHWINGUNG, *Z. Physik* **88**, 178.
214. J. Strong, PURE ROTATION SPECTRUM OF THE HCl FLAME, *Phys. Rev.* **45**, 887.
215. W. Waltersdorff, UNTERSUCHUNGEN ÜBER DIE DURCHLÄSSIGKEIT DÜN-NER METALLSCHICHTEN FÜR LANGWELLIGE ULTRAROTE STRAHLUNG UND IHRE ELEKTRISCHE LEITFÄHIGKEIT, *Z. Physik* **91**, 230.

1935

216. R. B. Barnes, THE PURE ROTATION SPECTRA OF NH_3 AND ND_3, *Phys. Rev.* **47**, 658.
217. R. B. Barnes, W. S. Benedict, and C. M. Lewis, THE FAR INFRARED ABSORPTION OF BENZENE, *Phys. Rev.* **47**, 129.
218. R. B. Barnes, W. S. Benedict, and C. M. Lewis, THE FAR INFRARED SPECTRUM OF H_2O, *Phys. Rev.* **47**, 918.
219. R. B. Barnes, R. R. Brattain, and R. S. Firestone, POWDER FILTERS FOR THE INFRARED, *Phys. Rev.* **47**, 792 (Abstract).
220. C. H. Cartwright, EXTREME INFRA-RED INVESTIGATION OF HINDERED ROTATION IN WATER, *Nature* **135**, 872.
221. C. H. Cartwright, EXTREME INFRA-RED ABSORPTION OF D_2O, ICE, AND D_2O IN DIOXANE, *Nature* **136**, 181.
222. C. H. Cartwright and J. Errera, ISOMÉRIE INTRAMOLÉCULAIRE DE L'α-PICO-LINE ÉTUDIÉE DANS L'INFRAROUGE LOINTAIN, *Compt. Rend.* **200**, 914 (1935).
223. C. H. Cartwright and J. Errera, POLARISATION ATOMIQUE ET ABSORPTION DE LIQUIDES DANS L'INFRAROUGE LOINTAIN, *Acta Physicochim. URSS* **3**, 649.

1936

224. R. B. Barnes and L. G. Bonner, FILTERS FOR THE INFRARED, *J. Opt. Soc. Am.* **26**, 428.
225. R. B. Barnes and L. G. Bonner, THE CHRISTIANSEN FILTER EFFECT IN THE INFRARED, *Phys. Rev.* **49**, 732.
226. R. B. Barnes and L. G. Bonner, THE EARLY HISTORY AND METHODS OF INFRARED SPECTROSCOPY, *Amer. Phys. Teacher* **4**, 181.
227. C. H. Cartwright, IONIC DISPERSION IN THE EXTREME INFRARED, *Phys. Rev.* **49**, 101.
228. C. H. Cartwright, ABSORBING AND REFLECTING POWERS OF H_2SO_4 SOLUTIONS IN THE FAR INFRARED, *J. Chem. Phys.* **4**, 413.
229. C. H. Cartwright, HINDERED ROTATION IN LIQUID H_2O AND D_2O, *Phys. Rev.* **49**, 470.
230. C. H. Cartwright and J. Errera, EXTREME INFRARED DISPERSION OF POLAR AND NONPOLAR LIQUIDS, *Proc. Roy. Soc.* (*London*) **154**, 138.
231. M. Lewitskaja, EINE NEUE QUELLE VON LANGWELLIGEN ULTRAROTEN STRAHLEN, *Phys. Z. Sowjetunion* **10**, 697.
232. A. H. Pfund, THE ELECTRIC WELSBACH LAMP, *J. Opt. Soc. Am.* **26**, 439.
233. J. H. Plummer, TRANSMISSION OF POWDER FILMS TO THE INFRARED SPECTRUM, *J. Opt. Soc. Am.* **26**, 434.

1937

234. R. B. Barnes and L. G. Bonner, A SURVEY OF INFRA-RED SPECTROSCOPY I, *J. Chem. Educ.* **14**, 564.

235. R. B. Barnes and L. G. Bonner, A SURVEY OF INFRA-RED SPECTROSCOPY II, *J. Chem. Educ.* **15**, 25.

236. C. H. Cartwright, ABSORBING AND REFLECTING POWERS OF ELECTRO-LYTES IN THE FAR INFRARED, *J. Chem. Phys.* **5**, 776.

237. K. H. Hellwege, ÜBER RASTERFÖRMIGE REFLEXIONSGITTER, *Z. Physik* **106**, 588.

238. H. W. Hohls, ÜBER DISPERSION UND ABSORPTION VON LITHIUMFLUORID UND NATRIUMFLUORID IM ULTRAROT, *Ann. Physik* **29**, 433.

239. M. Parodi, ÉTUDE DE QUELQUES BORATES ET DE QUELQUES OXYDES DANS L'INFRAROUGE LOINTAIN, *Compt. Rend.* **204**, 1111.

240. M. Parodi, ÉTUDE DE LA TRANSMISSION DE QUELQUES OXYDES DANS L'INFRAROUGE LOINTAIN, *Compt. Rend.* **204**, 1636.

241. M..Parodi, SUR LA TRANSMISSION DE QUELQUES OXYDES DANS L'INFRA-ROUGE LOINTAIN, *Compt. Rend.* **205**, 906.

242. H. M. Randall, D. M. Dennison, N. Ginsburg, and L. R. Weber, THE FAR INFRARED SPECTRUM OF WATER VAPOR, *Phys. Rev.* **52**, 106.

1938

243. P. Barchewitz and M. Parodi, SPECTRES D'ABSORPTION DES MONOSUB-STITUÉS DU BENZÈNE DANS L'INFRAROUGE LOINTAIN, DE 180 À 600 CM⁻¹ (17 À 55 μ), *Compt. Rend.* **207**, 903.

244. P. Barchewitz and M. Parodi, SUR UN SPECTROMÈTRE À FILS POUR L'ÉTUDE DE L'INFRAROUGE LOINTAIN, *Compt. Rend.* **206**, 1891.

245. M. Czerny and H. Röder, FORTSCHRITTE AUF DEM GEBIETE DER ULTRAROT-TECHNIK, *Ergeb. Exakt. Naturwiss.* **17**, 70.

246. W. M. Elsasser, NOTE ON ATMOSPHERIC ABSORPTION CAUSED BY THE ROTATIONAL WATER BAND, *Phys. Rev.* **53**, 768.

247. W. M. Elsasser, FAR INFRARED ABSORPTION OF ATMOSPHERIC WATER VAPOR, *Astrophys. J.* **87**, 497.

248. B. Koch, MESSUNGEN IM LANGWELLIGEN ULTRAROTSPEKTRUM DER QUARZQUECKSILBERLAMPE, *Ann. Physik* **33**, 335.

249. F. Matossi, ERGEBNISSE DER ULTRAROTFORSCHUNG, *Ergeb. Exakt. Naturwiss.* **17**, 108.

250. M. Parodi, SPECTRES DE QUELQUES DÉRIVÉS DU MÉTHANE DANS L'IN-FRAROUGE LOINTAIN, *Compt. Rend.* **207**, 1196.

251. M. Parodi, RECHERCHES DANS L'INFRAROUGE LOINTAIN PAR LA MÉ-THODE DES RAYONS RESTANTS, Herman et Cie, Paris.

252. H. M. Randall, THE SPECTROSCOPY OF THE FAR INFRARED, *Rev. Mod. Phys.* **10**, 72.

253. H. M. Randall and F. A. Firestone, A RECORDING SPECTROGRAPH FOR THE FAR INFRARED, *Rev. Sci. Instr.* **9**, 404.

254. J. Strong, PROCEDURES IN EXPERIMENTAL PHYSICS, Prentice-Hall, New York.

1939

255. P. Barchewitz and M. Parodi, SPECTRES D'ABSORPTION DANS L'INFRAROUGE LOINTAIN (20 À 60 MICRONS) DES DERIVÉS HALOGÉNÉS DU MÉTHANE ET DES DÉRIVÉS MONOSUBSTITUÉS DU BENZÈNE, *J. Phys. Rad.* **10**, 143.

256. P. Barchewitz and M. Parodi, ÉTUDE DE LA TRANSMISSION DE QUELQUES NITRILES DANS L'INFRAROUGE LOINTAIN, Compt. Rend. **209**, 30.

257. W. Dahlke, INTENSITÄT UND QUERSCHNITTSVERTEILUNG DER LANG-WELLIGEN ULTRAROT-STRAHLUNG (300 μ) DES QUECKSILBERHOCH-DRUCKBOGENS, Z. Physik **114**, 205.

258. W. Dahlke, DIE SPEKTRALE ENERGIEVERTEILUNG DER LANGWELLIGEN UR-EMISSION (300 μ) VON VERSCHIEDENEN HOCHDRUCKENTLA-DUNGEN, Z. Physik **114**, 672.

259. J. Duchesne and M. Parodi, STRUCTURE OF THE TETRACHLORETHYLENE MOLECULE, Nature **144**, 382.

260. T. Erb and H. Klumb, UNTERSUCHUNGEN ÜBER DIE HERSTELLBARKEIT LANGWELLIGER ULTRAROTER STRAHLUNG, Z. Physik **114**, 519.

261. N. Fuson, H. M. Randall, and D. M. Dennison, THE FAR INFRA-RED ABSORP-TION SPECTRUM AND THE ROTATIONAL STRUCTURE OF THE HEAVY WATER MOLECULE, Phys. Rev. **56**, 982.

262. K. H. Hellwege, UNTERSUCHUNGEN IM LANGWELLIGEN ULTRAROT ÜBER KOMBINATIONSSCHWINGUNGEN UND ÜBER DIE EXISTENZ VON METALLHYDRATKOMPLEXEN IN KRISTALLEN, Ann. Physik **34**, 521.

263. A. Kalugina, EFFECT OF GRAIN SIZE ON INTENSITY OF MASS RADIATORS, J. Exp. Theor. Phys. USSR **9**, 362.

264. O. Marr, SPEKTRALMESSUNGEN BEI 0.2–0.5 mm WELLENLÄNGE AN EINI-GEN HOCHFREQUENZISOLIERSTOFFEN UND OXYDEN, Z. Physik **113**, 415.

265. Z. I. Slawsky and D. M. Dennison, THE CENTRIFUGAL DISTORTION OF AXIAL MOLECULES, J. Chem. Phys. **7**, 509.

1940

266. W. Dahlke, DEUTUNG DER LANGWELLIGEN UR-EMISSION (300 μ) DES QUECHSILBERHOCHDRUCKBOGENS ALS TEMPERATURSTRAHLUNG, Z. Physik **115** 1.

267. H. Hopf, WASSERDAMPFABSORPTIONSLINIEN IM SPEKTRALGEBIET VON 0.15 BIS 0.5 mm WELLENLÄNGE, Z. Physik **116**, 310.

268. H. S. Seifert and H. M. Randall, TRANSMISSION AND REFLECTION OF PLAS-TICS AND METAL BLACKS IN THE FAR INFRA-RED, Rev. Sci. Instr. **11**, 365.

1941

269. H. M. Foley and H. M. Randall, FINE STRUCTURE IN THE FAR INFRARED SPECTRUM OF NH$_3$, Phys. Rev. **59**, 171.

1942

270. T. G. Cowling, THE ABSORPTION OF WATER VAPOR IN THE FAR INFRA-RED, Rept. Progr. Phys. **9**, 29.

271. L. Kellner, DISPERSION IN THE FAR INFRARED, Rept. Progr. Phys. **8**, 200.

1945

272. J. P. Cooley and J. H. Rohrbough, THE PRODUCTION OF EXTREMELY SHORT ELECTROMAGNETIC WAVES, *Phys. Rev.* **67**, 296.

1947

273. M. J. E. Golay, THEORETICAL CONSIDERATION IN HEAT AND INFRA-RED DETECTION WITH PARTICULAR REFERENCE TO THE PNEUMATIC DETECTOR, *Rev. Sci. Instr.* **18**, 347.
274. M. J. E. Golay, A PNEUMATIC INFRA-RED DETECTOR, *Rev. Sci. Instr.* **18**, 357.
275. E. K. Plyler, PRISM SPECTROMETRY FROM 24 TO 37 MICRONS, *J. Chem. Phys.* **15**, 885.
276. J. U. White, GRATINGS AS BROAD BAND FILTERS FOR THE INFRARED, *J. Opt. Soc. Am.* **37**, 713.

1948

277. R. J. Coates, A GRATING SPECTROMETER FOR MILLIMETER WAVES, *Rev. Sci. Instr.* **19**, 586.
278. N. Ginsburg, ADDITIONAL ROTATIONAL ENERGY LEVELS OF H_2O AND D_2O MOLECULES, *Phys. Rev.* **74**, 1052.
279. G. Hettner and G. Leisegang, DIE DISPERSION DER MISCHKRISTALLE TlBr–TlJ (KRS5) UND TlCl–TlBr (KRS6) IM ULTRAROT, *Optik* **3**, 305.
280. W. L. Hyde, THALLIUM BROMIDE–IODIDE PRISM SPECTROSCOPY IN THE FAR INFRARED, *J. Chem. Phys.* **16**, 744.
281. E. K. Plyler, VIBRATIONAL BANDS MEASURED WITH A THALLIUM BROMIDE–IODIDE PRISM, *J. Chem. Phys.* **16**, 1008.
282. E. K. Plyler, INFRARED PRISM SPECTROMETRY FROM 24 TO 40 MICRONS, *J. Res. Natl. Bur. Stand.* **41**, 125.
283. E. K. Plyler and J. J. Ball, INFRARED ABSORPTION OF DEPOSITED BLACKS, *J. Opt. Soc. Am.* **38**, 988.
284. V. Z. Williams, INFRARED INSTRUMENTATION AND TECHNIQUES, *Rev. Sci. Instr.* **19**, 135.

1949

285. D. M. Gates, AN EXPERIMENTAL AND THEORETICAL INVESTIGATION OF THE SKELETAL FREQUENCIES OF THE PARAFFIN HYDROCARBONS AND THE FAR INFRARED SPECTRUM OF CARBON TETRACHLORIDE, *J. Chem. Phys.* **17**, 393.
286. M. J. E. Golay, THE THEORETICAL AND PRACTICAL SENSITIVITY OF THE PNEUMATIC INFRA-RED DETECTOR, *Rev. Sci. Instr.* **20**, 816.
287. K. Lark-Horovitz and K. W. Meissner, THE OPTICAL PROPERTIES OF SEMI-CONDUCTORS. I. THE REFLECTIVITY OF GERMANIUM SEMI-CONDUCTORS, *Phys. Rev.* **76**, 1530.

288. J. Mevel, LES RÉSEAUX LAMELLAIRES ET LEUR APPLICATION A L'ÉTUDE DE L'INFRAROUGE LOINTAIN, *Rev. Opt.* **27**, 661.

289. J. Mevel, THE SPECTRAL DOMAIN BETWEEN THE INFRARED AND MICRO-WAVE REGIONS, *Rev. Gen. Sci. Pur. Appl.* **56**, 11.

290. E. K. Plyler, INFRA-RED ABSORPTION SPECTRA OF ORGANIC COMPOUNDS, *J. Chem. Phys.* **17**, 218.

291. L. W. Tilton, E. K. Plyler, and R. E. Stephens, REFRACTIVE INDICES OF THAL-LIUM BROMIDE–IODIDE CRYSTALS FOR VISIBLE AND INFRARED RADIANT ENERGY, *J. Res. Natl. Bur. Stand.* **43**, 81.

1950

292. T. G. Cowling, ATMOSPHERIC ABSORPTION OF HEAT RADIATION BY WATER VAPOR, *Phil. Mag.* **41**, 109.

293. T. K. McCubbin and W. M. Sinton, RECENT INVESTIGATIONS IN THE FAR INFRARED, *J. Opt. Soc. Am.* **40**, 537.

294. E. K. Plyler, INFRARED ABSORPTION SPECTRA OF TWELVE SUBSTITUTED BENZENE DERIVATIVES FROM 15 TO 40 MICRONS, *Disc. Faraday Soc.* **9**, 100.

295. G. Yamamoto and G. Onishi, ABSORPTION COEFFICIENT OF WATER VAPOR IN THE FAR-INFRARED REGION, *Sci. Rept. Tohoku Univ.* **1**, 5.

1951

296. E. K. Plyler and W. S. Benedict, INFRARED SPECTRA OF EIGHTEEN HALOGEN-SUBSTITUTED METHANES, *J. Res. Natl. Bur. Stand.* **47**, 202.

297. E. K. Plyler and F. P. Phelps, THE TRANSMITTANCE OF CESIUM BROMIDE CRYSTALS, *J. Opt. Soc. Am.* **41**, 209.

1952

298. S. S. Ballard, L. S. Combes, and K. A. McCarthy, A COMPARISON OF THE PHYSICAL PROPERTIES OF CESIUM BROMIDE AND THALLIUM BROMIDE–IODIDE, *J. Opt. Soc. Am.* **42**, 65.

299. J. P. Casey and E. A. Lewis, INTERFEROMETER ACTION OF A PARALLEL PAIR OF WIRE GRATINGS, *J. Opt. Soc. Am.* **42**, 971.

300. W. H. Eberhardt and T. G. Burke, AN ABSORPTION BAND OF NOCl AT 30 μ, *J. Chem. Phys.* **20**, 529.

301. R. M. Goody, A STATISTICAL MODEL FOR WATER VAPOR ABSORPTION, *Quart. J. Roy. Meteorol. Soc.* **78**, 165.

302. J. Lecomte, LE SPECTROMÉTRIE INFRAROUGE ET SES PROGRÈS RECENTS, *Cah. Phys.* **38**, 26.

303. R. C. Lord, FAR INFRARED TRANSMISSION OF SILICON AND GERMANIUM, *Phys. Rev.* **85**, 140.

304. R. C. Lord, R. S. McDonald, and F. A. Miller, NOTES ON THE PRACTICE OF INFRARED SPECTROSCOPY, *J. Opt. Soc. Am.* **42**, 149.

305. T. K. McCubbin, THE SPECTRA OF HCl, NH_3, H_2O, and H_2S FROM 100 TO 700 MICRONS, *J. Chem. Phys.* **20**, 668.

306. T. K. McCubbin and W. M. Sinton, A TWELVE-INCH FAR INFRARED GRATING SPECTROMETER, *J. Opt. Soc. Am.* **42**, 113.

307. R. A. Oetjen, W. H. Haynie, W. M. Ward, R. L. Hansler, H. E. Schauwecker, and E. E. Bell, AN INFRARED SPECTROGRAPH FOR USE IN THE 40–150 MICRON SPECTRAL REGION, *J. Opt. Soc. Am.* **42**, 559.

308. E. K. Plyler and N. Acquista, INFRARED PROPERTIES OF CESIUM BROMIDE PRIMS, *J. Res. Natl. Bur. Stand.* **49**, 61.

309. E. K. Plyler and J. J. Ball, FILTERS FOR THE INFRARED REGION, *J. Opt. Soc. Am.* **42**, 266.

310. E. K. Plyler and F. P. Phelps, GROWTH AND INFRARED TRANSMISSION OF CESIUM IODIDE CRYSTALS, *J. Opt. Soc. Am.* **42**, 432.

311. R. Q. Twiss, ON THE GENERATION OF MILLIMETER RADIATION, *SERL Tech. J.* **2**, 10.

312. A. Walsh, ECHELETTE ZONE PLATES FOR USE IN FAR INFRARED SPECTROSCOPY, *J. Opt. Soc. Am.* **42**, 213.

313. D. E. Williamson, CONE CHANNEL CONDENSER OPTICS, *J. Opt. Soc. Am.* **42**, 712.

314. H. Yoshinaga and Y. Yamada, CONSTRUCTION OF A FAR INFRARED SPECTROMETER WITH ECHELETTE GRATING, *Sci. Light (Japan)* **2**, 18.

315. H. Yoshinaga and Y. Yamada, CONSTRUCTION OF A FAR INFRARED SPECTROMETER WITH ECHELETTE GRATING, *J. Phys. Soc. Japan* **7**, 223.

                                    1953

316. N. Acquista and E. K. Plyler, INFRARED MEASUREMENTS WITH A CESIUM IODIDE PRISM, *J. Opt. Soc. Am.* **43**, 977.

317. S. S. Ballard, L. S. Combes, and K. A. McCarthy, THE PHYSICAL PROPERTIES OF CRYSTALLINE CESIUM IODIDE, *J. Opt. Soc. Am.* **43**, 975.

318. C. R. Bohn, N. K. Freeman, W. D. Gwinn, J. L. Hollenberg, and K. S. Pitzer, A SPECTROMETER FOR THE FAR INFRARED AND THE SPECTRUM OF 1,2-DICHLOROETHANE, *J. Chem. Phys.* **21**, 719.

319. E. Burstein, E. E. Bell, J. W. Davisson, and M. Lax, OPTICAL INVESTIGATIONS OF IMPURITY LEVELS IN SILICON, *J. Phys. Chem.* **57**, 849.

320. A. R. Downie, M. C. Magoon, T. Purcell, and B. Crawford, THE CALIBRATION OF INFRARED PRISM SPECTROMETERS, *J. Opt. Soc. Am.* **43**, 941.

321. A. Hadni, ANALYSE DE LA BANDE DU SULFURE DE CARBONE, À L'ÉTAT DE VAPEUR, VERS 25 μ, AU MOYEN D'UN SPECTROGRAPHE ENREGISTREUR A RÉSEAU, *Compt. Rend.* **236**, 1761.

322. A. Hadni, STRUCTURE FINE DES RAIES DE L'AMMONIAC DE $J'' = 14$ À $J'' = 18$, *Compt. Rend.* **237**, 317.

323. R. L. Hansler and R. A. Oetjen, THE INFRA-RED SPECTRA OF HCl, DCl, HBr, AND NH_3 IN THE REGION FROM 40 TO 140 MICRONS, *J. Chem. Phys.* **21**, 1340.

324. L. Harris and A. L. Loeb, CONDUCTANCE AND RELAXATION TIME OF ELECTRONS IN GOLD BLACKS FROM TRANSMISSION AND REFLECTION MEASUREMENTS IN THE FAR INFRARED, *J. Opt. Soc. Am.* **43**, 1114.

325. R. Meier, UNTERSUCHUNGEN IM GEBIET DER ZEHNTELMILLIMETERWELLEN, *Ann. Physik* **12**, 26.

326. J. K. O'Loane, SOME ABSORPTION BANDS IN THE FAR INFRARED, *J. Chem. Phys.* **21**, 669.

327. E. K. Plyler and N. Acquista, INFRARED SPECTROMETRY WITH A CESIUM IODIDE PRISM, *J. Opt. Soc. Am.* **43**, 212.
328. R. E. Stroup, THE ROTATIONAL SPECTRUM OF ND_3 BETWEEN 60 AND 200°K, *J. Chem. Phys.* **21**, 2072.
329. R. E. Stroup and R. A. Oetjen, PH_2D AND PHD_2 ROTATIONAL LINES IN THE REGION BETWEEN 50 AND 100°K, *J. Chem. Phys.* **21**, 2092.
330. R. E. Stroup, R. A. Oetjen, and E. E. Bell, THE ROTATIONAL SPECTRA OF REGULAR AND DEUTERATED PHOSPHINE AND ARSINE IN THE REGION BETWEEN 50 AND 200°K, *J. Opt. Soc. Am.* **43**, 1096.

1954

331. G. T. J. Alkemade, A PROPOSED METHOD OF INFRARED DETECTION BASED ON THERMAL CONVERSION OF RADIATION, *Physica* **20**, 433.
332. M. Born and K. Huang, DYNAMICAL THEORY OF CRYSTAL LATTICES, Clarendon Press, Oxford.
333. L. Genzel and W. Eckhardt, SPEKTRALUNTERSUCHUNGEN IM GEBIET UM 1 mm WELLENLÄNGE. I. KONSTRUKTION DES SPEKTROMETERS UND DESSEN GRUNDSPEKTREN, *Z. Physik* **139**, 578.
334. L. Genzel and W. Eckhardt, SPEKTRALUNTERSUCHUNGEN IM GEBIET UM 1 mm WELLENLÄNGE. II. DIE ROTATIONS-SPEKTREN VON HCN UND H_2S, *Z. Physik* **139**, 592.
335. A. Hadni, FILTRES POUR L'INFRAROUGE LOINTAIN ET SPECTRES DES MÉTHYLAMINES DE 30 À 43 μ, *J. Phys. Rad.* **15**, 317.
336. A. Hadni, SPECTRES DE QUELQUES MOLÉCULES SIMPLES DANS L'INFRA-ROUGE LOINTAIN, *J. Phys. Rad.* **15**, 417.
337. A. Hadni, ÉLIMINATION DE LA VAPEUR D'EAU À L'INTERIEUR D'UN SPECTROGRAPHE POUR L'INFRAROUGE LOINTAIN ET SPECTRE DE L'ALCOOL MÉTHYLIQUE, *Compt. Rend.* **238**, 573.
338. A. Hadni, DETERMINATION ET INTERPRÉTATION DES BASSES FRÉQUEN-CES D'ABSORPTION INFRAROUGE DE L'ÉTHER DIMÉTHYLIQUE. AP-PLICATION AU CALCUL DES FONCTIONS THERMODYNAMIQUES, *Compt. Rend.* **239**, 349.
339. A. Hadni, SPECTROMÈTRE À RÉSEAU POUR L'INFRAROUGE LOINTAIN, *Rev. Opt.* **33**, 576.
340. J. Lecomte, SPECTROPHOTOMÉTRIE DANS L'INFRAROUGE; SON IMPORT-ANCE EN PHYSIQUE ET EN CHIMIE, *Rev. Opt.* **33**, 533.
341. K. S. Pitzer and J. L. Hollenberg, CIS- AND TRANS-DICHLOROETHYLENES. THE INFRARED SPECTRA FROM 130–400 cm^{-1} AND THE THERMODYNAMIC PROPERTIES, *J. Am. Chem. Soc.* **76**, 1493.
342. E. K. Plyler, MÉSURES DANS L'INFRAROUGE AVEC UN PRISME EN IODURE DE CÉSIUM, *J. Phys. Rad.* **15**, 519.
343. E. K. Plyler and N. Acquista, INFRARED ABSORPTION OF LIQUID WATER FROM 2 TO 42 MICRONS, *J. Opt. Soc. Am.* **44**, 505.
344. H. M. Randall, INFRARED SPECTROSCOPY AT THE UNIVERSITY OF MICHIGAN, *J. Opt. Soc. Am.* **44**, 97.
345. V. Roberts, CALIBRATION OF INFRARED SPECTROMETERS IN THE WAVE-LENGTH REGION 15–25 μ, *J. Sci. Instr.* **31**, 226.

346. F. Rössler, VORSCHLAG ZUR DEUTUNG DER LÄNGSTWELLIGEN ULTRA-
     ROTEMISSION DER QUECKSILBERHOCHDRUCKENTLADUNG, Z. *Physik*
     139, 56.
347. W. M. Sinton and W. C. Davis, FAR INFRARED REFLECTANCES OF TlCl,
     TlBr, TlI, PbS, PbCl$_2$, ZnS, AND CsBr, *J. Opt. Soc. Am.* **44**, 503.

                                        1955

348. J. Gaunt, THE INFRARED SPECTRUM OF TeF$_6$ FROM 25–40 μ, *Trans. Faraday
     Soc.* **51**, 893.
349. L. Guilotto, SPECTRES DES CRISTAUX ET DES LIQUIDES DANS L'INFRA-
     ROUGE LOINTAIN, *Nuovo Cimento Suppl.* **2**, 571.
350. A. Hadni, SPECTROMÈTRE ENREGISTREUR À RÉSEAU POUR L'INFRA-
     ROUGE LOINTAIN, *Nuovo Cimento Suppl.* **2**, 710.
351. A. Hadni, SPECTRES D'ABSORPTION DE GAZ ET DE VAPEURS DANS
     L'INFRAROUGE LOINTAIN; APPLICATION AU CALCUL DES FONCTIONS
     THERMODYNAMIQUES, *Nuovo Cimento Suppl.* **2**, 710.
352. A. Hadni, STRUCTURE DES BANDES DE DÉFORMATION DU DICHLORO-
     DIFLUOROMÉTHANE VERS 447 cm^{-1}, *Compt. Rend.* **240**, 1702.
353. C. Haeusler, INSTRUMENTATION DANS L'INFRAROUGE LOINTAIN, *J.
     Phys. Rad.* **16**, 882.
354. W. Kaiser and H. Y. Fan, INFRARED ABSORPTION OF INDIUM ANTIMONIDE,
     Phys. Rev. **98**, 966.
355. J. A. A. Ketelaar, J. P. Colpa and F. N. Hooge, PURE ROTATIONAL ABSORPTION
     SPECTRUM OF HYDROGEN, *J. Chem. Phys.* **23**, 413.
356. J. A. A. Ketelaar, C. Haas, F. N. Hooge, and R. Brockhuijsen, FAR INFRARED
     EMISSION SPECTRUM OF FLAMES, ROTATIONAL SPECTRUM OF OH
     RADICALS, AND DETERMINATION OF FLAME TEMPERATURES, *Physica* **21**,
     695.
357. M. Lax and E. Burstein, INFRARED LATTICE ABSORPTION IN IONIC AND
     HOMOPOLAR CRYSTALS, *Phys. Rev.* **97**, 39.
358. J. Lecomte, RAPPORT SUR LA RÉFRACTOMÉTRIE DANS L'INFRAROUGE,
     Nuovo Cimento Suppl. **2**, 579.
359. R. P. Madden and W. S. Benedict, PURE ROTATION LINES OF OH, *J. Chem.
     Phys.* **23**, 408.
360. I. M. Mills and B. Crawford, TRANSMISSION OF GERMANIUM IN THE FAR
     INFRARED, *J. Opt. Soc. Am.* **45**, 489.
361. I. M. Mills, J. R. Scherer, B. Crawford, and M. Youngquist, CALIBRATION AND USE
     OF CESIUM IODIDE PRISM IN THE INFRARED, *J. Opt. Soc. Am.* **45**, 785.
362. E. D. Palik, THE PURE ROTATIONAL SPECTRA OF DBr, HI, AND DI IN THE
     SPECTRAL REGION BETWEEN 45 AND 170 MICRONS, *J. Chem. Phys.* **23**, 217.
363. E. D. Palik, THE PURE ROTATIONAL SPECTRUM OF H$_2$Se BETWEEN 50 AND
     250 cm^{-1}, *J. Chem. Phys.* **23**, 980.
364. E. K. Plyler and N. Acquista, SMALL GRATING SPECTROMETER FOR THE FAR
     INFRARED, *J. Chem. Phys.* **23**, 752.
365. E. K. Plyler and N. Acquista, THE USE OF CESIUM IODIDE PRISMS TO
     50 MICRONS, *Nuovo Cimento Suppl.* **2**, 629.
366. K. N. Rao and E. D. Palik, RESOLUTION OF v_{11}(E) OF ALLENE AT 28 μ, *J. Chem.
     Phys.* **23**, 2112.

367. W. M. Sinton, OBSERVATION OF SOLAR AND LUNAR RADIATION AT 1.5 MILLIMETERS, *J. Opt. Soc. Am.* **45**, 975.

368. W. G. Spitzer and H. Y. Fan, INFRARED ABSORPTION IN INDIUM ANTIMONIDE, *Phys. Rev.* **99**, 1893.

369. R. Waldron, INFRARED SPECTRA OF FERRITES, *Phys. Rev.* **99**, 1727.

370. H. Yoshinaga, ABSORPTION BANDS OF CCl_4 IN THE FAR INFRARED REGION, *J. Chem. Phys.* **23**, 2206.

371. H. Yoshinaga, REFLECTIVITY OF SEVERAL CRYSTALS IN THE FAR INFRARED REGION BETWEEN 20 AND 200 MICRONS, *Phys. Rev.* **100**, 753.

372. H. Yoshinaga and R. A. Oetjen, SAMPLING TECHNIQUE FOR TRANSMISSION MEASUREMENTS IN THE FAR INFRARED REGION, *J. Opt. Soc. Am.* **45**, 1085.

1956

373. E. Burstein, G. S. Picus, and H. A. Gebbie, CYCLOTRON RESONANCE AT INFRARED FREQUENCIES IN InSb AT ROOM TEMPERATURE, *Phys. Rev.* **103**, 825.

374. E. Burstein, G. S. Picus, B. W. Henvis, and R. F. Wallis, ABSORPTION SPECTRA OF IMPURITIES IN SILICON-I GROUP III ACCEPTORS, *J. Phys. Chem. Solids* **1**, 75.

375. H. Y. Fan, INFRARED ABSORPTION IN SEMICONDUCTORS, *Rept. Prog. Phys.* **19**, 107.

376. H. Y. Fan, W. G. Spitzer, and R. J. Collins, INFRARED ABSORPTION IN *n*-TYPE GERMANIUM, *Phys. Rev.* **101**, 566.

377. H. A. Gebbie and G. A. Vanasse, INTERFEROMETRIC SPECTROSCOPY IN THE FAR INFRARED, *Nature* **178**, 432.

378. L. Genzel, SPEKTRALUNTERSUCHUNGEN IM GEBIET UM 1 mm WELLENLÄNGE. III. DISPERSIONSMESSUNGEN AM LiF, *Z. Physik* **144**, 25.

379. L. Genzel, SPEKTRALUNTERSUCHUNGEN IM GEBIET UM 1 mm WELLENLÄNGE. IV. VAKUUM-SPEKTROMETER H_2S UND NH_3 ROTATIONSSPEKTREN, *Z. Physik* **144**, 311.

380. R. E. Glover and M. Tinkham, TRANSMISSION OF SUPERCONDUCTING FILMS AT MILLIMETER-MICROWAVE AND FAR INFRARED FREQUENCIES, *Phys. Rev.* **104**, 844.

381. C. Haas and J. A. A. Ketelaar, WIDTH OF INFRARED REFLECTION BANDS, *Phys. Rev.* **103**, 564.

382. A. Hadni, CONTRIBUTION À L'ÉTUDE THÉORIQUE ET EXPERIMENTALE DE L'INFRAROUGE LOINTAIN, *Ann. Physik* **1**, 234.

383. A. Hadni, SUR LE ROLE DE RÉCEPTEUR DANS LA DÉTERMINATION DES ÉLÉMENTS D'UN SPECTROMÈTRE À RÉSEAU, *Ann. Physik* **1**, 765.

384. A. Hadni, PETIT SPECTROMÈTRE À RESÉAU POUR L'INFRAROUGE LOINTAIN, *J. Phys. Rad.* **17**, 77.

385. A. Hadni, REMARQUES SUR L'INSTRUMENTATION DANS L'INFRAROUGE LOINTAIN, *J. Phys. Rad.* **17**, 311.

386. W. Klemperer, COMPARISON OF FORCE CONSTANTS AND MOLECULES. THE HALIDES OF ZINC, CADMIUM AND MERCURY, *J. Chem. Phys.* **25**, 1066.

387. S. Krimm, C. Y. Liang, and G. B. B. M. Sutherland, INFRARED SPECTRA OF HIGH POLYMERS. II. POLYETHYLENE, *J. Chem. Phys.* **25**, 549.

388. J. Lecomte, SPECTRES D'ABSORPTION INFRAROUGE OBTENUS AU LAB-
ORATOIRE DE RECHERCHES PHYSIQUES À LA SORBONNE AUX ENVIRONS
DE LA SIXIÈME OCTAVE, *Cah. Phys.* **73**, 26.

389. C. Y. Liang, S. Krimm, and G. B. B. M. Sutherland, INFRARED SPECTRA OF HIGH
POLYMERS. I. EXPERIMENTAL METHODS AND GENERAL THEORY, *J.
Chem. Phys.* **25**, 543.

390. R. Newman, CONCENTRATION EFFECTS ON THE LINE SPECTRA OF BOUND
HOLES IN SILICON, *Phys. Rev.* **103**, 103.

391. E. D. Palik and K. N. Rao, PURE ROTATIONAL SPECTRA OF CO, NO, and N_2O
BETWEEN 100 AND 600 μ, *J. Chem. Phys.* **25**, 1174.

392. G. S. Picus, E. Burstein, and B. W. Henvis, ABSORPTION SPECTRA OF IMPURITIES
IN SILICON-II GROUP V DONORS, *J. Phys. Chem. Solids* **1**, 75.

393. E. K. Plyler and N. Acquista, INFRARED MEASUREMENTS FROM 50 TO 125
MICRONS, *J. Res. Natl. Bur. Stand.* **56**, 149.

394. H. H. Theissing and P. J. Caplan, MEASUREMENT OF THE SOLAR MILLIMETER
SPECTRUM, *J. Opt. Soc. Am.* **46**, 971.

395. M. Tinkham, ENERGY GAP INTERPRETATION OF EXPERIMENTS OF
INFRARED TRANSMISSION THROUGH SUPERCONDUCTING FILMS, *Phys.
Rev.* **104**, 845.

396. N. G. Yaroslavski, B. A. Zheludov, and A. E. Stanevich, REGISTERING SPECTROM-
ETER FOR THE LONGWAVE INFRARED REGION, *Opt. i Spektr.* **1**, 507.

397. H. Yoshinaga and R. A. Oetjen, OPTICAL PROPERTIES OF INDIUM ANTIMON-
IDE IN THE REGION FROM 20 TO 200 MICRONS, *Phys. Rev.* **101**, 526.

1957

398. J. Bohdansky, ÜBER DIE EMISSIONSLINIEN DES Hg-HOCHDRUCKBOGENS
IM GESAMTEN ULTRAROT, *Z. Physik* **149**, 383.

399. W. S. Boyle and A. D. Brailsford, INFRARED RESONANT ABSORPTION FROM
BOUND LANDAU LEVELS IN InSb, *Phys. Rev.* **107**, 903.

400. R. E. Glover and M. Tinkham, CONDUCTIVITY OF SUPERCONDUCTING FILMS
FOR PHOTON ENERGIES BETWEEN 0.3 AND 4.0 kT_c, *Phys. Rev.* **108**, 243.

401. A. Hadni, LE SPECTRE INFRAROUGE AU DÉLÀ DE 20 MICRONS. II. SPECTRO-
GRAPHES À RÉSEAUX, *Cah. Phys.* **77**, 12 (1957).

402. H. Happ, W. Eckhardt, L. Genzel, G. Sperling, and R. Weber, DER KRISTALL-
DETEKTOR ALS EMPFÄNGER THERMISCHER STRAHLUNG IM GEBIET
VON 100–1000 WELLENLÄNGE, *Z. Naturforsch.* **12A**, 522.

403. J. T. Last, INFRARED ABSORPTION STUDIES ON BARIUM TITANATE AND
RELATED MATERIALS, *Phys. Rev.* **105**, 1740.

404. J. Lecomte, LE SPECTRE INFRAROUGE AU DÉLÀ DE 20 MICRONS. I.
SPECTROGRAPHES À PRISMES, *Cah. Phys.* **77**, 1.

405. R. C. Lord and T. K. McCubbin, INFRARED SPECTROSCOPY FROM 5 TO 200
MICRONS WITH A SMALL GRATING SPECTROMETER, *J. Opt. Soc. Am.* **47**,
689.

406. E. D. Palik and E. E. Bell, PURE ROTATIONAL SPECTRA OF THE PARTIALLY
DEUTERATED AMMONIAS IN THE FAR INFRARED SPECTRAL REGION,
J. Chem. Phys. **26**, 1093.

407. E. D. Palik and R. A. Oetjen, THE PURE ROTATIONAL SPECTRUM OF H_2Se
IN THE FAR INFRARED SPECTRAL REGION, *J. Mol. Spectry.* **1**, 223.

408. E. D. Palik and K. N. Rao, PURE ROTATIONAL ABSORPTION SPECTRA OF SOME METHYL HALIDES BETWEEN 100 AND 600 μ, *J. Chem. Phys.* **26**, 1401.

409. C. H. Palmer, RATIO RECORDING SYSTEM FOR THE JOHNS HOPKINS 100-FOOT ABSORPTION CELL, *J. Opt. Soc. Am.* **47**, 367.

410. K. N. Rao and E. D. Palik, RESOLUTION OF THE TYPE E BANDS OF SOME SYMMETRIC TOP MOLECULES IN THE REGION 25–35 MICRONS, *J. Mol. Spectry.* **1**, 24.

411. S. A. Rice and W. Klemperer, SPECTRA OF THE ALKALI HALIDES. II. THE INFRARED SPECTRA OF THE SODIUM AND POTASSIUM HALIDES, RbCl AND CsCl, *J. Chem. Phys.* **27**, 573.

412. W. G. Spitzer and H. Y. Fan, DETERMINATION OF THE OPTICAL CONSTANTS AND CARRIER EFFECTIVE MASS OF SEMICONDUCTORS, *Phys. Rev.* **106**, 882.

413. W. G. Spitzer and H. Y. Fan, INFRARED ABSORPTION IN n-TYPE SILICON, *Phys. Rev.* **108**, 268.

414. J. Strong, INTERFEROMETRY FOR THE FAR INFRARED, *J. Opt. Soc. Am.* **47**, 354.

415. N. G. Yaroslavski, METHODS AND APPARATUS FOR LONGWAVE INFRARED SPECTROSCOPY, *Uspekhi Fiz. Nauk* **62**, 159.

1958

416. F. Ables and J. P. Mathieu, CALCUL DES CONSTANTES OPTIQUES DES CRISTAUX IONIQUES DANS L'INFRAROUGE, À PARTIR DU SPECTRE DE RÉFLEXION, *Ann. Physik* **3**, 5.

417. F. F. Bentley, E. F. Wolfarth, N. Srp, and W. R. Powell, ANALYTICAL APPLICATIONS OF FAR INFRARED SPECTRA. I. HISTORICAL REVIEW, APPARATUS, AND TECHNIQUES, *Spectrochim. Acta* **13**, 1.

418. G. R. Bird, A. Danti, and R. C. Lord, PURE ROTATIONAL ABSORPTION OF NO_2 IN THE 50–200 MICRON REGION, *Spectrochim. Acta* **12**, 247.

419. W. S. Boyle, A. D. Brailsford, and J. K. Galt, DIELECTRIC ANOMALIES AND CYCLOTRON ABSORPTION IN THE INFRARED: OBSERVATIONS ON BISMUTH, *Phys. Rev.* **109**, 1396.

420. A. Büchler and W. Klemperer, INFRARED SPECTRA OF THE ALKALINE-EARTH HALIDES. I. BERYLLIUM FLUORIDE, BERYLLIUM CHLORIDE, AND MAGNESIUM CHLORIDE, *J. Chem. Phys.* **29**, 121.

421. J. Connes, DOMAINE D'UTILISATION DE LA MÉTHODE PAR TRANSFORMÉE DE FOURIER, *J. Phys. Rad.* **19**, 197.

422. F. A. Cotton, A. Danti, J. S. Waugh, and R. W. Fessenden, CARBON-13 NUCLEAR RESONANCE SPECTRUM AND LOW-FREQUENCY INFRARED SPECTRUM OF IRON PENTACARBONYL, *J. Chem. Phys.* **29**, 1427.

423. P. W. Davis and R. A. Oetjen, FAR INFRARED SPECTRA OF SEVERAL PYRAMIDAL TRIHALIDES, *J. Mol. Spectry.* **2**, 253.

424. H. A. Gebbie, SPECTRES D'ABSORPTION ATMOSPHÉRIQUE DANS L'INFRAROUGE LOINTAIN PAR INTERFÉROMÉTRIE À DEUX ONDES, *J. Phys. Rad.* **19**, 230.

425. L. Genzel and R. Weber, ZUR THEORIE DER INTERFERENZ-MODULATION FÜR ZWEISTRAHL-INTERFERENZ, *Z. Angew. Phys.* **10**, 127.

426. L. Genzel and R. Weber, SPEKTROSKOPIE IM FERNEN ULTRAROT DURCH INTERFERENZ-MODULATION, *Z. Angew. Phys.* **10**, 195.

427. A. F. Gibson, INFRARED AND MICROWAVE MODULATION USING FREE CARRIERS IN SEMICONDUCTORS, *J. Sci. Instr.* **35**, 273.

428. A. Hadni, SUR QUELQUES SPECTRES D'ABSORPTION AU DÉLÀ DE 18 μ, *Spectrochim. Acta* **11**, 632.

429. A. Hadni and E. Decamps, SPECTRES D'ABSORPTION ET DE RÉFLEXION ENTRE 100–300 μ, *Intern. Conf. Spectry. 7th, Liège 1958, Proc.*, p. 423.

430. A. Hadni, E. Decamps, and P. Delorme, POLARIZATION OF LIGHT BETWEEN 100 μ AND 300 μ BY ECHELETTE GRATINGS, *J. Phys. Rad.* **19**, 793.

431. G. Heilmann, DIE TEMPERATURABHÄNGIGKEIT DER OPTISCHEN KONSTANTEN VON LiF IM BEREICH DER ULTRAROT RESTSTRAHLENBANDEN, *Ann. Physik* **152**, 368.

432. H. J. Hrostowski and C. S. Fuller, EXTENSION OF INFRARED SPECTRA OF III–V COMPOUNDS BY LITHIUM DIFFUSION, *J. Phys. Chem. Solids* **4**, 155.

433. H. J. Hrostowski and R. H. Kaiser, INFRARED SPECTRA OF GROUP III ACCEPTORS IN SILICON, *J. Phys. Chem. Solids* **4**, 148.

434. H. J. Hrostowski and R. H. Kaiser, INFRARED SPECTRA OF HEAT TREATMENT CENTERS IN SILICON, *Phys. Rev. Letters* **1**, 199.

435. M. Klier, DIE TEMPERATURABHÄNGIGKEIT DER OPTISCHEN KONSTANTEN VON LITHIUMFLUORID UND NATRIUMFLUORID IM ULTRAROTEN, *Z. Physik* **150**, 49.

436. A. Mitsuishi, H. Yoshinaga, and S. Fujita, THE FAR INFRARED ABSORPTION OF SULFIDES, SELENIDES, AND TELLURIDES OF ZINC AND CADMIUM, *J. Phys. Soc. Japan* **13**, 1235.

437. A. Mitsuishi, H. Yoshinaga, and S. Fujita, THE FAR INFRARED ABSORPTION OF FERRITES, *J. Phys. Soc. Japan* **13**, 1236.

438. R. Newman, OPTICAL PROPERTIES OF *n*-TYPE InP, *Phys. Rev.* **111**, 1518.

439. R. C. Ohlmann, P. L. Richards, and M. Tinkham, FAR INFRARED TRANSMISSION THROUGH METAL LIGHT-PIPES, *J. Opt. Soc. Am.* **48**, 531.

440. E. K. Plyler and N. Acquista, TRANSMITTANCE AND REFLECTANCE OF CESIUM IODIDE IN THE FAR INFRARED REGION, *J. Opt. Soc. Am.* **48**, 668.

441. E. K. Plyler and L. R. Blaine, INFRARED MEASUREMENTS WITH SMALL GRATING FROM 100 TO 300 MICRONS, *J. Res. Natl. Bur. Stand.* **60**, 55.

442. H. Reimann, EIN VAKUUMGITTERSPEKTROMETER FÜR DAS FERNE INFRAROT, *Optik aller Wellenlängen*, Akademie-Verlag, Berlin, p. 110.

443. P. L. Richards and M. Tinkham, FAR INFRARED ENERGY GAP MEASUREMENTS IN BULK SUPERCONDUCTORS, *Phys. Rev. Letters* **1**, 318.

444. R. Rosenberg and M. Lax, FREE CARRIER ABSORPTION IN *n*-TYPE Ge, *Phys. Rev.* **112**, 843.

445. M. Sergent, ÉTUDE EXPÉRIMENTALE D'UN SPECTROGRAPH POUR L'INFRAROUGE LOINTAIN, *Rev. Opt.* **37**, 552.

446. J. Strong, CONCEPTS OF CLASSICAL OPTICS, Freeman, San Francisco.

447. J. Strong and G. A. Vanasse, MODULATION INTERFÉRENTIELLE ET CALCULATEUR ANALOGIQUE POUR UN SPECTROMÈTRE INTERFÉRENTIAL, *J. Phys. Rad.* **19**, 192.

448. A. E. Stanevich and N. G. Yaroslavski, SOME INFRARED RADIATION SOURCES WITHIN THE 20–110 μ RANGE, *Inzh.-Fiz. Zhur.* **1**, 49.

449. N. G. Yaroslavski and A. E. Stanevich, ROTATIONAL SPECTRUM OF WATER VAPOR IN THE 50 TO 1500 μ REGION, *Izvest. Akad. Nauk SSSR, Ser. Fiz.* **22**, 1145; *Bull. Acad. Sci. USSR* **22**, 1131.

450. N. G. Yaroslavski and A. E. Stanevich, ROTATIONAL SPECTRUM OF H_2O IN

THE LONG-WAVELENGTH INFRARED REGION 50–1500 μ, *Opt. i Spektr.* **5**, 384.

451. H. Yoshinaga, S. Fujita, S. Minami, A. Mitsuishi, R. A. Oetjen, and Y. Yamada, FAR INFRARED SPECTROGRAPH FOR USE FROM THE PRISM SPECTRAL REGION TO ABOUT 1 mm WAVELENGTH, *J. Opt. Soc. Am.* **48**, 315.

1959

452. S. I. Averkov and V. Y. Ryadov, INDICATION OF INFRARED RADIATIONS BY THERMAL FREQUENCY CONVERTERS, *Seriya Radiofizika* **2**, 697.

453. L. Beckmann, E. Funck, and R. Mecke, ENERGIEBEGRENZTE AUFLÖSUNG VON SPEKTROMETERN FÜR DAS MITTLERE UND FERNE INFRAROT, *Z. Angew. Physik* **11**, 207.

454. F. F. Bentley and E. F. Wolfarth, ANALYTICAL APPLICATIONS OF FAR INFRARED SPECTRA-II. SPECTRA-STRUCTURE CORRELATIONS FOR ALIPHATIC AND AROMATIC HYDROCARBONS IN THE CESIUM BROMIDE REGION, *Spectrochim. Acta* **15**, 165.

455. W. S. Boyle, FAR INFRARED MAGNETO-OPTIC EFFECTS FROM IMPURITIES IN GERMANIUM, *J. Phys. Chem. Solids* **8**, 321.

456. W. S. Boyle and K. F. Rodgers, DE HAAS-VAN ALPHEN TYPE OSCILLATIONS IN THE INFRARED TRANSMISSION OF BISMUTH, *Phys. Rev. Letters* **2**, 338.

457. W. S. Boyle and K. F. Rodgers, PERFORMANCE CHARACTERISTICS OF A NEW LOW-TEMPERATURE BOLOMETER, *J. Opt. Soc. Am.* **49**, 66.

458. R. Brout, SUM RULE FOR LATTICE VIBRATIONS IN IONIC CRYSTALS, *Phys. Rev.* **113**, 43.

459. D. G. Burkhard and D. M. Dennison, ROTATION SPECTRUM OF METHYL ALCOHOL, *J. Mol. Spectry.* **3**, 299.

460. R. S. Caldwell and H. Y. Fan, OPTICAL PROPERTIES OF TELLURIUM AND SELENIUM, *Phys. Rev.* **114**, 664.

461. R. J. Collins and D. A. Kleinman, INFRARED REFLECTIVITY OF ZINC OXIDE, *J. Phys. Chem. Solids* **11**, 190.

462. A. Danti and R. C. Lord, PURE ROTATIONAL ABSORPTION OF OZONE AND SULFUR DIOXIDE FROM 100 AND 200 MICRONS, *J. Chem. Phys.* **30**, 1310.

463. A. Danti and J. L. Wood, FAR INFRARED SPECTRUM AND BARRIER TO INTERNAL ROTATION IN 1,1,1,2-TETRAFLUOROETHANE, *J. Chem. Phys.* **30**, 582.

464. L. A. Duncanson, J. W. Eddel, M. B. Lloyd, and W. T. Moore, DIFFRACTION GRATINGS FOR THE MEASUREMENT OF SPECTRA IN THE 20–100 μ REGION, *Spectrochim. Acta* **15**, 64.

465. H. Y. Fan and P. Fisher, ABSORPTION SPECTRA OF GROUP V DONORS IN GERMANIUM, *J. Phys. Chem. Solids* **8**, 270.

466. P. Fisher and H. Y. Fan, OPTICAL AND MAGNETO-OPTICAL ABSORPTION EFFECTS OF GROUP III IMPURITIES IN GERMANIUM, *Phys. Rev. Letters* **2**, 456.

467. S. J. Fray and J. F. C. Oliver, PHOTOCONDUCTIVE DETECTOR OF RADIATION OF WAVELENGTH GREATER THAN 50 μ, *J. Sci. Instr.* **36**, 195.

468. K. D. Froome, A NEW MICROWAVE HARMONIC GENERATOR, *Nature* **184**, 808.

469. R. Geick, PHOTOCONDUCTIVE DETECTOR OF RADIATION OF WAVE-LENGTH GREATER THAN 50 μ, *Z. Naturforsch.* **14A**, 196.

470. L. Genzel, H. Happ, and R. Weber, EIN GITTERSPEKTROMETER FÜR DEN BEREICH DES FERNEN ULTRAROT UND DER MIKROWELLEN, *Ann. Physik* **154**, 1.

471. L. Genzel, H. Happ, and R. Weber, DISPERSIONSMESSUNGEN AN NaCl, KCl UND KBr ZWISCHEN 0.3 UND 3 mm WELLENLÄNGE, *Z. Physik* **154**, 13.

472. D. M. Ginsberg, P. L. Richards, and M. Tinkham, APPARENT STRUCTURE ON THE FAR INFRARED ENERGY GAP IN SUPERCONDUCTING LEAD AND MERCURY, *Phys. Rev. Letters* **3**, 337.

473. A. Hadni and E. Decamps, SPECTROMÈTRE À PETIT RÉSEAU POUR LES ONDES MILLIMÈTRIQUES 0.3–1.6 mm, *Compt. Rend.* **249**, 2048.

474. A. Hadni, E. Decamps, and C. Janot, SPECTROMÉTRIE DANS L'INFRAROUGE LOINTAIN (50–900 μ), (4th Meeting), Vol. 3, in *Advances in Molecular Spectroscopy*, Macmillan, New York, p. 130.

475. A. Hadni, C. Janot, and E. Decamps, CONTRIBUTION À L'ÉTUDE DE LA RÉFLEXION DE L'INFRAROUGE LOINTAIN PAR LES RÉSEAUX ECHELETTE, *J. Phys. Rad.* **20**, 705.

476. A. Hadni, C. Janot, and E. Decamps, SPECTROMÉTRIE DANS L'INFRAROUGE LOINTAIN (50 À 350 μ), *Rev. Opt.* **38**, 463.

477. V. Lorenzelli and K. D. Möller, SPECTROSCOPIE MOLÉCULAIRE, SPECTRE D'ABSORPTION DU TRICHLORURE DE PHOSPHORE DANS L'INFRAROUGE LOINTAIN, *Compt. Rend.* **248**, 1980.

478. F. A. Miller, G. L. Carlson, and W. B. White, INFRARED AND RAMAN SPECTRA OF NaCl, KCl, AND KBr CRYSTALS, *J. Phys. Soc. Japan* **14**, 110.

479. A. Mitsuishi, H. Yoshinaga, and S. Fujita, THE FAR INFRARED REFLECTIVITY OF NaCl, KCl, and KBr CRYSTALS, *J. Phys. Soc. Japan* **14**, 110.

480. G. S. Picus, E. Burstein, B. W. Henvis, and M. Hass, INFRARED LATTICE VIBRATION STUDIES OF POLAR CHARACTER IN COMPOUND SEMICONDUCTORS, *J. Phys. Chem. Solids* **8**, 282.

481. D. W. Robinson, EBERT SPECTROMETER FOR THE FAR INFRARED, *J. Opt. Soc. Am.* **49**, 966.

482. T. Sakurai, VARIABLE DEPTH ECHELETTE GRATINGS, *Sci. Rept. Tohoku Univ.* **11**, 352.

483. J. Strong and G. A. Vanasse, INTERFEROMETRIC SPECTROSCOPY IN THE FAR INFRARED, *J. Opt. Soc. Am.* **49**, 844.

484. G. A. Vanasse, J. Strong, and E. V. Loewenstein, FAR INFRARED SPECTRA OF H_2O AND H_2S TAKEN WITH AN INTERFEROMETRIC SPECTROGRAPH, *J. Opt. Soc. Am.* **49**, 309.

485. A. Yamaguchi, I. Ichishima, T. Shimanouchi, and S. Mizushima, THE FAR INFRARED SPECTRUM OF HYDRAZINE, *J. Chem. Phys.* **31**, 843.

486. N. G. Yaroslavski and A. E. Stanevich, LONG-WAVELENGTH INFRARED SPECTRUM OF H_2O VAPOURS AND THE ABSORPTION OF AIR IN THE REGION 20–2500 μ (500–4 cm^{-1}), *Opt. i Spektr.* **7**, 621; *Opt. Spectry.* **5**, 380.

487. N. G. Yaroslavski and A. E. Stanevich, ROTATIONAL SPECTRUM OF WATER VAPOR AND THE ABSORPTION OF HUMID AIR IN THE 40–2500 μ WAVELENGTH REGION, *Opt. i Spektr.* **6**, 799.

1960

488. J. R. Aronson, R. C. Lord, and D. W. Robinson, FAR INFRARED SPECTRUM AND STRUCTURE OF DISILOXANE, *J. Chem. Phys.* **33**, 1004.

489. H. Bilz, L. Genzel, and H. Happ, THE INFRARED DISPERSION OF ALKALI HALIDES. I. THE INTERPRETATION OF THE SPECTRA ACCORDING TO THE BORN-HUANG THEORY, Z. Physik 160, 535.
490. W. Boldt, UNTERSUCHUNGEN ÜBER DIE STRAHLUNGSERZEUGUNG IM WELLENLÄNGEN-BEREICH VON 100 BIS 1000 μm, Monatsber. Deut. Akad. Wiss. Berlin 212, 735.
491. W. Boldt and H. Reimann, ZUR LANGWELLIGEN ULTRAROTEMISSION VON GASENTLADUNGSLAMPEN, Ann. Physik 6, 293.
492. W. S. Boyle and A. D. Brailsford, FAR INFRARED STUDIES OF BISMUTH, Phys. Rev. 120, 1943.
493. S. I. Chan, J. Zinn, and W. D. Gwinn, DOUBLE MINIMUM VIBRATION IN TRI-METHYLENE OXIDE, J. Chem. Phys. 33, 295.
494. W. Culshaw, HIGH RESOLUTION MILLIMETER WAVE FABRY-PEROT INTER-FEROMETER, IRE Trans. Micr. Theor. Tech. MTT-8, 182.
495. A. Danti, W. J. Lafferty, and R. C. Lord, FAR INFRARED SPECTRUM OF TRI-METHYLENE OXIDE, J. Chem. Phys. 33, 294.
496. T. J. Dean, G. O. Jones, D. H. Martin, P. A. Mawer, and C. H. Perry, SUPER-CONDUCTING BOLOMETERS AND SPECTROMETRY IN THE FAR INFRARED, Optica Acta 7, 185.
497. W. Eckhardt, SPEKTRALUNTERSUCHUNGEN IM GEBIET UM 1 mm WEL-LENLÄNGE. V. EIN VERGLEICHSMASS FÜR OPTISCH SPEKTROSKOPISCHE SYSTEME, Z. Physik 159, 405.
498. W. Eckhardt, SPEKTRALUNTERSUCHUNGEN IM GEBIET UM 1 mm WELLEN-LÄNGE. VI. DIE KONSTRUCKTION EINES IMPULSMASSENSTRAHLERS UND SEINE EIGENSCHAFTEN ALS SPEKTROSKOPISCHE STRAHLUNGS-QUELLE, Z. Physik 160, 121.
499. P. Fisher and H. Y. Fan, ABSORPTION SPECTRA AND ZEEMAN EFFECT OF COPPER AND ZINC IMPURITIES IN GERMANIUM, Phys. Rev. Letters 5, 195.
500. S. J. Fray, F. A. Johnson, and R. H. Jones, LATTICE ABSORPTION BANDS IN INDIUM ANTIMONIDE, Proc. Phys. Soc. (London) 76, 939.
501. K. D. Froome, MILLIMETRE WAVES FROM MERCURY ARC HARMONIC GENERATOR, Nature 186, 959.
502. K. D. Froome, SUBMILLIMETRE WAVES BY HARMONIC GENERATOR FROM COLD CATHODE ARCS, Nature 188, 43.
503. H. A. Gebbie, N. B. W. Stone, and C. D. Walshaw, PURE ROTATIONAL ABSORP-TION OF OZONE IN THE REGION 125–500 MICRONS, Nature 187, 765.
504. R. Geick, THE REFRACTIVE INDEX OF CRYSTALLINE AND FUSED QUARTZ IN THE SPECTRAL REGION NEAR 100 μ, Z. Physik 161, 116.
505. L. Genzel, APERIODIC AND PERIODIC INTERFERENCE MODULATION FOR SPECTROGRAPHIC PURPOSES, J. Mol. Spectry. 4, 241.
505a. D. M. Ginsberg and M. Tinkham, FAR INFRARED TRANSMISSION THROUGH SUPERCONDUCTING FILMS, Phys. Rev. 118, 990.
506. M. Gottlieb, OPTICAL PROPERTIES OF LITHIUM FLUORIDE IN THE IN-FRARED, J. Opt. Soc. Am. 50, 343.
507. A. Hadni and E. Decamps, ÉTUDE DE SUBSTANCES TRANSPARENTES DANS L'INFRAROUGE LOINTAIN (50–2500 μ), Compt. Rend. 250, 1827.
508. A Hadni, E. Decamps, D. Grandjean, and C. Janot, ÉTUDE EXPERIMENTALE DE LA POLARIZATION DE LA LUMIÈRE PAR LES RÉSEAUX ECHELETTE ENTRE 1 ET 600 μ, Compt. Rend. 250, 2007.
509. A. Hadni and C. Janot, SPECTROMÈTRE A TRÈS GRANDE LUMINOSITÉ

POUR L'INFRAROUGE LOINTAIN. APPLICATION A L'ÉTUDE DE L'ABSORP-
TION DES CRISTAUX DE HALOGENURES ALCALINS ENTRE 250 ET 650 μ,
Rev. Opt. **39**, 451.

510. M. Hass, TEMPERATURE DEPENDENCE OF THE INFRARED SPECTRUM
OF SODIUM CHLORIDE, *Phys. Rev.* **117**, 1497.

511. M. Hass, FUNDAMENTAL LATTICE DISPERSION FREQUENCIES OF NaCl
AND KCl AT 82°K, *Phys. Rev.* **119**, 633.

512. D. Kleinmann, ANHARMONIC FORCES IN THE GaP CRYSTAL, *Phys. Rev.* **118**,
118.

513. D. A. Kleinmann and W. G. Spitzer, INFRARED LATTICE ABSORPTION OF
GaP, *Phys. Rev.* **118**, 110.

514. S. H. Koening and R. D. Brown, FAR INFRARED ELECTRON-IONIZED DONOR
RECOMBINATION RADIATION IN GERMANIUM, *Phys. Rev. Letters* **4**, 170.

515. H. Kondoh, ANTIFERROMAGNETIC RESONANCE IN NiO IN FAR-INFRARED
REGION, *J. Phys. Soc. Japan* **15**, 1970.

516. B. Lax, RESONANCE AND MAGNETO-ABSORPTION AT MILLIMETER AND
INFRARED FREQUENCIES, in *Solid State Physics in Electronics and Telecommunica-
tions*, Vol. 3, Academic Press, New York, p. 508.

517. E. V. Loewenstein, INTERFEROMETRIC SPECTRA OF AMMONIA AND
CARBON MONOXIDE IN THE FAR INFRARED, *J. Opt. Soc. Am.* **50**, 1163.

518. A. A. Maradudin and R. F. Wallis, LATTICE ANHARMONICITY AND OPTICAL
ABSORPTION IN POLAR CRYSTALS: I. THE LINEAR CHAIN, *Phys. Rev.*
120, 442.

519. A. Mitsuishi, Y. Yamada, S. Fujita, and H. Yoshinaga, POLARIZER FOR THE FAR
INFRARED REGION, *J. Opt. Soc. Am.* **50**, 437.

520. E. D. Palik, G. S. Picus, S. Teitler, and R. F. Wallis, INFRARED CYCLOTRON
RESONANCE ABSORPTION IN COMPOUND SEMICONDUCTORS, in *Proceed-
ings of the International Conference on Semiconductor Physics*, *Prague 1960*, Publishing
House of the Czechoslovak Academy of Sciences, Prague 1961, p. 587.

521. C. H. Palmer, EXPERIMENTAL TRANSMISSION FUNCTIONS FOR THE PURE
ROTATION BAND OF WATER VAPOR, *J. Opt. Soc. Am.* **50**, 1232.

522. E. K. Plyler and L. R. Blaine, TRANSMITTANCE OF MATERIALS IN THE FAR
INFRARED, *J. Res. Natl. Bur. Stand.* **64C**, **55**.

523. E. H. Putley, IMPURITY PHOTOCONDUCTIVITY IN *n*-TYPE InSb, *Proc. Phys.
Soc. (London)* **76**, 802.

524. P. L. Richards and M. Tinkham, FAR INFRARED ENERGY GAP MEASURE-
MENTS IN BULK SUPERCONDUCTING In, Sn, Hg, Ta, V, Pb, AND Nb, *Phys.
Rev.* **119**, 575.

525. D. W. Robinson and D. A. McQuarrie, FAR INFRARED SPECTRA OF SOME
SYMMETRIC TOP MOLECULES, *J. Chem. Phys.* **32**, 556.

526. G. Schaefer, DAS ULTRAROT SPEKTRUMS DES U-ZENTRUMS, *J. Phys. Chem.
Solids* **12**, 233.

527. J. Strong and G. A. Vanasse, LAMALLAR GRATING FAR-INFRARED INTER-
FEROMETER, *J. Opt. Soc. Am.* **50**, 113.

528. B. Szigeti, THE INFRA-RED SPECTRA OF CRYSTALS, *Proc. Roy. Soc. (London)*
A258, 377.

529. S. Takahashi, DISTRIBUTION OF THE RADIATION DIFFRACTED BY THE
VARIABLE DEPTH ECHELETTE GRATINGS, *Sci. Rept. Tohoku Univ.* **12**, 80.

530. T. Williams, DIRECT DETERMINATION OF LINE SHAPES OF ROTATIONAL
SPECTRA FROM INTERFEROMETRIC MEASUREMENTS, *J. Opt. Soc. Am.* **50**,
1159.

1961

531. D. Bloor, T. J. Dean, G. O. Jones, D. H. Martin, P. A. Mawer, and C. H. Perry, SPEC-TROSCOPY AT EXTREME INFRA-RED WAVELENGTHS. I. TECHNIQUES, *Proc. Roy. Soc. (London)* **260**, 510.

532. D. Bloor, T. J. Dean, G. O. Jones, D. H. Martin, P. A. Mawer, and C. H. Perry, SPECTROSCOPY AT EXTREME INFRA-RED WAVELENGTHS. II. THE LAT-TICE RESONANCES OF IONIC CRYSTALS, *Proc. Roy. Soc. (London)* **261**, 10.

533. D. Bloor and D. H. Martin, ANTIFERROMAGNETIC RESONANCE IN THE EXTREME INFRA-RED, *Proc. Phys. Soc. (London)* **78**, 774.

534. E. Burstein, D. N. Langenberg, and B. N. Taylor, SUPERCONDUCTORS AS QUAN-TUM DETECTORS FOR MICROWAVE AND SUBMILLIMETER-WAVE RADIA-TION, *Phys. Rev. Letters* **6**, 92.

535. D. Chin and P. A. Giguère, THE TORSIONAL OSCILLATION FREQUENCY OF H_2O_2, *J. Chem. Phys.* **34**, 690.

536. W. Cochran, S. J. Fray, F. A. Johnson, J. E. Quarrington, and N. Williams, LATTICE ABSORPTION IN GALLIUM ARSENIDE, *J. Appl. Phys. Suppl.* **32**, 2102.

537. J. Connes, RECHERCHES SUR LA SPECTROSCOPIE PAR TRANSFORMATION DE FOURIER, *Rev. Opt. Théor. Instrum.* **40**, 45, 116, 171, 231.

538. P. Delorme and A. Hadni, SPECTROMÈTRE À PETIT RÉSEAU POUR L'INFRA-ROUGE LOINTAIN (45–150 μ), *Compt. Rend.* **252**, 1299.

539. W. G. Fateley and F. A. Miller, TORSIONAL FREQUENCIES IN THE FAR INFRARED. I. MOLECULES WITH A SINGLE ROTOR, *Spectrochim. Acta* **17**, 857.

540. S. B. Field and D. H. Martin, THE SKELETAL MODES OF LONG-CHAIN MOLE-CULES, *Proc. Phys. Soc. (London)* **78**, 625.

541. S. J. Fray, F. A. Johnson, J. E. Quarrington, and N. Williams, THE OPTICAL ABSORP-TION OF GALLIUM ARSENIDE IN THE RESTSTRAHLEN BAND, *Proc. Phys. Soc. (London)* **77**, 215.

542. K. Frei and H. H. Günthard, DISTORTION OF SPECTRAL LINE SHAPES BY RECORDING INSTRUMENTS, *J. Opt. Soc. Am.* **51**, 83.

543. H. A. Gebbie, SUBMILLIMETER WAVE SPECTROSCOPY USING A MICHEL-SON INTERFEROMETER, in *Advances in Quantum Electronics*, Columbia Univ. Press, New York, p. 155.

544. R. Geick, ÜBER DAS ABSORPTIONSVERHALTEN VERSCHIEDENER QUARZ-GLASSORTEN IM SPEKTRALBEREICH VON 150 μ BIS 275 μ, *Z. Naturforsch.* **16A**, 1390.

545. R. Geick, DER BRECHUNGSINDEX VON KRISTALLINEM UND GESCHMOL-ZENEM QUARZ IM SPEKTRALBEREICH UM 100 μ, *Z. Physik* **161**, 116.

546. D. W. Goodwin and R. H. Jones, FAR INFRARED AND MICROWAVE DETECTORS, *J. Appl. Phys.* **32**, 2056.

547. A. Hadni, C. Henry, J. P. Mathieu, and H. Poulet, LES FRÉQUENCES FONDA-MENTALES DE VIBRATION DU CALOMEL, *Compt. Rend.* **252**, 1585.

548. D. Hadzi, FAR-INFRARED BANDS OF SOME CRYSTALS WITH STRONG HYDROGEN BONDS, *J. Chem. Phys.* **34**, 1445.

549. G. N. Harding, M. F. Kimmitt, J. H. Ludlow, P. Porteous, A. C. Prior, and V. Roberts, EMISSION OF SUBMILLIMETRE ELECTROMAGNETIC RADIATION FROM HOT PLASMA IN ZETA, *Proc. Phys. Soc. (London)* **77**, 1069.

550. T. C. Harman, A. J. Strauss, D. H. Dickey, M. S. Dresselhaus, G. B. Wright, and J. G. Mavoirdes, LOW ELECTRON EFFECTIVE MASSES AND ENERGY GAP IN $Cd_xHg_{1-x}Te$, *Phys. Rev. Letters* **7**, 403.

551. H. Happ and L. Genzel, INTERFERENZMODULATION MIT MONOCHROMA-TISCHEN MILLIMETERWELLEN, *Infrared Phys.* **1**, 39.

552. L. Harris, THE TRANSMITTANCE AND REFLECTION OF GOLD BLACK DEPOSITS IN THE 15 TO 100 MICRON REGION, *J. Opt. Soc. Am.* **51**, 80

553. L. Harris and P. Fowler, ABSORPTION OF GOLD IN THE FAR INFRARED, *J. Opt. Soc. Am.* **51**, 164.

554. G. Heilmann, THE OPTICAL CONSTANTS OF CALCIUM FLUORIDE IN THE INFRARED, *Z. Naturforsch.* **16A**, 714.

555. G. Heilmann, ÜBER DIE ANWENDUNG VON POLARISATOREN MIT KLEINEM POLARISATIONSGRAD IM ULTRAROTEN, *Optik* **18**, 440.

556. F. Keffer, A. J. Sievers, and M. Tinkham, INFRARED ANTIFERROMAGNETIC RESONANCE IN MnO, *J. Appl. Phys. Suppl.* **32**, 65.

557. B. P. Kozyrev and M. A. Kropotkin, STUDY OF SPECTRAL REFLECTION COEFFI-CIENTS OF THERMAL RADIATION RECEIVING COATINGS FOR THE WAVELENGTH RANGE FROM 10 TO 200 μ, *Opt. i Spektr.* **10**, 657; *Opt. Spectry.* **10**, 345.

558. D. C. Laine, A PROPOSAL FOR A TUNABLE SOURCE OF RADIATION FOR THE FAR INFRARED USING BEATS BETWEEN OPTICAL MASERS, *Nature* **191**, 795.

559. V. Lorenzelli and P. Delorme, SPECTRE DU FERROCYANURE DE POTASSIUM A L'ÉTAT CRISTALLIN DANS L'INFRAROUGE LOINTAIN (45–140 μ), *Compt. Rend.* **253**, 92.

560. F. J. Low, LOW TEMPERATURE GERMANIUM BOLOMETER, *J. Opt. Soc. Am.* **51**, 1300.

561. E. V. Loewenstein, OPTICAL PROPERTIES OF SAPPHIRE IN THE FAR INFRARED, *J. Opt. Soc. Am.* **51**, 108.

562. A. A. Maradudin and R. F. Wallis, LATTICE ANAHARMONICITY AND OPTICAL ABSORPTION IN POLAR CRYSTALS. II. CLASSICAL TREATMENT IN THE LINEAR APPROXIMATION, *Phys. Rev.* **123**, 777.

563. D. H. Martin and D. Bloor, THE APPLICATION OF SUPERCONDUCTIVITY TO THE DETECTION OF RADIANT ENERGY, *Cryogenics* **1**, 1.

564. F. A. Miller and G. L. Carlson, THE VIBRATIONAL SPECTRA AND STRUCTURE OF $Si(NCO)_4$ AND $Ge(NCO)_4$, *Spectrochim. Acta* **17**, 977.

565. A. Mitsuishi, THE OPTICAL PROPERTIES OF CADMIUM TELLURIDE IN THE FAR INFRARED REGION, *J. Phys. Soc. Japan* **16**, 533.

566. J. Neuberger and R. D. Hatcher, INFRARED OPTICAL CONSTANTS OF NaCl, *J. Chem. Phys.* **34**, 1733.

567. R. C. Ohlmann and M. Tinkham, ANTIFERROMAGNETIC RESONANCE IN FeF_2 AT FAR-INFRARED FREQUENCIES, *Phys. Rev.* **123**, 425.

558. D. C. Laine, A PROPOSAL FOR A TUNABLE SOURCE OF RADIATION FOR PROPERTIES OF ALKALI HALIDE AND THALLIUM HALIDE SINGLE CRYSTALS, *Rev. Phys. Chem. Japan* **31**, 1.

569. E. D. Palik, G. S. Picus, S. Teitler, and R. F. Wallis, INFRARED CYCLOTRON RESONANCE IN InSb, *Phys. Rev.* **122**, 475.

570. E. D. Palik, J. R. Stevenson, and R. F. Wallis, INFRARED CYCLOTRON RESON-ANCE ABSORPTION IN n-TYPE GaAs, *Phys. Rev.* **124**, 701.

571. E. D. Palik, S. Teitler, and R. F. Wallis, FREE CARRIER CYCLOTRON RESON-ANCE, FARADAY ROTATION, AND VOIGT DOUBLE REFRACTION IN COMPOUND SEMICONDUCTORS, *J. Appl. Phys. Suppl.* **32**, 2132.

572. E. D. Palik and R. F. Wallis, INFRARED CYCLOTRON RESONANCE IN n-TYPE InAs AND InP, *Phys. Rev.* **123**, 131.

573. G. S. Picus, RECOMBINATION PROCESSES IN FAR INFRARED PHOTO-CONDUCTORS, *J. Phys. Chem. Solids* **22**, 159.

574. E. H. Putley, IMPURITY PHOTOCONDUCTIVITY IN *n*-TYPE InSb, *J. Phys. Chem. Solids* **22**, 241.

575. P. Poinsot, X. Gerbaux, P. Strimer, and A. Hadni, SPECTROMÈTRE À RÉSEAU POUR L'INFRAROUGE LOINTAIN, ÉTUDE DU TITANATE DE BARYUM, DU QUARTZ ET DE L'EAU LIQUIDE, *Compt. Rend.* **253**, 2049.

576. H. Reimann, A VACUUM GRATING SPECTROMETER FOR THE FAR INFRARED, *Optik Aller Wellenlängen*, Akademie-Verlag, Berlin, p. 110.

577. P. L. Richards, ANISOTROPY OF THE SUPERCONDUCTIVITY ENERGY GAP IN PURE AND IMPURE TIN, *Phys. Rev. Letters* **7**, 412.

578. B. V. Rollin, DETECTION OF MILLIMETER AND SUBMILLIMETER WAVE RADIATION BY FREE CARRIER ABSORPTION IN SEMICONDUCTORS, *Proc. Phys. Soc. (London)* **77**, 1102.

579. A. J. Sievers and M. Tinkham, FAR-INFRARED EXCHANGE RESONANCE IN YTTERBIUM IRON GARNET, *Phys. Rev.* **124**, 321.

580. A. E. Stanevich and N. G. Yaroslavski, LOW FREQUENCY INFRARED HYDRO-GEN BOND ABSORPTION SPECTRUM OF LIQUID WATER AND CRYSTAL HYDRATES, *Dokl. Akad. Nauk. SSSR* **137**, 60; *Sov. Phys. Dokl.* **6**, 224.

581. A. E. Stanevich and N. G. Yaroslavski, ABSORPTION OF LIQUID WATER IN THE LONG WAVELENGTH PART OF THE INFRARED SPECTRUM (42–2000 MICRONS), *Opt. i Spektr.* **10**, 538; *Opt. Spectry.* **10**, 278.

582. J. Strong, IRIDESCENT KClO₃ CRYSTALS AND INFRARED REFLECTION FILTERS, *J. Opt. Soc. Am.* **51**, 853.

583. R. H. Wright and P. N. Daykin, A SPECTROMETER FOR THE FAR INFRARED, *Nature* **189**, 212.

584. A. Yamaguchi, I. Ichishima, T. Shimanouchi, and S. Mizushima, FAR INFRARED SPECTRUM OF HYDRAZINE, *Spectrochim. Acta* **16**, 1471.

1962

585. D. M. Adams, A SPECTROMETER FOR THE RANGE 20–300 μ, *Spectrochim. Acta* **18**, 1039.

586. R. E. Anacreon, C. C. Helms, and E. H. Siegler, A NEW DOUBLE-BEAM FAR-INFRARED SPECTROPHOTOMETER, in *Progress in Infrared Spectroscopy*, Vol. 1, Plenum Press, New York, p. 87.

587. A. Anderson, S. H. Walmsley, and H. A. Gebbie, FAR INFRA-RED SPECTRA OF CRYSTALLINE HYDROGEN CHLORIDE AND HYDROGEN BROMIDE, *Phil. Mag.* **7**, 1243.

588. A. S. Barker and M. Tinkham, FAR-INFRARED FERROELECTRIC VIBRATION MODE IN SrTiO₃, *Phys. Rev.* **125**, 1527.

589. H. Bilz and L. Genzel, ULTRAROTDISPERSION DER ALKALIHALOGENIDE. II. ZUR THEORIE DER ABSORPTION BEI TIEFEN TEMPERATUREN, *Z. Physik* **169**, 53.

590. L. R. Blaine, E. K. Plyler, and W. S. Benedict, CALIBRATION OF SMALL GRATING SPECTROMETERS FROM 166 TO 600 cm⁻¹, *J. Res. Natl. Bur. Stand.* **66A**, 223.

591. G. L. Carlson, THE VIBRATIONAL SPECTRUM OF Si(NCS)₄, *Spectrochim. Acta* **18**, 1529.

592. P. Delorme and V. Lorenzelli, TECHNIQUES D'UTILISATION DES ÉCHANTIL-LONS DANS L'INFRAROUGE LOINTAIN, *J. Phys. Rad.* **23**, 589.

593. T. Deutsch, POLARIZATION EFFECTS IN THE INFRARED LATTICE ABSORP-
TION BANDS OF CdS, *J. Appl. Phys.* **33**, 751.

594. J. R. Durig, R. C. Lord, W. J. Gardner, and L. H. Johnston, INFRARED TRANS-
MITTANCE and REFLECTANCE OF BERYLLIUM OXIDE, *J. Opt. Soc. Am.*
52, 1078.

595. W. G. Fateley and F. A. Miller, TORSIONAL FREQUENCIES IN THE FAR IN-
FRARED. II. MOLECULES WITH TWO OR THREE METHYL ROTORS, *Spectrochim. Acta* **18**, 977.

595a. P. Fisher, THE EFFECT OF CHEMICAL-SPLITTING ON THE LYMAN SERIES
OF ANTIMONY IMPURITY IN GERMANIUM, *J. Phys. Chem. Solids* **23**,
1346.

596. D. Fröhlich, DIE OPTISCHEN KONSTANTEN VON LiF IM GEBIET DER ULTRA-
ROTEN EIGENSCHWINGUNGEN, *Z. Physik* **169**, 114.

597. K. D. Froome, MICROWAVE HARMONIC GENERATOR CAPABLE OF FRE-
QUENCIES IN EXCESS OF 600 Gc/s, *Nature* **193**, 1169.

598. R. Geick, ZUR DISPERSION DES NaCl IM BEREICH SEINER ULTRAROTEN
EIGENSCHWINGUNG, *Z. Physik* **166**, 122.

599. X. Gerbaux and A. Hadni, ABSORPTION SPECTRUM OF SnI_4 IN THE FAR
INFRARED, *J. Phys. Rad.* **23**, 877.

600. A. K. Ghosh and S. D. Majumdar, SOME FORMULAE IN THE THEORY OF MOLE-
CULAR SPECTRA, *Spectrochim. Acta* **18**, 615.

601. D. M. Ginsberg and J. D. Leslie, FAR INFRARED ABSORPTION IN A LEAD-
THALLIUM SUPERCONDUCTING ALLOY, *IBM J. Res. Develop.* **6**, 55.

602. S. A. Golden, APPROXIMATE SPECTRAL ABSORPTION COEFFICIENTS FOR
PURE ROTATIONAL TRANSITIONS IN DIATOMIC MOLECULES, *J. Quant.
Spect. Rad. Transfer.* **2**, 201.

603. A. Hadni, J. Claudel, E. Decamps, X. Gerbaux, and P. Strimer, SPECTRES D'ABSORP-
TION DE MONOCRISTAUX DANS L'INFRAROUGE LOINTAIN (50–1600 μ), À
LA TEMPÉRATURE DE L'HÉLIUM LIQUIDE : IODURE DE CÉSIUM, QUARTZ,
GERMANIUM ET NITRATE DE NÉODYME, *Compt. Rend.* **255**, 1595.

604. H. Happ, H. W. Hofmann, E. Lux, and G. Seger, DISPERSIONSMESSUNGEN AN
CsBr UND CaF_2 IM FERNEN ULTRAROT UND BEI MILLIMETERWELLEN,
Z. Physik **166**, 510.

605. L. Harris and J. Piper, TRANSMITTANCE AND REFLECTANCE OF ALUMINUM
OXIDE FILMS IN THE FAR INFRARED, *J. Opt. Soc. Am.* **52**, 223.

606. M. Hass and B. W. Henvis, INFRARED LATTICE REFLECTION SPECTRA OF
III–V COMPOUND SEMICONDUCTORS, *J. Phys. Chem. Solids* **23**, 1099.

607. R. Heastie and D. H. Martin, COLLISION-INDUCED ABSORPTION OF SUB-
MILLIMETER RADIATION BY NON-POLAR ATMOSPHERIC GASES, *Can.
J. Phys.* **40**, 122.

608. S. Ikegami, I. Ueda, S. Kisaka, A. Mitsuishi, and H. Yoshinaga, FAR INFRARED
REFLECTIVITY OF $BaTiO_3$, *J. Phys. Soc. Japan* **17**, 1210.

609. C. Janot and A. Hadni, RÉFLEXION DE L'INFRAROUGE LOINTAIN SUR LES
RÉSEAUX ÉCHELETTE, *J. Phys. Rad.* **23**, 152.

610. W. Kaiser, W. G. Spitzer, R. H. Kaiser, and L. E. Howarth, INFRARED PROPERTIES
OF CaF_2, SrF_2, AND BaF_2, *Phys. Rev.* **127**, 1950.

611. R. W. Keyes, TRENDS IN THE LATTICE "COMBINATION BANDS" OF ZINC-
BLENDE-TYPE SEMICONDUCTORS, *J. Chem. Phys.* **37**, 72.

612. W. J. Lafferty and D. W. Robinson, FAR INFRARED SPECTRA OF SOLID
METHYL HALIDES, *J. Chem. Phys.* **36**, 83.

613. V. Lorenzelli, SPECTROMÈTRE À HAUTE RÉSOLUTION POUR L'INFRA-ROUGE LOINTAIN (20–180 μ), *Compt. Rend.* **254**, 1017.

614. D. H. Martin, FILLING THE SPECTROSCOPIC GAP BETWEEN MICROWAVES AND THE INFRA-RED I, II, *Contemp. Phys.* **4**, 139, 187.

615. R. S. McDowell and L. H. Jones, PURE ROTATIONAL SPECTRUM OF NT_3, *J. Mol. Spectry.* **9**, 79.

616. R. S. McDowell and L. H. Jones, LOW-FREQUENCY FUNDAMENTALS OF SOME METAL CARBONYLS, *J. Chem. Phys.* **36**, 3321.

617. S. S. Mitra, VIBRATION SPECTRA OF SOLIDS, in *Solid State Physics*, Vol. 13, Academic Press, New York, p. 1.

618. A. Mitsuishi, Y. Yamada, and H. Yoshinaga, REFLECTION MEASUREMENTS ON RESTSTRAHLEN CRYSTALS IN THE FAR-INFRARED REGION, *J. Opt. Soc. Am.* **52**, 14.

619. A. Mitsuishi and H. Yoshinaga, INFRARED ABSORPTION OF U-CENTERS IN KCl AND KBr, *Prog. Theoret. Phys. Suppl.* **23**, 241.

620. A. Mitsuishi, H. Yoshinaga, S. Fujita, and Y. Suemoto, VIBRATIONAL SPECTRA OF RUBY AND HAEMATITE IN THE INFRARED REGION, *Japan J. Appl. Phys.* **1**, 1.

621. J. M. Munier, J. Claudel, E. Decamps, and A. Hadni, POUVOIR RÉFLECTEUR DES RÉSEAUX ÉCHELETTE DANS L'ORDRE 1 EN LUMIÈRE POLARISÉE, *Rev. Opt.* **5**, 245.

622. V. N. Murzin and A. I. Demeshina, A SPECTROPHOTOMETER FOR THE FAR-INFRARED REGION, *Opt. i Spektr.* **13**, 826; *Opt. Spectry.* **13**, 467.

623. R. Papoular, SPECTROMÉTRIE INTERFÉRENTIELLE DES PLASMAS ENTRE 0.1 ET 10 mm, *J. Phys. Rad.* **23**, 185.

624. E. K. Plyler, D. J. C. Yates, and H. A. Gebbie, RADIANT ENERGY FROM SOURCES IN THE FAR INFRARED, *J. Opt. Soc. Am.* **52**, 859.

625. R. F. Potter and W. L. Eisenman, INFRARED PHOTODETECTORS: A REVIEW OF OPERATIONAL DETECTORS, *Appl. Opt.* **1**, 567.

626. R. C. Ramsey, SPECTRAL IRRADIANCE FROM STARS AND PLANETS, ABOVE THE ATMOSPHERE FROM 0.1 TO 100.0 MICRONS, *Appl. Opt.* **1**, 465.

627. K. N. Rao, W. W. Brim, V. L. Sinnett, and R. H. Wilson, WAVELENGTH CALIBRATIONS IN THE INFRARED. IV. USE OF A 1000-LINES-PER-INCH BAUSCH AND LOMB PLANE REPLICA GRATING, *J. Opt. Soc. Am.* **52**, 862.

628. R. L. Redington, THE INFRARED SPECTRUM AND BARRIER HINDERING INTERNAL ROTATION IN H_2S_2, CF_3SH, CF_3SD, *J. Mol. Spectry.* **9**, 469.

629. K. F. Renk and L. Genzel, INTERFERENCE FILTERS AND FABRY-PEROT INTERFEROMETER FOR THE FAR INFRARED, *Appl. Opt.* **1**, 643.

630. S. Roberts and D. D. Coon, FAR-INFRARED PROPERTIES OF QUARTZ AND SAPPHIRE, *J. Opt. Soc. Am.* **52**, 1023.

631. G. Rupprecht, D. M. Ginsberg, and J. D. Leslie, PYROLYTIC GRAPHITE TRANSMISSION POLARIZER FOR INFRARED RADIATION, *J. Opt. Soc. Am.* **52**, 665.

632. G. Seger and L. Genzel, ABSORPTIONMESSUNGEN AN LiF ZWISCHEN 0.3 UND 0.9 mm WELLENLÄNGE, *Z. Physik* **169**, 66.

633. J. E. Stewart and W. S. Gallaway, DIFFRACTION ANOMALIES IN GRATING SPECTROPHOTOMETERS, *Appl. Opt.* **1**, 421.

634. M. Tinkham, FAR INFRARED SPECTRA OF MAGNETIC MATERIALS, *J. Appl. Phys.* **33**, 1248.

635. W. J. Turner, INFRARED LATTICE BANDS OF ALUMINUM ANTIMONIDE, *Phys. Rev.* **127**, 126.

636. G. R. Wilkinson, S. A. Inglis, and C. Smart, FAR INFRARED SPECTROSCOPY, Institute of Petroleum Symposium on Spectroscopy, March 1962.

637. Y. Yamada, A. Mitsuishi, and H. Yoshinaga, TRANSMISSION FILTERS IN THE FAR-INFRARED REGION, *J. Opt. Soc. Am.* **52**, 17.

                                        1963

638. D. M. Adams and H. A. Gebbie, ABSORPTION SPECTRA OF SOME INORGANIC COMPLEX HALIDES BY FAR-INFRARED INTERFEROMETRY, *Spectrochim. Acta.* **19**, 925.

639. D. M. Adams, H. A. Gebbie, and R. D. Peacock, FAR INFRA-RED ABSORPTION SPECTRA OF SOME COMPLEX HALIDES OF MOLYBDENUM AND TUNGSTEN, *Nature* **199**, 278.

640. D. M. Adams, M. Goldstein, and E. F. Mooney, FAR-INFRA-RED SPECTRA OF SOME ANHYDROUS METAL HALIDES IN THE SOLID STATE, *Trans. Faraday Soc.* **59**, 2228.

641. A. Anderson, H. A. Gebbie, and S. G. Walmsley, FAR INFRA-RED SPECTRA OF MOLECULAR CRYSTALS, I. HYDROGEN AND DEUTERIUM HALIDES, *Mol. Phys.* **7**, 401.

642. S. I. Averkov, V. I. Anikin, V. Y. Ryadov, and N. I. Furashov, VACUUM SPECTROMETER FOR THE FAR INFRARED, *Pribory i Tekh. Eksper. USSR (Instruments and Experimental Techniques)* **1**, 708.

643. A. S. Barker and M. Tinkham, FAR-INFRARED DIELECTRIC MEASUREMENTS ON POTASSIUM DIHYDROGEN PHOSPHATE, TRIGLYCINE SULFATE, AND RUTILE, *J. Chem. Phys.* **38**, 2257.

644. D. W. Berreman, INFRARED ABSORPTION AT LONGITUDINAL OPTIC FREQUENCY IN CUBIC CRYSTAL FILMS, *Phys. Rev.* **130**, 2193.

645. L. R. Blaine, A FAR-INFRARED VACUUM GRATING SPECTROMETER, *J. Res. Natl. Bur. Stand.* **67C**, 207.

646. R. H. Bradsel, A HIGH SPEED POWER SWITCH FOR USE WITH THE NATIONAL PHYSICAL LABORATORY ULTRAMICROWAVE HARMONIC GENERATOR, *J. Sci. Instr.* **40**, 225.

647. M. A. C. S. Brown and M. F. Kimmitt, NARROW BANDWIDTH TUNABLE INFRA-RED DETECTORS, *Brit. Comm. and Elect.* **10**, 608.

648. M. R. Brown, GENERATION, DETECTION AND PROPERTIES OF COHERENT RADIATION OF WAVE-LENGTHS LESS THAN 1 mm, *Nature* **200**, 1270.

649. S. C. Brown, G. Bekefi, and R. E. Whitney, FAR-INFRARED INTERFEROMETER FOR THE MEASUREMENT OF HIGH ELECTRON DENSITIES, *J. Opt. Soc. Am.* **53**, 448.

650. C. Bouster, J. Claudel, X. Gerbaux, and A. Hadni, SPECTRES D'ABSORPTION DE LA CUPRITE ENTRE 30 ET 250 MICRONS À TEMPÉRATURE ORDINAIRE ET A TEMPÉRATURE DE L'HÉLIUM LIQUIDE, *Ann. Phys.* **8**, 299.

651. J. E. Chamberlain, J. E. Gibbs, and H. A. Gebbie, REFRACTOMETRY IN THE FAR INFRA-RED USING A TWO-BEAM INTERFEROMETER, *Nature* **198**, 874.

652. H. H. Claassen, C. L. Chernick, and J. G. Malm, VIBRATIONAL SPECTRA AND STRUCTURE OF XeF_4, *J. Am. Chem. Soc.* **85**, 1927.

653. K. Colbow, INFRARED ABSORPTION LINES IN BORON-DOPED SILICON, *Can. J. Phys.* **41**, 1801.

654. A. I. Demeshina and V. N. Murzin, ABSORPTION AND REFLECTION SPECTRA OF $BaTiO_3$ IN THE LONG-WAVELENGTH INFRARED REGION, *Fiz. Tverd. Tela* **4**, 2980; *Sov. Phys.-Solid State* **4**, 2185.

655. J. R. Durig, FAR-INFRARED ABSORPTION, VIBRATIONAL SPECTRA AND STRUCTURE OF BETA-PROPIOLACTONE, Spectrochim. Acta 19, 1225.

656. J. R. Durig and R. C. Lord, PURE ROTATIONAL ABSORPTION OF NITROSYL FLUORIDE AND NITROSYL CHLORIDE IN THE 80–250 μ REGION, Spectrochim. Acta 19, 421.

657. J. R. Durig and R. C. Lord, FAR INFRARED ABSORPTION, VIBRATIONAL SPECTRA, AND STRUCTURE OF TETRAFLUORO-1,3-DITHIETANE, Spectrochim. Acta 19, 769.

658. W. F. Edgell, C. C. Helms, and R. E. Anacreon, LOW-FREQUENCY INFRARED SPECTRUM OF $Fe(CO)_5$, J. Chem. Phys. 38, 2039.

659. J. T. Edmond, A. Anderson, and H. A. Gebbie, FAR INFRA-RED STUDY OF As_2Se_3-TYPE GLASSES, Proc. Phys. Soc. (London) 81, 378.

660. W. G. Fateley and F. A. Miller, TORSIONAL FREQUENCIES IN THE FAR INFRA-RED-III. THE FORM OF THE POTENTIAL CURVE FOR HINDERED INTERNAL ROTATION OF A METHYL GROUP, Spectrochim. Acta 19, 611.

661. O. K. Filippov and N. G. Yaroslavskii, TRANSMISSION OF LONG-WAVELENGTH INFRARED RADIATION (40–200 μ) BY HEATED CRYSTALLINE AND FUSED QUARTZ, Opt. i Spektr. 15, 558; Opt. Spectry. 15, 229.

662. H. A. Gebbie and N. W. B. Stone, MEASUREMENT OF WIDTHS AND SHIFTS OF PURE ROTATIONAL LINES OF HYDROGEN CHLORIDE PERTURBED BY RARE GASSES, Proc. Phys. Soc. (London) 82, 309.

663. H. A. Gebbie and N. W. B. Stone, COLLISION INDUCED ABSORPTION IN CARBON DIOXIDE IN THE FAR INFRARED, Proc. Phys. Soc. (London) 82, 543.

664. H. A. Gebbie, N. W. B. Stone, and D. Williams, AN INTERFEROMETRIC STUDY OF THE INFRA-RED SPECTRUM OF COMPRESSED NITROGEN, Mol. Phys. 6, 215.

665. H. A. Gebbie, G. Topping, R. Illsley, and D. M. Dennison, THE ROTATION SPECTRUM OF METHYL ALCOHOL FROM 20 cm^{-1} TO 80 cm^{-1}, J. Mol. Spectry. 11, 229.

666. J. H. S. Green, W. Kynaston, and H. A. Gebbie, FAR INFRARED SPECTROSCOPY OF BENZENE DERIVATIVES, Spectrochim. Acta 19, 807.

667. A. Hadni, POSSIBILITÉS ACTUELLES DE DÉTECTION DU RAYONNEMENT INFRAROUGE, J. Phys. 24, 694.

668. A. Hadni, FAR-INFRARED ABSORPTION OF RUBY AT LIQUID HELIUM TEMPERATURE, Appl. Opt. 2, 977.

669. A. Hadni, INSTRUMENTATION IN THE FAR INFRARED. APPLICATIONS TO MOLECULAR AND SOLID STATE PHYSICS, Spectrochim. Acta 19, 793.

670. A. Hadni, E. Decamps, and J. Munier, ÉMISSION PROPRE DE L'ARC À VAPEUR DE MERCURE DANS L'INFRAROUGE LOINTAIN, INTERPRÉTATION ET APPLICATIONS, Rev. Opt. 42, 584.

671. A. Hadni and P. Strimer, TRANSITIONS ÉLECTRONIQUES DANS L'INFRAROUGE LOINTAIN, Compt. Rend. 257, 398.

672. M. Hass, INFRARED LATTICE REFLECTION SPECTRA OF LiCl, LiBr, KF, RbF, AND CsF, J. Phys. Chem. Solids 24, 1159.

673. L. C. Hebel and P. A. Wolff, QUANTUM EFFECTS IN THE INFRARED REFLECTIVITY OF BISMUTH, Phys. Rev. Letters 11, 368.

674. G. Heilmann, MEASUREMENT OF THE OPTICAL CONSTANTS OF CaF_2 IN THE INFRARED RESTSTRAHLEN BAND, Z. Physik 176, 253.

675. C. C. Helms, H. W. Jones, A. J. Russo, and E. H. Siegler, DESIGN AND PERFORMANCE OF A DOUBLE-BEAM FAR-INFRARED SPECTROMETER, Spectrochim. Acta 19, 819.

676. P. J. Hendra, R. D. G. Lane, and B. Smethurst, AN INFRARED SPECTROMETER TO OPERATE OVER THE RANGE 8–100 MICROMETERS, *J. Sci. Instr.* **40**, 457.

677. R. Isaac, F. F. Bentley, H. Sternglanz, W. C. Coburn, C. V. Stephenson, and W. S. Wilcox, THE FAR INFRARED SPECTRA OF MONOSUBSTITUTED PYRIDINES, *Appl. Spectry.* **17**, 90.

678. C. Janot and A. Hadni, POLARISATION DE LA LUMIÈRE PAR LES RÉSEAUX ÉCHELETTES DANS L'INFRAROUGE LOINTAIN, *J. Phys.* **24**, 69.

679. M. F. Kimmitt and G. B. F. Niblett, INFRA-RED EMISSION FROM THE THETA PINCH, *Proc. Phys. Soc. (London)* **82**, 938.

680. M. A. Kinch and B. V. Rollin, DETECTION OF MILLIMETRE AND SUB-MILLI-METRE WAVE RADIATION BY FREE CARRIER ABSORPTION IN A SEMI-CONDUCTOR, *Brit. J. Appl. Phys.* **14**, 672.

681. B. P. Kozyrev and A. V. Mezenov, EXPERIMENTAL INVESTIGATION OF THE TRANSMISSION FUNCTIONS OF AIR IN THE REGION OF THE ROTATIONAL ABSORPTION BANDS OF WATER VAPOR, *Opt. i Spektr.* **15**, 549; *Opt. Spectry.* **15**, 293.

682. M. Křížek, POLARIZATION OF NON-PARALLEL RAYS OF INFRARED RADIA-TION BY REFLECTION, *Czech. J. Phys.* **13**, 599.

683. D. C. Laine, R. C. Srivastava, A. L. S. Smith, and J. E. Ingram, THE AMMONIA MASER OSCILLATOR AS A DETECTOR OF STIMULATED EMISSION IN THE FAR INFRARED, in *Proceedings of the Symposium on Optical Masers*, Vol. 13, Polytechnic Press, Brooklyn, N.Y., p. 617.

684. V. Lorenzelli, SPECTROMÈTRE POUR L'INFRAROUGE LOINTAIN, *Rev. Opt.* **42**, 129.

685. D. H. Martin and C. D. Stone, RELAXATION OF FERROELECTRIC PHOSPHATES (ADP AND KDP) IN THE EXTREME-INFRA-RED, *Phys. Letters* **5**, 26.

686. L. May and K. J. Schwing, THE USE OF POLYETHYLENE DISKS IN THE FAR INFRARED SPECTROSCOPY OF SOLIDS, *Appl. Spectry.* **17**, 166.

687. S. S. Mitra, PHONON ASSIGNMENTS IN ZnSe AND GaSb AND SOME REGULA-RITIES IN THE PHONON FREQUENCIES OF ZINCBLENDE-TYPE SEMI-CONDUCTORS, *Phys. Rev.* **132**, 986.

688. A. Mitsuishi, Y. Otsuka, S. Fujita, and H. Yoshinaga, METAL MESH FILTERS IN THE FAR INFRARED REGION, *Japan. J. Appl. Phys.* **2**, 574.

689. T. Miyazawa, K. Fukushima, and Y. Ideguchi, FAR INFRARED SPECTRA AND THEIR VIBRATIONAL ASSIGNMENTS OF ISOTACTIC POLYPROPYLENE, *J. Polymer Sci.* **1B**, 385.

690. K. D. Möller and R. V. McKnight, FAR-INFRARED-TRANSMISSION FILTER GRATINGS, *J. Opt. Soc. Am.* **53**, 760.

691. C. Nanney, INFRARED ABSORPTANCE OF SINGLE-CRYSTAL ANTIMONY AND BISMUTH, *Phys. Rev.* **129**, 109.

692. E. D. Palik and J. R. Stevenson, INFRARED CYCLOTRON RESONANCE IN *n*-TYPE InAs, *Phys. Rev.* **130**, 1344.

693. C. H. Plamer, NOTE ON THE USE OF GRATINGS AS BROAD-BAND INFRARED FILTERS, *J. Opt. Soc. Am.* **53**, 1005.

693a. E. H. Putley, THE DETECTION OF SUB-mm RADIATION, *Proc. IEEE* **51**, 1412.

694. K. N. Rao, R. V. deVore, and E. K. Plyler, WAVELENGTH CALIBRATIONS IN THE FAR INFRARED (30 to 1000 MICRONS), *J. Res. Natl. Bur. Stand.* **67A**, 351.

695. P. L. Richards, FAR-INFRARED MAGNETIC RESONANCE IN CoF_2, NiF_2, $KNiF_3$ AND YbIG, *J. Appl. Phys.* **34**, 1237.

696. D. W. Robinson, SPECTRA OF MATRIX-ISOLATED WATER IN THE "PURE ROTATION" REGION, *J. Chem. Phys.* **39**, 3430.

697. A. Sabatini and L. Sacconi, FAR-INFRARED SPECTRA OF SOME TETRA-CHLORO METAL (II) COMPLEXES, *Ric. Sci., Series 2, Part II-A,* **3**, No. 5, p. 755.

698. G. Sage and W. Klemperer, FAR-INFRARED SPECTRUM AND BARRIER TO INTERNAL ROTATION OF ETHYL FLUORIDE, *J. Chem. Phys.* **39**, 371.

699. R. B. Sanderson and N. Ginsburg, LINE WIDTHS AND LINE STRENGTHS IN THE ROTATION SPECTRUM OF WATER VAPOR, *J. Quant. Spect. Rad. Transfer* **3**, 435.

700. T. Shimanouchi, Y. Kyogoku, and T. Miyazaki, A DOUBLE-BEAM GRATING SPECTROPHOTOMETER WITH A CsBr FORE-PRISM AND WAVELENGTHS OF ABSORPTION LINES OF WATER VAPOR IN THE REGION FROM 16 TO 30 μ, *Spectrochim. Acta* **19**, 451.

701. E. H. Siegler, INTERNATIONAL SYMPOSIUM ON FAR INFRARED SPECTRO-SCOPY, Cincinnati, Ohio, August 21–24, 1962, in *Appl. Opt.* **2**, 663.

702. A. J. Sievers and M. Tinkham, FAR INFRARED ANTIFERROMAGNETIC RE-SONANCE IN MnO AND NiO, *Phys. Rev.* **129**, 1566.

703. A. J. Sievers and M. Tinkham, FAR INFRARED SPECTRA OF RARE-EARTH IRON GARNETS, *Phys. Rev.* **129**, 1995.

704. A. J. Sievers and M. Tinkham, FAR-INFRARED SPECTRA OF HOLMIUM, SAMARIUM, AND GADOLINIUM IRON GARNETS, *J. Appl. Phys.* **34**, 1235.

705. D. Steele, W. Kynaston, and H. A. Gebbie, THE FAR INFRA-RED SPECTRA OF *p*-DIFLUOROBENZENE AND *p*-DIFLUORODEUTEROBENZENE, *Spectrochim. Acta.* **19**, 785.

706. J. H. M. Thornley, FAR-INFRARED SPECTRA OF TWO CERIUM DOUBLE NITRATE SALTS, *Phys. Rev.* **132**, 1492.

707. R. Ulrich, K. F. Renk, and L. Genzel, TUNABLE SUBMILLIMETER INTERFERO-METERS OF THE FABRY-PEROT TYPE, *IEEE Trans. Microw. Theor. Techn.* **11**, 363.

708. S. Walles, TRANSMISSION OF SILICON BETWEEN 40 μm AND 100 μm. *Arkiv. Fysik* **25**, 33.

709. S. H. Walmsley and A. Anderson, FAR INFRA-RED SPECTRA OF MOLECULAR CRYSTALS. II. CHLORINE, BROMINE, AND IODINE, *Mol. Phys.* **7**, 411.

710. K. A. Wickersheim, OPTICAL AND INFRARED PROPERTIES OF MAGNETIC MATERIALS, *Magnetism*, G. T. Rado and H. Suhl, Eds., Academic Press, New York, p. 269.

711. H. A. Willis, R. G. J. Miller, D. M. Adams, and H. A. Gebbie, SYNTHETIC RESINS AS WINDOW MATERIALS FOR FAR INFRA-RED SPECTROSCOPY, *Spectrochim. Acta* **19**, 1457.

712. J. L. Wood, FAR-INFRARED SPECTROSCOPY, *Quart. Rev. (London)* **17**, 362.

713. L. A. Woodward, FAR INFRARED ABSORPTION OF DIMETHYL MERCURY, *Spectrochim. Acta* **19**, 1963.

714. K. Yoshihara, AN IDEA OF AN INTERFERENCE SPECTROMETER FOR THE FAR INFRARED, *Japan. J. Appl. Phys.* **2**, 818.

715. A. G. Zhukov, LONG-WAVELENGTH INFRARED SPECTROMETER, *Opt. i Spektr.* **14**, 422; *Opt. Spectry.* **14**, 225.

716. U. A. Zirnit and M. M. Suchchinskii, LOW-FREQUENCY VIBRATIONS OF METHYL-SUBSTITUTED CYCLOHEXANES, *Opt. i Spektr.* **15**, 190; *Opt. Spectry.* **15**, 102.

1964

717. A. Anderson, H. A. Gebbie, and S. H. Walmsley, FAR INFRARED SPECTRA OF MOLECULAR CRYSTALS, I. HYDROGEN AND DEUTERIUM HALIDES, *Mol. Phys.* **7**, 401.

718. A. Anderson and S. H. Walmsley, FAR INFRA-RED SPECTRA OF MOLECULAR CRYSTALS. III. CARBON DIOXIDE, NITROUS OXIDE, AND CARBONYL SULPHIDE, *Mol. Phys.* **7**, 583.

719. I. M. Arefyev, APPLICATION OF NEGATIVE LIGHT FLUX SPECTROSCOPY IN THE FAR-INFRARED REGION OF THE SPECTRUM, *Opt. i Spektr.* **17**, 300; *Opt. Spectry.* **17**, 157.

720. J. R. Aronson, H. G. McLinden, and P. J. Gielisse, LOW-TEMPERATURE FAR-INFRARED SPECTRA OF GERMANIUM AND SILICON, *Phys. Rev.* **135**, A785.

721. S. I. Averkov, V. I. Anikin, V. Y. Ryadov, and N. I. Furashov, ASTRONOMICAL STATION FOR FAR-INFRARED OBSERVATIONS, *Soviet Astron. A. J.* **8**, 432.

722. M. Balkanski, J. M. Besson, and R. LeToullec, DISPERSION CURVES OF PHONONS IN HEXAGONAL CADMIUM SULFIDE OBTAINED BY INFRARED SPECTROSCOPY, in *Proceedings of the Internationl Conference on the Physics of Semiconductors*, Paris, 1964, Dunod, Paris, p. 1091.

723. A. S. Barker, INFRARED LATTICE VIBRATIONS IN CALCIUM TUNGSTATE AND CALCIUM MOLBYDATE, *Phys. Rev.* **135**, A742.

724. A. S. Barker, TRANSVERSE AND LONGITUDINAL OPTIC MODE STUDY IN MgF_2 AND ZnF_2, *Phys. Rev.* **136**, A1290.

725. A. S. Barker and J. J. Hopfield, COUPLED-OPTICAL-PHONON MODE THEORY OF THE INFRARED DISPERSION IN $BaTiO_3$, $SrTiO_3$, AND $KTaO_3$, *Phys. Rev.* **135**, A1732.

726. J. A. Bastin, A. E. Gear, G. O. Jones, H. J. T. Smith, and P. J. Wright, SPECTROSCOPY AT EXTREME INFRA-RED WAVELENGTHS. III. ASTROPHYSICAL AND ATMOSPHERIC MEASUREMENT, *Proc. Roy. Soc. (London)* **278A**, 543.

727. E. E. Bell, EXPERIENCE WITH INTERFEROMETRIC TECHNIQUES IN THE FAR INFRARED, in *Proceedings of the Far Infrared Physics Symposium*, Riverside, Calif., 1964, NOL Corona Tech. Bull.

728. J. E. Bertie and E. Whalley, INFRARED-ACTIVE INTERCHAIN VIBRATION IN POLYETHYLENE, *J. Chem. Phys.* **41**, 575.

729. D. Bloor, FAR INFRARED SPECTRUM OF $K_3[Co(CN)_6]$ AND $K_3[Fe(CN)_6]$, *J. Chem. Phys.* **41**, 2573.

730. J. W. Brasch and R. J. Jakobsen, THE USE OF A POLYETHYLENE MATRIX FOR STUDYING DILUTION AND LOW-TEMPERATURE EFFECTS IN THE FAR INFRARED, *Spectrochim. Acta* **20**, 1644.

731. R. A. Buchanan, H. H. Caspers, and H. R. Marlin, INFRARED ABSORPTION SPECTRA OF LiOH FILMS IN THE 75 TO 650 cm^{-1} REGION, *J. Chem. Phys.* **40**, 1125.

732. E. Burstein, INTERACTION OF PHONONS WITH PHOTONS: INFRARED, RAMAN AND BRILLOUIN SPECTRA, in *Phonons and Phonon Interactions*, Benjamin, New York, p. 276.

733. H. H. Caspers, R. A. Buchanan, and H. R. Marlin, LATTICE VIBRATIONS OF LaF_3, *J. Chem. Phys.* **41**, 94.

734. W. S. C. Chang, and R. F. Rountree, PROPERTIES OF MATERIALS FOR SUB-MILLIMETER MASERS, in *Third International Congress on Quantum Electronics*, Columbia Univ. Press, New York, p. 677.

735. G. W. Chantry, A. Anderson, and H. A. Gebbie, FAR INFRARED SPECTRUM OF SULPHUR AND SELENIUM, *Spectrochim. Acta* **20**, 1223.

736. G. W. Chantry, A. Anderson, and H. A. Gebbie, FAR INFRARED SPECTRA OF NAPHTHALENE h_8 AND d_8, *Spectrochim. Acta* **20**, 1465.

737. A. Crocker, H. A. Gebbie, M. F. Kimmitt, and L. E. S. Mathias, STIMULATED EMISSION IN THE FAR INFRA-RED, *Nature* **201**, 250.

738. E. Decamps, A. Hadni, and J. M. Munier, SPECTRES D'ABSORPTION DE MOLÉ-CULES POLAIRES À L'ÉTAT GAZEUX ET LIQUIDE ENTRE 50 ET 700 MIC-RONS, *Spectrochim. Acta* **20**, 373.

739. P. Delorme, V. Lorenzelli, and M. Fournier, SPECTRE INFRAROUGE DES DÉRIVÉS HEXASUBSTITUÉS HALOGÉNÉS DU BENZÈNE (2000–70 CM^{-1}) ET ATTRIBU-TION DES VIBRATIONS FONDAMENTALES ACTIVES, *Compt. Rend.* **259**, 751.

740. H. Dötsch and H. Happ, TEMPERATURABHÄNGIGKEIT DER ABSORPTION VON NaCl ZWISCHEN 1 UND 3 mm WELLENLÄNGE, *Z. Physik* **177**, 360.

741. J. M. Dowling, DIRECT INTERPRETATION OF FAR-INFRARED INTERFERO-GRAMS WITH APPLICATION TO DIATOMIC AND LINEAR MOLECULES, *J. Opt. Soc. Am.* **34**, 663.

742. T. Dupius and V. Lorenzelli, CONTRIBUTION À L'ÉTUDE DE QUELQUES METASTANNATES DE MÉTAUX BIVALENTS PAR SPECTROMÉTRIE D'AB-SORPTION INFRAROUGE (2–15 μ), *Compt. Rend.* **259**, 4585.

743. W. G. Fateley, I. Matsubara, and R. E. Witkowski, VAPOR-LIQUID FREQUENCY SHIFTS IN THE LOW-FREQUENCY, INFRARED SPECTRUM, *Spectrochim. Acta* **20**, 1461.

744. E. Fatuzzo, FIELD DEPENDENCE OF THE ABSORPTION BANDS OF FERRO-ELECTRICS IN THE FAR INFRA-RED, *Proc. Phys. Soc. (London)* **84**, 709.

745. A. S. Filler, APODIZATION AND INTERPOLATION IN FOURIER-TRANSFORM SPECTROSCOPY, *J. Opt. Soc. Am.* **54**, 762.

746. O. K. Filippov and V. M. Pivovarov, RADIATION OF A PRK-4 LAMP IN THE FAR INFRARED REGION, *Opt. i Spektr.* **16**, 522; *Opt. Spectry.* **16**, 282.

747. A. O. Frenzel and J. P. Butler, STUDY OF THE 73 cm^{-1} BAND OF POLYETHY-LENE, *J. Opt. Soc. Am.* **54**, 1059.

748. H. A. Gebbie, F. D. Findlay, N. W. B. Stone, and J. A. Robb, INTERFEROMETRIC OBSERVATIONS ON FAR INFRA-RED STIMULATED EMISSION SOURCES, *Nature* **202**, 169.

749. H. A. Gebbie and N. W. B. Stone, A MICHELSON INTERFEROMETER FOR FAR INFRARED SPECTROSCOPY OF GASES, *Infrared Phys.* **4**, 85.

750. H. A. Gebbie, N. W. B. Stone, G. R. Bird, and G. R. Hunt, MAGNETIC EFFECTS IN THE FAR INFRA-RED SPECTRUM OF NITROGEN DIOXIDE (NO_2), *Nature* **200**, 1304.

751. H. A. Gebbie, N. W. B. Stone, and F. D. Findlay, A STIMULATED EMISSION SOURCE OF 0.34 MILLIMETRE WAVELENGTH, *Nature* **202**, 685.

752. R. D. Gillard, H. G. Silver, and J. L. Wood, THE FAR INFRARED SPECTRA OF TRIS-ACETYLACETONATO METAL III COMPLEXES, *Spectrochim. Acta* **20**, 63.

753. R. Geick, MEASUREMENT AND ANALYSIS OF THE FUNDAMENTAL LATTICE VIBRATION SPECTRUM OF PbS, *Phys. Letters* **10**, 5.

754. D. M. Ginsberg and J. D. Leslie, FAR INFRARED ABSORPTION IN SUPERCON-DUCTING LEAD ALLOYS, *Rev. Mod. Phys.* **36**, 198.

755. A. Hadni, L'ÉLARGISSEMENT ACTUEL DU DOMAINE DES APPLICATIONS DE L'INFRAROUGE LOINTAIN, *Ann. Physik* **9**, 9.

756. A. Hadni, FAR INFRARED ELECTRONIC TRANSITIONS IN IONS AND PAIRS OF IONS, *Phys. Rev.* **136**, A758.

757. A. Hadni, J. M. Munier, and E. Decamps, SPECTRE D'ABSORPTION DE MOLÉ-CULES POLAIRES À L'ÉTAT GAZEUX ET LIQUIDE ENTRE 50 ET 700 MICRONS, *Spectrochim. Acta* **20**, 373.

758. A. Hadni, G. Morlot, and P. Strimer, LES NIVEAUX D'ÉNERGIE DE BASSE FRÉ-QUENCE DU RUBIS CONCENTRÉ, *Compt. Rend.* **258**, 515.

759. A. Hadni and P. Strimer, TRANSITIONS ÉLECTRONIQUES DANS L'INFRA-ROUGE LOINTAIN: CHLORURES ANHYDRES DE PRASÉODYME ET DE SAMARIUM, *Compt. Rend.* **258**, 5616.

760. A. Hadni, B. Wyncke, P. Strimer, E. Decamps, and J. Claudel, MATÉRIAUX TRANS-PARENTS DANS L'INFRAROUGE LOINTAIN (50–1600 MICRONS), in *Third International Congress on Quantum Electronics*, Columbia Univ. Press, New York, p. 731.

761. R. K. Harris and R. E. Witkowski, INFRARED STUDIES OF ACYCLIC CONJU-GATED MOLECULES IN THE REGION 80 cm^{-1} TO 850 cm^{-1}, *Spectrochim. Acta* **20**, 1651.

762. M. Hass, LATTICE VIBRATIONS IN IONIC CRYSTALS OF FINITE SIZE, *Phys. Rev. Letters* **13**, 429.

763. M. Hass, SIZE EFFECTS IN INFRARED LATTICE ABSORPTION, in *Proceedings of the Far Infrared Physics Symposium*, Riverside, Calif., 1964, NOL Corona Tech. Bull.

764. G. R. Hunt, C. H. Perry, and J. Ferguson, FAR INFRARED REFLECTANCE AND TRANSMITTANCE OF POTASSIUM MAGNESIUM FLUORIDE AND MAG-NESIUM FLUORIDE, *Phys. Rev.* **134**, A688.

765. I. Iwahashi, K. Matsummoto, and M. Inaba, MODEL FIS-1 HITACHI FAR INFRA-RED SPECTROPHOTOMETER, *Hitachi Rev.* **SI 10**, 39.

766. R. J. Jakobsen and J. W. Brasch, FAR-INFRARED STUDIES OF INTERMOLECU-LAR FORCES, DIPOLE–DIPOLE COMPLEXES, *J. Am. Chem. Soc.* **86**, 3571.

767. A. V. Jones, THE TELLURIC EMISSION SPECTRUM IN THE RANGE 1 μ TO 3 mm, *Soc. Roy. Sci. Liège Men.* **9**, 289.

768. G. Jones and W. Gordy, EXTENSION OF SUBMILLIMETER WAVE SPECTRO-SCOPY BELOW A HALF-MILLIMETER WAVELENGTH, *Phys. Rev.* **135**, A295.

769. L. H. Jones, VIBRATIONAL SPECTRA, FORCE CONSTANTS, AND BONDING IN MIXED CYANIDE-HALIDE COMPLEXES OF GOLD, *Inorg. Chem.* **3**, 1581.

769a. R. L. Jones and P. Fisher, THE EFFECT OF UNIAXIAL STRESS ON THE EXCITA-TION SPECTRUM OF A GROUP III IMPURITY IN GERMANIUM, *Solid State Commun.* **2**, 369.

770. R. I. Joseph and B. D. Silverman, INFRARED ACTIVE MODES IN KTaO$_3$, *J. Phys. Chem. Solids* **25**, 1125.

771. A. Kahan, H. G. Lipson, and E. V. Loewenstein, INFRARED LATTICE VIBRATIONS OF MAGNESIUM STANNIDE, in *Proceedings of the International Conference on the Physics of Semiconductors*, Paris, 1964, Dunod, Paris, p. 1067.

772. B. Z. Katsenelenbaum, QUASIOPTICAL METHODS OF GENERATION AND TRANSMISSION OF MILLIMETER WAVES, *Uspekhi Fiz. Nauk* **83**, 81; *Soviet Phys.-Usp.* **7**, 385.

773. F. K. Kneubühl, J. F. Moser, and H. Steffen, A HIGH RESOLUTION SUBMILLI-METER SPECTROGRAPH FOR SOLID STATE RESEARCH, *Helv. Phys. Acta* **37**, 596.

774. R. Kopelman, FAR-IR SPECTRUM OF DIMETHYL ACETYLENE: INTERNAL ROTATION AND EVIDENCE FOR A D_{6h} EFFECTIVE SYMMETRY, *J. Chem. Phys.* **41**, 1547.

775. D. J. Kroon, EEN InSb FOTOGELEIDINGS-DETECTOR VOOR MILLIMETER EN SUBMILLIMETERSTRALING, *Ned. Natuurk.* **30**, 301.

776. M. A. Kropotkin and B. P. Kozyrev, STUDY OF THE REFLECTION SPECTRA OF NATURAL AND SYNTHETIC MATERIALS IN THE 0.7–100 μ RANGE, *Opt. i Spektr.* **17**, 259; *Opt. Spectry.* **17**, 136.

777. J. D. Leslie, R. L. Cappelletti, D. M. Ginsberg, D. K. Finnemore, F. H. Spedding, and B. J. Beaudry, FAR-INFRARED ABSORPTION IN SUPERCONDUCTING LANTHANUM, *Phys. Rev.* **134**, A309.

778. J. D. Leslie and D. M. Ginsberg, FAR INFRARED ABSORPTION IN SUPER-CONDUCTING LEAD ALLOYS, *Phys. Rev.* **133**, A362.

779. B. Lax and D. T. Stevenson, HIGH MAGNETIC FIELD RADIATION SOURCES, in *Proceedings of the Far Infrared Physics Symposium*, Riverside, Calif., 1964, NOL Corona Tech. Bull.

780. A. J. Lichtenberg, S. Sesnic, and A. W. Trivelpiece, MEASUREMENT OF THE SYNCHROTRON RADIATION SPECTRUM FROM A HOT PLASMA, *Phys. Rev. Letters* **13**, 387.

781. K. R. Loos, FAR-INFRARED SPECTRUM OF Fe(CO)$_5$, *J. Chem. Phys.* **40**, 3741.

782. R. Marshall and S. S. Mitra, OPTICALLY ACTIVE PHONON PROCESSES IN CdS AND ZnS, *Phys. Rev.* **134**, A1019.

783. L. E. S. Mathias and A. Crocker, STIMULATED EMISSION IN THE FAR INFRA-RED FROM WATER VAPOUR AND DEUTERIUM OXIDE DISCHARGES, *Phys. Letters* **13**, 35.

783a. R. A. McFarlane, W. L. Faust, C. K. N. Patel, and C. G. B. Garrett, NEON GAS MASER LINES AT 68.329 μ AND 85.047 μ, *Proc. IEEE* **52**, 318.

784. R. V. McKnight and K. D. Möller, FAR INFRARED SPECTRUM OF POLYPROPY-LENE, HIGH-DENSITY POLYETHYLENE, AND QUARTZ-CRYSTAL PLATES, *J. Opt. Soc. Am.* **54**, 132.

785. F. A. Miller and W. G. Fateley, THE INFRARED SPECTRUM OF CARBON SUB-OXIDE, *Spectrochim. Acta* **20**, 253.

786. S. S. Mitra and P. J. Gielisse, INFRARED SPECTRA OF CRYSTALS, in *Progress in Infrared Spectroscopy*, Vol. 2, H. A. Szymanski, Ed., Plenum Press, New York, p. 47.

787. S. S. Mitra and R. Marshall, TRENDS IN THE CHARACTERISTIC PHONON FREQUENCIES OF THE NaCl-, DIAMOND-, ZINC-BLENDE-, AND WURTZITE-TYPE CRYSTALS, *J. Chem. Phys.* **41**, 3158.

788. J. Murphy, H. H. Caspers, and R. A. Buchanan, SYMMETRY COORDINATES AND LATTICE VIBRATIONS OF LaCl$_3$, *J. Chem. Phys.* **40**, 743.

789. V. N. Murzin and A. I. Demeshina, TEMPERATURE STUDIES OF INFRARED REFLECTION SPECTRA OF BaTiO$_3$ AND SrTiO$_3$ IN THE 2–1000 μ REGION, *Fiz. Tverd. Tela* **5**, 2359; *Soviet Phys.-Solid State* **5**, 1716.

790. V. N. Murzin and A. I. Demeshina, EFFECT OF TEMPERATURE ON THE OSCIL-LATION SPECTRA OF POLYCRYSTALLINE BaTiO$_3$ AND SrTiO$_3$ IN A WIDE SPECTRAL RANGE, *Fiz. Tverd. Tela* **6**, 182; *Soviet Phys.-Solid State* **6**, 144.

791. T. P. Myasnikova and I. M. Arefyev, LOW-FREQUENCY ABSORPTION SPECTRA OF SOME FERROELECTRICS, *Opt. i Spektr.* **16**, 540; *Opt. Spectry.* **16**, 293.

792. E. D. Palik, FAR INFRARED MAGNETO-OPTICAL STUDIES OF SEMICON-DUCTORS, in *Proceedings of the Far Infrared Physics Symposium*, Riverside, Calif., 1964, NOL Corona Tech. Bull.

793. C. H. Palmer, DIFFRACTION ANOMALIES WITH MILLIMETRIC WAVE GRATINGS, in *Quasi-Optics*, Microwave Research Institute Symposium Series, Vol. XIV, Polytechnic Press, Brooklyn, N.Y., p. 151.

794. R. Papoular, FAR I.R. REFERENCE SOURCES FOR THE EVALUATION OF HIGH TEMPERATURES, *Infrared Phys.* **4**, 137.

794a. C. K. N. Patel, W. L. Faust, R. A. McFarlane, and C. G. B. Garrett, C. W. OPTICAL MASER ACTION UP TO 133 μ (0.133 mm) IN NEON DISCHARGES, *Proc. IEEE* **52**, 713.

795. C. K. N. Patel, W. L. Faust, R. A. McFarlane, and C. G. B. Garrett, LASER ACTION UP TO 57.355 μ IN GASEOUS DISCHARGES (Ne, He–Ne), *Appl. Phys. Letters* **4**, 18.

796. C. H. Perry, FAR INFRARED SPECTROSCOPY OF SOLIDS, in *Proceedings of the Far Infrared Physics Symposium*, Riverside Calif. 1964, NOL Corona Tech. Bull.

797. C. H. Perry, B. N. Khanna, and G. Rupprecht, INFRARED STUDIES OF PEROVSKITE TITANATES, *Phys. Rev.* **135**, A408.

798. E. H. Putley, THE ULTIMATE SENSITIVITY OF SUB-mm DETECTORS, *Infrared Phys.* **4**, 1.

799. E. H. Putley, FAR INFRA-RED PHOTOCONDUCTIVITY, *Phys. Stat. Sol.* **6**, 471.

800. C. M. Randall, R. M. Fuller, and D. J. Montgomery, INFRARED DISPERSION FREQUENCIES FOR ALKALI HALIDES, *Solid State Commun.* **2**, 273.

801. R. B. Reeves and D. W. Robinson, INFRARED SPECTRUM OF METHYLSILYL-d_3-ACETYLENE, *J. Chem. Phys.* **41**, 1699.

801a. J. H. Reuszer and P. Fisher, AN OPTICAL DETERMINATION OF THE GROUND-STATE SPLITTINGS OF GROUP V IMPURITIES IN GERMANIUM, *Phys. Rev.* **135**, A1125.

802. P. L. Richards, HIGH-RESOLUTION FOURIER TRANSFORM SPECTROSCOPY IN THE FAR-INFRARED, *J. Opt. Soc. Am.* **54**, 1474.

803. P. L. Richards, HIGH RESOLUTION FAR INFRARED INTERFEROMETRY WITH APPLICATIONS TO SOLID STATE PHYSICS, in *Proceedings of the Far Infrared Physics Symposium*, Riverside, Calif. 1964, NOL Corona Tech. Bull.

804. P. L. Richards and G. E. Smith, FAR-INFRARED CIRCULAR POLARIZER, *Rev. Sci. Instr.* **35**, 1535.

805. W. G. Rothschild, PURE ROTATIONAL ABSORPTION SPECTRUM OF HYDROGEN FLUORIDE BETWEEN 22 AND 250 μ, *J. Opt. Soc. Am.* **54**, 20.

806. V. Y. Ryadov and N. I. Furashov, MEASUREMENT OF ATMOSPHERIC TRANSPARENCY TO 0.87 mm WAVES, *Radio Eng. Electr. USSR* **9**, 773.

807. L. Sacconi, A. Sabatini, and P. Gans, INFRARED SPECTRA FROM 80 TO 2000 cm^{-1} OF SOME METAL-AMMONIA COMPLEXES, *Inorg. Chem.* **3**, 1772.

807a. R. Sehr, STIMULATED EMISSION OF 100-μ RADIATION FROM BI–Sb P–N JUNCTIONS, *Proc. IEEE* **52**, 725.

808. H. Shenker, W. J. Moore, and E. M. Swiggard, INFRARED PHOTOCONDUCTIVE CHARACTERISTICS OF BORON-DOPED GERMANIUM, *J. Appl. Phys.* **35**, 2965.

809. T. Shimanouchi and I. Haroda, FAR-INFRARED SPECTRA OF CYANURIC ACID, URACIL, AND DIKETOPIPERAZINE, *J. Chem. Phys.* **41**, 2651.

810. T. Shimanouchi and I. Nakagawa, INFRARED SPECTRA AND FORCE CONSTANTS OF AMMINE COMPLEXES, *Inorg. Chem.* **3**, 1805.

811. A. J. Sievers, FAR-INFRARED RESONANCE STATES IN SILVER-ACTIVATED POTASSIUM HALIDE CRYSTALS, *Phys. Rev. Letters* **13**, 310.

812. A. E. Stanevich, HYDROGEN BONDING AND LONG-WAVELENGTH INFRARED ABSORPTION SPECTRA OF SOME BENZENE DERIVATIVES, *Opt. i Spektr.* **16**, 781; *Opt. Spectry.* **16**, 425.

813. A. E. Stanevich, LONG-WAVELENGTH INFRARED ABSORPTION SPECTRA OF CARBOXYLIC ACIDS, *Opt. i Spektr.* **16**, 781; *Opt. Spectry.* **16**, 243.

814. R. G. Strauch, R. E. Cupp, M. Lichtenstein, and J. J. Gallagher, QUASI-OPTICAL TECHNIQUES IN MILLIMETER SPECTROSCOPY, in *Quasi-Optics*, Microwave Research Institute Symposia Series, Vol. XII, Polytechnic Press, Brooklyn, N.Y., p. 581.

815. P. Taimsalu and J. L. Wood, THE FAR INFRA-RED SPECTRA OF ALKYL TIN CHLORIDES, *Spectrochim. Acta* **20**, 1043.

816. P. Taimsalu and J. L. Wood, SOLVENT EFFECTS IN FAR INFRARED SPECTRA, *Spectrochim. Acta* **20**, 1357.

817. M. Tinkham, SPECTROSCOPY OF SOLIDS IN THE FAR-INFRARED, *Science* **145**, 240.

818. P. Vogel and L. Genzel, A METHOD FOR PERFORMING THE ANALOG FOURIER TRANSFORM FOR INTERFERENCE MODULATION SPECTROSCOPY, *Appl. Opt.* **3**, 367.

819. P. Vogel and L. Genzel, TRANSMISSION AND REFLECTION OF METALLIC MESH IN THE FAR INFRARED, *Infrared Phys.* **4**, 257.

820. S. Walles and S. Boija, TRANSMITTANCE OF DOPED SILICON BETWEEN 40 AND 100 μm, *J. Opt. Soc. Am.* **54**, 133.

821. S. H. Walmsley and A. Anderson, FAR INFRA-RED SPECTRA OF MOLECULAR CRYSTALS. II. CHLORINE, BROMINE, AND IODINE, *Mol. Phys.* **7**, 411.

822. R. Weber, SINGLE PHONON ABSORPTION IN NaCl DUE TO Ag^+ CENTERS, *Phys. Letters* **12**, 311.

823. W. Wetting and L. Genzel, APPLICATION OF FRUSTRATED TOTAL REFLEC-TION AS A FILTER FOR THE FAR INFRARED, *Infrared Phys.* **4**, 235.

824. E. Wiener and I. Pelah, INDICATION OF LOW FREQUENCY HYDROGEN MODES IN KH_2PO_4 FROM INFRARED MEASUREMENTS, *Phys. Letters* **13**, 206.

825. R. A. Williams and W. S. C. Chang, INTERFEROMETRIC WAVELENGTH SELEC-TION FOR SUBMILLIMETER RADIOMETRY, in *Quasi-Optics*, Microwave Research Institute Symposia Series, Vol. XIV, Polytechnic Press, Brooklyn, N.Y., p. 607.

826. W. J. Whitteman and R. Bleekrode, PULSED AND CONTINUOUS MOLECULAR FAR INFRARED GAS LASER, *Phys. Letters* **13**, 126.

827. P. A. Wolff, PROPOSAL FOR A CYCLOTRON RESONANCE MASER IN InSb, *Physics* **1**, 147.

828. H. R. Wyss, R. D. Werder, and H. H. Günthard, FAR INFRARED SPECTRA OF TWELVE ORGANIC LIQUIDS, *Spectrochim. Acta* **20**, 573.

829. D. J. C. Yates, ATTEMPTS TO OBSERVE THE FAR-INFRARED SPECTRA OF ADSORBED MOLECULES, *J. Chem. Phys.* **40**, 1157.

830. H. Yoshinaga, RECENT RESEARCH IN FAR INFRARED OPTICS IN JAPAN, *Appl. Opt.* **3**, 805.

831. H. Yoshinaga, S. Minami, I. Makino, I. Iwahashi, M. Inaba, and K. Matsumoto, AN OPTICAL NULL DOUBLE-BEAM FAR INFRARED SPECTROPHOTOMETER, *Appl. Opt.* **3**, 1425.

832. J. N. Zemel, TRANSVERSE OPTICAL PHONONS IN PbS, in *Proceedings of the International Conference on the Physics of Semiconductors*, Paris, 1964, Dunod, Paris, p. 1061.

833. A. G. Zhukov, LONG-WAVELENGTH INFRARED SPECTROMETER WITH ECHELETTES OF SMALL DIMENSIONS, *Opt. i Spektr.* **17**, 284; *Opt. Spectry.* **17**, 148.

834. U. A. Zirnit and M. M. Suchchinskii, LOW-FREQUENCY VIBRATIONS IN THE SPECTRA OF METHYL-SUBSTITUTED PENTANES, *Opt. i Spektr.* **16**, 902; *Opt. Spectry.* **16**, 489.

1965

835. D. M. Adams and D. M. Morris, FAR-INFRA-RED SPECTRA OF SOME SQUARE-PLANAR IONS, *Nature* **208**, 283.

835a. R. L. Aggarwal, P. Fisher, V. Mourzine, and A. K. Ramdas, EXCITATION SPECTRA OF LITHIUM DONORS IN SILICON AND GERMANIUM, *Phys. Rev.* **138**, A882.

836. F. Ambrosino, N. Neto, and S. Califano, THE INFRARED SPECTRUM IN POLARIZED LIGHT OF TRANS-TRANS-TRANS CYCLODODECATRIENE, *Spectrochim. Acta* **21**, 409.

837. A. Anderson and H. A. Gebbie, FAR INFRARED STUDY OF MOLECULAR CRYSTALS BY INTERFEROMETRIC METHODS, *Spectrochim. Acta* **21**, 883.

838. A. Anderson and S. H. Walmsley, FAR INFRA-RED SPECTRA OF MOLECULAR CRYSTALS. IV. AMMONIA, HYDROGEN SULPHIDE, AND THEIR FULLY DEUTERATED ANALOGUES, *Mol. Phys.* **9**, 1.

839. I. M. Arefeyev, THE CHOICE OF FILTERS AND DETERMINATION OF SPECTRAL PURITY OF RADIATION IN LONG-WAVE I. R. SPECTROSCOPY, *Zhur. Priklad. Spectrosk.* **11**, 462.

840. J. D. Axe, J. W. Gaglianello, and J. E. Scardefield, INFRARED DIELECTRIC PROPERTIES OF CADMIUM FLUORIDE AND LEAD FLUORIDE, *Phys. Rev.* **139**, 1211.

841. R. F. W. Bader and K. P. Huang, JAHN-TELLER EFFECT IN THE VIBRATIONAL SPECTRA OF PENTACHLORIDES, *J. Chem. Phys.* **43**, 3760.

842. A. S. Barker and J. A. Ditzenberger, INFRARED LATTICE VIBRATIONS IN CoF_2, *Solid State Commun.* **3**, 131.

843. E. E. Bell, MEASUREMENT OF SPECTRAL TRANSMITTANCE AND REFLECTANCE WITH A FAR INFRARED MICHELSON INTERFEROMETER, *Japan. J. Appl. Phys.*, *Suppl. 1*, **4**, 412.

844. R. J. Bell and T. E. Gilmer, A NEW RADIATION CHOPPER, *Appl. Opt.* **4**, 45.

845. L. V. Berman and A. G. Zhukov, OPTICAL PROPERTIES OF CaF_2 IN THE WAVELENGTH RANGE 170–600 μ, *Opt. i Spektr.* **19**, 783; *Opt. Spectry.* **19**, 443.

846. C. V. Berney, INFRARED, RAMAN, AND NEAR-ULTRAVIOLET SPECTRA OF CF_3COZ COMPOUNDS. II. HEXAFLUOROACETONE, *Spectrochim. Acta* **21**, 1809.

847. J. Besson, B. Philippeau, R. Cano, M. Mittiloli, and R. Papoular, LE DÉTECTEUR À ANTIMONIURE D'INDIUM DANS LA BANDE DE 0.1 À 10 mm (30 À 3000 GHz), *L'Onde Électrique* **45**, 107.

848. H. Bilz, K. F. Renk, and K. H. Timmesfeld, ROTATIONAL STRUCTURE OF THE LOCALISED MODE IN $KI:NO_2$, *Solid State Commun.* **3**, 223.

849. H. Bilz, D. Strauch, and B. Fritz, INFRARED ABSORPTION AND ANHARMONICITY OF THE U-CENTER LOCAL MODE, THEORY, AND DISCUSSION, *J. Phys. Suppl.* **C2**, 3.

850. R. Blinc and S. Svetina, LOW-FREQUENCY HYDROGEN MODES IN KH_2PO_4 TYPE FERROELECTRIC CRYSTALS, *Phys. Letters* **15**, 119.

851. D. Bloor, A RAPID SCANNING SPECTROMETER FOR THE FAR INFRARED (200–25 cm^{-1}), *Spectrochim. Acta* **21**, 595.

852. D. Bloor, THE FAR INFRA-RED SPECTRA OF THE ISOMORPHOUS NITRATES OF CALCIUM, STRONTIUM, BARIUM, AND LEAD, *Spectrochim. Acta* **21**, 133.

853. D. R. Bosomworth and H. P. Gush, COLLISION-INDUCED ABSORPTION OF COMPRESSED GASES IN THE FAR INFRARED, I., *Can. J. Phys.* **43**, 729.

854. D. R. Bosomworth and H. P. Gush, COLLISION-INDUCED ABSORPTION OF COMPRESSED GASES IN THE FAR INFRARED, II., *Can. J. Phys.* **43**, 751.

855. D. R. Bosomworth and H. P. Gush, FAR INFRARED SPECTROSCOPY OF COMPRESSED GASES USING A MICHELSON INTERFEROMETER, *Japan. J. Appl. Phys., Suppl. 1*, **4**, 588.

856. I. B. Bott, A POWERFUL SOURCE OF MILLIMETRE WAVELENGTH ELECTRO-MAGNETIC RADIATION, *Phys. Letters* **14**, 293.

857. E. B. Bradley, C. R. Bennett, and E. A. Jones, FAR-INFRARED ABSORPTION OF SULFUR MONOBROMIDE, *Spectrochim. Acta* **21**, 1505.

858. H. P. Broida, K. M. Evenson, and T. T. Kikuchi, COMMENTS ON THE MECHANISM OF THE 337-MICRON CN LASER, *J. Appl. Phys.* **36**, 3356.

859. M. A. C. S. Brown and M. F. Kimmit, FAR-INFRARED RESONANT PHOTO-CONDUCTIVITY IN INDIUM ANTIMONIDE, *Infrared Phys.* **5**, 93.

860. J. C. Burgiel and L. C. Hebel, FAR-INFRARED SPIN AND COMBINATION RE-SONANCE IN BISMUTH, *Phys. Rev.* **140**, 925.

861. J. C. Burgiel, H. Meyer, and P. L. Richards, FAR-INFRARED SPECTRA OF GAS MOLECULES TRAPPED IN β-QUINOL CLATHRATES, *J. Chem. Phys.* **43**, 4291.

862. P. S. Callahan, FAR INFRA-RED EMISSION AND DETECTION BY NIGHT-FLYING MOTHS, *Nature* **206**, 1172.

863. M. Camani, F. K. Kneubühl, J. F. Moser, and H. Steffen, CYANVERBINDUNGEN IM SUBMILLIMETERWELLEN-GASLASER, *Z. Angew. Math. Physik* **16**, 562.

864. J. E. Chamberlain, ON A RELATION BETWEEN ABSORPTION STRENGTH AND REFRACTIVE INDEX, *Infrared Phys.* **5**, 175.

865. J. E. Chamberlain, F. D. Findlay, and H. A. Gebbie, REFRACTIVE INDEX OF AIR AT 0.337-mm WAVE-LENGTH, *Nature* **206**, 4987.

866. J. E. Chamberlain and H. A. Gebbie, DETERMINATION OF THE REFRACTIVE INDEX OF A SOLID USING A FAR INFRA-RED MASER, *Nature* **206**, 602.

867. J. E. Chamberlain and H. A. Gebbie, SUB-MILLIMETRE DISPERSION AND ROTATIONAL LINE STRENGTHS OF THE HYDROGEN HALIDES, *Nature* **208**, 480.

868. G. W. Chantry, A. Anderson, D. J. Browning, and H. A. Gebbie, THE FAR INFRA-RED SPECTRUM OF ANTHRACENE, *Spectrochim. Acta* **21**, 217.

869. G. W. Chantry and H. A. Gebbie, SUB-MILLIMETRE WAVE SPECTRA OF POLAR LIQUIDS, *Nature* **208**, 378.

870. G. W. Chantry, H. A. Gebbie, and J. E. Chamberlain, A SUGGESTED MECHANISM FOR THE 337 μ CN MASER, *Nature* **205**, 377.

871. R. J. Clark and C. S. Williams, THE FAR INFRA-RED SPECTRA OF 2,2'-DIPY-RIDYL AND 1,10-PHENANTHROLINE COMPLEXES OF ALKYL TIN HALIDES, *Spectrochim. Acta* **21**, 1861.

872. G. A. Crowder, G. Gorin, F. H. Kruse, and D. W. Scott, TETRAMETHYLLEAD: FAR INFRARED SPECTRA, MOLECULAR VIBRATIONS, AND CHEMICAL THERMODYNAMIC PROPERTIES. RESOLUTION OF AN ENTROPY DIS-CREPANCY, *J. Mol. Spectry.* **16**, 115.

873. G. A. Crowder and D. W. Scott, LIQUID-VAPOR FREQUENCY SHIFTS AND TORSIONAL FREQUENCIES IN FAR INFRARED SPECTRA, *J. Mol. Spectry.* **16**, 122.

874. P. Datta and G. M. Barrow, SPECTRAL EVIDENCE FOR THE ROTATION OF MOLECULES IN THE LIQUID PHASE, *J. Chem. Phys.* **43**, 2137.

875. S. Deb, DETECTION, MODULATION, AND GENERATION OF OPTICAL AND SUB-MILLIMETRE WAVE RADIATION USING SEMICONDUCTING MA-TERIALS, *J. Sci. Indust. Res.* **24**, 398.

876. S. Deb and P. K. Chaudhuri, AMPLITUDE MODULATION OF INFRARED AND SUB-mm DIODES, *Proc. IEEE* **53**, 81.

877. J. W. Dees, V. E. Derr, J. J. Gallagher, and J. C. Wiltse, BEYOND MICROWAVES, *Intern. Sci. Tech.* **47**, 50.

878. A. N. Dellis, W. H. F. Earl, A. Malein, and S. Ward, FAR INFRARED FARADAY ROTATION IN A PLASMA, *Nature* **207**, 56.

879. P. Delorme, V. Lorenzelli, and A. Alemagna, VIBRATION SPECTRA OF THE BENZENE HALOGENS, *J. Chem. Phys.* **62**, 3.

879a. E. M. Dianov, N. A. Irisova, and N. V. Karlov, THE USE OF DIELECTRIC WAVE-GUIDES IN MILLIMETER SPECTROSCOPY, *Pribory Tekh. Eksper.* **4**, 144; *Instrum. Exper. Tech.* **4**, 885.

879b. E. M. Dianov, N. A. Irisova, and A. M. Prokhorov, APPARATUS FOR MEASURING THE REFLECTION AND TRANSMISSION COEFFICIENTS OF SUBSTANCES THAT OPERATES WITH MONOCHROMATIC SHORT-MILLIMETER AND SUBMILLIMETER RADIATION, *Pribory Tekh. Eksper.* **4**, 140; *Instrum. Exper. Tech.* **4**, 881.

880. S. I. Drasky and R. J. Bell, RADIATION CHOPPER FOR THE SUBMILLIMETER WAVELENGTH REGION, *Infrared Phys.* **5**, 137.

881. C. B. Farmer and P. J. Key, A STUDY OF THE SOLAR SPECTRUM FROM $7\,\mu$ TO $400\,\mu$, *Appl. Opt.* **4**, 1051.

882. W. G. Fateley, R. K. Harris, F. A. Miller, and R. E. Witkowski, TORSIONAL FREQUENCIES IN THE FAR INFRARED-IV. TORSIONS AROUND THE C–C SINGLE BOND IN CONJUGATED MOLECULES, *Spectrochim. Acta* **21**, 231.

883. J. R. Ferraro and A. Walker, COMPARISON OF THE INFRARED SPECTRA $(4000–70\,cm^{-1})$ OF SEVERAL HYDRATED AND ANHYDROUS SALTS OF TRANSITION METALS, *J. Chem. Phys.* **42**, 1278.

884. J. R. Ferraro and A. Walker, FAR-INFRARED SPECTRA OF ANHYDROUS METALLIC NITRATES, *J. Chem. Phys.* **42**, 1273.

885. J. R. Ferraro and W. R. Walker, INFRARED SPECTRA OF HYDROXY-BRIDGED COPPER (II) COMPOUNDS, *Inorg. Chem.* **4**, 1382.

886. J. H. Fertel and C. H. Perry, LONG-WAVE INFRARED SPECTRA OF ALKALI SALTS OF PLATINUM HALIDE COMPLEXES, *J. Phys. Chem. Solids* **26**, 1773.

887. A. Finch, I. J. Hyams, and D. Steele, FAR INFRARED SPECTRA OF BORON TRI-BROMIDE AND BORON TRI-IODIDE, *Trans. Faraday Soc.* **61**, 398.

888. A. Finch, R. C. Poller, and D. Steele, VIBRATIONAL SPECTRA OF SOME HETER-OCYCLIC TIN COMPOUNDS, *Trans. Faraday Soc.* **61**, 2628.

889. W. H. Fletcher and W. B. Barish, THE INFRARED SPECTRUM OF DIMETHYL KETENE, *Spectrochim. Acta* **21**, 1647.

890. S. G. Frankiss and F. A. Miller, INFRARED AND RAMAN SPECTRA AND STRUCTURE OF P_2Cl_4, *Spectrochim. Acta* **21**, 1235.

891. B. Fritz, U. Gross, and D. Baeuerle, INFRARED ABSORPTION AND ANHARMONICITY OF THE U-CENTRE LOCAL MODE. I. EXPERIMENTS, *Phys. Stat. Sol.* **11**, 231.

892. C. G. B. Garrett, FAR-INFRARED MASERS, *Intern. Sci. Tech.* **39**, 39.

893. H. A. Gebbie, N. W. B. Stone, F. D. Findlay, and E. C. Pyatt, ABSORPTION AND RE-FRACTIVE INDEX MEASUREMENTS AT A WAVE-LENGTH OF 0.34 mm, *Nature* **205**, 377.

894. L. Genzel, FAR INFRARED SPECTROSCOPY, *Japan. J. Appl. Phys. Suppl. 1*, **4**, 353.

895. L. Genzel, K. F. Renk, and R. Weber, CALCULATION OF THE IMPURITY-IN-DUCED LATTICE MODE ABSORPTION, *Phys. Stat. Sol.* **12**, 639.

896. J. E. Gibbs and H. A. Gebbie, CALCULATION OF THE POWER-SPECTRUM FROM AN INTERFEROGRAM SAMPLED AT POINTS NONE OF WHICH COINCIDES EXACTLY WITH ZERO PATH-DIFFERENCE, *Infrared Phys.* **5**, 187.

897. M. Goldstein, THE FAR-INFRA-RED SPECTRA AND STRUCTURE OF THE BIS (4-METHYLPYRIDINE) COMPLEXES OF CUPRIC HALIDES, *J. Inorg. Nucl. Chem.* **27**, 2115.

898. M. Goldstein, E. F. Mooney, A. Anderson, and H. A. Gebbie, THE INFRARED SPECTRA AND STRUCTURE OF COMPLEXES OF COPPER (II) HALIDES AND HETEROCYCLIC BASES, *Spectrochim. Acta* **21**, 105.

898a. W. Gordy, MICROWAVE SPECTROSCOPY IN THE REGION OF 4–0.4 MILLI-METRES, Eighth European Congress on Molecular Spectroscopy, Copenhagen 1965, Butterworth and Co., Ltd., London.

899. A. Hadni, J. Claudel, X. Gerbaux, G. Morlot, and J. Munier, SUR LE COMPORTE-MENT DIFFÉRENT DES CRISTAUX ET DES VERRES DANS L'ABSORPTION DE L'INFRAROUGE LOINTAIN (40–1500 μ) À LA TEMPÉRATURE DE L'-HÉLIUM LIQUIDE, *Appl. Opt.* **4**, 487.

900. A. Hadni, X. Gerbaux, G. Morlot, and P. Strimer, FAR INFRARED ELECTRONIC TRANSITIONS IN RARE EARTH AND TRANSITION ELEMENT IONS EM-BEDDED IN CRYSTALS, *Japan J. Appl. Phys., Suppl.* **4**, 574.

901. H. Hadni, Y. Henninger, R. Thomas, P. Vergnat, and B. Wyncke, SUR LES PROPRIÉTÉS PYROÉLECTRIQUES DE QUELQUES MATÉRIAUX ET LEUR APPLICATION À LA DÉTECTION DE L'INFRAROUGE, *J. Phys.* **26**, 345.

902. A. Hadni, Y. Henninger, R. Thomas, P. Vergnat, and B. Wyncke, ÉTUDE DES PRO-PRIÉTÉS PYROÉLECTRIQUES DE QUELQUES CRISTAUX ET DE LEUR UTILISATION À LA DÉTECTION DU RAYONNEMENT, *Compt. Rend.* **26**, 4186.

903. A. Hadni, G. Morlot, F. Brehat, and P. Strimer, SUR LE RÔLE DES IMPURÉTES POUR ACTIVER OPTIQUEMENT DANS L'INFRAROUGE LOINTAIN, LES ONDES ELASTIQUES DES CRISTAUX IONIQUES, *Compt. Rend.* **261**, 2605.

904. A. Hadni, G. Morlot, X. Gerbaux, D. Chanal, F. Brehat, and P. Strimer, ABSORPTION INDUITE DANS L'INFRAROUGE LOINTAIN PAR LES IMPURETÉS ET LES DÉFAUTS D'UN SOLIDE, *Compt. Rend.* **260**, 4973.

905. R. Hanna, THE STRUCTURE OF SODIUM SILICATE GLASSES AND THEIR FAR-INFRARED ABSORPTION SPECTRA, *J. Phys. Chem.* **69**, 3849.

906. M. Hass and M. O'Hara, SHEET INFRARED TRANSMISSION POLARIZERS, *Appl. Opt.* **4**, 1027.

907. P. J. Hendra and N. Sadasivan, THE FAR INFRA-RED SPECTRA OF IODINE COMPLEXES OF 1:4-DISELENAN AND 1:4-DITHIAN, *Spectrochim. Acta* **21**, 1127.

908. Y. Henninger, G. Morlot, and A. Hadni, FRÉQUENCES FONDAMENTALES D'ABSORPTION DES HALOGENURES CUIVREUX DANS L'INFRAROUGE LOINTAIN, ENTRE LA TEMPÉRATURE ORDINAIRE ET CELLE DE L'HÉLIUM LIQUIDE, *J. Phys.* **26**, 143.

909. K. Hisano, N. Ohama, and O. Matumura, FAR-INFRARED VIBRATION MODE IN $SrCl_2$, *J. Phys. Soc. Japan* **20**, 2294.

910. W. Honig, SUBMILLIMETER CERENKOV WAVE GENERATION IN SOLIDS, *Proc. IEEE* **53**, 182.

911. L. C. Hoskins, COMMENTS ON THE FAR-INFRARED SPECTRUM OF PF_5, *J. Chem. Phys.* **42**, 2631.

912. H. Hunziker and H. H. Günthard, VIBRATIONAL SPECTRA OF EIGHT

ISOTOPIC SPECIES, VALENCE FORCE CONSTANTS, AND ROTATIONAL BARRIERS OF 2-CHLOROPROPENE, *Spectrochim. Acta* **21**, 51.

913. R. H. Hunt, R. A. Leacock, C. W. Peters, and K. T. Hecht, INTERNAL-ROTATION IN HYDROGEN PEROXIDE: THE FAR-INFRARED SPECTRUM AND THE DETERMINATION OF THE HINDERING POTENTIAL, *J. Chem. Phys.* **42**, 1931.

914. B. B. van Iperen and W. Kuypers, EXPERIMENTAL C. W. KLYSTRON MULTIPLIER FOR SUBMILLIMETER WAVES, *Phillips Res. Rep.* **20**, 462.

915. R. J. Jakobsen and J. W. Brasch, FAR INFRARED STUDIES OF THE HYDROGEN BOND OF PHENOLS, *Spectrochim. Acta* **21**, 1753.

916. C. E. Jones, A. R. Hilton, J. B. Damrel, and C. C. Helms, THE COOLED GERMANIUM BOLOMETER AS A FAR INFRARED DETECTOR, *Appl. Opt.* **4**, 683.

917. G. O. Jones and J. M. Woodfine, DIRECT SPECTROSCOPIC OBSERVATION OF THE PHONON SPECTRUM OF SOLID ARGON, *Proc. Phys. Soc.* **86**, 101.

917a. R. L. Jones and P. Fisher, EXCITATION SPECTRA OF GROUP III IMPURITIES IN GERMANIUM, *J. Phys. Chem. Solids* **26**, 1125.

918. J. E. Katon and N. T. McDevitt, THE VIBRATIONAL SPECTRA OF PROPYNOIC ACID AND SODIUM PROPYNOATE, *Spectrochim. Acta* **21**, 1717.

919. J. E. Katon and N. T. McDevitt, THE FAR INFRARED SPECTRA OF BIPHENYL AND BIPHENYL-*d*-10, *J. Mol. Spectry.* **14**, 308.

919a. F. K. Kneubühl, J. F. Moser, H. Steffen, and W. Tandler, EIN SUBMILLIMETER-WELLEN-GASLASER, *Z. Angew. Math. Phys.* **16**, 560.

920. S. Krimm and M. I. Bank, ASSIGNMENT OF THE $71\,cm^{-1}$ BAND IN POLYETHYLENE, *J. Chem. Phys.* **42**, 4059.

921. K. Kudo and Y. Mochida, VERSATILE FAR INFRARED SPECTROPHOTOMETER, *Japan, J. Appl. Phys., Suppl. 1*, **4**, 372.

922. W. J. Lafferty, D. W. Robinson, R. V. St. Louis, J. W. Russel, and H. L. Strauss, FAR-INFRARED SPECTRUM OF TETRAHYDROFURAN: SPECTROSCOPIC EVIDENCE FOR PSEUDOROTATION, *J. Chem. Phys.* **42**, 2915.

923. L. N. Large and H. Hill, A COMPACT PULSED GAS LASER FOR THE FAR INFRARED, *Appl. Opt.* **4**, 625.

924. H. Levinstein, EXTRINSIC DETECTORS, *Appl. Opt.* **4**, 639.

925. R. Lochet and G. Nouchi, CONTRIBUTION À L'ÉTUDE DE L'ÉTAT TRIPLET DES COLORANTS: SPECTRES D'ABSORPTION DANS LE DOMAINE DES GRANDES LONGUEURS D'ONDE, INFRAROUGE ET PROCHE INFRA-ROUGE, *Compt. Rend.* **260**, 1897.

926. K. R. Loos and R. C. Lord, VIBRATIONAL SPECTRUM AND BARRIER TO INTERNAL ROTATION FOR CF_3CFO, *Spectrochim. Acta* **21**, 119.

927. V. Lorenzelli, F. Gesmundo, and J. Lecomte, NOUVELLE ÉTUDE DES SPECTRES D'ABSORPTION INFRAROUGE 2–150 μ DE SÉLÉNIATES MÉTALLIQUES, *J. Chim. Phys.* **62**, 320.

928. F. J. Low, PERFORMANCE OF THERMAL DETECTION RADIOMETERS AT 1.2 mm, *Proc. IEEE* **53**, 516.

929. I. Makino, T. Iwasaki, and I. Iwahashi, A FAR INFRARED SPECTROMETER DESIGNED MAINLY FOR USE IN SOLID STATE PHYSICS, *Japan. J. Appl. Phys., Suppl. 1*, **4**, 369.

930. D. Malz, INTERFERENCE PHENOMENA IN PLANE PARALLEL PLATES IN FOURIER SPECTROSCOPY AND THEIR EFFECT ON SPECTROSCOPIC INVESTIGATIONS ON SEMICONDUCTORS IN THE MIDDLE AND FAR INFRARED, *Exper. Tech. Phys. (Germany)* **13**, 257.

931. T. R. Manley and D. A. Williams, SCATTERING FILTERS IN THE FAR INFRARED, *Spectrochim. Acta* **21**, 737.

932. T. R. Manley and D. A. Williams, THE FAR INFRARED SPECTRA OF SOME GROUP Vb TRIHALIDES, *Spectrochim. Acta* **21**, 1773.

933. M. N. Markov, A BOLOMETER FOR THE FAR INFRARED, *Opt. i Spektr.* **18**, 119; *Opt. Spectry.* **18**, 60.

934. D. H. Martin, THE STUDY OF THE VIBRATIONS OF CRYSTAL LATTICES BY FAR INFRA-RED SPECTROSCOPY, *Advan. Phys.* **14**, 39.

935. M. D. Mashkovich and A. I. Demeshina, INVESTIGATION OF SOME INORGANIC DIELECTRIC MATERIALS IN THE FAR INFRARED, *Fiz. Tverd. Tela* **7**, 1634; *Soviet Phys.-Solid State* **7**, 1323.

936. L. E. S. Mathias, A. Crocker, and M. S. Wills, LASER OSCILLATIONS AT SUB-MILLIMETRE WAVELENGTHS FROM PULSED GAS DISCHARGES IN COMPOUNDS OF HYDROGEN, CARBON, AND NITROGEN, *Electronics Letters* **1**, 45.

937. N. T. McDevitt and A. D. Davidson, USE OF MERCURIC OXIDE AS A CALIBRATING MATERIAL FOR THE INFRARED REGION 700–50 cm^{-1}, *J. Opt. Soc. Am.* **55**, 1695.

938. N. T. McDevitt and A. D. Davidson, INFRARED STUDY OF Ag_2O IN THE LOW-FREQUENCY REGION, *J. Opt. Soc. Am.* **55**, 209.

939. N. T. McDevitt, A. L. Rozek, F. F. Bentley, and A. D. Davidson, INFRARED ABSORPTION SPECTRA OF CHLORO-, BROMO-, AND IODOALKANES IN THE 400 TO 100 cm^{-1} REGION, *J. Chem. Phys.* **42**, 1173.

940. F. A. Miller, S. G. Frankiss, and O. Sala, INFRARED AND RAMAN SPECTRA OF $P(CN)_3$ AND $As(CN)_3$, *Spectrochim. Acta* **21**, 775.

941. F. A. Miller, D. H. Lemmon, and R. E. Witkowski, OBSERVATION OF THE LOWEST BENDING FREQUENCIES OF CARBON SUBOXIDE, DICYANOACETYLENE, DIACETYLENE, AND DIMETHYLACETYLENE, *Spectrochim. Acta* **21**, 1709.

942. R. C. Milward, OBSERVATION OF A FAR INFRARED RESONANCE LINE IN ANTIFERROMAGNETIC CoO, *Phys. Letters* **16**, 244.

943. R. C. Milward and L. J. Neuringer, FAR-INFRARED ABSORPTION IN n-TYPE SILICON DUE TO PHOTO-INDUCED HOPPING, *Phys. Rev. Letters* **15**, 665.

944. S. Minami, H. Yoshinaga, and K. Matsunaga, AN APPLICATION OF DOUBLE CHOPPING SYSTEM TO FAR INFRARED SPECTROPHOTOMETER, *Japan. J. Appl. Phys., Suppl. 1*, **4**, 364.

945. S. Minami, H. Yoshinaga, and K. Matsunaga, A METHOD OF ELIMINATING ERROR DUE TO SAMPLE RADIATION IN DOUBLE-BEAM FAR INFRARED SPECTROPHOTOMETER, *Appl. Opt.* **4**, 1137.

946. A. Mitsuishi, H. Yoshinaga, K. Yata, and A. Manabe, OPTICAL MEASUREMENT OF SEVERAL MATERIALS IN THE FAR INFRARED REGION, *Japan. J. Appl. Phys., Suppl. 1*, **4**, 581.

947. K. D. Möller and R. V. McKnight, MEASUREMENTS ON TRANSMISSION-FILTER GRATINGS IN THE FAR INFRARED, *J. Opt. Soc. Am.* **55**, 1075.

948. K. D. Möller, V. P. Tomaselli, L. R. Skube, and B. K. McKenna, FAR-INFRARED VACUUM GRATING SPECTROMETER, *J. Opt. Soc. Am.* **55**, 1233.

949. W. J. Moore, EXCITED STATES OF NEUTRAL ZINC, COPPER, AND MERCURY IN PHOTOCONDUCTING GERMANIUM, *Solid State Commun.* **3**, 385.

950. W. J. Moore and H. Shenker, A HIGH-DETECTIVITY GALLIUM-DOPED GERMANIUM DETECTOR FOR THE 40–120 μ REGION, *Infrared Phys.* **5**, 99.

951. B. C. Murray and J. A. Westphal, INFRARED ASTRONOMY, *Sci. Amer.* **213**, 20.

952. V. N. Murzin, S. V. Bogdanov, and A. I. Demeshina, TRANSMISSION AND REFLEC-
     TION SPECTRA OF A GROUP OF TITANATES IN A WIDE INFRARED RANGE,
     Fiz. Tverd. Tela **6**, 3585; *Soviet Phys.-Solid State* **6**, 2869.

953. F. Y. Nad' and A. Y. Oleinikov, PHOTOCONDUCTIVITY OF *n*-TYPE INDIUM
     ANTIMONIDE IN THE LONG WAVELENGTH REGION OF THE SPECTRUM,
     Fiz. Tverd. Tela **6**, 2064; *Soviet Phys.-Solid State* **6**, 1629.

954. R. T. Neher and D. K. Edwards, FAR INFRARED REFLECTOMETER FOR
     IMPERFECTLY DIFFUSE SPECIMENS, *Appl. Opt.* **4**, 775.

955. E. D. Palik, A SOLEIL COMPENSATOR FOR THE FAR INFRARED, *Appl. Opt.*
     4, 1017.

956. C. Perchard, A. M. Bellocq, and A. Novak, VIBRATION SPECTRA OF IMIDAZOLE,
     IMIDAZOLE (d)-1, IMIDAZOLE (d$_3$)-2,4,5 AND IMIDAZOLE (d$_4$). II. THE
     REGION BETWEEN 1700–30 cm^{-1}, *J. Chim. Phys.* **62**, 1344.

957. Y. N. Petrov and A. M. Prokhorov, 75 MICRON LASER, *JETP Letters* **1**, 24.

957a. C. H. Perry, R. Geick, and E. F. Young, SOLID STATE STUDIES BY MEANS
      OF FOURIER TRANSFORM SPECTROSCOPY, *Appl. Opt.* **5**, 1171.

958. C. H. Perry, FAR INFRARED REFLECTANCE SPECTRA AND DIELECTRIC
     DISPERSION OF A VARIETY OF MATERIALS HAVING THE PEROVSKITE
     AND RELATED STRUCTURES, *Japan. J. Appl. Phys., Suppl. 1*, **4**, 564.

958a. E. I. Popov, SUBMILLIMETER-BAND RADIOMETERS USING INDIUM
      ANTIMONIDE DETECTORS, *Izv. VUS Radiofiz. USSR* **8**, 862; *Soviet Radiophys.*
      8, 611.

959. E. H. Putley, INDIUM ANTIMONIDE SUBMILLIMETER PHOTOCONDUCTIVE
     DETECTORS, *Appl. Opt.* **4**, 649.

960. K. F. Renk, ONE-PHONON LATTICE ABSORPTION AND LOCALIZED MODES
     IN THE PHONON GAP OF KI:NO$_2^-$, *Phys. Letters* **14**, 281.

961. P. L. Richards, HIGH RESOLUTION FAR INFRARED INTERFEROMETRY
     WITH APPLICATION TO SOLID STATE PHYSICS, *Japan. J. Appl. Phys., Suppl. 1*,
     4, 417.

962. W. G. Rothschild, FREQUENCY SHIFTS AND BAND CONTOURS OF VIBRA-
     TIONAL TRANSITIONS OF VAPORS TRAPPED IN HIGH-DENSITY POLY-
     ETHYLENE. THE v_3, v_5, v_6, $v_5–v_3$, AND $v_2–v_6$ BANDS OF CHCl$_3$ AND THE
     v_2, v_3 MODES OF CS$_2$, *J. Chem. Phys.* **42**, 694.

963. W. G. Rothschild, THE EFFECT OF POLYETHYLENE WINDOWS OF
     STANDARD ABSORPTION CELLS ON THE INFRARED SPECTRA OF VAPORS,
     Spectrochim. Acta **21**, 852.

964. H. N. Rundle, CONSTRUCTION OF A MICHELSON INTERFEROMETER FOR
     FOURIER SPECTROSCOPY, *J. Res. Natl. Bur. Stand.* **69C**, 5.

965. J. R. Rusk, LINE-BREADTH STUDY OF THE 1.64 mm ABSORPTION IN WATER
     VAPOR, *J. Chem. Phys.* **42**, 493.

966. J. W. Russell and H. L. Strauss, CZERNY-TURNER FAR INFRARED SPECTROM-
     ETER FOR THE 300–10 cm^{-1} REGION, *Appl. Opt.* **4**, 1131.

967. A. Sabatini and I. Bertini, INFRARED SPECTRA BETWEEN 100 AND 2500 cm^{-1}
     OF SOME COMPLEX METAL CYANATES, THIOCYANATES, AND SELENO-
     CYANATES, *Inorg. Chem.* **4**, 959.

968. T. Sakurai and S. Takahashi, FAR INFRARED SPECTROMETER WITH A VAR-
     IABLE DEPTH GRATING, *Japan. J. Appl. Phys., Suppl. 1*, **4**, 358.

969. R. B. Sanderson, FAR INFRARED OPTICAL PROPERTIES OF INDIUM ANTI-
     MONIDE, *J. Phys. Chem. Solids* **26**, 803.

970. K. Sato, INFRARED LATTICE VIBRATION SPECTRA OF CRYSTALLINE
     QUARTZ, *J. Phys. Soc. Japan* **20**, 795.

971. I. I. Shaganov, L. D. Kislovskii, and I. G. Rudyavskaya, ABSORPTION BY FREE CARRIERS IN SILICON IN THE 40–100 μ REGION, *Opt. i Spektr.* **18**, 318; *Opt. Spectry.* **18**, 174.

972. A. J. Sievers, FAR-INFRARED IMPURITY MODES IN POTASSIUM IODIDE, in *Proceedings of the Conference on Low Temperature Physics*, LT9 (Part B), Plenum Press, New York, p. 1170.

973. A. J. Sievers and C. D. Lytle, FAR-INFRARED ABSORPTION BY LOCALIZED LATTICE MODES IN KI:KNO_2, *Phys. Letters* **14**, 271.

974. A. J. Sievers, A. A. Maradudin, and S. S. Jaswal, INFRARED LATTICE ABSORPTION BY GAP MODES AND RESONANCE MODES IN KI, *Phys. Rev.* **138**, A272.

975. R. A. Smith, DETECTORS FOR ULTRAVIOLET, VISIBLE, AND INFRARED RADIATION, *Appl. Opt.* **4**, 631.

976. A. I. Stekhanov, A. A. Karamyan and N. I. Astaf'ev, INFRARED ABSORPTION SPECTRA OF PEROVSKITE-TYPE FERROELECTRIC CRYSTALS, *Fiz. Tverd. Tela* **7**, 157; *Soviet Phys.-Solid State* **7**, 199.

977. J. E. Stewart, FAR INFRARED SPECTROSCOPY, in *Interpretive Spectroscopy*, S. K. Freeman, Ed., Reinhold, New York, p. 131.

978. R. Stolen and K. Dransfeld, FAR-INFRARED LATTICE ABSORPTION IN ALKALI HALIDE CRYSTALS, *Phys. Rev.* **139**, A1295.

979. N. W. B. Stone and D. Williams, FAR INFRA-RED ABSORPTION IN LIQUID NITROGEN, *Mol. Phys.* **10**, 85.

980. H. Tadokoro, M. Kobayashi, M. Ukita, K. Yasufuku, S. Murahashi, and T. Torii, NORMAL VIBRATIONS OF THE POLYMER MOLECULES OF HELICAL CONFORMATION. V. ISOTACTIC POLYPROPYLENE AND ITS DEUTERO-DERIVATIVES, *J. Chem. Phys.* **42**, 1432.

981. P. Taimsalu and D. W. Robinson, THE FAR-INFRARED SPECTRA OF SOLID H_2S AND D_2S AT 5°K AND 77°K, *Spectrochim. Acta* **21**, 1921.

982. S. Takeno and A. J. Sievers, CHARACTERISTIC TEMPERATURE DEPENDENCE FOR LOW-LYING LATTICE RESONANT MODES, *Phys. Rev. Letters* **15**, 1020.

983. A. Vǎsko, OPTICAL CONSTANTS OF AMORPHOUS SELENIUM FROM ROENTGEN TO FAR-INFRARED REGION, *J. Opt. Soc. Am.* **55**, 894.

984. E. A. Vinogradov, E. M. Dianov, and N. A. Irisova, FABRY-PEROT INTERFEROMETER FOR THE SHORT MILLIMETER AND SUBMILLIMETER BANDS WITH METALLIC GRIDS HAVING PERIODS SMALLER THAN THE WAVELENGTH, *JETP Letters* **2**, 205.

985. E. A. Vinogradov, E. M. Dianov, and N. A. Irisova, MICHELSON INTERFEROMETER FOR MEASURING THE REFRACTIVE INDEX OF DIELECTRIC MATERIALS IN THE 2 mm WAVELENGTH RANGE, *Radiotekhnika i Elektronika USSR* **10**, 1804; *Radio Engng. Electr. Phys.* **10**, 1547.

986. A. Walker and J. R. Ferraro, INFRARED SPECTRA OF ANHYDROUS RARE-EARTH NITRATES FROM 4000–100 cm^{-1}, *J. Chem. Phys.* **43**, 2689.

987. W. H. Wells, PROPOSED GAS MASER PUMPING SCHEME FOR THE FAR INFRARED, *J. Appl. Phys.* **36**, 2838.

988. K. A. Wickersheim and R. A. Buchanan, SOME REMARKS CONCERNING THE SPECTRA OF WATER AND HYDROXYL GROUPS IN BERYL, *J. Chem. Phys.* **42**, 1468.

989. G. R. Wilkinson, LOW FREQUENCY INFRA-RED SPECTROSCOPY, in *Infra-Red Spectroscopy and Molecular Structure*, Elsevier, Amsterdam.

989a. W. Witte, CONE CHANNEL OPTICS, *Infrared Phys.* **5**, 179.

989b. W. J. Witteman and R. Bleekrode, PULSED AND CONTINUOUS MOLECULAR FAR INFRARED GAS LASER, *Z. Angew. Math. Phys.* **16**, 87.

990. H. Yoshinaga, RECENT TECHNIQUES IN FAR INFRARED SPECTROSCOPY, *Japan. J. Appl. Phys., Suppl. 1,* **4**, 420.

991. F. Zernike and P. R. Berman, GENERATION OF FAR INFRARED AS A DIFFERENCE FREQUENCY, *Phys. Rev. Letters* **15**, 999.

1966

991a. K. A. Aganbekyan, A. N. Vystavkin, V. N. Listvin, and V. D. Shtykov, RECEIVER WITH AN InSb DETECTOR FOR INVESTIGATING ABSORPTION SPECTRA IN THE SUBMILLIMETER WAVELENGTH RANGE, *Radiotekhnika i Elektronika* **11**, 1252; *Radio Engng. Electronic Phys.* **11**, 1093.

992. D. P. Akitt, W. Q. Jeffers, and P. D. Coleman, WATER VAPOR GAS LASER OPERATING AT 118 MICRONS WAVELENGTH, *Proc. IEEE* **54**, 547.

993. S. J. Allen, FAR-INFRARED SPECTRA OF HCl TRAPPED IN THE β-QUINOL CLATHRATE, *J. Chem. Phys.* **44**, 394.

994. V. Ananthanarayanan and A. Danti, LOW FREQUENCY VIBRATIONAL SPECTRUM OF TUTTON'S SALT, *J. Mol. Spectry.* **20**, 88.

995. A. Anderson and W. H. Smith, FAR-INFRARED SPECTRA OF CRYSTALLINE ACETYLENES C_2H_2 AND C_2D_2, *J. Chem. Phys.* **44**, 4216.

996. A. Anderson and S. H. Walmsley, FAR INFRA-RED SPECTRA OF MOLECULAR CRYSTALS. V. SULPHUR DIOXIDE, *Mol. Phys.* **10**, 391.

996a. F. Arams, C. Allen, B. Peyton, and E. Sard, MILLIMETER MIXING AND DETECTION IN BULK InSb, *Proc. IEEE* **54**, 612.

997. I. M. Aref'ev, P. A. Bazhulin, and T. V. Mikhal'tseva, LONG-WAVE INFRARED TRANSMISSION SPECTRA OF KH_2PO_4, *Fiz. Tverd. Tela* **7**, 2113; *Soviet Phys.-Solid State* **7**, 1948.

998. I. M. Aref'ev, P. A. Bazhulin, and I. S. Zheludev, LONG-WAVELENGTH INFRARED TRANSMISSION SPECTRA OF $NH_4H_2PO_4$, *Fiz. Tverd. Tela* **7**, 2834; *Soviet Phys.-Solid State* **7**, 2290.

998a. V. Y. Balakhanov and A. R. Striganov, INTERFERENCE FILTERS FOR THE MICROWAVE AND SUBMILLIMETRIC REGION OF THE SPECTRUM, *Zh. Priklad. Spektrosk.* **3**, 213.

999. A. S. Barker, TEMPERATURE DEPENDENCE OF THE TRANSVERSE AND LONGITUDINAL OPTIC MODE FREQUENCIES AND CHARGES IN $SrTiO_3$ AND $BaTiO_3$, *Phys. Rev.* **145**, 391.

1000. M. Baudler and G. Fricke, ON THE VIBRATION SPECTRUM AND MOLECULAR STRUCTURE OF P_2I_4, *Z. Anorg. Allgem. Chem.* **345**, 129.

1001. E. E. Bell, MEASUREMENT OF THE FAR INFRARED OPTICAL PROPERTIES OF SOLIDS WITH A MICHELSON INTERFEROMETER USED IN THE ASYMMETRIC MODE: PART I, MATHEMATICAL FORMULATION, *Infrared Phys.* **6**, 57.

1002. L. V. Berman and A. G. Zhukov, OPTICAL CONSTANTS OF CRYSTALLINE QUARTZ IN THE FAR INFRARED REGION, *Opt. i Spektr.* **21**, 735; *Opt. Spectry.* **21**, 401.

1003. D. W. Berreman, LOW PASS FILTERS FOR FAR INFRARED SPECTROSCOPY, *Rev. Sci. Instr.* **37**, 513.

1004. R. K. Bogens and A. G. Zhukov, OPTICAL CONSTANTS OF FUSED QUARTZ IN THE FAR INFRA-RED, *Zhur. Priklad. Spektrosk.* **4**, 68.

1005. T. R. Borgers and H. L. Strauss, FAR INFRARED SPECTRA OF TRIMETHYLENE SULFIDE AND CYCLOBUTANONE, *J. Chem. Phys.* **45**, 947.

1006. D. R. Bosomworth, THE FAR INFRARED OPTICAL PROPERTIES OF $LiNbO_3$, *Appl. Phys. Letters* **9**, 330.

1007. C. C. Bradley, H. A. Gebbie, A. C. Gilby, V. V. Kechin, and J. H. King, FAR INFRARED SPECTROSCOPY AT HIGH PRESSURES, *Nature* **211**, 839.

1008. F. Brehat, O. Evrard, A. Hadni, and J. Lambert, SPECTRES D'ABSORPTION DANS L'INFRAROUGE LOINTAIN DES OXYDES DE MAGNÉSIUM, DE FER, DE MANGANÈSE ET DE LEURS CRISTAUX MIXTES (FeO–MgO) ET (FeO–MnO), *Compt. Rend.* **263B**, 1112.

1009. J. I. Bryant, VIBRATIONAL SPECTRUM OF CESIUM AZIDE CRYSTALS, *J. Chem. Phys.* **45**, 689.

1010. R. A. Buchanan, H. E. Rast, and H. H. Caspers, INFRARED ABSORPTION OF Ce^{3+} IN LaF_3 AND CeF_3, *J. Chem. Phys.* **44**, 4063.

1011. W. J. Burroughs, E. C. Pyatt, and H. A. Gebbie, TRANSMISSION OF SUB-MILLIMETRE WAVES IN FOG, *Nature* **212**, 387.

1012. K. J. Button, H. A. Gebbie, and B. Lax, CYCLOTRON RESONANCE IN SEMICONDUCTORS WITH FAR INFRARED LASER, *IEEE J. Quant. Electr.* **QE2**, 202.

1013. E. G. Bylander and M. Hass, DIELECTRIC CONSTANT AND FUNDAMENTAL LATTICE FREQUENCY OF LEAD TELLURIDE, *Solid State Commun.* **4**, 51.

1014. S. I. Chan, T. R. Borgers, J. W. Russell, H. L. Strauss, and W. D. Gwinn, TRIMETHYLENEOXIDE, III. FAR-INFRARED SPECTRUM AND DOUBLE-MINIMUM VIBRATION, *J. Chem. Phys.* **44**, 1103.

1015. G. E. Campagnaro and J. L. Wood, TORSIONAL VIBRATION OF ACROLEIN, *Trans. Faraday Soc.* **62**, 263.

1016. G. L. Carlson, R. E. Witkowski, and W. G. Fateley, FAR INFRARED SPECTRA OF DIMERIC AND CRYSTALLINE FORMIC AND ACETIC ACIDS, *Spectrochim. Acta* **22**, 1117.

1016a. J. E. Chamberlain, G. W. Chantry, F. D. Findlay, H. A. Gebbie, J. E. Gibbs, N. W. B. Stone, and A. J. Wright, THE SPECTRAL TRANSMISSION AT INFRARED WAVELENGTHS OF MICHELSON INTERFEROMETERS WITH DIELECTRIC FILM BEAM DIVIDERS, *Infrared Phys.* **6**, 195.

1017. J. E. Chamberlain, G. W. Chantry, H. A. Gebbie, N. W. B. Stone, T. B. Taylor, and G. Wyllie, SUBMILLIMETRE ABSORPTION AND DISPERSION OF LIQUID WATER, *Nature* **210**, 790.

1018. J. E. Chamberlain and H. A. Gebbie, DISPERSION MEASUREMENTS ON POLYTETRAFLUORETHYLENE IN THE FAR INFRARED, *Appl. Opt.* **5**, 393.

1019. G. W. Chantry, H. A. Gebbie, P. R. Griffiths, and R. F. Lake, FAR INFRA-RED SPECTRA AND THE ROTATIONAL ISOMERISM OF SYMMETRICAL TETRACHLORO- AND TETRABROMO-ETHANE, *Spectrochim. Acta* **22**, 125.

1019a. S. Deb and P. K. Chaudhury, INFRARED AND SUBMILLIMETER WAVE MODULATION USING FREE CARRIER ABSORPTION IN p–n JUNCTION DIODES, *Solid State Electr.* **9**, 113.

1019b. E. D. Dianov, N. A. Irisova, and V. N. Timofeer, MEASUREMENT OF THE ABSORPTION COEFFICIENT OF GLASSES IN THE SUBMILLIMETER RANGE, *Fiz. Tverd. Tela* **8**, 2643; *Soviet Phys.-Solid State* **8**, 2113.

1020. E. M. Dianov and N. A. Irisova, DETERMINATION OF THE ABSORPTION COEFFICIENTS OF SOLIDS IN THE SHORT-WAVE PART OF THE MILLIMETRE BAND, *Zhur. Priklad. Spektrosk.* **5**, 251.

1021. J. M. Dowling and R. T. Hall, FAR INFRARED INTERFEROMETRY: UPPER LIMIT FOR LINE WIDTHS FOR THE PURE ROTATIONAL BAND OF CARBON MONOXIDE, *J. Mol. Spectry.* **19**, 108.

1022. D. A. Draegert, N. W. B. Stone, B. Curnutte, and D. Williams, FAR-INFRARED SPECTRUM OF WATER, *J. Opt. Soc. Am.* **56**, 64.

1023. J. R. Durig and R. C. Lord, FAR-INFRARED SPECTRUM OF FOUR-MEMBERED-RING COMPOUNDS. I. SPECTRA AND STRUCTURE OF CYCLOBUTANE, CYCLOBUTANONE-d_4, TRIMETHYLENE SULFIDE, AND PERFLUOROCY-CLOBUTANONE, *J. Chem. Phys.* **45**, 61.

1024. J. R. Durig, W. A. McAllister, J. N. Willis, and E. E. Mercer, FAR INFRA-RED SPECTRUM OF POTASSIUM AND AMMONIUM PENTAHALONITRO-SYLRUTHENATES, *Spectrochim. Acta* **22**, 1091.

1025. J. R. Durig and A. C. Morrissey, RAMAN AND INFRARED SPECTRA OF 3-METHYLENEOXETANE, *J. Chem. Phys.* **45**, 1269.

1026. W. G. Fateley, R. E. Witkowski, and G. L. Carlson, FAR-INFRARED TRANSMIS-SION OF COMMERCIALLY AVAILABLE CRYSTALS AND HIGH-DENSITY POLYETHYLENE, *Appl. Spectry.* **20**, 190.

1027. J. R. Ferraro, L. J. Basile, and D. L. Kovacic, THE INFRARED SPECTRA OF RARE-EARTH METAL CHLORIDE COMPLEXES OF 2,2'-BIPYRIDYL AND 1,10-PHENANTHROLINE FROM 650 TO 70 cm^{-1}, *Inorg. Chem.* **5**, 391.

1028. J. R. Ferraro and A. Walker, COMPARISON OF THE INFRARED SPECTRA OF THE HYDRATES AND ANHYDROUS SALTS IN THE SYSTEM $UO_2(NO_3)_2$ AND $Th(NO_3)_4$, *J. Chem. Phys.* **45**, 550.

1029. G. T. Flesher and W. M. Muller, SUBMILLIMETER GAS LASER, *Proc. IEEE* **54**, 543.

1030. M. L. Forman, W. H. Steel, and G. A. Vanasse, CORRECTION OF ASYMMETRIC INTERFEROGRAMS OBTAINED IN FOURIER SPECTROSCOPY, *J. Opt. Soc. Am.* **56**, 59.

1031. N. I. Furashov, FAR INFRARED ABSORPTION BY ATMOSPHERIC WATER VAPOR, *Opt. i Spektr.* **20**, 427; *Opt. Spectry.* **20**, 234.

1031a. C. G. B. Garrett, FAR INFRARED MASERS AND THEIR APPLICATIONS TO SPECTROSCOPY, *Physics of Quantum Electronics*, Edited by P. L. Kelley, B. Lax, and P. E. Tannenwald (McGraw-Hill Book Co. N.Y. 1966).

1031b. H. A. Gebbie, W. J. Burroughs, J. A. Robb, and G. R. Bird, OBSERVATIONS OF THE MAGNETIC DIPOLE ROTATION SPECTRUM OF OXYGEN, *Nature* **212**, 66.

1031c. H. A. Gebbie, N. W. B. Stone, W. Slough, J. E. Chamberlain, and W. A. Sheraton, SUB-MILLIMETRE MASER AMPLIFICATION AND CONTINUOUS WAVE EMISSION, *Nature* **211**, 62.

1032. H. A. Gebbie, N. W. B. Stone, W. Slough, and J. E. Chamberlain, SUB-MILLIMETRE MASER AMPLIFICATION AND CONTINUOUS WAVE EMISSION, *Nature* **211**, 5044.

1033. H. A. Gebbie, N. W. B. Stone, G. Topping, E. K. Gora, S. A. Clough, and F. X. Kneizys, ROTATIONAL ABSORPTION OF SOME ASYMMETRIC ROTOR MOLECULES. I. OZONE AND SULFUR DIOXIDE, *J. Mol. Spectry.* **19**, 7.

1033a. H. A. Gebbie and R. Q. Twiss, TWO-BEAM INTERFEROMETRIC SPECTRO-SCOPY, *Repts. Prog. Phys.* **29**, 729.

1034. R. Geick, W. J. Hakel, and C. H. Perry, TEMPERATURE DEPENDENCE OF THE FAR-INFRARED REFLECTIVITY OF MAGNESIUM STANNIDE, *Phys. Rev.* **148**, 824.

1035. R. Geick, C. H. Perry, and S. S. Mitra, LATTICE VIBRATIONAL PROPERTIES OF HEXAGONAL CdSe, *J. Appl. Phys.* **37**, 1994.
1036. R. Geick, C. H. Perry, and G. Rupprecht, NORMAL MODES IN HEXAGONAL BORON NITRIDE, *Phys. Rev.* **146**, 543.
1037. P. A. Giguère and C. Chapados, THE FAR INFRA-RED SPECTRA OF H_2O_2 AND D_2O_2 IN THE SOLID STATES, *Spectrochim. Acta* **22**, 1131.
1038. S. G. W. Ginn and J. L. Wood, INTERMOLECULAR VIBRATIONS OF CHARGE TRANSFER COMPLEXES, *Trans. Faraday Soc.* **62**, 777.
1039. C. C. Grimes, P. L. Richards, and S. Shapiro, FAR-INFRARED RESPONSE OF POINT-CONTACT JOSEPHSON JUNCTIONS, *Phys. Rev. Letters* **17**, 431.
1040. R. T. Hall and J. M. Dowling, COMMENTS ON A NOTE BY MOSER, STEFFEN, AND KNEUBÜHL, *Appl. Opt.* **5**, 1969.
1041. R. T. Hall and J. M. Dowling, PURE ROTATIONAL SPECTRUM OF NITRIC OXIDE, *J. Chem. Phys.* **45**, 1899.
1042. R. T. Hall, D. Vrabec, and J. M. Dowling, A HIGH-RESOLUTION FAR-INFRARED DOUBLE-BEAM LAMELLAR GRATING INTERFEROMETER, *Appl. Opt.* **5**, 1147.
1043. J. W. Halley, MICROSCOPIC THEORY OF FAR INFRARED TWO-MAGNON ABSORPTION IN ANTIFERROMAGNETS. I. PERTURBATION-THEORY SEARCH AND APPLICATION OF FOURTH-ORDER PROCESSES TO FeF_2, *Phys. Rev.* **149**, 423.
1044. J. W. Halley and I. Silvera, OBSERVATION AND THEORY OF NEW LINE IN FAR-INFRARED ABSORPTION IN ANTIFERROMAGNETIC FeF_2, *J. Appl. Phys.* **37**, 1226.
1045. J. Haraishi and T. Shimanouchi, LATTICE-VIBRATIONS AND THE FORCE FIELD OF K_2PtCl_4, K_2PdCl_4, AND K_2PtCl_6, *Spectrochim. Acta* **22**, 1483.
1046. R. E. Harris, R. L. Cappelletti, and D. M. Ginsberg, FAR-INFRARED TRANSMISSION THROUGH METAL LIGHT-PIPES WITH LOW THERMAL CONDUCTANCE, *Appl. Opt.* **5**, 1083.
1047. E. C. Heltemes, FAR-INFRARED PROPERTIES OF CUPROUS OXIDE, *Phys. Rev.* **141**, 803.
1047a. K. Hennerich, W. Lahmann, and W. Witte, THE LINEARITY OF GOLAY DETECTORS, *Infrared Phys.* **6**, 123.
1048. J. C. Hill and R. G. Wheeler, FAR-INFRARED SPECTRA OF ERBIUM, DYSPROSIUM, AND SAMARIUM ETHYL SULPHATE, *Phys. Rev.* **152**, 482.
1049. J. E. Hoffman and G. A. Vanasse, REAL-TIME SPECTRAL SYNTHESIS IN FOURIER SPECTROSCOPY, *Appl. Opt.* **5**, 1167.
1049a. R. M. Hornreich and S. Shtrikman, THEORY OF HARMONIC GENERATION USING ANTI-FERROMAGNETS IN THE FAR-INFRARED, *IEEE Trans.* **MAG-2**, 292.
1050. R. H. Hunt and R. A. Leacock, FAR-INFRARED SPECTRUM AND HINDERING POTENTIAL OF DEUTERIUM PEROXIDE, *J. Chem. Phys.* **45**, 3141.
1051. W. J. Hurley, INTERFEROMETRIC SPECTROSCOPY IN THE FAR INFRARED, *J. Chem. Educ.* **43**, 236.
1052. W. J. Hurley, I. D. Kuntz, and G. E. Leroi, FAR-INFRARED STUDIES OF HYDROGEN BONDING, *J. Am. Chem. Soc.* **88**, 3199.
1053. Y. Ishibashi, K. Siratori, and T. Nakamura, FAR INFRARED ABSORPTION OF THE $NaNO_2$ CRYSTAL, *J. Phys. Soc. Japan* **21**, 809.
1054. J. R. Jasperse, A. Kahan, J. N. Plendl, and S. S. Mitra, TEMPERATURE DEPENDENCE OF INFRARED DISPERSION IN IONIC CRYSTALS LiF AND MgO, *Phys. Rev.* **146**, 526.

1055. R. Kaplan, MAGNETO-OPTICAL STUDY OF HYDROGENIC DONOR IM-
      PURITY STATES IN InSb, in *Proceedings of the International Conference on the
      Physics of Semiconductors*, Kyoto, 1966, *J. Phys. Soc. Japan, Suppl.*, **21**, 249.
1056. L. F. Keyser and G. W. Robinson, INFRARED SPECTRA OF HCl AND DCl IN
      SOLID RARE GASES, *J. Chem. Phys.* **44**, 3225.
1057. P. H. Knapp and D. H. Martin, SUBMILLIMETER SPECTROSCOPY USING A
      FROOME HARMONIC GENERATOR (METAL-PLASMA JUNCTION), *Proc.
      IEEE* **54**, 528.
1058. F. K. Kneubühl, J. F. Moser, and H. Steffen, HIGH-RESOLUTION GRATING
      SPECTROMETER FOR THE FAR INFRARED, *J. Opt. Soc. Am.* **56**, 760.
1059. T. Kondo, K. Siratori, and H. Inokuchi, FAR-INFRARED ABSORPTION OF THE
      TCNQ-QUINOLINIUM COMPLEX, *J. Phys. Soc. Japan* **21**, 824.
1060. C. F. Krumm and G. I. Haddad, MILLIMETER- AND SUBMILLIMETER-WAVE
      QUANTUM DETECTORS, *Proc. IEEE* **54**, 627.
1061. M. Kubota, D. L. Johnston, and I. Matsubara, COMPLEXES OF SUCCINONITRILE
      WITH SILVER (I). A NITRATO-SILVER COMPOUND, *Inorg. Chem.* **5**, 386.
1062. R. F. Lake and H. W. Thompson, FAR INFRARED STUDIES OF HYDROGEN
      BONDING IN ALCOHOLS, *Proc. Roy. Soc. (London)* **A291**, 469.
1063. J. A. Lane, FAR-INFRARED SPECTRUM OF LIQUID WATER, *J. Opt. Soc. Am.*
      56, 1398.
1064. D. N. Langenberg, D. J. Scalapino, and B. N. Taylor, JOSEPHSON-TYPE SUPER-
      CONDUCTING TUNNEL JUNCTIONS AS GENERATORS OF MICROWAVE
      AND SUBMILLIMETER WAVE RADIATION, *Proc. IEEE* **54**, 560.
1065. A. J. Lichtenberg and S. Sesnic, ABSOLUTE RADIATION STANDARD IN THE
      FAR INFRARED, *J. Opt. Soc. Am.* **56**, 75.
1065a. T. M. Lifshits, F. Y. Nad, and V. I. Sidorov, IMPURITY PHOTOCONDUCTIVITY
      OF GERMANIUM DOPED WITH ANTIMONY, ARSENIC, BORON, OR
      INDIUM, *Fiz. Tverd. Tela* **8**, 3208; *Soviet Phys.-Solid State* **8**, 2567.
1066. H. G. Lipson and J. R. Littler, TUNING FORK CHOPPERS FOR INFRARED
      SPECTROMETERS, *Appl. Opt.* **5**, 472.
1067. E. V. Loewenstein, THE HISTORY AND CURRENT STATUS OF FOURIER
      TRANSFORM SPECTROSCOPY, *Appl. Opt.* **5**, 845.
1068. D. E. Mann, N. Acquista, and D. White, INFRARED SPECTRA OF HCl, DCl,
      HBr, AND DBr IN SOLID RARE-GAS MATRICES, *J. Chem. Phys.* **44**, 3453.
1069. D. E. McCarthy, TRANSMISSION OF IRTRAN MATERIALS FROM 50 μ TO
      300 μ, *Appl. Opt.* **5**, 472.
1070. N. T. McDevitt and A. D. Davidson, INFRARED LATTICE SPECTRA OF CUBIC
      RARE EARTH OXIDES IN THE REGION 700–50 cm^{-1}, *J. Opt. Soc. Am.* **56**, 636.
1071. R. S. McDowell, ON THE PURE ROTATIONAL SPECTRA OF SYMMETRIC TOP
      MOLECULES WITH UNRESOLVED K-STRUCTURE: REVISION OF THE
      ROTATIONAL CONSTANTS OF NT$_3$, *J. Mol. Spectry.* **19**, 239.
1072. Y. Mikawa, R. J. Jakobsen, and J. W. Brasch, FAR-INFRARED SPECTRA OF
      MERCURIC HALIDES AND THEIR DIOXANE COMPLEXES, *J. Chem. Phys.*
      45, 4528.
1073. S. S. Mitra, OPTICALLY-ACTIVE MULTIPHONON PROCESSES IN II–VI SEMI-
      CONDUCTORS, in *Proceedings of the International Conference on the Physics of
      Semiconductors*, Kyoto, 1966, *J. Phys. Soc. Japan, Suppl.*, **21**, 61.
1074. K. D. Möller, D. J. McMahon, and D. R. Smith, FAR INFRARED TRANSMISSION
      FILTERS FOR THE 300–18 cm^{-1} SPECTRAL REGION, *Appl. Opt.* **5**, 403.
1075. J. Morandat and J. Lecomte, ABSORPTION SPECTRA OF SOME METALLIC

CARBONATES IN THE FAR INFRARED. TRIAL INTERPRETATION, *Compt. Rend.* **263B**, 735.

1076. J. Morandat, V. Lorenzelli, and H. Paquet, ABSORPTION SPECTRA OF SOME METALLIC SULPHATES IN THE FAR INFRARED. TRIAL INTERPRETATION, *Compt. Rend.* **263B**, 697.

1077. T. Moriya, FAR INFRARED ABSORPTION BY TWO-MAGNON EXCITATIONS IN ANTIFERROMAGNETS, *J. Phys. Soc. Japan* **21**, 926.

1078. J. F. Moser, H. Steffen, and F. K. Kneubühl, A NOTE ON THE HIGH RESOLUTION SPECTROSCOPY IN THE FAR INFRARED, *Appl. Opt.* **5**, 1969.

1079. W. W. Müller and G. T. Flesher, CONTINUOUS WAVE SUBMILLIMETER OSCILLATION IN H_2O, D_2O, AND CH_3CN, *Appl. Phys. Letters* **8**, 217.

1080. I. Nakagawa and T. Shimanouchi, FAR INFRA-RED SPECTRA AND METAL-LIGAND FORCE CONSTANTS OF METAL AMMINE COMPLEXES, *Spectrochim. Acta* **22**, 759.

1081. I. Nakagawa and T. Shimanouchi, FAR INFRA-RED SPECTRA AND LATTICE VIBRATIONS OF HEXANITROCOBALT (III) COMPLEX SALTS AND AMMINE COMPLEX SALTS, *Spectrochim. Acta* **22**, 1707.

1082. T. Nakamura, SINGLE-OSCILLATOR MODEL OF THE "SOFT-MODE" LATTICE VIBRATION IN THE PEROVSKITE-TYPE FERROELECTRICS, *J. Phys. Soc. Japan* **21**, 491.

1083. L. Nemes, H. Ratajczak, and W. J. Orville-Thomas, MEASUREMENT OF REFRACTIVE INDEX IN THE FAR INFRA-RED, *Spectrochim. Acta* **22**, 156.

1084. L. J. Neuringer, R. C. Milward, and R. L. Aggarwal, FAR INFRARED RESONANT ABSORPTION IN n-TYPE SILICON, in *Proceedings of the International Conference on Physics in Semiconductors*, Kyoto, 1966, *J. Phys. Soc. Japan* **21**, 582.

1084a. S. L. Norman and D. H. Douglass, STRUCTURE IN THE PRECURSOR ABSORPTION IN SUPERCONDUCTING LEAD, *Phys. Rev. Letters* **17**, 875.

1085. E. L. Pace, A. C. Plaush, and H. V. Samuelson, VIBRATIONAL SPECTRA AND FREQUENCY ASSIGNMENTS FOR OCTAFLUOROPROPANE (C_3F_8) AND HEXAFLUOROACETONE (C_3F_6O), *Spectrochim. Acta* **22**, 993.

1086. C. H. Perry, R. Geick, and E. F. Young, SOLID STATE STUDIES BY MEANS OF FOURIER TRANSFORM SPECTROSCOPY, *Appl. Opt.* **5**, 1171.

1087. J. C. Picard and D. L. Carter, CYCLOTRON RESONANCE IN TELLURIUM AT SUBMILLIMETER WAVELENGTHS, in *Proceedings of the International Conference on Physics in Semiconductors*, Kyoto, 1966, *J. Phys. Soc. Japan Suppl.* **21**, 202.

1088. J. N. Plendl, A. Hadni, J. Claudel, Y. Henninger, G. Morlot, P. Strimer, and L. C. Mansur, FAR INFRARED STUDY OF THE COPPER HALIDES AT LOW TEMPERATURES, *Appl. Opt.* **5**, 397.

1089. H. Prask and H. Boutin, LOW-FREQUENCY MOTIONS OF H_2O MOLECULES IN CRYSTALS. III, *J. Chem. Phys.* **45**, 3284.

1089a. W. Prettl and L. Genzel, NOTES ON THE SUBMILLIMETER LASER EMISSION FROM CYANIC COMPOUNDS, *Phys. Letters* **23**, 443.

1089b. E. H. Putley, SOLID STATE DEVICES FOR INFRA-RED DETECTION, *J. Sci. Instr.* **43**, 857.

1090. K. Radcliff and J. L. Wood, INFLUENCE OF ENVIRONMENT ON TORSIONAL VIBRATIONS, *Trans. Faraday Soc.* **62**, 1678.

1091. K. F. Renk, LOCALISED GAP MODES IN THE FAR-INFRARED SPECTRUM OF $KI:OH^-$, *Phys. Letters* **20**, 137.

1092. K. F. Renk, NARROW TWO-PHONON DIFFERENCE BANDS IN KI, *Phys. Letters* **21**, 132.

1093. G. M. Ressler and K. D. Möller, FAR INFRARED TRANSMITTANCE OF IR-TRANS 1 TO 5 IN THE 250–10 cm^{-1} SPECTRAL REGION, *Appl. Opt.* **5**, 877.

1094. W. K. Rivers and A. P. Sheppard, HIGHLY ACCURATE NONSPHERICAL REFLECTORS FOR THE MILLIMETER AND SUBMILLIMETER WAVE-LENGTH, *Rev. Sci. Instr.* **37**, 195.

1095. D. W. Robinson and W. G. Von Holle, FAR-INFRARED SPECTRA OF MATRIX-ISOLATED HYDROGEN FLUORIDE, *J. Chem. Phys.* **44**, 410.

1096. W. G. Rothschild, GROUP FREQUENCIES IN CYCLOPROPANES. A CORRELA-TION DIAGRAM, *J. Chem. Phys.* **44**, 1712.

1097. W. G. Rothschild, ON THE EXISTENCE OF CONFORMERS OF CYCLOBUTYL MONOHALIDES, *J. Chem. Phys.* **44**, 2213.

1098. W. G. Rothschild, INFRARED AND RAMAN SPECTRUM OF BROMOCYCLO-PROPANE BETWEEN 3200 AND 240 cm^{-1}, *J. Chem. Phys.* **44**, 3875.

1099. W. G. Rothschild, ON THE EXISTENCE OF CONFORMERS OF CYCLOBUTYL MONOHALIDES. II. TEMPERATURE DEPENDENCE OF THE INFRARED SPECTRA OF BROMOCYCLOBUTANE AND CHLOROCYCLOBUTANE, *J. Chem. Phys.* **45**, 1214.

1100. W. G. Rothschild, ON THE EXISTENCE OF CONFORMERS OF CYCLOBUTYL MONOHALIDES. III. ASSIGNMENTS OF THE FUNDAMENTALS OF BROMO-CYCLOBUTANE AND CHLOROCYCLOBUTANE, *J. Chem. Phys.* **45**, 3599.

1101. I. G. Rudyavskaya, A. G. Kudryavtseva, and L. D. Kislovskii, TRANSMITTANCE OF ANTIREFLECTION-COATED SILICON IN THE FAR INFRARED, *Opt. i Spektr.* **21**, 476; *Opt. Spectry.* **21**, 266.

1102. E. E. Russell and E. E. Bell, MEASUREMENT OF THE FAR INFRARED OPTICAL PROPERTIES OF SOLIDS WITH A MICHELSON INTERFEROMETER USED IN THE ASYMMETRIC MODE: PART II, THE VACUUM INTERFEROMETER, *Infrared Phys.* **6**, 75.

1103. A. Sabatini and I. Bertini, FAR-INFRARED SPECTRA OF OXOCHLORO AND OXOBROMO COMPLEXES OF Nb(V), Mo(V), AND W(V), *Inorg. Chem.* **5**, 204.

1104. H. Sakai and G. A. Vanasse, HILBERT TRANSFORM IN FOURIER SPECTRO-SCOPY, *J. Opt. Soc. Am.* **56**, 131.

1105. H. Sakai and G. A. Vanasse, DIRECT DETERMINATION OF THE TRANSFER FUNCTION OF AN INFRARED SPECTROMETER, *J. Opt. Soc. Am.* **56**, 357.

1105a. V. I. Sidorov and T. M. Lifshits, PHOTOCONDUCTIVITY IN GERMANIUM DOPED WITH GROUP III IMPURITIES CAUSED BY OPTICAL EXCITATION OF THE IMPURITY LEVELS, *Fiz. Tverd. Tela* **8**, 2498; *Soviet Phys.-Solid State* **8**, 2000.

1106. A. J. Sievers, EXPLORING THE EXCITATION SPECTRA OF CRYSTALS USING FAR-INFRARED RADIATION, in *NATO Advanced Study on Elementary Excitations and their Interactions*, Cortina d'Ampezzo Italy, July 11–23.

1107. A. J. Sievers, R. W. Alexander, Jr., and S. Takeno, A RESONANT IMPURITY MODE AND THE MÖSSBAUER EFFECT OF EUROPIUM-DOPED MnF$_2$, *Solid State Commun.* **4**, 483.

1108. I. Silvera and J. W. Halley, INFRARED ABSORPTION IN FeF$_2$: PHENOMENO-LOGICAL THEORY, *Phys. Rev.* **149**, 415.

1109. D. R. Smith, B. K. McKenna, and K. D. Möller, FAR-INFRARED SPECTRA OF DIMETHYLSULFIDE, ISOBUTYLENE, AND ACETONE OBTAINED WITH A VACUUM-GRATING SPECTROMETER IN THE 240–10 cm^{-1} SPECTRAL RE-GION, *J. Chem. Phys.* **45**, 1904.

1110. W. H. Smith and G. E. Leroi, INFRARED AND RAMAN SPECTRA OF CARBON SUBOXIDE IN CONDENSED PHASES, *J. Chem. Phys.* **45**, 1767.

1111. O. M. Stafsudd, A METHOD FOR THE PRODUCTION OF FAR-INFRARED TRANSMISSION FILTER GRATINGS, *Appl. Phys.* **5**, 1957.

1112. A. E. Stanevich, THE LONG-WAVELENGTH INFRARED ABSORPTION SPECTRA OF DICARBOXYLIC ACIDS, *Opt. i Spektr.* **21**, 645; *Opt. Spectry.* **21**, 355.

1113. H. Steffen, J. Steffen, J. F. Moser, and F. K. Kneubühl, STIMULATED EMISSION UP TO 0.538 mm WAVELENGTH FROM CYANIC COMPOUNDS, *Phys. Letters* **20**, 20.

1113a. H. Steffen, J. Steffen, J. F. Moser, and F. K. Kneubühl, STIMULIERTE EMISSION VON ICN BIS ZU 0.774 mm WELLENLÄNGE, *Z. Angew. Math. Phys.* **17**, 472.

1114. H. Steffen, J. Steffen, J. F. Moser, and F. K. Kneubühl, COMMENTS ON A NEW LASER EMISSION AT 0.774 mm, *Phys. Letters* **21**, 425.

1115. H. Steffen, P. Schwaller, J. F. Moser, and F. K. Kneubühl, MECHANISM OF THE SUBMILLIMETER LASER EMISSIONS FROM THE CN-RADICAL, *Phys. Letters* **23**, 313.

1115a. J. Steffen, H. Steffen, J. F. Moser, and F. K. Kneubühl, UNTERSUCHUNG STIMU-LIERTER EMISSION IN SUBMILLIMETERWELLENBEREICH MIT EINEM KREUZGITTER FABRY-PEROT INTERFEROMETER, *Z. Angew. Math. Phys.* **17**, 470.

1116. D. L. Stierwalt, FAR INFRARED LATTICE BANDS IN INDIUM ANTIMONIDE, in *Proceedings of the International Conference on Physics in Semiconductors*, Kyoto, 1966, *J. Phys. Soc. Japan, Suppl.* **21**, 58.

1117. N. W. B. Stone and D. Williams, THE FAR INFRARED TRANSMISSION OF POLYETHYLENE AT REDUCED TEMPERATURES, *Appl. Opt.* **5**, 353.

1118. M. Takatsuji, THEORY OF FAR INFRARED DETECTION USING NONLINEAR OPTICAL MIXING, *Japan. J. Appl. Phys.* **5**, 389.

1118a. F. Varsanyi, FAR-INFRARED SOLID-STATE MASERS: A SPECULATIVE ACCOUNT WITH SOME RELATED EXPERIMENTS, Physics of Quantum Electronics, McGraw-Hill Book Co., New York, p. 370.

1119. H. W. Verleur and A. S. Barker, INFRARED LATTICE VIBRATIONS IN $GaAs_yP_{1-y}$ ALLOYS, *Phys. Rev.* **149**, 715.

1120. M. Veyssie, A. Hadni, and P. Strimer, SPECTRE D'ABSORPTION DANS L'INFRA-ROUGE LOINTAIN DU GALLATE DE DYSPROSIUM À STRUCTURE GRE-NAT, *J. Phys.* **27**, 43.

1121. H. Vogt and H. Happ, INFRARED SPECTRUM OF $NaNO_2$ BETWEEN 30 AND 150 μ IN THE FERRO- AND PARAELECTRIC PHASES, *Phys. Stat. Sol.* **16**, 711.

1121a. A. N. Vystavkin, V. N. Gubankov, and V. N. Listvin, EFFECT OF 0.2–8 mm ELEC-TROMAGNETIC RADIATION ON THE ELECTRICAL CONDUCTIVITY OF n-TYPE InSb AT HELIUM TEMPERATURES, *Fiz. Tverd. Tela* **8**, 443; *Soviet Phys.-Solid State* **8**, 350.

1122. V. Wagner, FAR-INFRARED ABSORPTION IN LIQUID AND SOLID BROMINE, *Phys. Letters* **22**, 58.

1123. W. H. Wells, MODES OF A TILTED-MIRROR OPTICAL RESONATOR FOR THE FAR INFRARED, *IEEE J. Quant. Electr.* **QE-2**, 94.

1124. R. G. Wheeler and J. C. Hill, SPECTROSCOPY IN THE 5 TO 400 WAVENUMBER REGION WITH THE GRUBB PARSONS INTERFEROMETRIC SPECTRO-METER, *J. Opt. Soc. Am.* **56**, 657.

1125. R. A. Williams and W. S. C. Chang, RESOLUTION AND NOISE IN FOURIER-TRANSFORM SPECTROSCOPY, *J. Opt. Soc. Am.* **56**, 167.

1126. R. A. Williams and W. S. C. Chang, OBSERVATION OF SOLAR RADIATION FROM 50 μ TO 1 mm, *Proc. IEEE* **54**, 462.

1127. R. H. Wright, ODOUR AND MOLECULAR VIBRATION, *Nature* **209**, 571.

1128. H. Yoshinaga, S. Fujita, S. Minami, Y. Suemoto, M. Inoue, K. Chiba, K. Nakano, S. Yoshida, and H. Sugimori, A FAR INFRARED INTERFEROMETRIC SPECTROMETER WITH A SPECIAL ELECTRONIC COMPUTER, *Appl. Opt.* **5**, 1159.

1967

1129. A. N. Aleksandrov, A. N. Sidorov, and N. G. Yaroslavskii, LONG-WAVELENGTH INFRARED ABSORPTION SPECTRA OF PHTHALOCYANINES, *Opt. i. Spektr.* **22**, 560; *Opt. Spectry.* **22**, 307.

1130. S. J. Allen, FAR-INFRARED SPIN-WAVE AND ANOMALOUS PHONON ABSORPTION IN ANTIFERROMAGNETIC URANIUM DIOXIDE, *J. Appl. Phys.* **38**, 1478.

1131. F. Arams, C. Allen, M. Wang, K. Button, and L. Rubin, FAR-INFRARED LASER POWER MEASUREMENTS, *Proc. IEEE* **55**, 420.

1132. K. Aring and A. J. Sievers, THERMAL CONDUCTIVITY AND FAR-INFRARED ABSORPTION OF UO_2, *J. Appl. Phys.* **38**, 1496.

1133. G. M. Arnold and R. Heastie, FAR INFRA-RED ABSORPTION IN EACH SOLID PHASE OF HCl AND DBr., *Chem. Phys. Letters* **1**, 51.

1134. J. D. Axe and G. D. Pettit, INFRARED DIELECTRIC DISPERSION OF SEVERAL FLUORIDE PEROVSKITES, *Phys. Rev.* **157**, 435.

1135. F. G. Baglin, S. F. Bush, and J. R. Durig, FAR-INFRARED SPECTRA AND SPACE GROUP OF CRYSTALLINE HYDRAZINE AND HYDRAZINE-d_4, *J. Chem. Phys.* **47**, 2104.

1136. M. Balkanski, P. Moch, and M. K. Teng, INFRARED LATTICE VIBRATIONS OF $KNiF_3$ FROM REFLECTION DATA AT 300° AND 90°K, *J. Chem. Phys.* **46**, 1621.

1137. J. M. Ballantyne, DOUBLE BEAM OPERATION OF A FOURIER SPECTROMETER, *Appl. Opt.* **6**, 587.

1138. A. S. Barker and R. Loudon, DIELECTRIC PROPERTIES AND OPTICAL PHONONS IN $LiNbO_3$, *Phys. Rev.* **158**, 433.

1139. A. S. Barker and H. W. Verleur, LONG WAVELENGTH OPTICAL PHONON VIBRATIONS IN MIXED CRYSTALS, *Solid State Commun.* **5**, 695.

1140. R. J. Bell and G. M. Goldman, OPTICAL CONSTANTS OF BLACK POLYETHYLENE, *J. Opt. Soc. Am.* **57**, 1552.

1141. R. J. Bell, S. I. Drasky, and W. L. Barnes, A VACUUM SUBMILLIMETER SPECTROMETER, *Infrared Phys.* **7**, 57.

1142. D. W. Berreman, ANOMALOUS RESTSTRAHL STRUCTURE FROM SLIGHT SURFACE ROUGHNESS, *Phys. Rev.* **163**, 855.

1143. D. R. Bosomworth, FAR INFRARED ABSORPTION MEASUREMENTS IN THE REGION OF THE NON-DEVONSHIRE LINES IN ALKALI HALIDES CONTAINING OH^- AND OD^- IMPURITIES, *Solid State Commun.* **5**, 681.

1144. D. R. Bosomworth, FAR-INFRARED OPTICAL PROPERTIES OF CaF_2, SrF_2, BaF_2, AND CdF_2. *Phys. Rev.* **157**, 709.

1145. D. R. Bosomworth and G. W. Cullen, ENERGY GAP OF SUPERCONDUCTING Nb_3Sn, *Phys. Rev.* **160**, 346.

1146. G. L. Bottger and A. L. Geddes, INFRARED ABSORPTION SPECTRUM OF CdTe, *J. Chem. Phys.* **47**, 4858.

1147. G. L. Bottger and A. L. Geddes, THE FAR-INFRARED SPECTRA OF SOME SILVER HALIDE COMPLEXES, *Spectrochim. Acta* **23A**, 1551.

1148. J. Bradbury, K. P. Forest, R. H. Nuttall, and D. W. A. Sharp, THE FAR INFRARED SPECTRA OF LIGAND-METAL-HALIDE COMPLEXES OF PSEUDOTETRAHEDRAL SYMMETRY, *Spectrochim. Acta* **23A**, 2701.

1149. C. C. Bradley, W. J. Burroughs, H. A. Gebbie, and W. Slough, OBSERVATION OF PRESSURE BROADENING EFFECTS IN D_2O USING A CN MASER, *Infrared Phys.* **7**, 129.

1150. E. Brannen, V. Sochor, W. J. Sarjeant, and H. R. Froelich, TIME DEPENDENCE OF THE POWER OUTPUT AT 119 μ IN A WATER VAPOR LASER, *Proc. IEEE* **55**, 462.

1151. M. O. Bulanin and M. V. Tonkov, A SMALL SPECTROMETER FOR LONG-WAVE INFRARED BAND, *Zh. Priklad. Spektrosk.* **6**, 713.

1152. R. Cano and M. Mattioli, FAR-INFRARED SPECTROSCOPY WITH A MICHELSON INTERFEROMETER AND AN InSb DETECTOR, *Infrared Phys.* **7**, 25.

1153. R. L. Cappelletti, D. M. Ginsberg, and J. K. Hulm, FAR-INFRARED ABSORPTION IN SUPERCONDUCTING NIOBIUM ALLOYS, *Phys. Rev.* **158**, 340.

1154. C. Carabatos, CONTRIBUTION À L'ÉTUDE DE LA BANDE FONDAMENTALE DE VIBRATION DU RÉSEAU DE LA CUPRITE À 146 cm^{-1}, *J. Phys. (France)* **28**, 825.

1155. D. L. Carter and J. C. Picard, RESONANCES IN SUBMILLIMETER ALFVÉN WAVE PROPAGATION IN BISMUTH, *Solid State Commun.* **5**, 719.

1156. J. E. Chamberlain, ON A CONVERGENCE CORRECTION FOR DISPERSIVE FOURIER SPECTROMETRY, *Appl. Opt.* **6**, 980.

1157. J. E. Chamberlain, A. E. Costley, and H. A. Gebbie, SUBMILLIMETRE DISPERSION OF LIQUID TETRABROMOETHANE, *Spectrochim. Acta* **23A**, 2255.

1158. J. E. Chamberlain, H. A. Gebbie, G. W. F. Pardoe, and M. Davies, MILLIMETER-WAVE FOURIER TRANSFORM SPECTROSCOPY FOR STUDIES OF THE LIQUID STATE, *Chem. Phys. Letters* **1**, 523.

1159. G. W. Chantry, H. A. Gebbie, and H. N. Mirza, FAR INFRA-RED SPECTRA AND THE MOLECULAR COMPLEXES OF CARBON TETRACHLORIDE AND CHLOROFORM WITH BENZENE, *Spectrochim. Acta*, **23A**, 2749.

1160. B. P. Clayman, I. G. Nolt, and A. J. Sievers, NEAR INSTABILITY OF LATTICE RESONANT MODES, *Phys. Rev. Letters* **19**, 111.

1161. N. D. Devyatkov and M. B. Golant, DEVELOPMENT OF ELECTRON DEVICES FOR THE MILLIMETER AND SUBMILLIMETER RANGES, *Radio Eng. Electr. Phys.* **12**, 1835.

1162. D. H. Dickey and J. O. Dimmock, EXCITATION SPECTRA OF GROUP III IMPURITIES IN GERMANIUM UNDER UNIAXIAL STRESS, *J. Phys. Chem. Solids* **28**, 529.

1163. D. H. Dickey, E. J. Johnson, and D. M. Larsen, POLARON EFFECTS IN THE CYCLOTRON-RESONANCE ABSORPTION OF InSb, *Phys. Rev. Letters* **18**, 599.

1164. J. M. Dowling and R. T. Hall, COMMENTS ON "HIGH-RESOLUTION GRATING SPECTROMETER FOR THE FAR INFRARED," *J. Opt. Soc. Am.* **57**, 269.

1165. H. D. Drew and A. J. Sievers, FAR INFRARED ABSORPTION IN SUPERCONDUCTING AND NORMAL LEAD, *Phys. Rev. Letters* **19**, 697.

1166. J. R. Durig, B. R. Mitchell, D. W. Sink, J. N. Willis, and A. S. Wilson, FAR INFRARED SPECTRA OF PALLADIUM COMPOUNDS—II. PYRIDINE AND 2,2'-BIPYRIDYL COMPLEXES OF PALLADIUM(II) AND PLATINUM(II), *Spectrochim. Acta* **23A**, 1121.

1167. S. F. Dyubko, V. A. Svich, and R. A. Valitov, SUBMILLIMETER CW GAS LASER, *Zh. ETP Pisma* **6**, 567 (1967); *JETP Letters* **6**, 80.

1168. E. Ellis, CRYSTAL FIELD SPECTRA OF RARE-EARTH SALTS IN THE FAR INFRARED (10–50 cm^{-1}), *Chem. Phys. Letters* **1**, 80.

1169. M. F. Farona, J. G. Grasselli, and B. L. Ross, LOW FREQUENCY INFRARED SPECTRA OF ACETONITRILE AND ACRYLONITRILE DERIVATIVES OF CHROMIUM, MOLYBDENUM, AND TUNGSTEN HEXACARBONYLS, *Spectrochim. Acta* **23A**, 1875.

1170. P. G. Frayne, VIDEO CRYSTAL DETECTION OF 337 μm MASER RADIATION, *Electr. Letters* **3**, 338.

1171. L. Frenkel, T. Sullivan, M. A. Pollack, and T. J. Bridges, ABSOLUTE FREQUENCY MEASUREMENT OF THE 118.6 μm WATER-VAPOR LASER TRANSITION, *Appl. Phys. Letters* **11**, 344.

1172. B. Frlec and H. H. Claassen, LONG-WAVELENGTH, INFRARED-ACTIVE FUNDAMENTAL FOR URANIUM, NEPTUNIUM, AND PLUTONIUM HEXAFLUORIDES, *J. Chem. Phys.* **46**, 4603.

1173. H. A. Gebbie, N. W. B. Stone, E. H. Putley, and N. Shaw, HETERODYNE DETECTION OF SUBMILLIMETER RADIATION, *Nature* **214**, 165.

1174. P. A. Giguère and K. Sathianandan, LATTICE VIBRATIONS IN KHF_2 AND $NaHF_2$, *Can. J. Phys.* **45**, 2439.

1175. R. G. J. Grisar, K. P. Reiners, K. F. Renk, and L. Genzel, IMPURITY-INDUCED FAR-INFRARED ABSORPTION IN THE PHONON GAP REGIONS OF KI AND KBr, *Phys. Stat. Sol.* **23**, 613.

1176. A. Hadni, ESSENTIALS OF MODERN PHYSICS APPLIED TO THE STUDY OF THE INFRARED, Pergamon Press, Oxford 1968.

1177. A. Hadni, J. Claudel, D. Chanal, P. Strimer, and P. Vergnat, OPTICAL CONSTANTS OF POTASSIUM BROMIDE IN THE FAR INFRARED, *Phys. Rev.* **163**, 836.

1178. A. Hadni, G. Morlot, and P. Strimer, FAR-INFRARED ELECTRONIC TRANSITIONS IN SOLIDS, *IEEE J. Quant. Electr.* **QE-3**, 111.

1179. A. Hadni and P. Strimer, ON THE ABSORPTION AND DICHROISM IN THE FAR INFRARED OF A MONOCRYSTAL OF PRASEODYMIUM FLUORIDE, *Compt. Rend.* **265B**, 811.

1180. A. Hadni, R. Thomas, and J. Weber, LASERS POUR L'INFRAROUGE LOINTAIN, *J. Chim. Phys. (France)* **64**, 71.

1181. R. T. Hall and J. M. Dowling, PURE ROTATIONAL SPECTRUM OF WATER VAPOR, *J. Chem. Phys.* **47**, 2454.

1182. J. W. Halley, MICROSCOPIC THEORY OF FAR-INFRARED 2-MAGNON ABSORPTION IN ANTIFERROMAGNETS, II. SECOND-ORDER PROCESS AND APPLICATION TO MnF_2, *Phys. Rev.* **154**, 458.

1183. I. Harada and T. Shimanouchi, FAR-INFRARED SPECTRA OF CRYSTALLINE BENZENE AT 138°K AND INTERMOLECULAR FORCES, *J. Chem. Phys.* **46**, 2708.

1184. G. C. Hayward and P. J. Hendra, THE FAR INFRA-RED AND RAMAN SPECTRA OF THE TRIHALIDE IONS IBr_2^- AND I_3^-, *Spectrochim. Acta* **23A**, 2309.

1185. L. O. Hocker and A. Javan, ABSOLUTE FREQUENCY MEASUREMENTS ON THE NEW CW HCN SUBMILLIMETER LASER LINES, *Phys. Letters* **25A**, 489.

1186. L. O. Hocker, A. Javan, D. R. Rao, L. Frenkel and T. Sullivan, ABSOLUTE FREQUENCY MEASUREMENT AND SPECTROSCOPY OF GAS LASER TRANSITIONS IN THE FAR INFRARED, *Appl. Phys. Letters* **10**, 147.

1187. L. O. Hocker, D. R. Ramachandra, and A. Javan, ABSOLUTE FREQUENCY MEA-SUREMENT OF THE 190 μ AND 194 μ GAS LASER TRANSITION, *Phys. Letters* **24A**, 690.

1188. W. F. Hoffman, N. J. Woolf, and C. L. Frederick, FAR INFRARED SURVEYS OF THE SKY, *Science* **157**, 187.

1189. H. Jacobs, G. Morris, and R. C. Hofer, INTERFEROMETRIC EFFECT WITH SEMICONDUCTORS IN THE MILLIMETER-WAVE REGION, *J. Opt. Soc. Am.* **57**, 993.

1190. R. J. Jakobsen, Y. Mikawa, and J. W. Brasch, FAR INFRARED STUDIES OF HY-DROGEN BONDING IN CARBOXYLIC ACIDS—I. FORMIC AND ACETIC ACIDS, *Spectrochim. Acta* **23A**, 2199.

1191. W. Q. Jeffers, SINGLE WAVELENGTH OPERATION OF A PULSED WATER-VAPOR LASER, *Appl. Phys. Letters* **11**, 178.

1192. W. Q. Jeffers and P. D. Coleman, THE FAR INFRARED STIMULATED EMISSION SPECTRUM OF D_2O, *Proc. IEEE* **55**, 1222.

1193. W. Q. Jeffers and P. D. Coleman, SPIKING AND TIME BEHAVIOR OF A PULSED WATER-VAPOR LASER, *Appl. Phys. Letters* **10**, 7.

1194. W. Q. Jeffers, C. F. Wittig, and P. D. Coleman, GAIN CHARACTERISTICS OF A WATER VAPOR LASER, *Proc. IEEE* **55**, 2163.

1195. D. R. Johnston and D. E. Burch, ATTENUATION BY ARTIFICIAL FOGS IN THE VISIBLE, NEAR INFRARED, AND FAR INFRARED, *Appl. Opt.* **6**, 1497.

1196. L. H. Jones and R. S. McDowell, REVISION IN ASSIGNMENT OF LOW-FRE-QUENCY VIBRATIONAL FUNDAMENTALS OF $Ni(CO)_4$, *J. Chem. Phys.* **46**, 1536.

1197. R. Kaplan, MAGNETOOPTICAL STUDIES OF SOLIDS USING FOURIER TRANSFORM SPECTROSCOPY, *Appl. Opt.* **6**, 685.

1198. R. Kaplan, E. D. Palik, R. F. Wallis, S. Iwasa, E. Burstein, and Y. Sawada, INFRARED ABSORPTION BY COUPLED COLLECTIVE CYCLOTRON EXCITATION-LONGITUDINAL OPTIC PHONON MODES IN InSb, *Phys. Rev. Letters* **18**, 159.

1199. B. Katz, A. Ron, and O. Schnepp, FAR-INFRARED SPECTRA OF HCl AND HBr IN SOLID SOLUTIONS, *J. Chem. Phys.* **46**, 1926.

1200. B. Katz, A. Ron, and O. Schnepp, FAR-INFRARED SPECTRUM OF HCl DIMERS, *J. Chem. Phys.* **47**, 5303.

1201. F. K. Kneubühl, J. F. Moser, and H. Steffen, NOTE ON A LETTER OF DOWLING AND HALL, *J. Opt. Soc. Am.* **57**, 271.

1202. F. K. Kneubühl and H. Steffen, RESONATOR MODES OF SUBMILLIMETER-WAVE LASER, *Phys. Letters* **25A**, 639.

1203. S. Kon, M. Yamanaka, J. Yamamoto, and H. Yoshinaga, EXPERIMENTS ON A FAR INFRARED CN LASER, *Japan. J. Appl. Phys.* **5**, 612.

1204. T. Kondow, K. Siratori, and H. Inokuchi, FAR-INFRARED SPECTRA OF TCNQ COMPLEXES: THEIR ELECTRONIC STRUCTURES, *J. Phys. Soc. Japan*, **23**, 98.

1205. K. Y. Kondratiev and Y. M. Timofeyev, ON THE ACCURACY OF APPROXIMATE METHODS OF THE CALCULATION OF THE TRANSMISSION FUNCTIONS FOR THE ROTATIONAL BAND OF WATER VAPOUR IN THE REAL ATMO-SPHERE, *Fiz. Atmosfer. i Okeana* **3**, 227.

1206. G. C. Kulasingam and W. R. McWhinnie, AN UNUSUAL FEATURE IN THE FAR INFRARED SPECTRUM OF DICHLORO DI-(2-PYRIDYL)-PHENYLAMINE-COBALT-(II), *Spectrochim. Acta* **23A**, 1601.

1207. J. Laane and R. C. Lord, FAR-INFRARED SPECTRA OF RING COMPOUNDS II. THE SPECTRUM AND RING-PUCKERING POTENTIAL FUNCTION OF CYCLOPENTENE, *J. Chem. Phys.* **47**, 4941.

1208. D. R. Lide, INTERPRETATION OF THE FAR INFRARED LASER OSCILLATION IN AMMONIA, *Phys. Letters* **24A**, 599.

1209. D. R. Lide and A. G. Maki, ON THE EXPLANATION OF THE SO-CALLED CN LASER, *Appl. Phys. Letters* **11**, 62.

1210. T. M. Lifshits and F. Y. Nad, IMPURITY PHOTOCONDUCTIVITY OF n-TYPE InSb IN STRONG MAGNETIC FIELDS, *Fiz. Tverd. Tela* **8**, 2149; *Soviet Phys.-Solid State* **8**, 1709.

1211. D. H. Martin, SPECTROSCOPIC TECHNIQUES FOR FAR INFRARED SUB-MILLIMETRE AND MILLIMETRE WAVES, North-Holland Publ. Co., Amsterdam 1967.

1212. L. E. S. Mathias, A. Crocker and M. S. Wills, PULSED LASER EMISSION FROM HELIUM AT 95 μm, *IEEE J. Quant. Elect.* **QE-3**, 170.

1213. L. E. S. Mathias, A. Crocker, and M. S. Wills, SPECTROSCOPIC MEASUREMENTS ON THE LASER EMISSION FROM DISCHARGES IN COMPOUNDS OF HYDROGEN, CARBON, AND NITROGEN, *S.E.R.L. Tech. J.* **17**, 15.

1214. D. E. McCarthy, BLACK POLYETHYLENE AS A FAR-INFRARED FILTER, *J. Opt. Soc. Am.* **57**, 699.

1215. L. Mertz, AUXILIARY COMPUTATION FOR FOURIER SPECTROMETRY, *Infrared Phys.* **7**, 17.

1216. R. C. Milward, ELECTRONIC EFFECTS IN THE FAR-INFRARED SPECTRA OF Dy, Ho, Tm, AND Yb ALUMINIUM GARNETS, *Physics Letters* **25A**, 19.

1217. F. A. Miller, W. G. Fateley, and R. E. Witkowski, TORSIONOL FREQUENCIES IN THE FAR INFRARED-V. TORSIONS AROUND THE C–C SINGLE BOND IN SOME BENZALDEHYDES, FURFURAL, AND RELATED COMPOUNDS, *Spectrochim Acta* **23A**, 891.

1218. A. Minoh, T. Shimizu, S. Kobayashi, and K. Shimoda, FAR INFRARED AND SUB-MILLIMETER MASER OSCILLATORS WITH H_2O AND D_2O, *Japan. J. Appl. Phys.* **6**, 921.

1219. K. D. Möller, A. R. DeMeo, D. R. Smith, and L. H. London, FAR-INFRARED TORSIONAL VIBRATION SPECTRA OF ONE-, TWO-, AND THREE-(CX_3) TOP MOLECULES, *J. Chem. Phys.* **47**, 2609.

1220. K. D. Möller and L. H. London, FAR-INFRARED TORSIONAL VIBRATION SPECTRUM OF CD_3CD_2Cl, *J. Chem. Phys.* **47**, 2505.

1221. J. E. Mooij, THIN FILM EMISSION OF KBr IN THE FAR INFRARED, *Phys. Letters* **24A**, 249.

1222. J. F. Moser, W. Zingg, H. Steffen, and F. K. Kneubühl, SUBMILLIMETREWAVE SPECTRA OF DOPED Al_2O_3 CRYSTALS, *Phys. Letters* **24A**, 411.

1223. W. M. Müller and G. T. Flesher, CONTINUOUS-WAVE SUBMILLIMETER OSCILLATION IN DISCHARGES CONTAINING C, N, AND H OR D, *Appl. Phys. Letters* **10**, 93.

1224. K. Nagasaka, Y. Oka, and S. Narita, PHOTOCONDUCTIVITY OF ANTIMONY-DOPED GERMANIUM IN THE FAR-INFRARED REGION, *Solid State Commun.* **5**, 333.

1225. S. Narita, H. Harada, and K. Nagasaka, OPTICAL PROPERTIES OF ZINC TELLURIDE IN THE INFRARED, *J. Phys. Soc. Japan* **22**, 1176.

1226. E. D. Nelson and J. Y. Wong, SPECTRAL PURITY FOR FAR INFRARED GRATING SPECTROSCOPY, *Appl. Opt.* **6**, 1259.

1227. E. D. Nelson, J. Y. Wong, and A. L. Schawlow, FAR INFRARED SPECTRA OF $Al_2O_3:Cr^{3+}$ AND $Al_2O_3:Ti^{3+}$, *Phys. Rev.* **156**, 298.

1228. L. J. Neuringer and R. C. Milward, HEAVILY DOPED SILICON AS A LOW TEMPERATURE TRANSMISSION FILTER FOR THE FAR INFRARED, *Appl. Opt.* **6**, 978.

1229. I. G. Nolt, R. A. Westig, R. W. Alexander, and A. J. Sievers, GAP MODES DUE TO Cl^- AND Br^- IN KI, *Phys. Rev.* **157**, 730.

1230. S. L. Norman and D. H. Douglass, NEW MEASUREMENTS OF THE ELECTROMAGNETIC ABSORPTION SPECTRUM OF SUPERCONDUCTING LEAD, *Phys. Rev. Letters* **18**, 339.

1231. G. P. O'Leary and R. G. Wheeler, FAR-INFRARED MEASUREMENT OF LONG-RANGE ORDER IN A DISPLACIVE PHASE TRANSITION, *Phys. Rev. Letters* **18**, 1054.

1232. E. D. Palik, D. G. Burkhard, and R. L. Wallis, THE FAR INFRARED SPECTRUM OF METHYL MERCAPTAN FROM $15\,cm^{-1}$ TO $45\,cm^{-1}$, *J. Mol. Spectry.* **23**, 425.

1233. D. N. Pande, DEUTERIUM ISOTOPIC SHIFTS IN THE SPECTRA OF URANYL SALTS, *Indian J. Pure Appl. Phys.* **5**, 339.

1234. C. H. Perry, D. P. Athans, E. F. Young, J. R. Durig, and B. R. Mitchell, FAR INFRARED SPECTRA OF PALLADIUM COMPOUNDS III. TETRAHALO, TETRAAMMINE, AND DIHALODIAMMINE COMPLEXES OF PALLADIUM (II), *Spectrochim. Acta* **23A**, 1137.

1235. C. H. Perry and E. F. Young, INFRARED STUDIES OF SOME PEROVSKITE FLUORIDES. I. FUNDAMENTAL LATTICE VIBRATIONS, *J. Appl. Phys.* **38**, 4616.

1236. R. Petit, SOME PROPERTIES OF METAL GRATINGS, *Optica Acta* **14**, 301.

1237. J. N. Plendl, L. C. Mansur, A. Hadni, F. Brehat, P. Henry, G. Morlot, F. Naudin, and P. Strimer, LOW TEMPERATURE FAR INFRARED SPECTRA OF SiO_2 POLYMORPHS, *J. Phys. Chem. Solids*, **28**, 1589.

1238. M. A. Pollack, T. J. Bridges, and A. R. Strand, CENTRAL TUNING DIP IN A SUBMILLIMETER MOLECULAR LASER, *Appl. Phys. Letters* **10**, 182.

1239. M. A. Pollack, T. J. Bridges, and W. J. Tomlinson, COMPETITIVE AND CASCADE COUPLING BETWEEN TRANSITIONS IN THE CW WATER VAPOR LASER, *Appl. Phys. Letters* **10**, 253.

1240. C. M. Randall, FAST FOURIER TRANSFORM FOR UNEQUAL NUMBER OF INPUT AND OUTPUT POINTS, *Appl. Opt.* **6**, 1432.

1241. C. M. Randall and R. D. Rawcliffe, REFRACTIVE INDICES OF GERMANIUM, SILICON, AND FUSED QUARTZ IN THE FAR INFRARED, *Appl. Opt.* **6**, 1889.

1242. R. D. Rawcliffe and C. M. Randall, METAL MESH INTERFERENCE FILTERS FOR THE FAR INFRARED, *Appl. Opt.* **6**, 1353.

1243. K. F. Renk, FAR INFRARED ABSORPTION THROUGH LATTICE VIBRATION IN DOPED KI, *Z. Physik* **201**, 457.

1244. G. M. Ressler and K. D. Möller, FAR INFRARED BANDPASS FILTERS AND MEASUREMENTS ON A RECIPROCAL GRID, *Appl. Opt.* **6**, 893.

1245. P. L. Richards, FAR-INFRARED ABSORPTION BY TWO-MAGNON EXCITATIONS IN ANTIFERROMAGNETS, *J. Appl. Phys.* **38**, 1500.

1246. J. N. A. Ridyard, A COMPLETE FOURIER SPECTROSCOPIC SYSTEM FOR THE FAR INFRARED, *J. Phys. (France) Suppl.* **28**, 63.

1247. E. E. Russell and E. E. Bell, OPTICAL CONSTANTS OF SAPPHIRE IN THE FAR INFRARED, *J. Opt. Soc. Am.* **57**, 543.

1248. R. B. Sanderson, MEASUREMENT OF ROTATIONAL LINE STRENGTHS IN HCl BY ASYMMETRIC FOURIER TRANSFORM TECHNIQUES, *Appl. Opt.* **6**, 1527.

1249. P. Schwaller, H. Steffen, and F. K. Kneubühl, RESONATORMODEN UND LINIEN-AUFSPALTUNGEN IM SUBMILLIMETERWELLEN GASLASER, *Z. Angew. Math. Phys.* **18**, 594.

1250. P. Schwaller, H. Steffen, J. F. Moser, and F. K. Kneubühl, INTERFEROMETRY OF RESONATOR MODES IN SUBMILLIMETER WAVE LASERS, *Appl. Opt.* **6**, 827.

1251. I. I. Shaganov, L. D. Kislovskii, and I. G. Rudyavskaya, ABSORPTION BY FREE CARRIERS IN SILICON IN THE 100–300 μ RANGE, *Opt. i Spektr.* **22**, 788; *Opt. Spectry.* **22**, 428.

1252. G. A. Slack, S. Roberts, and F. S. Ham, FAR-INFRARED OPTICAL ABSORPTION OF Fe^{2+} IN ZnS, *Phys. Rev.* **155**, 170.

1253. V. Sochor and E. Brannen, TIME DEPENDENCE OF THE POWER OUTPUT AT 337 μ IN A CN LASER, *Appl. Phys. Letters* **10**, 232.

1254. O. M. Stafsudd, F. A. Haak, and K. Radisavljević, FAR-INFRARED SPECTRUM OF CADMIUM TELLURIDE, *J. Opt. Soc. Am.* **57**, 1475.

1255. O. M. Stafsudd, F. A. Haak, and K. Radisavljević, LASER ACTION IN SELECTED COMPOUNDS CONTAINING C, N, AND H OR D, *IEEE J. Quant. Electr.* **QE-3**, 618.

1256. H. Steffen, B. Keller, and F. K. Kneubühl, RELATIONS BETWEEN MODES AND PULSE SHAPES OF SUBMILLIMETER LASER EMISSIONS, *Electr. Letters* **3**, 562.

1257. H. Steffen, J. F. Moser, and F. K. Kneubühl, RESONATOR MODES AND SPLITTING OF THE 0.337 mm EMISSION OF THE CN LASER, *J. Appl. Phys.* **38**, 3410.

1258. C. J. Summers, P. G. Harper, and S. D. Smith, POLARON COUPLING AND LINE WIDTH STUDIES OF CYCLOTRON RESONANCE ABSORPTION IN InSb IN THE FARADAY CONFIGURATION, *Solid State Commun.* **5**, 615.

1259. N. Takeuchi and S. Kobayashi, DETECTION OF PULSED FAR INFRARED LASER RADIATION BY A Ge:In DETECTOR, *Japan. J. Appl. Phys.* **6**, 1462.

1260. M. Tinkham and D. H. Martin, A PERIODIC LAMELLAR INTERFEROMETER FOR THE FAR INFRA-RED, *J. Phys. (France) Suppl.* **28**, 87.

1261. W. J. Tomlinson, M. A. Pollack, and R. L. Fork, ZEEMAN EFFECT STUDIES OF THE WATER-VAPOR LASER OSCILLATING ON THE 118.65 μm TRANSITION, *Appl. Phys. Letters* **11**, 150.

1262. V. I. Trifonov, BOLOMETER BASED ON THE HALL EFFECT, *Fiz. i Tekh. Poluprov.* **1**, 1342; *Soviet Phys.-Semicond.* **1**, 1117.

1263. E. H. Turner, OBSERVATION OF HCN MILLIMETER ABSORPTION IN CYANID LASER DISCHARGE, *IEEE J. Quant. Electr.* **QE-3**, 695.

1264. T. Ueda and T. Shimanouchi, RING-PUCKERING MOTION OF 2, 5-DIHYDRO-FURAN, *J. Chem. Phys.* **47**, 4042.

1265. R. Ulrich, EFFECTIVE LOW-PASS FILTERS FOR FAR INFRARED FREQUENCIES, *Infrared Phys.* **7**, 65.

1266. R. Ulrich, FAR-INFRARED PROPERTIES OF METALLIC MESH AND ITS COMPLEMENTARY STRUCTURE, *Infrared Phys.* **7**, 37.

1267. H. W. Verleur and A. S. Barker, LONG-WAVELENGTH OPTICAL LATTICE VIBRATIONS IN $Ba_y Sr_{1-y} F_2$ AND $Sr_y Ca_{1-y} F_2$, *Phys. Rev.* **164**, 1169.

1268. H. W. Verleur and A. S. Barker, OPTICAL PHONONS IN MIXED CRYSTALS OF $Cd Se_y S_{1-y}$, *Phys. Rev.* **155**, 750.

1269. A. N. Vystavkin, V. N. Gubankov, V. N. Listvin, and V. V. Migulin, INVESTIGATION OF THE INFLUENCE OF IMPURITIES AND A MAGNETIC FIELD ON THE DETECTION PROPERTIES OF n-TYPE InSb IN THE MILLIMETER AND SUBMILLIMETER RANGES, *Fiz. i. Tekh. Poluprov.* **1**, 844; *Soviet Phys.-Semicond.* **1**, 702.

1270. F. Watari, FAR INFRA-RED SPECTRA OF PYRIDINE-IODINE MONOCHLORIDE AND PYRIDINE-IODINE MONOBROMIDE COMPLEXES, *Spectrochim. Acta* **23A**, 1917.

1271. R. D. Werder, R. A. Frey, and H. H. Günthard, FAR-INFRARED MATRIX AND SOLUTION SPECTRA AND SOLID-STATE VIBRATIONAL SPECTRA OF NIOBIUMPENTACHLORIDE, *J. Chem. Phys.* **47**, 4159.

1272. E. Whalley and J. E. Bertie, THE FAR INFRARED SPECTRUM AND LONG-RANGE FORCES IN ICE, *J. Colloid Interface Sci.* **25**, 161.

1273. H. Yoshinaga, S. Kon, M. Yamanaka, and J. Yamamoto, FAR INFRARED GAS-LASERS, *Science of Light (Japan)* **16**, 50.

1274. E. F. Young and C. H. Perry, INFRARED STUDIES OF SOME PEROVSKITE FLUORIDES. II. MULTIPHONON SPECTRA, *J. Appl. Phys.* **38**, 4624.

1275. W. Zingg, J. F. Moser, and F. K. Kneubühl, SUBMILLIMETRE WAVE SPECTRA OF DOPED Al_2O_3 CRYSTALS, *Helv. Phys. Acta* **40**, 361.

1968

1276. I. G. Austin, B. D. Clay, C. E. Turner, and A. J. Springthorpe, NEAR AND FAR INFRARED ABSORPTION BY SMALL POLARONS IN SEMICONDUCTING NiO and CoO, *Solid State Commun.* **6**, 53.

1277. J. D. Axe, INFRARED DIELECTRIC DISPERSION AND APPARENT IONIC CHARGES IN SODIUM NITRITE, *Phys. Rev.* **167**, 573.

1278. W. Bagdade and R. Stolen, FAR INFRARED ABSORPTION IN FUSED QUARTZ AND SOFT GLASS, *J. Phys. Chem. Solids* **29**, 2001.

1279. L. J. Basile, D. L. Gronert, and J. R. Ferraro, INFRARED SPECTRA OF TERPYRIDYL-RARE EARTH HALIDE COMPLEXES (659–80 cm^{-1}), *Spectrochim. Acta* **24A**, 707.

1280. R. L. Bell and J. J. Uebbing, PHOTOEMISSION FROM InP-Cs-O, *Appl. Phys. Letters* **12**, 76.

1281. W. S. Benedict, IDENTIFICATION OF WATER VAPOR LASER LINES, *Appl. Phys. Letters* **12**, 170.

1282. J. E. Bertie, H. J. Labbé, and E. Whalley, FAR-INFRARED SPECTRA OF ICE II, V, and IX, *J. Chem. Phys.* **49**, 775.

1283. J. E. Bertie, H. J. Labbé, and E. Whalley, INFRARED SPECTRUM OF ICE VI IN THE RANGE 4000–50 cm^{-1}, *J. Chem. Phys.* **49**, 2141.

1284. C. L. Bertin and K. Rose, RADIANT ENERGY DETECTION BY SUPERCONDUCTING FILMS, *J. Appl. Phys.* **39**, 2561.

1285. G. Birnbaum and A. Rosenberg, COLLISION INDUCED ABSORPTION IN GASEOUS CH_4 AND CD_4 IN THE FAR INFRARED REGION, *Phys. Letters* **27A**, 272.

1286. S. G. Bishop, INFRARED ABSORPTION BY COUPLED COLLECTIVE CYCLOTRON EXCITATION-LONGITUDINAL-OPTICAL MODES IN PbTe, *Proc. IX Intern. Conf. Phys. Semicond.*, Nauka, Moscow 1968, p. 301.

1287. H. D. Bist, THE FAR INFRARED SPECTRA OF URANYL SALTS, *J. Mol. Spectry.* **27**, 542.

1288. R. Blinc, M. Mali, and A. Novak, FAR INFRARED SPECTRA OF FERROELEC-
TRIC SbSI, *Solid State Commun.* **6**, 327.

1289. W. Block, D. Keune, and H. Sievering, MATERIAL MEASUREMENT SCHEMES
FOR THE FAR INFRARED, *Appl. Opt.* **7**, 2319.

1290. J. Boettcher, K. Dransfeld, and K. F. Renk, FAR INFRARED HIGH RESOLUTION
SPECTROSCOPY OF EPR LINES WITH A LASER, *Phys. Letters* **26A**, 146.

1291. D. R. Bosomworth, R. S. Crandall, and R. E. Enstrom, DONOR SPECTROSCOPY IN
GaAs, *Phys. Letters* **28A**, 320.

1292. J. W. Brasch, R. J. Jakobsen, W. G. Fateley, and N. T. McDevitt, THE EFFECT OF
HYDROGEN BOND STRENGTH ON THE FREQUENCY OF A HYDROGEN
BOND STRETCHING VIBRATION, *Spectrochim. Acta* **24A**, 203.

1293. A. Bree and R. A. Kydd, INFRARED SPECTRUM OF ANTHRACENE CRYSTALS,
J. Chem. Phys. **48**, 5319.

1294. C. Brot, B. Lassier, G. W. Chantry, and H. A. Gebbie, FAR INFRARED SPECTRA
OF LIQUID AND CRYSTALLINE TERTIARY BUTYL CHLORIDE $(CH_3)_3CCl$,
Spectrochim. Acta **24A**, 295.

1295. D. E. Burch, ABSORPTION OF INFRARED RADIANT ENERGY BY CO_2 AND
H_2O. III. ABSORPTION BY H_2O BETWEEN 0.5 AND 36 cm^{-1} (278 μ–2 cm),
J. Opt. Soc. Am. **58**, 1383.

1296. M. J. Campbell, R. Grzeskowiak, and M. Goldstein, THE FAR INFRARED SPECTRA
(450–80 cm^{-1}) OF SOME COMPLEXES FORMED BY SEMICARBAZIDE, THIO-
SEMICARBAZIDE, ACETONE SEMICARBAZONE, AND 1,2,4-TRIAZOLE WITH
CUPRIC HALIDES AND OXYACID SALTS, *Spectrochim. Acta* **24A**, 1149.

1297. I. F. Chang and S. S. Mitra, APPLICATION OF A MODIFIED RANDOM-ELE-
MENT-ISODISPLACEMENT MODEL TO LONG-WAVELENGTH OPTIC
PHONONS OF MIXED CRYSTALS, *Phys. Rev.* **172**, 924.

1298. T. Y. Chang, N. Van Tran, and C. K. N. Patel, ABSOLUTE MEASUREMENT OF
SECOND ORDER NONLINEAR COEFFICIENT FOR OPTICAL GENERATION
OF MILLIMETER WAVE DIFFERENCE FREQUENCIES IN GaAs, *Appl. Phys.
Letters* **13**, 357.

1299. I. Darmon, C. Brot, G. W. Chantry, and H. A. Gebbie, FAR INFRA-RED SPECTRA
AND THE MOLECULAR LIBRATIONS IN SOME CRYSTALLINE HEXA-
SUBSTITUTED CHLORO-METHYL BENZENES, *Spectrochim. Acta* **24A**, 1517.

1300. P. Datta and G. M. Barrow, ROTATIONLIKE MOTION OF MOLECULES IN
LIQUIDS: FAR-INFRARED ABSORPTION OF POLAR MOLECULES IN AN
INERT SOLVENT, *J. Chem. Phys.* **48**, 4662.

1301. M. Davies, G. W. F. Pardoe, J. Chamberlain, and H. A. Gebbie, THE CHARACTER
OF THE FAR INFRA-RED ABSORPTIONS IN NON-POLAR LIQUIDS, *Chem.
Phys. Letters* **2**, 411.

1302. G. W. Day, O. L. Gaddy, and R. J. Iverson, DETECTION OF FAST INFRARED
LASER PULSES WITH THIN FILM THERMOCOUPLES, *Appl. Phys. Letters* **13**,
289.

1303. J. Dowling, THE ROTATION-INVERSION SPECTRUM OF AMMONIA, *J. Mol.
Spectry.* **27**, 527.

1304. D. A. Draegert and D. Williams, FAR-INFRARED ABSORPTION SPECTRA OF
AQUEOUS SOLUTIONS OF STRONG ELECTROLYTES, *J. Chem. Phys.* **48**, 401.

1305. J. R. Durig, D. J. Antion, and F. G. Baglin, FAR-INFRARED AND RAMAN SPEC-
TRA OF PHOSPHONIUM IODIDE AND PHOSPHONIUM IODIDE-d_4, *J. Chem.
Phys.* **49**, 666.

1306. J. R. Durig, G. L. Coulter, and D. W. Wertz, FAR-INFRARED SPECTRA AND

STRUCTURE OF SMALL RING COMPOUNDS. ETHYLENE CARBONATE, γ-BUTYROLACTONE, AND CYCLOPENTANONE, *J. Mol. Spectry.* **27**, 285.

1307. J. R. Durig and K. L. Hellams, LOW-FREQUENCY VIBRATIONS OF SOME ORGANOTRICHLORO-SILANES, *Appl. Spectry.* **22**, 153.

1308. J. R. Durig and D. W. Wertz, FAR-INFRARED SPECTRA AND STRUCTURE OF SMALL RING COMPOUNDS VIII. PSEUDOROTATION IN 1,3-DIOXOLANE, *J. Chem. Phys.* **49**, 675.

1309. S. F. Dyubko, V. A. Svich, and R. A. Valitov, HYDROGEN AS A BUFFER GAS FOR SUB-MILLIMETRE WATER VAPOR LASERS, *Zhur. Tekh. Fiz.* **38**, 1988.

1310. S. F. Dyubko, V. A. Svich, and R. A. Valitov, SO_2 SUBMILLIMETER LASER GENE-RATING AT WAVELENGTHS 0.141 AND 0.193 mm, *Zh. ETF Pisma* **7**, 408; *JETP Letters* **7**, 320.

1311. K. M. Evenson, H. P. Broida, J. S. Wells, R. J. Mahler, and M. Mizushima, ELECTRON PARAMAGNETIC RESONANCE ABSORPTION IN OXYGEN WITH THE HCN LASER, *Phys. Rev. Letters* **21**, 1038.

1312. O. K. Filippov and E. V. Ukhanov, ON THE EMISSION OF HIGH PRESSURE XENON LAMPS IN THE FAR IR REGION OF THE SPECTRUM, *Opt. i Spektr.* **24**, 648; *Opt. Spectry.* **24**, 347.

1313. J. B. Flynn and J. J. Schlickman, CROSS-FIELD FREE CARRIER FAR-INFRARED MODULATION IN GERMANIUM, *Proc. IEEE* **56**, 322.

1314. P. G. Frayne, FAR INFRA-RED FARADAY EFFECT IN RARE-EARTH IRON GARNETS, *Brit. J. Appl. Phys.* **1**, 741.

1315. I. P. French, MICROWAVE AND FAR-INFRARED ABSORPTION AS A DIAG-NOSTIC FOR THE WAKE OF HYPERSONIC PROJECTILES, *J. Quant. Spectr. Rad. Transfer* **8**, 1655.

1316. R. M. Fuller, D. G. Rathbun, and R. J. Bell, TRANSMITTANCE OF CER-VIT GLASS-CERAMIC IN THE ULTRAVIOLET, VISIBLE, INFRARED, AND SUB-MILLIMETER WAVELENGTH REGIONS, *Appl. Opt.* **7**, 1243.

1317. W. H. Green and A. B. Harvey, RING-PUCKERING VIBRATION OF 2,5-DIHY-DROTHROPHENE, *J. Chem. Phys.* **49**, 177.

1318. P. R. Griffiths and H. W. Thompson, FAR INFRARED SPECTRA OF PROLATE AND OBLATE SYMMETRIC TOPS, *Spectrochim. Acta* **24A**, 1325.

1319. C. C. Grimes, P. L. Richards, and S. Shapiro, JOSEPHSON-EFFECT FAR- INFRA-RED DETECTOR, *J. Appl. Phys.* **39**, 3905.

1320. A. Hadni, F. Brehat, J. Claudel, and P. Strimer, FAR-INFRARED-ACTIVE PHONON PROCESSES IN CuCl, *J. Chem. Phys.* **49**, 471.

1321. A. Hadni, D. Charlemagne, and R. Thomas, POLARIZATION COMPETITION BETWEEN MODES AND MAGNIFICATION OF SOME LINES IN THE EMIS-SION OF A PROPIONITRILE LASER NEAR 337 μ, *Compt. Rend.* **266B**, 1230.

1322. A. Hadni, J. Claudel, G. Morlot, and P. Strimer, TRANSMISSION AND REFLEC-TION SPECTRA OF PURE AND DOPED POTASSIUM IODIDE AT LOW TEM-PERATURES, *Appl. Opt.* **7**, 161.

1323. A. Hadni, J. Claudel, and P. Strimer, FAR INFRARED REFLECTION SPECTRA OF AgCl, AgBr, and AgI AT LOW TEMPERATURES, *Appl. Opt.* **7**, 1159.

1324. A. Hadni, E. Decamps, and J. P. Herbeuval, FAR-INFRARED ABSORPTION SPECTRA OF MOLECULAR CRYSTALS OF TETRAHEDRAL OR PYRAMIDAL MOLECULES, *J. Chim. Phys.* **65**, 959.

1325. J. Hiraishi, I. Nakagawa, and T. Shimanouchi, FAR INFRA-RED SPECTRA AND FORCE CONSTANTS OF AMINE COMPLEXES OF Pt(IV), Pt(II), AND Pd(II), *Spectrochim. Acta* **24A**, 819.

1326. T. M. Hard and R. C. Lord, A DOUBLE-BEAM HIGH-RESOLUTION SPECTRO-METER FOR THE FAR INFRARED, *Appl. Opt.* **7**, 589.

1327. W. N. Hardy, I. F. Silvera, K. N. Klump, and O. Schnepp, OPTICAL PHONONS IN SOLID HYDROGEN AND DEUTERIUM IN THE ORDERED STATE, *Phys. Rev. Letters* **21**, 291.

1328. B. Hartman and B. Kleman, ON THE ORIGIN OF THE WATER VAPOR LASER LINES, *Appl. Phys. Letters* **12**, 168.

1329. P. J. Hendra and P. J. D. Park, THE VIBRATIONAL SPECTRA OF SULPHUR AND SELENIUM MONOHALIDES, *J. Chem. Soc.* A, 908.

1330. A. Hezel and S. D. Ross, THE FAR INFRA-RED SPECTRA OF SOME DIVALENT METAL PYROPHOSPHATES, *Spectrochim. Acta* **24A**, 131.

1331. L. O. Hocker and A. Javan, ABSOLUTE FREQUENCY MEASUREMENTS OF NEW CW DCN SUBMILLIMETER LASER LINES, *Appl. Phys. Letters* **12**, 124.

1332. L. O. Hocker and A. Javan, LASER HARMONIC FREQUENCY MIXING OF TWO DIFFERENT FAR INFRARED LASER LINES UP TO 118 μ, *Phys. Letters* **26A**, 255.

1333. L. O. Hocker, D. R. Sokoloff, V. Daneu, A. Szoke, and A. Javan, FREQUENCY MIXING IN THE INFRARED AND FAR INFRARED USING A METAL-TO-METAL POINT CONTACT DIODE, *Appl. Phys. Letters* **12**, 401.

1334. R. G. Howell and D. J. Newman, HIGH RESOLUTION STUDY OF CRYSTAL FIELD LINES IN THE FAR INFRARED, *Phys. Stat. Sol.* **29**, 697.

1335. W. Q. Jeffers, AN EXPERIMENTAL TEST OF PROPOSED WATER-VAPOR LASER TRANSITION ASSIGNMENTS, *Appl. Phys. Letters* **13**, 104.

1336. W. Q. Jeffers and P. D. Coleman, RELAXATION PHENOMENA IN THE WATER VAPOR LASER, *Appl. Phys. Letters* **13**, 250.

1337. W. Q. Jeffers and C. J. Johnson, SPECTRAL RESPONSE OF THE Ge:Ga PHOTO-CONDUCTIVE DETECTOR, *Appl. Opt.* **7**, 1859.

1338. C. J. Johnson, A PULSED BREWSTER WINDOW WATER VAPOR LASER OPER-ATING BETWEEN 20 AND 120 μ, *IEE J. Quant. Electr.* **QE-4**, 701.

1339. K. C. Johnson and A. J. Sievers, A STUDY OF THE Hg-Xe ARC AS A SOURCE FOR THE FAR INFRARED, *Appl. Opt.* **7**, 2430.

1340. R. Kaplan, MAGNETO-OSCILLATORY EXCITATION SPECTRA OF SHALLOW ACCEPTOR IMPURITIES IN InSb AND Ge, *Phys. Rev. Letters* **20**, 329.

1341. R. Kaplan, S. G. Bishop, and B. D. McCombe, MAGNETO-OSCILLATORY AB-SORPTION AND PHOTOCONDUCTIVITY DUE TO THE EXCITATION OF ACCEPTOR IMPURITIES IN GERMANIUM, *Proc. IX Intern. Conf. Phys. Semicond.*, Nauka, Moscow 1968, p. 317.

1342. R. Kaplan and R. F. Wallis, IMPURITY-STATE-OPTICAL-PHONON COUPLING IN A MAGNETIC FIELD IN InSb, *Phys. Rev. Letters* **20**, 1499.

1343. T. Kasuya, A. Minoh, and K. Shimoda, A NEW LASER EMISSION FROM DEU-TERIUM OXIDE VAPOR, *J. Phys. Soc. Japan* **25**, 1201.

1344. T. Kasuya and K. Shimoda, NOTES ON THE IDENTIFICATION OF WATER LASER LINES, *Japan. J. Appl. Phys.* **7**, 782.

1345. T. Kawamura, A. Mitsuishi, and H. Yoshinaga, FAR-INFRARED SPECTRA OF KH_2PO_4 AND $NH_4H_2PO_4$, *Japan. J. Appl. Phys.* **7**, 1303.

1346. M. A. Kinch, HETERODYNE DETECTION AT mm WAVELENGTHS USING A ROLLIN InSb BOLOMETER, *Appl. Phys. Letters* **12**, 78.

1347. R. D. Kirby, A. E. Hughes, and A. J. Sievers, FAR INFRARED ABSORPTION IN NcCl: OH$^-$ AT LOW TEMPERATURES, *Phys. Letters* **28A**, 170.

1348. R. D. Kirby, I. G. Nolt, R. W. Alexander, and A. J. Sievers, FAR INFRARED PRO-

PERTIES OF LATTICE RESONANT MODES. I. ISOTOPE SHIFTS, *Phys. Rev.* **168**, 1057.

1349. M. V. Klein and H. F. Macdonald, FAR-INFRARED ABSORPTION INDUCED BY ISOTOPES IN NaCl AND LiF, *Phys. Rev. Letters* **20**, 1031.

1350. M. V. Klein, LOCALIZED MODES AND RESONANCE STATES IN ALKALI HALIDES, *Phys. Color Cent.* **1968**, 429, Acad. Press, New York.

1351. S. Kon, M. Otsuka, M. Yamanaka, and H. Yoshinaga, SPATIALLY RESOLVED MEASUREMENT OF PLASMA DENSITY USING A 337 μ CYANIDE LASER, *Japan. J. App. Phys.* **7**, 434.

1352. J. P. Kotthaus, HIGH POWER OUTPUT FROM A SUBMILLIMETER CW GAS LASER, *Appl. Opt.* **7**, 2422.

1353. B. Krakow, R. C. Lord, and G. O. Neely, A HIGH-RESOLUTION FAR-INFRARED STUDY OF ROTATION IN HN_3, HNCO, HNCS, AND THEIR DEUTERIUM DERIVATIVES, *J. Mol. Spectry.* **27**, 148.

1354. A. F. Krupnov, V. A. Skvortsov, and L. A. Sinegubko, SUBMILLIMETER SEALED-OFF $H_2O + H_2$ LASER, *Izv. VUZ Radiofiz.* **11**, 778; *Soviet Radiophys.* **11**, 700.

1355. J. Laane and R. C. Lord, FAR-INFRARED SPECTRA OF RING COMPOUNDS III. SPECTRUM, STRUCTURE, AND RING-PUCKERING POTENTIAL OF SILACYCLOBUTANE, *J. Chem. Phys.* **48**, 1508.

1356. R. F. Lake and H. W. Thompson, FAR INFRARED SPECTRA OF CHARGE-TRANSFER COMPLEXES BETWEEN BROMINE AND SUBSTITUTED PYRIDINES, *Spectrochim. Acta* **24A**, 1321.

1357. T. M. Lifshitz, N. P. Likhtman, and V. I. Sidorov, PHOTOELECTRIC SPECTROSCOPY OF IMPURITIES IN SEMICONDUCTORS, *Zh. ETF Pisma* **7**, 111; *JETP Letters* **7**, 84.

1358. T. M. Lifshitz, N. I. Likhtman, and V. I. Sidorov, TEMPERATURE DEPENDENCE OF PHOTOTHERMAL IONIZATION AND THE PROBABILITY OF HOLE TRAPPING BY EXCITED STATES OF HYDROGEN-LIKE ACCEPTORS, *Fiz. i Tekh. Poluprov.* **2**, 782; *Soviet Phys.-Semicond.* **2**, 652.

1359. A. S. Lvova, L. M. Sabirov, I. M. Arefev, and M. M. Sushchinskii, THE $100 \, cm^{-1}$ ABSORPTION BAND OF P-AZOXYPHENETOLE IN THE SOLID, LIQUID-CRYSTAL, AND LIQUID STATES, *Opt. i Spektr.* **24**, 613; *Opt. Spectry.* **24**, 322.

1360. A. G. Maki, ASSIGNMENT OF SOME DCN AND HCN LASER LINES, *Appl. Phys. Letters* **12**, 122.

1361. V. Y. Maleev and A. E. Stanevich, LONG-WAVELENGTH INFRARED ABSORPTION SPECTRA OF NUCLEOSIDE CRYSTALS, *Opt. i Spektr.* **24**, 72; *Opt. Spectry.* **24**, 36.

1362. W. S. Martin and M. Tinkham, MAGNETIC FIELD DEPENDENCE OF FAR-INFRARED ABSORPTION IN THIN-FILM SUPERCONDUCTORS, *Phys. Rev.* **167**, 421.

1363. L. E. S. Mathias, A. Crocker, and M. S. Wills, SPECTROSCOPIC MEASUREMENTS ON THE LASER EMISSION FROM DISCHARGES IN COMPOUND OF HYDROGEN, CARBON, AND NITROGEN, *IEEE J. Quant. Electr.* **QE-4**, 205.

1364. J. F. Moser, H. Steffen, and F. K. Kneubühl, SUBMILLIMETERWELLEN-TECHNIK, *Helv. Phys. Acta* **41**, 607.

1365. J. F. Moser, H. Steffen, and F. K. Kneubühl, AUSTAUSCHWECHSELWIRKUNGEN, INTRABANDSCHWINGUNGEN UND Ti^{3+}-KRISTALLFELDAUFSPALTUNG IN DOTIERTEM Al_2O_3, *Phys. Kond. Mat.* **7**, 261.

1365a. F. Y. Nad, PHOTOCONDUCTIVITY OF n-TYPE InSb DUE TO CYCLOTRON RESONANCE, *Fiz. i Tekh. Poluprov.* **2**, 1380; *Soviet Phys.-Semicond.* **2**, 1159.

1366. K. Nagata and Y. Tomono, FAR-INFRARED ABSORPTION IN THE FERRI-MAGNETIC SPIRAL MnCr$_2$O$_4$, *J. Phys. Soc. Japan* **24**, 1397.

1367. G. O. Neely, INTERPRETATION OF EXTREME CENTRIFUGAL DISTORTION IN HNCO, HNCS, AND THEIR DEUTERIUM DERIVATIVES, *J. Mol. Spectry.* **27**, 177.

1368. I. G. Nolt and A. J. Sievers, FAR-INFRARED PROPERTIES OF LATTICE RESONANT MODES II. STRESS EFFECTS, *Phys. Rev.* **174**, 1004.

1369. S. L. Norman, FAR-INFRARED ABSORPTION SPECTRA OF THICK SUPERCONDUCTING FILMS, *Phys. Rev.* **167**, 393.

1370. Y. Oka, K. Nagasaka, and S. Narita, FAR-INFRARED GERMANIUM DETECTORS, *Japan. J. Appl. Phys.* **7**, 611.

1371. E. D. Palik, B. W. Henvis, J. R. Stevenson, and S. Iwasa, REFLECTIVITY MEASUREMENTS OF COUPLED COLLECTIVE CYCLOTRON EXCITATION-LONGITUDINAL OPTICAL PHONON MODES IN POLAR SEMICONDUCTORS, *Solid State Commun.* **6**, 721.

1372. E. D. Palik, R. Kaplan, B. W. Henvis, J. R. Stevenson, S. Iwasa, and E. Burstein, MAGNETOOPTICAL STUDIES OF COUPLED COLLECTIVE CYCLOTRON EXCITATION-LONGITUDINAL OPTICAL PHONON MODES IN POLAR SEMICONDUCTORS, *Proc. IX Intern. Conf. Phys. Semicond.*, Nauka, Moscow 1968, p. 263.

1373. L. H. Palmer and M. Tinkham, FAR-INFRARED ABSORPTION IN THIN SUPERCONDUCTING FILMS, *Phys. Rev.* **165**, 588.

1374. C. Perchard and A. Novak, FAR-INFRARED SPECTRA AND HYDROGEN-BOND FREQUENCIES OF IMIDAZOLE, *J. Chem. Phys.* **48**, 3079.

1375. A. J. Perkins, PURE ROTATION SPECTRUM OF DEUTERIUM FLUORIDE VAPOR, *Spectrochim. Acta* **24A**, 285.

1376. M. A. Pollack, T. J. Bridges, and W. J. Tomlinson, COMPETITIVE AND CASCADE COUPLING BETWEEN TRANSITIONS IN THE CW WATER VAPOR LASER, *Appl. Phys. Letters* **10**, 253.

1377. M. A. Pollack, L. Frenkel, and T. Sullivan, ABSOLUTE FREQUENCY MEASUREMENT OF THE 220 μm WATER VAPOR LASER TRANSITION, *Phys. Letters* **26A**, 381.

1378. M. A. Pollack and W. J. Tomlinson, MOLECULAR LEVEL PARAMETERS AND PROPOSED IDENTIFICATIONS FOR THE CW WATER VAPOR LASER, *Appl. Phys. Letters* **12**, 173.

1379. C. Postmus, J. R. Ferraro, and S. S. Mitra, PRESSURE DEPENDENCE OF INFRARED EIGENFREQUENCIES OF KCl AND KBr, *Phys. Rev.* **174**, 983.

1380. C. M. Randall and R. D. Rawcliffe, FAR INFRARED OPTICAL PROPERTIES OF PRESSED CdTe, *Appl. Opt.* **7**, 213.

1381. J. H. Reuszer and P. Fisher, EXCITATION SPECTRA OF ARSENIC AND ANTIMONY IMPURITIES IN GERMANIUM UNDER UNIAXIAL COMPRESSION, *Phys. Rev.* **165**, 909.

1382. A. Rosenberg and G. Birnbaum, FAR-INFRARED ABSORPTION IN GASEOUS CF$_4$, *J. Chem. Phys.* **48**, 1396.

1383. W. G. Rothschild, ROTATIONAL MOTION IN SOLUTION: HYDROGEN HALIDES IN CYCLOHEXANE, *J. Chem. Phys.* **49**, 2250.

1384. W. G. Rothschild, MOBILITY OF SMALL MOLECULES IN VISCOUS MEDIA. I. ROTATIONAL MOTION OF METHYLENE CHLORIDE MOLECULES IN POLYSTYRENE BY FAR-INFRARED SPECTROSCOPY, *Macromolecules* **1**, 43.

1385. H. Sakai, G. A. Vanasse, and M. L. Forman, SPECTRAL RECOVERY IN FOURIER SPECTROSCOPY, *J. Opt. Soc. Am.* **58**, 84.

1386. K. Sakai, Y. Nakagawa, and H. Yoshinaga, LOW TEMPERATURE PROPERTIES OF THE RESTSTRAHLEN POWDER FILTERS IN THE FAR-INFRARED REGION, *Japan. J. Appl. Phys.* **7**, 792.

1387. T. Shimizu, K. Shimoda, and A. Minoh, STARK EFFECTS OF THE ROTATIONAL LINES OF AMMONIA OBSERVED BY A FAR-INFRARED LASER, *J. Phys. Soc. Japan* **24**, 1185.

1388. A. L. Smith, LOW FREQUENCY VIBRATIONAL SPECTRA OF GROUP IV-A PHENYL COMPOUNDS, *Spectrochim. Acta* **24A**, 695.

1389. D. R. Smith, R. L. Morgan, and E. V. Loewenstein, COMPARISON OF THE RADIANCE OF FAR-INFRARED SOURCES, *J. Opt. Soc. Am.* **58**, 433.

1390. R. A. Smith, THE DEVELOPMENT AND EXPLOITATION OF THE VERY-FAR INFRA-RED REGION OF THE SPECTRUM, *Helv. Phys. Acta* **41**, 6.

1391. R. A. Smith, S. Zwerdling, S. N. Dermatis, and J. P. Theriault, ABSORPTION OF COMPENSATED GERMANIUM IN THE VERY-FAR INFRARED, *Proc. IX Intern. Conf. Phys. Semicond.*, Nauka, Moscow 1968, p. 149.

1392. V. Sochor, CALCULATION OF THE FAR-INFRARED SPECTRUM OF A WATER-VAPOR MOLECULAR LASER, *Czech. J. Phys.* **18**, 60.

1393. O. Stafsudd and V. Stevens, THERMOPILE PERFORMANCE IN THE FAR INFRARED, *Appl. Opt.* **7**, 2320.

1394. H. Steffen and F. K. Kneubühl, RESONANT INTERFEROMETRY OF PULSED SUBMILLIMETER-WAVE LASERS, *IEEE J. Quant. Electr.* **QE-4**, 992.

1395. G. E. Stillman, C. M. Wolfe, I. Melngailis, C. D. Parker, P. E. Tannenwald, and J. O. Dimmock, FAR-INFRARED PHOTOCONDUCTIVITY IN HIGH-PURITY EPITAXIAL GaAs, *Appl. Phys. Letters* **13**, 83.

1396. R. W. Stimets and B. Lax, KRAMERS-KRONIG ANALYSIS OF MAGNETO-PLASMA-PHONON INTERACTIONS IN InAs, *Phys. Letters* **28A**, 321.

1397. N. W. B. Stone, H. A. Gebbie, D. W. E. Fuller, A. R. Lott, and C. C. Bradley, INTERFERENCE OF SUB-MILLIMETRE WAVES OVER 450 m PATH DIFFERENCE, *Nature* **217**, 1042.

1398. C. J. Summers, R. B. Dennis, S. D. Smith, and C. W. Litton, INFRA-RED MAGNETO-OPTICAL STUDIES OF LINE-WIDTH ANOMALIES CAUSED BY RESONANT POLARON COUPLING TO MAGNETIC STATES IN POLAR SEMICON-DUCTORS, *Proc. IX Intern. Conf. Phys. Semicond.*, Nauka, Moscow 1968, p. 1029.

1399. C. J. Summers, R. B. Dennis, B. S. Wherrett, P. G. Harper, and S. D. Smith, RESONANT-POLARON-COUPLING INVESTIGATION BY A STUDY OF LINE-WIDTHS, STRENGTHS, AND FREQUENCIES OF CYCLOTRON RESONANCE AND MAGNETIC-IMPURITY ABSORPTION IN InSb, *Phys. Rev.* **170**, 755.

1400. C. J. Summers, P. G. Harper, and S. D. Smith, POLARON COUPLING AND LINE WIDTH STUDIES OF CYCLOTRON RESONANCE ABSORPTION IN InSb IN THE FARADAY CONFIGURATION, *Solid State Commun.* **5**, 615.

1401. K. Tamagake, M. Tsuboi, and A. Y. Hirakawa, INTERNAL ROTATION SPECTRA OF METHYLAMINES. I. CH_3NH_2 AND CH_3ND_2, *J. Chem. Phys.* **48**, 5536.

1402. R. Turner, A. K. Hochberg, and T. O. Poehler, MULTIPLE PULSE EMISSION FROM A HCN LASER, *Appl. Phys. Letters* **12**, 104.

1403. R. Turner and T. O. Poehler, FAR-INFRARED LASER INTERFEROMETRY FOR ELECTRON DENSITY MEASUREMENTS, *J. Appl. Phys.* **39**, 5726.

1404. R. Ulrich, INTERFERENCE FILTERS FOR THE FAR INFRARED, *Appl. Opt.* **7**, 1987.

1405. H. J. van Daal, THE STATIC DIELECTRIC CONSTANT OF SnO_2, *J. Appl. Phys.* **39**, 4467.

1406. A. N. Vystavkin, Y. S. Gal'pern, and V. N. Gubankov, ABSORPTION OF MILLI-
METER AND SUBMILLIMETER RADIATION IN n-TYPE InSb AT HELIUM
TEMPERATURES, *Fiz. i Tekh. Poluprov.* **1**, 1735; *Soviet Phys.-Semicond.* **1**, 1439.

1407. G. H. Wegdam, R. Bonn, and J. Van der Elsken, FAR INFRARED SPECTRA OF
SOME SOLID AND FUSED ALKALIMETAL NITRATES, *Chem. Phys. Letters*
2, 182.

1408. R. G. Wheeler, F. M. Reames, and E. J. Wachtel, STATUS OF EXPERIMENTS ON
FAR-INFRARED ABSORPTION SPECTROSCOPY IN MAGNETIC MATER-
IALS, *J. Appl. Phys.* **39**, 915.

1409. J. Y. Wong, FAR-INFRARED SPECTRA OF IRON-DOPED MgO, *Phys. Rev.*
168, 337.

1410. J. Y. Wong, M. J. Berggren, and A. L. Schawlow, FAR-INFRARED SPECTRUM OF
$Al_2O_3:V^{4+}$, *J. Chem. Phys.* **49**, 835.

1411. T. Yajima and K. Inoue, SUBMILLIMETER-WAVE GENERATION BY OPTICAL
DIFFERENCE FREQUENCY MIXING OF RUBY R_1 AND R_2 LASER LINES,
Phys. Letters **26A**, 281.

1412. J. Yamamoto and H. Yoshinaga, InSb PHOTOCONDUCTIVE DETECTORS IN
THE FAR INFRARED, *Japan. J. Appl. Phys.* **7**, 498.

1413. J. Yamamoto, H. Yoshinaga, and S. Kon, ENHANCEMENT OF THE RESPONSE
IN FAR INFRARED InSb DETECTOR BY NEAR INFRARED IRRADIATION,
Japan. J. Appl. Phys. **7**, 957.

1414. M. Yamanaka, H. Yoshinaga, and S. Kon, TIME BEHAVIOR OF PULSED OUTPUT,
TRANSVERSE MODE PATTERN, AND POLARIZATION CHARACTERISTICS
IN FAR INFRARED CN LASER, *Japan. J. Appl. Phys.* **7**, 250.

1415. M. Yamanaka, S. Kon, J. Yamamoto, and H. Yoshinaga, MECHANISM AND TIME
BEHAVIOR OF THE 538 μm LASER EMISSION FROM ICN, *Japan. J. Appl.
Phys.* **7**, 554.

1416. S. Zwerdling, R. A. Smith, and J. P. Theriault, A FAST HIGH-RESPONSIVITY BOLO-
METER DETECTOR FOR THE VERY-FAR INFRARED, *Infrared Phys.* **8**, 271.

1417. S. Zwerdling, J. P. Theriault, and H. S. Reichard, A FAR INFRARED BOLOMETER
PRE-AMPLIFIER WITH LOW-NOISE PERFORMANCE AT HIGH IMPEDANCE,
Infrared Phys. **8**, 135.

1418. S. Zwerdling and J. P. Theriault, THE LOW TEMPERATURE TRANSMITTANCE
OF TWO FAR INFRARED LWP FILTERS, *Appl. Opt.* **7**, 209.

                                        1969

1419. D. P. Akitt and W. Q. Jeffers, CORRELATION EFFECTS IN H_2O AND D_2O LASER
TRANSITIONS, *J. Appl. Phys.* **40**, 429.

1420. C. Allen, F. Arams, M. Wang, and C. C. Bradley, INFRARED-TO-MILLIMETER
BROADBAND SOLID-STATE BOLOMETER DETECTORS, *Appl. Opt.* **8**, 813.

1421. W. S. Benedict, M. A. Pollack, and W. J. Tomlinson, THE WATER-VAPOR LASER,
IEEE J. Quant. Electr. **QE-5**, 108.

1422. F. A. Benson, MILLIMETRE AND SUBMILLIMETRE WAVES, Iliffe Books,
London.

1423. C. V. Berney, SPECTROSCOPY OF CF_3COZ COMPOUNDS. III. VIBRATIONAL
SPECTRUM AND BARRIER TO INTERNAL ROTATION OF TRIFLUORO-
ACETALDEHYDE, *Spectrochim. Acta* **25A**, 793.

1424. J. R. Birch, W. J. Burroughs, and R. J. Emery, OBSERVATION OF ATMOSPHERIC

ABSORPTION USING SUBMILLIMETRE MASER SOURCES, *Infrared Phys.* **9**, 75.

1425. S. G. Bishop and B. W. Henvis, MAGNETOREFLECTION STUDIES OF COUPLED PLASMON-PHONON MODES IN PbTe, *Solid State Commun.* **7**, 437.

1426. T. G. Blocker, M. A. Kinch, and F. G. West, FAR-INFRARED ABSORPTION IN SOLID ALPHA OXYGEN, *Phys. Rev. Letters* **22**, 853.

1427. E. Brannen, M. Hoeksema, and W. J. Sarjeant, LINEARLY POLARIZED MONO-CHROMATIC RADIATION FROM A WATER VAPOR LASER AT 118.6 MICRONS, *Can. J. Phys.* **47**, 597.

1428. M. Brith and A. Ron, FAR-INFRARED SPECTRA OF CRYSTALLINE ETHYLENE C_2H_4 AND C_2D_4, *J. Chem. Phys.* **50**, 3053.

1428a. F. Brown, PROPOSED FAR-INFRARED DIFFERENCE FREQUENCY GEN-ERATION USING THE LORENTZ NONLINEARITY OF A SOLID-STATE PLASMA, *IEEE J. Quant. Electr.* **QE-5**, 586.

1429. L. A. Carreira and R. C. Lord, FAR-INFRARED SPECTRA OF RING COM-POUNDS. IV. SPECTRA OF COMPOUNDS WITH AN UNSYMMETRICAL POTENTIAL FUNCTION FOR RING INVERSION, *J. Chem. Phys.* **51**, 2735.

1430. J. Chamberlain, A. E. Costley, and H. A. Gebbie, THE SUB-MILLIMETRE DIS-PERSION ROTATIONAL LINE STRENGTHS AND DIPOLE MOMENT OF GASEOUS AMMONIA, *Spectrochim. Acta* **25A**, 9.

1431. G. Chanin and M. Herse, A LIQUID HELIUM-COOLED MONOCHROMATOR AS A FAR INFRARED FILTER, *Appl. Opt.* **8**, 1739.

1432. G. W. Chantry, H. M. Evans, J. Chamberlain, and H. A. Gebbie, ABSORPTION AND DISPERSION STUDIES IN THE RANGE 10–1000 cm^{-1} USING A MODULAR MICHELSON INTERFEROMETER, *Infrared Phys.* **9**, 85.

1433. G. W. Chantry, H. M. Evans, J. W. Fleming, and H. A. Gebbie, TPX, A NEW MATERIAL FOR OPTICAL COMPONENTS IN THE FAR INFRA-RED SPECTRAL REGION, *Infrared Phys.* **9**, 31.

1434. D. Charlemagne and A. Hadni, THE BIREFRINGENCE AND ROTATORY POWER OF QUARTZ IN THE FAR INFRARED AT LIQUID NITROGEN AND ROOM TEMPERATURES, *Opt. Acta* **16**, 53.

1435. B. P. Clayman, I. G. Nolt, and A. J. Sievers, FAR-INFRARED ABSORPTION SPEC-TRUM OF NaI:NaCl, *Solid State Commun.* **7**, 7.

1436. P. E. Clegg, R. A. Newstead, and J. A. Bastin, MILLIMETRE AND SUBMILLIM-ETRE ASTRONOMY, *Phil. Trans.* **A264**, 293.

1436a. V. J. Corcoran, FREQUENCY LOCK OF THE HYDROGEN CYANIDE LASER TO A MICROWAVE FREQUENCY STANDARD, *J. Quant. Electr.* **QE-5**, 424.

1437. V. J. Corcoran, R. E. Cupp, and J. J. Gallagher, FREQUENCY LOCK OF THE HY-DROGEN CYANIDE LASER TO A MICROWAVE STANDARD, *IEEE J. Quant. Elect.* **QE-5**, 424.

1438. M. R. Daniel and A. P. Cracknell, MAGNETIC SYMMETRY AND ANTIFERRO-MAGNETIC RESONANCE IN CoO, *Phys. Rev.* **177**, 932.

1439. H. D. Drew and A. J. Sievers, A ^{3}He-COOLED BOLOMETER FOR THE FAR INFRARED, *Appl. Opt.* **8**, 2067.

1440. J. R. Durig, K. K. Lau, G. Nagarajan, M. Walker, and J. Bragin, VIBRATIONAL SPECTRA AND MOLECULAR POTENTIAL FIELDS OF MERCURIOUS CHLORIDE, BROMIDE, AND IODIDE, *J. Chem. Phys.* **50**, 2130.

1441. D. W. Faries, K. A. Gehring, P. L. Richards, and Y. R. Shen, TUNABLE FAR-IN-FRARED RADIATION GENERATED FROM THE DIFFERENCE FREQUENCY BETWEEN TWO RUBY LASERS, *Phys. Rev.* **180**, 363.

1442. P. D. Feldman and D. P. McNutt, FAR INFRARED NIGHTGLOW EMISSION FROM ATOMIC OXYGEN, *J. Geophys. Res. Space Phys.* **74**, 4791.

1443. P. D. Feldman and D. P. McNutt, A ROCKET-BORNE LIQUID HELIUM COOLED INFRARED TELESCOPE. II: PHOTOCONDUCTIVE DETECTORS, *Appl. Opt.* **8**, 2205.

1444. J. M. Fertel and C. H. Perry, OPTICAL PHONONS IN $KCl_{1-x} Br_x$, AND $K_{1-x}Rb_xI$ MIXED CRYSTALS, *Phys. Rev.* **184**, 874.

1445. R. Foglizzo and A. Novak, LOW-FREQUENCY INFRARED AND RAMAN SPECTRA OF HYDROGEN-BONDED PYRIDINIUM HALIDES, *J. Chem. Phys.* **50**, 5366.

1446. P. G. Frayne, REPETITIVE Q-MODULATED HCN GAS MASER, *J. Phys. (B)* (*Atom. Molec. Phys.*) **2**, 247.

1446a. M. J. French, D. E. H. Jones, and J. L. Wood, A DOUBLE-BEAM EBERT VACUUM GRATING SPECTROMETER FOR THE $10-200 \text{ cm}^{-1}$ REGION OF THE SPECTRUM, *J. Sci. Instr.* **2**, 664.

1447. J. N. Gaitskell, R. A. Newstead, and J. A. Bastin, SUBMILLIMETRE SOLAR RADIATION AT SEA LEVEL, *Phil. Trans. Roy. Soc. (London)* **A264**, 195.

1448. H. A. Gebbie, FOURIER TRANSFORM VERSUS GRATING SPECTROSCOPY, *Appl. Opt.* **8**, 501.

1449. H. A. Gebbie and W. J. Burroughs, SUBMILLIMETRE ASTRONOMY: RESULTS AND PROSPECTS, *Phil. Trans. Roy. Soc. (London)* **A264**, 307.

1450. W. H. Green, RING-PUCKERING VIBRATION OF 2,3-DIHYDROFURAN, *J. Chem. Phys.* **50**, 1619.

1451. J. A. Greenhouse and H. L. Strauss, SPECTROSCOPIC EVIDENCE FOR PSEUDOROTATION. II. THE FAR-INFRARED SPECTRA OF TETRAHYDROFURAN AND 1,3-DIOXOLANE, *J. Chem. Phys.* **50**, 124.

1452. T. M. Hard, SULFUR DIOXIDE SUBMILLIMETER LASER, *Appl. Phys. Letters* **14**, 130.

1453. T. M. Hard, F. A. Haak, and O. M. Stafsudd, SIMULTANEOUS HCN AND H_2O LASER EMISSION IN MIXTURES CONTAINING H, C, N, AND O, *IEEE J. Quant. Electr.* **QE-5**, 132.

1454. M. Harwit, J. R. Houck, and K. Fuhrmann, ROCKET-BORNE LIQUID HELIUM COOLED TELESCOPE, *Appl. Opt.* **8**, 473.

1455. J. C. Hassler and P. D. Coleman, FAR-INFRARED LASING IN H_2S, OCS, AND SO_2, *Appl. Phys. Letters* **14**, 135.

1456. W. Hayes and D. R. Bosomworth, ABSORPTION OF OXYGEN IN SILICON IN THE FAR INFRARED, *Phys. Rev. Letters* **23**, 851.

1457. M. Hoeksema, W. J. Sarjeant, and E. Brennen, FAR-INFRARED GAS LASERS AS SOURCES OF LINEARLY POLARIZED RADIATION, *IEEE J. Quant. Elect.* **QE-5**, 477.

1458. K. Itoh and T. Shimanouchi, FAR-INFRARED SPECTRA OF ACETANILIDE, *Spectrochim. Acta* **25A**, 290.

1459. I. Iwahashi, K. Matsumoto, S. Matsudaira, S. Minami, and H. Yoshinaga, IMPROVED DOUBLE BEAM VACUUM FAR INFRARED SPECTROPHOTOMETER, *Appl. Opt.* **8**, 583.

1460. P. Jacquinot, INTERFEROMETRY AND GRATING SPECTROSCOPY: AN INTRODUCTORY SURVEY, *Appl. Opt.* **8**, 497.

1461. R. J. Jakobsen, Y. Mikawa, and J. W. Brasch, FAR INFRARED STUDIES OF HYDROGEN BONDING IN CARBOXYLIC ACIDS-II. THE *n*-ALKYL ACIDS, PROPANOIC TO UNDECANOIC, *Spectrochim. Acta* **25A**, 839.

1462. C. J. Johnson, G. H. Sherman, and R. Weil, FAR INFRARED MEASUREMENT OF THE DIELECTRIC PROPERTIES OF GaAs and CdTe AT 300°K AND 8°K, *Appl. Opt.* **8**, 1667.

1463. R. G. Jones, C. C. Bradley, J. Chamberlain, H. A. Gebbie, N. W. B. Stone, and H. Sixsmith, TRANSIENT PHENOMENA IN THE 337 μm MASERS, *Appl. Opt.* **8**, 701.

1464. R. R. Joyce and P. L. Richards, FAR-INFRARED SPECTRA OF Al_2O_3 DOPED WITH Ti, V, AND Cr, *Phys. Rev.* **179**, 375.

1465. T. Kasuya, K. Shimoda, N. Takeuchi, and S. Kobayashi, UNDULATION IN THE OUTPUT OF A PULSE WATER-VAPOR LASER, *Japan. J. Appl. Phys.* **8**, 478.

1466. T. Kasuya, N. Takeuchi, A. Minoh, and K. Shimoda, CORRELATION BETWEEN FAR-INFRARED LASER OSCILLATIONS IN DEUTERIUM OXIDE VAPOR, *J. Phys. Soc. Japan* **26**, 148.

1467. F. Kneubühl, DIFFRACTION GRATING SPECTROSCOPY, *Appl. Opt.* **8**, 505.

1468. J. A. Koningstein, A COMPARISON OF FAR INFRARED AND RAMAN SPECTRA OF SOME RARE EARTH GARNET SINGLE CRYSTALS, *Chem. Phys. Letters* **3**, 303.

1469. I. V. Kucherenko, PHOTOIONIZATION OF ACCEPTOR IMPURITIES IN UNI-AXIALLY COMPRESSED GERMANIUM, *Fiz. Tekh. Poluprov.* **2**, 1069; *Soviet Phys.-Semicond.* **2**, 899.

1470. J. Laane, FAR-INFRARED SPECTRA AND THE RING-PUCKERING POTEN-TIAL FUNCTION OF SILACYCLOPENT-3-ENE AND SILACYCLOPENT-3-ENE-1, 1-d_2, *J. Chem. Phys.* **50**, 776.

1471. J. Laane, FAR-INFRARED SPECTRUM AND THE BARRIER TO PSEUDORO-TATION OF SILACYCLOPENTANE, *J. Chem. Phys.* **50**, 1946.

1472. B. Lassier, C. Brot, G. W. Chantry, and H. A. Gebbie, FAR INFRARED ABSORP-TION SPECTRA OF METHYL-CHLOROFORM IN ITS PLASTIC AND TRULY CRYSTALLINE PHASES, *Chem. Phys. Letters* **3**, 96.

1473. J. S. Levine and A. Javan, FAR-INFRARED CONTINUOUS-WAVE LASER OSCIL-LATION IN PURE HELIUM, *Appl. Phys. Letters* **14**, 348.

1474. E. V. Loewenstein and D. C. Newell, RAY TRACES THROUGH HOLLOW METAL LIGHT-PIPE ELEMENTS, *J. Opt. Soc. Am.* **59**, 407.

1475. R. C. Lord and W. C. Pringle, A SEARCH FOR PURE ROTATIONAL ABSORPTION IN ALLENE, *J. Chem. Phys.* **50**, 565.

1476. S. J. Martinich, C. J. Johnson, and D. P. Akitt, MEASUREMENT OF FAR-INFRARED LASER POWER, *Rev. Sci. Instr.* **40**, 359.

1477. D. P. McNutt, K. Shivanandan, and P. D. Feldman, A ROCKET-BORNE LIQUID HELIUM COOLED INFRARED TELESCOPE. I: THE DEWAR AND OPTICS, *Appl. Opt.* **8**, 2199.

1477a. I. Melngalis, G. E. Stillman, J. O. Dimmock, and C. M. Wolfe, FAR-INFRARED RECOMBINATION RADIATION FROM IMPACT-IONIZED SHALLOW DON-ORS IN GaAs, *Phys. Rev. Letters* **23**, 1111.

1478. M. Mikami, I. Nakagawa, and T. Shimanouchi, INFRARED SPECTROSCOPIC INVESTIGATION OF PYRIDINE-2-CARBOXAMIDE CHELATE OF BIVALENT METALS-II. FAR INFRARED SPECTRA AND NORMAL COORDINATE TREATMENTS, *Spectrochim. Acta* **25A**, 365.

1479. R. C. Milward, A SMALL LAMELLAR GRATING INTERFEROMETER FOR THE VERY FAR-INFRARED, *Infrared Phys.* **9**, 59.

1480. J. E. Mooij, W. B. Van den Bunt, and J. E. Schrijvers, THIN FILM EMISSION OF KBr, RbCl, RbBr, and NaCl IN THE FAR INFRARED, *Phys. Letters* **28A**, 573.

1481. A. Murai, LONG-TIME OPERATION OF HCN LASER BY PULSED DISCHARGES IN A SEAL-OFF TUBE, *Phys. Letters* **28A**, 540.

1482. K. Nagasaka and S. Narita, TWO TYPES OF FAR-INFRARED PHOTOCONDUCTIVITY IN ANTIMONY-DOPED GERMANIUM, *Solid State Commun.* 7, 467.

1483. R. W. Noyes, THE SOLAR CONTINUUM IN THE FAR INFRARED AND MILLIMETRE REGIONS, *Phil. Trans. Roy. Soc. (London)* **A264**, 205.

1484. J. F. Parrish, R. P. Lowndes, and C. H. Perry, FAR INFRARED ELECTRONIC AND VIBRONIC TRANSITIONS OF Nd^{3+} IN TYSONITE LANTHANIDE FLUORIDES, *Phys. Letters* **29A**, 133, 270.

1485. C. K. N. Patel and N. Van Tran, PHASE-MATCHED NONLINEAR INTERACTION BETWEEN CIRCULARLY POLARIZED WAVES, *Appl. Phys. Letters* **15**, 189.

1486. J. E. Pearson, D. T. Llewellyn-Jones, and R. J. Knight, WATER VAPOR ABSORPTION NEAR A WAVELENGTH OF 0.79 mm, *Infrared Phys.* **9**, 53.

1487. C. H. Perry and N. E. Tornberg, OPTICAL PHONONS IN MIXED SODIUM-POTASSIUM TANTALATES, *Phys. Rev.* **183**, 595.

1488. J. Petzelt, FAR INFRARED REFLECTIVITY OF SbSI, *Phys. Stat. Sol.* **36**, 321.

1489. C. M. R. Platt, A BROAD-BAND THERMAL RADIOMETER AT 1.2 mm USING A PUTLEY DETECTOR, *Infrared Phys.* **9**, 1.

1489a. M. A. Pollack, FAR-INFRARED LASER GAIN RESULTING FROM ROTATIONAL PERTURBATIONS, *IEEE J. Quant. Electr.* **QE-5**, 558.

1490. P. L. Richards and S. A. Sterling, REGENERATIVE JOSEPHSON EFFECT DETECTOR FOR FAR-INFRARED RADIATION, *Appl. Phys. Letters* **14**, 394.

1491. W. H. Robinette and R. B. Sanderson, DIPOLE MOMENT OF HBr FROM FAR INFRARED DISPERSION MEASUREMENTS, *Appl. Opt.* **8**, 711.

1492. R. Ruppin, SIZE EFFECTS OF PLASMON-PHONON MODES IN POLAR SEMICONDUCTORS, *J. Phys. Chem. Solids* **30**, 2349.

1493. R. V. St. Louis and O. Schnepp, ABSOLUTE FAR-INFRARED INTENSITIES OF α-NITROGEN, *J. Chem. Phys.* **50**, 5177.

1494. K. Sakai, T. Fukui, Y. Tsunawaki, and H. Yoshinaga, METALLIC MESH BANDPASS FILTERS AND FABRY-PEROT INTERFEROMETER FOR THE FAR INFRARED, *Japan. J. Appl. Phys.* **8**, 1046.

1495. B. D. Saksena, D. R. Pahwa, M. M. Pradhan, and K. Lal, REFLECTION AND TRANSMISSION CHARACTERISTICS OF WIRE GRATINGS IN THE FAR INFRARED, *Infrared Phys.* **9**, 43.

1496. A. E. Salomonovich, SOME PROBLEMS AND INSTRUMENTAL FEATURES OF SUBMILLIMETRE ASTRONOMY, *Phil. Trans. Roy. Soc. (London)* **A264**, 283.

1497. Y. A. Schwartz, A. Ron, and S. Kimel, FAR-INFRARED SPECTRUM AND PHASE TRANSITION OF ACETYLENE, *J. Chem. Phys.* **51**, 1666.

1498. O. M. Stafsudd and Y. C. Yeh, THE CW GAIN CHARACTERISTICS OF SEVERAL GAS MIXTURES AT 337 μ, *IEEE J. Quant. Electr.* **QE-5**, 377.

1499. R. H. Stolen, FAR-INFRARED ABSORPTION IN HIGH RESISTIVITY GaAs, *Appl. Phys. Letters* **15**, 74.

1499a. R. J. Strauch, TECHNIQUE FOR MEASUREMENT OF HCN-LASER LINEWIDTH, *Electr. Letters* **5**, 246.

1500. R. G. Strauch, D. A. Stephenson, and V. E. Derr, REFRACTIVE INDEX OF D_2O AT THE HCN LASER FREQUENCY, *Infrared Phys.* **9**, 137.

1501. T. Timusk and R. W. Ward, OBSERVATION OF SINGULAR POINTS IN DEFECT-INDUCED FAR-INFRARED SPECTRA OF KBr, *Phys. Rev. Letters* **22**, 396.

1502. R. Ulrich, PREPARATION OF GRIDS FOR FAR INFRARED FILTERS, *Appl. Opt.* **8**, 319.

1503. J. T. Vallin, G. A. Slack, S. Roberts, and A. E. Hughes, NEAR AND FAR INFRARED ABSORPTION IN Cr DOPED ZnSe, *Solid State Commun.* 7, 1211.

1504. N. Van Tran and C. K. N. Patel, FREE-CARRIER MAGNETO-OPTICAL EFFECTS IN FAR-INFRARED DIFFERENCE FREQUENCY GENERATION IN SEMI-CONDUCTORS, *Phys. Rev. Letters* 22, 463.

1505. S. P. Varma and K. D. Möller, FAR INFRARED BAND-PASS FILTERS IN THE 400–16 cm^{-1} SPECTRAL REGION, *Appl. Opt.* 8, 2151.

1506. S. P. Varma and K. D. Möller, FAR INFRARED INTERFERENCE FILTERS, *Appl. Opt.* 8, 1663.

1507. A. N. Vystavkin, Y. S. Gal'pern, and V. N. Gubankov, INVESTIGATION OF MILLI-METER AND SUBMILLIMETER RADIATION ABSORPTION IN *n*-TYPE InSb AT HELIUM TEMPERATURES, *Fiz. i Tekh. Poluprov.* 2, 1651; *Soviet Phys.-Semicond.* 2, 1373.

1508. A. N. Vystavkin and V. N. Gubankov, INVESTIGATION OF THE TEMPERATURE DEPENDENCE OF THE DETECTION PROPERTIES OF *n*-TYPE InSb IN THE MILLIMETER AND SUBMILLIMETER BANDS, *Fiz. i Tekh. Poluprov.* 2, 1158; *Soviet Phys.-Semicond.* 2, 968.

1509. J. L. Walter and R. J. Hooper, FAR INFRARED SPECTRA OF ISOMERIC GLY-CINE CHELATES, *Spectrochim. Acta* 25A, 647.

1510. N. J. Woolf, W. F. Hoffman, C. L. Frederick, and F. J. Low, A FAR INFRARED SKY SURVEY, *Phil. Trans. Roy. Soc. (London)* A264, 267.

1511. T. Yajima and K. Inoue, SUBMILLIMETER-WAVE GENERATION BY DIF-FERENCE-FREQUENCY MIXING OF RUBY LASER LINES IN ZnTe, *IEEE J. Quant. Electr.* QE-5, 140.

1512. F. Zernike, TEMPERATURE-DEPENDENT PHASE MATCHING FOR FAR-INFRARED DIFFERENCE-FREQUENCY GENERATION IN InSb, *Phys. Rev. Letters* 22, 931.

Author Index to Bibliography

Index
